International Conference on Applications in Computational Engineering and Sciences (IConACES 2020)

IOP Conference Series: Materials Science and Engineering
Volume 1128

Online
30-31 October 2020

Editors:

K. Annamalai
S. Jeyanthi
Davidson Jebaseelan
M. Lenin Babu

ISBN: 978-1-7138-4611-6
ISSN: 1757-8981

Printed from e-media with permission by:

Curran Associates, Inc.
57 Morehouse Lane
Red Hook, NY 12571

Institute of Physics
Dirac House, Temple Back
Bristol BS1 6BE UK

Phone: 44 1 17 929 7481
Fax: 44 1 17 920 0979

techtracking@iop.org

Additional copies of this publication are available from:

Curran Associates, Inc.
57 Morehouse Lane
Red Hook, NY 12571 USA
Phone: 845-758-0400
Fax: 845-758-2633
Email: curran@proceedings.com
Web: www.proceedings.com

TABLE OF CONTENTS

IOP Conf. Series: Materials Science and Engineering 1128 (2021) 011001 doi:10.1088/1757-899X/1128/1/011001

PREFACE

The 1st International Conference on "Applications in Computational Engineering & Sciences "ICONACES 2020, was organized by the School of Mechanical Engineering at Vellore Institute of Technology, Chennai on 30th and 31st Oct 2020. However, due to the COVID 19 pandemic the conference was held Online. The main objective of ICONACES 2020 was intended to bring together the computational experts in Engineering and Sciences in the fields of computational fluid dynamics, finite element methods, boundary element methods, optimisation methods, computational material modelling etc.

The conference focused on applications of engineering and sciences in computations domain in Automotive, Bio-engineering, Energy & Environment, Materials, and Automation in which VIT Chennai has an in-house expertise. Further, given the current situation of COVID'19, all major universities and relevant institutes, companies are into developing tools, solutions, products etc to protect mankind from the virus. Hence this conference has included a special track on Computational Aspects related to COVID'19, to provide a platform to share the information and knowledge in this field.

IConACES2020 also provided an opportunity to exchange research evidence and innovative ideas and also the issues related to computational engineering and science. This proceeding will include presentations on the latest research areas such as Automotive, Bio-engineering, Energy & Environment, Materials, and Automation.

The highlight of the conference was the excellent group of keynote speakers from around the globe on various themes of the conference. The keynote speakers included,

Dr. Yukio Tamura,Tokyo Polytechnic University,Japan.
DATO DR. IR. TS. Mohd Abdul,Karim Bin Abdullah,CEO, Serba Dinamik Holdings Berhad,Malaysia
Dr.Kalaikovan Antony,SerbaDinamik Holdings Berhad,Malaysia
Dr. AnjaliAwasthi,Concordia University,Canada
Dr. MVJJ Suresh,General Electric,India
Dr.ElammaranJayamani,Swinburne University,Malaysia
Dr.Lucas Hof,ETS-Canada
Dr. David Hughes,TEESIDE University,UK
Dr.Dong Ruan,Swinburne University,Australia
Dr. G.Saravana Kumar,IITMadras,India
Dr. Jorge Posada Velásquez,Vicomtech,Spain
Dr. Suman Chakraborty,IIT Kharagpur,India
Dr. N.M. Anoop Krishnan,IIT Delhi,India
Dr. Geeske Scholz,University of Osnabruck,Germany
Dr. Saurabh S Diwan,IISc Bangalore,India
Dr Chiara Bisagni,TU Delft,The Netherlands
Dr. Deepak Subramani,IISc Bangalore,India
Dr. Kambiz Ebrahimi,Labourough University,UK
Dr. Rod Giles,Royal Enfield,UK
Dr.Satyaprasad,Askhok Leyland,
Dr. Zahra Motamed,McMaster University,Canada
Dr. Mark Taylor,Flinders University,Australia
Dr. Christian M. Puttlitz,Colorado State University,USA

IOP Conf. Series: Materials Science and Engineering 1128 (2021) 011001 doi:10.1088/1757-899X/1128/1/011001

Call for Papers & Registration

This year, ICONACES 2020 had more than 300 participants from around the globe. A total of 120 research articles were submitted and based on the double-blind peer review process 51 full length papers were accepted to be published with IOP Conference Series: Materials and Engineering after presentations. The evaluation of all the papers is based on the review reports from various experts and authors revised their original submissions accordingly.

We would like to take this opportunity to deeply express our heartfelt appreciation to our management for providing us all the support required for the successful completion of the conference. Our special thanks to our partnership University, Middle East University, Turkey, Industry partners, SERBA, ST Advanced Composites, Asimptote, Keynote Speakers, Session Chairs, all the Committee Members and Reviewers involved in the technical evaluation of conference papers without whose support the conference would not have been successful.

We look forward to the IConACES 2021 edition !

Best Regards,
Dr K Annamalai, Dr.S.Jeyanthi, Dr Davidson Jebaseelan, Dr M. Lenin Babu
Guest Editors

IOP Conf. Series: Materials Science and Engineering 1128 (2021) 011002 doi:10.1088/1757-899X/1128/1/011002

IConACES 2020

Peer review declaration

All papers published in this volume of IOP Conference Series: Materials Science and Engineering has been peer reviewed through processes administered by the Editors. Reviews were conducted by expert referees to the professional and scientific standards expected of a proceedings journal published by IOP Publishing.

Type of peer review: Double-blind review

Conference submission management system: Through email, iconaces2020@vit.ac.in

Number of submissions received: 120 full length papers

Number of submissions sent for review: 120 papers

Number of submissions accepted: 51 papers

Acceptance Rate (Number of Submissions Accepted / Number of Submissions Received X 100): 42.5%

Average number of reviews per paper: 2 Reviewers

Total number of reviewers involved: 40 Reviewers

Any additional info on review process:

The first level review was received from the session chairs of various tracks during the IConACES2020 conference who had given their recommendation for further publication of the work based on the conference presentation and the extended abstract proceedings.

The received full papers were sent to two reviewers and the reviewers were asked to review :

1. Adherence to the IOP format and usage of standard English language throughout the manuscript https://publishingsupport.iopscience.iop.org/author-guidelines-for-conference-proceedings/
2. Plagiarism to be within 15%
3. Detailed technical comments: Problem statement/definition, methodology, results with respect to the methodology with discussion and existing literature & conclusion

Decision of the reviewer:

1. Accepted with minor corrections - (corrected manuscript need not be sent back to the reviewer)

IOP Conf. Series: Materials Science and Engineering 1128 (2021) 011002 doi:10.1088/1757-899X/1128/1/011002

2. Accepted with major corrections - (corrected manuscript to be sent back to the reviewer)
3. Rejection with reasons

The reviewer comments were sent to the authors again and a response to the comments were received with the revised paper. The editors had made sure the comments have been incorporated before it is sent for the publication process.

Contact person for queries:

Dr.K Annamalai
Dr S.Jeyanthi
Dr Davidson Jebaseelan
Dr M.C. Leninbabu
Editors -ICONACES-2020
School of mechanical Engineering
VIT CHENNAI
Vandalur-Kelambakkam road
Chennai, amil nadu,INDIA-600127
kannamalai@vit.ac.in
Jeyanthi.subramanian@vit.ac.in
davidson.jd@vit.ac.in
Lenin.babu@vit.ac.in
Phone no-91-9840477440,91-9445122485,91-7598622404

IOP Conf. Series: Materials Science and Engineering 1128 (2021) 012001 doi:10.1088/1757-899X/1128/1/012001

Geo located alert via bottom up IoT model in times of COVID-19 and other disasters

H O Lekshmy[1], T R Krishnaprasad[2], G Vishwanath Gupta[3] and K Namitha[4]

[1-4] Dept of Computer Science and Engineering, Amrita School of Engineering, Amrita Vishwa Vidyapeetham, Amritapuri, Kerala, India

* Corresponding author: vishwanathg2202@gmail.com

Abstract: Various disasters either man made or natural have necessitated people around the world to adopt ad-hoc practices in managing or mitigating the aftermath of the disaster. In times of a pandemic or prolonged flood or war it is even more pertinent that people do not detach up on the precautionary/ advisory measures. This paper proposes a mechanism integrating a mobile app with other IoT devices, using which the public can participate in issuing a broadcast alert in their locality.

1. Introduction

Wars, epidemics, etc. brings about a situation where the routine of a day-to-day life is disrupted and certain strategies are recommended by the authorities for enhanced safety of people. In the current scenario of a disaster such as the COVID-19 pandemic continues to result in a high number of health-related effects and increasing mortality rates, thus grinding the world economy into a turmoil. To contain the proliferation of the disease, many countries around the world have enforced various preventive measures such as locking down distinct locales depending upon the concentration of COVID-19 patients, limiting transportation, exhorting citizens to adhere to maintaining social distancing etc. [1]. While many laws are implemented to attenuate the effects of any disaster, information technology and communication applications continue to provide a major role in studying as well as in mitigating the adverse effects of crisis or disasters [2][3]. This paper seeks to provide a mechanism which can forewarn the local citizenry about any disaster and can act as a tool in mitigating the aftermath of a disaster. The proposed work integrates a mobile application with Raspberry-pi, connected with other devices in issuing an effective response in a bottom-up manner so that people can take adequate precaution in times of a disaster.

2. Motivation

If prudent precautionary measures are broadcast to people, several lives can be saved in times of a disaster. Very often, there is a shortfall of precise information collection and dissipation to promulgate an immediate response in managing the disaster and thus saving lives and livelihood. Sometimes, maintaining social distancing is required in times of COVID-19 and technology can improve the maintenance of social distancing. Hence, the proposed model brings forth a participatory framework based on a ground-up rather than top-down approach of information flow in which the local citizenry relays information to the area which would be affected, without waiting for the government to help. Thus, this model strives to reduce the workload of the government as well as to democratize disaster management.

3. Related Works

This paper [4] provides a mobile application, developed on Android using Java, for first responders to a disaster, like fire rescue workers. The back end is provided by EC2 instance of Amazon and with Twilio being the choice for communication related purposes. The application enables the team leader and the front-line rescue workers to communicate by using alerts and notifications. In the paper [5], an automated solution that detects a disaster and displays a prospective route to the disaster site from entities that can provide succor to people affected by the disaster is provided. The model proposes using Arduino, Raspberry Pi to receive sensor data from various clusters via wireless transmission. The data is processed by the server, which in turn communicates with the application whose GUI displays an alert along with route of travel to the user in the event of a disaster.

This paper [6] proposes a crowdsourced geographic information system where common citizens feed in reports to the system about the occurrence of a disaster from a mobile GUI application. The data is then fed to a server along with many other parameters. The server runs a simple additive weighting method for measuring the accuracy of information from a given area as well as to prioritize responses. The data of different regions is then made available on a dashboard.

In the paper [7] the authors propose a crowd sourced disaster relief mobile and web application where the victims of a disaster can issue requests for various requirements like medical, food, evacuation etc. These requests are tagged based on their locations and made available via a map-based interface. The application also provides data integrity, data privacy and data security.

The authors of the paper [8] have proposed a wearable device that helps the wearer in with an alert mechanism in ensuring social distancing in times of a pandemic like COVID-19. The model uses Passive Infrared Motion (PIR) sensor, piezoelectric buzzers, NodeMCU amongst other hardware components. If a user comes within a certain distance then infrared radiation is detected by the sensor and connected NodeMCU triggers the alarm. The module can also be used with Bluetooth and WiFi. In paper [9] the authors propose a model which can connect the medical personnel with the people who have symptoms of COVID-19. The application is developed using Java and Firebase and has GUI with location-based data, with an option to contribute as a volunteer.

In our proposed application, we are using alarms connected to Arduino to warn the residents of a locality about any threat regarding an approaching disaster. Our application can also warn people about overcrowding in times of a situation like COVID-19 using location-based services and image processing algorithms.

4. Proposed model

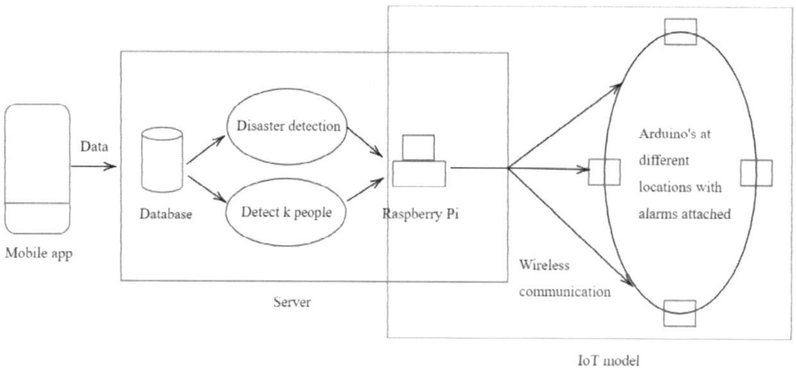

Figure 1. Overview of the proposed model

Our model has been categorized into three modules, namely, the mobile application module, image processing module in the server and the IoT module. An overview of the proposed model is in the diagram given in figure 1.

IOP Conf. Series: Materials Science and Engineering 1128 (2021) 012001 doi:10.1088/1757-899X/1128/1/012001

4.1 Mobile application

The mobile application provides the user with the option to take a picture and upload it to the server. The user need not provide any other information regarding the place from which the picture is uploaded. The user interface of the application also has an option for the user to enter the type of picture that is sent to the server. The data sent by the user is processed by the server. Server has dual objectives, to determine the type of disaster and to determine whether there are too many people on the road so that an alert can be issued to lessen the spread of diseases like COVID-19. The UI of the mobile application is as shown in figure 2. Due to the travel restriction, we could not take a photo of the street and hence have taken a sample picture from Unsplash.com by Koushik Das [10]. Once the 'Send' button is clicked the image is uploaded onto Firebase. We use Pyrebase to download the image from Firebase to send to the server.

4.2 Server-side image processing for detection of k people

Depending upon the text sent by the user, one of the two purposes of the image processing module in the server is to identify whether the image sent by the user has more than K number of persons in close proximity with each other. The server receives the GPS coordinates of the user along with the picture and the subtext.

Figure 2. Mobile UI Figure 3. Result

Once the coordinates and images are obtained, the server uses HaarCascade [12] algorithm to identify the number of people in a picture. The output of our algorithm, in the given context is either 0 or 1, depending on whether there are less than K or more than K number of people in the image, respectively. If more than K people are detected, then that information is passed into locally installed Raspberry Pi in that particular region. Pi creates alert messages and passes it into speakers.

4.3 Server-side image processing for disaster detection:

Figure 4. A sample data set

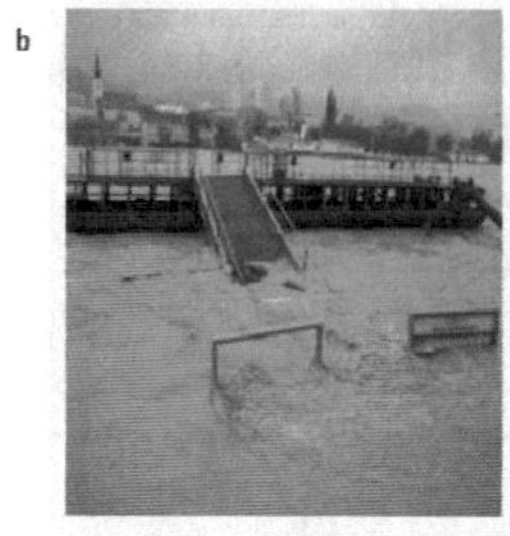

Figure 5. Test case

Once the image of disaster (say flood) is sent by the user through the mobile application, both the image and the GPS coordinates are received by the server. The server runs a machine learning algorithm to determine whether the image is of a disaster like a flood. The classification is made using a Bag of Features (BoF) model. The BoF model creates a multiplicity of features of an image using Speed Up Robust Features (SURF) technique. K-means algorithm is used to reduce this huge multiplicity of features into a much lesser number. Thus, the feature vector, from each category is given as input into a multi class linear Support Vector Machine (SVM) classifier for classification. The server also needs to evaluate the location of the user from which the image has been sent so that the appropriate alarm can be triggered.

Once the image is classified as a 'flood' or any other disaster, then the server transmits the GPS location from which the image has been sent to Raspberry Pi. We are using a dataset mentioned in [11]. From the images in the dataset, we are training the model using 328 images which are in the 'flood' category and 275 images of the 'pollution' category to balance the images in both the categories. The average accuracy of the model is 0.86. The figure 4 represents a sample data set and while figure 5 represents the test data respectively.

4.4 Raspberry Pi connection with Arduino

Raspberry Pi -3 is a credit card sized minicomputer that can perform all functions of a normal computer. It is widely used in IoT projects due its low cost and portability. Raspberry Pi -3 features a Broadcom system on a chip (SoC) with an integrated ARM-compatible CPU and on-chip GPU. Raspberry Pi-3 board contains 40 general purpose input-output pins, two 5-volt pins, two 3-volt pins and several ground pins. Different versions of Raspberry Pi s are available in the market. In our project we are using Raspberry Pi-3. The Raspberry Pi and each Arduino UNO is connected to nRF24L01. A circuit of the connection between Raspberry Pi and nRF24L01 is given in figure 6.

nRF24L01 is a single chip wireless radio transceiver module available in the market. This module can send and receive data. nRF24L01 operates on 2.4 GHz-2.5GHz ISM band. This module works with the help of SPI communications. Normally nRF24L01 covers a range of 100 m. We can increase its range by using antennas and powerful batteries. Nowadays more advanced boards are available in the market. In this proposed model we are using 1100-meter-long distance nRF24L01+PA+LNA wireless module with antenna. This module can provide a range over 1100 m.

4.5 Arduino connection with nRFVL01

Arduino UNO is an open-source microcontroller. This board incorporates fourteen input output pins that can connect to various external elements. Apart from this, Arduino UNO incorporates six analog pins. Arduino UNO is inexpensive when compared with Raspberry Pi. Each Arduino is connected to a speaker. These speakers having a dimension of 70 * 30 * 16 can produce loud stereophonic sound.

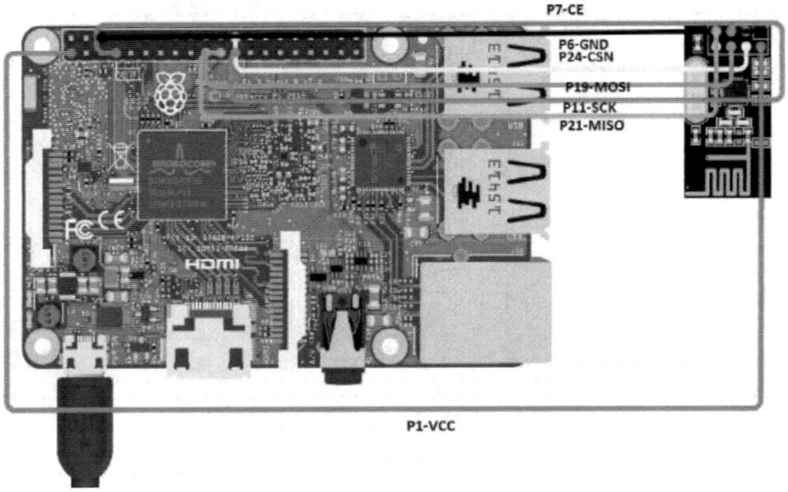

Figure 6. Raspberry-Pi connection with nRF24L01

The Arduino-nRF24L01 connection circuit diagram is given in figure.7. Power cable of nRF24L01 is connected to VCC in Arduino UNO to provide power. Similarly, the ground pin of nRF24L01 is connected with ground pin in Arduino UNO. MISO from nRF24LO1 is connected to pin no twelve of Arduino UNO. Similarly, MOSI in nRF24L01 is connected to pin number eleven. Chip Set (CS) from nRF24L01 is connected to pin number ten of Arduino UNO and SCK is connected to pin number thirteen of Arduino UNO. This is the connection between our Arduino UNO and nRF24L01. After that we connect the stereo speaker with the circuit. Stereo speakers usually have two terminals - A POS terminal and ground. The POS terminal produces sound in a specified direction. POS terminal of the stereo speaker is connected to pin number three of Arduino UNO. Ground of the speaker is connected to the ground pin in Arduino. We can use a breadboard for proper connection.

Each Raspberry Pi is connected with multiple Arduino UNO. This connection can be formed by using the principle of master- slave circuit. Here nRF24L01 connected in Raspberry Pi is considered as Master. While nRF24L01 connected with an Arduino UNO can be considered as a slave. By this method we can connect up to 5 slaves /nRF24L01 in the same wireless network. For example: Let us consider our central server is placed in the district capital, then our local server Raspberry Pi (local server) is placed in each Panchayat, while our Arduino modules are placed in each ward of

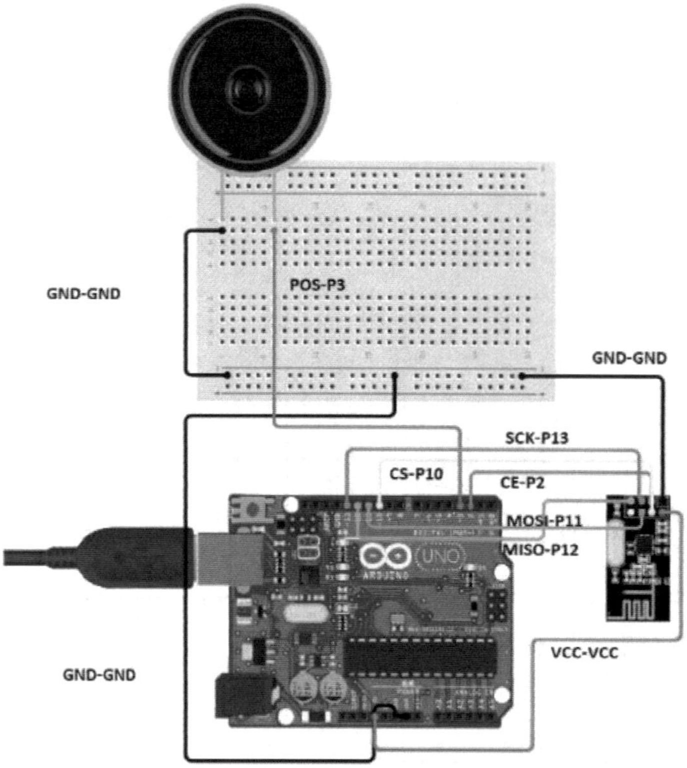

Figure 7. Arduino-nRF24L01 connection with speaker

the Panchayat. This way we can divide our entire region into small clusters. By this, communication becomes easy in every corner of the district.

4.6 Wireless communication between Raspberry-Pi-nRFVL01 and Arduino-nRFVL01

Here we are creating a wireless network (as is visualized in figure.1) using the nRF24L01 module. RF24 network library helps to form an Arduino wireless network. Now we define a byte array that contains the pipe address, through which nRF24L01 modules communicate, we will assign these addresses in octal format. Here we are using the concept of a master-slave circuit to form the wireless network. nRF24L01 connected with Raspberry is considered as master, while those connected with Arduino are treated as slaves. The GPS location of each Arduino module is mapped with the address that is already stored in Raspberry. When the server sends GPS location to Raspberry. Raspberry finds Arduino located near to the corresponding location, then it retrieves the receiver side address of nRF24L01 connected with that Arduino. After that Raspberry starts transmitting messages through the master nRF24L01 transceiver module. Slave nRF24L01 transceiver module in Arduino sides listens to these transmitted messages. Arduino connected with this module analyses the incoming signals and takes necessary actions like issuing alarm, sending messages to local authorities, etc. To avoid the local populace from panicking upon hearing an alarm, the alarm will be followed by a customized message. The message to be relayed via the alarm will be decided by the Arduino from a set of preexisting messages depending upon the incoming data.

5. Conclusion

Image captured using a mobile phone and its location is sent to the server using a mobile app, Server runs face recognition algorithm and determines whether the image contains more than k people or not. If the output of the image classification algorithm indeed states that there are more than k people, then the server labels the picture as crowded and passes information to Raspberry Pi-3 in corresponding GPS coordinates. If the image is detected as that of an approaching disaster like a flood, then that information is sent to Raspberry Pi-3 along with the GPS coordinates. Here Raspberry-Pi acts as a local server. This is usually placed in places that have an internet connection. Information received in Raspberry-pi is passed to Arduino UNO through nRF24L01 transceiver connected on both sides. Arduino generates auto generated voice messages or alarms to warn people. In case of emergency messages are sent to Raspberry Pi, Raspberry pi uses Google text to speech API to produce high quality audio. This audio file is saved in Raspberry pi. Raspberry locates Arduino UNO located near corresponding GPS coordinates. And passes the audio file to Arduino Uno through nRF24L01. This sound is reflected through 3W Stereo Speaker.

References

[1] Lakshmi Priyadarsini S, Suresh M. Factors influencing the epidemiological characteristics of pandemic CoVID- 19: A TISM approach. Int J Healthc Manag 2020;13:89–98. https://doi.org/10.1080/20479700.2020.1755804.

[2] Geetha M, Manoj M, Sarika AS, Mohan M, Rao SN. Detection and estimation of the extent of flood from crowd sourced images. Proc 2017 IEEE Int Conf Commun Signal Process ICCSP 2017 2018;2018-Janua:603–8. https://doi.org/10.1109/ICCSP.2017.8286429.

[3] Mclain ML, Babu SK, Bijlani K, Jayakhrishnan R, Bhavani RR. Collaborative Game Based Learning of Post-Disaster Management. 2016 IEEE 8th Int Conf Technol Educ Collab 2016:80–7. https://doi.org/10.1109/T4E.2016.23.

[4] Shaghaghi N, Patel S, Pabari B, Francis M. Implementing Communications and Information Dissemination Technologies for First Responders. 2019 IEEE Glob. Humanit. Technol. Conf. GHTC 2019, 2019. https://doi.org/10.1109/GHTC46095.2019.9033107.

[5] Goyal A, Meena K, Kini K, Parmar P, Zaveri M. Real Time Collaborative Processing for Event Detection and Monitoring for Disaster Management in IoT Environment. 2019 10th Int Conf Comput Commun Netw Technol ICCCNT 2019 2019:4–10. https://doi.org/10.1109/ICCCNT45670.2019.8944580.

[6] Puspita IA, Soesanto RP, Muhammad F. Designing Mobile Geographic Information System for Disaster Management by Utilizing Wisdom of the Crowd. 2019 IEEE 6th Int. Conf. Ind. Eng. Appl. ICIEA 2019, 2019. https://doi.org/10.1109/IEA.2019.8715031.

[7] Guntha R, Rao SN, Ramesh M V. Architectural Considerations for Building a Robust Crowdsourced Disaster Relief Application. 2020 Int. Conf. Commun. Syst. NETworkS, COMSNETS 2020, 2020. https://doi.org/10.1109/COMSNETS48256.2020.9027394.

[8] Raghav S, Vijay G, Harika PS, Rao AV, Gopinath A, Sai Shibu NB, et al. Suraksha: Low Cost Device to Maintain Social Distancing during CoVID-19 2020:1476–80. https://doi.org/10.1109/iceca49313.2020.9297503.

[9] Saxena T, Anuragi P, Shinde G, Yadav N, Digalwar M. COWAR: An Android Based Mobile Application to Help Citizens and CoVID-19 Warriors, 2021. https://doi.org/10.1109/cict51604.2020.9312073.

[10] Unsplash n.d.

[11] Barz B, Schröter K, Münch M, Yang B, Unger A, Dransch D, et al. Enhancing Flood Impact Analysis using Interactive Retrieval of Social Media Images 2019.

[12] Viola P, Jones M. Rapid object detection using a boosted cascade of simple features. Proc IEEE Comput Soc Conf Comput Vis Pattern Recognit 2001;1:11–8. https://doi.org/10.1109/cvpr.2001.990517.

CFD simulation for pedestrian comfort and wind safety in VIT campus for wind resource management

S A Prabhaharan[1], KB Rajasekarababu[2], G Vinayagamurthy[3*], R Sivakumar[4]

[1]PG Student, [2]PhD Student, [3*]Associate Professor, [4]Professor
[2, 3*] Centre for Innovation and New Product Development.
[1,2,3*,4,] School of Mechanical Engineering, VIT Chennai, India.
saprabhaharan@gmail.com[1], kb.rajasekarababu2014@vit.ac.in [2], vinayagamurthy.g@vit.ac.in[3*], sivakumar.r@vit.ac.in[4]

Abstract
Recent researches on Computational Wind Engineering (CWE) studies widely focus on wind pressure coefficients and wind flow field for the aerodynamic design of buildings, urban planning and dispersion studies. Nowadays, CWE had been widely used for identifying the critical wind locations on the field for warning the pedestrians as well as harnessing the wind. Hence, highly functional regions like universities are currently equipping micro wind turbines and roof-mounted wind turbines to meet their demands. This study focused on exploiting the wind flow conditions around the premises and wind force characteristics of the buildings situated at VIT Chennai premises for effective utilization of wind energy and determining the hot spots of unfavourable wind. By introducing CFD simulation earlier into the design process, it is easy to assess the wind resource conditions, pedestrian safety and comfort. However, the accuracy and reliability of CFD simulations can easily be compromised. For this reason, several sets of best practice guidelines have been developed in the past decades. Based on the best practices, this work presents the CFD simulation and framework for evaluating pedestrian comfort and possible areas to commission small wind turbines in the VIT campus. The simulations for assessing the wind flow conditions are carried out based on the *k-ε* realizable turbulent model with high-quality grid-based convergence. The preliminary results can ensure wind comfort, safety and wind resources with CFD and add to enriched wind environmental quality in living.
Keywords: *Computational Wind Engineering, Wind resource management, micro-scaled wind turbines, Pedestrian safety.*

1. Introduction

Urban planning and sustainability deal with the utilization of available natural resources in framing our use of land and built-in environment. Notably, universities and colleges implement the process in a full-fledged manner. Wind resource is one of the main stakeholders in power production and sustainable use of resources. Hence universities involve commissioning various micro-scaled, building-integrated wind turbines, small scale wind turbines to meet their electricity requirement to a certain extent. While concentrating on utilizing the wind resources, analyzing the appropriate wind conditions, considering the wind loads on buildings, and metrological data in the study field is essential. We need to trace out the precise locations where favourable conditions to harness wind energy are available. Ample researches have proven the reliability and efficiency of these computational studies. Hong et al. [1] Gaining valuable insights towards flow patterns around buildings, specifically in urban areas through CFD, was much facile than on-field testing. H. Montazeri et al. [2] demonstrated the accuracy and reliability of CFD studies over medium-rise buildings based on 3D RANS, involving multiple facades in the building architecture. Another influencing factor in the urban planning process is pedestrians' comfort and safety in the living environment. Since the location involved in this study has many high rise buildings, aligning with wind safety standards, it becomes necessary to analyze wind safety towards pedestrian comfort. The work focuses on developing a simulation framework to determine the best regions to utilize the wind resources and examine wind comfort and safety inside the VIT campus located in Chennai. Although evident studies prove the credibility of numerical simulations, still they remain questionable without validations. Consequently, our study aimed at providing detailed insights for further in-field testing and wind tunnel experiments. The simulation procedure is followed based on AIJ guidelines by (Tominaga et al. [3]) (Best Practise Guidelines for practical applications of CFD to pedestrian wind environment around buildings) developed by earlier researchers. In that way, the study can be graded as an accountable and standard one.

2. Methodology

The simulation procedure was carried out based on the standard wind assessment studies from the literature given by previous researchers in this field of study. Based on that, workflow is defined and are followed for the research. The process is initiated with the mesoscale inspection and CAD development of the entire campus. Then the necessary wind conditions are assessed, and suitable wind flow conditions are set up for the CFD analysis. The results are extracted based on best practice guidelines and are correlated with wind comfort standards to ensure the feasibility of deploying wind turbines inside the campus.

2.1 CAD model development

VIT University Chennai campus [refer Fig.1 (a)] is located in the south-west portion of Chennai city in Tamilnadu, India. It is situated about 10 km away from the east coast region. The premises consist of 15 buildings and divided into 3 clusters (as shown in Table 1). The highest building height is about 82 m (H_{max}) is shown in Fig.1(b). The whole premises is modelled using Solid Works 19 shown in Fig.1(c). The initial plan layout and orientation of buildings in the campus are extracted with the help of Google maps imported to the CAD platform and is scaled for the known dimensions. With the help of essential building constructional data, every building's topology (involving many approximations) is designed based on explicit modelling accordingly and altered to suit the needs of the computational study. Once the CAD data is finalized, the geometry details are mapped to suitable standard geometry interchange formats for further computational platforms.

a)VIT Chennai campus

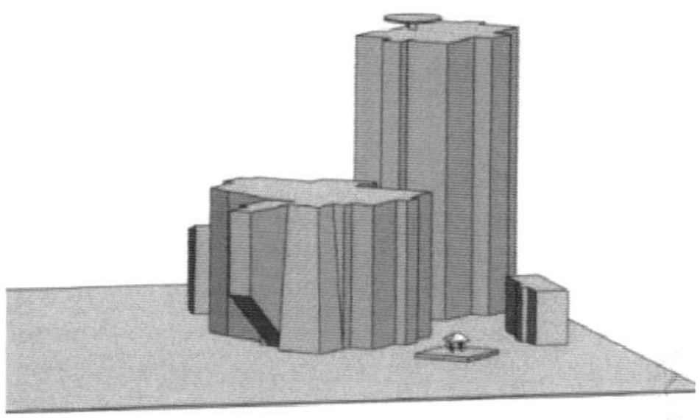

b)Developed CAD model

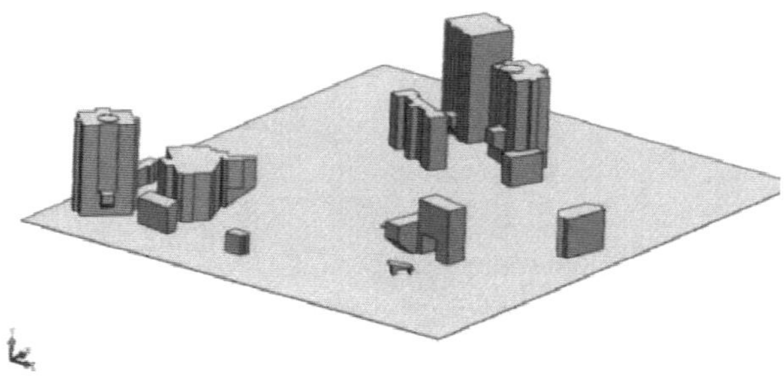

c)Full Mesoscale geometry of campus in Isometric View

Figure 1: Details of the campus constructions

IOP Conf. Series: Materials Science and Engineering 1128 (2021) 012002 doi:10.1088/1757-899X/1128/1/012002

2.2 CFD best practise guidelines

The numerical study is followed based on best practice guidelines developed by various researchers. The detailed studies performed in computational wind engineering and guidelines provided by institutes such as AIJ are the fundamentals for the simulation procedure discussed in the paper. The extensive wall function formulations presented by Tominaga et al. [3], Franke et al. [4], Blocken et al. [5] and Richards et al. [6] are utilized. This study is on the mesoscale; the domain must be vast enough to capture the flow patterns. Hence hexagonal domain modelling was highly recommended (Y. Toparlar et al. [7]).

- **Computational domain:** Over the past researches, CFD simulations involved computations in various shapes of domains such as circular, rectangular, pentagonal etc. Nevertheless, those domains are sufficient for a single model-based study, such as aerofoils, low-rise, tall, setback buildings (Rajasekarababu et al. [8]). This study is on the mesoscale; the domain must be vast enough to capture the flow patterns. Hence hexagonal domain modelling was highly recommended (Y. Toparlar et al. [7]).
- **The shape of the computational domain:** Since the study's geometry includes mesoscale models, it is necessary to simulate all possible wind conditions based on weather data. Ultimately it involves simulating the model for multiple wind flow conditions with different wind directions. Y. Toparlar et al. [7] adopted hexagonal domain modelling to analyze mesoscale models for micro-climate studies.
- **Size of the computational domain:** Our area of interest in this study, as mentioned earlier, comprises three clusters of building collections. They involve all types of building structures ranging from low rise, mid-rise to high elevation. Specific high rise structures also involve unconventional with purely contemporary design characteristics. Wind load parameters tend to be predicted, for which steady-state RANS simulations are undertaken. Franke et al. [4] defined the minimum size for the computational domain, which has to be based on the highest building height in the area of interest.
- **Discretization scheme:** Wind assessment-related computational studies involving a first-order discretization scheme resulted in dissipation errors, and hence to rectify those errors, it was recommended to follow appropriate second-order discretization schemes.

3. Pre-processing for the computational study

The pre-processing phase, during the simulation, goes through four important stages. Initially, the process starts with the selection of an appropriate computational domain. The hexagonal domain setup is adapted to suit the mesoscale assessments. Followed by the computational domain setup, the grid is generated with sufficient quality clubbed with grid independence tests. The later stages involve the choice of inflow boundary conditions and solver setups based on the wind characteristics associated with the campus location.

3.1 Computational domain

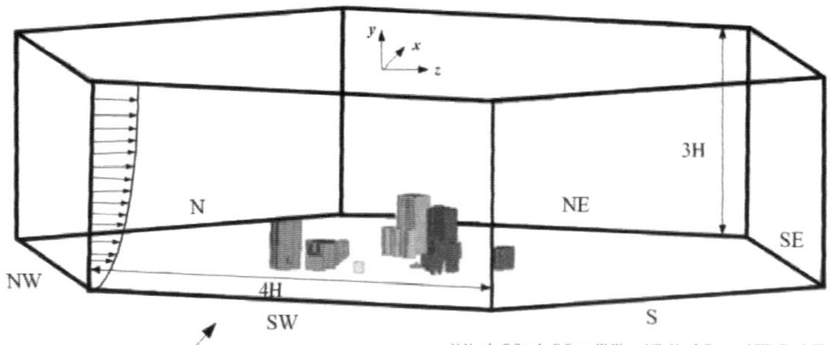

a) The model fitted inside the computational domain

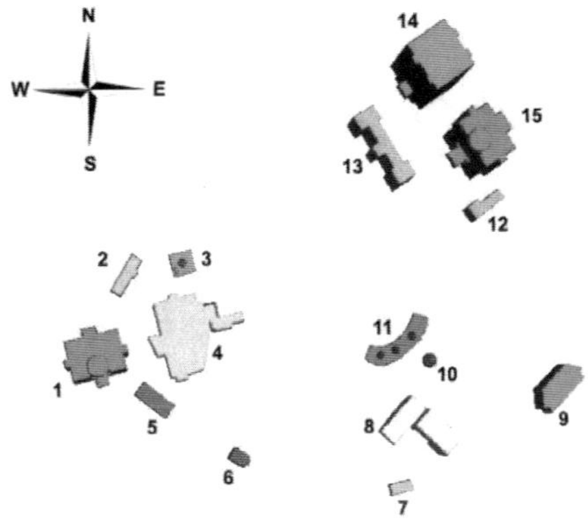

b) Top view of the geometries inside the domain

Figure 2: Computational domain setup

All the building structures inside the computational domain (which affect the flow potentially) are modelled, as shown in Fig.2 (a). Considering the buildings' position and orientation are classified into 15 numbers, shown in Fig.2 (b), and the buildings are categorized into three clusters, as mentioned in Table 1. The computational domain is shown in Fig.2. Since the study is highly sensitive in predicting pressure and velocity fields on and around the buildings, constructing a flow vicinity produces reliable quality (Revuz, Julia & Hargreaves., 2012[9]). The simulations are done with six different wind directions; suitable upstream and downstream distance is fixed (to meet the criteria as per best practice guidelines) as mentioned earlier since the proper downstream length must be maintained to aid in accurate flow prediction wake region. Also, these parameters influence the convergence, and grid sensitivity analysis is studied. The domain is quantified with the tallest building height: 14 in cluster C as H_{Max}. Both upstream & downstream of the domain measured to 8H and height of the domain is 3H,

3.2 Computational Grid

The computational grid is constructed with a high-resolution adaptive mesh technique, where the locations of adverse gradients are to be monitored near the buildings. This will lead to increased computational efficiency without compromising the accuracy of the study. The grid comprises only tetrahedral cells with prism layers generated in areas of interest, which can be seen in Fig.3. Following the best practice guidelines, grid sensitivity analysis is being performed with minimum tetrahedral elements. Following that, the grid is subjected to further refinement with a medium and fine level mesh resolution. A grid independence study is undergone for the mean pressure distribution around one of the tallest building, number 13, and the C_p plot results are provided in Fig.4. The plot exhibits a reliable result prediction when the mesh quality transits from medium quality to finer one. Hence the utmost fine quality meshing is adopted for the study and further wind direction simulations.

Table 1: Buildings situated inside the university

They are building no.	Cluster	Name	H_{max} (m)
1	A	Boy's Hostel, A block	69
2		Campus Bank	13.5
3		North Square	6.30
4		Academic Block 1	36
5		Gym Khanna	22.50
6		Guest House	15
7	B	Campus main gate	8.36
8		Admin block & Library	41
9		Auditorium	30
10		Sit-out	6.10
11		Gazebos	4.79
12	C	Health Centre block	22.50
13		Academic block 2	40
14		Boy's Hostel, C block	82
15		Girl's Hostel	69

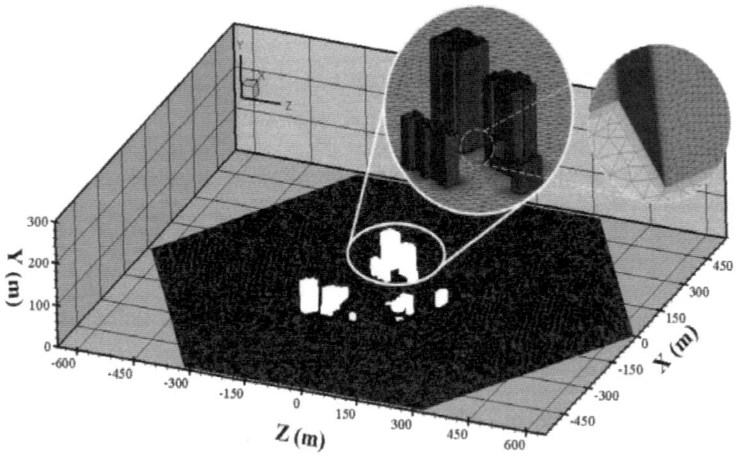

Figure 3: Computational Grid

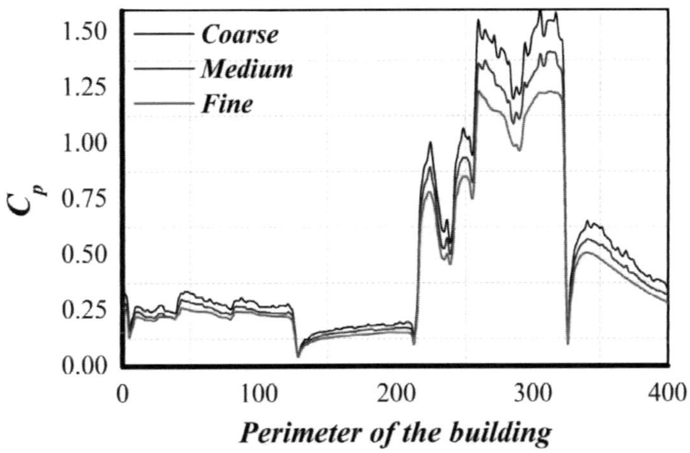

Figure 4: Grid Independence Study

3.3 Inflow boundary conditions

The wind flows are simulated under the open terrain condition. Simulated mean wind and turbulence profiles are validated from previous literature (Rajasekarababu, K.B et al., 2019 [10]). The profiles are shown in Fig.5. The boundary conditions in Table 3 are defined using the User Defined Function (UDF) in FLUENT 19.2. The open terrain velocity profile follows the logarithmic law and turbulence properties such as turbulence intensity (I), turbulent kinetic energy (k) and turbulent dissipation rate (ε) in the inlet. These equations are used to execute inflow boundary conditions in this simulation.

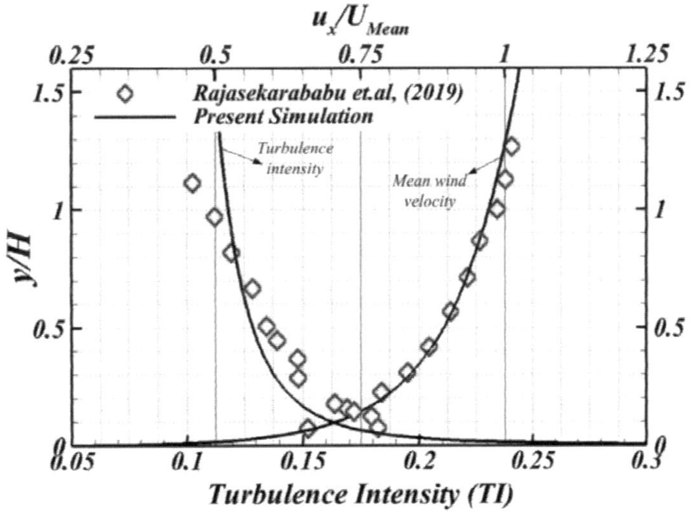

Figure 5: Turbulence Intensity and mean wind velocity at inlet

Table 2: In-flow Boundary Conditions

Inflow boundary (Inlet)	$(U_y) = \frac{u_{ABL}*}{\kappa} \ln\left(\frac{y+y_o}{y_o}\right)$ $\varepsilon(y) = \frac{u*^3}{\kappa(y+y_0)}$ $I(y) = \frac{1}{\ln\left(\frac{y}{y_0}\right)}$
Outlet	$\frac{\partial}{\partial x}(U,V,W,k,\omega)=0$ (pressure outlet)
Sky (top)	$U = U_{ABL}, k = k_{ABL}, \omega = \omega_{ABL}, W = 0, \frac{\partial}{\partial x}(U,V,W,k,\omega)=0$
Wall (sidewall)	$V = 0, \frac{\partial}{\partial x}(U,V,W,k,\omega)=0$
Ground wall	$Ks = \frac{9.793 y_0}{C_s}$

3.4 Solver Settings

The commercial CFD code ANSYS Fluent 19.2 is used to perform the isothermal simulations. The 3D steady RANS equations are solved in combination with the Realizable *k-ε* Model. The SIMPLE algorithm is used for pressure-velocity coupling, pressure interpolation is second order, and second-order discretization schemes are used for convection terms and the viscous terms of the governing equations. Convergence is assumed to be obtained when all the scaled residuals level off and reach a minimum of 10^{-6} for *x, y* momentum, 10^{-5} for *z* momentum and 10^{-4} for*k, ε* and continuity. As also observed by Rajasekarababu et al., [11 and 12], the simulations show oscillatory convergence.

4. Results And Discussion

The CFD simulations of the *3D RANS (k - ε) model* are obtained for six wind directions: *South, South, West, South East, North, North West* and *North East.* To assess the wind loads affecting the buildings, mean C_p values from six wind directions are shown in Fig.6. From the mean pressure coefficient, the tall buildings on the campus experienced high positive pressures for all six wind directions. Adverse effects of suction pressures are noticed towards the leeward side of these buildings. This occurs because of upwind stagnations of inlet wind and resulting recirculation followed by wake production in those areas. Due to the cascading effect of the high wind loads, the zones amidst comprising the cluster A and C (refer table-1) confronts high amounts of suction $-1.2 < C_p < -0.6$. These zones are not advisable to install small scale wind turbines. On observing the tall buildings' rooftops, (Building no: 1, 14 and 15) due to wind shear high amount of turbulence in wind flow are characterized. Hence installing roof-mounted wind turbines will not feasible in those locations, as it could affect the smooth function of the turbine and affect its performance. In contrast with the tall buildings, the roofs of medium-sized buildings inside the campus were not experienced by such adverse effects as tall buildings. Hence the medium-sized buildings and the surrounding locality proves to be permissible for the SWT installations.

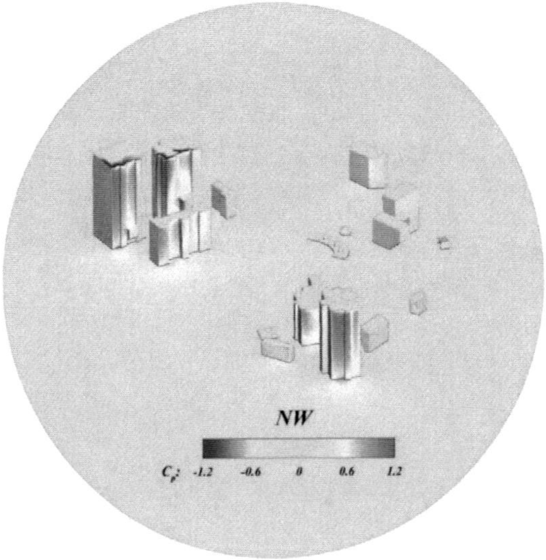
NW
C_p: -1.2 -0.6 0 0.6 1.2

(a)

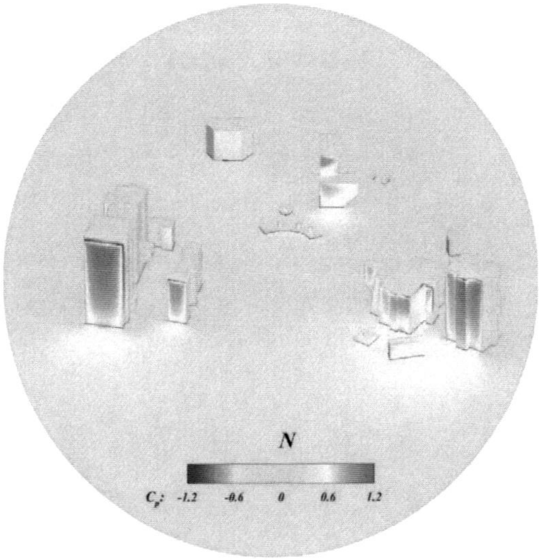
N
C_p: -1.2 -0.6 0 0.6 1.2

(b)

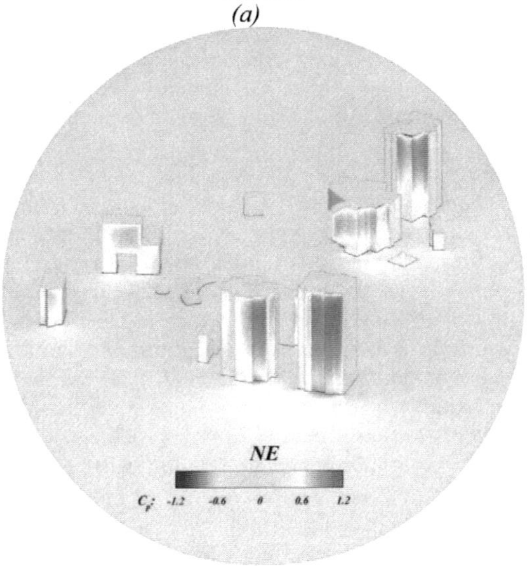
NE
C_p: -1.2 -0.6 0 0.6 1.2

(c)

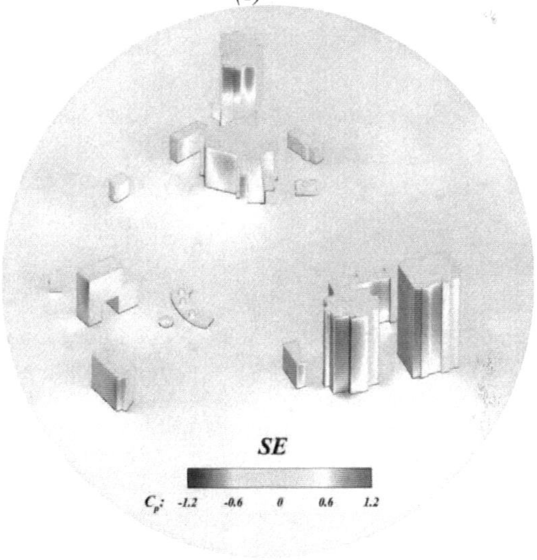
SE
C_p: -1.2 -0.6 0 0.6 1.2

(d)

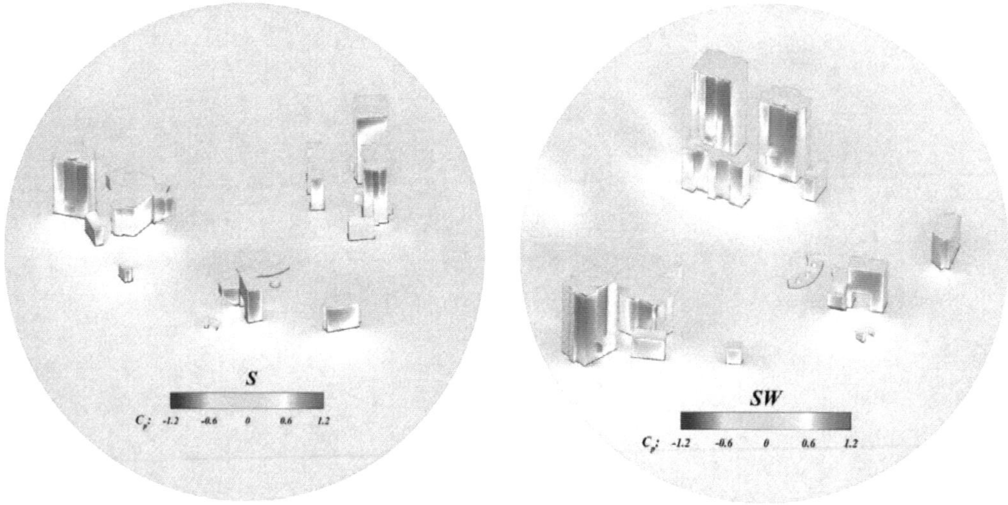

(e) (f)

Figure 6: C_p contours for all six wind directions *(a)* NW, *(b)* N, *(c)* NE, *(d)* SE, *(e)* S and *(f)* SW

4.1 Streamline velocity contours

In the view of harvesting the wind energy around the buildings, it is necessary to access the flow patterns in all possible areas since previous studies showed that the shape and structure of the surrounding buildings directly affect the performance of the wind turbines (Calautit, Katrina & Aquino [13]). The main idea of capturing these wind patterns is to assess wind velocity utilization by amplifying the flow due to building arrays. Hong Zhou, Yujie Lu [1] analyzed the wind acceleration due to low – rise buildings in improving wind turbines' operation. They infer that building shapes and orientations play a vital role in accelerating the wind flow. Therefore it is our due focus to determine the possibilities to increase the wind turbines' operating efficiency. The mean velocity of 7m /s, an average wind speed in the university's location, is set at a suitable height of 10m. The patterns for six wind directions are shown in Fig.7. From the streamline patterns, we can understand that wind speeds surrounding the building structures are low, between 2.6 m/s to 5.2 m/s. Also, in certain zones inside the campus, the building's shape had significantly improved the wind speed, around 7 to 7.8 m/s. Approximately 40% of the increase in wind speed is observed in those regions (see Fig.7). Fig.7 (a) & (b) shows, the wind speed increase is noticed in the east direction. Accordingly, the wind accelerations have developed due to the venturi effect between building 14 and 15 towards its leeward side.

Similarly, wind amplification in the passage contained in building 2. Due to wind shear along the building walls, certain zones experienced wind accelerations as well. In Fig.7 (c), Wind shear alongside building no. 13, leading to an increase in wind speed and the space between buildings no.13, 14 and 15 can be observed. Similar phenomena happen in building no.13, also as shown in Fig.7 (d).

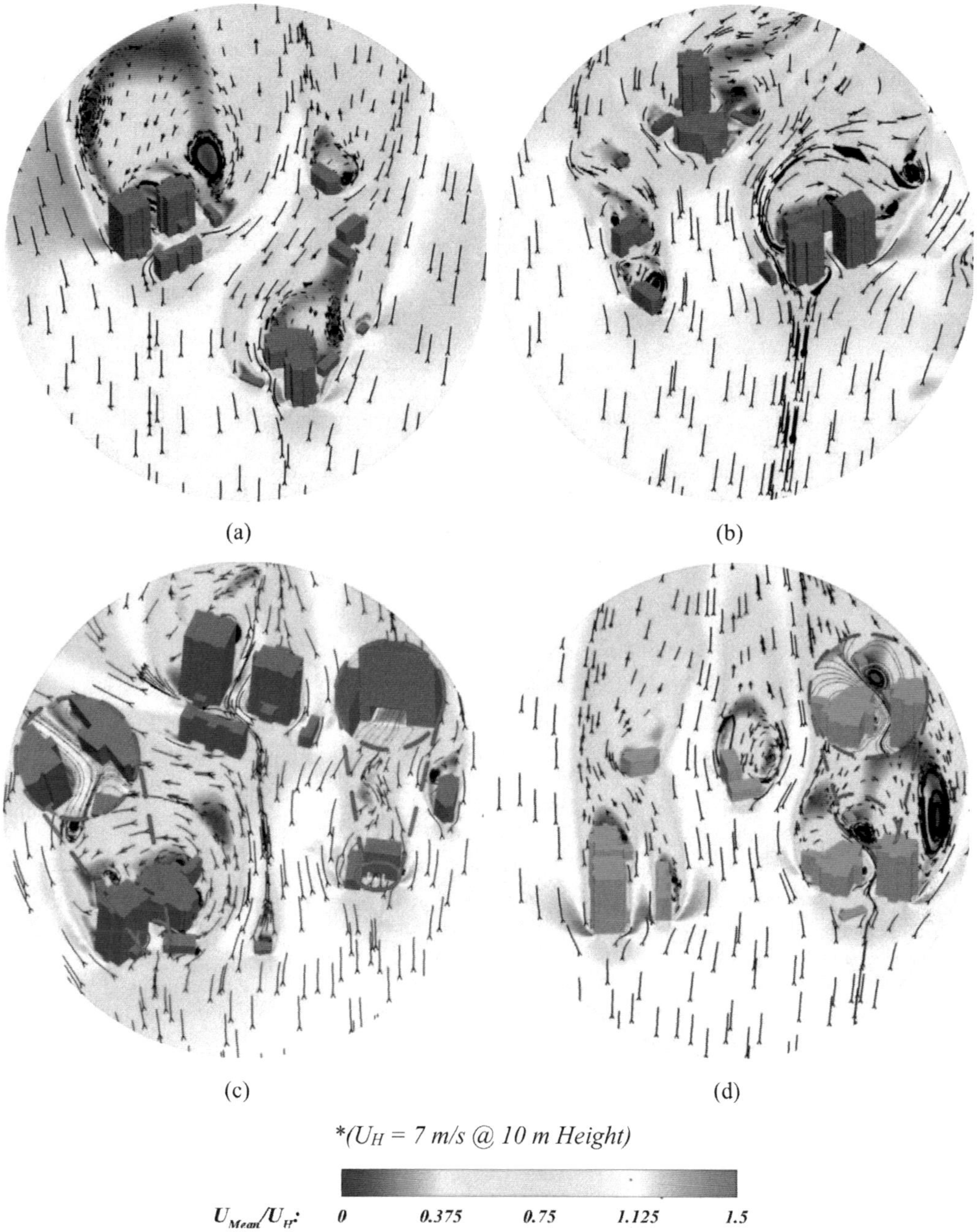

Figure 7: Streamline velocity contours for wind directions (a) SE (b) NE (c) NW and (d) N

4.2 Wind Comfort and safety factors

Wind comfort and safety factors are assessed in this study. Wind comfort is a function of many parameters such as weather, temperature, humidity and predominantly wind speed. Simulations for the mean inlet wind speed for one direction is not satisfactory, precisely for wind assessments. Hence, with weather data, the wind assessment for six different wind directions, as discussed in the previous section. To illustrate the process, the wind patterns are captured from the simulation, extracted at the height of 10m above the ground level. Examining the velocity streamlines, adverse velocity gradients are developed in accommodating the building structures' corner. Therefore, three critical regions were deduced, which are potentially unsafe for pedestrians under extreme wind conditions are situated inside the campus perimeter, as shown in Fig.8. Locations include 1) The pathway behind the building no. 4A and 3A, 2) Passage at the centre of building no.8B and 3) Area behind building no.13C (See Table.1 for building clusters and numbers). These locations are presented in the following figures comparing with the corresponding C_p value in those locations. The figures explain the venturi effect in critical areas due to pressure drop and wind amplifications.

The Dutch wind nuisance standard formulated the wind comfort and danger criteria comprising pedestrians' activities in certain wind conditions. Following the NEN8100 [14] standard, they are emphasized from the following table [4]. After a long computational study, the wind regions in the campus concerning the wind nuisance standard formulations, pedestrian comfort areas are sorted out (marked in red lines) in Fig.9. The traced out regions comply according to the quality classes A & B of the NEN8100 [14]. Pathways, sit-outs, gardens can be set out in these areas since it is entirely safe even in adverse weather conditions.

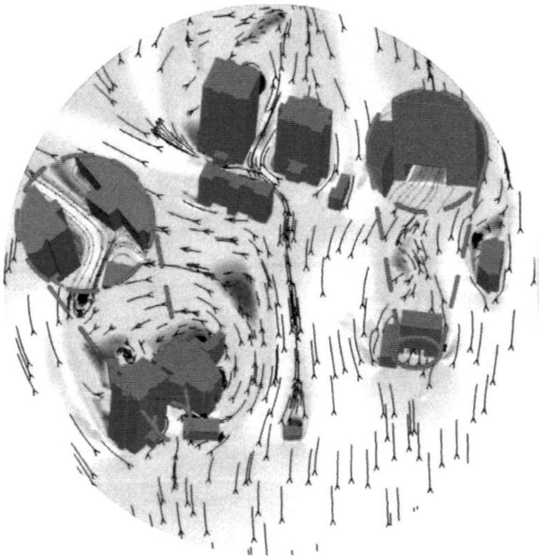

(a)

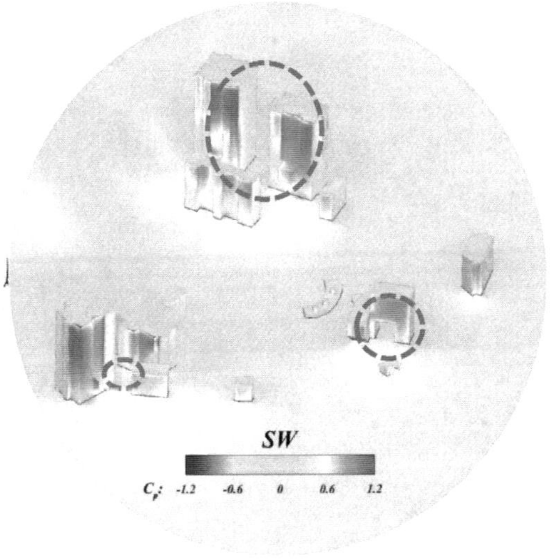

(b)

Figure 8: Critical areas in adverse wind conditions (a) Streamlines (b) C_p

Table 3: Dutch wind comfort criterion

PU_{THR}>5 m/s	Quality class	Activity		
		Traversing	Strolling	Sitting
<2.5	A	Good	Good	Good
2.5- 5.0	B	Good	Good	Moderate
5.0-10	C	Good	Moderate	Poor
10-20	D	Moderate	Poor	Poor
>20	E	Poor	Poor	Poor

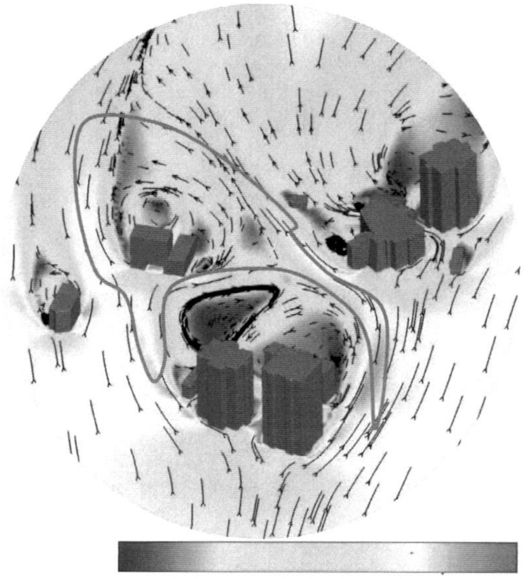

Figure 9: Favorable regions for wind comfort & safety

5. Conclusion:

CFD simulations performed using 3D RANS equations along with k-ε Realizable turbulence model for the mesoscale, explicitly modelled campus geometry. Following the best practice guidelines developed for complex environment studies, this study accessed the campus wind conditions for six wind directions. Turbulent parameters, along with mean velocities, are accessed for each simulation cases, and critical zones in the campus are charted out. Following the NIWE & IEC regulations, suitable locations for SWTs are traced out from the CFD study results. Comparing the intensities from all the cases, the adjacent zones to building no.11B, the pathway from building no.9B to building no.12C are more suitable for harnessing wind energies that receive nominal/acceptable levels of turbulence (≃ 5%) suited for SWT installation. To ensure safe transit and comfortable living conditions, wind comfort analysis is performed. Based on the obtained conclusions, three critical pedestrian discomfort locations are emphasized, where they are alarmingly dangerous in extreme wind conditions. About the NEN criteria, the best regions for public transit are derived out.

Nomenclature:

- ASME - American Society of mechanical engineers
- AIJ – Architectural Institute of Japan
- AIAA - American Institute of Aeronautics and Astronautics
- AOI – Angle of incidence
- BPG - Best practise guidelines
- NAFEMS - National Agency for Finite Element Methods and Standards
- NIWE – National Institute of Wind Energy
- RANS - Reynolds's Averaged NavierStoke's Equation
- SWT – Small Wind Turbines.
- TKE – Turbulence Kinetic Energy

References:

[1] Zhou, Hong, Yujie Lu, Xiaodong Liu, Ruidong Chang, and Bo Wang. "Harvesting wind energy in low-rise residential buildings: Design and optimization of building forms." *Journal of Cleaner Production* 167 (2017): 306-316.

[2] Montazeri, Hamid, and Bert Blocken. "CFD simulation of wind-induced pressure coefficients on buildings with and without balconies: validation and sensitivity analysis." *Building and Environment* 60 (2013): 137-149.

[3] Tominaga, Yoshihide, Akashi Mochida, Ryuichiro Yoshie, Hiroto Kataoka, Tsuyoshi Nozu, Masaru Yoshikawa, and Taichi Shirasawa. "AIJ guidelines for practical applications of CFD to pedestrian wind environment around buildings." *Journal of wind engineering and industrial aerodynamics* 96, no. 10-11 (2008): 1749-1761.

[4] Franke, J., and A. Hellsten. "Schl? nzen, H., Carissimo, B.(Eds.). Best practice guideline for the CFD simulation of flows in the urban environment. COST Office Brussels." (2007).

[5] Blocken, B., Stathopoulos, T. and Carmeliet, J., 2007. CFD simulation of the atmospheric boundary layer: wall function problems. Atmospheric environment, 41(2), pp.238-252.

[6] Richards, P. J., and S. E. Norris. "Appropriate boundary conditions for computational wind engineering models revisited." Journal of Wind Engineering and Industrial Aerodynamics 99, no. 4 (2011): 257-266.

[7] Toparlar, Yasin, Bert Blocken, P. V. Vos, G. J. F. Van Heijst, W. D. Janssen, Twan van Hooff, Hamid Montazeri, and H. J. P. Timmermans. "CFD simulation and validation of urban microclimate: A case study for Bergpolder Zuid, Rotterdam." *Building and environment* 83 (2015): 79-90.

[8] Rajasekarababu, K. B., G. Vinayagamurthy, and S. Selvi Rajan. "Experimental and computational investigation of outdoor wind flow around a setback building." In *Building Simulation*, vol. 12, no. 5, pp. 891-904. Tsinghua University Press, 2019.

[9] Revuz, J., D. M. Hargreaves, and J. S. Owen. "On the domain size for the steady-state CFD modelling of a tall building." *Wind and structures* 15, no. 4 (2012): 313.

[10] Rajasekarababu, K. B., and G. Vinayagamurthy. "Experimental and computational simulation of an open terrain wind flow around a setback building using hybrid turbulence models." *J. Appl. Fluid Mech* 12 (2019): 145-154.

[11] Rajasekarababu, K. B., and G. Vinayagamurthy. "Computational Simulation of Wind Flow Behavior Around a Building Structure." In *Innovative Design, Analysis and Development Practices in Aerospace and Automotive Engineering (I-DAD 2018)*, pp. 325-331. Springer, Singapore, 2019.

[12] Rajasekarababu, K. B., and G. Vinayagamurthy. "CFD validation of wind pressure distribution on a tall building under the influence of upstream terrain." *Progress in Computational Fluid Dynamics, an International Journal* 20, no. 5 (2020): 284-298.

[13] Calautit, Katrina, Angelo Aquino, John Kaiser Calautit, Payam Nejat, Fatemeh Jomehzadeh, and Ben Richard Hughes. "A review of numerical modelling of multi-scale wind turbines and their environment" *Computation* 6, no. 1 (2018): 24.

[14] Willemsen, Eddy, and Jacob A. Wisse. "Design for wind comfort in The Netherlands: Procedures, criteria and open research issues." Journal of Wind Engineering and Industrial Aerodynamics 95, no. 9-11 (2007): 1541-1550.

IOP Conf. Series: Materials Science and Engineering 1128 (2021) 012003 doi:10.1088/1757-899X/1128/1/012003

Controlling virus droplets diffusion in an isolated room using CFD

N.V.Chandra Shekar,B.Abhinay, C.Raj Pavan,R.Harish*
School of Mechanical Engineering, Vellore Institute of Technology, Chennai, Tamil Nadu-600127, India.
*Corresponding author Email: harish.r@vit.ac.in

Abstract. This paper presents an isolation room model with a patient carrying acute air born disease with an effective ventilation system being functioned. Different models have been simulated with different vent position and the results have been observed. With the unruly rise in the spread of the COVID-19, it is paramount to restrain the virus propagation within the hospital premises. The purpose of the research is to control the virus dissemination caused by sneezing, by altering the position of the ventilation i.e. air inlet and outlet, by using flow across the room to direct it towards the outlet with maintaining a negative pressure. The negative strain helps in confining the air-borne transmission of the deadly virus from spreading across the room and not letting them permeate outside the isolated region. An isolation room model has been studied using computational fluid dynamics, by setting up a discrete phase model by using injection spray modelling to observe the permeation of the virus droplets. The behaviour of these aerosol droplets are studied using simple-semi implicit method for the equation associated with pressure by specifying the droplet size.By altering inlet and outlet locations we are able to minimise the spread of these harmful droplets by using the flow from the air inlet to go against the diffusing droplets. These paper aides well for a sudden isolation room setup anywhere with peruse dimensions of the inlet and outlet height at most optimum position.

1. Introduction

Over the last decade there have been raised concerns on the air borne diseases spread across the globe. Various trials have been carried out by profuse organisations in successfully containing it. An isolation room is one such major aspect, while allowing the sick to develop antibodies and not allowing it to transmit to others. An isolation room of a hospital has exceptional conditions for successful containment of the contagion. Conditions like position of air vents, pressure variations, direction of flow of air inside the room are among required for effective containment. Through years it has been observed how the doctors or the nurses who have regular visit to the infected even with proper safety measures followed end up having got a virus contact. This ends up worse with contact spreading throughout the hospital premises and the worst case is the hospital turning out to be virus spread hotspot. The reason being that they when enter the patients room,they are exposed to air present in the room,the infected coughs/sneezes which in turn the virus which in form of droplets diffuses in the room and makes its presence throughout the room. With the increased awareness of the consistency of indoor air and its association with ACMV(air-conditioning and mechanical ventilation) system with proper design. The knowledge and need for the air change per hour become necessary and a major improvement in containing the air-borne transmissions.Then with the advent of the isolation room major tests were conducted resulting fruitful omission of the spread of the transmission was achieved. By directing the air flow-across the room the spread can be swayed and ensure safety. Negative pressure addition plays a key role in ensuring safety environment. The room is generally given negative strain in order to trap the air flowing out of the room via doors or windows. The negative pressure can be created using vents and with the help of the air flow we can prevent the spread of the contagion. Negative pressure prevents movement of air outside the isolatedenvironment. Computational fluid dynamics (CFD) is a field of science based on the solving of numerical problems, most commonly incompressible, viscous flow, including multi component gas mixtures. It is very powerful and authentic tool for investigating the variable airflow patterns regarding the ventilation across its inlet and outlet including variable flow patterns and behaviours. It is considered highly cost effective, explicit way to do analysis. To solve the equations we need some boundary conditions which are vital no-slip boundary condition, outlet, inlet, periodic boundary condition.This paper evaluates the air flow patterns through various setups of ventilation arrangements, finding out the appropriate ventilation arrangement for minimizing of virus droplet diffusion.Using computational fluid dynamics (CFD), we observe in detail where an infected person coughs in an isolated room and how it will diffuses. In our simulations, the minimization of virus droplets is achieved by using infection spray modelling,

To our knowledge, the studying of virus diffusion has never done through infection spray modelling. Many researchers investigated the similar issue and literature stated in their respective papers that are relevant to this study has been mentioned. H.Qian and Y.Li [1] explained that the existing ventilation system design based on the Centres for Disease Control and Prevention (CDC) is ineffective in eliminating neither fine nor large respiratory related particles. Thus, they built an isolation chamber with three separate-type exhaust system designs which supplying downwards.The results showed that exhausts at ceiling-level are more systematic at eliminating large amount of air particles situated in an isolated space.Kao and Yang [2] constructed cough model in order to perform arithmetic simulations related to virus droplets diffusion in an isolated room for various types of ventilation arrangements, not only the analysis related to droplet resulted but also the time of dilution of virus propagation for different-type ventilation arrangement.The analysis results proven to be airflow pattern with parallel directional ventilation setup are the most productive method for restraining virus propagation.

Adamset al. [3] have properly designed an Airborne infectious isolation room (AIIR) with anteroom-equipped at varying differential pressures under the conditions of chosen no supplier movement and also with high supplier movement for studying containment effectiveness of AIIR.Results shown that containment effectiveness with increasing negative pressure differential has resulted in increased effectiveness and with rising supplier movement the effectiveness decreased.Shelly et al. [4] has designed a negative pressure isolation ward with 30 bed capacity introduced with varying ventilation system. The flow introduced has been characterized throughout the room the pressure difference is observed constantly at each and every location inside the ward. The results showed that the room is slightly becoming neutral or positively pressurized when the safety personnel or other staff enters the room, this design has become effective in increasing the capacity in the hospital.

Bhattacharyya et al. [5]evaluation stated that the efficacy of conditioned air emitted from air-conditioning systems to be combined with aerosol sanitizer was examined in order to enter every section of the isolated space and destroy corona-virus. Contemplating factors influencing aerosol sanitizer transporting structures, such as flow dynamics, temperature and turbulent kinetic energy, analysis was done. The results of the analysis show that high turbulent fields in an isolated space are an effective way to distribute sanitizer to eliminate corona-virus. Cheong [7] examined in an isolated room containing negative pressure the patterns of airflow and pollutant diffusion by using CFD modelling based on the implementation of three ventilation systems. The results show that outlet at low level location have better pollutant removal efficiency than at the ceiling level outlet. The study also shows that ventilation techniques and arrangement of furniture have a significant influence on the patterns of airflow and pollutant scattering in an isolated room.

Julian et al. [8]explained the impact of air flow with different types of doors that are used in negative pressure isolated rooms. Single hinged, double hinged, single slide and double slide doors where examined and observed that the double hinged doors while opened have high risk factors of getting the leakage of airflow into or out of the space and sliding door have better efficiency in maintain the required airflow rate. Tung et al. [9] worked on the model of mock-up isolated room with different ventilation arrangement rates and also with pressure differentials of negative values for implementing better ventilation efficiency. Quantitative analysis of ventilation effect is based on two types of indexes namely EI and QI.The numerous experimental [10-13] and analytical models [14-16] developed to estimate the mass flow rates of bidirectional air exchanges in open enclosures [17-19] and rooms.Results obtained that a ventilation system arrangement with a negative pressure differential of value -15.0 Pa in an isolated chamber gives the better possible ventilation efficacy to extract the contaminants out of the isolated system.

Table 1. Different parameters with respect to other researchers.

Researcher name	Boundary conditions and other parameters
H.Qian and Y.Li [1]	Supply ventilation rate = 12ACH Exhaust rate = 13.2ACH Emission particles air flow rate = 6.1/min Exhale temperature rates = 32 ± 5°C Mode particle size = 80 μm
Kao and Yang [2]	Cough model conditions = Time ≤ 0.5s, Velocity=8m/s and 0.5s< Time ≤1s,Velocity= 10e(−7t)m/s Constant inlet velocity flow = 0.4m/s Outlet pressure = -8Pa Room differential pressures = 2.5 to 30Pa
Bhattacharyya et al. [5]	Inlet velocity at ceiling = 3.91m/s Inlet temperature at ceiling = 24°C Mass flow rate = 29.97kg/s Inlet velocity of sanitizing machine = 1.5m/s Inlet temperature of sanitizing machine = 30°C Mass flow rate of sanitizing machine = 1.854kg/s Isolation room measurements = 9.144m x 6.096m x 3.658m
Cheong [7]	Temperature at air supplied =18°C Total air exchange rate = 29.9ACH Temperature of all other walls =24°C Emission source sphere diameter = 38mm Emission species flow rate from source = 0.631/min Mass flow rate at extract = 62.6 E-6 kg/s. Isolation room measurements =3.35 m x 4.8m x 2.5m Air-gap underneath the door = 0.9m x 0.012m
Tung et al. [9]	Room ventilation supply = 24ACH Volume flow rate of air ranges as 85 to 3400m³/hr Room differential pressures = 0 to 62.5Pa Isolation room measurements = 2.4m x 2.4m x 2.4m

2. Modelling

A full-Scale isolation room present in most of the hospitals is considered. Using Solid works software and ANSYS Space claim we had built a CAD model which includes patient lying on a bed and considering a room with 9ft x 9ft x 8ft. To make sure the effective control of virus containment in an isolated room, we had considered a room with a rectangular inlet and a squared outlet and no other openings so that the air flow can be kept in a stable state. The inlet is a natural vent where we positioned it in such a way that ventilation air directly sweeps the patient's head and trunk regions, and moves to the ventilation outlet and the outlet was placed on the wall right above the patient's head. The inner surface of the patient's human mouth is differentiated by the inlet-mouth boundary condition, because this surface is assumed as the reference boundary of discrete phase corona virus release.

Table 2. Dimensions of CAD model description.

S.NO	Part Name	X Direction Length in m	Y Direction Length in m	Z Direction Length in m
1.	Room	1.97m	3m	3.3m
2.	Rectangular inlet vent	1.4m	0.25m	0.1m
3.	Square outlet vent	0.75m	0.5m	0.1m
4.	Bed	0.96m	1.05m	1.9m

5.	Human body	0.59m	0.07m	1.4m

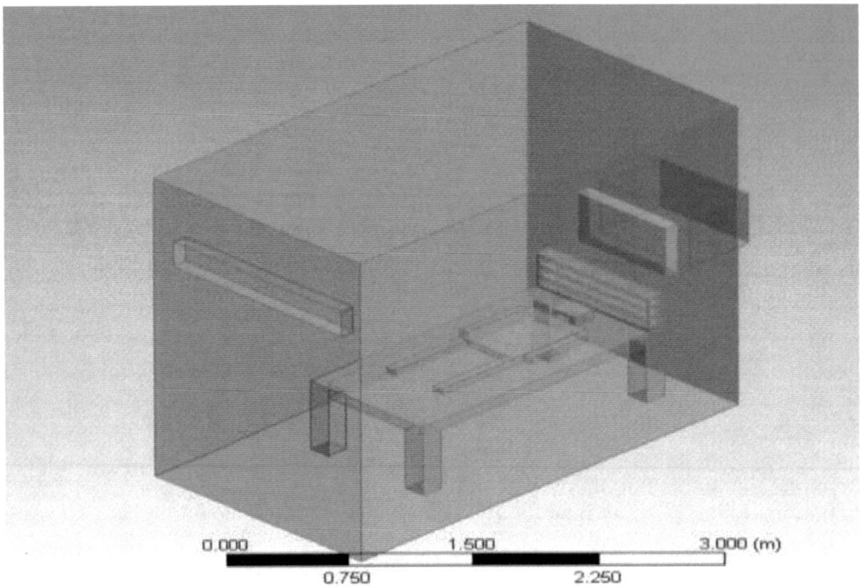

Figure 1. Cad model representing patient with room.

3. Meshing

Meshing has been carried out through Ansys workbench and used Fluent as the solver preference. Geometry clean up has been carried out and checked for free edges, scar lines, duplicate surfaces, small fillets and holes and intersection of parts. The element size has been set to 0.03m and the mesh type is unstructured. We have used edge sizing and face sizing for the human body and patient's bed to attain the good mesh quality (in terms of skewness and orthogonal quality) and for less computational time. The element number is 458943.

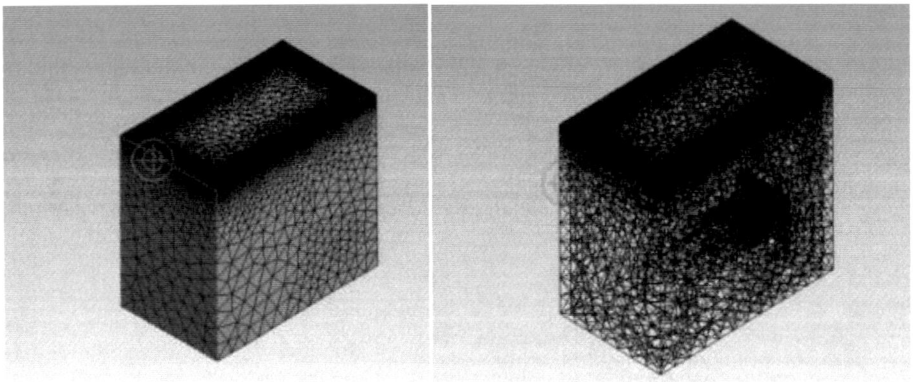

Figure 2. Meshing of the room in ansys workbench.

4. Analysis & methodology

We had performed a pressure-based solver. For particle tracking related to discrete phase over time we had made simulation as a transient. The gravity effect on the fluid is given as 9.81m.s-2. We had used viscous k-epsilon model with RNG and near wall treatment as standard wall function. The discrete phase model (DPM) is used, since this model allows us to study a mass of particles discretely in a continuous fluid space. The wet particles of the corona virus secreted from the patient's mouth are considered as a discrete phase and the air flow transmitted through the elevator ventilation valves is considered as a continuous phase inside the interior of the room. The physical models of discrete particles defined in this simulation include two-way turbulence coupling meaning the two-way interaction between continuous and discrete phase. Stochastic collision is activated which means irregular droplets collide with each other. Coalescence is activated means droplets merge with each other, and breakup is activated for collapse of the droplets. The type of discrete phase behaviour will be time dependent and with a time step of 0.001 s by activating the unsteady particle tracking mode. After activating the discrete phase model, the injection process is defined, which determines the type and quality of discrete particles injected into the model. We had defined injection particles as droplet; thus, water is defined as droplets and water vapour is defined as an evaporating gas species. The injection is performed superficially and through the inner Surface of the patient's mouth. According to the definition of injection, human cough virus particles are physically expelled from the patient's mouth by water droplets that are evaporating in space. These virus droplets have a temperature of 310 K, a velocity of 31.85 m.s-1, and a mass flow of 0.018 kg.s-1, which are emitted at intervals of 0s to 0.1s. The particle diameter of the virus is not constant during propagation and the rosin-rammler-logarithmic distribution method is considered for the size of the diameters. We had activated the species transport to enable the drop mode. The boundary conditions of the discrete phase model are defined as particles at the boundaries of the patient's mouth and the inlet and outlet of the airflow with Escape mode, for the particles movement through these boundaries. For the boundary of the human body the Reflect mode is applied for the reflection of particles that collide with this boundary, and at the boundary of the side walls of the elevator cabin, the Trap mode is used, which enables that particles are trapped and accumulate in this boundary. Also the inlet ventilation is such that fresh air flow enters the room as a continuous fluid with a speed equal to 2ms-1 and a temperature equal to 291K.The air flow exits from the outlet to the outside environment at a pressure equivalent to atmospheric pressure. We had considered that the air entering the room has a oxygen mass fraction equal to 0.18. The simulation process is performed with a time step size equal to 0.0025s.

Table 3. Boundary conditions used in fluent analysis.

MODELS		
Viscous		K-epsilon
	K-epsilon model	RNG
	Near-wall treatment	Standard wall function
Species model		Species transport
	Number of volumetric species	3(H_2O,O_2,N_2)
Discrete phase model		on
	Interaction	Interaction with continuous phase
	Particle treatment	Unsteady particle tracking
	Physical models	Two way turbulence coupling Stochastic collision Coalescence breakup
Injection		Active
	Injection type	Droplet
	Release from surfaces	Inlet-nose
	Material	Water liquid
	Evaporating species	H_2O
	Diameter distribution	Rosin-rommler-logarithmic
	Point properties	Temperature – 310K Velocity – 31.85ms^{-1} Total flow rate – 0.018 kg.s^{-1}

Energy		on
Boundary conditions		
Floor		Wall
	Wall motion	Stationary wall
	Heat flux	0 $W.m^{-2}$
	Discrete phase conditions	reflect
Walls		wall
	Wall motion	Stationary wall
	Heat flux	0 $W.m^{-2}$
	Discrete phase conditions	trap
Inlet-Air		Velocity inlet
	Velocity magnitude	2 ms^{-1}
	Temperature	291K
	Discrete phase conditioning	Escape
	H_2O mass fraction	0
	O_2 mass fraction	0.18
Outlet-Air		Pressure outlet
	Gauge pressure	0 Pascal
	Discrete phase conditions	escape
Inlet-Nose		Wall
	Wall motion	
	Heat flux	0 $W.m^{-2}$
	Discrete phase condition	escape
Human body		Wall
	Wall motion	
	Heat flux	0 $W.m^{-2}$
	Discrete phase condition	reflect

5. Results and discussion

The stop-convergence criterion was selected to be 1*10^-4 with independent of the results obtained to the mesh density less than 5% contrast to the reference mesh taken. The analysis has been carried out at constant pressure assuming the room at negative pressure. Using ANSYS Fluent post processing vectors, contours and stream line flow across the room has been observed. It has been observed that the flow from the inlet are directed towards to the outlet placed at back of the patients head and the sneeze droplets from the human body are directed directly towards the outlet without affecting the air present in the room. Results of the room temperature have been observed. Figure 3 indicates the streamline flow patterns inside the isolated room and flow patterns originating from the inlet and human nose. Figure 4 indicates the wall shear distribution and maximum values are seen distributed on the upper and lower walls of the room. The time variation of the local inlet velocity, shows spikes whenever the velocity has abrupt changes. Additionally, these spikes are in phase and have the same sign as the wall shear stress. This means that wall shear stress variation during the transient event can be partially described when sudden changes of mean velocity are observed. Figure 5 and 6 shows the streamline pattern and temperature distribution inside the isolated room and it is observed that the temperature distribution is higher near the doorways.

Figure 3. Stream line flow from inlet and human nose.

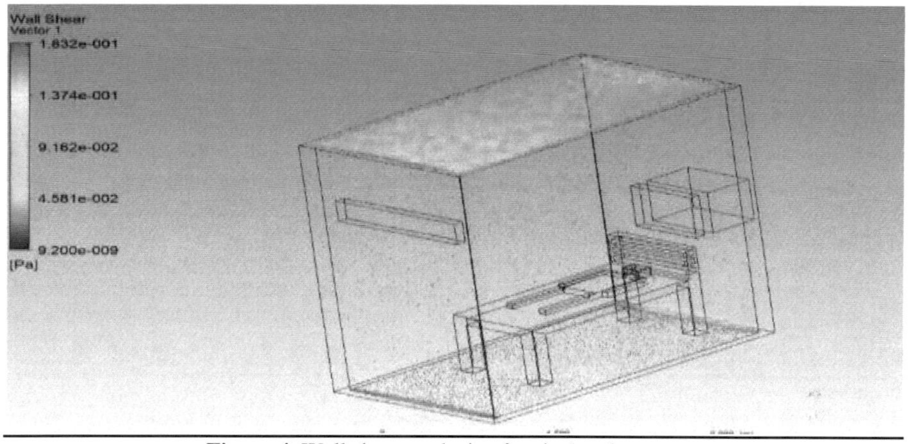

Figure 4. Wall shear analysis of an isolated room

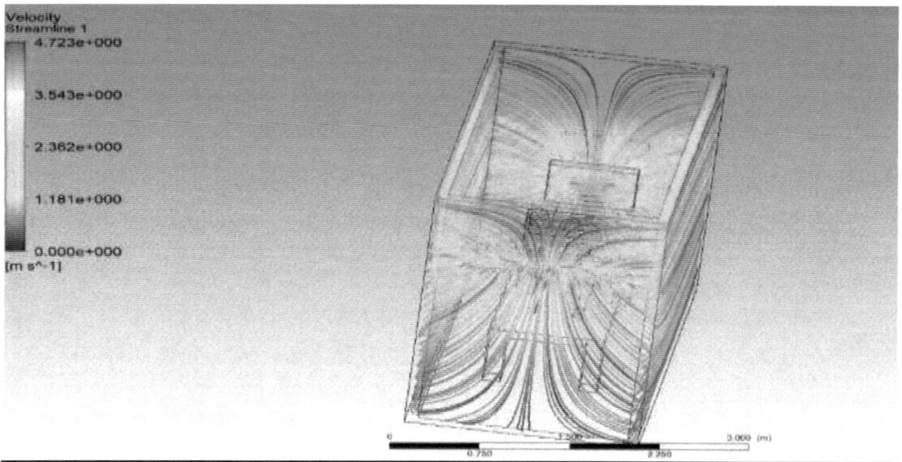

Figure 5. Streamline flow from the inlet directing towards the outlet.

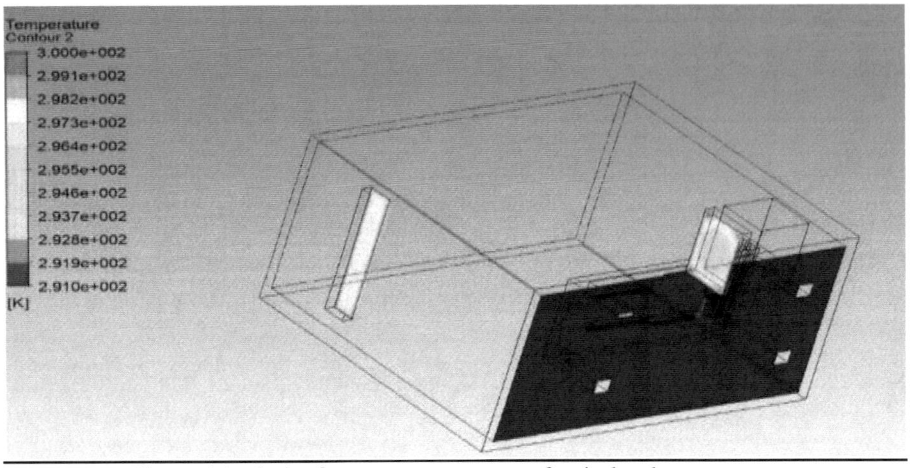

Figure 6. Analysis of temperature contour of an isolated room.

6. Conclusions

This paper investigates an isolation room model with a patient carrying acute air born disease with a effective ventilation system being functioned. It was found that the SST k-ω model gave the minimum error comparing with the wall shear stress simulated from other turbulent models, i.e. STD k-ε, RNG k-ε and Realizable k-ε models. Different models have been simulated with different vent position and the results have been studied. The purpose of the research is to contain the spread of the virus caused by sneezing, by altering the position of the ventilation i.e. air inlet and outlet, by using flow across the room to direct it towards the outlet with maintaining a negative pressure. The present study identified that the negative strain helps in confining the air-borne transmission of the deadly virus from spreading across the room and not letting them permeate outside the isolated region. It was also observed that by altering the position of the inlet and outlet the spread and diffusion of harmful droplets were significantly reduced.

References

1) Kao PH and Yang R J2006 Virus diffusion in isolation rooms*Journal of Hospital Infection*.**62(3)** 338-345.

2) Adams NJ, Johnson D L and Lynch R A 2011 The effect of pressure differential and care provider movement on airborne infectious isolation room containment effectiveness*Am J Infect Control*. **39** 91-7.

3) Cheong K W DandPhua S Y2006 Development of ventilation design strategy for effective removal of pollutant in the isolation room of a hospital *Building and Environment*. **41** 1161-70.

4) Shelly L Miller PhD, Nicholas Clements PhD and Steven A Elliott 2017 Implementing a Negative Pressure Isolation Ward for a Surge in Airborne Infectious Patients *AJIC Journal*.**45** P652-659.

5) Yun-Chun Tung, Shih-Cheng Hu , Tung-I Tsai and I-Lang Chang 2009 An experimental study on ventilation efficiency of isolation room *Building and Environment*.**44** 271-79.

6) Qian H and Li Y 2010 Removal of exhaled particles by ventilation and deposition in a multibed airborne infection isolation room *John Wiley & Sons A/S*.**4** 284-97.

7) Yang-Cheng Shih, Cheng-Chi Chiu and Oscar Wang 2007 Dynamic airflow simulation within an isolation room *Building and Environment*.**42** 3194–209.

8) Eames, Shoaib D,Klettner C A and Taban V 2009 Movement of airborne contaminants in a hospital isolation room *J. R. Soc. Interface*. **6** S757–766.

9) Suvanjan Bhattacharyya, Kunal Dey, Akshoy Ranjan Paul and Ranjib Biswas 2020 Oct A novel CFD analysis to minimize the spread of COVID-19 virus in hospital isolation room *Elsevier Public Health Emergency Collection - Chaos Solitons Fractals.* **139** 110294.

10) Harish. R, Venkatasubbaiah. K, Mathematical modeling and computation of fire induced turbulent flow in partial enclosures, Applied Mathematical Modelling, vol.37, pp: 9732-9746, 2013.

11) Harish. R, Venkatasubbaiah. K, Transport phenomena of turbulent fire spread through compartment connected to vertical shaft in tall building, Fire Safety Journal, vol.61, pp: 160-174, 2013.

12) Harish. R, Venkatasubbaiah. K,Effects of buoyancy induced roof ventilation systems for smoke removal in tunnel fires, Tunnelling and Underground Space Technology, vol.42, pp: 195-205, 2014.

13) Harish. R, Venkatasubbaiah. K, Numerical investigation of instability patterns and nonlinear buoyant exchange flow between enclosures by variable density approach, Computers & Fluids,vol.96, pp: 276-287, 2014.

14) Harish. R, Venkatasubbaiah. K,Large Eddy Simulation of thermal plume behavior in horizontally partitioned dual enclosure, Building Simulation, vol.8(2), pp: 137-148, 2015.

15) Harish. R, Venkatasubbaiah. K, Numerical study of water spray interaction with fire plume in dual chambers connected to tall shaft, Fire Safety Journal, vol. 74, pp: 1-10, 2015.

16) Harish. R, Venkatasubbaiah. K, Non-Boussinesq approach for turbulent buoyant flows in enclosure with horizontal vent and forced inlet port, Applied Mathematical Modelling vol. 40, pp: 927-941, 2016.

17) Harish. R, Venkatasubbaiah. K, Numerical simulation of turbulent plume spread in ceiling vented enclosure, European Journal of Mechanics-B/Fluids, vol.42,pp:142-158, 2013.

18) Harish. R,Effect of heat source aspect ratio on turbulent thermal stratification in a naturally ventilated enclosure,Building and Environment,143,pp:473-486, 2018.

19) Harish. R, Buoyancy driven turbulent plume induced by protruding heat source in vented enclosure,International Journal of Mechanical Sciences,148,pp:209-222, 2018.

Computational study of Corona Virus diffusion in a closed environment

S.Harish, G.N.Nihaarikha, R.Harish*

School of Mechanical Engineering, Vellore Institute of Technology, Chennai, Tamil Nadu 600127, India
*Email address: harish.r@vit.ac.in

Abstract : The biggest challenge that is faced by the human race after World war 2 is the Covid-19, which has affected more than 27.8 million people. The Covid 19 pandemic has entirely changed the livelihood of the people. It is not only seen as a global health crisis but also a Socio-economic issue that has disrupted the Governments worldwide. There is situation where normality is being questioned in the post pandemic era. Many regulations such as Quarantine, self-isolation, social-distancing, and travel restrictions have been advised to prevent the spread of covid-19. After several months of quarantine, the world is now slowly pacing its way back to normality as the vaccine has been invented and brought into use. It is expected that the educational institutions, work places and other public arenas will be opened for the use. However, the regulation of wearing mask and following social distancing are mandatory in public places. Considering the above situation as a need of the hour, this article mainly focuses on the Computational study of the Corona Virus diffusion for a confined space such that it can be applied to places with high population mass. The numerical model was simulated using ANSYS fluent and the results obtained are in excellent accordance with the literature.

Keywords: Covid 19; CFD Simulation; virus diffusion

1. Introduction

The world is now witnessing a global pandemic which has led to the fall of many economies. Throughout the entire world, schools, colleges, factories, and all other public areas have been ghosted due to this outbreak. Many countries are now going through the stage of recovery and people are moving towards following their normal routines with precautionary measures. These measures majorly include the acts of wearing a mask and following social distancing. Many studies have been conducted since the onset of the outbreak regarding the morphology the virus, its spreading pattern, and the preventive measures to be undertaken to avoid spread and infection. One such study analyzed the flow pattern in an isolation room, a room where the virus affected people are isolated in a hospital. It was concluded that the most effective way of controlling the spread of droplets (containing the virus) is by implementing a parallel directional airflow pattern in the isolation rooms.[1] Another study was conducted experimentally and numerically to analyze and understand the physics behind the human sneeze. The sneeze action was modeled numerically to study the droplet motion and its flow behavior. The experiment gave a realistic data which was similar to the numerical data obtained from the CFD model which helped in understanding the sneeze dynamics.[2] Another study was conducted to analyze the respiratory droplet transmission in an office room with only two people inside the room. Distinct types of ventilation strategies like personal ventilation, mixing ventilation and displacement ventilation were examined in this study. It was concluded that the droplet behavior can vary if the sneeze or cough velocity is very high in the case of personal ventilation.[3] This study opened doors for many ideas on the ventilation systems that can be used in our model. The research done by Suvanjan Bhattacharyyaa et al.[4] investigated the effective mixing of air from an air-conditioner mixed with an aerosol sanitizer in an isolation room. The airflow pattern was analyzed such that the sanitizer droplets that are sprayed could reach all the corners of the isolation room. This will prevent the infection of people who work in the isolation rooms such as nurses, doctors, health care workers, etc. The behavior of cough model was developed by Particle Image Velocimetry (PIV) and Hot Wire Anemometry experiments conducted in different conditions. The study by Dudalski et al.[5] assessed the transmission of airborne viruses using bio-aerosol sampling through a distance of 1m, which was later validated with CFD results. This research inferred the quantification of air movements caused from coughs to determine the safe distance that has to be maintained between two people to mitigate viral infection. One of the researches was, analyzing the droplet fallout and diffusion or dilution time for a bacterium in the coughed region. The transient analysis simulation provided the path and direction of the particle and its concentration was traced to check the formation zones of the pathogen [6]..Another study stated that the most effective way to prevent pathogen contagion is to cater proper ventilation. Ventilation systems should be designed such that it will fulfill the needs of the respective environment which will be best way in diluting the pathogen concentration.[7] . One another research by Ji-Xiang Wang et al.[8] simulated the sneezing process to analyze the virus carrier flow pattern and life span once it is released from patient's mouth. The mobility of the pathogen and the cross-infection probability were evaluated with the parameters, total maximum time, and overall particle concentration, when a proper outlet facility is provided, the overall particle concentration can be reduced to 95%and the total maximum time by 60% .Another study found that displacement ventilation in a room lowered the chances of exposure to virus from a normal respiration activity. But the probability of cross-infection was high when there was an action of sneeze or cough in a situation when two persons stand facing each other[9].The studies which were published recently in the domain of Covid virus transmission were mostly in agreement to the '2m distance' requirement to prevent transmission of virus from one person to another. But the current information shows that the virus particle containing droplets can channel through air to another person who is at a distance greater than 2m. Masks are an effective way to prevent the spread. In addition to wearing a mask, maintaining social distance of 2m would create a fall in transmission rates. When an infected person is not wearing a mask, there is a high chance of spread to people who are within a radius of up to 10m [10]. There are also numerous experimental [11-13] and analytical models [14-17] developed to estimate the mass flow rates of bidirectional air exchanges in open enclosures [18-20] and rooms.

This study applies the concepts of computational fluid dynamics to analyze motion and behavior of velocity streamlines and to calculate the distance which people must maintain to avoid the spread of the virus considering the ventilation conditions. Two classrooms are modeled using SOLIDWORKS. The dimensions of first room are taken as 5m x 8m x 3.5m and that of second is taken as 10m x 11.5m x 7m. ANSYS Fluent software was used to create a cough and sneeze model and to evaluate the results.A steady state, 3-D turbulent, multi phase flow using volume of fluids was generated.The plots were created using Tecplot software using the data obtained from Fluent. The findings of this study can be applied to various public places such as school and college classrooms, offices, and many other closed room environments.

Table1: Thermophysical properties of multiphase materials

Materials	ρ [kg/m^3]	μ [kg/m-s]	Cp[KJ/kg K]	k[W/m-K]
Air	1.225	0.000017	1	0.024
Water	998.2	0.001003	4.182	0.6

2. Methodology

2.1. Geometry description

A practical and real-life example of an educational institution (School class room) is taken for the numerical study. The rooms are in the shape of cuboid with dimensions 5m x 8m x 3.5m and 10m x 11.5m x 7m. Room 1 consists of 6 students, 1 faculty, an entrance and exhaust whereas Room 2 consists of 24 students, 4 fans, 2 entrances and 3 windows. Two locations in room 2 are considered for this study, location 1 is opposite to entrance and location 2 is adjacent to entrance. A numerical simulation is carried out separately for room1 and room 2 to analyze the virus diffusion when people cough or sneezed by considering factors such as ventilation and results are obtained.

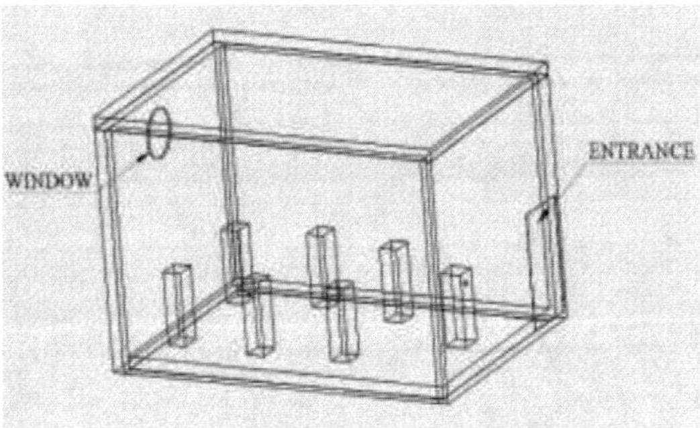

Figure1: Schematic diagram of classroom 1

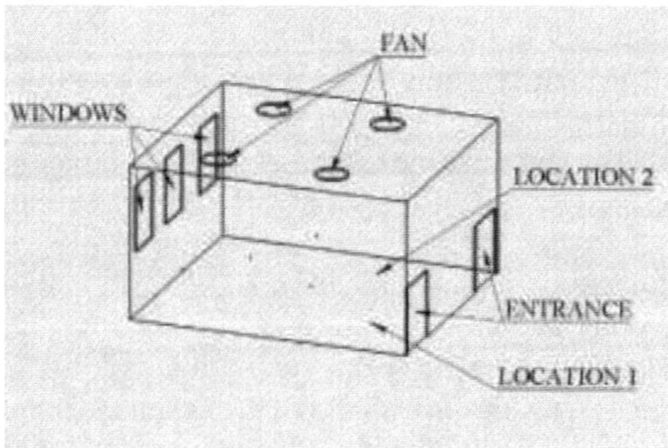

Figure 2: Schematic diagram of classroom 2

2.2. Initial and boundary conditions

A pressure based steady state analysis was carried out by neglecting the effect of gravity. The multi phase, flow- turbulence models are turned on during the iterations. Air is chosen as primary phase and water as secondary phase. Turbulence intensity is taken to be 5% with viscosity ratio as 10 for the turbulence model. The solution initialization is hybrid with time step size 0.01 and the reference frame is taken to be 'relative to cell zone.' Table 2 and 3 denotes the various initial and boundary conditions of two rooms used to run the simulation.

For room1 the no-slip boundary condition was given to walls with occupant mouth,nose and room entrance were taken to be velocity-inlet and exhaust as pressure-outlet whereas occupant mouth and room entrance were given velocity-inlet, fans as fan-inlet and windows as exhaust-outlet for room2.

Table 2: Boundary conditions of Room 1

	No. of models	Type	Magnitude
Mouth-Cough	6	Velocity-inlet	1m/s
Nose-Sneeze	6	Velocity-inlet	2 m/s
Entrance	1	Velocity-inlet	3 m/s
Exhaust	1	Pressure-inlet	0

Table 3: Boundary conditions of Room 2

	No. of models	Type	Magnitude
Mouth-Cough	2	Velocity-inlet	2 m/s
Fans	4	fan-inlet	2 m/s
Entrance	2	Velocity-inlet	5 m/s
Window	3	exhaust-outlet	NA

3. Results and discussions

3.1. Results of room 1:

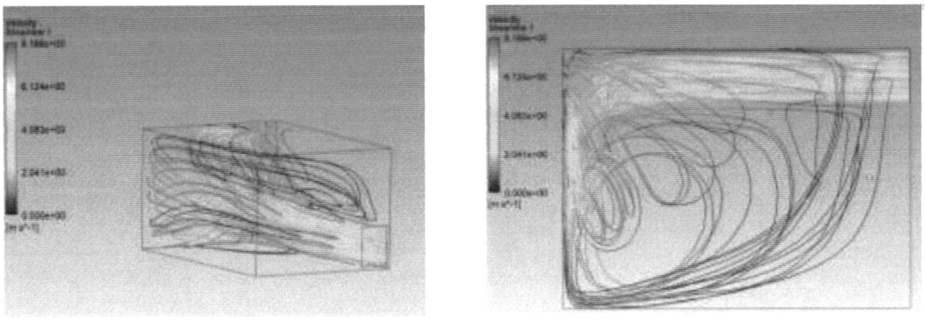

Figure 3:Velocity streamlines of the class entrance of room1

The above figures 3 represents the velocity flow from the entrance of the classroom. The trend seen is that the velocity decreases gradually and then increases as it gets closer to the exhaust.

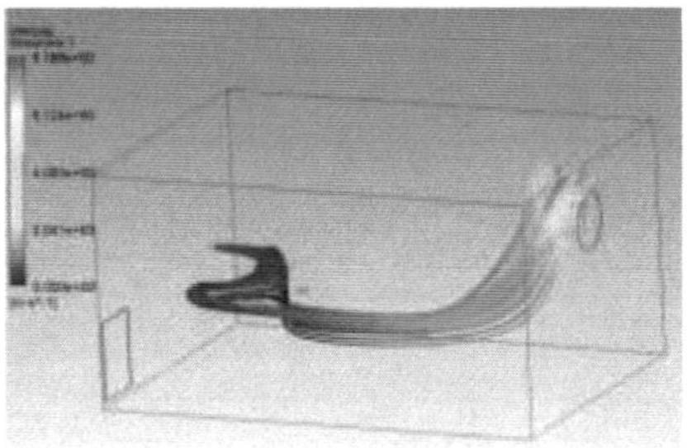

Figure 4: Velocity streamlines of cough

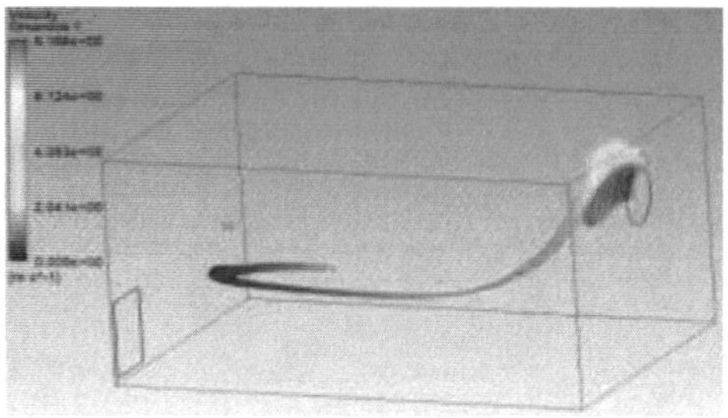

Figure 5: Velocity streamlines of sneeze

The cough is generated from mouth and sneeze from nose. Figure 5 shows the flow of velocity caused by the cough. The velocity streamlines from mouth hits the person who is at a distance of 2m. Figure 4 represents the velocity streamlines due to sneeze. It is seen that initially velocity decreases but as it crosses the midsection of the room the velocity increases since it approaches the exhaust.

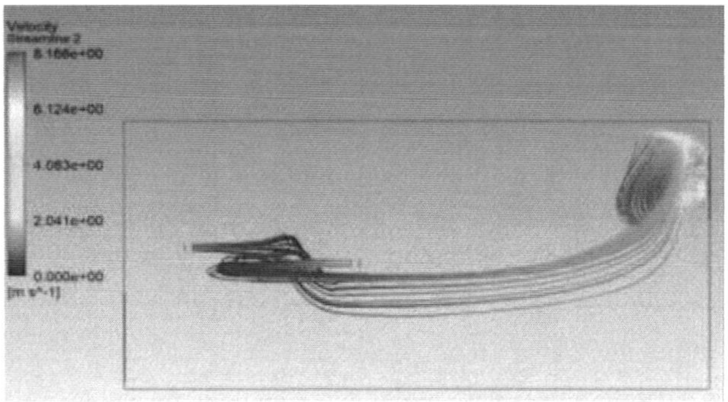

Figure 6: Velocity streamlines of cough and sneeze

Figure 5 and 6 shows the velocity streamlines of both cough and sneeze at the same time. It can be observed that the velocity flow pattern of cough and sneeze intersect each other when they are generated from opposite direction. It is also noted that streamlines of sneeze travels more distance with its initial value compared to that of cough before their velocities decreases because the initial velocity of sneeze is greater than that of cough.

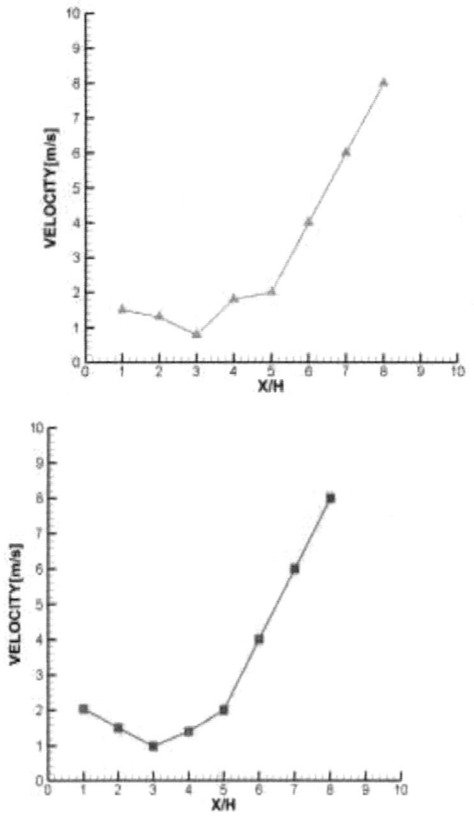

Figure 7: Plots of cough and sneeze

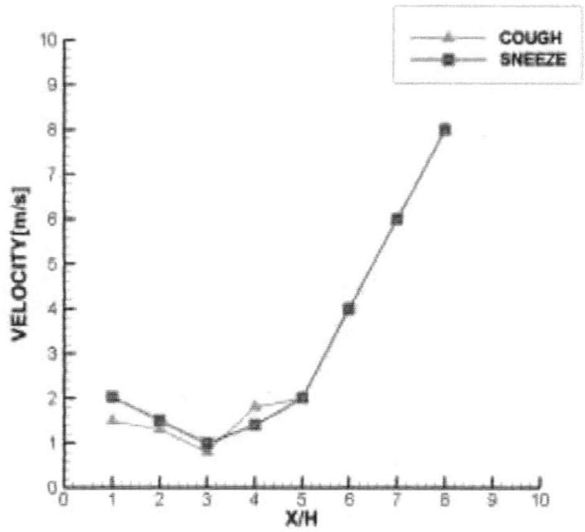

Figure 8: Comparison plots of cough and sneeze

Figure 7 and 8 shows the comparison plots between cough and sneeze models. It can be seen that sneeze begins with a higher velocity compared to cough. The velocity flow of both cough and sneeze decreases upto one-third of the room and then increases. The maximum velocity is observed in the vent region for both the models.

3.2. Results of Room 2:

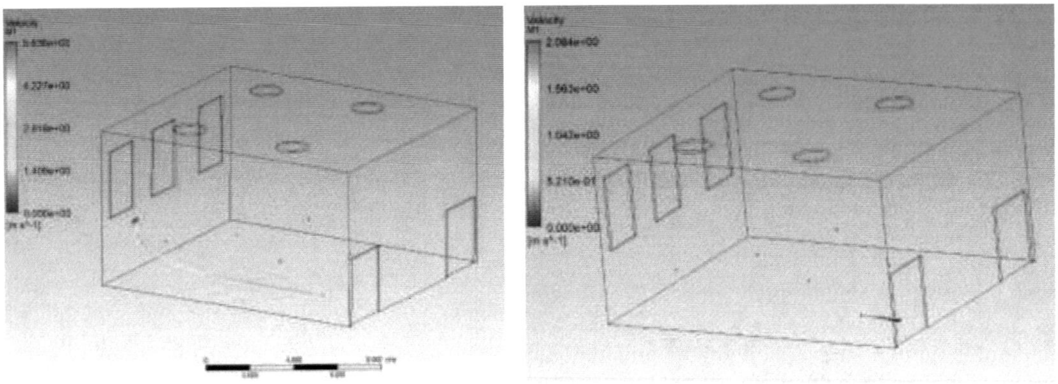

Figure 9: Cough streamlines at Location 1 with and without ventilation

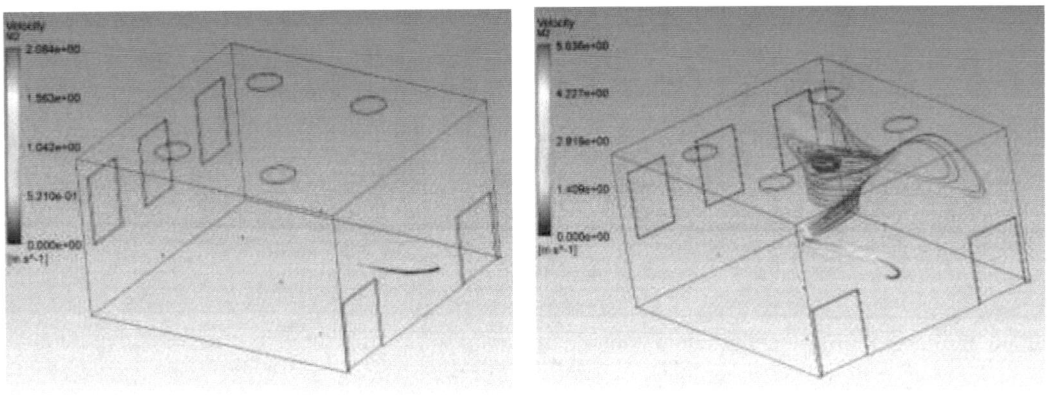

Figure 10: Cough streamlines at Location 2 with and without ventilation

The above figures represent cough streamlines in presence and absence of ventilation. Figures 9 represent the velocity streamlines in presence of ventilation while Figure 10 denotes velocity streamlines in absence of ventilation. It can be noticed that the streamlines travel in forward direction when there is no ventilation . In presence of ventilation, the streamlines travel in forward direction for a small distance and travels back in reverse direction due to opposing air velocity from the entrance, thereby approaching the exhaust. Location 2 in Figures 10 is not situated in front of the entrance unlike Location 1 in Figures 9. A similar trend in latter is seen as that of former in absence of ventilation. The velocity streamlines initially travel in forward direction but due to presence of ventilation source by its side (entrance) and the action of fan, it appears to be a swirl pattern.

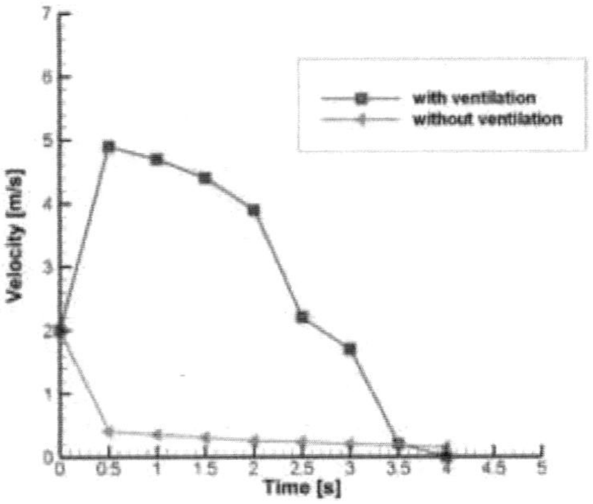

Figure 11: Comparison plots of location 1

Figure 11 represents the comparison plots of velocity vs time at location 1 which is in front of entrance. The velocity increases from its initial value when entrance and fans are given some velocity magnitude. It increases suddenly for a particular period due to the air flow from parallel direction (entrance) and then decreases gradually, whereas in case of zero ventilation velocity decreases gradually with time.

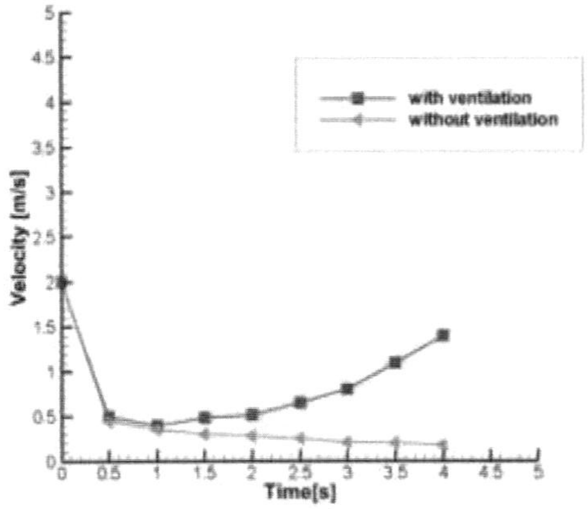

Figure 12: Comparison plots of location 2

The above figure represents the comparison of velocity with time in presence and absence of ventilation for location 2 which is not in front of entrance(ventilation source). It can be seen that the velocity decreases for both the cases for a specific time interval after which it increases gradually in presence of ventilation. In the absence of ventilation , the velocity decreases from its initial value as time increases which is similar to Figure 12 without ventilation. Results from room 2 indicate that the virus spreads faster and travel in all directions in presence of ventilation and is tabulated in Table 4.

Table 4: Results of room 2

S.NO	LOCATION	WITH VENTILATION	WITHOUT VENTILATION
1	Opposite to entrance	Increase by 59.18% Decrease by 96%	Decrease by 92.5%
2	Adjacent to entrance	Decrease by 76% Increase by 68.4%	Decrease by 92%

4. Conclusions

Numerical investigation is carried out to analyze the COVID virus diffusion in a classroom with fan, exhaust, teacher, and students. The streamlines of velocities are reported at various positions in the room. The results from room 1 indicates that all velocity particles of sneeze, cough and entrance first decreased as they move away from the inlets and then increased as they approach the exhaust. It is also seen that streamlines of sneeze travels more distance with its initial velocity value compared to that of cough before their velocities decreases. All the particles move towards exhaust or vent to equalize the inside and outside pressure i.e., to make the gauge pressure zero. It is also seen that velocity from entrance decreases gradually but velocities from nose and mouth decrease suddenly. The intersection of cough and sneeze velocity streamlines indicates that a minimum social distancing of 2m should be followed to prevent the spread of virus. In the presence of ventilation, the velocity increases and then decreases for location 1(opposite to entrance) whereas, it initially decreases and then increases for location 2 (adjacent to entrance). In absence of ventilation, the velocities at both the locations dropped from their initial values. The velocity increases instantaneously by 59.18 % and then drops by 96% at location 1 when there is ventilation, whereas it decreases by 92.5 % when there is no ventilation. For location 2, velocity decreases by 76% and then increases by 68.4 % in presence of ventilation, whereas it decreases by 92 % from its initial value in absence of ventilation. The total world is adversely affected due to COVID-19. As time passes people have adapted themselves to live with it. Many research studies are going on to eradicate the virus and to bring life back to normal. This paper concentrates on the velocity flow pattern of cough and sneeze in a confined room and insist the mankind to follow the precautionary measures such as social distancing, wearing masks etc. to fight against this deadly virus.

References

[1] Kao, P. H., & Yang, R. J. (2006). Virus diffusion in isolation rooms. Journal of Hospital Infection, 62(3), 338–345. doi:10.1016/j.jhin.2005.07.019.
[2] Busco, G., Yang, S. R., Seo, J., & Hassan, Y. A. (2020). Sneezing and asymptomatic virus transmission. Physics of Fluids, 32(7), 073309. doi:10.1063/5.0019090
[3] He, Q., Niu, J., Gao, N., Zhu, T., & Wu, J. (2011). CFD study of exhaled droplet transmission between occupants under different ventilation strategies in a typical office room. Building and Environment, 46(2), 397–408. doi:10.1016/j.buildenv.2010.08.003
[4] Bhattacharyya, S., Dey, K., Paul, A. R., & Biswas, R. (2020). A novel CFD analysis to minimize the spread of COVID-19 virus in hospital isolation room. Chaos, Solitons & Fractals, 139, 110294. doi:10.1016/j.chaos.2020.110294
[5]Dudalski, N., Mohamed, A., Mubareka, S., Bi, R., Zhang, C., & Savory, E. (2020). Experimental Investigation of Far Field Human Cough Airflows from Healthy and. Indoor Air. doi:10.1111/ina.12680
[6]Balocco, C. Hospital ventilation simulation for the study of potential exposure to contaminants. Build. Simul. **4,** 5–20 (2011). https://doi.org/10.1007/s12273-011-0019-6
[7]Peng, S., Chen, Q., & Liu, E. (2020). The role of computational fluid dynamics tools on investigation of pathogen transmission: Prevention and control. Science of The Total Environment, 142090. doi:10.1016/j.scitotenv.2020.142090
[8]Ji-Xiang Wang, Xiang Cao, Yong-Ping Chen, An air distribution optimization of hospital wards for minimizing cross-infection, Journal of Cleaner Production, Volume 279,2021,123431,ISSN 0959-6526,https://doi.org/10.1016/j.jclepro.2020.123431.
[9]Gao, N., & Niu, J. (2006). Transient CFD simulation of the respiration process and inter-person exposure assessment. Building and Environment, 41(9),1214–1222. doi:10.1016/j.buildenv.2005.05.014
[10]Setti, L., Passarini, F., De Gennaro, G., Barbieri, P., Perrone, M. G., Borelli, M., Miani, A. (2020). Airborne Transmission Route of COVID-19: Why 2 Meters/6 Feet of Inter-Personal Distance Could Not Be Enough. International Journal of Environmental Research and Public Health, 17(8), 2932. doi:10.3390/ijerph17082932
[11] Harish. R, Venkatasubbaiah. K(2013). Mathematical modeling and computation of fire induced turbulent flow in partial enclosures, Applied Mathematical Modelling, vol.37, pp: 9732-9746.https://doi.org/10.1016/j.apm.2013.05.011.
[12] Harish. R, Venkatasubbaiah. K(2013).Transport phenomena of turbulent fire spread through compartment connected to vertical shaft in tall building, Fire Safety Journal, vol.61, pp:160-174.https://doi.org/10.1016/j.firesaf.2013.09.005
[13] Harish. R, Venkatasubbaiah. K(2014) Effects of buoyancy induced roof ventilation systems for smoke removal in tunnel fires, Tunnelling and Underground Space Technology, vol.42, pp:195-205.https://doi.org/10.1142/9789814635165_0024
[14] Harish. R, Venkatasubbaiah. K(2014). Numerical investigation of instability patterns and nonlinear buoyant exchange flow between enclosures by variable density approach, Computers & Fluids,vol.96, pp: 276-287.https://doi.org/10.1016/j.compfluid.2014.03.026
[15] Harish. R, Venkatasubbaiah. K(2015). Large Eddy Simulation of thermal plume behavior in horizontally partitioned dual enclosure, Building Simulation, vol.8(2), pp: 137-148.https://doi.org/10.1007/s12273-014-0198-z

[16] Harish. R, Venkatasubbaiah. K(2015).Numerical study of water spray interaction with fire plume in dual chambers connected to tall shaft, Fire Safety Journal, vol. 74, pp: 1-10.https://doi.org/10.1016/j.firesaf.2015.03.007
[17] Harish. R, Venkatasubbaiah. K(2016). Non-Boussinesq approach for turbulent buoyant flows in enclosure with horizontal vent and forced inlet port, Applied Mathematical Modelling vol. 40,pp: 927-941.https://doi.org/10.1016/j.apm.2015.05.013
[18] Harish. R, Venkatasubbaiah. K(2013). Numerical simulation of turbulent plume spread in ceiling vented enclosure, European Journal of Mechanics-B/Fluids, vol.42,pp:142-158.https://doi.org/10.1016/j.euromechflu.2013.06.001
[19] Harish. R(2018).Effect of heat source aspect ratio on turbulent thermal stratification in a naturally ventilated enclosure, Building and Environment,143,pp:473-486.https://doi.org/10.1016/j.buildenv.2018.07.043
[20] Harish. R(2018). Buoyancy driven turbulent plume induced by protruding heat source in vented enclosure,International Journal of Mechanical Sciences,148,pp:209-222.https://doi.org/10.1016/j.ijmecsci.2018.09.001

Analyzing social distancing policy effectiveness using Computational Fluid Dynamics inside a bus to prevent COVID-19 airborne transmission

A.Pavansai, P.Deepak, S.Ragul Hari, R.Harish*, M.B. Shyam Kumar

School of Mechanical Engineering, Vellore Institute of Technology, Chennai, Tamil Nadu-600127, India
*Corresponding author E-mail: harish.r@vit.ac.in

Abstract:

Coronavirus disease (COVID-19) is a kind of disease that transmits from one body to another through air by the moist particles caused during sneezing and coughing. As a result, to reduce the virus spreading accordingly, 1.83 m of social distancing has been advised to be followed among the humans. Also, it is not well studied that whether the ambient wind and relative humidity (RH) will cause COVID-19 laden cough droplets to transport farther in the air, and make the current social distancing practice ineffective. In this study, computational fluid dynamics simulations are carried out to analyze the transient transport, condensation/evaporation, and deposition of COVID laden droplets emitted by coughs, with the different environmental conditions, cough velocities and Relative Humidity using the developed bus model of size 12 m x 3.5 m x 3.2 m. Different conditions of sneezing and coughing from the human's i.e., the laminar and the turbulent flows of the laden droplets in the air inside the bus were also examined. The distance between the two virtual humans is set as 6 m for a 12 m bus length in order to track the distance covered by cough particles. The facial covering effect on reducing the airborne transmission of the cough droplets has also been evaluated. It is found that due to the ambient air and humidity, the generation of secondary laden droplets occur which travels far and accumulates on the ground or any other third human being around and causes a strong potential risk to their health. The secondary droplets are transforming into large droplets due to high humidity and the hygroscopic effect is evaluated. The 6 feet social distancing is found to be ineffective for halting the spread of viruses among human beings because the micro-laden droplets caused during the sneezing and coughing are influenced by the convection effects and transport from one body to another within 5 s. It is thus recommended to wear masks for both infected and healthy humans to reduce the airborne cough droplets.

1.Introduction:

The outbreak of the COVID-19 virus creates a massive impact and destruction in all aspects of human well-being. Even the most powerful countries in the world suffer to tackle the effects of the pandemic as there is no proper medication officially released. Governing bodies and United Nations have introduced different measures to protect humans from being prone to the virus, using facemask and sanitizer has been an essential form of living [1]. Different protective measures for individuals like social distancing between humans are being the new normal [2]. A tentative distance of 1.8 m from each other is being advised, to avoid being contagious to viral infection. So, it creates a tough environment for industries working with huge manpower to function with their routine practices. The transportation sector, tourism, and many other sectors are being restricted and affected by the social distancing policy. As people couldn't afford to be idle for a longer run, they are bound to work for their survival despite the pandemic. This paper will focus on analyzing the effectiveness of the social distancing policy by carrying out Computational Fluid Dynamics (CFD) simulation on a bus model. A detailed study has been carried out to track the path and the distance traveled by the airborne cough droplets. A transportation model of the bus has been considered, primarily a bluff body design of the bus is created and also placing a victim of COVID-19 virus inside the closed body[3]. By considering different scenarios such as bus being stagnant or moving with different velocities and also including many influencing factors such as inlet air, sneeze, cough velocity, etc. path tracking analysis of pathogens has been obtained and mapped it to graphs and figures.

The research done by Bhattacharyyaa et al. [1] studied on the mixing of air from an air-conditioner with an aerosol sanitizer in an isolation room. The airflow pattern was analyzed such that the sanitizer droplets that are sprayed could reach all the corners of the isolation room. There are numerous studies carried out to analyze the air flow patterns in different ventilation systems in isolated rooms and enclosures [2-4]. This will prevent the infection of people who work in the isolation rooms such as nurses, doctors, health care workers, etc. The study by Dudalski et al. [5] assessed the transmission of airborne viruses using bio-aerosol sampling through a distance of 1m, which was later validated with CFD results. Empirical models and numerical methods were used along with experiments to study the diffusion of droplets in enclosures and are reported in literature [6,7]. There are also numerous experimental [8-10] and analytical models [11-13] developed to estimate the mass flow rates of bidirectional air exchanges in open enclosures [14-17] and rooms.

2.Methodology:

A rectangular passenger bus model, designed in SOLIDWORKS is taken for the analysis. The dimensions and specifications of the bus model are given below.

2.1 Bus dimensions and specifications:

The designed bus model (Figure 1) took has dimensions of 12 m x 3.5 m x 3.2 m and it consists of four windows of dimensions 1.5 m x 0.94 m on both sides of the bus. The windows are separated by a distance of 4.5 m on both sides. Also, the windows are located at a height of 1.96 m from the base of the bus. The material of the bus taken for the analysis is aluminum. The bus compartment is modeled as an empty rectangular box; in actual buses, it is furnished with seats, partitions, and luggage racks, as well as passengers. The doors are closed; in reality, doors are often left open. The effect of wheels is not considered. There are three passengers inside the bus, the infected passenger is at the rear and the other two normal passengers are at the front as shown in the Figure 1. The distance between the infected passenger and the middle standing passenger is 6m and the front passenger is 10m from the infected passenger. The height of the passenger is 1.82m. The windows are symmetrically located and the distance of the center of each window from the front side of the bus is given in Table-1.

Table 1: The distance between center of windows and front end of bus

Description	*Front windows (Win 1, Win 2)*	*Rear windows (Win 3, Win 4)*
Distance of center of window from the front end of the bus	3m	9m

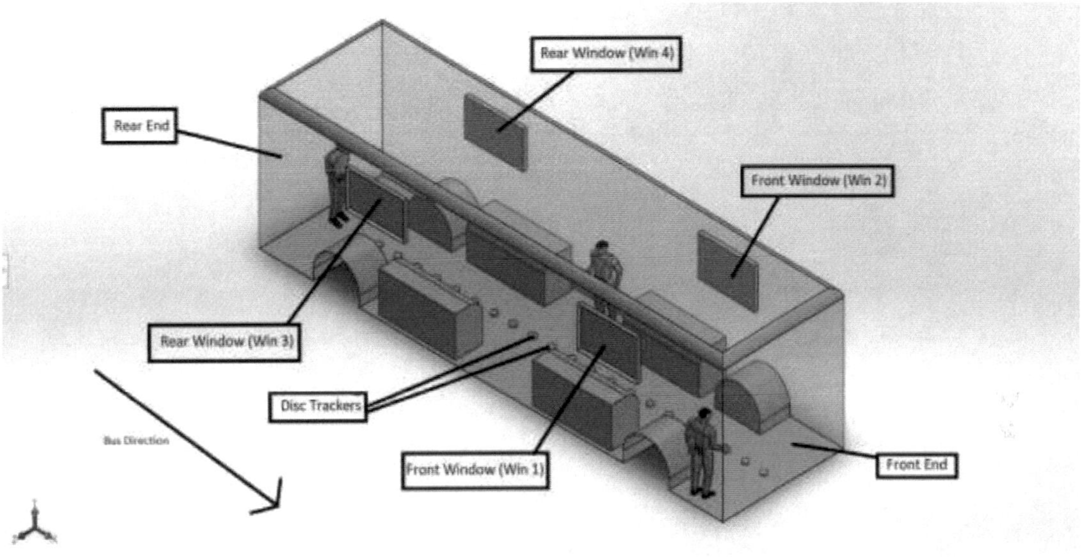

Figure 1: Bus model.

The whole simulation was done based on turbulent conditions of the air. When the bus cruises at different speeds, the air inlet velocity through windows are different and they follow a different path. To understand this, a simulation similar to a wind tunnel experiment using ANSYS FLUENT under turbulent wind conditions was carried out. The different velocities, with which the bus was traveling are given below in the Table-2.

Table 2: Velocities with which the bus cruises

S.NO	*Velocity of bus (in kmph)*	*Velocity of bus (in m/s)*
1	0	0
2	20	5.556
3	30	8.334
4	40	11.112

The velocities were taken up to 40kmph because there is not much change in the experiment beyond this velocity.

3.Flow visualization of window inlet:

To begin the simulation, the below steps are to be followed:

3.1 Pre-Processing:

This is the primary step of CFD simulation which helps in describing the geometry. A fluid domain is created which encapsulates the bus model. The domain of interest filled with air is then further divided into smaller tetrahedral segments known as the mesh generation. Boolean is used to separate the bus body from the fluid domain.

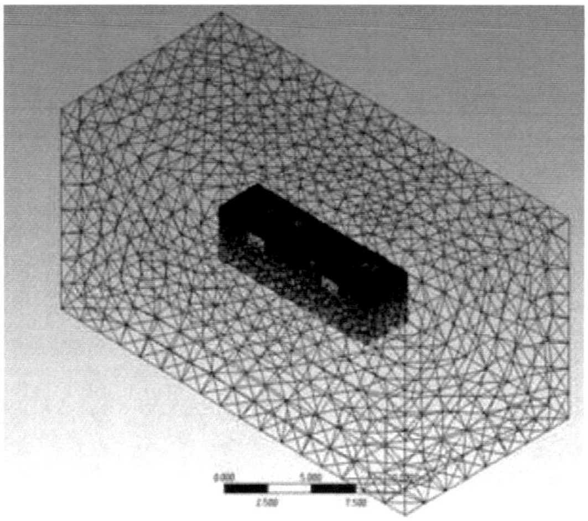

Figure 2*:* Schematic showing the model after meshing

After meshing, the different sections are named as inlet, outlet, walls, and body which in our case is the bus. As the bus travels, the air touches the bus through the inlet and exits through the outlet of the fluid domain. In this process, the air enters into the windows at different velocities which needs to be determined.

3.2 Solver:

After meshing, the model is exported to ANSYS FLUENT and the governing equations, materials and boundary conditions are set. Standard k-epsilon with enhanced wall treatment is taken for the best possible output. Second-order accurate discretization scheme was used for pressure, whereas for momentum discretization, the second-order upwind scheme was used keeping a time step of 0.1 s. The solver was initiated with the different vehicle speeds throughout the domain with 5% longitudinal turbulence intensity. Various boundary conditions employed are as follows:

***Inlet**:* Velocity inlet of 0, 5.556 m/s, 8.334 m/s, 11.112 m/s (With 5% turbulent intensity and turbulent viscosity ratio of 10 and also, the velocities are taken with an interval of 10 kmph to have uniformity as a boundary condition.)

Outlet: Pressure outlet (ambient pressure with 5% backflow intensity)

Domain walls: Velocity same as bus velocity

Bus walls: Stationary

Body surfaces: With open windows (to let the air in)

3.3 Post-Processing:

After the simulation, the results can be observed and analyzed with different contour plots of velocity, temperature, pressure, density, and vector plot of streamlines. Given below are the velocity contours after simulation at different bus speeds. Figure 3 (a), 3 (b) and 3 (c) shows the velocity of inlet air at the windows. A YX plane Z=1 bisects the YZ plane of bus at center represents the velocity distribution across the bus.

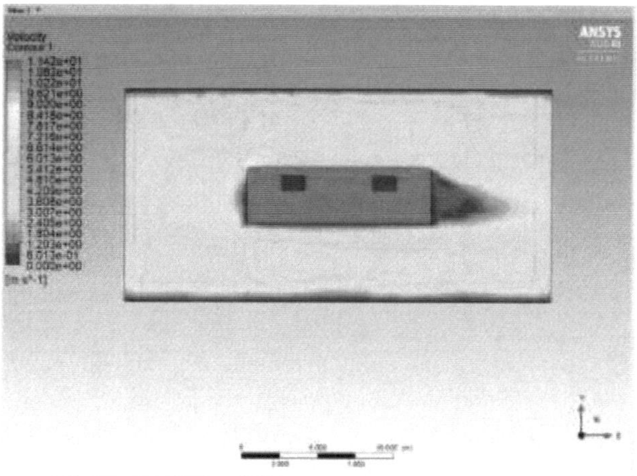

Figure 3 : Velocity contour at 20kmph (5.556m/s)

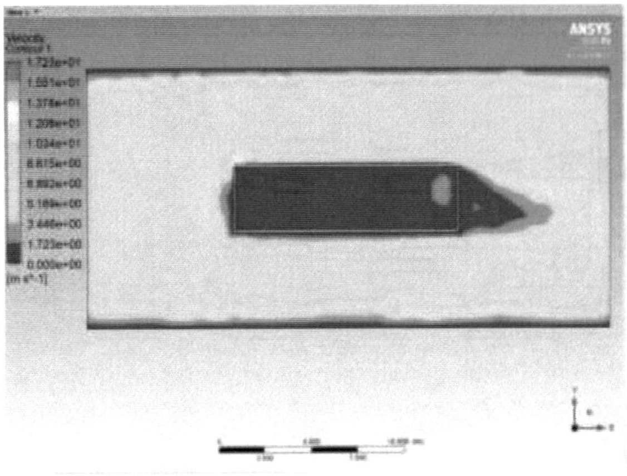

Figure 4 : Velocity contour at 30 kmph (8.334 m/s)

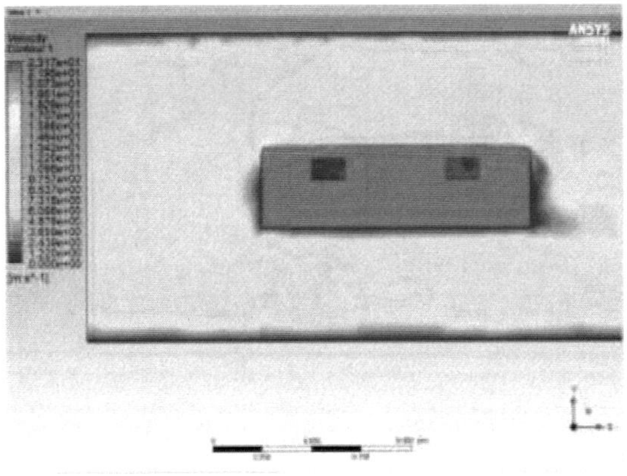

Figure 5: Velocity contour at 40 kmph (11.112 m/s)

The above figures are taken from the simulation of bus model enclosed in a fluid domain with air inside as fluid. It represents the condition of a moving bus. It is similar to wind tunnel experiment but in simulated form. Here, the velocity of air entering into the bus from windows have been measured. Although the exact velocity cannot be found, a non-uniform velocity distribution of air through bus windows have been taken from the given velocity range provided by the velocity contour for study purpose.

In the above Fig 3, the velocity of inlet air through all the windows is the same with velocity in between 0.4-0.86 m/s, in fig 4, the velocity of air through all windows is the same with a range in between 1-1.7 m/s, in fig 5 the velocity of air in the front windows is in between 0-1.22 m/s and the rear windows is

1.22-2.43 m/s. A table has been provided with values that have been taken from velocity contour after the simulation in ANSYS FLUENT.

Table 3: Boundary conditions for velocity inlets of bus

S.NO	***Velocity of bus (kmph)***	***Velocity of bus (m/s)***	***Velocity of air through windows as per results (m/s)***	
			Front windows (Win 1, Win 2)	***Rear windows (Win 3, Win 4)***
1	0	0	0.2	0.2
2	20	5.556	0.7	0.7
3	30	8.334	1.5	1.5
4	40	11.112	1	2.2

4.Discrete particle method:

For the simulation a human being model with a height of 1.82m and the natural velocities at which cough and sneeze particles get released are considered. Sprays can be a little difficult to track, but some scientists have managed to make a rough estimate like naturally, cough particles can travel fast up to 50 mph and can execute almost 3,000 droplets in just one go, and coming to Sneeze particles, they can travel up to 100 mph and creates roughly 100,000 droplets. By considering average cough and sneeze particle size, it varies from 0.5-12 μm [3]. In these simulations, a standard value of 4 μm is considered. To show the flow of sneeze and cough particles, injections are given from the injected person's inlet mouth. Cough and sneeze flows enforce high velocities of low viscosity gas through narrow openings; therefore, they are turbulent flows. The present study uses standard k-epsilon turbulence model with enhanced wall treatment. This model gives better results for problems involving ventilation, air quality, and air-borne infection risk in the health care facilities. Since the discrete phase method is employed, its approach is to ensemble large sneeze and cough particles, which are injected in the continuum, and equations for heat and mass transfer are solved simultaneously. Neglecting radiation, the mechanisms for droplet mass and temperature change are convection and evaporation. Sherwood correlations relate mass transfer coefficient to the Reynolds and Schmidt numbers.

$$m_p c_p (dT_p/dt) = hA_p(T_\infty - T_p) + (dm_p/dt)h_{fg} \quad (1)$$

IOP Conf. Series: Materials Science and Engineering 1128 (2021) 012005 doi:10.1088/1757-899X/1128/1/012005

The trajectory of a discrete phase droplet can be determined by integrating the force balance written in the Lagrangian reference frame. This force balance equates the particle inertia with the forces acting on the droplet. The component form of this equation in the Cartesian coordinate can be written as follows [4].

$$du_p/dt = F_D (u-u_p) + g(\rho_p-\rho)/\rho + F \quad (2)$$

The boundary conditions for the discrete phase are as follows:

Diameter of particles: 4 μm

Material taken: water- liquid (typically what saliva made of)

Velocity of the particles: 20, 30, 40 m/s is taken.

Time period of sneeze: 4seconds

Temperature of particles: 338 K

Flow rate: 2.51e-7, 3.77e-7, 4.96e-7 kg/s (for 20, 30, 40 m/s respectively)

Injections: 8 identical injections are taken for flow visualization of particles.

A simple pressure-velocity coupling is considered along with the least-squared cell-based gradient. Second-order discretization is used for pressure and second-order upwind is considered for momentum, turbulent kinetic energy and dissipation rates. Here the gravity is taken into consideration as the particles will reach the surface of the bus. After performing the hybrid initialization with all the boundary conditions given, the calculation is performed by taking 700 numbers of time steps and 0.1 s as the time step size. Also, the maximum iteration/time step is taken as 20. As per the boundary conditions, the particles come out from capsule shaped human mouth at different velocities. Similarly, the bus travels at different velocities which in turn have different velocities of air inlet through windows thus altering the trajectory of human cough and sneeze particles.

5.Results and Discussion:

The main aim of the paper is to track the safest distance inside a bus that a human being needs to maintain. To track the distance, disc-shaped trackers are installed inside a bus with a range of 1.5 m from the rear bus wall and 0.5 m spacing in between two adjacent discs. There are a total of 21 discs that have been installed in the bus. The number of iterations taken is 700 after which, there is no activity observed inside the bus. Next, the results of the simulations for different scenarios are discussed.

Figures 6 and 7 are conditions at rest i.e., bus velocity is equal to 0m/s. The inlet air through windows is taken as 0.2m/s because when the bus is at rest, there will be air entering the windows with some velocities due to disturbances in the surroundings.

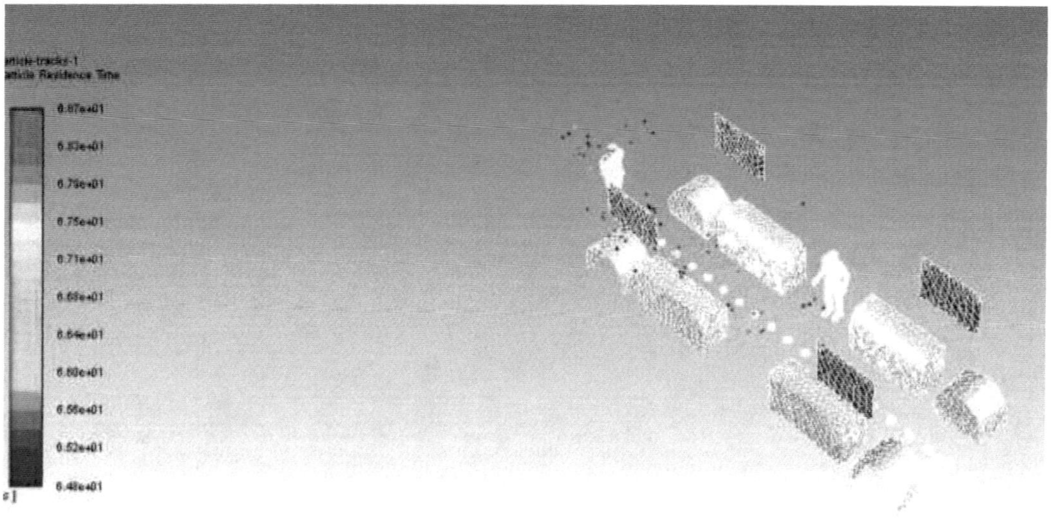

Figure 6: Particle residence time at 20 m/s cough velocity and bus at rest

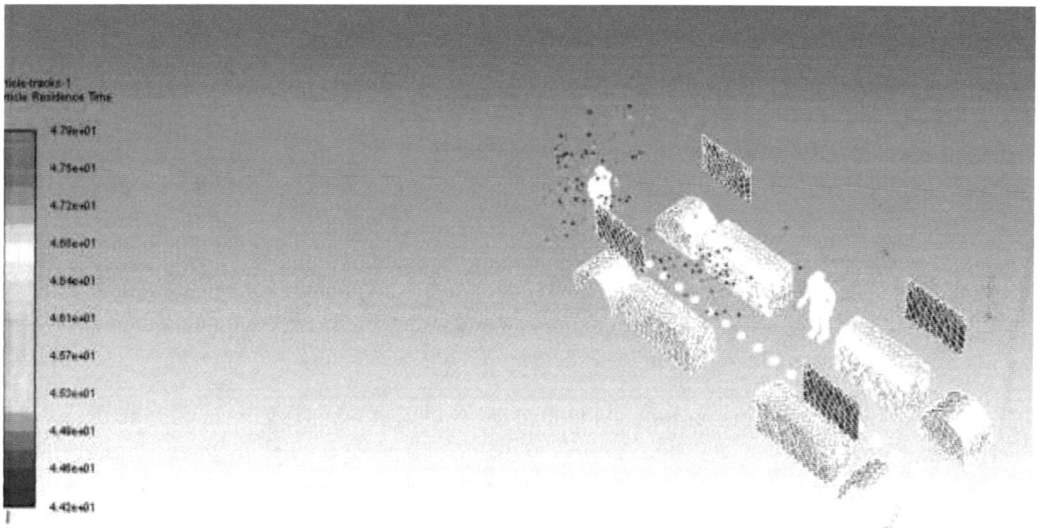

Figure 7: Particle residence time at 30 m/s cough velocity and bus at rest

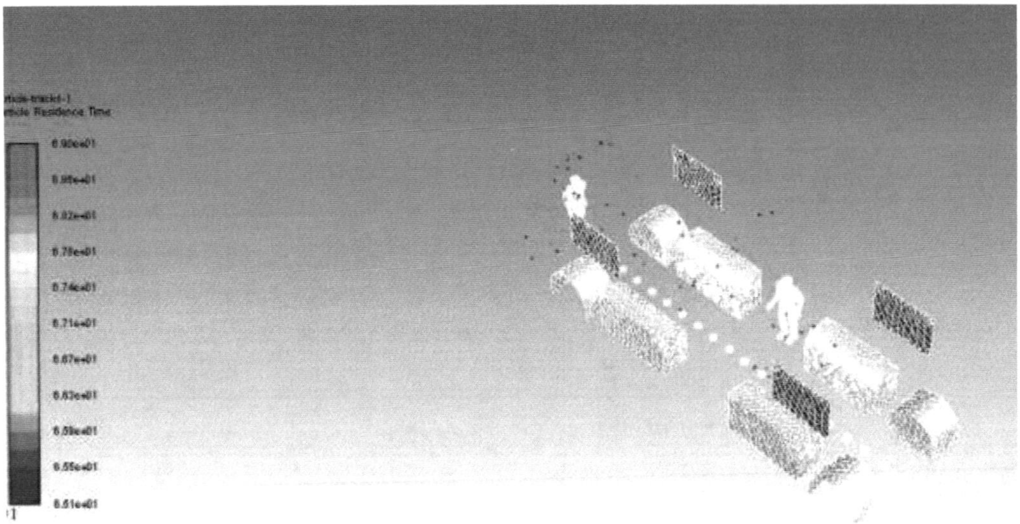

Figure 8: Particle residence time at 40 m/s cough velocity and bus at rest

Since there are disc trackers on the bus floor equally spaced at a distance of 0.5 m, the characteristics of particle tracking have been analyzed from animation video in ANSYS FLUENT and distance travelled by the particles have been observed w.r.t the disc trackers and the distance has been measured by how many discs the particles crossed. From the above figures, the distance traveled by the particles inside the bus due to sneezing and coughing is 6.0m, 6.8m, and 7.5m for the cough velocities of 20m/s, 30m/s and 40m/s respectively. The human cough velocity has been taken with an interval of 10 m/s starting from 20 m/s. The maximum human sneeze velocity or the velocity with which a human being can emit disease causing droplet particles goes up to 44 m/s and any number of simulations can be done within the range of 44 m/s. Because to have uniformity and study purpose, these values has been considered.

In the above case, since it is a stationary condition there is less disturbance from outside and the particles travel farther and many people will have a chance of getting affected. Figures 8, 9 and 10 represents the bus velocity equal to 0 m/s. The air inlets through rear and front windows are taken as 0.2 m/s which we got from previous simulation for the bus. As the human cough velocity increases, the distance travelled by the particles increases.

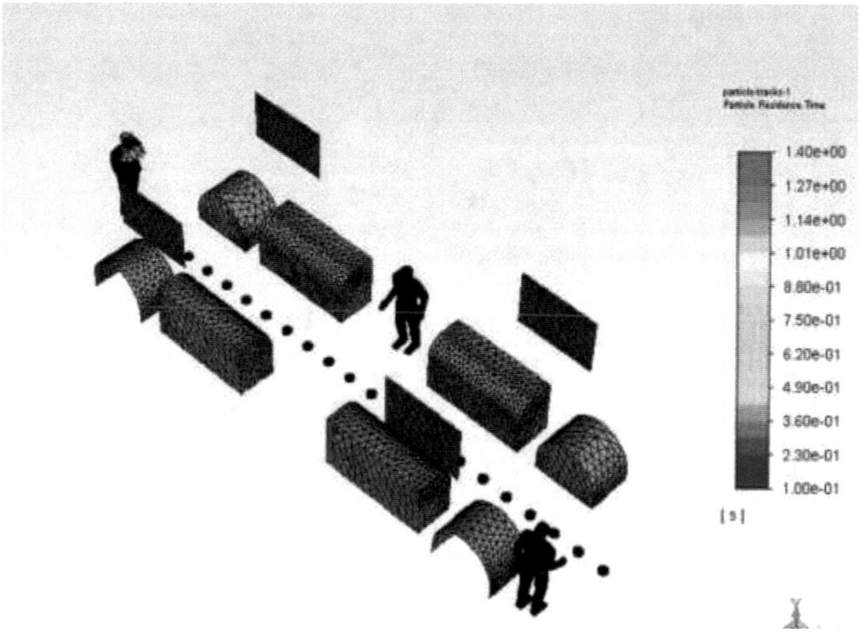

Figure 9: Particle residence time at 20 m/s cough velocity and bus at 40kmph

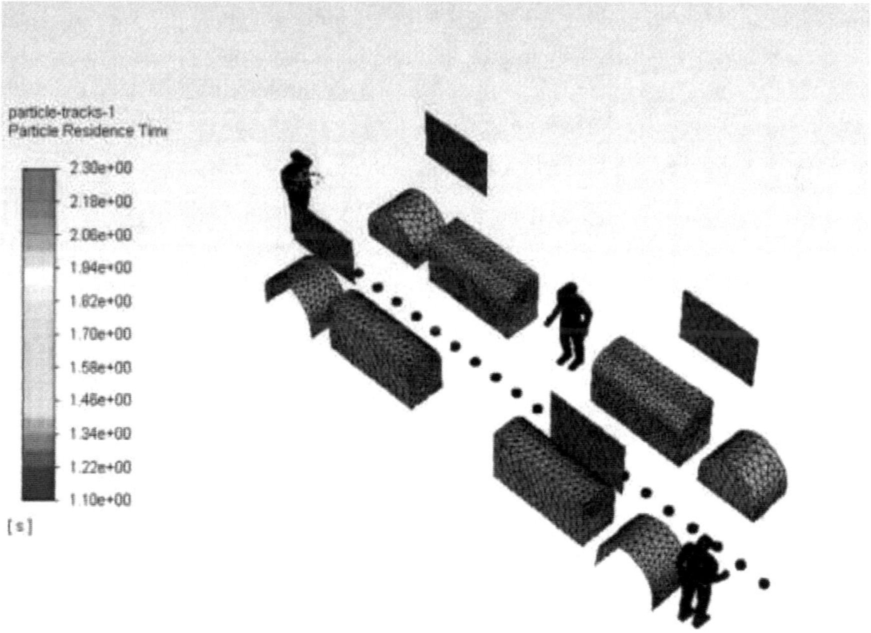

Figure 10: Particle residence time at 30 m/s cough velocity and bus at 40kmph

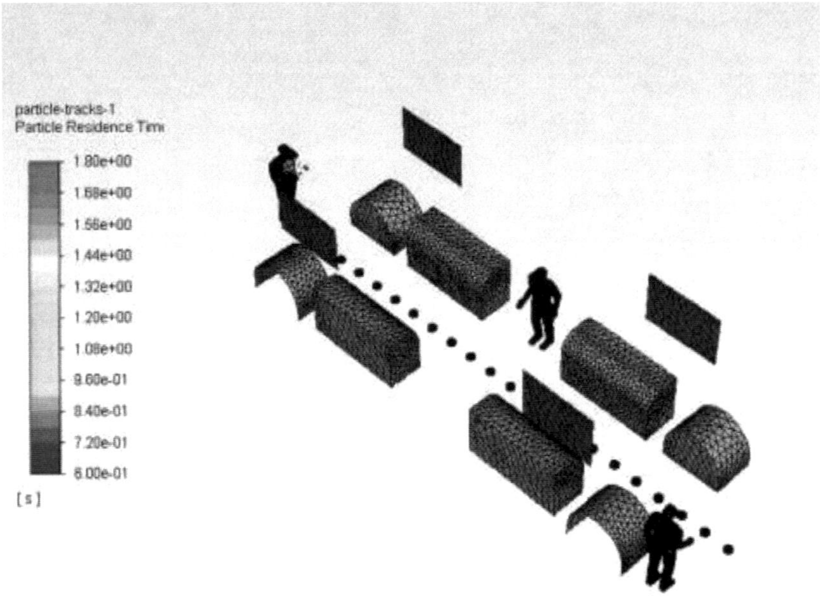

Figure 11: Particle residence time at 40 m/s cough velocity and bus at 40kmph

Similarly, the distance has been found with the help of disc trackers and animation video for the particles when the bus is at 40 kmph. From the above figures 11-13, the distance travelled by particles is less than 0.5 m for the human cough velocities of 20 m/s and 30 m/s. Also, for human cough velocity of 40 m/s, the distance found was 0.5 m. This is mainly due to the outside disturbance of air which flows through the windows, resulting in low distance. Here the velocity of air from front and rear windows is taken as 1 m/s and 2.2 m/s which we obtained from previous simulation for bus model.

Also, the distance travelled by particles at 20 kmph and 30 kmph bus velocity has been tracked and displayed in the below table along with previous readings. These are the total simulations done for different bus and human cough velocities in order to find distance covered by disease causing particles.

Table 4: This table represents the Distance w.r.t velocities

S.NO	***Velocity of bus(km/hr)***	***Velocity of bus(m/s)***	***Velocity of human cough(m/s)***	***front inlet (Win 1, Win 2)***	***back inlet (Win 3, Win 4)***	***Distance (m)***
1.	0	0	20	0.2	0.2	6.0
2.	0	0	30	0.2	0.2	6.8
3.	0	0	40	0.2	0.2	7.5
4.	20	5.56	20	0.7	0.7	1.3
5.	20	5.56	30	0.7	0.7	1.5
6.	20	5.56	40	0.7	0.7	2
7.	30	8.34	20	1.5	1.5	0.5
8.	30	8.34	30	1.5	1.5	0.65
9.	30	8.34	40	1.5	1.5	0.70
10.	40	11.12	20	1	2.2	<0.5
11.	40	11.12	30	1	2.2	<0.5
12.	40	11.12	40	1	2.2	0.5

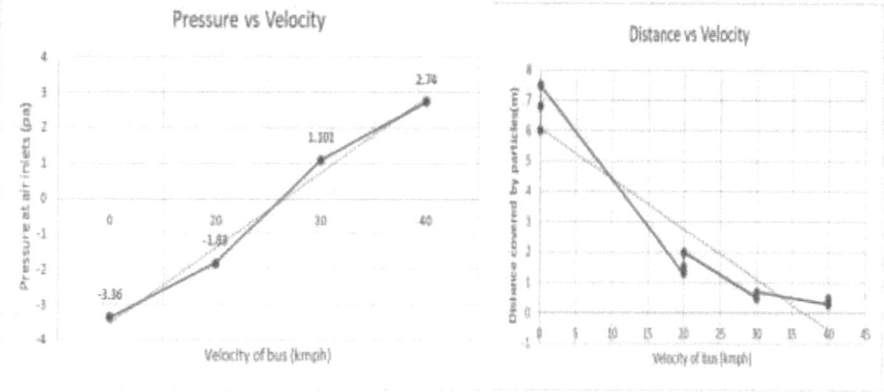

Figure 12: variation of velocity of the bus w.r.t to pressure inside the bus (image to the left) and distance travelled by the particles(image to the right) distributions

The Figures 13,14 and 15, illustrates the streamline flow in the bus. The streamline flow patterns are plotted and analyzed to determine the particle distribution inside the bus.

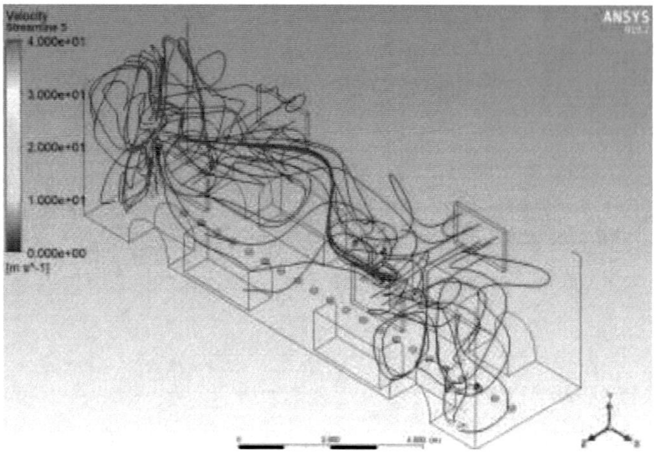

Figure 13: stream line flow when bus at rest

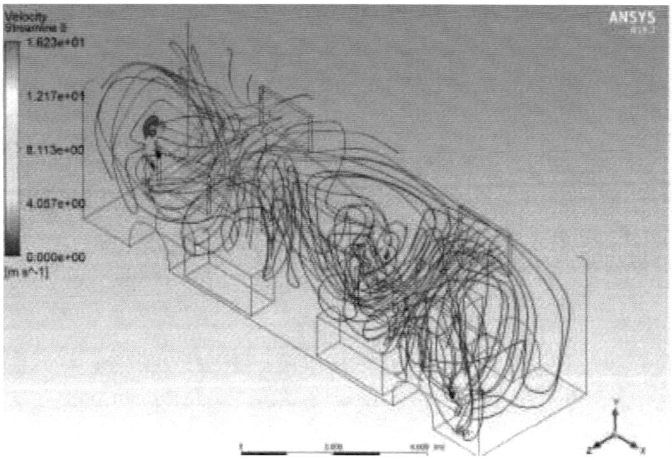

Figure 14: stream line flow when bus is at 40kmph

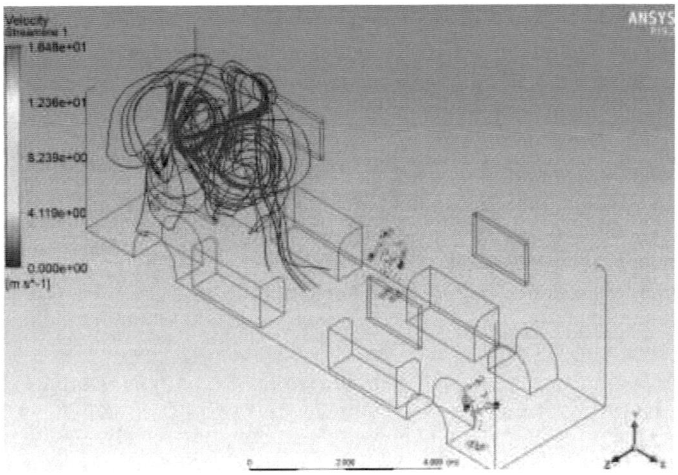

Figure 15: stream line flow when windows are closed.

6.Conclusion:

Hence the distance that the COVID-19 particles cover inside a moving bus with air disturbance has been measured. To achieve this, a simulation for the bus has been carried out in order to find the velocity of air through the windows of the bus. After finding out the inlet air velocity, the values have been taken to analyze the particle tracking inside the bus with cough particles coming out of human mannequin at different velocities to find the distance travelled by the particles with the help of trackers. Also, it is found that, the safest distance which the humans shall maintain in order to prevent the spread of COVID-19 inside a bus is 7.5m. When the bus velocities are greater than 40kmph the particles will not be able to travel farther due to the high disturbances created by the air through windows which pushes the particles backward, only the passengers sitting or standing behind the infected passenger will get affected. Over this velocity, the safest distance will be around 1 m. When the bus is at rest or having a halt in a bus station, the spread of COVID-19 is greater covering a larger distance and affecting the passengers in that region significantly. So, it is better to reduce halts and set a limited number of bus stations, and is, even more, better to have a single starting and ending destination. It is more hazardous to close the windows and doors when the bus is at rest as the COVID-19 will travel farther due to no disturbances from the air outside.

References:

1)Bhattacharyya, S., Dey, K., Paul, A. R., & Biswas, R. (2020). A novel CFD analysis to minimize the spread of COVID-19 virus in hospital isolation room. Chaos, Solitons & Fractals, 139, 110294. doi:10.1016/j.chaos.2020.110294

2) Jianjian Wei PhD, Yuguo Li PhD, Airborne spread of infectious agents in the indoor environment, American Journal of Infection Control , 44 (2016) S102-S108.

3) Naiping Gao , Jianlei Niu , Transient CFD simulation of the respiration process and inter-person exposure assessment, Building and Environment 41 (2006) 1214–1222

4)Ji-Xiang Wang, Xiang Cao, Yong-Ping Chen, An air distribution optimization of hospital wards for minimizing cross-infection, Journal of Cleaner Production, Volume 279,2021,123431,ISSN 0959-6526,https://doi.org/10.1016/j.jclepro.2020.123431.

5)Dudalski, N., Mohamed, A., Mubareka, S., Bi, R., Zhang, C., & Savory, E. (2020). Experimental Investigation of Far Field Human Cough Airflows from Healthy and. Indoor Air. doi:10.1111/ina.12680

6) Deepthi Sharan Thatiparti, Urmila Ghia, Kenneth R. Mead, Assessing Effectiveness of Ceiling-Ventilated Mock Airborne Infection Isolation Room in Preventing Hospital-Acquired Influenza Transmission to Health Care Workers, ASHRAE Trans. 2016 ; 122(2): 35–46

7) Yang J, Sekhar SC, Cheong KW, et al. Performance evaluation of a novel personalized ventilation-personalized exhaust system for airborne infection control. Indoor Air 2015;25:176-87
8) Harish. R, Venkatasubbaiah. K, Mathematical modeling and computation of fire induced turbulent flow in partial enclosures, Applied Mathematical Modelling, vol.37, pp: 9732-9746, 2013.

9) Harish. R, Venkatasubbaiah. K, Transport phenomena of turbulent fire spread through compartment connected to vertical shaft in tall building, Fire Safety Journal, vol.61, pp: 160-174, 2013.

10) Harish. R, Venkatasubbaiah. K,Effects of buoyancy induced roof ventilation systems for smoke removal in tunnel fires, Tunnelling and Underground Space Technology, vol.42, pp: 195-205, 2014.

11) Harish. R, Venkatasubbaiah. K, Numerical investigation of instability patterns and nonlinear buoyant exchange flow between enclosures by variable density approach, Computers & Fluids, vol.96, pp: 276-287, 2014.

12) Harish. R, Venkatasubbaiah.K, Large Eddy Simulation of thermal plume behavior in horizontally partitioned dual enclosure, Building Simulation, vol.8(2), pp: 137-148, 2015.

13) Harish. R, Venkatasubbaiah. K, Numerical study of water spray interaction with fire plume in dual chambers connected to tall shaft, Fire Safety Journal, vol. 74, pp: 1-10, 2015.

14) Harish. R, Venkatasubbaiah. K, Non-Boussinesq approach for turbulent buoyant flows in enclosure with horizontal vent and forced inlet port, Applied Mathematical Modelling vol. 40, pp: 927-941, 2016.

15) Harish. R, Venkatasubbaiah. K, Numerical simulation of turbulent plume spread in ceiling vented enclosure, European Journal of Mechanics-B/Fluids, vol.42,pp:142-158, 2013.

16) Harish. R, Effect of heat source aspect ratio on turbulent thermal stratification in a naturally ventilated enclosure, Building and Environment,143,pp:473-486, 2018.

17) Harish. R, Buoyancy driven turbulent plume induced by protruding heat source in vented enclosure, International Journal of Mechanical Sciences,148,pp:209-222, 2018.

IOP Conf. Series: Materials Science and Engineering 1128 (2021) 012006 doi:10.1088/1757-899X/1128/1/012006

The Effect of Ramp Location in a Strut Based Scramjet Combustor under non reacting flow field

A. Antony Athithan[1], S. Jeyakumar [1,2*] and R. Sivakumar[3]

[1] Faculty of Engineering, Lincoln University College, Malaysia.
[2] Aeronautical Engineering, Kalasalingam Academy of Research and Education, India
[3] School of Mechanical Engineering, VIT University, Chennai, India

* sjeyakumar1974@gmail.com

Abstract. The scramjet engine is considered to be the high-speed propulsive system for hypersonic air-breathing vehicles. Researchers focus on various geometries of strut injectors to enhance the mixing and combustion efficiencies of the scramjet combustor. This paper reveals the non-reacting flow characteristics of various ramp locations in a scramjet combustor with strut injection. The ramps are symmetrically placed at three different locations upstream of the strut within the combustor. Air enters into the combustor inlet with Mach number 2, and the fuel is injected at the sonic velocity from the strut. The 2D supersonic flow characteristics are numerically investigated using ANSYS 18.0 software. The steady-state flow simulations are performed in this study. The flow field is modelled using RANS equations, and the k-ω SST turbulence model and default constants are chosen to investigate the flow characteristics within the combustor. The flow characteristics of the ramp based DLR scramjet combustors are compared with the baseline DLR scramjet model. The computational results are validated with the experimental data. The shock wave interaction from the ramps enhances the distribution of hydrogen in the lateral direction of the flow than the baseline strut configuration. The ramps create recirculation regions that could enhance fuel-air mixing. The ramp induced strut based DLR scramjet engine provides increased total pressure loss of the baseline DLR scramjet.

Keywords: Strut injection, ramp, hydrogen jet, computational fluid dynamics (CFD), scramjet, static pressure, total pressure loss.

1. Introduction

High-speed transportation depends on supersonic and hypersonic flights with a good understanding of fuel and air mixing performance inside the supersonic combustor is essential. Mixing of fuel with air and ignition and flame holding are the fundamental constrain in the design of the scramjet engine[1,2]. The main focus is on the fuel injection system because the fuel's residence time with air in the combustor is less than 1ms. Researchers have suggested different fuel injectors like flush wall injection[3–8], cavity-type injection[9–15], strut based injection[16–21], and their blends[22–27], etc., to augment mixing and flame holding techniques. The strut injectors of specific geometry enhance mixing and combustion efficiencies with optimum total pressure loss than other injection and flame holding configurations.

The experiments on strut based DLR scramjet combustor were conducted with a hydrogen fuel injection system by Waidmann et al.[28]. The numerical investigation on a two-dimensional scramjet combustor with a flamelet model [29] is validated with the reported experimental results of the DLT strut scramjet combustor. Kummitha et al. [30] investigated the supersonic combustor with an innovative design of strut injector, which involves the modified double arrow strut and rocket models shows. The results show that the impinging of the first oblique shock is created from the strut leading edge that is very small when compared to the basic strut. Thus, multiple reflections of shock waves help the fuel to mix early with the supersonic airstream. Rocket strut fuel injector causes high-pressure rise due to impingement of shock wave occurs at the same location. Consequently, the pressure downstream of the strut is high in rocket and double strut-type fuel injector when compared to basic strut. Due to the inclined fuel injector, the combustion efficiency is significantly higher in the case of both double and rocket strut models. Therefore, the decrease in ignition delay is noted in double and rocket strut models than basic DLR scramjet models.

The study conducted by Kumaran and Babu [31] numerically simulated the hydrogen-fueled supersonic combustor employing a multistep chemistry model and compared it with the single-step reaction model to evaluate the performance of the combustor. The study findings disclosed that a multistep chemistry model could be an exercise to evaluate the insight properties of the combustion reaction like heat release rate, ignition delay, etc. Conversely, the single-step model can offer better results for the combustor's overall performance with a decrease in computational cost. Gerlinger and Bruggemann [32] have studied the mixing of hydrogen jets supplied from a strut injector under cold supersonic airflow conditions. It is indicated that the thickness of mixing layerand the total pressure loss increases by raising the strut lip thickness that is mainly owing to the increased diffusivity of the hydrogen at the outer strut wall and the more robust shock wave formation. Huang et al. [33] executed the numerical simulation studies on hydrogen-air reaction mechanism, the injection pressure and temperature variations of a strut-type scramjet combustor. Their study proved that shocks are formed from the strut is pushed out of the duct with the subsonic flow for increasing the temperatures and injection pressure.

Choubey and Pandey [34] executed the numerical simulation analysis on two strut configuration in a scramjet combustor model by changing the strut's angle of attack and asserted that zero angles of attack make a surge in mixing and combustion efficiencies. There is another work by the researchers [20], and it deals with the effect of altering the strut geometry and orientation in the combustor from the inlet. Furthermore, it is disclosed that the optimum lip height and position of the strut functions in an essential role in improving the combustion efficiency. Three strut positioning in a scramjet combustor was computationally examined by Kumar et al. [35]. It was identified that the maximum combustion efficiency and thrust had been attained by Pareto-optimal optimization studies accordingly position the struts in the combustor.

The numerical study of Athithan A A et al. [40] on the effect of double ramp configurations in a strut based scramjet combustor shows the distribution of hydrogen is enhanced in the double ramp combustor along spatial direction than normal DLR strut combustor. The study concluded that deceleration of flow at downstream of the strut due to the shocks and its interactions from the double ramps and strut, furthermore leads to more ignition delay.

The above literature reveals the performance of strut based DLR combustor with various active methods. The effect of ramps in a strut based scramjet combustor is not detailed in the open literature. The present study influences the performance of wall-mounted ramps in a DLR scramjet model under a non-reacting steady supersonic field. The ramps are located at the bottom and top walls of the duct towards the upstream of the strut injector. The numerical investigation is carried out using the two dimensional RANS equation with k-ω SST turbulence model. The flow shock patterns and overall performance of the 2D combustor provide acceptable results as that of 3D analysis [39]. The effect of the ramps is accessed based on shock generation and interactions with the shear layers, wall pressures of the combustor, and the stagnation pressure loss.

2. Computational Method

2.1. Numerical Modelling

The numerical studies of the scramjet model are accomplished by solving the two-dimensional momentum, mass, and energy conservative equations. In the current approach, the compressible Reynolds averaged Navier-Stokes equations (RANS) are solved with the SST k-ω turbulence model [36], which offers a good prediction of jet flows[37]. The flow governing equations [18] are discretized by the finite volume method framework of ANSYS18.0. The density-based solver with the implicit formulation and advection upstream splitting methods are adopted in this work [38]. The governing equations, i.e., mass, momentum, and energy equations written for the total enthalpy are expressed as:

Continuity equation

$$\frac{\partial \rho}{\partial t} + \frac{\partial}{\partial x_i}(\rho u_i) = 0 \quad (1)$$

Momentum equation

$$\frac{\partial}{\partial t}(\rho u_i) + \frac{\partial}{\partial x_i}(\rho u_i u_j) + \frac{\partial P}{\partial x_i} = \frac{\partial}{\partial x_i}(\tau_{ij}) \quad (2)$$

Energy equation

$$\frac{\partial}{\partial t}(\rho H) + \frac{\partial}{\partial x_i}(\rho H u_i) = -\frac{\partial}{\partial x_i}(\tau_{ij} u_i) + \frac{\partial q_i}{\partial x_i} \quad (3)$$

The turbulence kinetic energy, k, and the specific dissipation rate, ω is obtained from the following transport equations:

$$\frac{\partial}{\partial x_i}(\rho k u_i) = \frac{\partial}{\partial x_j}\left(\Gamma_k \frac{\partial k}{\partial x_j}\right) + G_k - Y_k + S_k \quad (4)$$

and

$$\frac{\partial}{\partial x_j}(\rho \omega u_j) = \frac{\partial}{\partial x_j}\left(\Gamma_\omega \frac{\partial \omega}{\partial x_j}\right) + G_\omega - Y_\omega + D_\omega + S_\omega \quad (5)$$

The terms, Gk represents the production of turbulent kinetic energy and Gω as the generation of ω, Γk and Γω represent the effective diffusivity of k and ω respectively, Yk and Yω represent the dissipation of k and ω due to turbulence, Dω represents the cross-diffusion terms and Sk and Sω as the user-defined source terms

2.2. The Computational domain

Figure 1 shows the configuration of the supersonic model combustors that have been numerically investigated under non-reacting field using commercial code ANSYS 18.0. The channel has an inlet height of 50mm an inlet Mach number of 2.0 and an inlet static temperature of 300k. The DLR scramjet model experimented with by Waidmann [28] is shown in Figure 1. In the DLR scramjet combustor, a wedge-shaped strut is located at 77mm downstream of the combustor entrance and 25 mm away from the bottom wall. The strut has a length of 32mm and a height of 6mm. The hydrogen jet is injected through 15 orifices of 1.0mm in diameter. The ramps are placed at the bottom and top walls of the combustor at three axial locations, 77mm, 50mm, and 33mm from the combustor inlet and are denoted as Case 2, Case3, and Case 4, respectively. The ramp dimensions are also detailed in Figure 1, and the performance of the ramp locations in the strut type scramjet model is compared with the baseline DLR model (Case 1). The incoming air at the combustor inlet is at Mach number 2, and the hydrogen fuel is injected parallel to the airstream at the sonic condition. The flow is considered compressible and two dimensional. The operating conditions of the scramjet combustor are identical for all the cases.

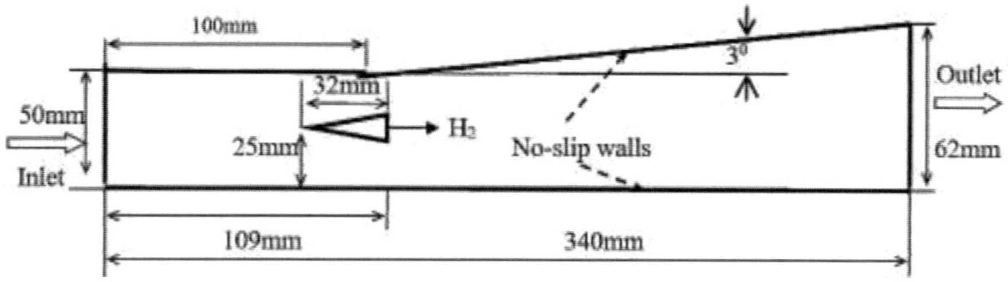

(a) Case 1

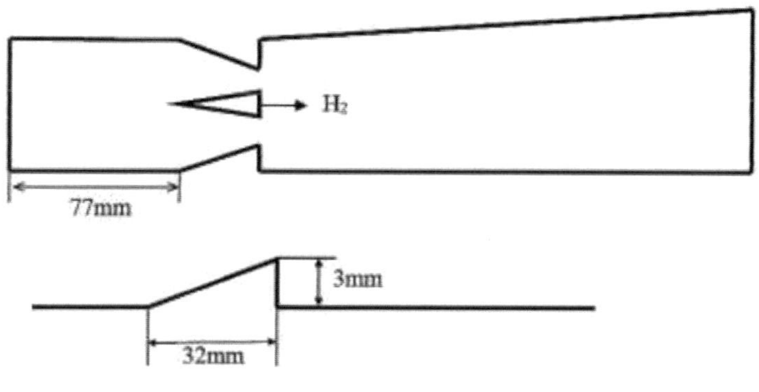

(b) Case 2

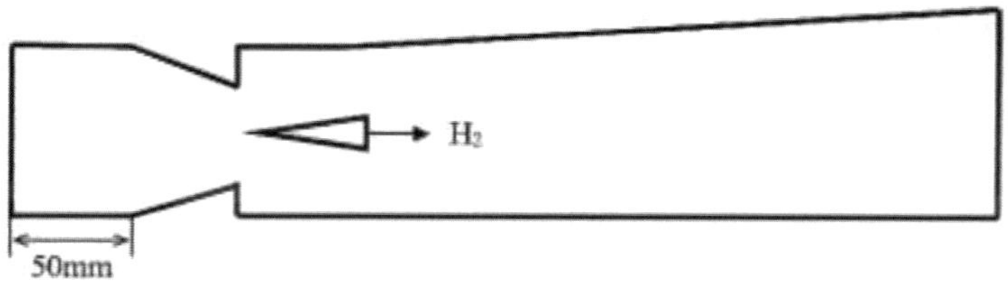

(c) Case 3

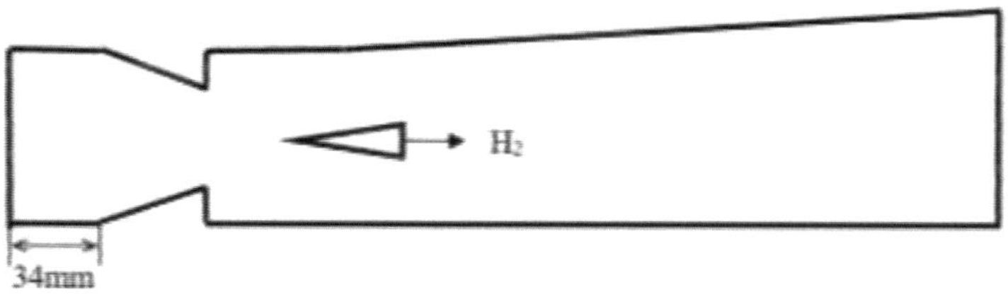

(d) Case 4

Figure1. Geometric details of scramjet Combustor for different configurations

2.3. Boundary condition

The boundary conditions are essential to solve the numerical governing equations. The air inlets are defined as supersonic conditions with specified variables of Mach number 2.0, and the hydrogen jet is injected from the strut at the sonic condition parallel to the flow direction. Dirichlet and Neumann boundary conditions are used as inlet and outlet boundary conditions, respectively. A fixed wall with no-slip boundary condition is chosen for the walls of the combustor. The details of the inflow conditions are detailed in Table 1.

Table 1. Inflow conditions of air and fuel.

Variable	Air	Hydrogen
Mach Number	2	1
Velocity (m/s)	706	1240
Temperature(K)	340	250
Pressure (bar)	1.0	1.0
Density (kg/m^3)	1.002	0.097
Mass fraction of O_2	0.232	0
Mass fraction of H_2	0	1
Mass fraction of H_2O	0.032	0

2.4. *Grid independence study*

The numerical solution accuracy to the problem mainly depends on grid size, hence for this investigation grid independence study has been performed. Three different grids, namely coarse (146146), medium (191607), and fine (290112) meshes, are employed to optimize the grid resolution for the convergence analysis. For the entire flow field, the y+ value is less than 1.0 (6.1e-7), which corresponds to the first-row cell height specified at 0.001 mm. Figure 4 shows the grid independence study, and it is found that the distribution of static pressure at the combustor's bottom wall for all the mesh size shows the variance of less than 1%. The medium and fine meshes show the same result, hence to reduce the computational time, the medium-sized mesh is considered for the analysis. The convergence criteria for the numerical simulation is considered based on the variation of the net mass flux of the flow across the combustor falls below 0.001kg/s, i.e., is less than 0.1% of the fuel mass flow.

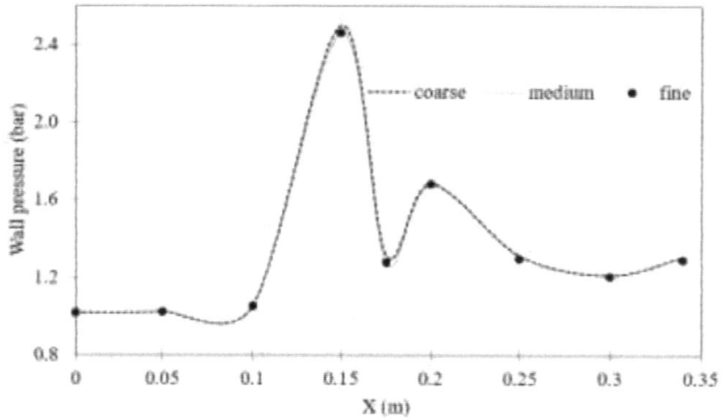

Figure 2. Grid independence study

3. Validation

The computational study is validated with the experimental results of Waidmann et al. [28] in terms of shadowgraph image and wall pressure plot and is shown in Figures 3 and 4, respectively. The inflow conditions for the experimental and numerical investigations are identical. The inlet flow to the combustor is Ma=2.0. The inflow parameters are presented in Table 1. The numerical result shows the shock generated from the strut, jet stream from the injector, and shock reflection pattern from the combustor wall is similar to that of the experimental result. The wall static pressure values at the bottom wall of the DLR scramjet model are presented in figure 4 for the validation at the steady-state condition. The simulation results of the wall static pressures along the axis are well-matched with the experiment result are shown in figure 4. The maximum wall static pressure is observed at X = 0.12m, for both experimental and numerical studies. It is noted that numerical results are in qualitative agreement with the experimental results of the reference [28].

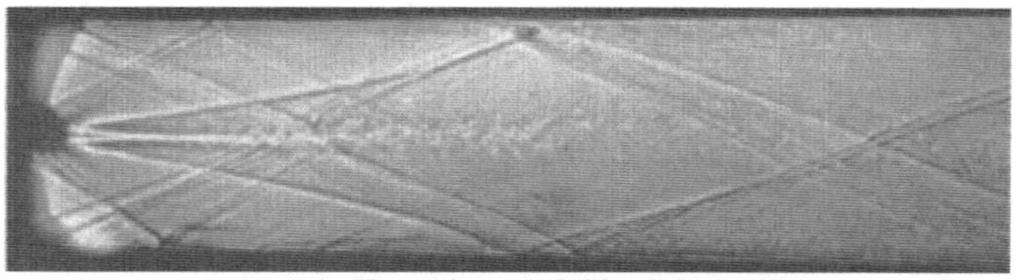

(a) Experimental Shadowgraph

(b) Numerical shadowgraph

Figure 3. Comparison of (a) Experiment result with (b) numerical simulation

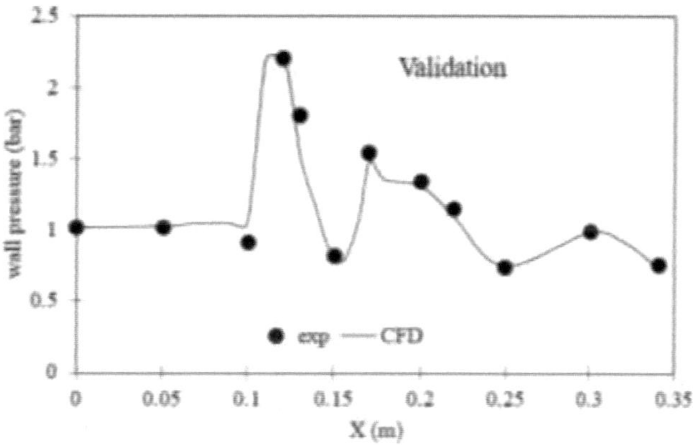

Figure 4. Validation of wall static pressure with experimental results

4. Result and discussion

The numerical investigations on the flow characteristics of ramps positioned on the top and bottom walls of a DLR scramjet combustor under a non-reacting flow field are discussed and compared with the typical DLR model. Figure 5 shows the shadowgraph images of the different geometry profiles that provide the details about the shocks and shear layer patterns inside the duct. The leading edge oblique shock waves from the strut in the scramjet combustor undergo multiple shock reflections from the combustor walls that are observed from Case 1. These reflected shock waves interact with the trailing edge shock from the strut and impinge on the fuel stream shear layer. For case 2, additional shockwaves are observed at the leading edge of the ramps that interact with the oblique shock wave emanated from the strut that decelerates the flow downstream. The shock waves emanated from the trailing edges of the ramps and strut interact with the fuel stream downstream that may enhance mixing of the fuel-air stream downstream with the lesser supersonic stream than the case 1. The ramp related shocks interact with leading-edge shock from strut upstream of strut injector as the ramps are located upstream of the strut (case 3). These reflected shocks from the internal walls of the combustor overlap at different locations downstream of the strut with lesser intensity than case 2. In case 4, the ramp-related shock from the leading edge of ramps interacts upstream of the strut, and the leading edge shock waves from the strut impinge the trailing edge shocks of the ramp which results in deceleration of the flow downstream of the strut. Moreover, the upstream positioning of ramps creates the flow separation downstream of the ramp base due to the shock boundary layer interactions. The shock interactions and the flow separations in the supersonic flow field are perceived from Figure 6.

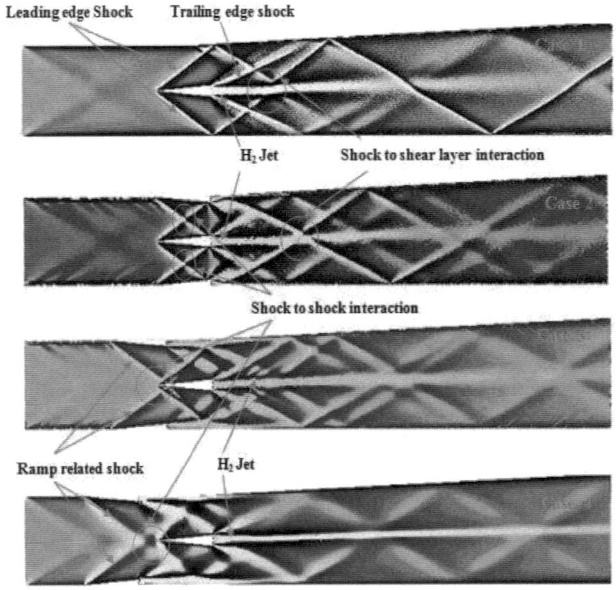

Figure 5. Shadowgraph image of combustor model with ramps at the bottom and top wall of duct in different positions

Figure 6 shows the Mach number contour comparison of different scramjet combustor profile. It is seen that the oblique shock waves are created at the leading edge of the strut, which gets reflected into the strut from the bottom and top wall of the combustor. The flow separations are observed downstream of the ramps due to the shock boundary layer interactions. The flow separation increases as the ramps are moved toward the combustor inlet. The recirculation regions are formed at the trailing edge of ramps, and these recirculation regions will act as flame holders and enhance fuel-air mixing. Figure 7 shows the recirculation region formed inside the combustors. For case 1, small vortices are noticed near the trailing edge of the strut. However, for case 2, 3 and 4, the additional recirculation regions are formed near the top and bottom wall of ramps. The large vortices are observed downstream of the ramps for case 3 compared to other ramp cases. It is well known that recirculation will be formed at the downstream of a ramp or a backward facing step in a supersonic flow field. For case 3, the shock boundary layer interaction at the base of the ramps creates a flow separation downstream of the ramp. The flow separation regime is comparatively larger for case 3 compared to the other cases where the recirculation region is formed through the separation region.

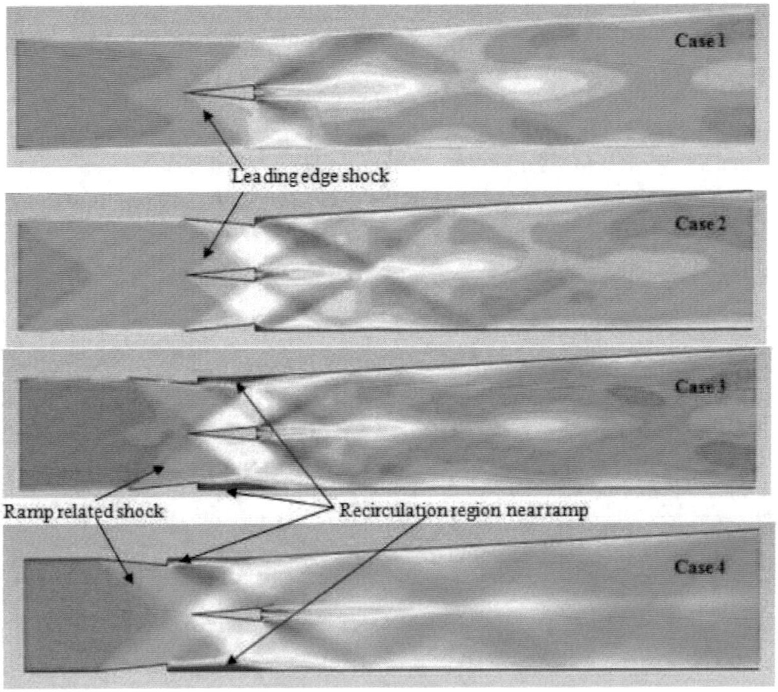

Figure 6. Mach number contour comparison of scramjet combustor with strut ramp at different position

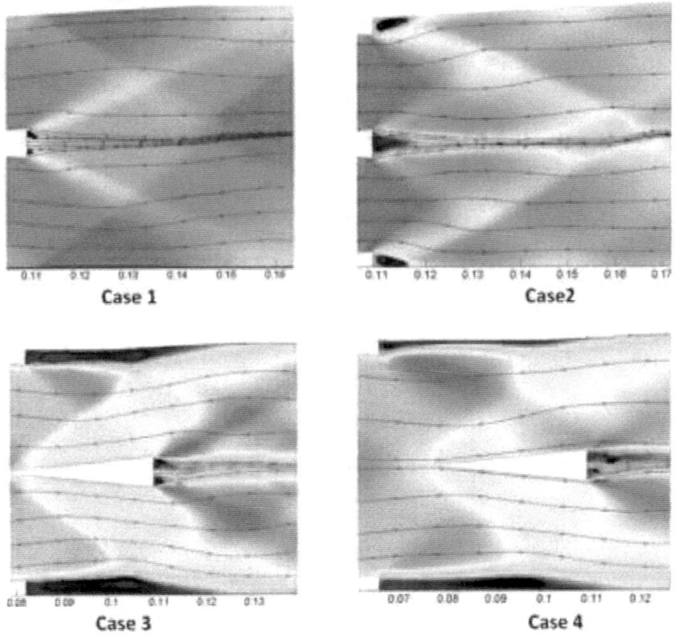

Figure 7. Streamline for various axial locations of the ramps in the DLR scramjet model

4.1. Wall static pressure

The wall pressure variations along the bottom wall and centreline of the combustor are shown in

Figure 8 (a, b). The wall pressure distribution is uniform for the typical DLR scramjet model till x=0.1m, subsequently, a wavy pressure profile is noted. The wavy nature of the pressure profile is due to the shock waves generated from the strut edges and its reflections from the combustor walls. An increase in static pressure profiles is observed for case 2 and case 3 from x = 0.05m due to the shock generated from the edges of the ramps. A peak pressure value is observed for case 4, at x = 0.1m due to the shock interactions and flow separation, which creates a compressive zone till the strut injection regime. It is notable that maximum static pressure from the bottom wall and midline of the duct at point x=0.1m to 0.15m, respectively. It is due to the intense shock wave reflections and flow separations inside the combustor.

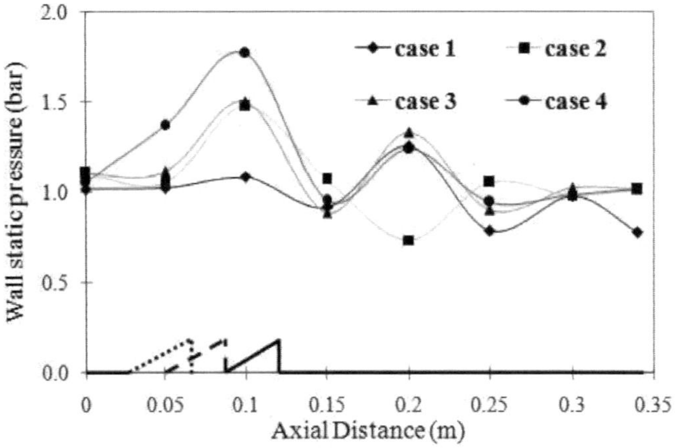

(a)

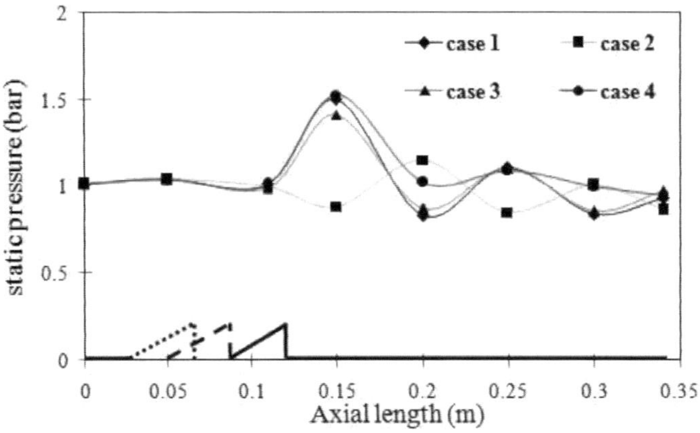

(b)

Figure 8. Pressure variation at (a) bottom wall and (b) centreline of the combustor

4.2. Mass fraction of H_2

The hydrogen mass fraction distribution at the three axial locations viz, X=0.15m, 0.2m, and 0.275m across the combustor are shown in Figure 9. At X = 0.15m, the distribution of hydrogen in the spatial direction is enhanced for case 2 than other configurations that may improve the combustion zone. Case 1 and case 4 show higher hydrogen mass fraction at X=0.2m, indicates that the shock waves' interaction with the fuel stream is less intense than case 2 and case 3. The hydrogen mass fraction is almost uniform for all the cases at X = 0.275 m, indicating that the fuel-air stream mixing is identical for all the configurations. The distribution of hydrogen near the walls of the combustor is negligible for all the cases. Further investigations on the reacting flow studies will reveal the effectiveness of the mixing and hydrogen distribution in the scramjet combustor.

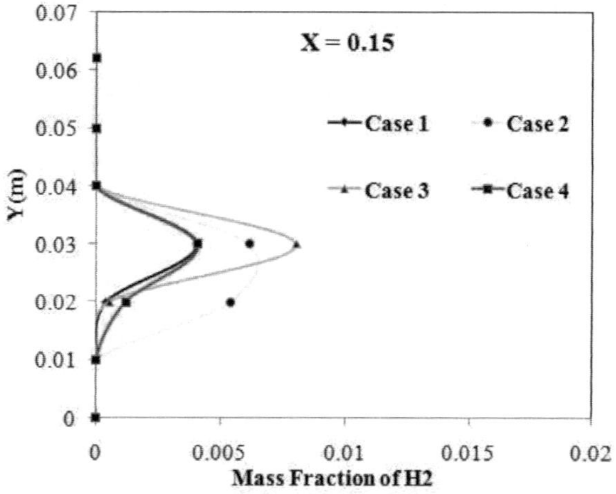

(a)

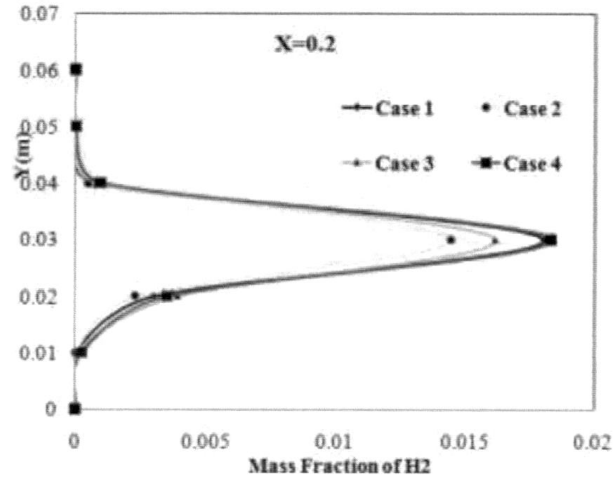

(b)

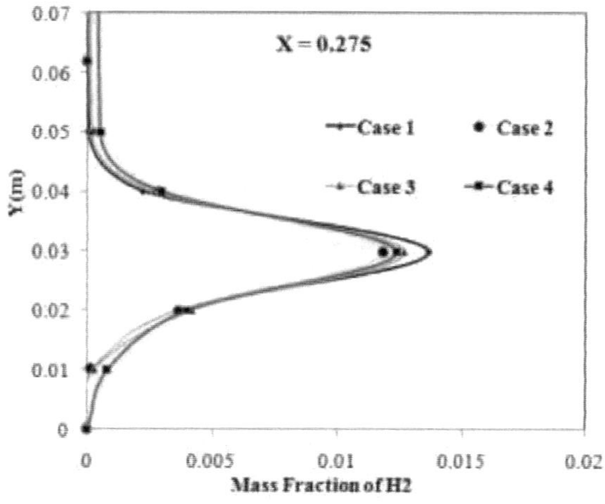

(c)

Figure 9. Mass fraction of hydrogen at (a) X=0.15m (b) X=0.2m and (c) X=0.275m.

4.3. *Total pressure loss*

The process of mixing with the shock waves is the irreversible process and causes the generation of entropy with total pressure losses. Generally, the more pressure losses occurs with an increase in mixing enhancement due to more oblique shock waves than other cases and their interaction with the shear mixing layer. The total pressure loss is calculated as,

$$\eta_t = 1 - \frac{\int_A P_o \rho_u dA}{\int_A P_{oinl} \rho_u dA} \tag{4}$$

Figure 12 shows the total pressure loss across the combustor at different locations. The higher pressure loss is noticed for the combustor with double ramps, due to the more shock wave interactions which decelerate the flow downstream and enhances total pressure loss compared to the DLR scramjet model (case1). It is observed that the maximum pressure loss of about 18% is observed for case 4 due to the shock waves interaction upstream of the strut, shock boundary layer interaction and flow separations create a compressive zone that decelerates the flow to the low supersonic condition compared to other ramp cases. For case 1, the total pressure loss of about 7%, which is comparatively lower than other cases with double ramps.

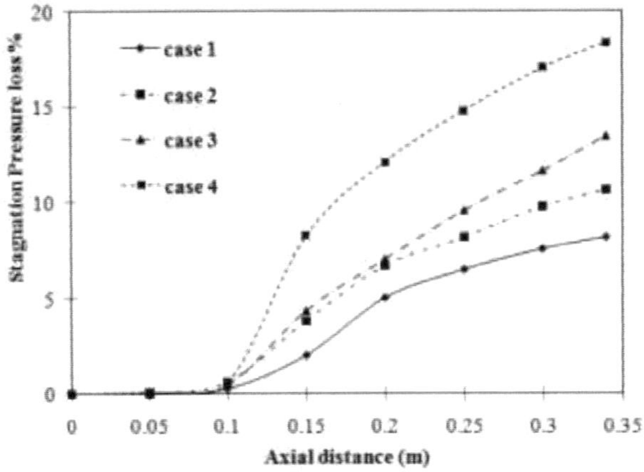

Figure 10. Total pressure loss for various geometry profile of the combustor

5. Conclusion

The numerical investigations of the normal DLR strut type scramjet combustor with upstream double ramps at the bottom and top walls of the combustor duct are carried out with non-reacting flow field conditions using the RANS equation. The effectiveness of the ramps locations along with the strut injection is analyzed based on the key parameters, such as wall static pressures, hydrogen mass fraction at various cross-sections, and total pressure loss across the combustor. The numerical results are validated with the reported experimental data, which shows higher correlations. The shadowgraph images show that more shock is formed for the double ramp DLR combustor than the typical DLR combustor model.

- The shock boundary layer interactions and flow separation resulting in higher wall static pressure for case 4 are compared to other double ramp profiles.
- The hydrogen distribution near the strut injector region enhances in the lateral direction as the ramps are positioned towards the combustor inlet.
- The shockwaves from the ramps decelerate the supersonic airflow downstream of the strut compared with the DLR scramjet model.

- In addition, more recirculation regions are formed for a double ramp combustor profile than the DLR scramjet model that acts as flame holders.
- The total pressure loss of the flow increases compared to the baseline scramjet model due to the more shock waves interaction arises from the ramps. The reacting flow studies will reveal the effectiveness of ramps in the strut based combustor is considered for future investigations.

Appendix

Nomenclature

Ma	Mach number
Po	total pressure
P	Static pressure
ρ	density
u	velocity
$\dot{m}$	mass flow rate
k	turbulence kinetic energy
ω	specific dissipation rate
G_k	production of turbulent kinetic energy
G_ω	generation of ω
Γ_k and Γ_ω	effective diffusivity of k and ω
Y_k and Y_ω	dissipation of k and ω
D_ω	cross-diffusion terms

References

[1] J. Kurian, K.. Habeeb, Cavities for Thermal and Momentum Mixing of Dual Supersonic Streams, AIAA 2002-4277. (2002).

[2] R.S. Fry, A Century of Ramjet Propulsion Technology Evolution, J. Propuls. Power. 20 (2004) 27–58. https://doi.org/10.2514/1.9178.

[3] K. Ambe Verma, K. Murari Pandey, M. Ray, K. Kumar Sharma, Effect of transverse fuel injection system on combustion efficiency in scramjet combustor, Energy. 218 (2021). https://doi.org/10.1016/j.energy.2020.119511.

[4] S.H. Won, I.S. Jeung, B. Parent, J.Y. Choi, Numerical investigation of transverse hydrogen jet into supersonic crossflow using detached-eddy simulation, AIAA J. 48 (2010) 1047–1058. https://doi.org/10.2514/1.41165.

[5] E. Erdem, K. Kontis, Numerical and experimental investigation of transverse injection flows, Shock Waves. 20 (2010) 103–118. https://doi.org/10.1007/s00193-010-0247-1.

[6] Y. Zhang, W. Liu, B. Wang, Effects of oblique and transverse injection on the characteristics of jet in supersonic crossflow, Acta Astronaut. 115 (2015) 356–366. https://doi.org/10.1016/j.actaastro.2015.06.004.

[7] W. Huang, Transverse jet in supersonic crossflows, Aerosp. Sci. Technol. 50 (2016) 183–195. https://doi.org/10.1016/j.ast.2016.01.001.

[8] A. Anazadehsayed, M. Barzegar Gerdroodbary, Y. Amini, R. Moradi, Mixing augmentation of transverse hydrogen jet by injection of micro air jets in supersonic crossflow, Acta Astronaut. 137 (2017) 403–414. https://doi.org/10.1016/j.actaastro.2017.05.007.

[9] S. Jeyakumar, S.M. Assis, K. Jayaraman, Experimental study on the characteristics of axisymmetric cavity actuated supersonic flow, Proc. Inst. Mech. Eng. Part G J. Aerosp. Eng. 231 (2017) 2570–2577. https://doi.org/10.1177/0954410016667149.

[10] S.M. Assis, S. Jeyakumar, K. Jayaraman, The Effect of Transverse Injection Upstream of an Axisymmetric Aft Wall Angled Cavity in a Supersonic Flow Field, in: J. Phys. Conf. Ser., 2019. https://doi.org/10.1088/1742-6596/1276/1/012019.

[11] S. Jeyakumar, K.A. Yuvaraj, K. Jayaraman, F. Cardona, M.T.H. Sultan, Effect of cavity fore wall modifications in supersonic flow, in: IOP Conf. Ser. Mater. Sci. Eng., 2016. https://doi.org/10.1088/1757-899X/152/1/012002.
[12] S. Jeyakumar, V. Venkateshwaran, N. Surjith, A.K. Raja, G.S. Samy, Experimental investigations on aft ramp cavities with fore wall modifications in scramjet combustors, 2017. https://doi.org/10.1007/978-81-322-2743-4_114.
[13] S. Jeyakumar, K. Jayaraman, Effect of finite width cavity in axisymmetric supersonic flow field, Proc. Inst. Mech. Eng. Part G J. Aerosp. Eng. 232 (2018) 180–184. https://doi.org/10.1177/0954410016674036.
[14] S. Jeyakumar, S.M. Assis, K. Jayaraman, Effect of Axisymmetric Aft Wall Angle Cavity in Supersonic Flow Field, Int. J. Turbo Jet Engines. 35 (2018) 29–34. https://doi.org/10.1515/tjj-2016-0027.
[15] S. Jeyakumar, P. Balachandran, S. Indira, Experimental Investigations on Supersonic Stream Past Axisymmetric Cavities, J. Propuls. Power. 22 (2006) 1141–1144. https://doi.org/10.2514/1.21024.
[16] K. Wu, P. Zhang, W. Yao, X. Fan, Computational realization of multiple flame stabilization modes in DLR strut-injection hydrogen supersonic combustor, Proc. Combust. Inst. 37 (2019) 3685–3692. https://doi.org/10.1016/j.proci.2018.07.097.
[17] C.N.R. Chung-jen Tam, Kuang-Yu Hsu, Mark R. Gruber, Fuel / Air Mixing Characteristics of Strut Injections for Scramjet Combustor Applications, AIAA Pap. 2008-6925. (2008).
[18] L. Suneetha, P. Randive, K.M. Pandey, Numerical investigation on influence of diamond shaped strut on the performance of a scramjet combustor, Int. J. Hydrogen Energy. 44 (2019) 6949–6964. https://doi.org/10.1016/j.ijhydene.2019.01.187.
[19] K. Kumaran, P.R. Behera, V. Babu, Numerical Investigation of the Supersonic Combustion of Kerosene in a Strut-Based Combustor, J. Propuls. Power. 26 (2010) 1084–1091. https://doi.org/10.2514/1.46965.
[20] G. Choubey, K.M. Pandey, Effect of parametric variation of strut layout and position on the performance of a typical two-strut based scramjet combustor, Int. J. Hydrogen Energy. 42 (2017) 10485–10500. https://doi.org/10.1016/j.ijhydene.2017.03.014.
[21] C. Li, X. Chen, Y. Li, O. Musa, L. Zhu, Numerical investigation on the performance of scramjet combustor with a novel strut configuration, Appl. Therm. Eng. 159 (2019) 113894. https://doi.org/10.1016/j.applthermaleng.2019.113894.
[22] L. Suneetha, P. Randive, K.M. Pandey, Numerical investigation on implication of dual cavity on combustion characteristics in strut based scramjet combustor, Int. J. Hydrogen Energy. 44 (2019) 32080–32094. https://doi.org/10.1016/j.ijhydene.2019.10.064.
[23] J. Miao, Y. Fan, T. Liu, Influence of strut on cavity at subsonic speeds: Ignition characteristics, Proc. Inst. Mech. Eng. Part G J. Aerosp. Eng. 0 (2020) 1–11. https://doi.org/10.1177/0954410020904832.
[24] Y. Zong, W. Bao, J. Chang, J. Hu, Q. Yang, J. Song, M. Wu, Effect of fuel injection allocation on the combustion characteristics of a cavity-strut model scramjet, J. Aerosp. Eng. 28 (2015) 1–8. https://doi.org/10.1061/(ASCE)AS.1943-5525.0000374.
[25] K.-Y. Hsu, C.D. Carter, M.R. Gruber, T. Barhorst, S. Smith, Experimental Study of Cavity-Strut Combustion in Supersonic Flow, J. Propuls. Power. 26 (2010) 1237–1246. https://doi.org/10.2514/1.45767.
[26] L. Suneetha, P. Randive, K.M. Pandey, Numerical investigation on mixing behavior of fuels in reacting and non-reacting flow condition of a cavity-strut based scramjet combustor, Int. J. Hydrogen Energy. 44 (2019) 16718–16734. https://doi.org/10.1016/j.ijhydene.2019.04.262.
[27] G. Choubey, K.M. Pandey, Effect of different wall injection schemes on the flow-field of hydrogen fuelled strut-based scramjet combustor, Acta Astronaut. 145 (2018) 93–104. https://doi.org/10.1016/j.actaastro.2018.01.034.
[28] W. Waidmann, F. Alff, U. Brummund, M. Bohm, W. Clauss, M. Oschwald, Experimental

Investigation of the Combustion Process in a Supersonic Combustion Ramjet (Scramjet), Hardthausen, Ger. DLR Inst. Chem. 470 Propuls. Eng. (1994).
[29] M. Oevermann, Numerical investigation of turbulent hydrogen combustion in a SCRAMJET using flamelet modeling, Aerosp. Sci. Technol. 4 (2000) 463–480. https://doi.org/10.1016/S1270-9638(00)01070-1.
[30] O.R. Kummitha, K.M. Pandey, R. Gupta, Numerical analysis of hydrogen fueled scramjet combustor with innovative designs of strut injector, Int. J. Hydrogen Energy. 45 (2020) 13659–13671. https://doi.org/10.1016/j.ijhydene.2018.04.067.
[31] K. Kumaran, V. Babu, Investigation of the effect of chemistry models on the numerical predictions of the supersonic combustion of hydrogen, Combust. Flame. 156 (2009) 826–841. https://doi.org/10.1016/j.combustflame.2009.01.008.
[32] P. Gerlinger, D. Br-uacute, Ggemann, Numerical Investigation of Hydrogen Strut Injections into Supersonic Airflows, J. Propuls. Power. 16 (2000) 22–28. https://doi.org/10.2514/2.5559.
[33] W. Huang, Z.G. Wang, S. Bin Luo, J. Liu, Parametric effects on the combustion flow field of a typical strut-based scramjet combustor, Chinese Sci. Bull. 56 (2011) 3871–3877. https://doi.org/10.1007/s11434-011-4823-2.
[34] G. Choubey, K.M. Pandey, Effect of variation of angle of attack on the performance of two-strut scramjet combustor, Int. J. Hydrogen Energy. 41 (2016) 11455–11470. https://doi.org/10.1016/j.ijhydene.2016.04.048.
[35] S. Kumar, S. Das, S. Sheelam, Application of CFD and the Kriging method for optimizing the performance of a generic scramjet combustor, Acta Astronaut. 101 (2014) 111–119. https://doi.org/10.1016/j.actaastro.2014.04.003.
[36] M.C. Banica, T. Scheuermann, J. Chun, B. Weigand, J. Von Wolfersdorf, Numerical Study of Supersonic Combustion Processes with Central Strut Injection, J. Propuls. Power. 26 (2010) 869–874. https://doi.org/10.2514/1.43599.
[37] S. Aravind, R. Kumar, Supersonic combustion of hydrogen using an improved strut injection scheme, Int. J. Hydrogen Energy. 44 (2019) 6257–6270. https://doi.org/10.1016/j.ijhydene.2019.01.064.
[38] W. Huang, L. Yan, Numerical investigation on the ram-scram transition mechanism in a strut-based dual-mode scramjet combustor, Int. J. Hydrogen Energy. 41 (2016) 4799–4807. https://doi.org/10.1016/j.ijhydene.2016.01.062.
[39] Wu, K., Zhang, P., Yao, W. and Fan, X., Numerical Investigation on Flame Stabilization in DLR Hydrogen Supersonic Combustor with Strut Injection, Combust. Sci. Technol. 189 (2017), 2154–2179.
[40] Athithan, A. A., Jeyakumar, S., Sczygiol, N., Urbanski, M., & Hariharasudan, A., The Combustion Characteristics of Double Ramps in a Strut-Based Scramjet Combustor, Energies, 14(4), (2021), 831. https://doi.org/10.3390/en14040831.

Aerodynamic Performance Improvement by Streamlining The Front Fairing of a Racing Vehicle

Ferdaus F[1,*], N Raghukiran[1], Vijaykumar[2], and Karthikeyan[3]

[1]Centre for Automation, School of Mechanical Engineering, [1]Vellore Institute of Technology, Chennai Campus, India. [2]School of Computer Science, New South Wales University, NSW, Australia, [3]School of Mechanical Engineering, Kings Cornerstone International College, Chennai, India.

*Corresponding author, E-mail: Fadyfithu@gmail.com

Abstract: The paper helps to explore and demonstrate the weight reduction of a component with the help of efficient aerodynamics. This can help to reduce fuel consumption and improve the performance of a racing vehicle. The vehicle used is a KTM RC200 2015 which was tested on various factors such as aerodynamics, weight, fuel consumption, and performance. The results were eventually validated with appropriate simulation and experimental trials. To improve and sustain the performance of the engine, it is necessary to explore various factors such as weight reduction through material selection and aerodynamically streamlining the vehicle design. The main component that was tested is the front fairing of the bike. The front fairing plays a very important role in a bike's performance. Therefore, the stock front fairing component is to be replaced with a modified component which also needs to meet all the regulatory safety standards. In the present work, two different front fairing components were used and one of them is made up of plastic representing the stock component while the second one is the customized and improved component made up of glass fiber reinforced epoxy. The results had shown promising improvements to the performance of the vehicle and it was observed that the material density and strength played a factor in the bike's performance and fuel efficiency.

Keywords: Aerodynamics, fuel efficiency, weight optimization, performance, computational fluid dynamics, front fairing, drag reduction, and motorcycle.

NOMENCLATURE:

1. Computer Aided Design	- CAD
2. Kraftfahrzeuge Trunkenpolz Mattighofen	- KTM
3. Bayerische Motoren Werke	- BMW
4. Digital holographic microscopy	- DHM
5. Electromyography	- EMG
6. Rapid Upper Limb Assessment	- RULA
7. Rapid Entire Body Assessment	- RABE
8. Ovako Working posture Assessment System	- OWAS
9. Quick Exposure Checklist	- QEC
10. 2-Dimensional	- 2D

11. 3-Dimensional - 3D
12. Calorific value - C_v
13. Density - ρ
14. Pressure - P
15. Temperature - T
16. Mechanical efficiency - η_m
17. Area - A

1. INTRODUCTION

While driving motorcycles, especially in open places such as highways, the most important resistance we feel is the wind. This resistance has a significant impact on not only the stability of and safety of the ride but also the fuel economy, reduction of noise, etc. This major retarding force that affects the movement of objects is aerodynamics drag [1].

The design of fairing for motorcycles should be considered based on the aerodynamics of its body. It is important to understand the various sources, effects, and impact of drag, then calculate their power which helps in designing fairings to reduce the drag effect [2]. The design of fairing should be such a way that it covers the rider as well as the major portion of the motorcycle body with a key objective of reducing the aerodynamic drag. The whole journal in precise states the significant role of the aerodynamic effects and performance by streamlining the front fairing of a racing vehicle supported by CATIA and CFD simulations along with its material selection, weight optimization, fuel efficiency, and also the comparison of custom side and stock side is compared and stated with an appropriate application. Also, reasons have been given on why custom side stock is mostly preferred by everyone.

Aerodynamics is characterized as the cooperation among airflow and the development of strong bodies through the air. A quick-moving motorcycle pushing its way through an air mass will meet a degree of resistance [3] which increments as the motorcycle's speed increments. This air resistance is all the more ordinarily known as drag. A machine with an enormous frontal zone will deliver more prominent degrees of drag than a motorcycle with a little frontal zone, and this connection between frontal territory and drag is legitimately relative. For instance, a motorcycle with a frontal region of one square meter, including the rider, creates twice as much drag as a motorcycle with just a 0.5 square meter frontal surface.

Auto industries emphasize a lot regarding the design of their vehicles since the movement has a significant impact due to aerodynamics and such efforts are generally called as "flow Analysis". Usual parameters such as speed, pressure, density, and temperature with respect to the moving object are analyzed in the flow analysis [3]. The flow analysis is performed under two different categories. One of them is an external analysis where the airflow is analyzed around the motorcycle. The external analysis is performed at different parts: flow in the front of the motorcycle, flow around the body and passenger and flow behind the motorcycle.

In large automobiles such as a car, bus, etc., the flow analysis underneath the body of the vehicle (between body and the ground) is also performed while that is minimal for motorcycle and is usually neglected [3]. The second type of analysis is internal to the body of the vehicle. The sudden disturbances caused by heavy crosswinds and the effect of air circulation while passing big vehicles are well established in the literature. Such air streams will flow through the radiator fins, which helps in cooling the engine [4]. Fairing design has more impact on external analysis and hence the internal analysis is not a vital part of the present study. To improve the flow path, boundary-layer separation happens to be very important. The boundary layer is the sheet of air adjacent to the body. While moving at slow speeds, motorcycles with fairings and smooth surface, this layer is thin and smooth. This becomes turbulent and thick layer if the surface is rough [4, 5].

The forces such as drag, lift, and side forces are the significant inertia forces that we have to deal with for effective aerodynamic effect. Drag is the most prominent force amongst these as this is the major obstruction for the forward movement of the motorcycle [5]. To calculate the drag, it is important to understand the airflow during the motorcycle ride. The drag is not consistent during the airflow from the front to rear of the motorcycle [6]. The air that hits the front of the motorcycle immediately forms a boundary layer and gets separated like a fierce stream. With fairing and smooth surface, this effect can be minimized. Continuing its journey towards the body of the motorcycle, once the stream is split, during the middle part of the motorcycle (around the body & passenger), the stream becomes unstable and oscillate.[6] Finally, it loses energy sufficiently to separate from the body surface (boundary layer). The separated flow continues to travel behind the motorcycle, and it forms a turbulent wake. The sudden hit we feel when a truck crossed us is due to this turbulence. The surface variations around the body of motorcycles are the major cause of such changes in aerodynamic forces [7]. Identifying various parts of the airflow and boundary layer, especially where it becomes turbulent and the distance it traveled over the body, and flow speed are important factors for analysis.

Vehicle design plays a significant role in flow separation. This is mainly in the body as well as If the object is flat and straight, then the separation is consistent throughout the body causing a significant drag effect at the end [8,9]. Any uneven surface will cause an additional drag component. Hence, one of the design considerations of fairing should be such that the air deflection towards the top should be higher than the passenger level. Otherwise, there will be an additional drag. If the body of the motorcycle is designed in a concave manner, then the boundary layer will stick close to the body and cause significant drag. A convex design (one that deflects air away from the motorcycle), makes easy separation and, hence reducing the drag [10].

The design patterns and considerations are to reduce the drag effect and force of the motorcycle during movement. The main objective of design considerations is initial approaches were taken at the overall motorcycle itself shaped in a curve, smooth and tapered [11]. Later, the design went to the level of components also. The reason for the curved fuel tank and every component comes with a tapered edge and curved shapes are primarily due to drag. Many design ideas were borrowed by airplane and ship development. Similarly, the design of the frontal part of the motorcycle should be curved and convex (deflect boundary layer away from the motorcycle body). The faring on the motorcycle should be angled as well rather than a straight vertical one. An angle of 8° to 11° gives a drag coefficient reduction of up to 8% [12].

This work mainly attempts to showcase the correlation of various factors in a racing bike. Firstly, the effect of a better shape for the front fairing on the drag is observed using CFD analysis, and subsequently, the material selection to make the aftermarket part is explored to make it lighter than the stock part. Subsequently, the effect of the combination of both materials used and better aerodynamic shape on the performance of the vehicle is highlighted.

2. METHODOLOGY

2.1 Material selection

Material selection plays a very important role on a motorcycle as they have to be as light as possible at the same time strong and sturdy. Material selection helps achieve this, the motorcycle industry and the racing industry have always been ahead on this in finding new composite materials that offer strength and lightweight components. Material selection affects all the performance factors if done correctly. The only downside is that some materials can be very expensive to acquire. There are various and countless composite materials but very few materials meet the specific need in the racing industry. Materials like aluminum, fiberglass, epoxy, plastics, resins, etc. are very common materials used in the lower end of the racing industry as they are cheap and strong at the same time provided the components are made correctly and accurately [13].

2.2. Modelling

Initially, a conceptual design has been created for the custom fairing in order to measure the angle using multiple views. Figure 1(a) represents the side view and top view of the conceptual design for the custom fairing. Moreover, CATIA V5 was used to create the basic 2D models which are the precursors for the study (Figures 1).

Using CATIA we will design both the stock side and custom side model. The dimensions of the CATIA model for the stock side (fig 1 c and e) are that the baseline is 74 cm, the length is 45 cm and the breadth is 15 cm. On the other hand, the dimensions for the custom side model (fig 1 d and f) are that the baseline is 64 cm and the length is 50cm and the breadth is 20cm. The angle that we have measured for the bends is 50 cm and 30 cm respectively. Figure 1shown below is the completed picture of the CATIA diagram shown with the above-given dimensions. The basic methodology to create these diagrams starts with opening the CATIA workbench where which we select and move through the mechanical design and part design. Once we have clicked part design. In fig 1 image the PAD operation has been used to produce a geometry. The stock top diagram dimensions are 28mm and for the custom top diagram, the dimensions are 30mm.

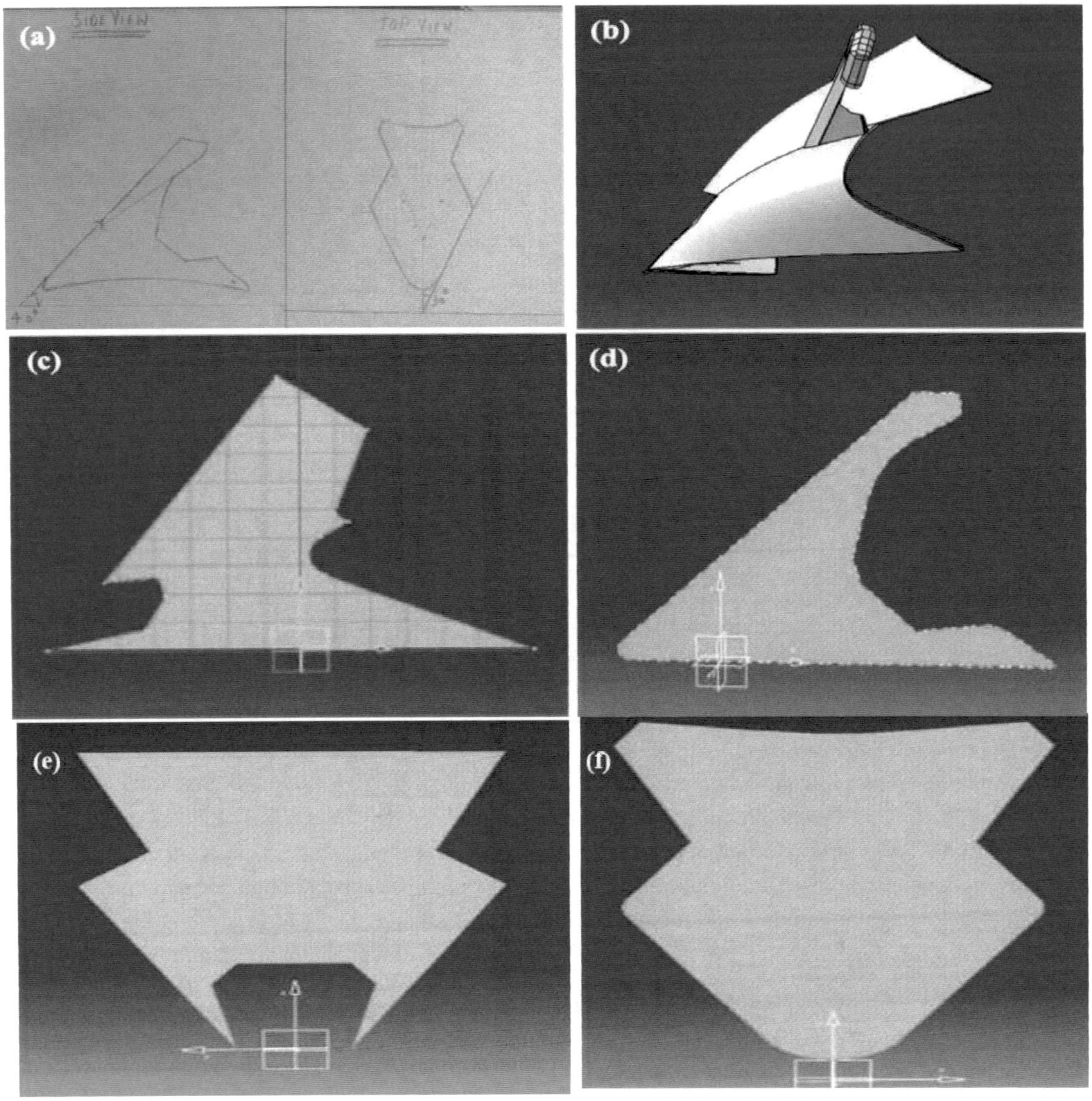

Figure 1: Design of faring: (a) conceptual design of the custom fairing, (b) 3d of the custom fairing, (c) stock side view of the fairing, (d) custom side view of the fairing, (e) stock top view of the fairing and (f) custom top view of the fairing

The figure 2 a and b depict the stock side and top view of the meshing. Figure number 2 c, and d depict the custom side and top view of the meshing. In the meshing, there are two different types of the element have been used to maintain a flow. One quadrilateral element and the other in a triangular mesh.

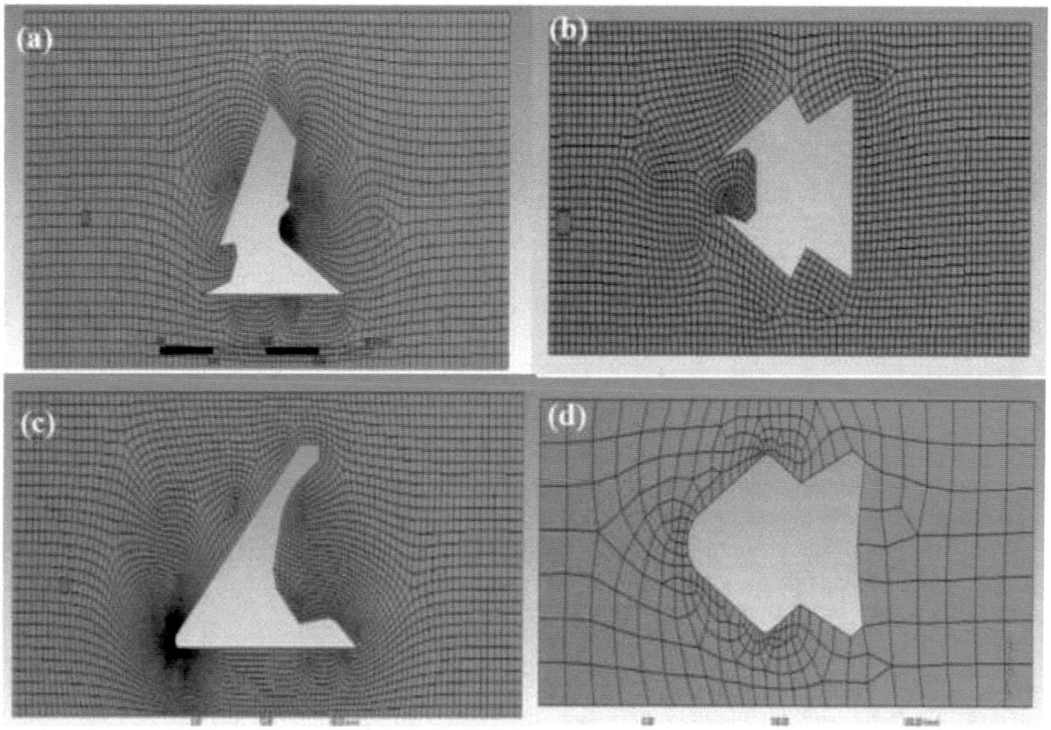

Figure 2: Meshes of fairing: (a) stock side view of the meshing, (b) stock top view of the meshing, (c) custom side view of the meshing and, (d) custom top view of the meshing

Table 1 depicts the meshing criteria and parameters for the mesh. The mesh criteria play a major role in the meshing. It's mandatory to use the meshing criteria before generating mesh. In order to maintain the mesh flow, the targeted length was 5, the minimum and maximum lengths are 3 and 7. Also to maintain a right angle on the plane, the given aspect ratio, warping and skewness are 1, 45^0, 10.

Table 1 Meshing criteria and parameters for the mesh

Sl. No	Mesh Criteria	Value
1	Aspect Ratio	1
2	Skewness	45^0
3	Warping	10
4	Taper	0.35
5	Jacobian	0.7
6	Minimum length	3
7	Maximum length	7
8	Targeted Length	5
9	Minimum angle quadrilateral	45^0
10	Maximum angle quadrilateral	135^0
11	Minimum angle triangle	30^0
12	Maximum angle triangles	120^0

13	Mesh distortion	4

3. RESULTS AND DISCUSSION

Consequently, as a result, given in table 2, shows and discusses about the boundary condition of the existing models and proposed model. Its shows various parameters of the computational domain and the solver such as materials, solver, density, viscosity, etc.

Table 2 The Computational detail and solver detail for fairing

Sl .No	Computational Domain	Conditions
1	Processing	Series
2	Solver	Pressure Based
3	Timer	Steady State
4	Dimensional	2D
5	Models	Energy/Viscous –
6	Material	Air
7	Viscous Model	K Epsilon - Standard
8	C_p Specific Heat	1006.43 j/kg - K
9	Density	1.225 Kg/m^3
10	Thermal Conductivity	0.0242 w/ m -k
11	Viscosity	1.789 e^{-5} Kg/m -s
12	Boundary Condition	Velocity Inlet
13	Reference temperature:	300 K
14	Inlet total Pressure:	0 Pa (assumed)
15	Outlet Static Pressure:	0 Pa
16	Wall	no Slip conditions

3.1 Stock Results

Aerodynamics velocity results of stock side velocity, stock top velocity stock side view pressure, custom side velocity, and custom top velocity are depicted in Figures 3 to 6. Figure 3 a and b depicts the stock side velocity and the stock top velocity. These figs 4 a and b show that the stock fairing is less streamlined and the stock fairing has a lot of turbulent flows and pressure spots, Compared to the custom side. The faring stock available and its performance from both sides and the top view is shown in the respective diagrams with the results on the left.

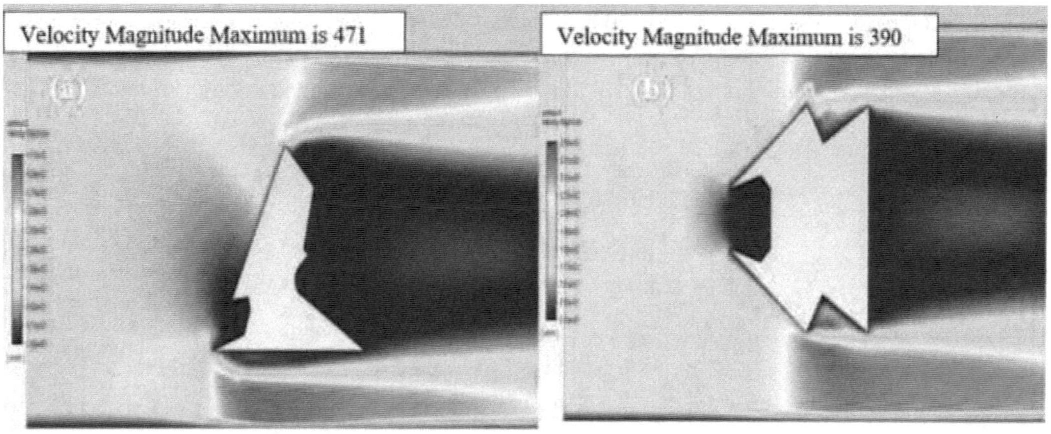

Fig 3: A stock side view and top view velocity magnitude (km/hr.) contours: (a) fairing side view and (b) fairing top view.

Figure 4 (a, b) depicts the stock side view and top view turbulence. The turbulence shown here represents the turbulence kinetic energy. Mostly the turbulence describes the turbulent fluctuations of the velocities. The turbulence is calculated with the help of the Reynolds number.

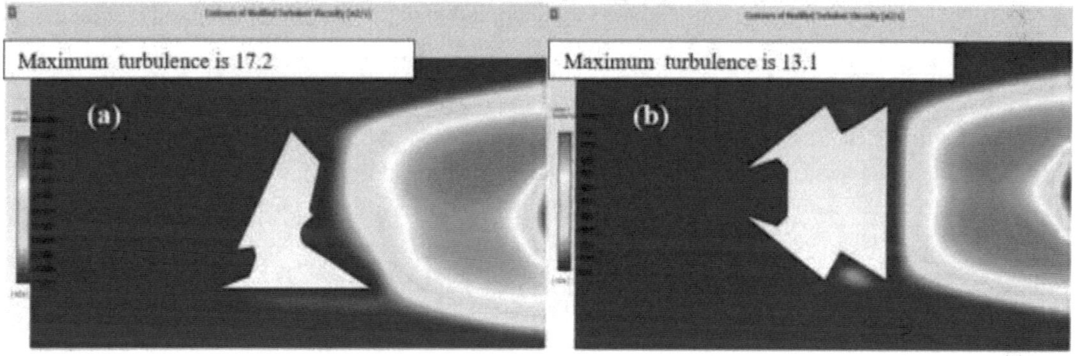

Fig 4: A stock side view and top view turbulence ($m^2/s.$) contours: (a) fairing side view and (b) fairing top view.

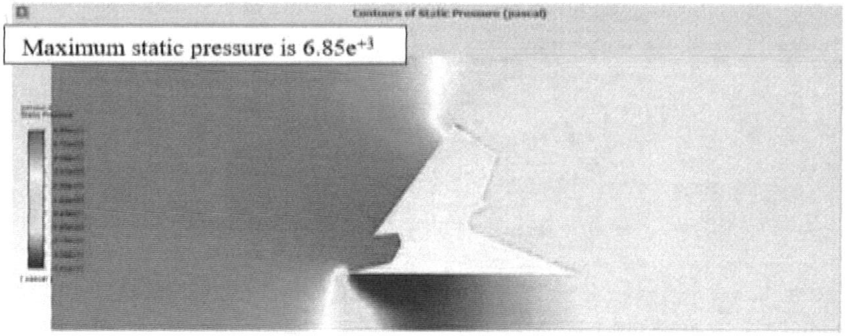

Fig 5 stock side view pressure distribution

IOP Conf. Series: Materials Science and Engineering 1128 (2021) 012007 doi:10.1088/1757-899X/1128/1/012007

Figure 5 depicts the stock side view pressure. When we consider the weight as a major factor the pressure plays a major role which supports deformation and helps in creating stress in the entire model and simulation till it reaches the equilibrium state. In other words, this is also called a static structural model. According to our simulation Figure 6 states the simulation image of the static pressure.

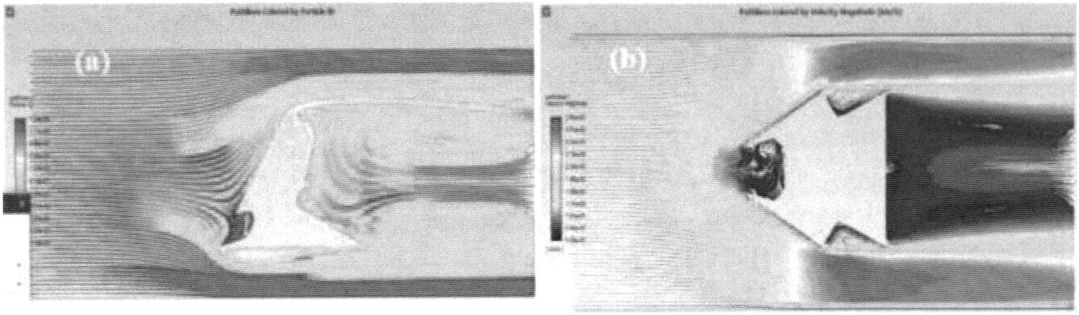

Fig 6: A stock side view and top view particles path lines: (a) fairing side view and (b) fairing top view

Figure 6 a and b represent particle pathlines of the stock side and top view. In this figure, it can clearly see the flow path line are understandable with color codes. With the help of path lines, the air flows, turbulent flow, and velocity flows can be verified.

3.2 Custom Fairing Results

Fig 7 a and b depict the custom side velocities. It can be noticed that with respect to the differences in the above Fig 5 a and b, where the stock fairing has a lot of turbulent flows and pressure spots. However, in the custom fairing, it can be seen that the flow is much cleaner and more streamlined compared to the stock fairing. This means the custom fairing has lesser drag and more velocities which was the objective of these tests. Custom results and simulation would be much sharper because the fairing has been modified and kept accordingly to the user's comfort. Hence observing from the below picture, we can find the velocities are sharper and faster. The overall custom side velocity result is 650 km/hr. that is observed from the simulation for custom fairing.

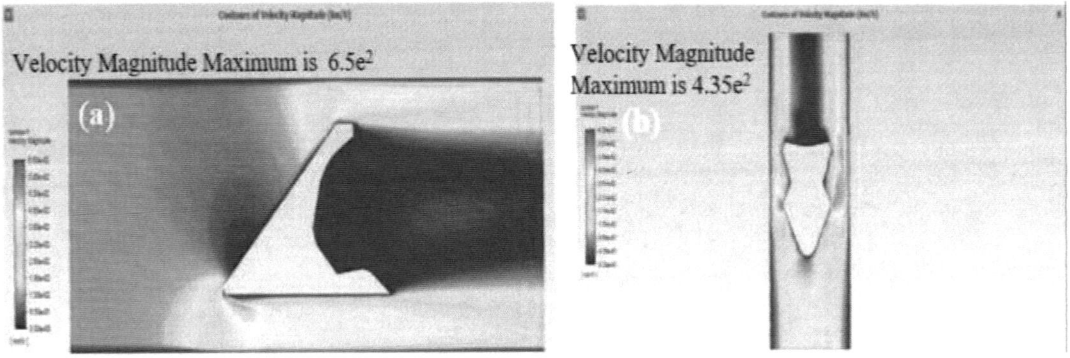

Fig 7: A custom side view and top view velocity magnitude (km/hr.) contours: (a) fairing side view and (b) fairing top view

Figure 7 a and b depict the custom side velocities. It can be noticed that with respect to the differences in the above Fig 9 (a) and (b), where the stock fairing has a lot of turbulent flows and pressure spots. However, in the custom fairing, it can be seen that the flow is much cleaner and more streamlined compared to the stock fairing.

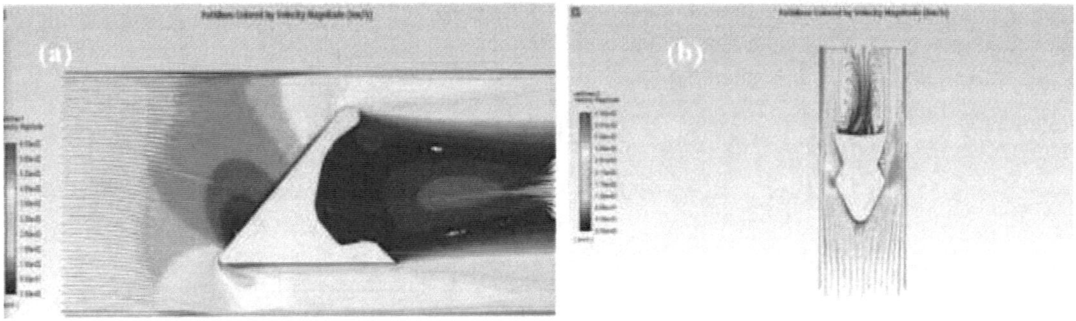

Fig 8: A custom side view and top view particles path lines: (a) fairing side view and (b) fairing top view

Figure 8 a and b represent the custom side view particle path line. These path lines are the lines that are traveled by the neutrally buoyant particles in equilibrium with the fluid motion. These particle path lines are also called as the dimensional flow. From the image, we can see that how the air flows are passing towards the top and side view.

Figure 9 a and b depict the custom turbulence and the solution converged graph. Mostly the turbulence describes the turbulent fluctuations of the velocities. The graph tells us the final result is converged this means that the solution we have obtained is correct. Now that we know that the boundary of separation and its design has a huge impact on aerodynamic drag. If the force is huge around the boundary layer and if the separation is not smooth, the drag will be more. So, it is important to custom design the faring and makes sure the boundary layer separation and flow throughout the body of the vehicle is important. The type of turbulence stated here is the turbulence kinetic energy.

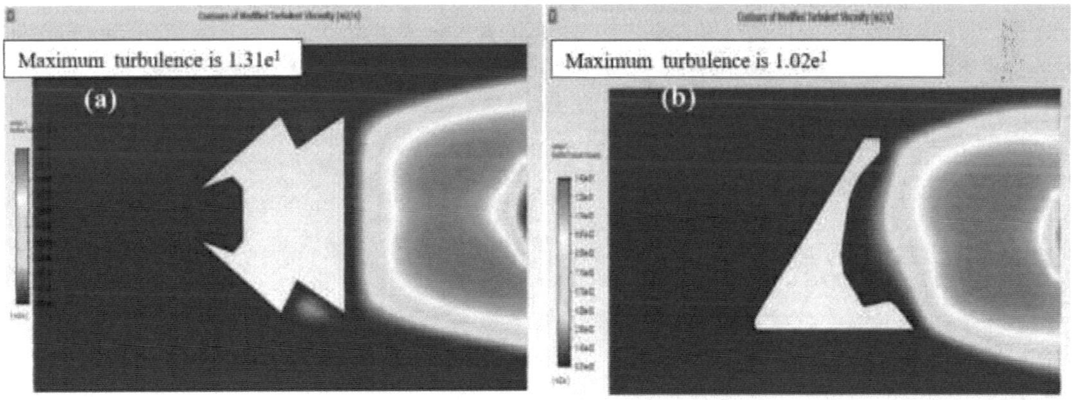

Fig 9: A custom side view and top view turbulence (m²/s.) contours: (a) fairing side view and (b) fairing top view.

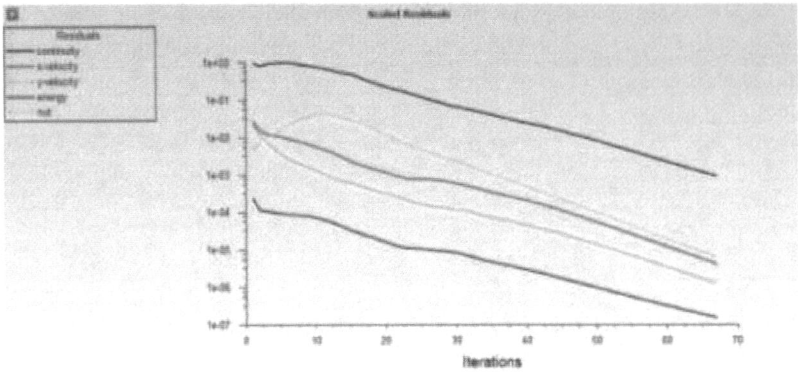

Fig 10 Result Iteration Graph

Figure 10 indicated the converged solution graph. The graph has plotted with 5 different colors. The black color indicates continuity, the blue color indicated energy, blue and green indicated x and y velocity.
In this graph, x-direction indicates the iterations, and y-direction indicates e. From the graph we can see that continuity increasing compare to others, the x and y velocity increasing the iterations and decreasing the e. We can conclude that the converged solution is based on continuity, energy, x, and y velocity.

3.3 Comparison of Stock and Custom Fairing

The stock weight represents the actual weight of the product based on and brought out by its manufacturing. In Fig 10 (a) if we observe the picture of the stock weight the weight can be observed which 2.215kg. The stock weight is also represented as the default weight of an object. On the other hand, when we look at 10 (b) custom weight can be observed. Custom weight represents the modified weight of an object as per the user's interest. For example, taking the below pictures as an example it can be observed that the look of the fairing is modified and the custom weight has been drastically reduced by 1.15 kg.

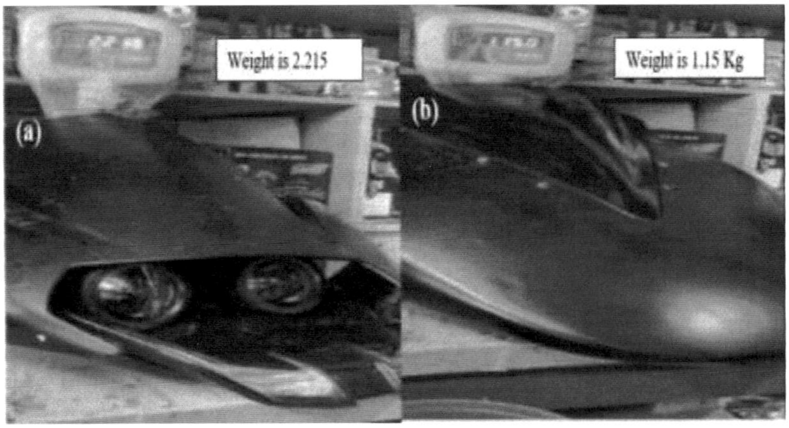

Fig 11: Weight of the fairing in kg. (a) stock fairing and (b) custom fairing

With the newly modified component in place, the bike was run for 5 km at an average speed on the same roads, and the mileage was recorded. The difference in the performance was very profound and proved that the custom fairing made a huge difference in the fuel consumption rates as seen in Fig 12 (a) and (b).

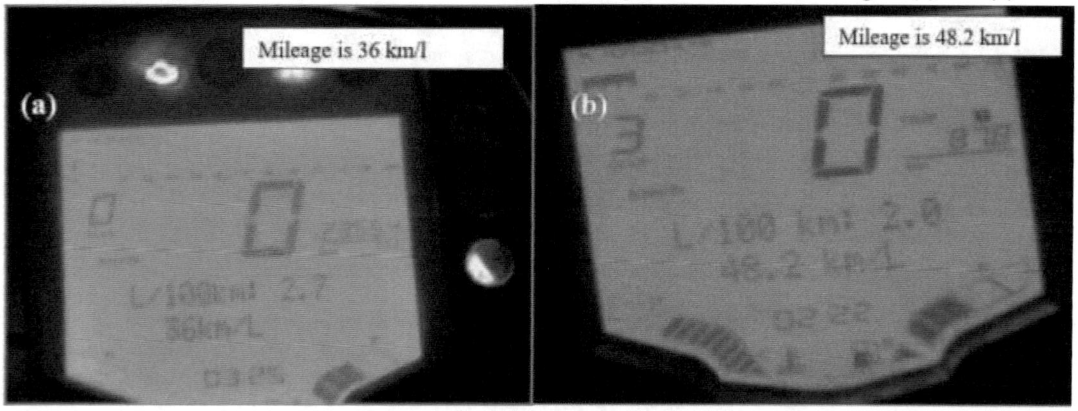

Fig 12: Mileage of the vehicles in km/l. (a) stock mileage and (b) custom mileage

The two components that were used are made of plastic (stock) and glass fiber reinforced epoxy (custom) respectively. Therefore, the influence of material weight on the bike's performance and fuel efficiency is to be inferred.

Using Catia, a solid block of dimensions 10 cm length, 5 cm breadth, 0.2 cm depth was modeled to depict a part of the front fairing panel. The component is then attributed with two different materials mentioned in the previous sections. Boundary conditions are applied to test them under 10 N force with two sides fixed. The simulation was done for both of the materials (the plastic and the GFRE) separately (Figure 13 (a) and (b)).

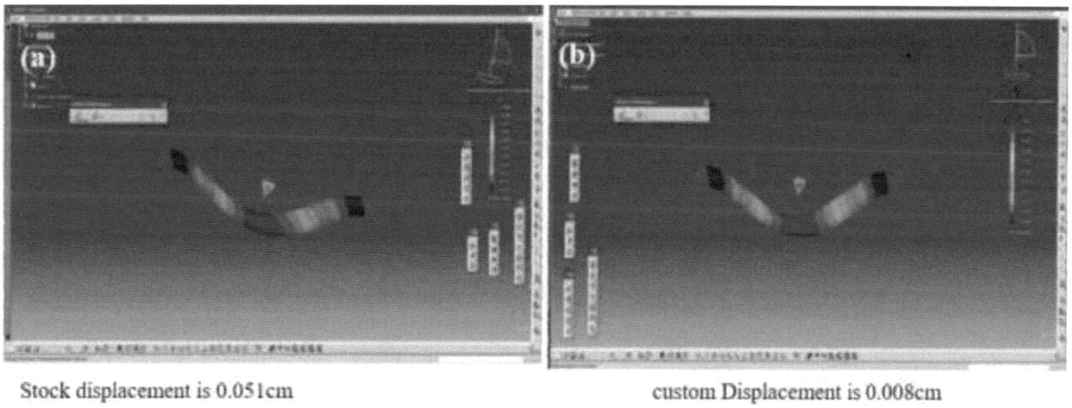

Figure 13: Material displacement: (a) displacement for the stock fairing, (b) displacement for the custom fairing

The stock of RC200 has a stock speed of 134 km/h. On the other hand, the custom fairing has a speed of 137 km/h as shown in fig 14. Looking at aspects like the weight and fuel efficiency the performance and advantages were totally supportive for the custom fairing. Since we have customized the fairing the difference in the performance was very profound and it proved that the custom fairing made a huge difference in the fuel consumption rates. On the contrary, as we have modified and customized it ourselves the weight has decreased. Since the weight has decreased the motorbike now has more aerodynamic, speed, less weight, and more fuel efficiency. Overall, the performance of the custom fairing has been good and more effective compared to the stock fairing.

Fig 14 custom top speed

Table 3 denotes the stock and custom design comparison, and the custom design provides a high efficiency compared to the existing one. Similarly, the top speed, weight, and velocity also obtained better variation and results. It can see that custom results have increased efficiency. The table gives the detailed values of each aspect and differentiating and stating how the custom design is mostly preferred and efficient

Table 3 Comparison of stock and custom results

Sl.No	Parameters	Stock	Custom
1	Velocity	471 km/h	650 km/h
2	Material Displacement	0.051cm	0.008cm
3	Weight	2.215kg	1.150kg
4	Fuel Efficiency	36km/l	48km/l
5	Top Speed	134 km/h	137km/h

Figure 15 a) represents the custom fairing for the race bike and b) represents the stock (existing) fairing for the race bike. It shows vehicle comparison of custom fairing and stock. In conclusion, custom fairing has more aerodynamics compare to stock.

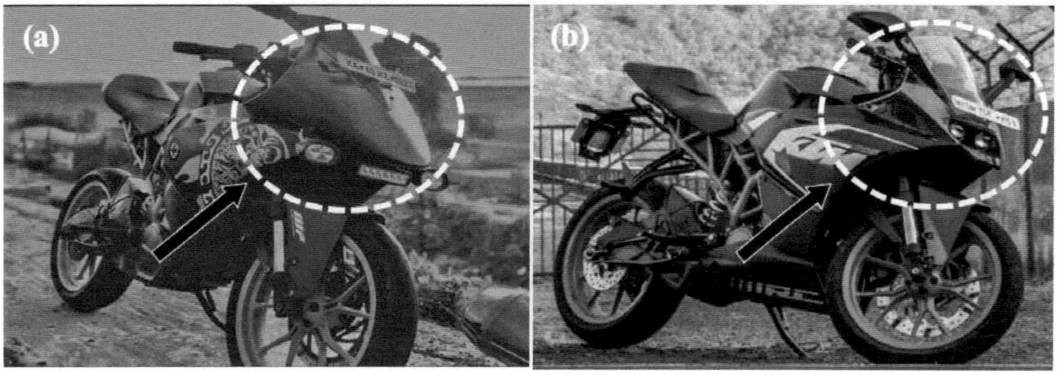

Fig 15: The fairing of a racing motorcycle: (a) custom fairing for a race bike and (b) stock fairing for a race bike.

4. CONCLUSIONS

From the results and outcomes from various analysis and tests from the present work, it can be clearly seen how material selection and relatively better aerodynamic component helps in reducing the weight of the vehicle and with resulting enhanced performance of a motorcycle, the fuel consumption will be reduced. The objective of this work was to test and analyses various factors and incorporate them into a single component and see how each of these factors complement each other in practical conditions. Hence, an aftermarket component can increase fuel efficiency and performance and reduce weight, and suitably changing the material and shape of a component.

6. REFERENCE:

[1] de Rome, L., Taylor, E. A., Croft, R. J., Brown, J., Fitzharris, M. and Taylor, N. A. S. (2015) 'Thermal and Cardiovascular Strain Imposed by Motorcycle Protective Clothing Under Australian Summer Conditions', Ergonomics, 139(June), pp. 1–29. doi:10.1080/00140139.2015.1082632.
[2] Fakir M. CFD study on aerodynamic effects of a rear wing/spoiler on a passenger vehicle.
[3] Zhu, S., Murakami, S. and Nishimura, H. (2012) 'Internal analysis of a motorcycle taking into account the rider's effects', Vehicle System Dynamics, 50(8), pp. 1225–1245.
[4] Scappaticci L, Risitano G, Battistoni M, Grimaldi C. Drag optimization of a sport motorbike. SAE Technical Paper; 2012 Apr 16.
[5] Gilkeson C, Toropov V, Thompson H, Wilson M, Foxley N, Gaskell P. Aerodynamic shape optimization of a low drag fairing for small livestock trailers. In12th AIAA/ISSMO multidisciplinary analysis and optimization conference 2008 Sep (p. 5903).
[6] Motorcycle Handling and Chassis Design – Tony Foale Design and Testing of a Large Composite Asymmetric Payload Fairing - Tomoya Ochinero, Ph.D.,1 and Thomas Deiters, P.E.2 ATA Engineering, Inc., Los Angeles, CA, 90245
[7] Scott, G. B., and Lambe, N. R. (1996) 'Working practices in a perchery system, using the OVAKO Working Posture Analyzing System (OWAS)', Applied Ergonomics, 27(4), pp. 281–284. doi: 10.1016/0003-6870(96)00009-9.

[8] SAE J 243:2006 Motorcycle Classifications and testing

[9] Malizia F, Blocken B. Bicycle aerodynamics: History, state-of-the-art, and future perspectives. Journal of Wind Engineering and Industrial Aerodynamics. 2020 May 1;200:104134.

[10] Bruzzone F, Scorrano M, Nocera S. The combination of e-bike-sharing and demand-responsive transport systems in rural areas: A case study of Velenje. Research in Transportation Business & Management. 2020 Sep 29:100570.

[11] Epema HK, Van Den Brand S, Gregoor W, Kooijman JD, Pereboom HP, Wielemaker DC, Van der Zweep CJ. Bicycle Design: A different approach to improving on the world human-powered speed records. Procedia Engineering. 2012 Jan 1;34:313-8

[12] Holeman AS, Holeman HD, inventors; Novigo Designs LLC, assignee. Protective cover for motorcycle fairing. United States patent application US 29/500,531. 2015 Aug 18.

[13] Kamble A, Bhorge R, Janunkar R, Karamkar B. A Study on Effect of Air Resistance on Motorcycle.

[14] Cakir M. CFD study on aerodynamic effects of a rear wing/spoiler on a passenger vehicle.

[15] Sedlak V. Motorcycle cornering improvement: An aerodynamically approach based on flow interference.

[16] Gilkeson C, Toropov V, Thompson H, Wilson M, Foxley N, Gaskell P. Aerodynamic shape optimization of a low drag fairing for small livestock trailers. In12th AIAA/ISSMO multidisciplinary analysis and optimization conference 2008 Sep (p. 5903).

[17] Reiman J, Heppel P, inventors; Aerotail LLC, assignee. Deployable vehicle fairing structure. United States patent US 6,799,791. 2004 Oct 5

[18] Abe KH, Gevert KV, inventors; Bayerische Motoren Werke AG, assignee. Fairing for motorcycles. United States patent application US 07/316,685. 1991 Feb 19.

[19] Selvamuthu T, Thangadurai M. Motorcycle Drag Reduction using a Streamlined Object Ahead of the Rider. Journal of The Institution of Engineers (India): Series C. 2019 Oct 1;100(5):801-10

[20] Ribaldone, E., Puri D., Cogoti, I., Pizzoni, R., 2011. Optimizing the external shape of vehicle at the concept stage: integration of aerodynamics and ergonomics, SAE International Journal of engines. (4) 2:2622-2628.

Design of 3D Printed Aircraft Seat Structure using Latticing in combination with Topology Optimization and Generative Design

Wedyn Noronha[1], Vedant Joshi[1], Dr. S Jeyanthi[1]
[1]Vellore Institute of Technology, Chennai
[1]Vellore Institute of Technology, Chennai
[1]Vellore Institute of Technology, Chennai
nwedyn999@gmail.com, vedantjoshi@gmail.com, Jeyanthi.subramanian@vit.ac.in

corresponding author: nwedyn999@gmail.com

Abstract: The aviation industry is responsible for 2% of all human-induced CO_2 emissions. According to Iata (a trade body), the number of air passengers is predicted to touch 16 billion by 2050. Minimizing weight is central in achieving an ideal balance between payload and range of an aircraft and consequently, low fuel consumption. For instance, in a Boeing 787, a 20% weight savings will result in 10 to 12% superior fuel efficiency. Besides reducing carbon footprint, aircraft performance improvements like better acceleration, higher structural strength and stiffness, and increased safety could also be achieved by lightweight design. In this research paper, we propound the design of an aircraft seat structure which can be fabricated from Polyetherimide resin by 3D Printing. The structure is designed using topology optimization, generative design, and latticing. The lattice structures are made and tested on nTopology™. The seat's ergonomics are evaluated on Catia V5. The seat design meets all functional requirements without compromising on passenger comfort. The design can be used for standard commercial jets, and the weight savings are predicted to reduce fuel consumption drastically, resulting in lower costs and lower emissions.

1. INTRODUCTION

Light-weighting is associated with the use of advanced materials on structures that are optimized to have minimal mass. Light weighting can be achieved through the computational and numerical development of structures. Conventional numerical methods are sizing, shaping, finite element analysis, and topology optimization [1]. At the same time, latticing and generative design are popular computational techniques. Six factors are considered while choosing materials for seating in the aircraft industry: mechanical strength, ductility, low density, corrosion resistance, machinability, and cost [2]. According to studies conducted by Francis Froes [3], aluminum and seat cushion foam contribute 43% and 8% to seat weight, respectively. This accounts for the seat's structure and the cushioning. Therefore, this paper aims to reduce these components' weight, specifically since they account for 51% of the total seat weight.

Latticing converts solid-designed components into structures consisting of a network of nodes that are interconnected by beams or struts. Latticing reduces the weight of components significantly without compromising on structural integrity. In fact, the structure's interconnected nature enables it to adjust to loads from different angles, making it more resilient. Lattice structures are widely used because they are associated with mechanical properties, which standard bulk materials cannot achieve. This can be achieved by changing the position of nodes or altering beam/strut dimensions. Latticing gives designers control over how structures will interact with external loads in the form of shock, vibrations, or fatigue loading. Lattices can be incorporated in designs to provide higher strength to weight ratio and better energy absorbance. For this paper, the main concerning load was the compressive load on the seat. Studies by An Hou and Kurt Gramol [4] showed that, under compression, triangle lattices are weaker due to bridging, which causes delamination. Hexagon structures are hence preferred.

Before proceeding with design optimization and analysis, several factors such as safety, comfort, and ergonomics must be analyzed to finalize design contours. Accordingly, several research studies have been previously published discussing the optimum seat designing parameters for maximum comfort. Ishant Gupta et al. [5] worked on simulating comfort and analyzing several design parameters of aircraft seats in the civil aviation industry by deploying the Delmia Human v6 software. It was inferred from this research study that the 95th percentile of the human population portrayed the most favorable results of a backrest angle of 105° and a seat height of 440mm. On the contrary, the 5th percentile population depicts the optimum design parameters of a seat height of 330mm and a backrest angle of 90°. Other techniques such as pressure mapping and the practical implementation of *Ideal Pressure Distribution* were carried out by Gabriel H. Campos et al. [6]. A morphed seat design was presented with an optimized seat curvature. Decentralization of the pressure distribution on the backrest was successful, and a similar contour was adopted in this current research study. Further research on biomechanical analysis on aircraft seats was done by R.H.M. Goossens et al. [7], where various factors, including seat height, depth, backrest height, width, armrest length, width, and the backrest inclination was analyzed for best comfort. These values and calculations were further employed in this research study. Light-weight techniques were done by S. Dangal et al. [8] by incorporating spring-foam technology to model a comfortable aircraft seat cushion. An overall weight reduction of 20% was achieved accordingly. Throwing light on this area, numerous other light-weighting techniques can be employed in industrial products.

Topology Optimisation is a commonly known technique by which a particular component can be optimized based on the numerous loads. Topology Optimization works iteratively, removing mass in low-stress concentration areas and reserving high-stress concentration regions, thereby generating an optimized structure. Each iteration removes a certain amount of mass from the total component according to the forces acting on it and the mesh. Generative Design, on the other hand, is also a flourishing technique for design optimization. Generative Design, as its name suggests, adds material based on the stress concentration in specific areas. The method requires creating preserve regions and obstacles as boundary conditions. Preserve regions are areas where mass cannot be compromised, and obstacle regions are specific locations on which mass cannot be added due to real-life constraints. This paper compares topology optimization and generative design of the supporting members on an aircraft seat and creates a low-weight design. Bearing this in mind, the following research study works on an alternative solution to creating light-weight aircraft seats for maximizing fuel efficiency, thereby minimizing fuel consumption.

2. METHODOLOGY

2.1 THEORETICAL CALCULATIONS

The leg of an aircraft seat is one of the most load-bearing components among the entire seat assembly. Hence, a rigorous analysis must be done to ensure reliability and optimization. This paper has focussed mainly on two computational techniques, i.e., topology optimization and generative design, to optimize the part with minimum weight and maximum performance. Before performing computational structural analysis on the leg, the forces acting on it were analyzed. A free-body diagram of the leg was created, and we subsequently concluded that the two forces acting on it were as follows:

1. Force exerted by the weight of the passenger as well as the remaining components of the seat (Vertically downward force) – The maximum weight of a passenger was considered to be 125kg, and the remaining weight of the seat as 10kg for a total of 135kg (worst-case scenario). Since the acceleration due to gravity is $9.81 m/s^2$, the total force exerted on 2 legs will be 1324.35N. As a result, the force exerted on a single leg will be around 660N.
2. The pseudo force exerted on the leg due to acceleration and deceleration of the aircraft (Horizontal forward/backward force) – An aircraft in motion does not travel at a constant velocity, and hence it is essential to consider the force on the leg, which acts in a horizontally backward direction. The average maximum velocity attained by an aircraft was taken to be

285kmph by analyzing several aircraft speeds. Based on a research study, the ground roll distance before take-off was taken as 2314ft, which nearly equals 705.3072m. Using the 3rd kinematic equation of motion i.e., $(v^2 - u^2 = 2 \times a \times s)$, the maximum acceleration was calculated to be 4.426m/s². Hence, the pseudo force acting on one leg will be approximately 300N.

VERTICALLY DOWNWARD FORCE

$$w_{person} = weight\ of\ person = 125kg$$
$$w_{seat} = weight\ of\ seat = 10kg$$
$$g = graviational\ constant = 9.81ms^{-2}$$
$$W_{Total} = w_{person} + w_{seat} = 125 + 10 = 135kg$$
$$F_{Vertical} = W_{Total} \times g = 135 \times 9.81 = 1324N$$
$$F_{Vertical\ on\ 1\ leg} = \frac{F_{Vertical}}{2} = \frac{1324.35}{2} = 662.175 \sim \mathbf{660N}$$

HORIZONTALLY BACKWARD FORCE

$$v^2 - u^2 = 2 \times a \times s$$
$$v = Final\ Velocity = 285\ kmph = 79ms^{-1}$$
$$u = Initial\ Velocity = 0\ ms^{-1}$$
$$s = ground\ roll\ distance = 2314ft = 705.3072m$$
$$a = acceleration = 4.426ms^{-2}$$
$$F_{Total} = Total\ Horizontal\ Force = m \times a$$
$$= 135\ x\ 4.426 \sim 597.51N$$
$$= 600N$$
$$F_{1\ leg} = Horizontal\ Force\ on\ 1\ leg = \frac{F_{Total}}{2} = \frac{600}{2} = \mathbf{300N}$$

2.2 TOPOLOGY OPTIMISATION

After the forces acting on the leg were calculated, a solid model was created using Solidworks and was further imported to Ansys Workbench Design Modeller. A highly refined mesh sized 1mm was generated, and a mesh refinement of 3x was applied on the three holes where beams will be attached. This was done to further refine the mesh in areas subjected to a more significant amount of stress and deformation. Subsequently, the three holes and an offset material of 10mm were kept around it as a preserve region. The remaining volume was kept solid to allow the software to optimize that region based on stress experienced and total deformation. The design optimization of the leg was done using numerous Aluminium grades using topological methods. For this, Aluminium 2, 6 and 7 series materials were chosen. Al 7075- T6511, Al- 7050-T7651, Al 6061-T89 and Al 2024-T861 showed the best results, as shown in Table 1.

2.3 GENERATIVE DESIGN FOR ADDITIVE MANUFACTURING

Generative design is an iterative design process that uses a program to generate multiple-segmented outputs that meet certain constraints. Further, an engineer can refine the optimal results by selecting specific output or changing input values, ranges and distribution. The novel generative design feature in Autodesk Fusion 360 was deployed for an alternative optimization of the aircraft seat leg. Instead of importing a solid model into the design software, a smaller model consisting of preserve and obstacle geometry was used. Generative design is thus a method by which the software adds material, not interfering with the obstacle geometry and maintaining the preserved geometries. The generative design studies were mainly conducted using the Aluminum grades 7075 T6 and 6061 T6. Four outcome studies were run for each of the two grades based on the manufacturing method required for the leg.

2.4 LATTICING

Lattice structures are generally created using one of two popular techniques, i.e., a uniform voxel-based approach or the conformal latticing approach. The uniform voxel technique divides the component into unit volumes and maps lattice topology into those volumes. [9] In conformal latticing, the unit cells are structured along the surfaces that are to be reinforced. Their deformation occurs such that it conforms to the component surface. Since the lattices are mapped into the unit volumes in the uniform voxel technique, many lattice unit cells are broken or disconnected at the surfaces, making them weak. These discontinuities result in fracture of the structure when loaded in compression or tension. In contrast, the conformal latticing technique creates hexahedral meshes on the component surface, and the unit cells are then mapped onto these meshes. This methodology eliminates the incomplete cells as observed in the uniform vortex technique. This makes conformally latticed structures more resilient. Compressive strength varies based on whether the lattice shows bending-dominated or stretch-dominated behavior. Stretch-dominated behavior shows greater strength as compared to bending-dominated. [10] However, it must be noted that conformal latticing ensures that lattice structures in the same layer are fully connected and provide strength in the direction of the connections. If loads come from other directions, the structure could be weakened. So, engineers should accurately determine the loading directions before orienting the lattices within the structure for optimal performance. In this paper, we have latticed the backrest, cushion, and hand rest of the aircraft seat. The conformal latticing technique was used on topology PlatformTM software. To compare the different lattice unit cells, we replicated them along with material properties on Solidworks and tested them in Ansys. The CAD model of the components was designed on Solidworks. The CAD files were imported onto the nTopology platform and then converted to implicit bodies. A volume mesh had to be created to lattice the structure. While latticing the edge length, the lattice cell shape was altered in different iterations. The edge length was set as 25mm. The Triangular Honeycomb (THC), Octet, Body Centred Cubic (BCC), Face Centred Cubic (FCC), Diamond and Kelvin Cell lattices were tested.

2.5 MATERIAL SELECTION

The rigid aromatic rings in high-temperature thermoplastics help them withstand high temperatures. The aromatic rings hinder the backbone chain's movement since two chemical links have to break for a chain break to occur. This is better than aliphatic structures since they require only one link to be broken for chain breaks. Thus, mechanical properties, high-temperature capability, and chemical resistance are greatly enhanced and even better than crosslinked, thermosetting polymers.

ULTEM™ 9085 resin is a Polyetherimide resin that is a flame-retardant, high-performance thermoplastic. It has a high strength-to-weight ratio, excellent impact strength, and high heat resistance. In addition, it has an optimal flame, smoke, and toxicity (FST) properties. ULTEM™ 9085 resin CG meets stringent test criteria and retains traceability required by the aerospace industry. Polyetherimide (PEI) is also a widely used high-temperature thermoplastic in the aircraft industry. PEI is selected for internal aircraft applications for its inherent flame retardancy and low smoke emissions. It also has excellent chemical resistance to fuels and fluids used in the aircraft industry. Its heat deflection temperature (HDT) is approximately 153° C. ULTEM 9085 has a flexural strength of 16200 psi, making it resistant to fatigue failure. Furthermore, ULTEM-9085 is compatible with FDM 3D printing technology, making it an appropriate choice since 3D printing is majorly used in the seat's manufacturing.

3.RESULTS AND DISCUSSION

3.1 *ANSYS WORKBENCH – TOPOLOGY OPTIMIZATION SETUP*

Topology optimization in Ansys Workbench was carried out in 4 iterations to reduce mass: 10%, 15%, 20%, and 25% mass retained. The results can be seen in Fig 1. It can be inferred that optimizing the

IOP Conf. Series: Materials Science and Engineering 1128 (2021) 012008 doi:10.1088/1757-899X/1128/1/012008

geometry with 10% and 15% mass retention resulted in several imperfections around the upper holes. The iteration with 25% mass retention shows a promising structure with minor imperfections. However, this iteration does not efficiently reduce weight to a minimal amount. The iteration with 20% mass retainment generated nearly 500g of weight reduction compared to the iteration with 25% mass retained while maintaining mesh consistency, strength, and minor imperfections. Hence, further optimization was executed to reduce additional weight. However, before optimizing the geometry further, the Aluminium grades 7075-T76511, 7050-T7651, 6061-T89, and 2024-T861 were finalized among 15 grades due to various factors, including density, yield strength, and availability.

3.1.1 POST PROCESSING AND STRUCTURAL ANALYSIS IN ANSYS SPACECLAIM

Before analyzing the four grades in Ansys, the optimized geometry was post-processed in the Ansys Spaceclaim modeler. The generated design was smoothened, shrink-wrapped and auto-fixed to make it compatible with additive manufacturing requirements, as shown in figure 2(b). Figure 2(a) shows the aircraft seat leg's structural analysis with the four materials mentioned earlier. Three factors, mainly total deformation, maximum elastic equivalent strain and maximum equivalent stress, were analyzed. The four materials showed similar results since their structural and mechanical properties nearly the same with minor differences. It was inferred that Al 7075-T76511 showed the minimum deformation while Al 6061-T89 experienced minimum equivalent stress. Consequently, Al 6061-T89 was finalized attributed to lower density (2700kg/m^3) than the other three grades. Based on the mentioned results, the grade Al 6061-T89 was finalized.

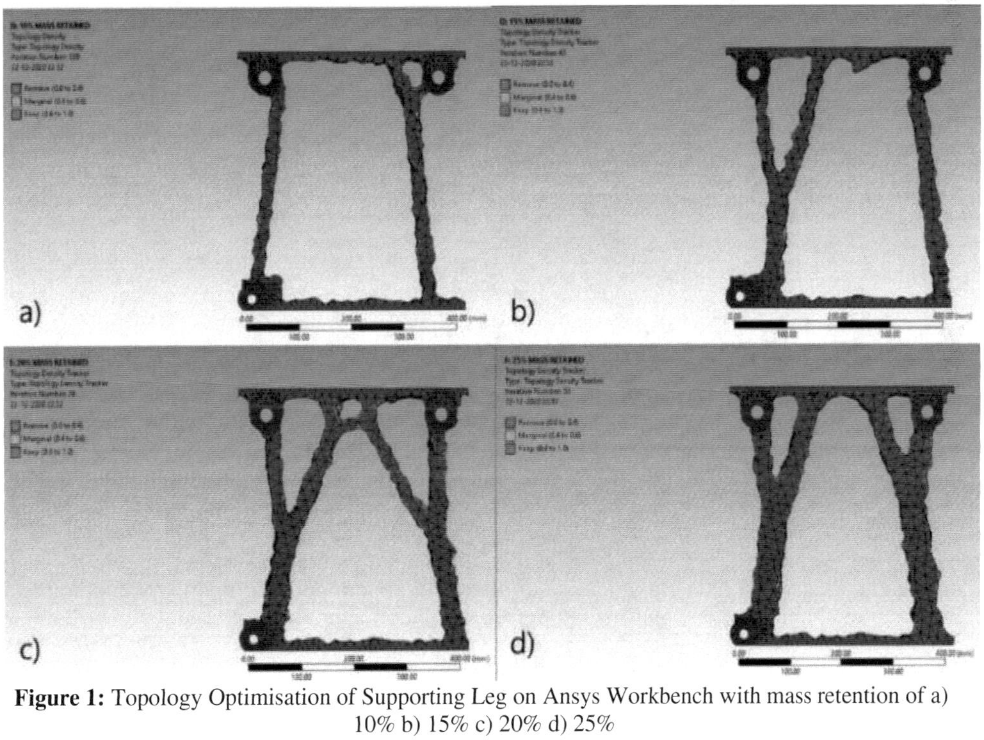

Figure 1: Topology Optimisation of Supporting Leg on Ansys Workbench with mass retention of a) 10% b) 15% c) 20% d) 25%

IOP Conf. Series: Materials Science and Engineering 1128 (2021) 012008 doi:10.1088/1757-899X/1128/1/012008

Table 1: Comparison of mechanical behaviour between several materials for the aircraft seat leg

Material	Total Deformation (mm)	Maximum Equivalent Elastic Strain	Maximum Equivalent Stress (MPa)	Yield Strength (MPa)
Al 7075-T6511	0.026453	8.5567E-05	5.9799716	510
7050-T7651	0.026851	8.5567E-05	5.9799716	500
6061-T89	0.027479	8.7547E-05	5.8791732	370
2024-T861	0.026705	8.5081E-05	5.8891731	490

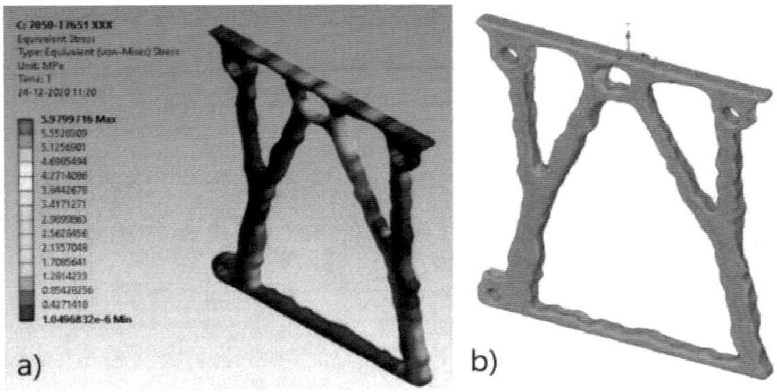

Figure 2: (a) Structural analysis of the topology optimised structure (b) Post processing of the topology optimised structure

3.1.2 TOPOLOGY OPTIMIZATION IN AUTODESK FUSHION 360

The leg was further optimized using the shape optimization tool in Autodesk Fusion 360 software. Instead of importing a solid model, the resultant topology optimized model in Ansys was used for subsequent analysis and optimization. The four iterations of simulations based on load path criticality in Fusion 360 can be seen in figure 3. It can be inferred that in all 4 cases, material from the middle portion of the leg has been removed since that particular area experiences minimal stress as compared to other portions, thereby reducing more weight. After carefully analyzing the equivalent stress in every region of the optimized design, it was concluded that the iteration with a 0.8 load path criticality would best suit optimization. This version was post-processed in Solidworks, and fillets were added to the edges for weight reduction and minimizing stress concentration around sharp corners.

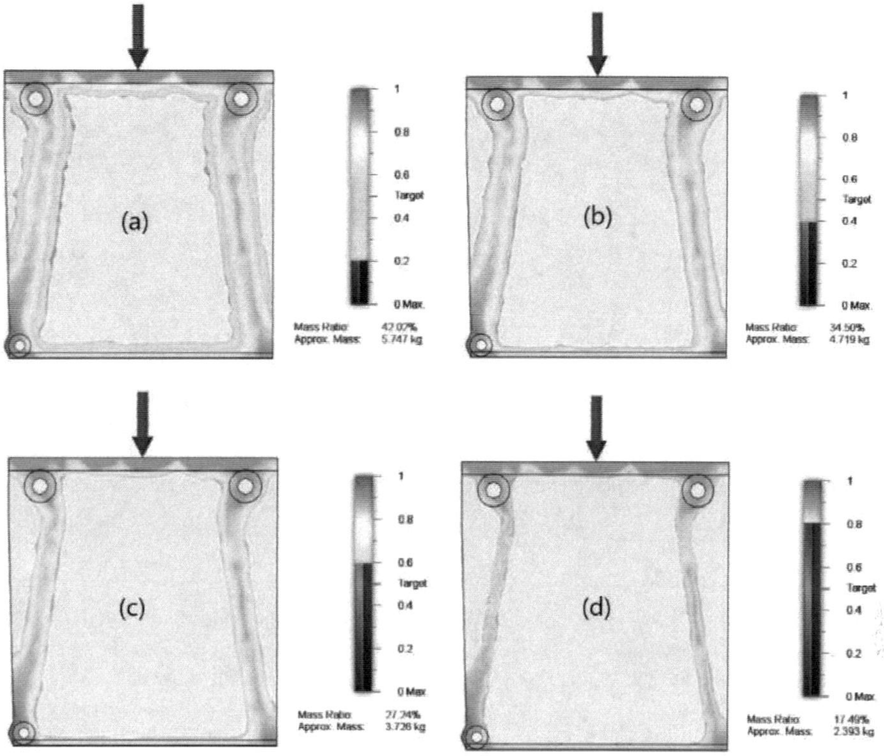

Figure 3: Topology Optimization iterations based on load path criticality of (a) 0.2 (b) 0.4 (c) 0.6 (d) 0.8

3.1.3 STATIC STRUCTURAL ANALYSIS IN ANSYS WORKBENCH

Afterward, the thickness of the leg was varied between 0.5 and 1 inches for further optimization. The results of the structural analyses of the leg variations of 3 thicknesses are shown in Table 2. It can be inferred that the leg with thicknesses 1in and 0.75in are overdesigned. Despite having a thickness of 1in, the maximum equivalent stress experienced by it is roughly 10MPa less than that experienced by the leg having a thickness of 0.5 inches. It is also evident that the leg with a thickness of 0.5 inches weighs around 1.2 kg lesser than the leg with a thickness of 1in, proving that it is a valid, optimized design while maintaining structural strength and durability. It also shows a considerable deformation value coming out to be 0.54081mm. Hence, after a series of simulations and further modifications, the final optimized design is shown in Figure 4(a).

Table 2. Mechanical behaviour of aircraft seat leg iterations with different thicknesses

Thickness of Leg (in)	Total Deformation (mm)	Maximum Equivalent Elastic Strain	Maximum Equivalent Stress (MPa)	Factor of Safety	WEIGHT (kg)
1	0.28149	2.9578E-04	20.991	17.62744164	2.37
0.75	0.37116	2.9545E-04	21.175	17.47343566	1.79
0.5	0.54081	4.3066E-04	30.876	11.98341754	1.21

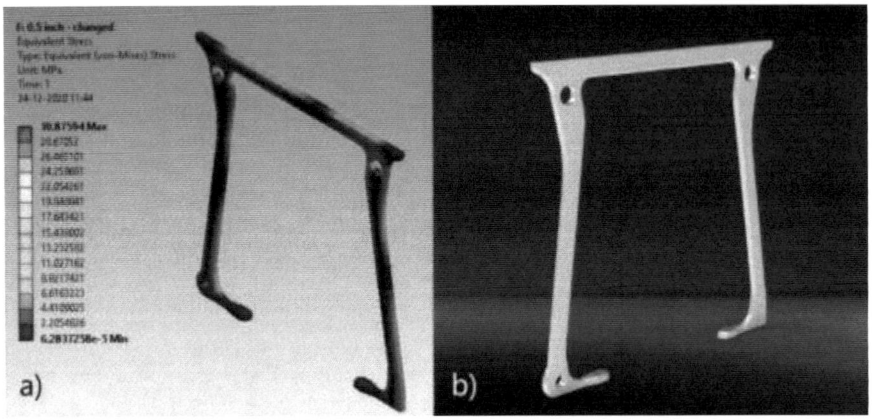

Figure 4: (a) Structural analysis of the topology optimized structure (b) Post processing of the topology optimized structure

3.2 FINDINGS AND INFERENCES FROM 8 GENERATIVE DESIGN OUTCOMES

Numerous conclusions can be drawn after analyzing the structure, durability, reliability, and other structural results. A comprehensive comparison of various characteristics of the eight iterations is listed in Table 3. According to the results listed in Table 3, it can be inferred that the designs for Al 7075 T6 and Al 6061 T6(unrestricted manufacturing method) show the least global displacement among all other iterations. They also exhibit a high and reliable factory of safety. On the other hand, optimized designs (b) and (f) show the maximum global deformation of 0.54mm and 0.98mm, respectively, amongst all other outcomes, which prove unreliable and easily deformable. These designs also have a minimum factor of safety of 2 since their orientations for additive manufacturing are incompatible with the loads experienced. Hence (b) and (f) cannot be post-processed further for 3D printing.

3.2.1 *COMPARATIVE ANALYSIS OF THE RESULTS*

One iteration each (for AL7075 T6 and 6061 T6), as shown in figure 5 (a and e), was performed irrespective of the required manufacturing method required. The remaining six designs were generated based on a motive for subsequent 3D printing in the X, Y, and Z directions. Figure 5(b c d) and figure 5(f g h) show the design optimization for additive manufacturing in the X, Y, and Z direction for Al7075 T6 and Al6061 T6, respectively. Furthermore, designs (b), (c), (f), and (g) show geometry inconsistencies in load-bearing areas, such as the two holes on the top and one hole at the bottom. Although designs (c) and (g) show a moderate deformation of 0.5mm and 0.17mm, they lack the geometry to be manufactured without post-processing irregularities and eliminated from the others. Concerning maximum weight reduced, designs (e) and (h) show the most promising results having a weight of 1.566kg and 1.571kg, respectively. The software's designs were tailored to be 3D printed in the Z+ direction concerning their isometric view. On the other hand, iterations (a) and (d) also resulted in considerable weight reduction, with the final mass of each outcome being 1.632kg and 1.636kg, respectively. It is also evident from the diagram that iterations tailored to be 3D printed subsequently in the Z+ direction showed the most optimized and safe results along with geometry compatibility. However, while comparing outcomes (a), (d), (e), and (h), it is clear that Al 6061 T6, which is to be further 3D printed in the Z+ direction, experienced the maximum equivalent stress of 8.4MPa as compared to the other three outcomes. Finally, after finally narrowing down to the parameters of weight, maximum equivalent stress, global deformation, and the generative design tool's percentage recommendation, iteration (e) was finalized and considered for further post-processing.

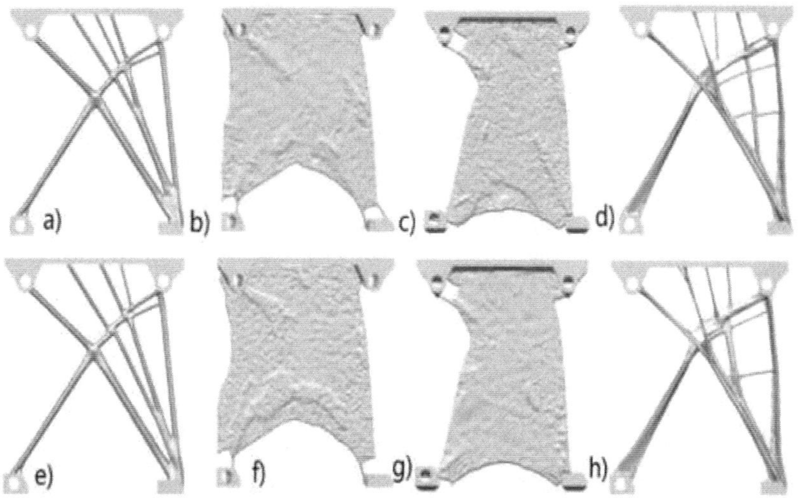

Figure 5: Generative Design outcomes based on material properties and manufacturing methods

Table 3. Comparison of Generative Design outcomes based on material properties and manufacturing techniques

Outcome	Material	Manufacturing Method	Orientation	Iterations	Mass (kg)	Maximum Von Mises Stress (MPa)	Minimum Factor of Safety	Maximum Global Displacement (mm)	Recommendation by Autodesk Fusion 360 Software (%)
a	Al 7075 T6	Unrestricted	-	36	1.632	4.4	32.96	0.04	91
b	Al 7075 T6	Additive Manufacturing	X+	41	2.15	72.5	2	0.54	43
c	Al 7075 T6	Additive Manufacturing	Y+	39	2.214	50.3	2.88	0.50	19
d	Al 7075 T6	Additive Manufacturing	Z+	45	1.636	5.4	26.85	0.08	70
e	Al 6061 T6	Unrestricted	-	42	1.566	4.8	57.02	0.04	92
f	Al 6061 T6	Additive Manufacturing	X+	40	2.055	137.5	2	0.98	49
g	Al 6061 T6	Additive Manufacturing	Y+	39	2.113	40.9	6.72	0.17	41
h	Al 6061 T6	Additive Manufacturing	Z+	46	1.571	8.4	32.61	0.09	71

3.3 LATTICE STRUCTURE COMPARISONS ON ANSYS AND NTOPOLOGY PLATFORM

The tables below depict the analysis results for the compared lattice structures. The weights of the latticed components are also shown.

Table 4. Comparison of Lattice structure performances under service conditions

Lattice type	Thickness	Total Deformation (mm)	Maximum Equivalent Elastic Strain	Maximum Equivalent Stress (MPa)	Factor of Safety
BCC	2mm	60.552	1.5455	2832.7	0.04
	3mm	0.12003	0.0039757	737.06	0.33
	5mm	0.1346	0.000848	129.59	1.92
FCC	2mm	112.12	1.4746	2913.1	0.04
	3mm	22.08	0.56431	1130.1	0.12
	5mm	2.956	0.14598	303.33	0.46
OCTET	2mm	2.9362	0.1747	281.38	0.49
	3mm	1.2985	0.094068	197.02	0.71
	5mm	0.41029	0.040426	70.432	1.98
TETRAHEDRAL	2mm	9.7797	0.25217	544.69	0.25
	3mm	3.9588	0.15568	333.2	0.42
	5mm	1.5688	0.09556	208.54	0.7
THC	2mm	1.7115	0.078826	160.38	0.87
	3mm	0.65104	0.040347	81.851	1.71
	5mm	0.20946	0.012388	23.305	6
DIAMOND	2mm	1.0311	3.92E-02	75.33	1.85
	3mm	0.7564	3.62E-02	71.456	1.97
	5mm	0.32575	3.24E-02	67.178	2.08

Table 5. Comparison of Lattice Structure Weights

Lattice type	Thickness	Mass of Seat Cushion (g)	Mass of Back Rest (g)	Mass of Hand-rest (g)
BCC	2mm	841.20554	1308.23	155.4292
	3mm	1819.9124	2831.59	
	5mm	2814.1625	7,392.48	
FCC	2mm	127.9261	1597.39	184.4612
	3mm	227.96	3459.95	
	5mm	401.4139	8817.99	
OCTET	2mm	106.402	3076.95	335.0241
	3mm	225.354	6598.9	
	5mm	489.2857	16124.2	
TETRAHEDRAL	2mm	1234.38	1877.49	151.992
	3mm	2634.62	4029.84	
	5mm	6532.97	10241.1	
THC	2mm	844.972	11738.6	646.852
	3mm	1136.27	24421.8	
	5mm	1819.91	37002.3	
DIAMOND	2mm	278.3425	1289.48	156.6279
	3mm	558.3638	2853.48	
	5mm	662.6248	7,316.89	

The seat cushion bears the most load amongst all the latticed components. The best iterations for the seat cushion in terms of Factor of Safety (FoS) were Octet (1.98), Diamond (1.85), and Triangular Honeycomb (1.71). Thus, we considered these lattices for the seat structure. The THC lattice structure had a comparable Factor of Safety but was heavier than Octet and Diamond structures, so it was discarded. The diamond iteration of 2mm beam thickness had the lowest weight (278.3425g). However, we selected the Octet Lattice due to its lower deformation (0.41029 mm) than Diamond (1.0311mm).

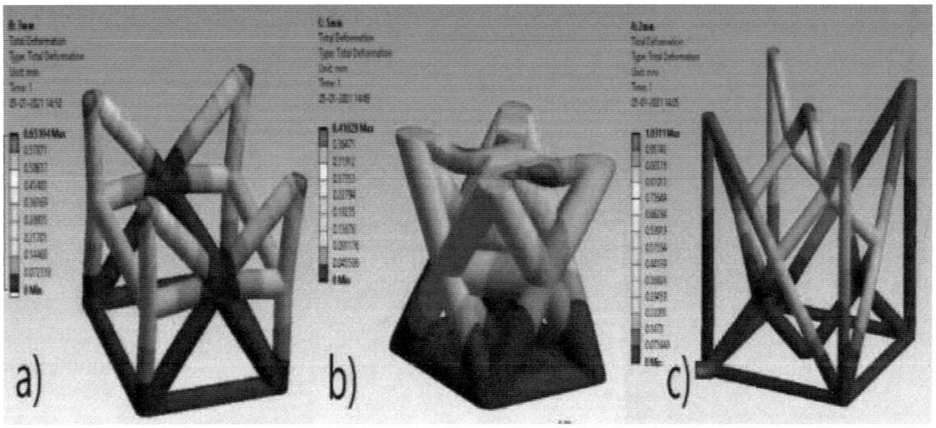

Figure 6: Total deformation on (a) Triangular Honey Comb (b) Octet and (c) Diamond Lattices for Seat Cushion.

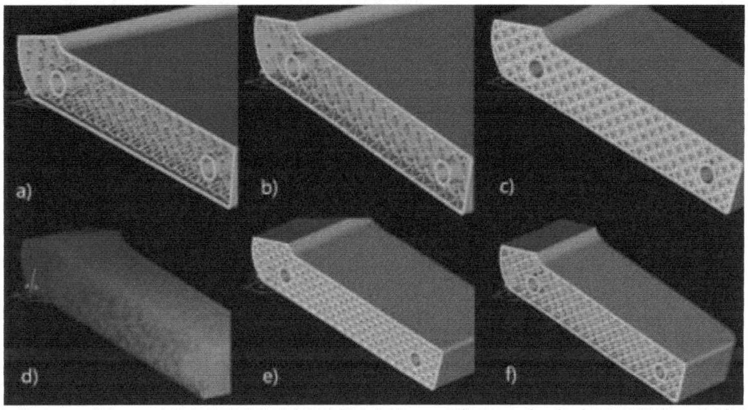

Figure 7: Seat cushion with (a) BCC (b) FCC (c) Octet (d) Tetrahedral (e) THC and (f) Diamond lattice.

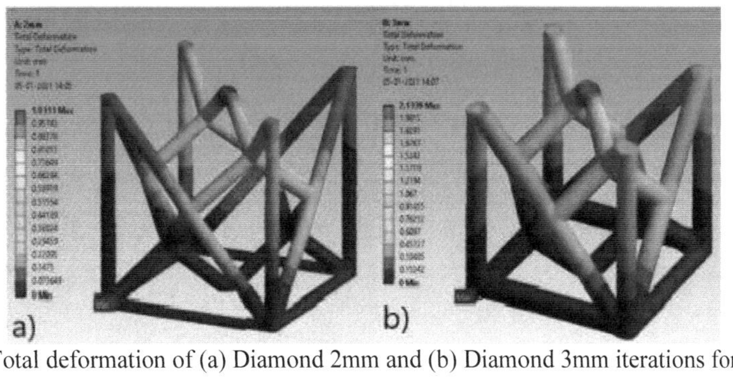

Figure 8: Total deformation of (a) Diamond 2mm and (b) Diamond 3mm iterations for back-rest.

The Diamond 2mm and 3mm beam thickness iterations had a safety factor of 1.85 and 1.97, respectively. The 2mm iteration was selected for the back-rest latticing on account of its lower weight.

The hand rests of aircrafts seats are usually about 2 inches wide; however, seat manufacturers now make them 1.6 inches wide. Our design is 1.6 inches wide and has been latticed with a 2mm thick shell. The

BCC, FCC, Tetrahedral, and Diamond Lattices had comparable weights of 155.4292g, 184.4612g, 151.992g, and 156.6279g, respectively. The BCC and FCC structures had very high deformation, so these lattices were rejected. The tetrahedral design had 9.7797 mm deformation while the diamond lattice showed just 1.0311 mm deformation. Therefore, it was selected.

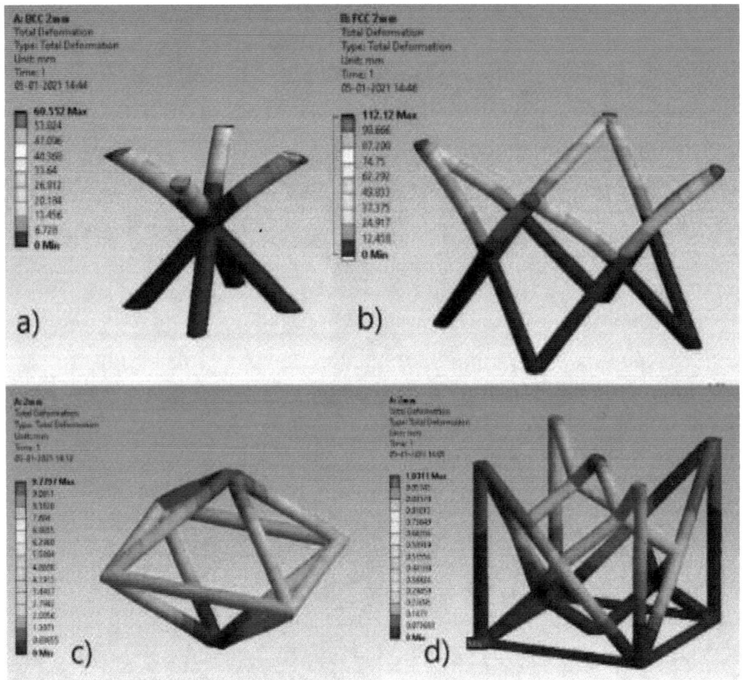

Figure 9: Total deformation of (a) BCC (b) FCC (c) Tetrahedral (d) Diamond lattices for hand rest.

The final latticed components are shown below.

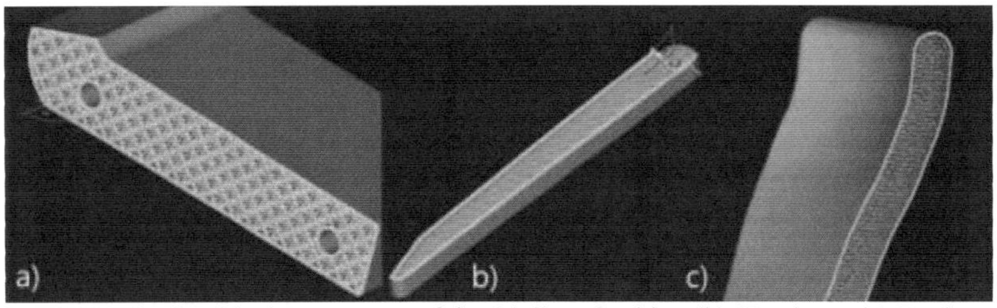

Figure 10: Finalised lattice structure of the (a) Seat Cushion (b) Hand-rest (c) Back-rest.

The seat cushion was latticed with an octet lattice structure of 5 mm thickness, while the hand rest and backrest were latticed using the diamond lattice structure of 2mm thickness.

3.4 ERGONOMICS EVALUATION ON CATIA V5

We tested our design on Catia V5 software to obtain a RULA score. For the evaluation, we considered a passenger that is a 95 percentile American man whose weight is 125 Kg. A RULA score of 3 is preferred, while scores below 3 are okay, and scores above 3 are considered inferior. The figure below shows the results of the ergonomic evaluation. The seat design obtained a score of 3.

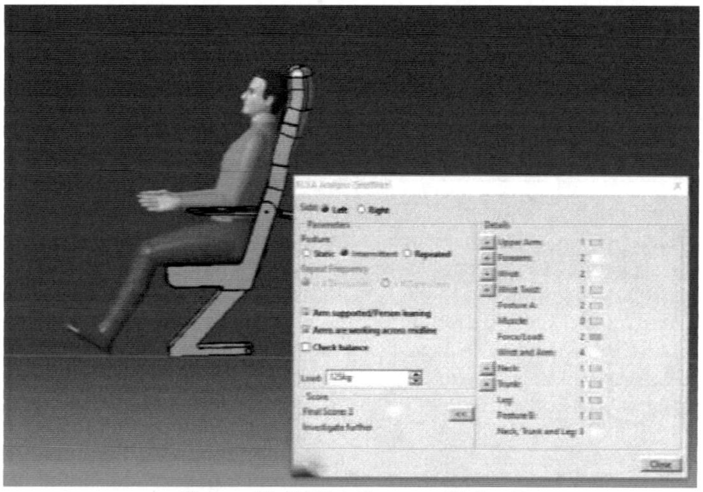

Figure 11: RULA Score on Catia v5

4. COMPARISON WITH STANDARD AIRCRAFT SEATS

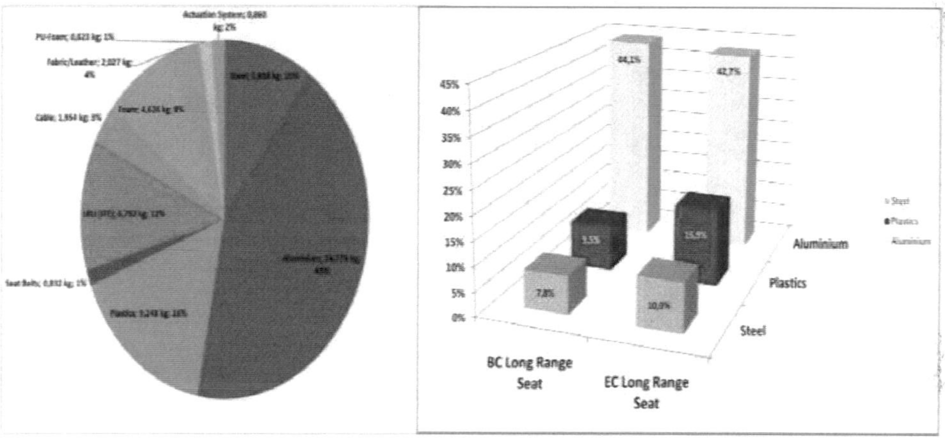

Figure 12: Weight Distribution of Materials in Aircraft Seat

Standard Aircraft seats weigh about 15 kg on average. The figure above shows that 43% of the seat weight is from aluminum components. Therefore, aluminum components account for about 6.45 kgs of the seat's weight. At the same time, plastic foam contributes to about 3.3 kg of the seat weight.

Table 6: Weight of Optimised Components.

COMPONENT	WEIGHT
Supporting Leg	1.21 Kg
Actuating Mechanism Housing	1.8 Kg
Hand rest	156.6 g
Seat Back Rest	1289.48 g
Bottom Cushion	489.28 g
Serving Tray	330.45 g

The table above shows that the optimized designs show a 3.29 kg weight saving for Aluminium components and a 1.1 kg weight reduction for plastic and foam components. They are effectively reducing each seat weight by approximately 4.4 Kg.

5. CONCLUSION

Weight reduction in aircraft components is imperative to improve fuel efficiency. Each seat contributes significantly to the aircraft weight; therefore, we decided to reduce weight on aircrafts seats using computational techniques like latticing, topology optimization, and generative design simulations on the seat hand-rest, supporting leg, backrest, housing mechanism, serving tray and bottom cushion. This yielded positive results in the form of a 4.4 kg weight reduction per seat. Since, on average, there are 180 seats within a single passenger aircraft, the total amount of weight reduced will be approximately 800 kg. To summarise, combining different and simple computational & numerical techniques can help design and develop very light aircraft seats. This concept can be extended to other components in the aircraft's interior as well.

6. REFERENCES

[1] L. Zhu, N. Li, P.R.N. Childs, Light-weighting in aerospace component and system design, Propulsion and Power Research, Volume 7, Issue 2, 2018 ISSN 2212-540X https://doi.org/10.1016/j.jppr.2018.04.001
[2] Information found "The evolution of the Airline Seat" in web, at 10-02-13: http://www.cntraveler.com/features/2011/05/From-Wicker-to-Wow
[3] Francis Froes, 19 - Combining additive manufacturing with conventional casting and reduced density materials to greatly reduce the weight of airplane components such as passenger seat frames, Editor(s): Francis Froes, Rodney Boyer, Additive Manufacturing for the Aerospace Industry, Elsevier, 2019, Pages 419-425, ISBN 9780128140628, https://doi.org/10.1016/B978-0-12-814062-8.00021-2.
[4] Compressive Strength of Composite Lattice Structures An Hou, Gramoll, Kurt 1998/03/01 1998 1998/03/01 doi: 10.1177/073168449801700505 Journal of Reinforced Plastics and Composites
[5] Ishant Gupta, Parveen Kalra, Puneet Chawla, Jagjit Singh, Evaluation of Pilot's Seat Design of Civil Aircraft for Indian Anthropometric Data by using Delmia Human Software, Procedia Manufacturing, Volume 26, 2018, Pages 70-75, ISSN 2351-9789, https://doi.org/10.1016/j.promfg.2018.07.009.
[6] Gabriel H. Campos, Fengfeng (Jeff) Xi, Pressure sensing of an aircraft passenger seat with lumbar control, Applied Ergonomics, Volume 84, 2020, 103006, ISSN 0003-6870, https://doi.org/10.1016/j.apergo.2019.103006.
[7] R.H.M Goossens, C.J Snijders, T Fransen, Biomechanical analysis of the dimensions of pilot seats in civil aircraft, Applied Ergonomics, Volume 31, Issue 1, 2000, Pages 9-14, ISSN 0003-6870, https://doi.org/10.1016/S0003-6870(99)00028-9.
[8] S. Dangal, M. Smulders, P. Vink, Implementing spring-foam technology to design a lightweight and comfortable aircraft seat-pan, Applied Ergonomics, Volume 91, 2021, 103174, ISSN 0003-6870, https://doi.org/10.1016/j.apergo.2020.103174.
[9] Park, Sang-in & Rosen, David & Duty, Chad. (2014). Comparing mechanical and geometrical properties of lattice structure fabricated using Electron Beam Melting. 1.
[10] Tobias Maconachie, Martin Leary, Bill Lozanovski, Xuezhe Zhang, Ma Qian, Omar Faruque, Milan Brandt, SLM lattice structures: Properties, performance, applications and challenges, Materials & Design, Volume 183, 2019, 108137, ISSN 0264-1275, https://doi.org/10.1016/j.matdes.2019.108137.

Rupture Mechanism of Cerebral Carotid Aneurysm Using Computational Methods

Vrushabh Sawant[1], Rohit Chaudhari[1], Sreeja V[2], Albert Eienstein[2], Davidson Jebaseelan[1], Jebaraj C[1]

[1]School of mechanical engineering, VIT University Chennai Campus, Chennai-600127, Tamil Nadu, India.
Email : vrushabh.sawant@gmail.com

[2]EinNel Technologies-Research & Development Centre
Chennai-600073, Tamil Nadu, India.

Abstract.
This study helps to predict the rupture mechanism of an Inter Carotid Artery (ICA) for saccular aneurysm study through the Finite Element Method (FEM) combined with Discrete Element Method (DEM) of LS DYNA and also compared with a Fluid Solid Interaction (FSI) study. Also the thickness of an artery wall was found by using James-Guth Stress Strain Relation with validation. It was concluded that the Finite Element Method with Discrete Element Method was more efficient than FSI as the computation time was reduced by 51.55% with an error of 0.08%.

Keywords: discrete element method (DEM); James-Guth stress strain relation; fluid solid interaction (FSI); internal carotid artery (ICA); saccular aneurysm; rupture mechanism ; finite element method (FEM)

1. Introduction

An aneurysm is a localized inflation of an artery due to various factors such as stress, smoking, drug abuse and trauma etc. It's one among the cerebral vascular disorder. Ignorance to it may cause hemorrhage and might even prove to be fatal. There are three types of aneurysm shown in Figure 1 as saccular, fusiform and berry. Also they are classified based on the diameter of the aneurysm as small, medium and large aneurysms. Takato Morioka et.al. (1949) proved that the odds of predicting the existence of aneurysm based on the symptoms are very slim. Based on the previous analyzed data, from all the patients examined, two third of the patient failed to show any symptoms at all. But not only predicting the presence of an aneurysm is important but also it is equally important to estimate its failure criteria. Hence understanding the rupture mechanism of an aneurysm is of at most important. Many patients live long normally even though an aneurysm exist in their body as the rate of aneurysm inflation is very slow, but this statement can be contradicted based on the past cases encountered by the doctors in which the inflation rate has amplified and ultimately causing death of the patient due to hemorrhage.

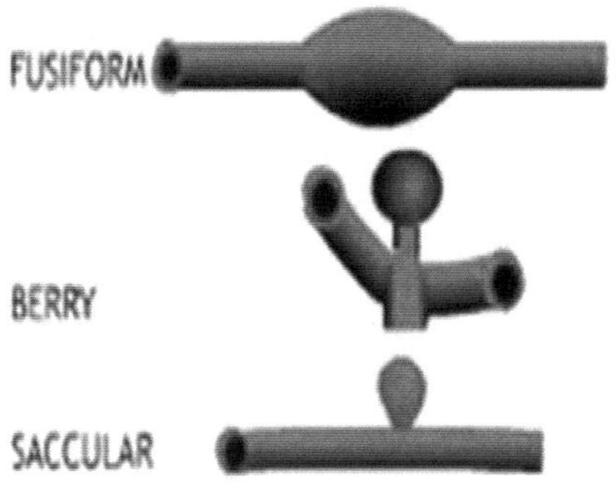

Figure 1 Types of Aneurysm

Rupture Mechanism studies using CAE approach

Many studies have been conducted in the past like Joseph F. Polak et.al. (1999) in order to understand the rupture mechanism of an aneurysm and also to predict its failure criteria. An orthodox method of doing so is by taking a CT scan or a MRI of a patient and based on its dimensional specifications various parameters are found and these parameters have a tolerance limit which is decided from past experiments and theoretical studies. Parameter such as intima media thickness, Wall Shears Stress & Aspect Ratio etc. There are many other parameters related to hemodynamic study but in this study we will be concerned with the above mentioned parameters.

Knowing the significance and precision of the results obtained from the CAE analysis, various CFD based studies have been conducted in the past. Chien et al (2009) conducted similar such hemodynamic study on an internal carotid artery-ophthalmic artery aneurysm. It was observed that aneurysm having more complex flow pattern in the aneurysm were more prone to rupture. Also in the rupture aneurysm the WSS distribution was comparatively more homogeneous then the unruptured ones. Another such similar analysis was done on a giant internal carotid artery aneurysm. The motive of this study was to use CFD approach to analyze the surgical management along with is clinical course of a giant carotid aneurysm. For this the two models were analyzed which were taken before and after the treatment. A CFD analysis was conducted on both the models to simulate the flow and the WSS distribution. It was observed that the hemodynamic features have improved than the previous intervention, thus concluded that CFD study to be helpful in such cases which were similar to Russin eta al (2015).

Literature review shows that most of the studies have used CFD approach in order to carry out analysis related to rupture mechanism Feng et al (2007) have also conducted such studies.

IOP Conf. Series: Materials Science and Engineering 1128 (2021) 012009 doi:10.1088/1757-899X/1128/1/012009

Also, there is less literature available on numerical study of internal carotid artery (ICA). Also the CFD studies takes high computational time and hence taking in to account all the above factors it was decided to study the rupture mechanism of ICA using computational methods as a topic of our study.

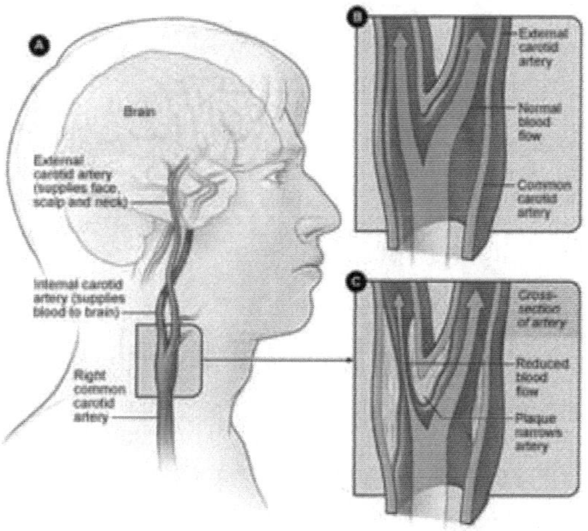

Figure 2 Anatomy of Carotid Artery

In this study aneurysm of internal carotid artery saccular aneurysms considered. The anatomy of carotid artery is given in Figure 2. Although the previous studies done on aneurysm rupture were convincing but there was limited approach to do the study. Use of CAE in the aneurysm study reduced the error caused by humans and also aids to draw a relevant inference based on the flow simulation obtained from a CFD study which would be more precise. But this approach has a significant drawback of large computation time and it follows the frequently used CFD approach to obtain sound results, hence the approach to do the rupture mechanism study was limited. The motive of this study was to find an alternative computationally efficient approach to simulate and analyze rupture mechanism of ICA. Finite Element Method coupled with Discrete Element Method has been a new field of study in the recent times and due to its ability to obtain appropriate solution in fewer periods of time then the Finite Element Method which has been a trend from a very long time in studying various aspects using CAE approach, FEM/DEM has been an alternative approach used in this study. Displacement and other parameters will be obtained from FEM but it will not be enough to replace unless this method facilitates flow simulation. Hence in order to accomplish this limitation it was combined with the Discrete Element Method which would be responsible for flow simulation of the blood. The key aspect of this study was to simulate the flow of blood in the form of particles. Russin et al (2015) proved that it helps to predict the complex flow inside the aneurysm which will in turn be helpful for predicting the effectiveness of treatments on a patient possessing ICA.

This method to study rupture mechanism of an ICA needs validation, and that would be achieved by the convention method involved in the CAE method which would be CFD approach (Fluid Solid Interaction model). Stress parameter obtained by both the approaches would be compared and the precision would be checked. Based on this the credibility of the new rupture mechanism study approach would be tested and concluded. An additional objective of this study was to find the algorithm to find the thickness of the aneurysm at various stages, right from the time of its formation till its rupture.

2. Methodology

The analysis for the rupture mechanism consisted of two fundamental approaches Finite Element Method (FEM) and Discrete Element Method (DEM). The motive of using these approaches was to reduce computation time for the analysis and to simulate the pulsating flow of blood without performing any CFD (FSI) analysis. This need of computation time reduction can be achieved by the use of high-performance system but that will prove to be expensive. Another alternative method to achieve this would be by reducing the number of elements involved in the simulation. FEM would not serve the purpose as the mesh involved in the CFD study would be volume mesh. But on the contrary FEM/DEM needs the 2D mesh. And the utilization of the DEM would reduce the elements used to model the fluid and the aid to simulate the flow would also be achieved which could replace the FSI study or prove to be an alternative with less computation time. More specifically this method would be more helpful where only the structural parameters would be of more significance.

The aneurysm considered in our study is an Internal Carotid Artery aneurysm; it is classified as a small aneurysm as its diameter is 8.1 mm and its type is saccular. The aneurysm is having an aspect ratio of 1.5, height ratio of 1.2 and an inflow angle of 850. Table 1 consists the specifications of the ICA. A subsection

Table 1 Specification of Blood and Artery Wall

Parameter	**Units**	**Value**
Blood		
Density (kg/m^3)	1100	
Specific Heat (J/kg-K)	4186	
Reynolds Number	176 (laminar flow)	
Artery Wall		
Density (kg/m^3)	0.92	
Wall Thickness (m)	0.5	
Poisson's Ratio	0.49	

Conventionally to model a artery the entire volume inside (blood) and outside (artery wall) would be discretised in to solid elements. The pressure volume relation and the total volume would be the contribution of the addition of all the elements in the entire system. But an artery is not a simple geometry to model; it has many twists and turns. As a result, such geometries experience the limitation of the FEM mesh which is not suitable for such sharp curves and irregular surfaces. As a result, the computation cost would hike up due to these solid elements. Hence there was a need for an alternative approach to replace volume mesh modeling to surface mesh. This could be achieved by modeling the artery as a control volume. In other way it can be interpreted as the entire volume of the artery would be covered by an envelope, which is also termed as the domain. The domain is modeled and meshed by using shell element or membrane mesh. A specific thickness would be assigned to it along with appropriate material for the domain. As there is an existing method in which the surface area, orientation and the position of the membrane or shell element can be computed and stored at every time step. This property of the control shell element is utilized to calculate the volume of the artery. This control surface area can be related to the control volume of the artery by using the Green's Theorem. The first two integrals are the integrals over the control volume (dv = dxdydz) and the last integral would be the integration of the surface enclosing the control volume, n_x is the direction cosine between the surface normal and the 'x' direction corresponding to the x-partial derivative.

$$\iiint \phi \frac{\partial \psi}{\partial x} dxdydz = -\iiint \psi \frac{\partial \phi}{\partial x} dxdydz + \oint \phi \psi n_x d\Gamma \quad (1)$$

The volume integral can be written as,

$$V = \iiint dxdydz \quad (2)$$

Comparing the equation (1) and (2) a volume integral can be obtained. Assuming an arbitrary function '$\varphi = 1$' and '$\psi = x_x$'

$$V = \iiint dxdydz = \oint x n_x d\Gamma \quad (3)$$

To obtain the volume of the entire control volume the surface integral so obtained from equation (3) can be further added for all the elements of the artery.

$$\oint x n_x d\Gamma \cong \sum_{i=1}^{N} \bar{x}_i n_{ix} A_i \quad (4)$$

Where the x_i is the average x coordinate n_{ix} is the direction cosine between the surface normal and the x direction and A_i is the surface area of the element.

The force equation for the DEM element can be written as,

$$\hat{f}^{i+1} = \hat{f}^{i} + \Delta\hat{f} \quad (5)$$

whereas the subscript i+1 indicates the time increment and the superpose caret (ˆ) indicates the force in the local coordinate system which is along the axis of the element.

Considering the case in which the orientation vector is not used, the global component of the forces for the discrete element are defined by the use of the direction cosine of the elements,

$$\begin{Bmatrix} F_x \\ F_y \\ F_z \end{Bmatrix} = \frac{\hat{f}}{l} \begin{Bmatrix} \Delta l_x \\ \Delta l_y \\ \Delta l_z \end{Bmatrix} = \hat{f} \begin{Bmatrix} n_x \\ n_y \\ n_z \end{Bmatrix} = \hat{f} \underline{n} \quad (6)$$

Where, $l$is the length

$$\underline{\Delta l} = \begin{Bmatrix} \Delta l_x \\ \Delta l_y \\ \Delta l_z \end{Bmatrix} = \begin{Bmatrix} x_2 - x_1 \\ y_2 - y_1 \\ z_2 - z_1 \end{Bmatrix} \quad (7)$$

$$l= \sqrt{\Delta l_x^2 + \Delta l_y^2 + \Delta l_z^2} \tag{8}$$

The (x_i, y_i, z_i) are the global coordinates for the nodes of the spring elements. The force obtained in equation (6) would be added and subtracted from first and second node respectively. If the node is attached to the ground the approach used will be same but instead of (x_2, y_2, z_2) the initial coordinates of the node 1 would be used as (x_0, y_0, z_0).

$$\begin{Bmatrix} F_x \\ F_y \\ F_z \end{Bmatrix} = \frac{\hat{f}}{l} \begin{Bmatrix} x_0 - x_1 \\ y_0 - y_1 \\ z_0 - z_1 \end{Bmatrix} = \hat{f} \begin{Bmatrix} n_x \\ n_y \\ n_z \end{Bmatrix} \tag{9}$$

The force-displacement relation woul be used in order to determine the increase of element force. There are nine type of force-displacement/velocity relation, among which our study will be concerned with the 'linear elastic' force-displacement/velocity relationship. The relation is given by,

$$\hat{f} = K\Delta l \tag{10}$$

whereas the 'K' is the element stiffness and the' Δl' is the change in length of the element. The force-velocity relation can be defined for the linear viscous element which is given by,

$$\hat{f} = C\frac{\Delta l}{\Delta t} \tag{11}$$

where 'C' is the viscous damping parameter and 'Δt ' is the time step increase. In order to consider the strain rate effect, the forces are scaled which are based on relative velocities of all the springs. The static force computed from all the springs are amplified by multiplying a factor which results to obtain the dynamic force.

$$F_{dynamic} = \left(1 + k_d \frac{V}{V_0}\right) F_{static} \tag{12}$$

k_0 is the user defined input value, V is the absolute relative velocity and V_0 is the dynamic test velocity.

The deformation limit of the tension- compression of the body is set according to the application. Once the limit of compression-tension is reached the computation of momentum conservation is done and the common acceleration is computed.

$$\hat{a}_{Common} = \frac{\hat{f}_1 + \hat{f}_2}{m_1 + m_2} \tag{13}$$

Table 2 Stages of DEM

IOP Conf. Series: Materials Science and Engineering 1128 (2021) 012009 doi:10.1088/1757-899X/1128/1/012009

DISCRETE ELEMENT METHOD (DEM)	
STAGES	***ACTIVITIES***
1	Particle Initialization
2	Collision Detection
3	Contact Force Calculation
4	Computing Acceleration Vector
5	Compute Velocity and Position of All Bodies
6	Output Data

The various stages of the DEM are summarized in Table 2. This approach would be implemented in the study for that the process to setup a simulation goes as follows. A C.T scan of a patient is considered suffering from a ICA aneurysm. This data is then provided to an imaging tool in order to convert the data in to a BEM model. The segmentation of the geometry is done prior to converting in to BME model. The pre-processing of this model is done by importing the model to LS PrePost 4.2 and ANSA V15.0.2. The type of mesh used in this study is the triangular mesh, with a mesh count of 29074 and an aspect ratio of 4.5. After the meshing process is complete it is then imported to LS DYNA R8 which is used as a solver.

The output file is then imported to LS PrePost for further processing. In these various parameters were extracted from the output file such as Von Mises Stress, Average Displacement and Pressure Distribution.

The Fluid Solid Interaction model is a type of analysis in which the computational fluid dynamics and the structural analysis are coupled and the results are linked with each other in order to obtain appropriate results. In this study the FEM model is imported to the ANSYS WORKBENCH 14.5. The Artery model is used for the Transient Structural Analysis shown in Figure 3 and the Blood model is used for Computational Fluid Analysis shown in Figure 4.

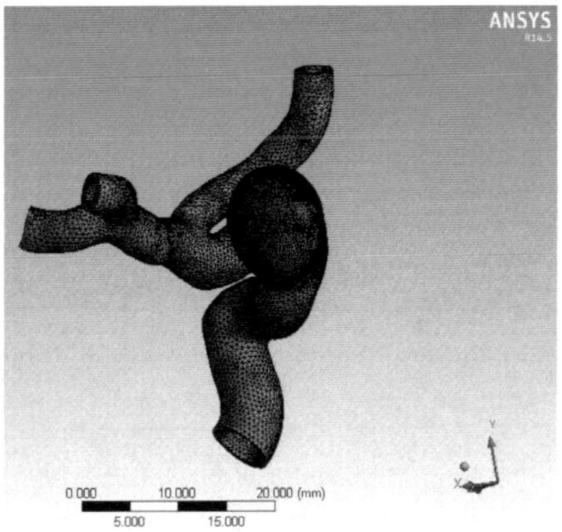

Figure 3 Artery Wall for FSI

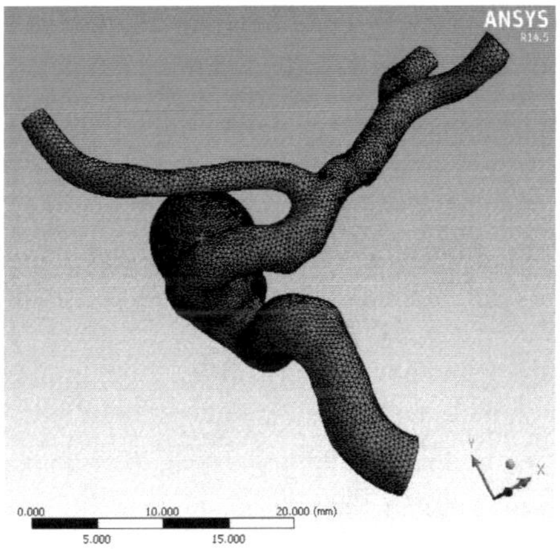

Figure 4 Fluid Volume for FSI

Both the bodies are meshed in ANSYS 14.5 and type of mesh used was tetra mesh. Mesh count of the Artery and the Blood body obtained was 322332 and 557778 respectively. The boundary conditions used in this analysis is shown in Table 3.

IOP Conf. Series: Materials Science and Engineering 1128 (2021) 012009 doi:10.1088/1757-899X/1128/1/012009

Table 3 Boundary Conditions for FSI Model

Boundary Conditions for FSI Model		
Artery	One inlet and Three outlets	Fixed Support
Blood	Inlet	Velocity Inlet (Pulsating Flow)
	Outlet 1	Pressure Outlet
	Outlet 2	Pressure Outlet
	Outlet 3	Pressure Outlet

The data file so obtained from the analysis of the secular aneurysm is then imported to CFD POST 14.5 and further processed. Since the main objective of this study was to validate the FEA/DEM based analysis and also to find the WSS. Hence two parameters were extracted during the post processing Wall Shear Stress (WSS) and Von Mises Stress. As it can be see the WSS and Von Mises Stress were obtained from a FSI study for validation of the FEM/BEM method studied earlier, a comparison will help to conclude the similarity of both the methods. This will aid to conclude as the FEM/BEM method can be a parallel approach of studying the aneurysm-based studies.

The aneurysm is of non-uniform thickness hence it is very essential to determine the thickness distribution of the artery wall. This is essential to analyze various phenomenon of the artery using CAE approach. Till date there were many efforts done to find the approximate thickness distribution. Efforts were made by inserting a dye inside the blood stream and the intensity of the MRI would say the area of minimum and maximum thickness though less accurate. Till date the thickness of the artery wall for the modeling of the artery wall in FEA analysis is considered to be of uniform thickness. This is an assumption made results in inaccurate results.

Considering the similarities in the inflation of the aneurysm and the inflation of the balloon one can obtain the relation of the James-Guth Stress Strain Relation which was introduced by him to find the nature of inflation of the ideal rubber balloon inflation. According to this relation the thickness and the pressure gradient parameters are related for the thickness variation. Also the profile of inflation can be determined from the relation derived from this relation by providing appropriate boundary conditions. Consider a non-uniform inflated domain given in Figure 5.

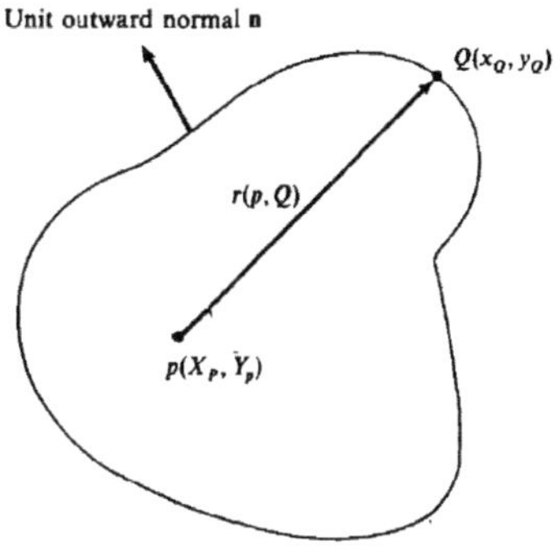

Figure 5 Non-uniformly Inflated Domain

Now consider a point 'p' in the center of the domain which is fixed and is equidistant from all the nodes of the domain and the inflation of the domain is in the direction of the unit normal 'n'. Consider the pressure outside the domain to be atmospheric P_{out} and the inside pressure to be P_{in}. As this condition will be considered to be the initial condition after a particulate period of time't' seconds the inflation of the domain occurs and the point 'Q_0' further displaced to 'Q'. The distance between the two points increases from 'r_0' to 'r'.

- Application of James-Guth Stress Strain Relation for an Ideal Rubber.

$$f_i = \frac{1}{L_i[kKT\left(\frac{L_i}{L_o}\right)^2 -pV]} \text{ (James-Guth Stress Strain Relation)} \quad (14)$$

In this algorithm the pressure gradient at various stages are considered. The gradient is considered between the interior of the aneurysm and its exterior. This equation if then derived further to obtain

$$P_{in} - P_{out} = \frac{f_t}{\pi r^2} = \frac{C}{r_o^2 r[1-\left(\frac{r_o}{r}\right)6]} \quad (12)$$

Using the boundary conditions the constant 'C' can be obtained. From this equation we can further derive the relation between thickness and distance of the aneurysm from the fixed center.

$$\frac{t}{t_o}=\left(\frac{r_o}{r}\right)^2 \tag{13}$$

(Relation between thickness and radial distance of a node from a common point)

Now to prove the above explained method a case study will be considered in which the theoretical and the simulation results will be compared. The comparison is shown in Table 4.

3. Results & Discussions

The case study done to validate the James-Guth stress strain relation by results obtained in both simulation and the theory calculations are very close. This algorithm can be put in to a computer programming code and thus predicting thickness would be easy. The database of the pressure gradients at various stages can be prepared and the database can be linked to the work which would possess the logic as the relation of thickness and distance between the two points of consideration, in the CAE study it would be the nodes which would be considered. This method can be further studied as a research field for better understanding.

Table 4 Comparative Study of James-Guth Stress Strain Relation

CASE STUDY		
Pressure	Initial Pressure (2.409E-06 Pa)	Final Pressure (0.2409E-06 Pa)
Shell Thickness (m)	0.0415 (simulation value)	0.0376
Linear Extension Of a Node From a Fixed Centre Point Of Saccular Aneurysm (m)	14.6939 (simulation value)	15.1708

Till date rupture mechanics of an aneurysm was conducted using the FSI method. The analysis took 900 minutes to complete where as the FEM/DEM simulation using took 126 minutes to converge. If WSS is to be obtained, then an internal flow CFD analysis with a pulsating flow was done with a computation time of 310 minutes. Thus both analyses together tool 436 minutes. Thus there is a percentage computation time reduction of 51.55 %. The curiosity of an individual would be ignited to find the reason of such a drastic reduction of computation time. The answer to that would be, CFD study requires volume mesh whereas FEM/DEM uses surface mesh. If an FSI approach is to be considered, in addition to the artery wall volume mesh there is a volume mesh of the flowing fluid. But if the FME/DEM approach is considered the artery wall would be shell meshed and there is no need to provide a volume mesh for the flowing fluid. The use of DEM in LS-DYNA eliminated the use of volume mesh. Thus the number of elements is reduced. As number of element is directly proportional to the computational time, the time reduction is achieved. Figure 6 shows the meshed model of an ICA saccular aneurysm.

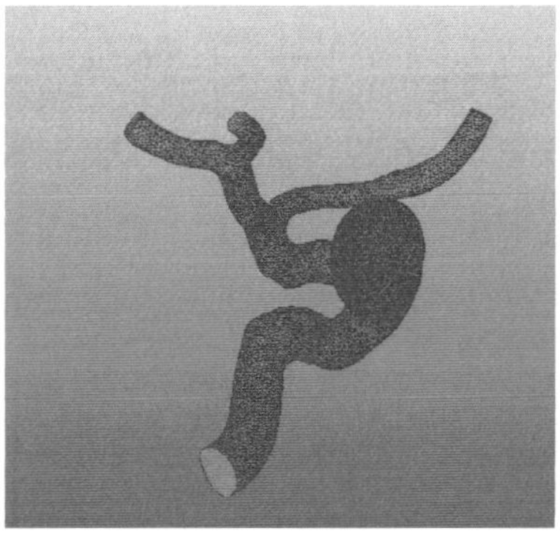

Figure 6 Shell Mesh Model

As we have seen that based on an observational study done by Jonathan Russin eta al (2015) of the blood flow pattern doctors can predict the effectiveness of the treatment at every intervention. Thus appropriate measures can be taken in order to decide the further method of treatment.

The WSS value obtained from the CFD analysis was 3.5 Pa. According to the study done by Masaaki Shojima eta al (2007) this value is considered to be in the safe range according to the previous study. Figure 7 shows the WSS contour.

Figure 7 Wall Shear Stress: 3.5 Pa

Aspect ratio obtained from the geometric specifications of the aneurysm was found to be 1.5. This aspect ratio is considered to be safe and the odds of rupture phenomenon to occur are very less. The ground on which the statement proves to be true is the previous studies done on similar topic. According to Weir B eta al (2003) the aneurysm up to AR of 1.8 is likely to be unruptured but as the aneurysm AR reaches 3.4 the chances of it getting rupture is maximum.

The Inflow Angle has a greater significance to the cause of rupture it also influences the WSS.It was observed that as the IA increases the peak velocity and the shear jet zone of the aneurysm increases. The shear jet zone is the zone near the neck of the aneurysm where maximum shear action occurs due to the blood flow at high pressure. In our case the IA was found to be 85°,according to Merih leta al (2010) the influence of IA would be moderate as the angle lies between 60^{0} to 140^{0}.This range of IA was studies in a previous study and hence taking under consideration this study we can say the aneurysm rupture chances are less due to IA of 85°. Figure 8 provides the displacement contour which is having the influence of the IA as the maximum displacement was found to be at the tip of the aneurysm. As greater the IA angle the horizontal component of fluid force increases causing greater displacement to compensate for the increase in the kinetic energy of the fluid.

In the past there were many efforts done by Shang EK eta al (2013) to study the significance of the variable thickness of the aneurysm body. Due to this there is a non-homogeneous inflation in an aneurysm in most of the cases. Till date in order to simulate the rupture mechanism of an aneurysm the thickness of the wall is assumed to be uniform. Even though the simulations are

IOP Conf. Series: Materials Science and Engineering 1128 (2021) 012009 doi:10.1088/1757-899X/1128/1/012009

very precise, there will always be an error factor that would be the huddle for any study to obtain the precise results.

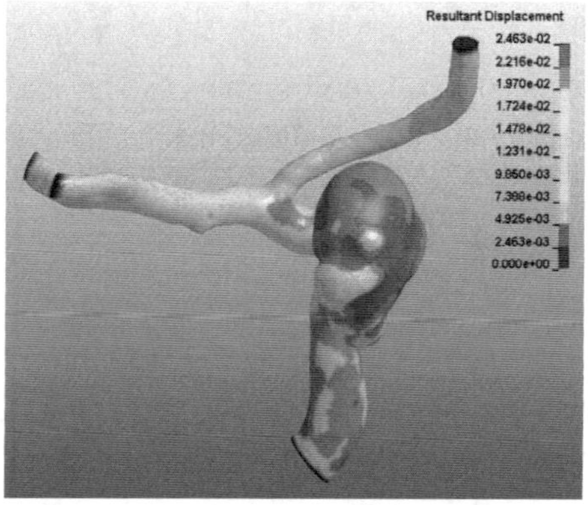

Figure 8 Max displacement 2.46e^{-05} m

But using the James-Guth's Stress Strain relation for the ideal rubber not only helps to predict the variable thickness distribution at any point on the aneurysm but also aids us to predict the appropriate inflation profile of the aneurysm at every stages of inflation with respect to the pressure gradient between inside and outside of the aneurysm. This can be achieved by putting the equation as an algorithm in any programming language and the series of pressure gradient data would be acquired from the simulation itself. This method can be considered as one of the significant scope of this study. Figure 9 shows the pressure distribution contour inside the aneurysm.

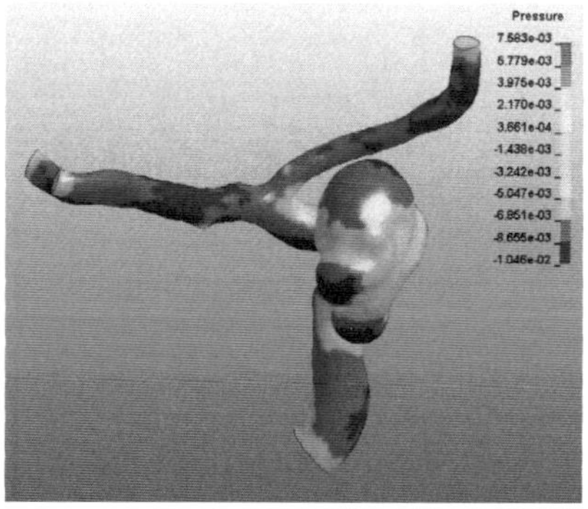

Figure 9 Max Pressure is 7.583e^{-09} Pa

Maximum pressure obtained was 7.583e-09 Pa. Taking into consideration the mesh parameters of the BEM study and the CFD study the validation process was successfully achieved as the Von Mises Stress obtained from FEM/DEM and CFD analysis was found to be 0.32 Pa and 0.4 Pa. respectively. Since the difference was of 0.08 Pa the result was satisfactory. Figure 10 shows the contour of Von Mises Stress obtained from the CFD study

Figure 10 Simulation Results of FEM and DEM: Von Mises Stress 0.318 Pa

The simulation so obtained by the FEM/DEM simulation is very helpful for the prediction of aneurysm rupture and also to study the flow pattern. This can be seen in the simulation result shown in Figure 10

4. Conclusion

The process of FEM/DEM analysis was comparatively simpler then the CFD process. The Von Mises Stress obtained from both the study after comparison was found to have the difference of 0.08 Pa.

But due to the reduction in the computation time for the objective to achieve by both the approach had a significant variation. The use of FME/DEM approach along with the regulars CFD approach instead of the conventional CFD study using the FSI model would reduce the computational time by 51.55%.

Thus the advantage of less computation time by using BME process for the rupture mechanism study outwits the minor difference of the Von Mises stress. Also using the Corpuscular method, a person can even study the flow pattern of the blood, this further adds to the advantages of the FEM/DEM study.

Based on the case study done to validate the use of James-Guth's Stress Strain relation for ideal rubber the values of thickness and the distance of a node from a common center distance was found to be 10.37% and 3.16%. Assuming the error in the CAE analysis to be +10%. The results acquired are satisfactory and precise. Hence use of the James-Guth's equation will enhance the future studies related to aneurysm.

References

[1] Daniel H.O'Leary, M.D, Joseph F. Polak, M.D., M.P.H., Richard A Kronmal, Ph.D., Teri A. Manolio, M.D., M.H.S., Gregory L.Burke, M.D., M.S., and Sidney K. Wolfson, Jr., M.D.,"Caroted Artery Intima and Media Thickness as a Risk Factor For Myocardial"For the Cardiovascular Health Study Collaborative Research Group, The new England Journal of Medicine, 1999 Jan 7;340(1):14-22.

[2] Luca Antiga Ph.D., Prof. Andrea Remuzzi, Prof. Robert Fumero, Ing. Alberto Redaelli, Prof. Giancarlo Ferrigno, Patient-Specific Modeling of Geometry and Blood Flow in Large Arterie, Politecnico Di Milano, XV- Ciclo, 1999-2002.

[3] Takato Morioka MD, Shunji Nishio MD, Yutaka Kimura MD, Seiichi Kashimura MD, Masashi Fukui MD. Rupture of a giant middle Cerebral Aneurysm after blunt head injury. Journal of Clinical Neuroscience (1999) 6 (2), 175-177, 1949 Harcourt Brace & Co Ltd, 1948.

[4] David A. Steinman, Jonathan B. Thomas. Hanif M. Ladak, Jaques S. Milner, Brian K. Rutt, and J David Spence, "Reconstruction of Carotid Bifurcation Hemodynamic and Wall Thicknes Using Computational Fluid Dynamics and MRI", Magnetic Resonance in Medicine 2002 47: 149-159.

[5] Bryce Weir, M.D., Christian Amidei, M.S.N ., J. Max Findlay, M.D., Ph.D., F.R.C.S.(C), Neal F. Kassel, M.D., Lanting Dai, M.S., and Theodore G. Karrison , Ph. D.,'The Aspect Ratio (dome/neck) of Ruptured and Unruptured Aneurysm', Journal Of Neurosurgery, 2003 Vol.99, No.3, Page 447-451.

[6] Weir B, Amidei C, Kongable G, Findlay JM, Kassell NF, Kelly J, Dai L, Karrison T.G., "'The Aspect Ratio (dome/neck) of Ruptured and Unruptured Aneurysm", Journal Of Neurosurgery, 2003 99(3) 447-451.

[7] John O. Hallquist, "LS DYNA Theory Manual", March 2006.

[8] Bin Feng, Bai-Yan Li, Eric A. Nauman and John H. Schild, "Theoretical and electrophysiological evidence for axial loading about aortic baroreceptor nerve terminals in rat", Am J Physiol Heart Circ Physiol 293, H3659-H3672, 2007.

[9] Masaaki Shojima, M.D., Marie Oshima, Ph.D., Kiyoshi Takagi, M.D., Ph.D. Ryo Torii, PhD, Motoharu Hayakawa, M.D PhD., Kazuhiro Kataga, M.D PhD, Akio Morita, M.D. PhD. Takaaki Kirino, MD. PhD. Method for Asessing the Ruptur risk of an Aneurysm of a Patient and Associated System, German Application Number 02007026519.2, 2007.

[10] Sujan Dhara, B.E., Markus Tremmel, Ph.D., J Mocco, M.D., M.S.,Minsuok Kim, Ph.D., Junichi

Yamamoto, M.D., Ph.D., Adnan H. Siddiqui, M.D., Ph.D, L.Nelson Hopkins, M.D., "Morphology Parameter for Intracranial Aneurysm Rupture Risk Assesment", NIH-PA authors manuscript, 2008.

[11] Aiche Chien, Ph.D., Satoshi Tateshima, M.D, James Sayre, Ph.D., Marcelo Castro, Ph.D., Juan Cebral, Ph.D., Fernando Vinuela, M.D., Patient-Specific Hemodynamic Analysis of Small Internal Carotid Artery-Ophthalmic Artery Aneurysm, Surgical neurology. 2009 Nov 30;72(5):444-50.

[12] Merih I, Baharoglu, M.D., Clemens M. Schirmer, M.D., Daniel A. Hoit, M.D, M.P.H., Bu-Lang Gao, M.D., Adel M. Malek, M.D. Ph.D., "Aneurysm Inflow Angle as a Discriminant for Rupture in Sidewall Cerebral Aneurysms, Morphometric and Computational Fluid Dynamics', Stroke, 2010, 41, 1423-1430.

[13] Acropolis Convention Center, 'Simulation of Hemodynamics in a Patient-Specific Aortic Coarctation Model', The 15th International Conference on Medical Image Computing and Computer Assisted Intervention, Sacom 2012.

[14] Feng Gao and Hiroshi Okada, "Hemodynamic Changes in Coronary Artery After Stent Implantation Based on Patient- Specific Model", Journal of Medical and Bioengineering, June 2013 Vol. 2(2).

[15] Luisa C. Sousa, Catarina F., Castro, Carlos C. Antonio, Rui Chaves, Rosa Santos, Pedro Castro and Elsa Azevedo, "Hemodynamics of A Stenosed Carotid Bifurcation", VI international Conference on Adaptive Modeling and Simulation", ADMOS 2013.

[16] Shang EK, Nathan DP, Sprinkle SR, Fairman RM, Bavaria JE, Gorman RC, Gorma JH 3rd, Jackson BM,"Impact Of Wall Thickness and Saccular Geometry on the Computational Wall Stress of Descending Thoracic Aortic Aneurysm", US National Library of Medicine National Institute Of Health, 2013 128 (11 suppl 1).

[17] Jonathan Russin, Haithem Babiker, Justin Ryan, Leonardo Rangel-Castilla, David Fraker, Peter Nakaji. Computational Fluid Dynamics to Evaluate the Management of a Giant Internal Carotid Artery Aneurys, World neurosurgery, 2015 Jun 30;83(6):1057-65.

REVIEWING THE CHALLENGES IN MAINTAINING THE RELIABILITY AND ACCURACY OF IOT SYSTEMS FOR REMAINING USEFUL LIFE PREDICTION

Kalaiarasan Sekar[1, 2], Shahani Aman Shah[2], A Antony Athithan[1]

[1] Faculty of Engineering, Lincoln University College, Kuala Lumpur, Malaysia.
[2] UniKL MIAT, Universiti Kuala Lumpur, Kuala Lumpur, Malaysia

*Shahani@unikl.edu.my

Abstract. Internet of things (IoT) has been implemented in aviation predictive maintenance in recent years for the enhancement of better maintenance prediction, to reduce downtime, unnecessary maintenance actions, increase safety, increase system readiness, and refine the management process and to improve component design. The IoT system in predictive maintenance is very optimistic in gathering and analysing, predicting the component failures and to determine the remaining useful life of a systems. Since Remaining useful life of an system is defines as the length from the current time to the end of its useful life. Due to its futuristic increasing demand of IoT in aviation maintenance, the biggest challenge is to ensuring the reliability and accuracy of any specific IoT system allotted for monitoring aircraft components in the near future. Hence, this review paper clearly explains the challenges associated with IoT systems on predicting Remaining useful life.

1. Introduction

Internet of things and Artificial intelligence plays a vital role in Aircraft predictive maintenance. Internet of things (IoT) has been implemented in aviation predictive maintenance in recent years for the enhancement of better maintenance prediction, to reduce downtime, unnecessary maintenance actions, increase safety, increase system readiness, and refine the management process, and improve the component design. The Internet of things has heterogeneous applications in the aviation industry. Hence, ensuring the reliable outcome and performance of the IoT systems when synchronizing with complex computational devices in the aircraft components is necessary to reduce the prediction challenges due to false negatives and false positives data sets. Hence the same can impact the accuracy in prediction of the remaining useful life of an aircraft component. The Remaining useful life prediction for aircraft systems and subsystems can be measured using the data-driven and model-driven approaches. The theoretical methods and prognosis algorithms can be developed for predicting the remaining useful life of an aircraft component, but which has a major challenge of validation to ensure the accuracy of the predictions [1]. Similarly, structural health monitoring with IoT platforms can be used to predict the endurance of the damage and damage evaluation. The structural monitoring can be developed by obtaining real-time data which could be achieved through reliable high-speed internet and wireless sensor networks. Although, the huge challenge lies in providing a low-cost computational system amid growing maturity in IoT systems [2, 3]. The complexity also being faced while measuring the QoS (Quality of service) of heterogeneous IoT systems, through Model-driven approach verification.[4].The quality of data is vital in predicting the remaining useful life of a system and structure since the proposed system will have to predict less or no false negatives with a critical operating environment[5]. The limitations in the IoT system include security challenges, quality of service, constraints in obtaining the real-time data, and also difficulty in collecting remaining useful life data from a particular machine type [6]. This section reveals the complexity faced by the researchers on various techniques proposed with the IoT platform for remaining useful life prediction with high-quality reliable data.

2. Challenges in IoT System reliability

2.1. The IoT System Reliability

The major challenge lies with the heterogeneous devices from small low power to high range systems is implementing a multilayer security. Since, these heterogeneous networks are more vulnerable security attacks and providing with the fault data. So, any IoT system would contain a standard mechanism within itself to indicate the redundancy during the malfunction and security attack [7]

The IoT system with the heterogeneous multilayered infrastructure has a greater challenge in obtaining a reliable data which eventually leads to false Remaining useful life predictions. The vulnerability in the system leads to anomalous data being generated and sent which in most severe cases affect human lives [8]

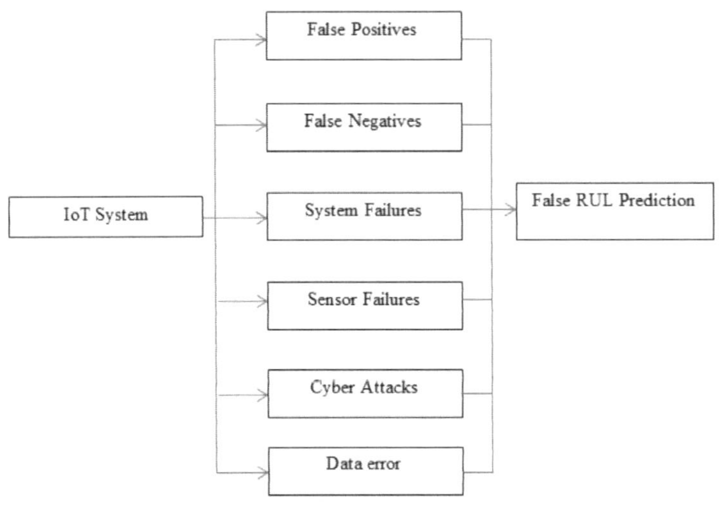

Figure 1. IoT fault syndromes leading to false RUL Prediction

2.2. Challenges in Anomaly detection

IoT systems working on heterogeneous Platform are in need to generate huge amount of data which the large computation is necessary to be adopted. When it comes to handling of huge data with large computational systems, anomaly detection is largely needed for identifying the misbehaving data with normal data sets [9].The Key issues restricting the anomaly detection and the possible causes for those issues are shown in Table 1.

Table 1. Key Issues and Possible Causes in Anomaly Detection

Key Issues	Possible Causes
Incomplete Data Points	Incomplete External data from Environment
Data Error	Device Failure
Encrypted Data	Protected Data
Sensor Error	Multilayered Sensors
Data Noise	Transmission system failure
Data Surge	Overload of Data

2.3. Equipment Reliability Challenges

The Reliability of the equipment makes possible to meet the requirements expected by the manufacturers and Maintenance personals [10, 11]. Optimization of the IoT devices and equipment's during management of large data becomes a challenge [12].The computational and Mathematical models developed with generated data rely highly on equipment efficiency. The Quality of the data will be compromised upon reliability affected on those equipment's which in turn leads to False Prediction of Remaining Useful life of a components [13].

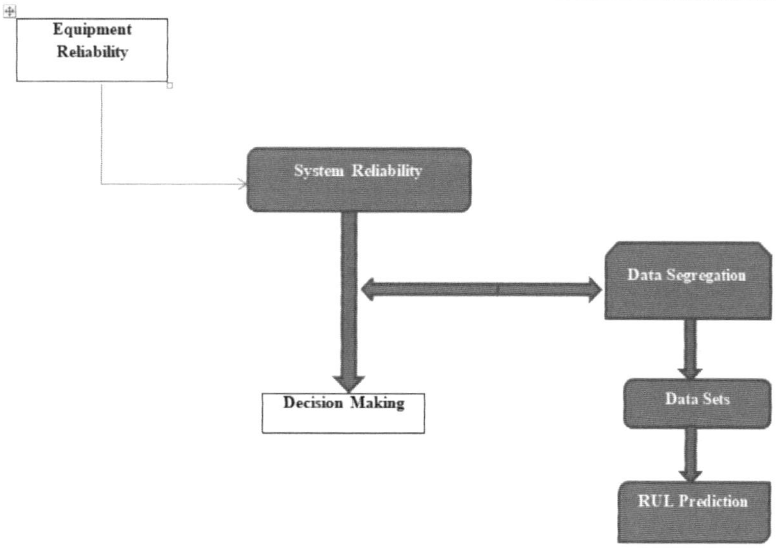

Figure 2. Scope of Equipment Reliability where coloured boxes focus mainly on this review

2.4. Challenges in IoT Architecture

Due to multilayered architecture of IoT system, it is imperative that the system produces reliable output throughout its mission cycle[14-16].In Terms of architectural challenges for IoT system itself, four major layers are considered to prove its reliability for providing the output.[17,18].In Other words, the common multilayered IoT architecture contains Service layer, support, communication and Perception layers .Each layers on the architecture poses different failure conditions on functioning which questions the reliability and leads to false Prediction.[19,20].The service layer in the multilayered architecture on aircraft engine components will use smart sensors to measure engine parameters like Exhaust gas temperature(EGT),N1 compressor speed, whereas the support layer intended to work on FDEP(Functional dependency),service switches, trigger switches , and in the modes of MTBF and MTTR, through where the availability of the system is measured[21,22,23].

The Communication layers poses failures on wireless communication ,noisy data ,attenuation of signals and perception layer provides challenges in reliable monitoring in terms of sensor nodes failures to determine measurements like temperatures and Humidity, which all provides False output or no output condition. Figure 3 shows the IoT architectural layers and possible failure modes leading to false RuL Prediction.

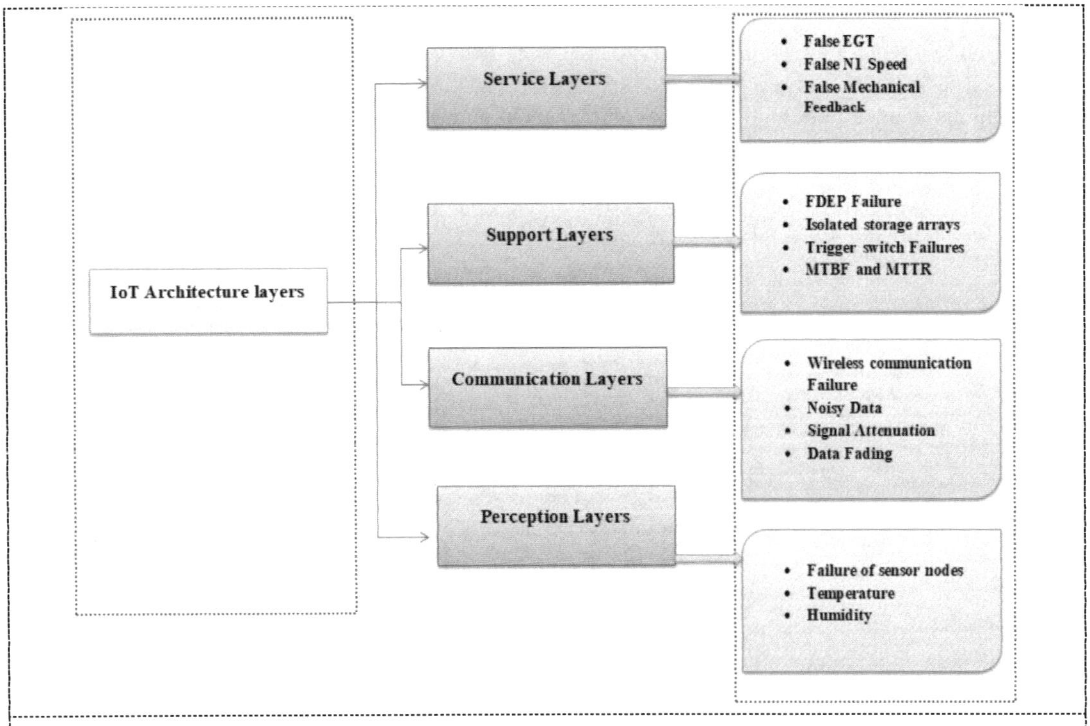

Figure 3. IoT Architecture and Possible failures

2.5. Performance Challenges

The Heterogeneity of an IoT system will have the complexity and constraints on hardware and software which requires massive computational system which leads to noticeable degradation on the performance parameters in terms of High throughout, latency of the system, and accuracy of the data

[24]. Specifically, the high accuracy requirement in the IoT system may affect the control aspects in case of unmanned air vehicle which affects the Ultra, Low and End to End latencies. Also the entire system would liable to provide unique complex challenges in terms of sensors [25, 26].Table 2 Shows the specific possible causes which affects the performance efficiency of any Multilayered IoT Systems.

Table 2. Performance Affecting Parameter and Causes

Possible Causes	Performance affecting Parameters
Infeasible Raw Data	High Throughputs and frames
Communication Delay	Low Latencies
High accuracy Requirement	Control failure

2.6. Challenges on Data Registration
The Future challenges on Data registration and Data generation need to incorporate rich sensors like LiDAR [26] and high computational systems for maximum reliability. The complexity is defined in data segregation, data extraction and categorisation of data in timely manner [27]

Considering the aircraft systems, the connected IoT systems may use both Operational data and BITE Data for Flight warning systems(FWS) and Central fault display unit(CFDS) for indication and creating a maintenance reports using cloud platforms. Since the fault syndromes developed in the sensors would eventually leads to Non-Reliable maintenance indications to the crew and leads to false prediction. The review points out the possible challenges from data registration to remaining useful life prediction.

Especially on the studies conducted for data registration over years, major precise fault happens due to sensor overlapping at the close proximity regions [28, 29]. This is highly susceptible in heterogeneous systems. Figure 4 Shows the Overview of Challenges in Multi-layered CFDS system on RuL Prediction and data extraction

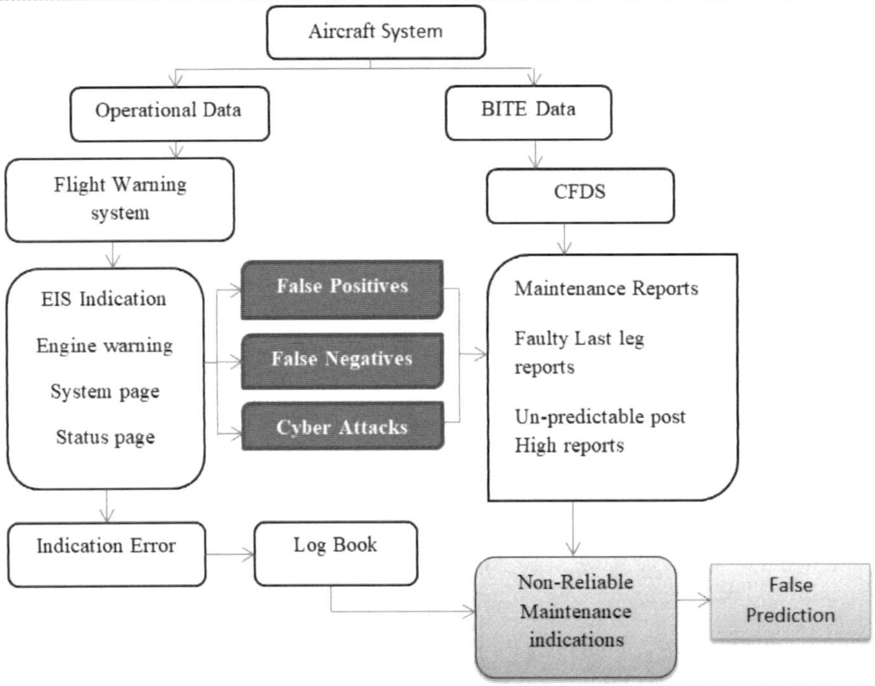

Figure 4. Overview of Challenges in Multi-layered CFDS system on RuL Prediction and data extraction

2.7. *Hardware Reliability Challenges*

The Hardware non-reliability on the IoT system is highly susceptible due to non-quantification and evaluation of physical materials in the connected system. So the whole challenges create the necessity for prediction methodologies for assessing the hardware reliability. The Common methods are Physics of Failure (PoF) Prediction [30].

The Physics of Failure method is commonly used method which provides potential results for accurate prediction of RuL and mode of failure. Figure 5 Shows the steps involved in Physics of Failure (PoF).

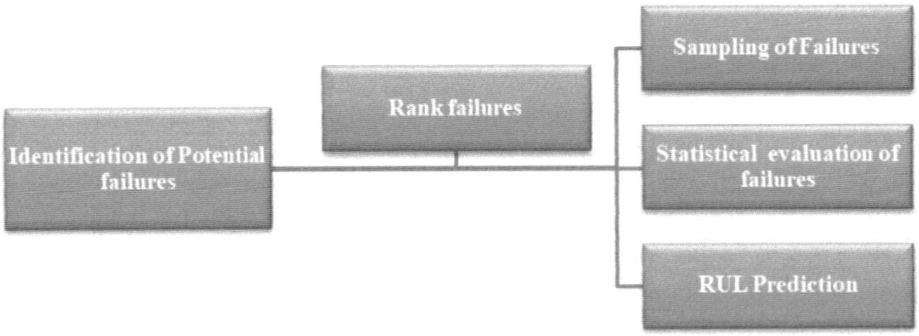

Figure 5. Steps involved in PoF Prediction

2.8. Challenges in Network Reliability

The Major challenge in maintaining network reliability in the IoT Systems is very crucial where importantly Assessing QoS (Quality of Service) and Continuous Quantification should be considered. So always the user-friendly assessment and prediction technique should be assigned to evaluate the network efficiency of the system [30, 31]. Quantification of delay throughputs for QoS metric analysis is carried out to provide sufficient information on reliability of end to end IoT systems [32].The QoS Profile generation which is linked with various components in the multi-layered system has been proposed for determination of latency and bandwidth [33].The Statistical Modelling approach is carried out to calculate the QoS metrics like time consuming, time of response, and Repair times [34]. The Redundancy models were studied the infrastructure of Gateway and ISP redundancy [35]. The Various findings have been carried out by past researchers on assessing network reliability on the IoT systems. The Previous works carried out on Network reliability assessment will make a pathway for future researchers for selecting suitable and appropriate method for multi-layered systems.

2.9. System Security Challenges

The heterogeneous Multilayered IoT System will have more vulnerability for security attacks. To address this issues the IoT system design must be optimized to have important factors which includes Perfect Physical coupling, Communication, security, Scalability and Privacy requirements [36].Especially various types of threats have been identified by previous researchers. Table 3 Shows the Summarized Literature review showing contribution of each works related to security attacks on the IoT System.

Table 3. Summarized Literature review showing contribution of Each Works related to Security Attacks on the IoT System

Contribution	Work	Findings
Cyber-attacks	P. McDaniel et al.(2009)[37] A.O.Otuoze et al.(2018)[38] S.Goel et al.(2015)[39] V. Delgado-Gomes et al[40]	Several Potential Cyber-attacks have been discussed through this works where Active and Passive attacks poses significant threats based on spy, eavesdrop and DoS
Spoofing Attacks	P. Pradhan et al.(2016)[41] P. Risbud et al.(2018)[42]	The Major Challenge in the IoT system is that susceptibility to the Spoofing attacks where GPS spoofing is due to high strength incorrect signals and ARP Spoofing is due to false messages linkage to MAC address of the hackers. The control protocol is affected which may mislead the network operating systems.
Replay Attacks	J. Zhao et al.(2016)[43] T. Tran el al.(2013)[44]	The Authenticity of the Information is highly intercepted due to replay attacks in the IoT systems. Those Incorrect information may lead

		to False RuL Prediction.
Smart Meter DoS Attacks	P. Yi et al.(2014)[45] C. Bekara et al.(2014)[46] Y. Guo et al.(2015)[47]	The Denial of Service attacks will large amount of replies and request packets which may leads to total system failure. The corrective action is achieved through integration of IoT devices in to Smart Grid.
Malware Attacks	E. Modiri Dovom et al.(2017)[48] P. Eder-Neuhauser et al.[49]	The malicious software is injected to the system which may cause interruptions or No service. The Communication layer of the IoT system is more prone to these attacks which may have to be Integrated for prevention.

3. Validation and Prediction Challenges

3.1. Overview of Consideration and Challenges in IoT System Architecture design

This Section Provides the Summary of Literature review for various Modern Tool Validation Approaches in Providing Safety Aspect and Statistical prediction including Machine learning and Deep Learning Approaches. Especially, while concerning the importance of security in the IoT systems various challenges and Considerations have been put forth to design and Validate the system. The Consideration includes Better interaction of the IoT system with the Physical World [50], Constraints in the available resources [51], heterogeneity [52], Large Scale data registration [53] and segregation, Security breaches, Privacy settings [54] and trust management [55].

Table 4. Summarized Literature review showing contribution of Each Works related to Security Attacks on the IoT System

Consideration	Challenges
Physical World Integration	• OS Update • Compatibility • Coupling with Physical and Cyber world
Heterogeneity	• Continuous Monitoring needed • Reliable Detection Mechanism for Abnormal Behaviour • Reliability issues over Traditional Mechanisms • Reliable Security

	Algorithms
Constraints in the Resources	• Memory space allocation for large amounts of data • Availability of the system • Detection of Intrusion Points
Privacy	Firewall Reliability and Intrusion Detection
Large Scale Distribution	Scalable
Trust Management Administration	Reliable

3.2. RUL Prediction approaches and drawbacks

The section provides the summary of different modern prediction approaches which includes Deep learning and Machine learning models used for RUL Prediction using IoT Systems. RUL Prediction derived from the concept of Prognostics where future of the entire system is predicted using observations, statistical and mathematical models [55].Any RUL Prediction is defined as observing the functional range of the entire system or component before it reaches the fatigue or failure range [56].

The RUL Prediction can be predicted and analysed through various approaches which includes traditional supervised and unsupervised machine learning, and deep learning models[57,58].The Modern prognostic approaches contains various challenges in RUL Prediction process. Table 4 shows the summary of review of challenges and drawbacks considered on the various RUL Prediction approaches.

Table 5. Shows the summary of review of Challenges considered on Various RUL Prediction Approaches

RUL Prediction Approach	Work	Challenges
Physics Based Approach	A. Cubilo et al.(2016)[59] H.M Elattar et al.(2016)[60]	• Intense Computations Requirement • High fidelity • Complexity on modelling the defect • Reusable limits are less • Complex mechanical systems • Difficulty on identifying the fault

Hybrid based Approach	M. Schwabacher (2005)[61] Youdao Wang et al(2020)[58]	• Noisy data • Inaccuracy because of Noisy data • Both Model and Data Required
Data-Driven Approaches	X .S. Si et al(2011)[62] Youdao Wang et al(2020)[58]	• Intense and Large algorithms required • Short Prediction ranges • Inadequate and shortage of data for new systems
Supervised Machine learning	Brownlee et al(2016)[63] Kushal .R. D(2020)[57]	• Labelling of Data • Huge Labelling requires Intense training of the models • Identification of Inputs and Outputs
Unsupervised Machine Learning	Kushal .R. D(2020)[57] Shanthamallu et al.(2017)[64]	• Large Volume of Clustering • High Monitoring needed for Unlabelled Information

4. Conclusion

To the best of our knowledge, this paper will be able to reveal the challenges associated with remaining useful life prediction of components or system using the IoT platform, and recommends the pre-requisites and requirements to be considered for any IoT infrastructure if in case of false data sets with sensor and system failure. Also, the literature review briefs the specific challenges obtained from the heterogeneous IoT network. This includes system reliability challenges, challenges in anomaly detection using IoT systems, challenges associated with constructing the IoT architecture, challenges in data registration and data segregation, challenges associated with hardware reliability, and security challenges. Also, one of the other major challenge lies with validation of the selected IoT model in consideration with different real time factors as discussed in chapter 3 from this paper. The IoT model which is constructed for Remaining useful life prediction basically perform on the three approaches which includes physics based , hybrid and data-driven approaches. The summary on table 4 describes the drawbacks encountered by the researchers on testing the model with these three different approaches. The challenges are summarised to provide the knowledge on the machine learning approaches which incorporates IoT system for remaining useful life prediction.

We believe this review will be able to provide the insights for researchers to identify the adoptability of IoT systems on aviation Predictive maintenance with reliable and cost-effective computational systems. The outcome of this review is to provide the understanding of the challenges associated with multi-layered IoT systems and leads to consider the consequences developed on the specific conditions in the system prior to construction of the IoT Model. So we strongly believe, these consideration and knowledge about the system construction and performance of the particular model would eventually reduce the downtime, enhance the cost saving in the aviation predictive maintenance.

References

[1] Xiongzi, C., Jinsong, Y., Diyin, T. and Yingxun, W., 2011, August. Remaining useful life prognostic estimation for aircraft subsystems or components: A review. In *IEEE 2011 10th International Conference on Electronic Measurement & Instruments* (Vol. 2, pp. 94-98). IEEE.

[2] Abdelgawad, A. and Yelamarthi, K., 2017. Internet of things (IoT) platform for structure health monitoring. *Wireless Communications and Mobile Computing, 2017.*
[3] Angadi, A., Dias, R. and Bagali, M.U., 2016. An Aircraft Health Monitoring System using IOT. *Indian Journal of Science and Technology*, *9*(33), pp.1-5.
[4] Costa, B., Pires, P.F., Delicato, F.C., Li, W. and Zomaya, A.Y., 2016, August. Design and analysis of IoT applications: a model-driven approach. In *2016 IEEE 14th Intl Conf on Dependable, Autonomic and Secure Computing, 14th Intl Conf on Pervasive Intelligence and Computing, 2nd Intl Conf on Big Data Intelligence and Computing and Cyber Science and Technology Congress (DASC/PiCom/DataCom/CyberSciTech)* (pp. 392-399). IEEE.
[5] The 5 step Roadmap to IoT-based Predictive Maintenance, how to leverage sensor data predictive analytics & Machine learning for more intelligent maintenance, *Article Published by XM Pro, Inc.Agile Industrial IoT, Version 2.0.*
[6] Calabrese, M., Cimmino, M., Fiume, F., Manfrin, M., Romeo, L., Ceccacci, S., Paolanti, M., Toscano, G., Ciandrini, G., Carrotta, A. and Mengoni, M., 2020. SOPHIA: An event-based IoT and machine learning architecture for predictive maintenance in industry 4.0. *Information*, *11*(4), p.202.
[7] Khan, M.A. and Salah, K., 2018. IoT security: Review, blockchain solutions, and open challenges. *Future Generation Computer Systems*, *82*, pp.395-411..
[8] Moore, S.J., Nugent, C.D., Zhang, S. and Cleland, I., 2020. IoT reliability: a review leading to 5 key research directions. *CCF Transactions on Pervasive Computing and Interaction*, pp.1-17.
[9] Sharma, B., Sharma, L. and Lal, C., 2019, December. Anomaly Detection Techniques using Deep Learning in IoT: A Survey. In *2019 International Conference on Computational Intelligence and Knowledge Economy (ICCIKE)* (pp. 146-149). IEEE.
[10] Bianchini, A., Pellegrini, M. and Rossi, J., 2019. Maintenance scheduling optimization for industrial centrifugal pumps. *International Journal of System Assurance Engineering and Management*, *10*(4), pp.848-860.
[11] Ustundag, A. and Cevikcan, E., 2017. *Industry 4.0: managing the digital transformation*. Springer.
[12] Lee, C.K.M., Zhang, S.Z. and Ng, K.K.H., 2017. Development of an industrial Internet of things suite for smart factory towards re-industrialization. *Advances in manufacturing*, *5*(4), pp.335-343.
[13] SCHEER, A.W., 2019. Enterprise 4.0-From disruptive business model to the automation of business processes.
[14] Kempf, J., Arkko, J., Beheshti, N. and Yedavalli, K., 2011, March. Thoughts on reliability in the internet of things. In *Interconnecting smart objects with the Internet workshop* (Vol. 1, pp. 1-4).
[15] Han, C., Jornet, J.M., Fadel, E. and Akyildiz, I.F., 2013. A cross-layer communication module for the Internet of Things. *Computer Networks*, *57*(3), pp.622-633.
[16] Palattella, M.R., Dohler, M., Grieco, A., Rizzo, G., Torsner, J., Engel, T. and Ladid, L., 2016. Internet of things in the 5G era: Enablers, architecture, and business models. *IEEE Journal on Selected Areas in Communications*, *34*(3), pp.510-527.
[17] Burhan, M., Rehman, R.A., Khan, B. and Kim, B.S., 2018. IoT elements, layered architectures and security issues: A comprehensive survey. *Sensors*, *18*(9), p.2796.
[18] Darwish, D., 2015. Improved layered architecture for Internet of Things. *Int. J. Comput. Acad. Res.(IJCAR)*, *4*, pp.214-223.
[19] Frustaci, M., Pace, P., Aloi, G. and Fortino, G., 2017. Evaluating critical security issues of the IoT world: Present and future challenges. *IEEE Internet of things journal*, *5*(4), pp.2483-2495.
[20] Xing, L., 2020. Reliability in Internet of Things: Current status and future perspectives. *IEEE Internet of Things Journal*, *7*(8), pp.6704-6721.
[21] Gopalakrishnan, M., Skoogh, A., Salonen, A. and Asp, M., 2019. Machine criticality assessment for productivity improvement. *International Journal of Productivity and*

Performance Management.
[22] Quijada Fumero, P. and Salunkhe, O., 2017. *Demonstration of real-time criticality assessment using a test-bed* (Master's thesis).
[23] Souza, M.L.H., da Costa, C.A., de Oliveira Ramos, G. and da Rosa Righi, R., 2020. A survey on decision-making based on system reliability in the context of Industry 4.0. *Journal of Manufacturing Systems*, *56*, pp.133-156.
[24] Bagchi, S., Abdelzaher, T.F., Govindan, R., Shenoy, P., Atrey, A., Ghosh, P. and Xu, R., 2020. New Frontiers in IoT: Networking, Systems, Reliability, and Security Challenges. *IEEE Internet of Things Journal*, *7*(12), pp.11330-11346.
[25] Yick, J., Mukherjee, B. and Ghosal, D., 2008. Wireless sensor network survey. *Computer networks*, *52*(12), pp.2292-2330.
[26] Wang, Q., Zhu, Y. and Cheng, L., 2006. Reprogramming wireless sensor networks: challenges and approaches. *IEEE network*, *20*(3), pp.48-55.
[27] Habib, A., Ghanma, M., Morgan, M. and Al-Ruzouq, R., 2005. Photogrammetric and LiDAR data registration using linear features. *Photogrammetric Engineering & Remote Sensing*, *71*(6), pp.699-707.
[28] Bellekens, B., Spruyt, V., Berkvens, R., Penne, R. and Weyn, M., 2015. A benchmark survey of rigid 3D point cloud registration algorithms. *Int. J. Adv. Intell. Syst*, *8*, pp.118-127.
[29] Liu, S., Tong, X., Chen, J., Liu, X., Sun, W., Xie, H., Chen, P., Jin, Y. and Ye, Z., 2016. A linear feature-based approach for the registration of unmanned aerial vehicle remotely-sensed images and airborne LiDAR data. *Remote Sensing*, *8*(2), p.82.
[30] Ahmad, M., 2014, November. Reliability models for the internet of things: A paradigm shift. In *2014 IEEE International Symposium on Software Reliability Engineering Workshops* (pp. 52-59). IEEE.
[31] Maalel, N., Natalizio, E., Bouabdallah, A., Roux, P. and Kellil, M., 2013, May. Reliability for emergency applications in internet of things. In *2013 IEEE International Conference on Distributed Computing in Sensor Systems* (pp. 361-366). IEEE.
[32] Kamyod, C., 2018, February. End-to-end reliability analysis of an IoT based smart agriculture. In *2018 International Conference on Digital Arts, Media and Technology (ICDAMT)* (pp. 258-261). IEEE.
[33] Brogi, A. and Forti, S., 2017. QoS-aware deployment of IoT applications through the fog. *IEEE Internet of Things Journal*, *4*(5), pp.1185-1192.
[34] Li, S. and Huang, J., 2017, June. GSPN-based reliability-aware performance evaluation of IoT services. In *2017 IEEE International Conference on Services Computing (SCC)* (pp. 483-486). IEEE.
[35] Sinche, S., Polo, O., Raposo, D., Femandes, M., Boavida, F., Rodrigues, A., Pereira, V. and Silva, J.S., 2018, June. Assessing redundancy models for IoT reliability. In *2018 IEEE 19th International Symposium on" A World of Wireless, Mobile and Multimedia Networks"(WoWMoM)* (pp. 14-15). IEEE.
[36] Sha, K., Wei, W., Yang, T.A., Wang, Z. and Shi, W., 2018. On security challenges and open issues in Internet of Things. *Future Generation Computer Systems*, *83*, pp.326-337.
[37] McDaniel, P. and McLaughlin, S., 2009. Security and privacy challenges in the smart grid. *IEEE Security & Privacy*, *7*(3), pp.75-77
[38] Ghansah, I., 2012. *Smart Grid Cyber Security Potential Threats, Vulnerabilities and Risks: Interim Project Report*. California Energy Commission.
[39] Goel, S. and Hong, Y., 2015. Security challenges in smart grid implementation. In *Smart grid security* (pp. 1-39). Springer, London.
[40] Brito, R., Carvalho, A. and Gericota, M., 2015, June. A new three-phase voltage sourced converter laplace model. In *2015 9th International Conference on Compatibility and Power Electronics (CPE)* (pp. 160-166). IEEE.
[41] Pradhan, P., Nagananda, K., Venkitasubramaniam, P., Kishore, S. and Blum, R.S., 2016,

October. GPS spoofing attack characterization and detection in smart grids. In *2016 IEEE Conference on Communications and Network Security (CNS)* (pp. 391-395). IEEE.
[42] Risbud, P., Gatsis, N. and Taha, A., 2018. Vulnerability analysis of smart grids to GPS spoofing. *IEEE Transactions on Smart Grid*, *10*(4), pp.3535-3548.
[43] Gao, Y.L., An, X.H. and Liu, J.M., 2008, December. A particle swarm optimization algorithm with logarithm decreasing inertia weight and chaos mutation. In *2008 international conference on computational intelligence and security* (Vol. 1, pp. 61-65). IEEE.
[44] Tran, T.T., Shin, O.S. and Lee, J.H., 2013, January. Detection of replay attacks in smart grid systems. In *2013 International Conference on Computing, Management and Telecommunications (ComManTel)* (pp. 298-302). IEEE.
[45] Yi, P., Zhu, T., Zhang, Q., Wu, Y. and Li, J., 2014, June. A denial of service attack in advanced metering infrastructure network. In *2014 IEEE International Conference on Communications (ICC)* (pp. 1029-1034). IEEE.
[46] Bekara, C., 2014. Security issues and challenges for the IoT-based smart grid. *Procedia Computer Science*, *34*, pp.532-537.
[47] Guo, Y., Ten, C.W., Hu, S. and Weaver, W.W., 2015, February. Modeling distributed denial of service attack in advanced metering infrastructure. In *2015 IEEE power & energy society innovative smart grid technologies conference (ISGT)* (pp. 1-5). IEEE.
[48] Dovom, E.M., Azmoodeh, A., Dehghantanha, A., Newton, D.E., Parizi, R.M. and Karimipour, H., 2019. Fuzzy pattern tree for edge malware detection and categorization in IoT. *Journal of Systems Architecture*, *97*, pp.1-7.
[49] Eder-Neuhauser, P., Zseby, T. and Fabini, J., 2018. Malware propagation in smart grid monocultures. *e & i Elektrotechnik und Informationstechnik*, *135*(3), pp.264-269.
[50] Boyer, S.A., 1999. *SCADA: supervisory control and data acquisition* (Vol. 3). Research Triangle Park: Isa.
[51] Sha, K., Wei, W., Yang, T.A., Wang, Z. and Shi, W., 2018. On security challenges and open issues in Internet of Things. *Future Generation Computer Systems*, *83*, pp.326-337.
[52] Morrow, B., 2012. BYOD security challenges: control and protect your most sensitive data. *Network Security*, *2012*(12), pp.5-8.
[53] Yu, T., Sekar, V., Seshan, S., Agarwal, Y. and Xu, C., 2015, November. Handling a trillion (unfixable) flaws on a billion devices: Rethinking network security for the internet-of-things. In *Proceedings of the 14th ACM Workshop on Hot Topics in Networks* (pp. 1-7).
[54] Sha, K., Alatrash, N. and Wang, Z., 2016. A secure and efficient framework to read isolated smart grid devices. *IEEE Transactions on Smart Grid*, *8*(6), pp.2519-2531.
[55] Yan, Z., Zhang, P. and Vasilakos, A.V., 2014. A survey on trust management for Internet of Things. *Journal of network and computer applications*, *42*, pp.120-134.
[56] Salunkhe, T., Jamadar, N.I. and Kivade, S.B., 2014. Prediction of Remaining Useful Life of mechanical components-a Review. *International Journal of Engineering Science and Innovative Technology (IJESIT)*, *3*(6), pp.125-135.
[57] Dalal, K.R., 2020, July. Analysing the Role of Supervised and Unsupervised Machine Learning in IoT. In *2020 International Conference on Electronics and Sustainable Communication Systems (ICESC)* (pp. 75-79). IEEE.
[58] Wang, Y., Zhao, Y. and Addepalli, S., 2020. Remaining Useful Life Prediction using Deep Learning Approaches: A Review. *Procedia Manufacturing*, *49*, pp.81-88.
[59] Cubillo, A., Perinpanayagam, S. and Esperon-Miguez, M., 2016. A review of physics-based models in prognostics: Application to gears and bearings of rotating machinery. *Advances in Mechanical Engineering*, *8*(8), p.1687814016664660.
[60] Elattar, H.M., Elminir, H.K. and Riad, A.M., 2016. Prognostics: a literature review. *Complex & Intelligent Systems*, *2*(2), pp.125-154.
[61] Schwabacher, M., 2005. A survey of data-driven prognostics. In *Infotech@ Aerospace* (p. 7002).

[62] Si, X.S., Wang, W., Hu, C.H. and Zhou, D.H., 2011. Remaining useful life estimation–a review on the statistical data driven approaches. *European journal of operational research*, *213*(1), pp.1-14.
[63] Brownlee, J., 2016. *Master Machine Learning Algorithms: discover how they work and implement them from scratch*. Machine Learning Mastery.
[64] Shanthamallu, U.S., Spanias, A., Tepedelenlioglu, C. and Stanley, M., 2017, August. A brief survey of machine learning methods and their sensor and IoT applications. In *2017 8th International Conference on Information, Intelligence, Systems & Applications (IISA)* (pp. 1-8). IEEE.

Failure Analysis of Drive Axle Shaft failed under Torsional Stress

P Maurya[1], N Mulani[1], C Michael[1] and D Jebaseelan[1]

[1]School of Mechanical Engineering, Vellore Institute of Technology, Chennai, India
E-mail: punitmaurya@hotmail.com

Abstract. Drive axles are an essential component for power transmission in any automobile. It transfers the power from the differential to the hub of the wheels which drives the vehicle. The axles are subjected to torsional stresses due to the torque from the differential and the resisting torque present due to the frictional resistance from the ground on the wheels. Hence, they are prone to torsional failures under fatigue with and without shock loading. In this paper, a comprehensive engineering failure analysis of the failed drive axle shaft of material EN 24-T, custom designed for a 200kg BAJA SAE all-terrain vehicle was performed. Destructive, non-destructive testing and material analysis along with finite element simulations using Ansys were carried out to better understand the nature and cause of the failure. Failure analysis data hints at the flaws in design considerations and scanning electron microscopy (SEM) images reveal the formation of bainitic microstructures within the component due to improper heat treatment which lowers the ultimate tensile strength of the material.

1. Introduction

The movement of a vehicle happens when the torque produced by the prime mover (here engine) is transferred to the wheels in the desired amount via the transmission. This transfer and modification of torque in a vehicle is called power transmission. It has always been a field of challenges and innovations. Most on-road automobiles today use rigid drive axles to deliver power from a transmission to the wheels but an off-road automobile like an All-terrain Vehicle (ATV) requires a variable length drive axle to accommodate for the vertical wheel travel due to rough terrain.

Most of Original Equipment Manufacturer (OEM) drive axles comprise of either two Constant Velocity (CV) joints or two tripod joints. CV joints give maximum angularity of 40° with no plunge whereas tripod joints have maximum angularity of 23° with a plunge of about 50 mm to account for variable length of drive axle and to accommodate side thrust impacts during cornering. The desire to use a cross-groove joint which is a hybrid of a CV and a tripod joint was to get maximum angularity of 31° along with a plunge of 50 mm. But due to limited track width requirements of the buggy, the wheel side cross-groove joint was custom designed to work as a wheel hub while enclosing the joint within itself. Also, due to the in-board mounting of rear brake rotor, the drive axle shafts varied in length and had to be custom designed and manufactured. These axle rods are susceptible to torsional, shock and fatigue loads which greatly affect the component's life.

The present work does a systematic failure analysis for a failed drive axle rod, which was a critical failure in terms of the ATV buggy [1]. Bipin's [3], Bhagoria's [6] and Zhao's [7] papers on failure analysis of axle shaft of an automobile helped us in defining the steps for this comprehensive failure analysis.

2. Methods

Figure 1 shows the powertrain assembly of the buggy consisting of a Briggs and Stratton© OHV 10 HP Engine, CVTech™ continuously variable transmission (CVT), a custom designed constant reduction gearbox which connects with the two drive axles integrated with the wheel hub. The buggy was designed according to the rules published by the Society of Automotive Engineers (SAE) in the rulebook for BAJA all-terrain vehicle championships which take place across the globe.

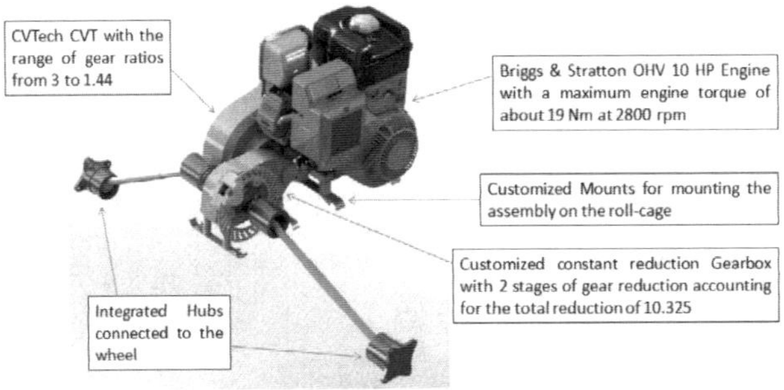

Figure 1. Powertrain assembly of the ATV buggy.

Figure 2 shows the hub integrated drive axle design consisting of an OEM cross groove joint that fits on the splines of the gearbox output shaft. The other end of the axle is another cross groove joint sitting in a customized hub, machined out of Aluminium 7071 block and bolted to the wheel rim.

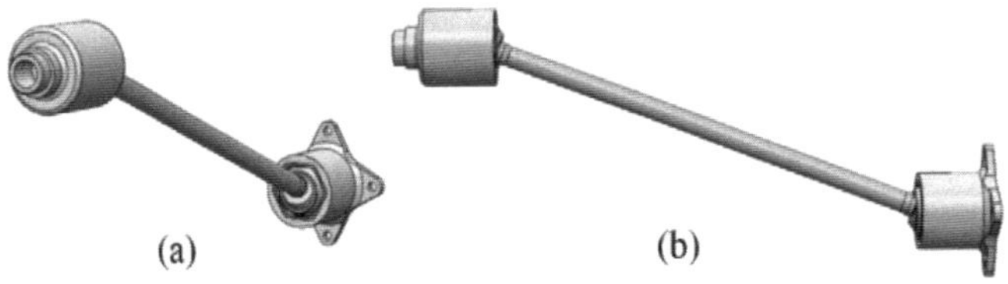

Figure 2. (a) Isometric and (b) front view of complete drive axle assembly.

Figure 3 shows an exploded view of the integrated cross-groove drive axle assembly which helps us determine the loading points on the axle rod at the splined areas.

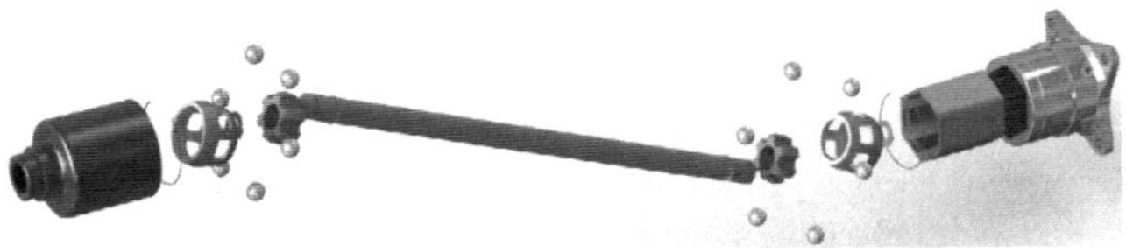

Figure 3. Exploded view of complete drive axle assembly.

Detailed engineering draft of the drive axle rod in figure 4 shows its key features which helps us understand the sequence of manufacturing processes the rod must have gone through.

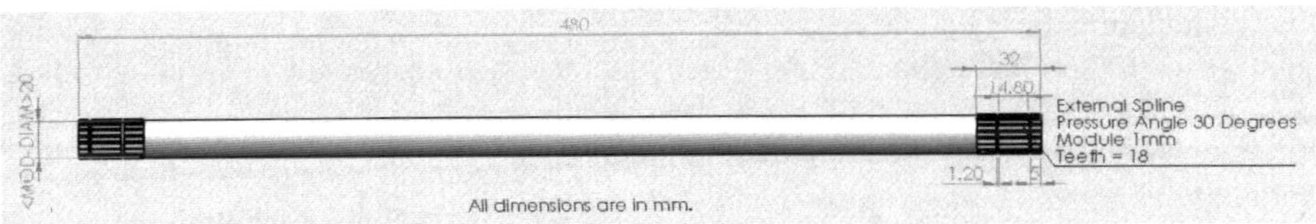

Figure 4. CAD model of the custom designed drive axle.

EN 24-T was chosen as the material for the drive axle rod due to its high tensile strength and other mechanical properties as seen in table 1 [2]. Tempering of the axle rod was carried out by heating it up to 660°C and holding at heat for two hours per inch of total thickness which increases the toughness while reducing brittleness and internal stresses.

Table 1. Mechanical properties of EN 24-T steel.

Condition	Tensile Strength (N/mm^2)	Yield Strength (N/mm^2)	Elongation (%)	Brinell Hardness
T	850 - 1000	650	13	248 - 302

EN24-T has the following chemical composition as shown in table 2 [2].

Table 2. Chemical composition of EN 24-T steel.

	C	Si	Mn	P	S	Mo	Cr	Ni
Min.	0.36	0.10	0.45	-	-	0.20	1.00	1.30
Max.	0.44	0.35	0.70	0.035	0.04	0.35	1.40	1.70

The 20 mm diameter EN 24-T billet was cut in the desired length. 1 mm module splines with 30° pressure angle were splined on both the ends of the rod. After splining, 2 circlip grooves were machined on both the splined areas to hold the cross-groove cage in its place.

The rod was heated to its austenization temperature using oxygas flames for flame hardening. It was performed to achieve a surface hardness of around 55-60 HRC and case depth between 0.127 mm and 6.35 mm.

The maximum torque condition coming on the drive axle will be at the moment when the vehicle takes a jump start or cornering at high speeds [3].

- Maximum torque from the Briggs and Stratton© engine (E_{max}) = 19 N-m
- Maximum gear reduction of the CVTech™ CVT (CVT_{max}) = 2.64
- Constant gear reduction of the gearbox ($gb_{reduction}$) = 10.325
- Diameter of the axle rod (d) = 20 mm
- Length of the axle rod (l) = 486 mm
- Shear modulus of steel (G) = 80,000 MPa
- Tensile strength of EN 24-T ($\tau_{tensile}$) = 850 N/mm^2

$$T_{max} = E_{max} \cdot CVT_{max} \cdot gb_{reduction} \tag{1}$$

Maximum torque transferred to the drive axles is given by equation (1) and comes out to be 518 N-m.

$$\tau_{permissible} = 0.7 \cdot \tau_{tensile} \tag{2}$$

Maximum permissible stress is given by equation (2) and it comes out to be 595 MPa.

$$\tau_{max} = \frac{16 \cdot T_{max}}{\pi \cdot d^3} \tag{3}$$

Maximum torque transferred to the drive axle rod is given by equation (3) and it comes out to be 330 MPa.

$$FOS = \frac{\tau_{\text{permissible}}}{\tau_{\text{max}}} \tag{4}$$

Factor of safety of the axle rod is given by equation (4) and it comes out to be 1.8.

$$\theta_{\text{max}} = \frac{584 \text{ x } T_{\text{max}} \text{ x } l}{\text{G x } d^4} \tag{5}$$

Maximum angular deflection of axle rod under maximum torque is given by equation (5) and it comes out to be 0.0115 rad or 0.66°.

After the flame hardening of the axle rod no rusting was observed anywhere for the complete duration of service. The un-splined areas of the shaft usually remained in open and often dust, dirt and mud would accumulate over it which would be easily cleaned with a spray of water or some rugged piece of cloth. The splined area is covered and sealed by a rubber boot or 'bellow' which protects the joint and the spline from outside dirt particles and moisture. A grease lubricant of standard OEM grade was filled into the bellows to provide lubrication to the joint and also help in heat dissipation. It was constantly being replenished after about 10 hours of running as the bellows use to either crack and leak or slide open from the joint on the hub side. The axles used to operate for an average of 8 hours after which the axle rods used to break at the same circlip groove location. Axle rods usually used to break after an aggressive start, a jump down from a high location or at attempts of very steep turns at high speeds.

The sample was collected from the testing field where the drive axle failed and covered in dirt, grease and mud. It was gently washed off with water and a piece of cloth. Later detergent water was used to remove the grease. Small samples of the fractured end were cut off by an angle grinder for detailed analysis under optical microscope.

Ansys© – Mechanical was used for the finite element analysis of the axle rod. For the design under consideration, finite element mesh is generated using tetrahedral mesh type taking fine size to 1mm and minimum edge length as 0.1mm with 20690 nodes.

3. Results and discussion

As seen in figure 5(a), to replicate the loading condition, maximum torque coming on the axle rod of 518 N-m was applied at one splined end of the drive axle while the other splined end was fixed [3]. Maximum deformation of about 2.34 mm was observed in figure 5(b) under the applied torque.

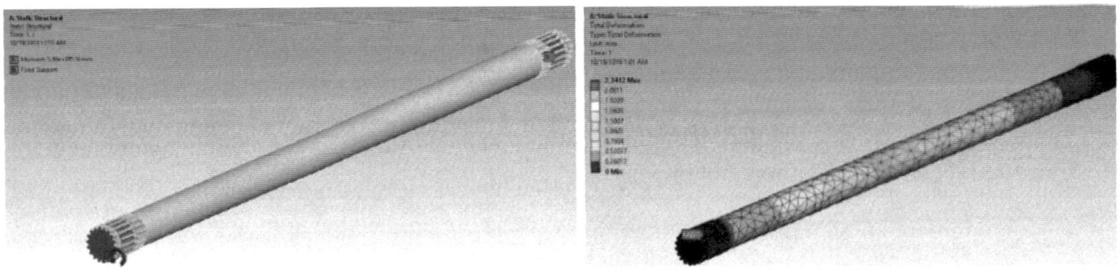

(a) (b)

Figure 5. (a) Loading of the axle rod and (b) maximum deflection under the load.

In figure 6(a) and 6(b), it can be observed that the regions of maximum stresses of about 5.57GPa are found around the circlip groove area on the edges of the start of the splines and it prevails over to the bottom part of the spline. It is clearly observed that this is the exact point of the plane from where the fracture takes place.

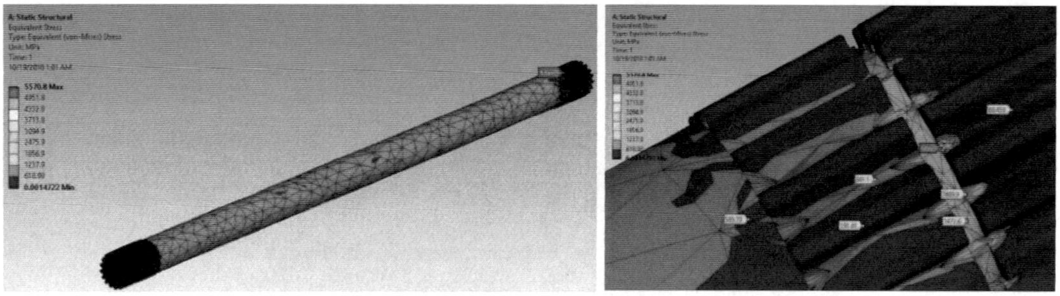

(a) (b)

Figure 6. (a) Equivalent (von-misses) stress plot of drive axle, (b) enlarged view.

From figure 7(a) it is visible that the major part of the drive axle (the non-splined part) has an FOS between 1 and 5. It is least around the circlip groove (about 0.12) and it continues to be low till the bottom of the splines as seen in figure 7(b).

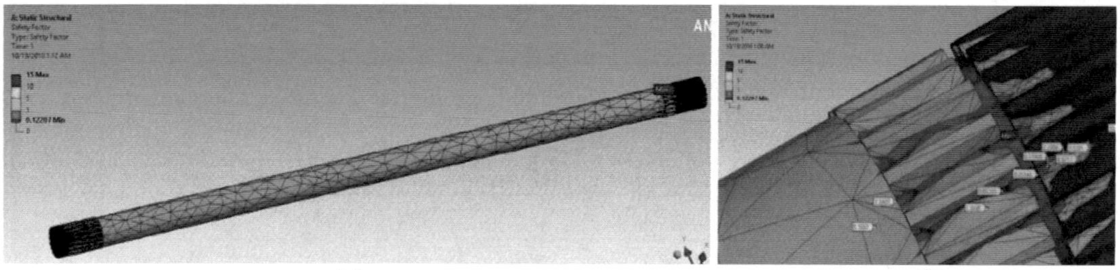

(a) (b)

Figure 7. (a) Factor of safety plot of drive axle, (b) enlarged view close to the spline.

Observation of the failed component

The fractured drive axle was recovered from the vehicle and there was still a lot of grease and dirt particles on it. The smaller portion of the drive axle that broke out was not found. Observed closely in figure 8(a), a slight bend in the drive axle was discovered and after thorough investigation it was found that this bend was observed after the heat treatment process which suggests that flame hardening was not the best heat treatment process for this component. This bend might have caused irregular stresses in the component over its duration of service.

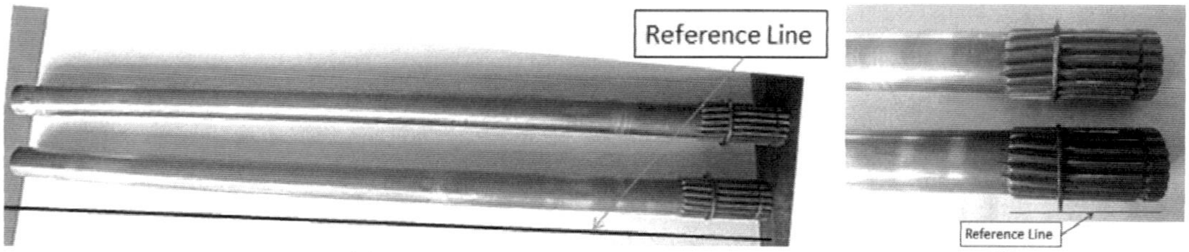

(a) (b)

Figure 8. (a) Axle rod found to be bent, (b) torsional deflection seen in the splines.

When the other side of the spline was viewed closely in figure 8(b), it was visible how the splines had deformed under the torsional forces.

When the fracture surface was photographed and is shown in figure 9 (a) and (b), the deformations due to torsional stresses were clearly visible which originated from the outer periphery with different points of fracture initiation and ended up somewhere in between. The shiny and rough patches of the surface indicate towards both tensile and brittle nature of the fracture [1].

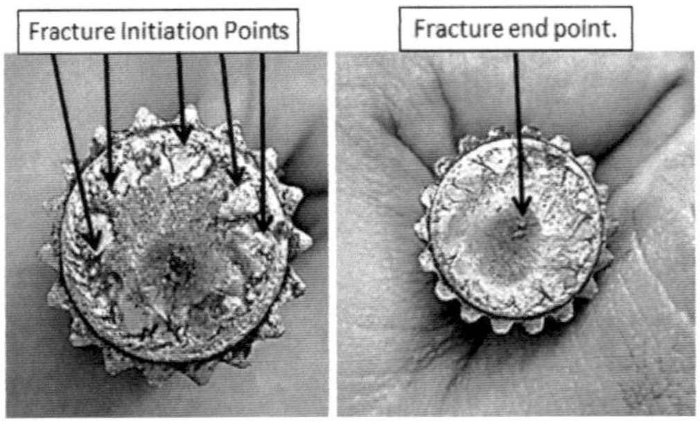

(a) (b)

Figure 9. (a) Fracture initiation points (b) fracture end point.

Rockwell hardness test was conducted to check for the hardness of the component at two different locations and the EN 24-T hardness ranged from 248-302 BHN (table 1) and will be helpful to know the quality of the heat treatment process.

Table 3. Hardness values from rockwell hardness test.

Hardness at the centre	Hardness at the periphery
30 HRC = 286 BHN	54 HRC = 543 BHN

The hardness value from table 3 is well within known range of hardness of the material and also helped to verify the application of the heat treatment processes along with the material hardness characteristics.

A tensile test was conducted on a specimen machined from the centre region of the bar based on ASTM E8 Standards with a strain rate of 4×10^{-4} S^{-1}.

Table 4. Tensile test values

Property	Values from test	EN 24-T, flame hardened [2]
Yield Strength	665 MPa	1110 MPa
Ultimate Tensile Strength	916 MPa	1294 MPa
Elongation	12.8 %	14 %
Reduction of Area	52.6 %	50 %

The values in table 4 reveal the imperfection in heat treatment processes applied on the axle rod as it could not achieve the desired values as per design considerations.

Optical microscopic analysis of the component

As seen in figure 10, optical microscopy at about 3000X magnification revealed the formation of bainitic microstructure in the material sample. Hence it was decided to go for Scanning Electron Microscopy for more understanding [4].

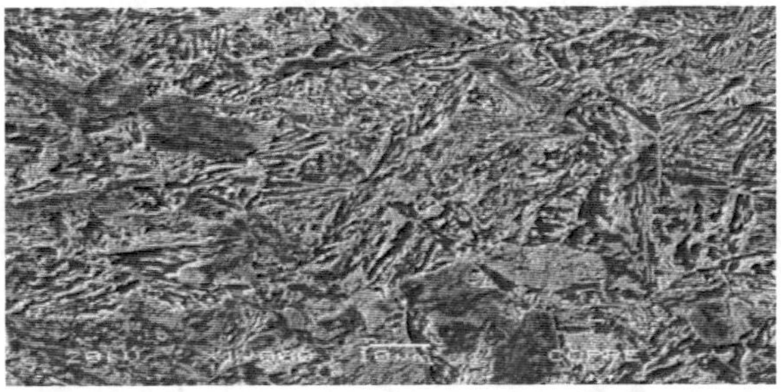

Figure 10. Presence of bainitic microstructure detected.

SEM analysis of the component

Detailed observation of the fracture surface was made by scanning electron microscopy (SEM), as shown in Figure 11. The images were made at specific points: close to the edge, at a quarter-radius from the edge, mid-radius and at centre of the fracture surface.

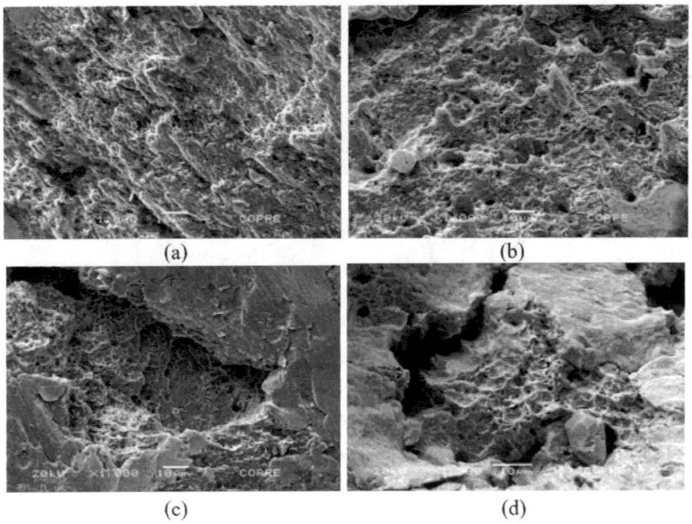

Figure 11. Fracture morphology: (a) close to the edge, (b) at a quarter-radius from the edge, (c) midway between edge and centre and (d) at the centre.

It was observed from the scanning electron microscopy images in figure 11that part of the surface was damaged due to rubbing. However, a significant part of the fracture was preserved and dimple morphology was evidenced [5]. The elongated dimples close to the surface, as shown in figure 11(a) are observed by and associated with the larger shear deformations in that region. Besides the elongated dimples, secondary cracks were observed.

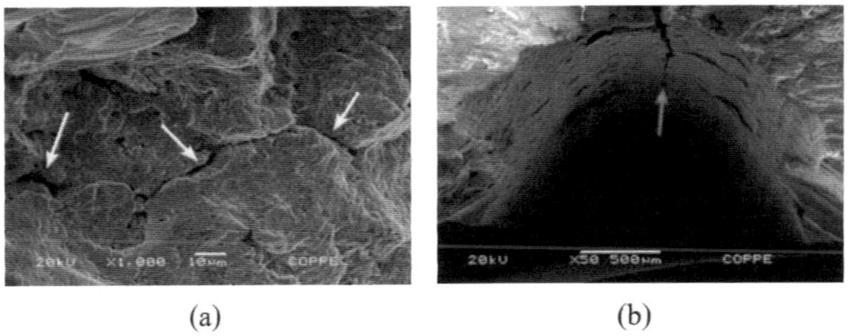

Figure 12. (a) Secondary cracks observed on the fracture surface close to the edge (yellow arrows) (b) spline region with intense transverse cracking and a longitudinal crack at the base of the spline root (red arrow).

Figure 12 shows a lateral image of the spline close to the fracture surface revealing several small transverse cracks and a large longitudinal crack emanating from the spline root.

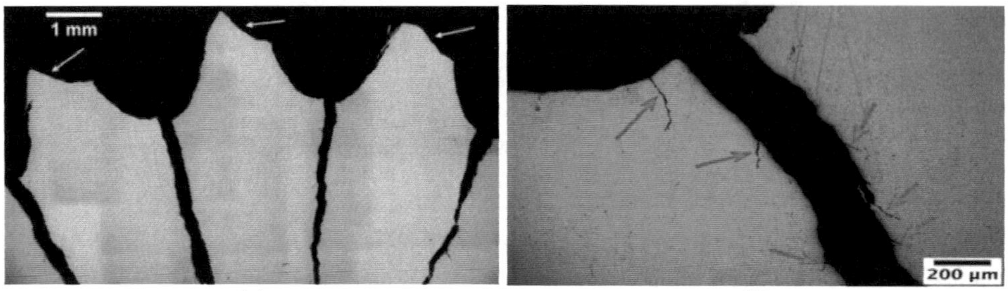

(a) (b)

Figure 13. Transverse view of the drive-axle close to the fracture surface (a) large cracks emanate from the roots of the splines and the wear of the splines (yellow arrows), (b) secondary cracks (red arrows) observed inside a crack.

It is possible to see in figure 13(a) that the tips of the splines treads were worn away from the component. It's also evidenced from figure 13 that cracks emanating from the spline roots are associated to fatigue process due to torsional loading [5]. Similar results can be corroborated by another study too [7]. Bibin et al study points out the effect of bainitic structure and the same was observed in the current failure analysis too. The FE study helped in identifying the design changes that are required to avoid future failure.

4. Conclusion

The above conducted failure analysis of the fractured drive-axle reveals the following:

- The design considerations did not take into account of the circlip groove which was a high stress concentration area and the primary reason for fracture.
- The material EN24 T selected for the application was appropriate but the heat treatment process applied on the drive axles lead to bainitic microstructure which lowers the yield and ultimate tensile strength of the material. Also, improper heating and clamping lead to slight deformations in the component.

A detailed failure analysis of the drive axle rod was conducted using destructive and non-destructive testing as well as microscopic analysis was done. It showed that there were some flaws in the design considerations and improper heat treatment processes were applied which led to the formation of bainitic microstructure. Due to which the brittleness of the component increased and its tensile strength decreased leading to the failure.

References

[1] Muhammad W and Deen K M 2010 Failure Analysis of Water Pump Shaft, *ASM International*, 10, pp. 161-166.

[2] Smiths, S EN24 STEEL (817M40T). [Online]. Available from: https://www.smithmetal.com/en24.

[3] Bipin W, Prashanta C and Tejpal P N 2016 Failure Analysis Of Automotive Front Wheel Drive Shaft, *Journal Of Information, Knowledge And Research In Mechanical Engineering*, 4(1), Pp. 707-713.

[4] Khodamorad A and Hassan S 2012 Effect of Bainite Morphology on Mechanical Properties of the Mixed Bainite-martensite Microstructure in D6AC Steel, *Journal of Materials Science & Technology* Pages 336-342, 28(4), pp. 336-342.

[5] Lassen T and Spagnoli A 2013 Fatigue Crack Paths in Shafts Subjected to Bending and Torsion. *CP 2013*.
[6] Bhagoria P, John ST, Patangia P, Purohit R. Failure Analysis of the Axle Shaftof an Automobile. *Materials Today: Proceedings*. 2017 Jan 1;4(4):5398-407.
[7] Zhao LH, Xing QK, Wang JY, Li SL, Zheng SL. Failure and root cause analysis of vehicle drive shaft. *Engineering Failure Analysis*. 2019 May 1;99:225-34.

Design verification through tolerance stack up analysis of mechanical assembly for 3d printed components

R Jafrin Beryl[1], A N Saravanan[2], Annamalai K[3] and K Janardhan Reddy[4]

[1,2] PG student, School of Mechanical Engineering, VIT University, Chennai, India.
[3,4] Professor, School of Mechanical Engineering, VIT University, Chennai, India.

E-mail: [1]jafrinberyl.r2020@vitstudent.ac.in, [2]saravanan.an2020@vitstudent.ac.in, [3] [3]kannamalai@vit.ac.in and [4]kjanardhanreddy@vit.ac.in

Abstract. Automobile sector encounters new challenges and hurdles every day. New design trends and day-by-day technology advancement from research, push companies to develop new competitive models. Automobile manufacturing is one of the most competitive business domains where every organization will thrive to prove themselves in the society and market. In order to increase the production rate of the company, automobile manufactures can increase the efficiency of their research and development process enabling them to get their product in the market with less time and better quality. Additive manufacturing is being used in the prototyping, tool and dies making process due to less time, labour efforts and cost. But the hurdles faced by company in additive manufacturing are poor dimensional accuracies and tolerances. Automotive industries find difficult to get their exact dimensional tolerance values resulting in failure of the component. In this paper authors demonstrated tolerances stack-up analysis through a case study of critical 3D printed components and its assembly used in automobile. This paper provides a review of role and future scope of additive manufacturing in automobile industries.

1. Introduction

Design verification is an essential step in the development of any product, it makes sure that the final product designed matches the intended product, but many design projects do not complete thorough design qualification resulting in products that do not satisfy customer expectations and then requires a very expensive design modification.

Tolerance Stack-Ups plays a vital role in addressing mechanical fit and performance requirements. Fit(mechanical) simply answers the question, “Do the parts that make up the assembly always go together?” Mechanical performance requirements would include the performance of mechanisms, like latches and actuators. Some other performance requirements are accurate optical alignments and efficiency (of drives).

Additive manufacturing (AM) is a technique of mixing materials by either fusion, binding, or solidifying materials such as liquid resin and powders. It fabricates in a layer-by-layer manner using 3D Cad modelling. The terms such as 3D printing (3DP), direct digital manufacturing (DDM), rapid prototyping (RP), and rapid manufacturing (RM) can be used to describe this process.

Additive manufacturing processes prints components using STL (Standard Tessellation Language) files, which contain information pertaining to the geometry of the object. This manufacturing technique is very useful production volumes is low, presence of high design entanglement, and recurrent design changes are required. This technique helps to fabricate complex parts which will solve the shortcomings of design constraints of conventional manufacturing methods. This manufacturing technique has many gains, but its applications are still narrow because of its poor accuracy and long print times compared to CNC machines. But it does not have the same limitation as CNC machining because it separates the part in cross sections with a resolution equal to that of the process. Nevertheless, the shortcomings such as accuracy and build time can be improved by engaging suitable part orientation. Optimized part orientation can improve the accuracy and the building time and support volume, which in turn reduces the part production cost.

Furthermore, Additive manufacturing consists of additional controllable process parameters and higher active interaction between the material properties and process parameters. There are different kinds of Additive manufacturing processes depending on the material preparation, material type, phase change phenomenon, layer generation technique and application.

In this paper author attempt to do tolerance stack up analysis for 3D printed components and its assembly. Two approaches have been used viz: Worst case (WC) and Root Sum Square (RSS). As a case study, front wheel assembly of an All-terrain vehicle (ATV) has been designed in SOLIDWORKS 2018 and the prototype has been fabricated using Dreamer NX flash forge machine and the tolerance stack analysis were carried out and design verification done for dimensional accuracy.

Based on the result obtained, design modification is recommended to statically distribute appropriate tolerance to fulfil the functionality of the individual components and the whole assembly at product level.

2. Literature review

He and Gibson (1992) [1] developed a computerized trace method which determines the relationship between geometrical tolerances and manufacturing dimensions and tolerances.

Ngoi (1997) [2] presented a refined method by deploying the 'Quickie' technique for tolerance stack up analysis for geometrical tolerances.

Ngoi, B.K.A., (1999) [3] also presented a straightforward graphical approach known as the "Catena" method for tolerance stack up, involving geometric characteristics in form control – flatness, straightness, circularity and cylindricity.

Ngoi, B.K.A., (2010) [4] discussed the stack up of geometrical tolerances using generic capsule method.

Mansuy (2011) [5] presented a method that helps to solve problems for the case of serial assembly (stacking) without any clearances. This is based on the utilization of influence coefficients to get the relationship between the functional tolerance and the tolerances associated with the geometry of the mechanism's interface surfaces.

Chetan H (2013) [6], in their research., their aim was to redesign an existing passenger car side door latch to improve the manufacturability using Design for Manufacture (DFM). Tolerance stack up analysis was used to find the clearance or interference between two features on a part and their assembly variation.

A.K Sahani (2014) [7], in their research, aimed towards the systematic solution of tolerance stack up problem involving geometric characteristics. In their work, angularity tolerance has been considered for illustration of the methodology. Two approaches viz. Worst Case (WC) and Root Sum Square (RSS) had been used. An example of dovetail mounting mechanism was taken for purpose of stack up of angularity.

Ahmed Jawad Qureshi (2018) [8] a new geometric deviation modelling is presented based on analytical modelling of the process and specifying its effects on GD&T features. The geometric deviation and compensation model based on the new methodology is being developed to quantify and removes the geometric form and shape error as much as possible.

IOP Conf. Series: Materials Science and Engineering 1128 (2021) 012012 doi:10.1088/1757-899X/1128/1/012012

3. Methodology

3.1. 3D Modelling of Main Assembly

A case study of front wheel assembly as shown in figure 1 of an All-terrain vehicle (ATV) has been taken for stack up analysis. This assembly consists of components like knuckle, disc, hub (rotor), M6 bolt, bearings and circlip. The individual parts of the knuckle assembly are created using Solidworks and its assembly is shown in figure 2.

Figure 1. Arrangement of the knuckle in all-terrain vehicle.

Figure 2. Knuckle assembly.

Dimensional details and tolerance information for each subassembly is given in the table 1. There are four subassemblies are considered for the tolerance stack-up analysis viz: rotor and bolt, knuckle and bearing, rotor and bearing and disc and bolt.

Table 1. Dimensions and Tolerances for Subassemblies

(a) Rotor and Bolt

Tolerance (mm)	Basic Dimension (mm)	Rotor (mm)	Bolt (mm)	Tolerance (mm)	Limits (mm)	Fit	Clearance (mm)
0.5	6	6.5[a]	M6	0.25	6.25[a]	Clearance fit	0.75
0.5	6	5.5[b]	M6	-0.25	5.75[b]		

[a] Upper limit, [b] Lower limit

(b) Knuckle and Bearing

Tolerance (mm)	Basic Dimension (mm)	Knuckle (mm)	Bearing (mm)	Tolerance (mm)	Limits (mm)	Fit	Clearance (mm)
0.5	25	25.5[a]	25	0	25[a]	Transition fit	0.62
0.5	25	24.5[b]	25	-0.12	24.88[b]		

(c) Rotor and Bearing

Tolerance (mm)	Basic Dimension (mm)	Rotor (mm)	Bearing (mm)	Tolerance (mm)	Limits (mm)	Fit	Clearance (mm)
0.5	47	47.5[a]	47	0	47[a]	Transition fit	0.62
0.5	47	46.5[b]	47	-0.12	46.88[b]		

(d) Disc and Bolt

Tolerance (mm)	Basic Dimension (mm)	Disc (mm)	Bolt (mm)	Tolerance (mm)	Limits (mm)	Fit	Clearance (mm)
0.5	6	6.5[a]	M6	0.25	6.25[a]	Clearance fit	0.75
0.5	6	5.5[b]	M6	-0.25	5.75[b]		

[a] Upper limit
[b] Lower limit

3.2. Positional tolerance.

It is required to estimate the position tolerance for each assembly feature for implement the tolerance stack analysis. In this case study positional tolerance for main and subassembly feature are calculated and given in the table 2,3 and 4

Table 2. Positional tolerance for disc bolt and rotor.

Bolt outer diameter (mm)	Tolerance type	Tolerance (mm)	Positional tolerance (mm)	Rotor inner diameter (mm)
5.99	Fixed	0.498	3.302	9.296

Table 3. Positional tolerance for rotor and bearings.

Bearing outer diameter (mm)	Clearance (mm)	Rotor inner diameter (mm)
47	3.302	50.3174

Table 4. Positional tolerance for bearing and shaft

Shaft outer diameter (mm)	Clearance (mm)	Bearing inner diameter (mm)
25	3.302	28.321

3.3. Assembly shift

The main reason for finding the assembly shift is to avoid the linear movement of the component and allowing the rotational movement the component.

3.3.1. Assembly shift between Rotor (hole) and Bearing (shaft).

In this assembly, there are pairs of hole and shaft. The main function of this assembly is to provide rotational movement instead of linear movement in order to arrest linear movement. Assembly shift is calculated as follows

Hole = 47mm (maximum = 47.5mm)
Bearings = 47mm
Clearance = Hole value - Shaft value
Clearance = 0.5mm
Both sides = 0.5/2= 0.25mm

In this assembly we have two bearings, so 0.25x2= 0.5mm is the clearance value provided for both the bearings.

3.3.2. Assembly shift between Bearing (hole) and knuckle (shaft).

Here, the assembly shift is calculated between bearing (hole) and knuckle (shaft), where the bearing is placed over the knuckle shaft which provides rotational movement.

Bearings inner diameter =25 mm
Shaft diameter =24.5
Max bearings diameter =25mm
Clearance = Bearing (Hole) - Knuckle (Shaft)
Clearance = 25mm-24.5mm = 0.5mm
Per axis = 0.5/2= 0.25mm
Clearance = 0.5mm

In this assembly we have two bearings so, 0.25x2 = 0.5mm is the clearance value provided for both the bearings.

3.4. Design of Knuckle assembly

In the existing knuckle assembly, the knuckle is attached with two bearings and over this rotor is placed for providing rotational movement of the vehicle. During assembly due to improper tolerance valve the bearing is coming out from the knuckle and which leads to improper function of assembly.

To avoid this, design modification for the knuckle-bearing assembly is proposed as shown in figure 3, which will support the bearing and it also control the assembly.

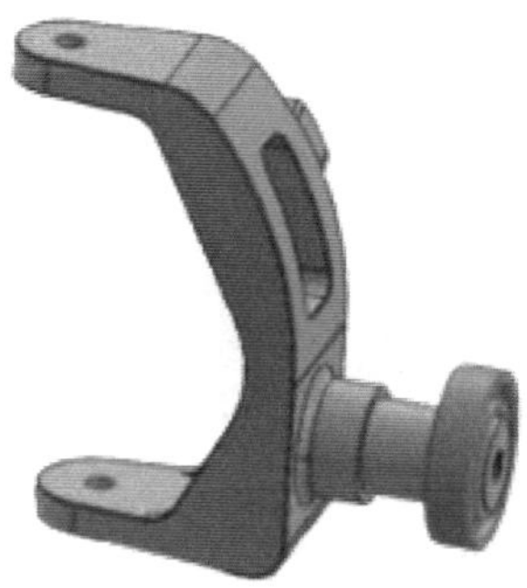

Figure 3. Knuckle and bearing misalignment.

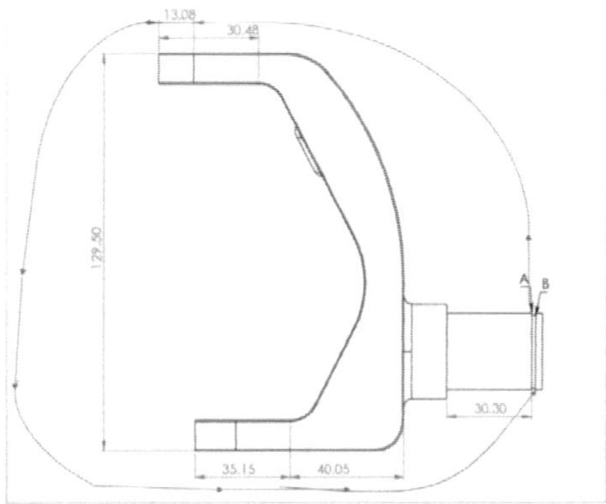

Figure 4. Knuckle dimensions.

3.4.1. Individual tolerance values

Tolerance values for individual dimensions of knuckle – bearing sub assembly are calculated and shown in the table 5.

Table 5. Knuckle tolerance calculation

Positive value (mm)	Negative value (mm)	Tolerance (mm)
---	13.02	0.05
---	30.48	0.05
---	30.30	0.01
---	64.75	0.01
64.75	---	0.05
40.05	---	0.03
35.15	---	0.05
Total= 139.95	138.55	0.29

In order to arrest the linear movement of bearing and to provide a proper assembly, tolerance stack up analysis is carried out for the knuckle. And the knuckle dimensions are separated into positive and negative columns (Table 5).

Two points are considered in the knuckle i.e., point A and point B (Figure 4). The arrow mark pointing away from point A is considered as a negative value and the arrow mark pointing towards the point B is considered as a positive value.

Now, the difference between the positive and negative value in mm are

Positive value = 139.95
Negative value = 138.55
Nominal distance = 1.4
Clearance = +/- 0.29

From the above calculation, the maximum distance is 1.69 mm and 1.11mm is the minimum distance. In order to arrest linear bearing movement and providing rotational movement, a component is selected based on the above-mentioned value. As it an interference fit, we are considering a constant dimension circlip as 1.2mm thickness which will arrest the bearing linear movement and provide proper assembly.

3.4.2. Circlip thickness

In section 3.4.1 we have calculated the thickness (of circlip) for bearing assembly. The placement of circlip is also very important in order to get proper assembly. The circlip and bearing distance should be as minimum as possible. Distance AC* (Figure 5) is considered from the right end of the knuckle and the distance is calculated for the circlip to fit in the groove.

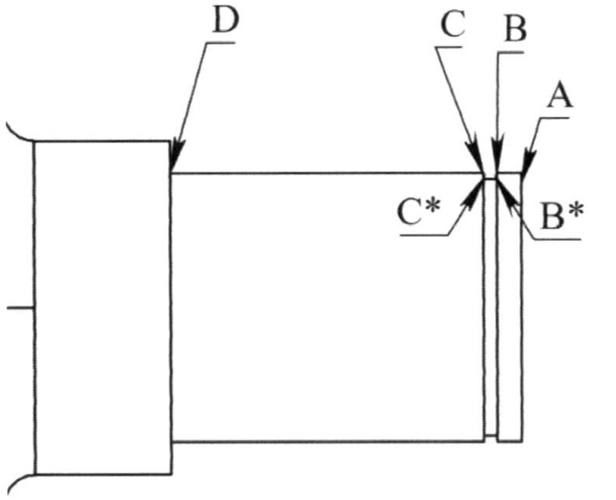

Figure 5. Knuckle shaft dimensions with circlip groove.

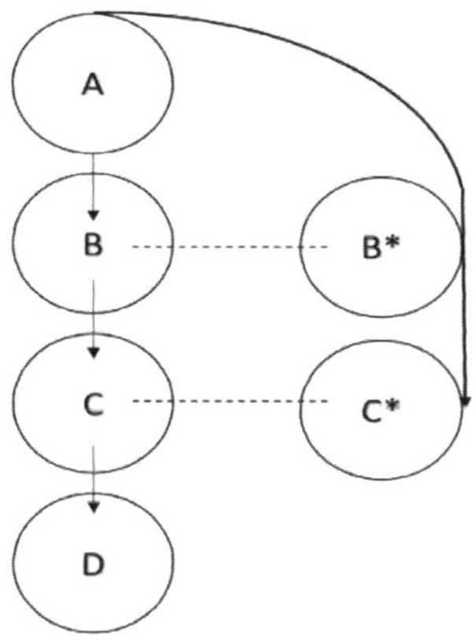

Figure 6. GD&T model for knuckle shaft.

WORST CASE METHOD

AC* is the distance where the circlip will be placed.

C*C is the vertical distance from the dept.

The dimensions are separated into positive and negative values. The clockwise dimensions are considered to be positive value and the anti-clockwise are considered to be negative value.

Now,

AC*+C*C-CB-BA=0

X+(0.5543±0.02)-(1.3±0.01)-(2.4±0.01)=0

$X = 3.15 \pm 0.04$

$X_{MAX} = 3.19$

$X_{MIN} = 3.11$

ROOT SUM SQUARE METHOD (RSS) METHOD

$X = 3.15 \pm \sqrt{((0.02*0.02)+(0.01*0.01)+(0.01*0.01))}$

$X = 3.15 \pm (0.024)$

$X_{MAX} = 3.174$

$X_{MIN} = 3.13$

From calculation it can be found that 3.2mm is the distance where we need to place the circlip. Now the problem is to fit circlip in the respective depth. But as per the design one end of the circlip is placed but the other end of the circlip is not locating exactly.

3.4.3. Circlip gap

To achieve the proper assembly of circlip we have to measure the distance between AB1 and AB*. Both C1 and C* are combined together but B1 and B* are not joined there is a gap between these two points. In order to reduce the gap, we need to find the difference between AB1 and AB* which are explained geometrically in figures 7 and 8

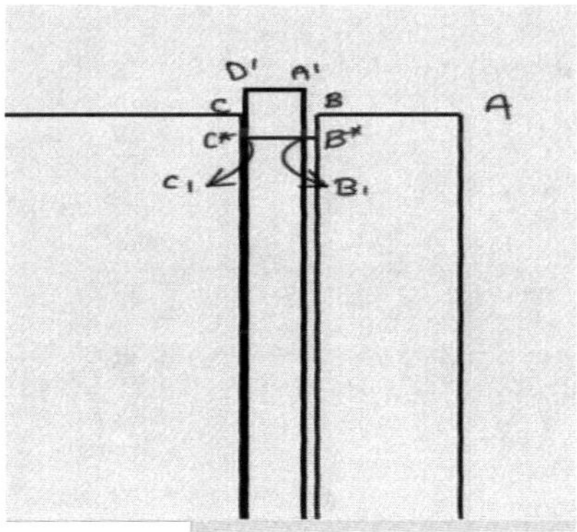

Figure 7. GD&T model for knuckle shaft.

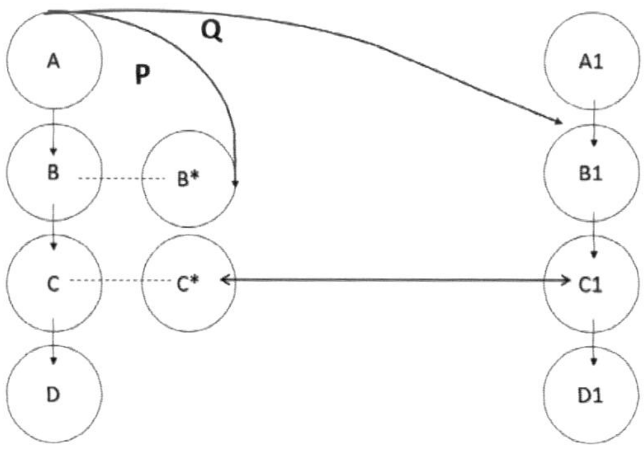

Figure 8. GD&T model assembly.

Let's consider AB* as P and AB1 as Q

TO FIND P

AB*+BB*-BA=0

WORST CASE METHOD

$P+(0.5543\pm0.02) - (2.4\pm0.01) = 0$
$P = 1.8457\pm0.03$
$P_{MAX} = 1.87$
$P_{MIN} = 1.82$

ROOT SUM SQUARE METHOD

$1.8457\pm\sqrt{((0.02*0.02) + (0.01*0.01))}$
1.8457 ± 0.224
$P_{MAX} = 1.87$
$P_{MIN} = 1.83$

TO FIND Q

AB1+B1C1+C1C*+CC*-CB-BA=0

WORST CASE METHOD

$Q+(1.2\pm0.01) + (0) + (0.5543\pm0.02) - (1.3\pm0.01) - (2.4\pm0.01) = 0$
$Q = 1.95\pm0.5$
$Q_{MAX} = 2.00$
$Q_{MIN} = 1.90$

ROOT SUM SQUARE METHOD

$1.95+\sqrt{((0.01*0.01) +(0.02*0.02) + (0.01*0.01) + (0.01*0.01))}$
1.95 ± 0.0265
$Q_{MAX} = 1.97$
$Q_{MIN} = 1.92$

TO CALCULATE CLEARANCE

WORST CASE METHOD GIVES

$Q_{MAX} = 2.00$
$P_{MIN} = 1.82$

MAX CLEARENCE

$R_{MAX} = Q_{MAX} - P_{MIN}$
$R_{MAX} = 0.18$

MIN CLEARENCE

$R_{MIN} = Q_{MIN}-P_{MAX}$
$Q_{MIN} = 1.90$
$P_{MAX} = 1.87$
$R_{MIN} = 1.90 - 1.87$
$R_{MIN} = 0.03$

ROOT SUM SQUARE METHOD GIVES

MAX CLEARENCE

$R_{MAX} = Q_{MAX} - P_{MIN}$
$Q_{MAX} = 1.97$
$P_{MIN} = 1.83$
$R_{MAX} = 0.14$

MIN CLEARENCE

$R_{MIN} = Q_{MIN} - P_{MAX}$
$Q_{MIN} = 1.92$
$R_{MAX} = 1.87$
$R_{MIN} = 1.92 - 1.87$
$R_{MIN} = 0.05$

4. Results

Tolerance values obtained from Worst Case (WC) and Root Sum Square (RSS) approach for individual parts and their assembly are shown in table 6. The result of the stack-up analysis shows that the maximum clearance value and minimum clearance value are positive. From the calculation we got 0.05 as a clearance value and this value engages the point B1 and B* as shown in the figure 7.

Table 6 Dimensional and tolerance values

		Approach			
		Worst Case		Root Sum Square	
Part Name	Alphabet	Maximum Value	Minimum Value	Maximum Value	Minimum Value
Knuckle	X	3.19	3.11	3.17	3.13
Knuckle and Circlip Assembly	P	1.87	1.82	1.87	1.83
	Q	2	1.9	1.97	1.92
	R	0.18	0.03	0.14	0.05

Design verification:

After completing the tolerance calculation, 3D printing is done using polylactic acid (PLA) material in the dreamer NX flash forge machine which is shown in the figure 9. Tolerance verification was done using SolidWorks 2016 and the complete assembly is shown in figure 10. The tolerance values shown in the result (Table 6) is applied to assemble the 3D printed component. The final assembled view of the 3D printed component is shown in the figure 11. The calculated tolerance values are used for both CAD model assembly and working of the 3D printed component assembly.

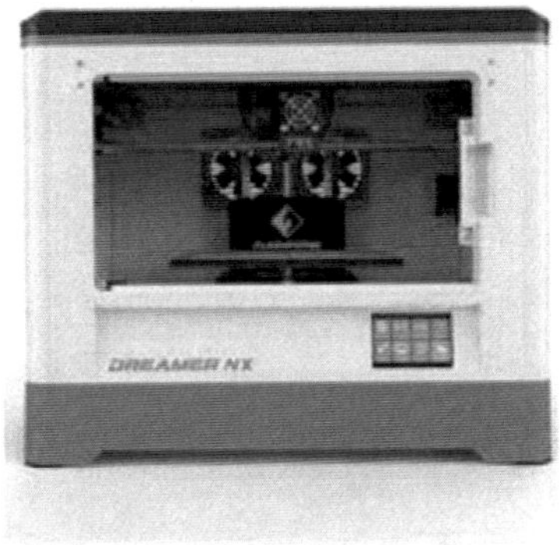

Figure 9. Dreamer NX flash forge machine.

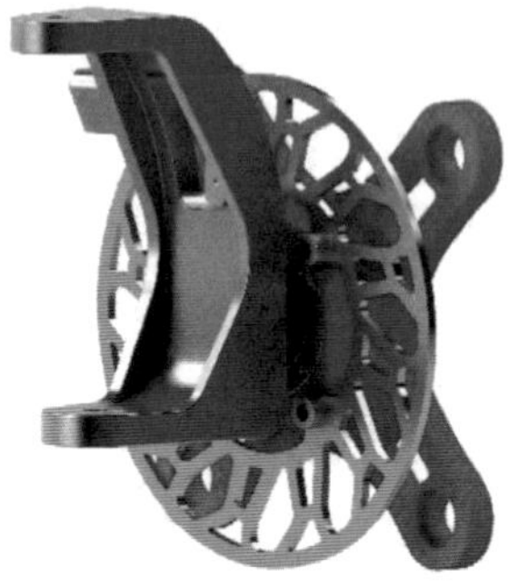

Figure 10. Assembly after verification.

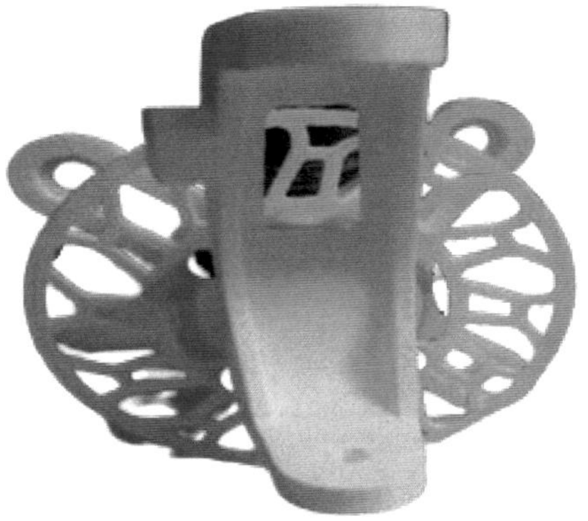

Figure 11. Assembled view of the 3D printed component.

5. Conclusions

In this paper, a graphical approach is demonstrated for tolerance stack up analysis. As a case study knuckle assembly used in All Terrain Vehicle is chosen. Dimensional details are physically measured and assembly tolerances are estimated. A graphical approach is employed to determine individual tolerance values for the all the components in the assembly and corresponding tolerance values are obtained using Worst Case (WC) analysis and Root Mean Square (RMS) values methods. The result is validated through 3D printed components fabricated using polylactic acid (PLA) material in the dreamer NX flash forge machine which shows that maximum clearance and minimum clearance value is positive as found in the graphical approach.

References

[1] He, J.R. & Gibson P.R. (1992). Computer Aided Geometrical Dimensioning and Tolerancing for Process-Operation Planning and Quality Control. *Int Adv Manuf Technol*, 7, pp. 11-20

[2] Ngoi, B.K.A., Tan, C.S., Goh, L.C., 1997. Graphical Approach to Assembly Tolerance Stack Analysis—the 'Quickie' GDT Method", Proc Institute of Mechanical Engineers **Volume 211** Part B, pp. 463-472.

[3] Ngoi, B.K.A., Lim, L.E.N., Ang, P.S., Ong, A.S., 1999. Assembly Tolerance Stack Analysis for Geometric Characteristics in Form Control – the "Catena" Method, *International Journal of Advance Manufacturing Technology*, **Vol.15**, pp. 292–298.

[4] Ngoi, B.K.A., Agrawal, A.M., Chua, C.S., 2010. The Generic Capsule Approach to Tolerance Stack Analysis, *International Journal of Production Research*, 36:12, pp. 3273-3293.

[5] Mansuy, M., Giordano, M., Hernandez, P., September 2011. A New Calculation Method for The Worst Case Tolerance Analysis and Synthesis in Stack-Type Assemblies, *Computer-Aided Design*, **Volume 43**, Issue 9, pp. 1118-1125.

[6] Chethan H, 2 Naveen Udupa, 3Ramakrishna Hegde, 4Girish kumar.R, Design and Tolerance Stack up Analysis of Car Side Door Latch. *IJRMET* **Vol. 3**, Issue 2, May - Oct 2013 ISSN : 2249-5762 (Online)

[7] Sahani, A.K., Sharma, A.K., Jain, P.K., Sharma, Satish C., Bajpai, J.K., 2012. Review of Assembly Tolerance Stack up Analysis Techniques.

[8] Baltej Singh Rupal, Ahmed Jawad Qureshi, Geometric Deviation Modeling and Tolerancing in Additive manufacturing: A GD&T Perspective, 1st Conference of NSERC Network for Holistic Innovation in Additive Manufacturing (HI-AM).

[9] Sivakumar, K., Balamurugan, C., Ramabalan, S., February 2011. Simultaneous Optimal Selection of Design and Manufacturing Tolerances with

[10] Singh, P.K., Jain, P.K., Jain, S.C., 2009. Important Issues in Tolerance Design of Mechanical Assemblies (Part 1: tolerance analysis), *Proc.ImechE, J. Engineering Manufacture*, **Vol. 223** Part B, pp. 765-778.

[11] Singh, P.K., Jain, P.K., Jain, S.C., 2009. Important Issues in Tolerance Design of Mechanical Assemblies (Part 2: tolerance synthesis), *Proc.ImechE, J. Engineering Manufacture*, **Vol. 223** Part B, pp. 1249-1287.

[12] Chase, K.W., Greenwood, W.H., Loosli, B.G., Loren, H.F., 1990. Least Cost Tolerance Allocation for Mechanical Assemblies with Automated Process Selection, *Manufacturing Review*, **Vol. 3, No. 1**, pp. 49-59.

[13] Ahluwalia, R.S. & Karolin, A.V. (1984). CATC – A computer aided tolerance control system. *Journal of manufacturing system*, 3(2), pp.153-160

[14] Drake, Paul J. Jr. (1999). *Dimensioning & Tolerancing Handbook*, Mc Graw Hill, ISBN 0- 07-018131-4, New York

[15] Swift, K.G.; Raines M. & Booker, J.D. (1999). Tolerance optimization in assembly stacks based on capable design. Proceedings of the Institution of Mechanical Engineers, *Journal of Engineering Manufacture*, Part B, pp. 677-693

[16] Guo N, Leu MC. Additive manufacturing: technology, applications and research needs. *Frontiers of Mechanical Engineering*. 2013 Sep 1;8(3):215-43.

[17] Zuowei ZH, Keimasi S, Anwer N, Mathieu L, Lihong QI. Review of Shape Deviation Modeling for Additive Manufacturing. InAdvances on Mechanics, *Design Engineering and Manufacturing 2017* (pp. 241-250). Springer, Cham.

[18] Rebaioli L, Fassi I. A review on benchmark artifacts for evaluating the geometrical performance of additive manufacturing processes. *The International Journal of Advanced Manufacturing Technology*. 2017 Nov 1;93(5-8):2571-98.

[19] P.G.Maropoulos, D Ceglarek,"Design Verification and Validation in Product Lifecycle", *CIRP Annals – Manufacturing Technology*, Vol. 54, Issue 2, pp. 607-622, 2010

[20] T. Hussain, Z. Ali, J. Larik,"A Study On Tolerance Representation, Variation Propagation Analysis and Control In Mechanical Assemblies", *Sindh University Research Journal (Science Series)*, **Vol. 44**, Issue 3, pp. 427 – 432, 2012.

IOP Conf. Series: Materials Science and Engineering 1128 (2021) 012013 doi:10.1088/1757-899X/1128/1/012013

Design Study of an Electric Motorcycle Chassis Obtained using Topology Optimization

Mohammad Irfan Ahmad[1] and Pankaj Dorlikar[2]

[1] PG Student, Department of Mechanical Engineering, Army Institute of Technology, Pune-411015, India
[2] Assistant Professor, Department of Mechanical Engineering, Army Institute of Technology, Pune-411015, India

E-mail: [1] irfanahmad_18987@aitpune.edu.in, [2] pvdorlikar@aitpune.edu.in

Abstract. Rising Carbon emission and associated pollution forcing government of India to make stringent emission norms and policies. Hence, it is need of the hour to switch to alternative fuels such as electricity. About 60% of the petrol consumption in India is attributed to the motorcycles and there hasn't been significant research in the electric mobility in motorcycles. This work is based on study of an electric motorcycle chassis with integrated battery-pack space that has been obtained as a conceptual geometry using topology optimization. The layout of the chassis for this work has been obtained by applying topology optimization on a design domain representing a space obtained on the basis of presently available dimensional measurements of commuter segment motorcycles in India. The chassis has been subjected to linear static and modal analysis for further enhancing the strength, stiffness and natural frequency using the design study approach on critical region.

1. Introduction

Chassis, also known as frame is the support framework that bears static and dynamic loads acting on the vehicle. Its design plays critical role in vehicle performance, rider comfort, steering quality, handling etc. Kurdi et al. used FEM to analyse stresses in truck chassis [1]. They have stressed on predicting fatigue life of the chassis and durability loading in its designing so as to verify safety during its use in real life situations. Also, they have shown that the critically stressed zones in the chassis due to loading must be located using FEM because such regions are the first points of fatigue failure propagation. Jonathan Hastie et al. have studied the front and side impacts and loading due to shock mounts on SAE BAJA vehicle [2]. It helped them in modifying the chassis by adding three key structural components to bear the loading conditions. It was also found out that the design failure occurred in roll over and hence an integrated solution from the beginning could be proposed using the FEA methodology.

Four repetitions of analysis have been studied by Bennett et al. for designing a frame [3]. They applied simple load cases to various frames and the one with highest factor of safety (FOS) was considered for further analysis such as side impact, drop test etc. In order to reaffirm ability of the vehicle to endure extreme loading conditions, Raina et al. have analyzed the frame with considerable factor of safety (FOS) [4]. This was done so as to validate the frame to withstand loading and minimize the chances of failure and the risks arising out from it.

Topology Optimization is an optimization technique that helps in finding the most efficient material distribution in a design domain, that can withstand externally applied loads. The last few decades have seen significant application of this technique in addressing structural problems using FEA, especially in the automotive industry. Topology optimization and topography optimization have been combined and has been used to design a contact start-stop suspension system subject to various frequencies such as torsional, bending and sway modes by Kilian et al. [5]. Chiandussi et al. have used topology optimization on automotive suspension [6]. In it, the suspension system was subjected to three static loads under volume constraint and minimum compliance. Similarly, Chang and Lee have applied topology optimization on a bracket used in automotive air conditioner subject to minimum compliance design and natural frequency constraints [7]. Wang et al. optimised the thickness and distribution of material reinforcements and thereby improved the stiffness of an automobile body using size and topology optimizations [8]. Pederson et al. in his work studied a crushed structure and its energy absorption history using topology optimization [9-10]. Jeyapandiarajan et al. have done a comparison between various materials by comparing stress, deformation and FOS to design the chassis for an electric two-wheeler using FEA [11]. They have shown a compromise in terms of material strength and cost of manufacturing for the chassis. Gazaly has observed the suitability of common FEA tools in studying stress analysis [12]. He has analysed stress and deformation of a truck chassis using four FEA tools viz., Ansys, Abaqus, Hyperview and Nastran. He has also found out that Ansys and Abaqus to be most frequently used FEA packages in such studies.

Furthermore, the vibration characteristics including the mode shapes and natural frequencies of a truck chassis have been studied by Fui et al. [13]. Observations regarding displacement and stress distribution of the chassis under different boundary conditions has been made using finite element technique. Also observation that road irregularity causing excitation to the truck chassis is the main disturbance to the frame and that various mounting locations of the components can be altered based on mode shape response. He has also suggested some modifications in order to reduce vibration and improve stiffness of the chassis. A hybrid prototype of an electric motorcycle using an effective and compact powertrain has been designed by Stuecke et al. [14]. The prototype has been designed by altering the original machine with minimal changes, such that high performance is achieved without increasing the overall weight of the chassis significantly. For example, the electric components have been mounted in such a way that original shape and aerodynamics of the vehicle has been retained.

The development in electric vehicle has gained tremendous pace in the past few years, especially due to reduction in the prices of power storage battery packs and various incentives offered by the governments across the globe. Topology optimization as a tool to design automotive components has also been used extensively in the recent past, but Stefanos et al. has applied nested topology optimization to design the chassis for an electric two-wheeler, combining the structural applicability of the tool with the positioning of battery pack space within the design domain [15]. Also, recently Airbus APWorks developed an electric motorcycle by combining the study and the manufacturing process (additive manufacturing) [16]. The frame weighed as low as 6kg and the entire motorcycle weighed less than 35kg. This work has taken inspiration from the research work that has been under progress in designing innovative and novel chassis for a motorcycle segment that will cater to largest proportion of masses in the two-wheeler industry.

2. Methodology of Study

The conceptual geometries were obtained using topology optimization (level-set algorithm) applied to the design domain developed for commuter segment motorcycles presently available in India. The geometries obtained from topology optimization were used to generate 15 possible conceptual design of the chassis. The mode shapes have been utilised to find the optimal parameters of the chassis such as pipe diameter, thickness, tube width and height, keeping wheelbase constant for all purposes. The analysis has been done by subjecting the model to three different loading conditions i.e., for maximum braking at front axle, maximum acceleration at rear wheels and lateral load when motorcycle is on curve within limits of pneumatic adherence.

The work has involved FEA to assess the validity of the chassis when used on real life electric motorcycle. The preliminary stage involved data collection of the current commuter segment motorcycles in India. Various dimensions of the motorcycles such as length, width, seat height, ground clearance etc. were utilised to design the control volume (design domain) as shown in Figure 1, on which topology optimization was performed using Autodesk's Fusion 360™. The volume of the design domain was such that it could accommodate as much space as required to cover the chassis of the electric motorcycle. Boundary conditions based on which topology optimization was performed is beyond the scope of this paper. The results obtained by topology optimization have been used as the basis of the design of the chassis.

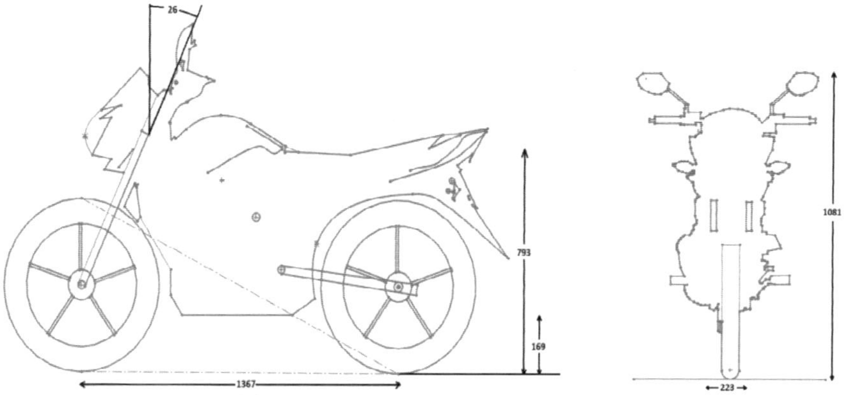

(a) (b)

Figure 1. Side view and Front view of the commuter motorcycle and associated dimensions.

Table 1 shows the technical specifications and geometric dimensions based on which the control volume for the chassis has been developed. These dimensions have been taken as average value of individual dimensions of the commuter segment motorcycles.

Table 1. Geometric dimensions and technical specs for design domain development

Sr. No.	Parameter	Specification (mm)
1	Wheelbase	1367
2	Ground clearance	169
3	Seat height	793
4	Length	2013
5	Width	748
6	Overall Height	1081

The CAD model has been prepared keeping in mind manufacturability using Solidworks2019®. The CAD model of the chassis was analyzed for stresses and mode shapes using Altair HyperWorks and further design study was carried out. The Figure 2 represents the flowchart of the process applied in the study.

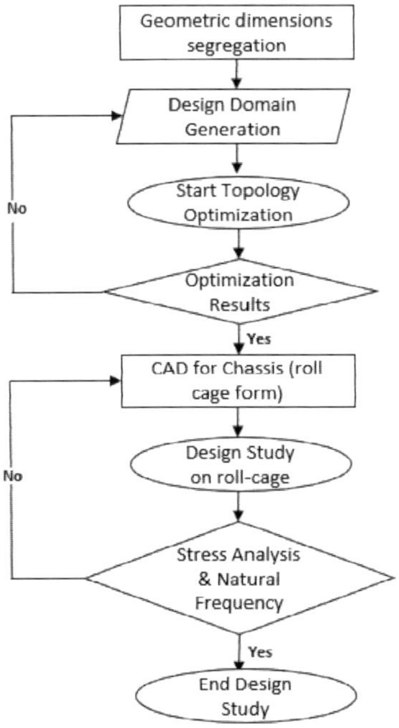

Figure 2. Flowchart for the design process.

2.1. Design Domain generation and Meshing

The design domain in this study represents a volumetric space of material that covers the entire region occupied by the chassis of commuter segment motorcycles. Hence, for the purpose, the dimensions for the design domain have been drawn from the existing motorcycle chassis and meshed with 34471 solid Hexahedron elements of size 10mm for application of the topology optimization. The volume occupied by the battery space has been cut out from the design domain and has been geometrically restricted in order to avoid any optimization in that region. Figure 3 depicts the process for the same.

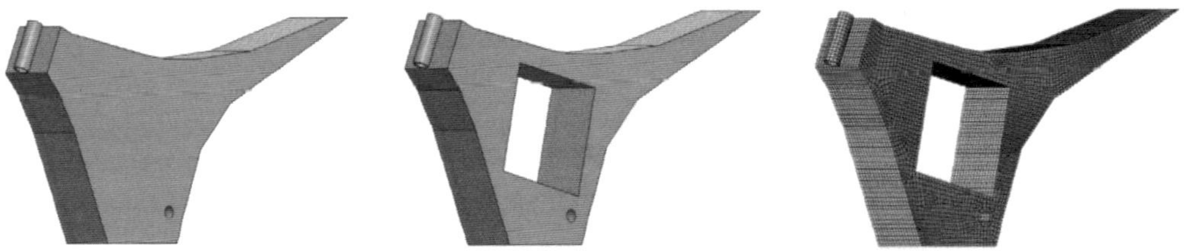

Figure 3. Design domain and meshing for topology optimization.

2.2. Topology Optimization results and development of roll-cage type manufacturable chassis

The results of topology optimization have been used as the basis for designing the chassis. Topology Optimisation has been performed using targets for 80% volume reduction and stiffness maximisation using level-set algorithm and stereolithographic file (.stl) was promoted to generate the layout of the frame as shown in Figure 4. Using the final layout, CAD model was developed using tubular sections such that manufacturability improves.

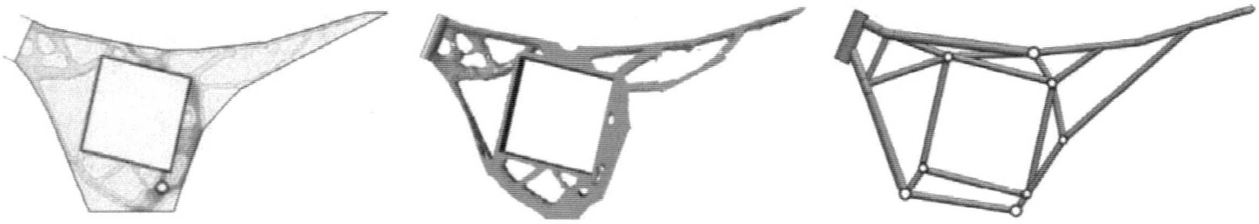

Figure 4. Design process based on topology optimization result using .stl file and CAD.

3. Design Study based on Stress Analysis Results

3.1 Boundary Conditions and Mesh

As loads are calculated for three situations viz. for extreme braking at front wheels, extreme acceleration at rear wheels and lateral load under pneumatic adherence limits on the curves, for the purpose of design study load due to extreme braking have been considered as the values are much higher than that of the other cases. Single point constraint is applied at the swing arm pivot hole of the chassis and point loads act on the headstock as shown in Figure 5. The chassis has been meshed on the mid-surface with 2D elements of size 5mm and quads and trias totaling 37723 elements in numbers.

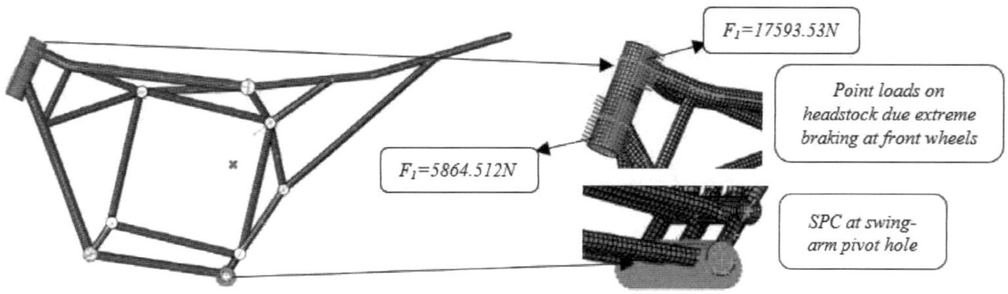

Figure 5. Forces and constraint on the chassis.

3.2 Mode Shapes, Natural Frequency and Stress analysis (1st iteration)

Natural frequency has been used as the criteria to inspect the feasibility to manufacture the chassis. Deflection result has been used to add elements around the stressed zones. Figure 6 to Figure 9 represent the results obtained from the first iteration of the design study.

3.2.1 Mode Shapes

(a) First order mode, 2.33Hz (b) Second order mode, 4.62 Hz (c) Third order mode, 6.93Hz

Figure 6. Natural frequency and modal analysis results upto third order mode.

(a) Fourth order mode, 12.56 Hz (b) Fifth order mode, 14.92 Hz (c) Sixth order mode, 19.73Hz

Figure 7. Natural frequency and modal analysis results from fourth to sixth order mode.

3.2.2 Maximum stress

(a) Max braking at front axle (b) Max acceleration at rear axle (c) Pneumatic adherence limit curve

Figure 8. Stress induced under three loading situations.

3.2.3 Maximum Deflection

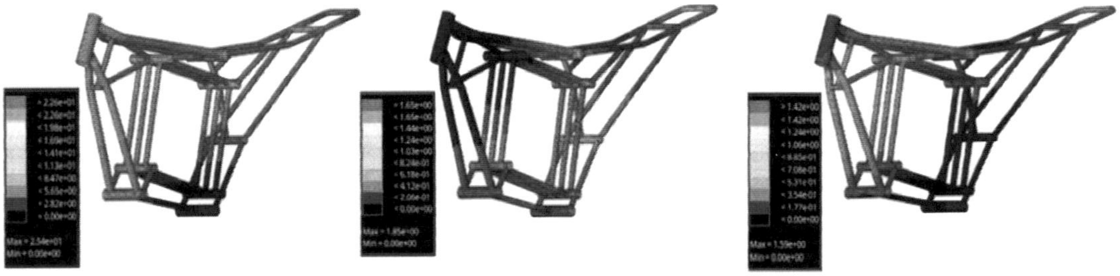

(a) Max braking at front axle (b) Max acceleration at rear axle (c) Pneumatic adherence limit curve

Figure 9. Deflection of chassis under loading condition.

A brief observation leads to the understanding that the natural frequency of the chassis for the first six mode shapes are way too low to actually build one. In fact, the natural frequency in the second and third modes coincides with human body's natural frequency and that would cause discomfort to the rider in the real life scenarios. In order to improve its performance, most stressed region is identified at the bottom portion joint and two parallel tubes are added to it. The modified chassis is again analyzed for its behavior and have been shown in Figure 10 to Figure 13 as the results obtained from the second iteration of the design study.

3.3 Mode shapes, Natural frequency and Stress analysis (2nd iteration)

3.3.1 Mode shapes and corresponding natural frequencies

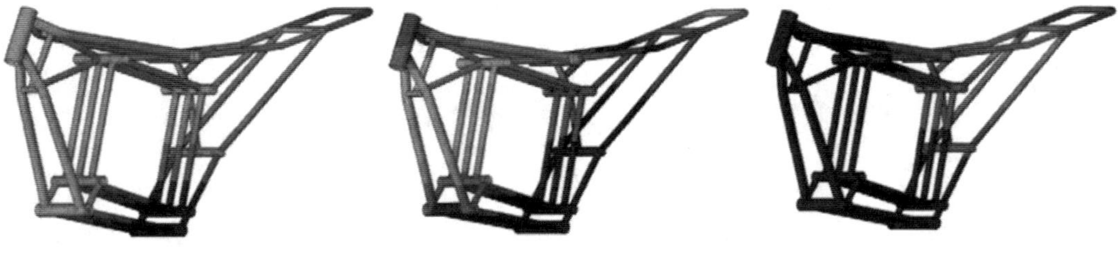

(a) First order mode, 29.44Hz (b) Second order mode, 58.02Hz (c) Third order mode, 69.83Hz

Figure 10. Natural frequency and modal analysis results up to third order mode.

(a) Fourth order mode, 133.46Hz (b) Fifth order mode, 182.45Hz (c) Sixth order mode, 205.6Hz

Figure 11. Natural frequency and modal analysis results from fourth to sixth order mode.

3.3.2 Maximum stress

(a) Max braking at front axle (b) Max acceleration at rear axle (c) Pneumatic adherence limit curve

Figure 12. Stress induced under three loading situations.

3.3.3 Maximum deflection

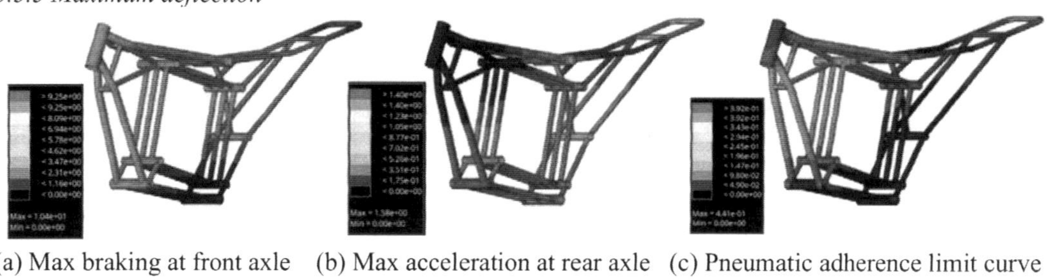

(a) Max braking at front axle (b) Max acceleration at rear axle (c) Pneumatic adherence limit curve

Figure 13. Deflection of chassis under loading condition.

As it can be seen that there is significant improvement in natural frequencies, that is now on the acceptable levels of manufacturability (approximately four-fold above that of human body). But the deformation needs to be constrained to lower values. Hence, an additional third tube is added as a connection to the swing-arm pivoting hole tube with increased thicknesses of all the three elements. The whole procedure is re-analyzed as shown in Figure 14 to Figure 17 as results obtained from the third iteration of the design study.

3.4 Mode shapes, Natural frequency and Stress analysis (3rd iteration)

3.4.1 Mode shapes and corresponding natural frequencies

(a) First order mode, 38.59Hz (b) Second order mode, 61.89Hz (c) Third order mode, 70.60Hz

Figure 14. Natural frequency and modal analysis results upto third order mode.

(a) Fourth order mode, 133.46Hz (b) Fifth order mode, 182.45Hz (c) Sixth order mode, 205.6Hz

Figure 15. Natural frequency and modal analysis results from fourth to sixth order mode.

3.4.2 Maximum Stress

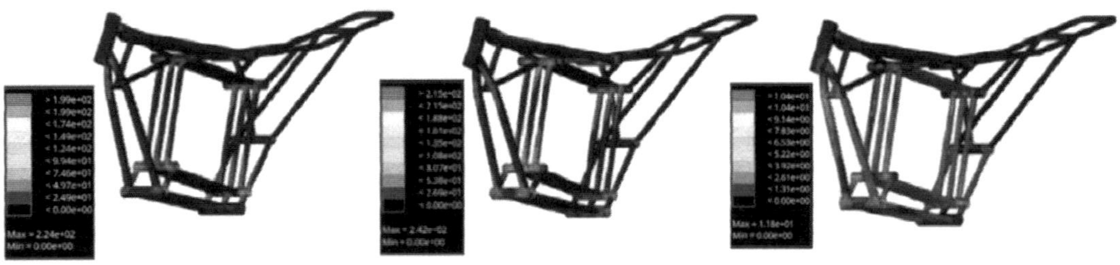

(a) Max braking at front axle (b) Max acceleration at rear axle (c) Pneumatic adherence limit curve

Figure 16. Stress induced under three loading situations.

3.4.3 Maximum Deflection

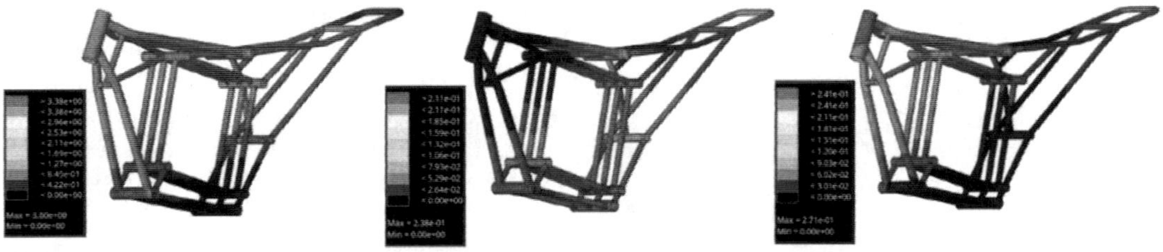

(a) Max braking at front axle (b) Max acceleration at rear axle (c) Pneumatic adherence limit curve

Figure 17. Deflection of chassis under loading condition.

Addition of parallel tubes on the lower region of the chassis helped in enhancing the stiffness as well as increased the natural frequency to permissible levels which are safe for the riders use. Table 2 shows a comparison of the natural frequencies vis-à-vis mode numbers.

Table 2. Natural frequency and corresponding mode numbers for the three iterations.

Mode Number	Natural Frequency (Hz) (First Iteration)	Natural Frequency (Hz) (Second Iteration)	Natural Frequency (Hz) (Third Iteration)
1	2.33	29.44	38.59
2	4.62	58.02	61.89
3	6.93	69.83	70.60
4	12.56	133.46	138.04
5	14.92	182.445	187.18
6	19.73	364.68	208.30

For the sake of clarity, a graphical representation for the stress and deflection for the three load cases, i.e. when maximum braking is applied to front wheels, when maximum acceleration occurs at rear wheels and when the vehicle is on curves subject to pneumatic adherence limits has been presented in Figure 18. The maximum stress and deflection have reduced as the stiffness improved.

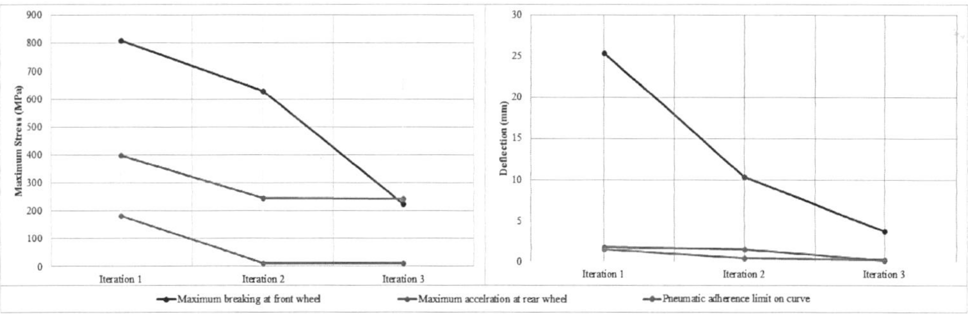

(a) Variation in stress (b) Variation in deflection

Figure 18. Stress and deflection for the three load cases.

4. Conclusion

This study has been conducted to design a novel chassis for an electric motorcycle using numerical methods and CAD tools. The mode shapes and corresponding natural frequencies helped modifying the chassis in such a way that its stiffness increased significantly thereby reducing deflections at critical points. It was witnessed that addition of two parallel tubular elements at the base region of the chassis helped increase the first natural frequency from 2.33Hz to 29.44Hz. Further addition of a parallel element at the central portion just above the swing-arm pivot hole with increased pipe thickness led to improved natural frequency to 38.39Hz. As normal sustainable human body natural frequency lies in the range of 5Hz to 7Hz with the present situation is very safe and comfortable for the riders. Also the maximum stress and deflection reduced from 810 MPa to 181 MPa and 25.4mm to 1.85mm respectively for the first load case of maximum braking at front wheels, from 628 MPa to 12 MPa and from 10.4mm to 1.58mm respectively in the second load case of maximum acceleration at rear wheels and from 224 MPa to 11.8 MPa and 3.8mm to 0.271mm respectively in the third load case of pneumatic adherence limit on curves. With the considerably reduced induced stresses and deflection, the electric motorcycle chassis is safe and easily for fabrication. This study has revealed the importance of design study before fabrication of final chassis designs. It primarily enhances the safety of the rider due to increased stiffness and secondarily as a compromise between weight and stiffness, as further addition of elements would only make the chassis heavier.

References

[1] Kurdi O, Rahman A R, 2010 Finite Element Analysis of Road Roughness Effect on Stress Distribution of Heavy Duty Truck Chassis *International Journal of Technology*, Vol **1**, No 1 pp 57-64

[2] Hastie J, 2005 Mini Baja Vehicle Design Optimization Honors Thesis Department of Mechanical, Industrial and Manufacturing Engineering *College of Engineering, Northeastern University* Boston MA

[3] Bennett C, Lockwood E, McClinton A, McRee R, Pemberton C 2013 SAE BAJA mini frame analysis Project Report *Department of Mechanical Engineering Northern Arizona University* Flagstaff AZ

[4] Raina D, Gupta R D, Phanden R K 2015 Design and Development for Roll Cage of All Terrain Vehicles *International Journal for Technological Research in Engineering* Vol **2** Issue 7

[5] Kilian S, Zander U, Talke F E 2003 Suspension modeling and optimization using finite element analysis *Tribol Int* Vol **36** pp 317-324

[6] Chang J W, Lee Y S 2008 Topology optimization of compressor bracket *Journal of Mechanical Science and Technology* Vol **22** pp 1668-1676

[7] Chiandussi G, Gaviglio I, Ibba A 2004 Topology optimisation of an automotive component without final volume constraint specification *Adv Eng Softw* Vol **35** pp 609-617

[8] Wang L, Basu P K, Leiva J P 2004 Automobile body reinforcement by finite element optimization *Finite Elem Anal Des* Vol **40** pp 879-893

[9] Pedersen C B W 2003 Topology optimization design of crushed 2d frames for desired energy absorption history *Struct Multidisc Optim* Vol **25** pp 368-382

[10] Pedersen C B W 2004 Crashworthiness design of transient frame structures using topology optimization *Comput Methods Appl Mech Eng* **193** pp 653-678

[11] Jeyapandiarajan P, Kalaiarassan G, Joel J, Rutwesh S, Fastin F T 2018 Design and Analysis of Chassis for an Electric Motorcycle *Materials Today: Proceedings* Vol **5** pp 13563-13573

[12] Gazali N M 2014 Application of Finite Element Stress Analysis of Heavy Truck Chassis: Survey and Recent Development *Journal of Mechanical Design and* Vibration Vol. **2** No. 3 pp 69-73

[13] Fui T H, Rahman R A 2007 Statics and Dynamics Structural Analysis of a 4.5 Ton Car Chassis *Jurnol Mekanikal* No **24** pp 56-57

[14] Stuecke P, Fethke S, Foken W, Jentsch M, Lehmann A, Kloetzner K, Reissmann D 2011 Development of a Racing Motorbike with Electric Power Train *SAE Technical Papers* Small Engine Technology Conference & Exposition pp 1-6

[15] Stefanos E M, Lagaros N D, Nassiopoulos E 2019 Nested Topology Optimization Methodology for Designing Two-Wheel Chassis *Front Built Environ* Vol **5** No 34

[16] Airbus Group, APWorks 2018 Inspired by Nature: Electric Motorcycle Goes 3D–Combining Topology Optimization, New Materials, and Additive Manufacturing in the Development of the Airbus APWorks Light Rider Results in a Revolutionary LightweightDesign *Altai*

IOP Conf. Series: Materials Science and Engineering 1128 (2021) 012014 doi:10.1088/1757-899X/1128/1/012014

Design study of electric vehicle roll-cage obtained using modal analysis

Shubham Arunrao Katait [1] and Pankaj Dorlikar [1]

[1] Department of Mechanical Engineering, Army Institute of Technology, Pune, India
shubhamkatait@gmail.com, pvdorlikar@aitpune.edu.in

Abstract. Understanding the dynamic behaviour of structures has become increasingly essential in the design process of any mechanical system. Modal factors like natural frequencies and mode shapes characterize the dynamic response of the structure and can be used for determination of the possible bending and torsion phenomenon. The Computer-Aided Drafting (CAD) model plays an important role in the design and development phases of an electric vehicle. For doing Finite Element Analysis (FEA) using software CAD model has to be imported so that after conversion Finite Element (FE) model can be solved. In the present case the CAD model is imported in HyperMesh. The FE model is generated and solved by using OptiStruct. Ignoring the first six natural frequencies and mode shapes due to its association with six degrees of freedom, subsequent six natural frequencies and mode shapes are proposed for an electric vehicle roll-cage. We can learn even more from CAD geometry and mode shapes in every direction about its stiffness and weight. Once we know our design baseline, we can use a software simulation to use a design study and improve its performance. We have considered the pipe diameter, thickness, tube width and height as main design parameters which needs to be optimized. In this paper, we optimized all these design parameters to make driver comfortable during driving by increasing natural frequency of the roll-cage structure by 3.34 times the human body threshold.

1. Introduction

Electric vehicles are grabbing everybody's attention as an important solution for enhancing air quality to achieve zero pollution [1]. Today, many of the researchers are actively working on to improve the fields of safety, range, acceleration, economy as well as energy, power density and recharging time of the batteries, which are the main drawbacks of the electric vehicles [2]. Due to availability of FEM analysis capability of the analysis software, we are using it as a model simulator for the manufacturing end products besides a design process tool.

Multi-tubular space frame, frequently suggested to as roll-cage goes about as a structural symbolize for different sorts of automotive vehicles. The roll-cage needs to adjust to the principles indicated by SAE guidelines. The roll-cage is exposed to the different dynamic and static load experienced during front, side, roll-over, torsional impact. The frame of roll-cage ought to be adequately hardened to withstand against all possible loads so that we get with adept strength to weight ratio [3]. To meet these rules it is critical to consider different boundaries engaged with the plan of a roll-cage, directly from the material used for the fabrication and exposing it to the impacts and forces it may experience. Factor of safety 2 was established, which gave a moderate design. In a modal analysis, the weakest parts of our design expose themselves as low frequency resonant modes [4].The main objective of the present work is to find the optimum pipe diameter, thickness, tube width and height of roll-cage pipe at the very first step after the CAD geometry creation. This is followed by the further static as well as dynamic analysis of the roll-cage.

2. Dimensions and Material properties

The design of the roll-cage in any condition should consistently guarantee sufficient clearances from roll-cage members for the safety of driver along with his/her free movement. A minimum roll-cage provision that can accommodate two individuals with an arrangement to place lead-acid battery and small cargo (230L) in front to place bags is considered. It can also handle two in-wheel electric motors on the rear suspension frame. Outer dimensions of the roll-cage are presented in Table 1.

.

Table 1. Outer dimensions of roll-cage

Parameter	Length	Width	Height	Gross vehicle weight
Dimension	2696mm	1558mm	1540mm	570kg

From the availability of huge variety of alloys steel, AISI 4130 chromoly steel was picked to be the reasonable material due to its less weight per meter length when compared with AISI 1018. Steel pipes from AISI-4130 were used as a material for the construction of roll-cages because of high tensile and yield strength. Besides these, the risks of welding capability are also minimal. The Material properties are displayed in Table 2.

Table 2. Material properties of AISI 4130

Property	Values
Initial Density	7890 kg/m^3
Young's Modulus	210 GPa
Poisson's Ratio	0.3
Yield Stress	180MPa
Hardening Parameter	230MPa
Hardening Exponent	0.2
Maximum Stress	300MPa
Ultimate Tensile Strength	670MPa
Yield Tensile Strength	440MPa

3. Research methodology

First we had developed the CAD geometry in SolidWorks. Then the CAD geometry is imported to HyperMesh. The FE model is generated and solutions are obtained using OptiStruct solver. Then, we found out the first feasible mode shapes and then use HyperStudy, SolidWork Simulation is also used to optimize the pipe thickness. As the thickness of our pipe adds more weight than stiffness, we need to find out the critical point in our design. We have to optimize the thickness and maximize the first mode resonant frequency. The detail methodology is shown in Figure 1.

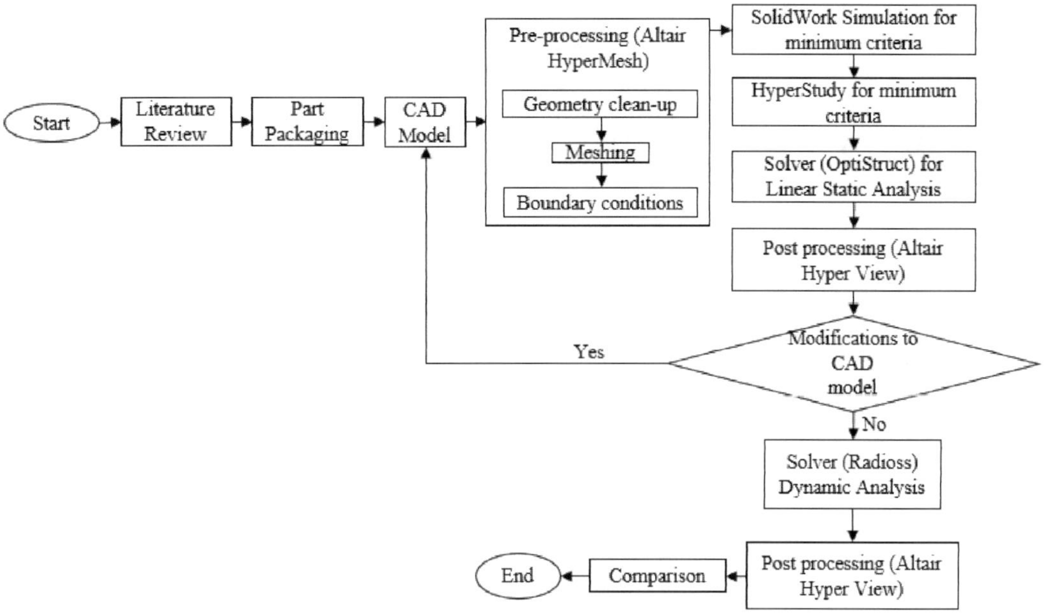

Figure 1. Detailed research methodology

4. CAD Modelling

In CAD modelling, we made one roll-cage and added different parts like the seat, the battery, the mannequin. Then the reach of mannequin is checked to make it comfortable for the human users while driving. We made the final roll-cage in SolidWork with the given specification. Detailed packaging of the roll-cage parts is illustrated in the Figure 2.

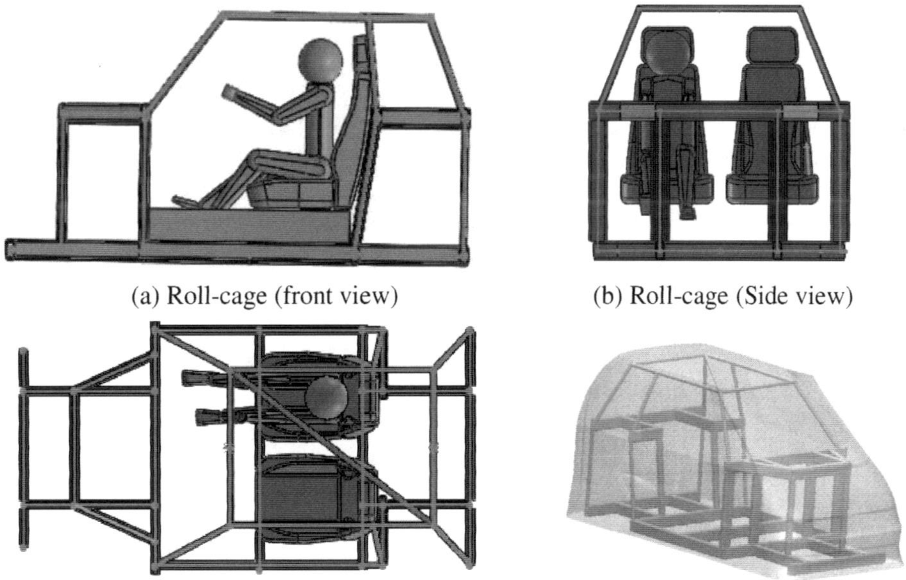

(a) Roll-cage (front view) (b) Roll-cage (Side view)

(c) Roll-cage (Top view) (d) Roll-cage outer packaging

Figure 2. Part packaging of roll-cage

IOP Conf. Series: Materials Science and Engineering 1128 (2021) 012014 doi:10.1088/1757-899X/1128/1/012014

5. Modal analysis optimization

Modal analysis is a method used to study the vibrational excitement of a system. The most basic form of analysis is modal analysis and is typically the starting point of further, more detailed analysis of dynamics. Modal analysis is used to classify the vibrational features of a system, including natural frequencies and mode shapes. Modal analysis helps design to prevent resonant vibrations or to vibrate at a certain frequency and allows engineers to see how the design can respond to various complex load types.

Mesh size is a key consideration when doing Finite Element Analysis. The mesh size is defined as the size of the smallest uniform element to be analyzed and thus has a crucial role to play in any analysis procedure. Mixed shell meshing with tria and quad element with 7mm fine mesh element size is used to perform different types of analysis for the consideration of accurate and very low level of assumption. Firstly, we take the circular pipe of a diameter of 25mm and a thickness of 3mm, rectangular pipe of 101.6mm x 50.8mm and a thickness of 3mm, rectangular plate of a thickness of 2.5 mm, is used in roll-cage. The final meshed roll-cage model is having 183652 total elements with 123.5 Kg mass of roll-cage. The final meshed roll-cage is shown in Figure 3.

Figure 3. FE model

5.1 Free-free modal Analysis FE model

Modal analysis of the roll-cage was performed using solver OptiStruct from finite element software HyperWork. Due to the advantages of reasonably accurate visualization and simulation of the mode shapes, finite element analysis software is used. Also the deformations that occur in the roll-cage could be precisely located. The different mode shapes of the FE model are displayed in Figure 4. The six consecutive mode shape frequencies are summarized in Table 3.

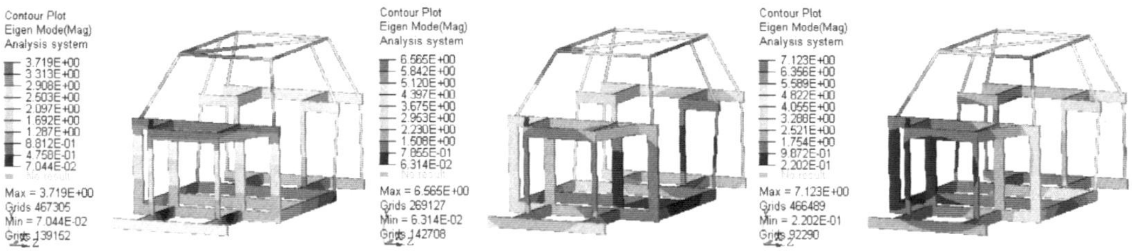

(a) F7: 24.971 Hz (b) F8: 28.510 Hz (c) F9: 28.671 Hz

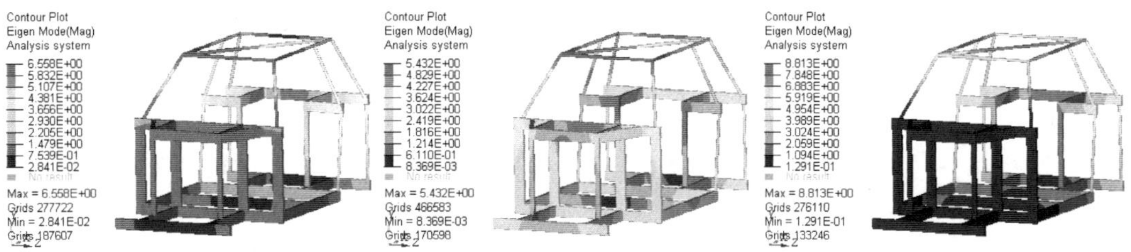

(d) F10: 33.231 Hz (e) F11: 37.003 Hz (f) F12: 46.727 Hz

Figure 4. Free-free modal Analysis of FE model

Table 3. Free-free modal Analysis of FE model

Frequency	Free-free(Hz)
F7	24.971
F8	28.510
F9	28.671
F10	33.231
F11	37.003
F12	46.727

5.2 HyperStudy optimization

HyperStudy is a multi-disciplinary software design study that allows engineers to study the performance of their products and optimize their stability. Optimization begins after the execution of a clean basic analysis (e.g. sliding energies and negligible hourglassing). New design variables or further optimization can be restarted conveniently from a new position onto the response surface.

First, we must choose optimizing parameter in HyperStudy. Then we have to include the range in the module as input values of software that is the optimizing parameter's upper and lower bound, which is inferred from the modal analysis. Then we make one of the OptiStruct linear static analysis input files with all set boundary conditions. This file will be imported at the time of HyperStudy optimization and solve in HyperStudy software. After running solver, we get the optimized result of every parameter shown in Table 4.

Table 4. HyperStudy optimization result table

Optimizing parameter	Lower Bound	Upper Bound	Optimal
Thickness rectangle(mm)	4	6	4.928
Width (mm)	25.4	76.2	48.7934
Height (mm)	50.8	152.4	102.1334
Thickness pipe (mm)	1.5	3	1.758
pipe diameter (mm)	12.7	50.8	24.9428

5.3 SolidWorks Simulation for optimization

A DoE-base optimization approach is used in SolidWorks Simulation. In order to run a problem, the engineer sets the maximum and minimum values for his dimensional design variables, and then chooses the "Standard" or "High Quality" optimization. The Standard approach assumes that the objective response curve between the limit values is linear and only calculates the response to those values. High-quality optimization takes into account the possibility of a second-order response between limits and evaluates both the mid-value and the extremes.

We may also find somewhat the same nature of conclusion through other design analysis (however with less colorful pictures). SolidWork Simulation is providing us with the optimized solution (as not every result must be released), through setting our analysis to increase or decrease the pipe or rectangle tube's thickness by iterating, thus optimizing the first mode of resonance. Here we can even get more accurate results by reducing the increments to just 1inch. Optimized parameter design study inputs window of SolidWork Simulation shown in Figure 5.

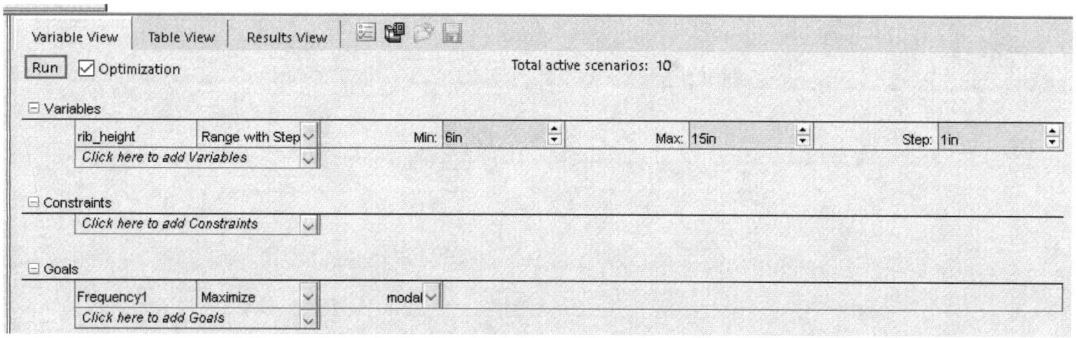

Figure 5. Optimized parameter design study input window

We get the final optimized parameter table from SolidWork Simulation as shown in Table 5.

Table 5. SolidWork Simulation result table

Optimizing parameter	Current	Optimal	Scenario-1	Scenario-2	Scenario-3
Rectangle Thickness (mm)	4	5	3	4.5	6
Width (mm)	38.1	50.8	25.4	63.5	76.2
Height(mm)	76.2	101.6	50.8	127	152.4
Pipe thickness (mm)	2.5	2	1.5	1.9	3
pipe diameter (mm)	19.05	25.4	12.7	31.75	38.1
Frequency(Hz)	24.32	26.11	21.55	21.65	25.65

5.4 Final optimized FE Roll-cage

The result of HyperStudy optimization (Table 4) and SolidWork simulation (Table 5) finalizes the best possible design parameter values as shown Table 6, which are the final optimized FE roll-cage parameters for further design study.

Table 6. Optimized FE roll-cage parameter

Optimizing parameter	Optimal
Rectangle Thickness (mm)	5
Width (mm)	50.8
Height(mm)	101.6
Pipe Thickness (mm)	2
pipe diameter (mm)	25.4

5.5 Free-free modal Analysis optimize FE roll-cage

After final optimization of the parameter check with free-free modal analysis. We get to maximize the first shape of the roll-cage mode as displayed in Figure 6 and are summarized in Table 7.

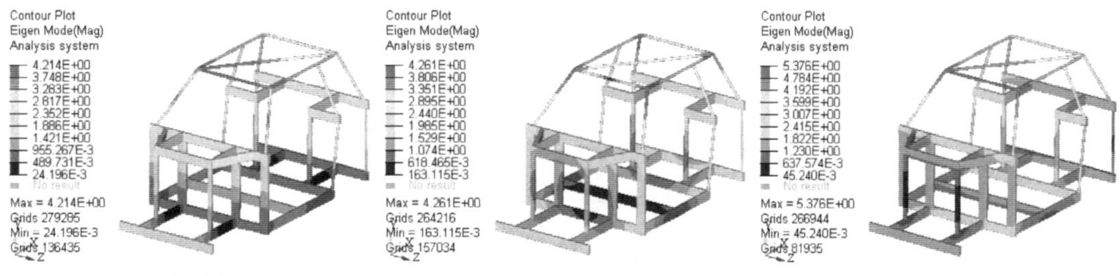

(a) F7: 2.5024 Hz (b) F8: 2.9186 Hz (c) F9: 30.958 Hz

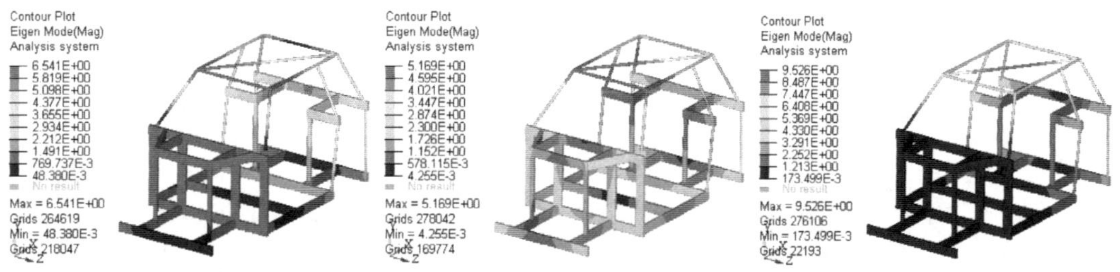

(d) F10: 33.562 Hz (e) F11: 38.481 Hz (f) F12: 47.38 Hz

Figure 6. Free-free modal Analysis of optimize FE roll-cage

Table 7. Free-free modal Analysis of optimize FE roll-cage

Frequency	Free-free(Hz)
F7	25.024
F8	29.186
F9	30.958
F10	33.562
F11	38.481
F12	47.38

The natural frequency of a human body in standing posture is about 7.5 Hz and in sitting posture is 4-6 Hz [10]. After referring table 7, we can easily conclude that operator can drive this roll-cage based vehicle comfortably, since the FEA optimised frequency is approximately 3.34 times higher than maximum human body bearable frequency.

6. Design Study of optimize FE roll-cage

The roll-cage strength is such that the working conditions are moderately permanent deflected. To ensure the study of the finite element analysis and to calculate stress levels. The strength of the [roll-cage tubes] frame is set at a value that exceeds the stress value reached in the study.

G-force is used for static analyzes, according to the Newton second movement law.

$$F = \text{mass} \times \text{acceleration} = 570 \times 9.81 \approx 5592$$

Where, F = Force exerted on roll-cage in N; m = mass of vehicle in kg, a = acceleration in m/s^2

To assess its feasibility and structural integrity according to various static and dynamic load tests.

Static Structural Analysis

- Front Impact
- Rear Impact

Explicit Dynamic Analysis

- Front Impact
- Rear Impact

IOP Conf. Series: Materials Science and Engineering 1128 (2021) 012014 doi:10.1088/1757-899X/1128/1/012014

6.1 Static Structural Analysis

Static analysis is performed to check the integrity of the roll-cage and the strength to withstand forces during the actual simulation. Boundary conditions for static structural analysis as shown in Figure 7.

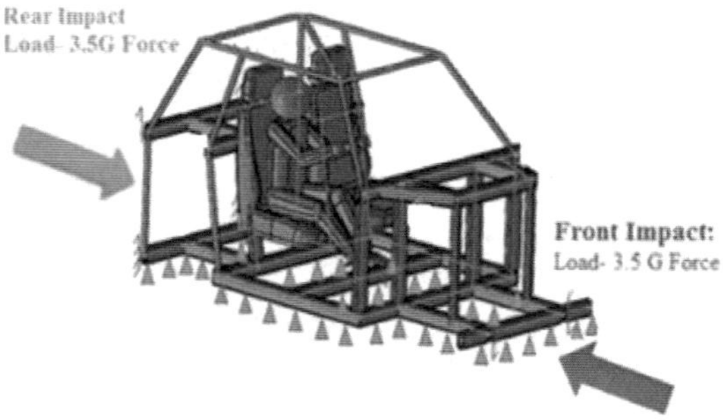

Figure 7. Boundary condition for static structural analysis of optimize FE roll-cage

- Front Impact

Front impact analysis is carried out to monitor the integrity of the roll-cage and resist the impact force on the front. 3.5G Nodal strength on the front part nodes shall be equally distributed. The nodes on the rear side of the roll-cage in all directions are fixed to simulate this condition. Front impact result are shown in figure 8.

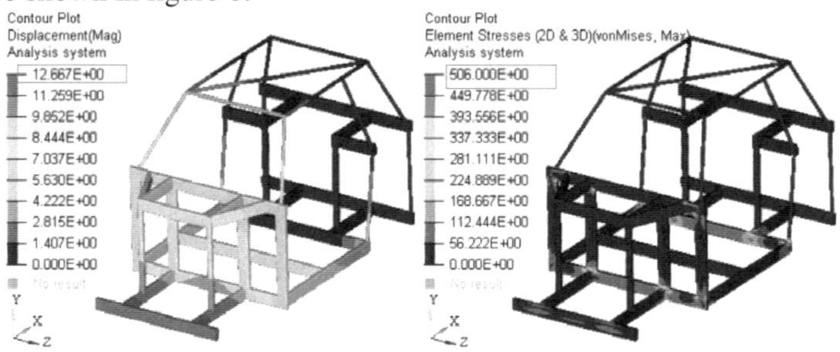

Figure 8. Static structural analysis of roll-cage under frontal impact

- Rear Impact

Rear impact analysis is carried out to monitor the integrity of the roll-cage and resist the impact force on the rear. 3.5G Nodal strength on the front part nodes shall be equally distributed. The nodes on the front side of the roll-cage in all directions are fixed to simulate this condition. Rear impact result are shown in figure 9.

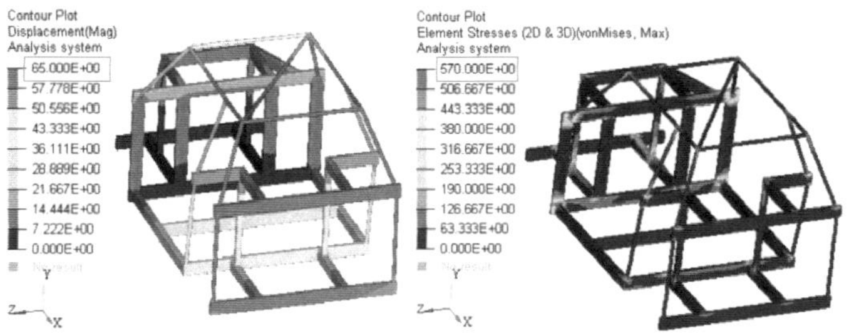

Figure 9. Static structural analysis of roll-cage under rear impact

6.2 Explicit Dynamic Analysis

Dynamic analysis is performed to check the performance of the roll-cage under dynamic loading conditions. This simulation is conducted in the software package of HyperWork. Boundary conditions for explicit dynamic analysis as illustrated in Figure 10.

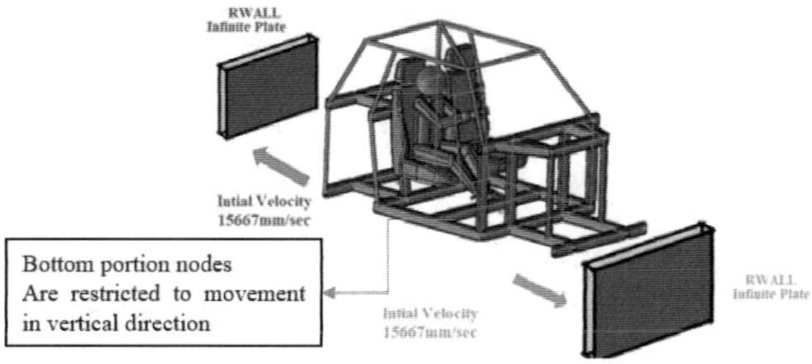

Figure 10. Boundary condition for explicit dynamic analysis of optimize FE roll-cage

- Front Impact

The Front Impact Analysis was conducted on the HyperWork RADIOSS 2017. For the analyses, the limit conditions used include an infinite wall (RWALL) which is fixed in front of the roll-cage. The initial speed of 56 kmph according to SAEINDIA is the rules for roll-cage. For better graph tracing, ENG-RUN time (Tstop) will be used as 0.1. As the boundary condition is exposed in figure 10.

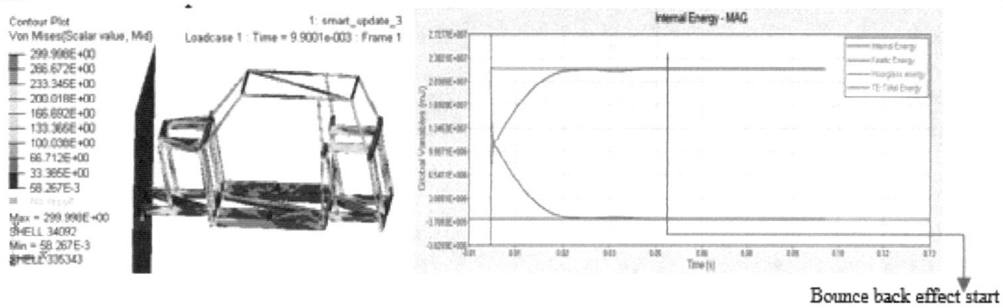

Figure 11. Explicit Dynamic analysis of roll-cage under frontal impact

- Rear Impact

The Rear Impact Analysis was conducted on the HyperWork RADIOSS 2017. Same rigid wall, initial speed and ENG-RUN time like front impact is used in rear impact. As the boundary condition is exposed in figure 10.

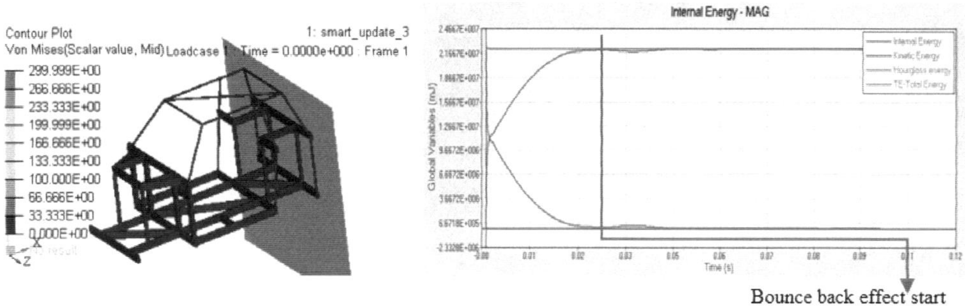

Figure 12. Explicit Dynamic analysis of roll-cage under rear impact

7. Result

After studying the outputs from figure 8, 9, 11 and 12, the results of FE optimized roll-cage are summarized as shown in table 8.

Table 8. Design Study result of optimize FE roll-cage

Analysis	Loading	Maximum Deformation (mm)	Max. Deformation (mm) (Matlab)	Max. Stress (MPa)	Max. Stress (MPa) (Matlab)	FOS
Front Impact	3.5G	1.046	1.24	195	187	3.43
Front Impact (Dynamic)	-	-	-	299.97	-	2.23
Rear Impact	3.5G	17.94	18.10	253.63	254.10	2.64
Rear Impact (Dynamic)		-	-	353	-	1.86

We have obtained the optimum design parameter (Thickness, width, Height, diameter) of the pipe as well as tubular section by modal analysis. We have also written a Matlab code for validation. As seen from the Table 8, the roll-cage is found secure when loaded statically and dynamically with a factor of safety for the roll-cage reaches around 2.

Free-free modal Analysis of optimize FE roll-cage after Ignoring the first six natural frequencies and mode shapes due to its association with six degrees of freedom, subsequent six natural frequencies and mode shapes are proposed for an electric vehicle roll-cage range between 25.024hz to 47.38hz.

8. Conclusion

This paper provide the optimum design parameter (Thickness, width, Height, diameter) of a roll-cage considering the modal analysis. We come up with different static structural and explicit dynamic analysis from that we can figure out the Table 7 that the roll-cage is secure when we loaded statically and dynamically. The factor of safety for the roll-cage reaches 2, which is safe from design point of view. The naturally produced first bending mode will meet the dynamic frame performance requirements under different conditions. The frequency of the first mode is higher than the frequency of road excitation in the most common operating conditions and is also higher than the frequency of the electric motor, thus trying to avoid these common resonance problems. For improving the comfort level of driver it is necessary to either reduce or increase the natural frequency of roll-cage. Since the FEA optimised frequency is approximately 3.34 times higher than maximum human body bearable frequency.

References

[1] Bonisoli E, Delprete C, Esposito M and Mottershead J 2011 Structural dynamics with coincident eigenvalues: modelling and testing. *Modal Analysis Topics, Volume 3, Proceedings of the 29th IMAC, 2011.*

[2] Klerk D, Rixen D and Voormeeren S 2008 General Framework for Dynamic Substructuring : History, Review and Classification of Techniques *AIAA Journal, Vol. 46, No. 5, (2008), pp. 1169–1181.*

[3] Heylen W and Sas P 2006 Modal Analysis Theory and Testing *Katholieke Universteit Leuven, Departement Werktuigkunde.*

[4] Richardson M H 1978 Measurement and Analysis of the Dynamics of Mechanical Structures, *Hewlett-Packard Conference for Automotive and Related Industries* Detroit MI pp 1-46

[5] Lee D, Choi H and Han S 2006 Design of Automotive Body Structure Using Multicriteria Optimization *Journal of Structural and Multidisciplinary Optimization, Vol. 32, 2006, pp. 161-167.*

[6] Lee N and Nikravesh E 1994 Steady State Analysis of Multibody Systems with Reference to Vehicle Dynamics *Journal of Nonlinear Dynamics, Vol. 5, 1994, pp. 181- 192.*
[7] Riley W and George A 2002 Design, Analysis and Testing of a Formula SAE Car Chassis *SAE TECHNICAL PAPER SERIES 2002-01-3300.*
[8] Balmes E 2018 Shapes & DOF: on the use of modal concepts in the context of parametric non-linear studies *ISMA, Sep 2018, Leuven, Belgium, hal-01912722*
[9] Balaguru, Natarajan E, Ramesh and Muthuvijayan 2019 Structural and modal analysis of scooter frame for design improvement *International Conference on Advances in Materials, Manufacturing and Applied Sciences, 2214-7853*
[10] Anas M, Nandu N, Anjai K, Nairc A and Sreedharand P 2017 Design, Analysis, Fabrication and Testing of a Formula Car Chassis *International Conference on Advances in Materials and Manufacturing Applications [IConAMMA 2017], 24944–24953.*
[11] Ren W, Peng B,Shen J,Li Y, and Yu Y 2018 Study on Vibration Characteristics and Human Riding Comfort of a Special Equipment Cab *Journal of Sensors Volume 2018, Article ID 7140610, 8 pages*
[12] Lombard M 2018 *Mastering SolidWorks* (Inc., Indianapolis, Indiana, Canada)
[13] Gokhale N, Deshpande S, Bedekar S and Thite A 2008 *Practical Finite Element Analysis*(India)

The thermal and stress analysis of disc brake

Alampally Sainath[1], Prathamesh Mahesh Dehadray[2], P. Bharath[3], Lokavarapu Bhaskara Rao[4, *]

[1,2,3] School of Mechanical Engineering, Vellore Institute of Technology, Chennai Campus, Vandalur- Kelambakkam Road, Chennai, Tamil Nadu - 600127, India.
[1]E-mail: sainath.sai8197@gmail.com, [2]E-mail: prathamesh.dehadray@gmail.com,
[3]Email: bharathpappu01@gmail.com, [4, *]Email: bhaskarbabu_20@yahoo.com (Corresponding author)

Abstract: Braking system has huge importance in vehicles. The disc brake is one of the most important amongst the brake rotors. Brake rotors generate opposing torque on the shaft of the wheel thus converts the kinetic energy of the wheel to heat. The braking components are used to decelerate the vehicle by friction developed on to the wheel when brake is applied. The heat is produced during friction, which will cause negative effects on the brake assembly. This study is to analyze the motor cycle disc brake, which affects the surface of the disc. In order to get those effects on the disc surface the thermal stress analysis and thermal deformation is analyzed using ANSYS. This analysis is used to avoid the damage of the disc due to heat generation. In this study, different materials are compared to choose optimum one for the better performance to deformation, temperature and stresses.

Keywords: Disc brake, Friction, Deformation, Materials, Thermal analysis

INTRODUCTION:

Motor cycle plays vital role in our day-to-day life. If some obstacle comes in the path of moving vehicle, braking system is used to stop or decelerate the vehicle to avoid collision. But failure of braking system can cause accidents, hence study of braking system is mandatory. Sung Soo Kang and Seong - Keun Cho [10], studied geometry of vents in motorcycle disk brake. The thermal analysis is carried out in ANSYS. They compared the results for ventilated disc and solid disc for maximum temperature and concluded the importance of ventilated disc. Thilak [12], studied the importance of disc brake and materials required to make it. Cast iron is widely used for disc brakes but now a days those discs are made up from carbon-carbon composites (CCC) or ceramic matrix composites (CMC). He performed transient thermo-elastic analysis for repetitive braking and compared the results obtained. Sarip [11], modelled an assembly of disc brake with brake pads and wheel hub and fixed with nut bolt fitment. He applied pressure to carry out the braking operation by means of friction. The friction is necessary to stop a vehicle, these brake pads acts as the friction material to decelerate the vehicle speed. Guru Murthy et al [4], studied the operation of disc brake. They modelled the disc brake in CATIA and analysis is carried out in ANSYS. They took carbon ceramic composite material for study and calculated normal force, shear force, piston force and braking distance. Thermal and modal analysis is also carried out to obtain heat flux and deflection.

During continuous braking, the temperature of the brake increases. As the heat increases continuously, the effectiveness of the brake reduces and it may lead to damage of the brake rotor [1-3]. The critical performance requirements for a brake rotor includes good thermal conductivity, low wear rate, long durability, low vibration, heat resistant and high cooling rate [5-8]. For this purpose, disc brake is widely used. The best combination of the discs like the disc material, pad thickness and vent diameter are chosen. The brakes should be very strong and flexible to stop the vehicle immediately in fraction of seconds. The rider must have proper control on the vehicle during braking. The rider must control the vehicle not to have skidding of the vehicle during braking. The effectiveness should not be reduced up to its usage. Brakes should have very less wear property.

METHODOLOGY:

Modelling

It is difficult to model the exact disc brake for transient thermoelastic behavior hence in this study, the geometry of disc brake is modelled using some assumptions. These assumptions are developed to reduce the complexity in the mathematical calculations. The assumptions are made depending on the accuracy required during modeling [7]. The disc material is homogenous and isotropic. The disc should be stress free before the application of brake. Brakes should be applied on both the wheels. The thermal conductivity of the vehicle should be uniform throughout the disc. The disc brake is modelled in CATIA. The design is developed by using the disc of the Hero Honda Ambition bike of 135 cc as shown in figure 1.

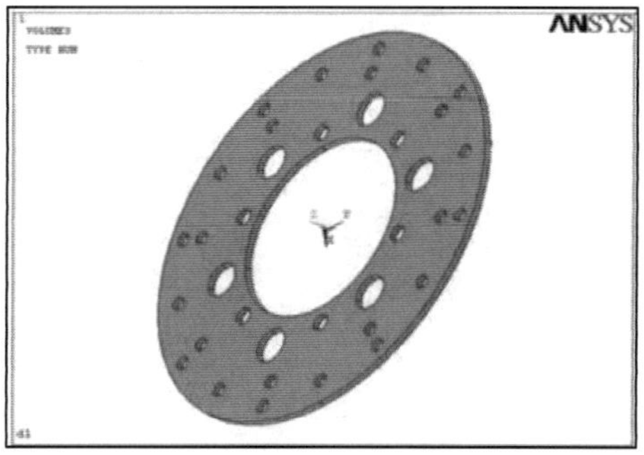

Figure. 1. Solid model of 240mm disc brake

Meshing

The finite element analysis is the basic concept, which is used to analyze the structure. It is very difficult to analyze the entire structure once is difficult so in order to reduce the complexity we use meshing. The meshing consists of nodes and elements. The combined network of nodes and elements is called mesh. Foe this study, tetra mesh is used as shown in figure 2.

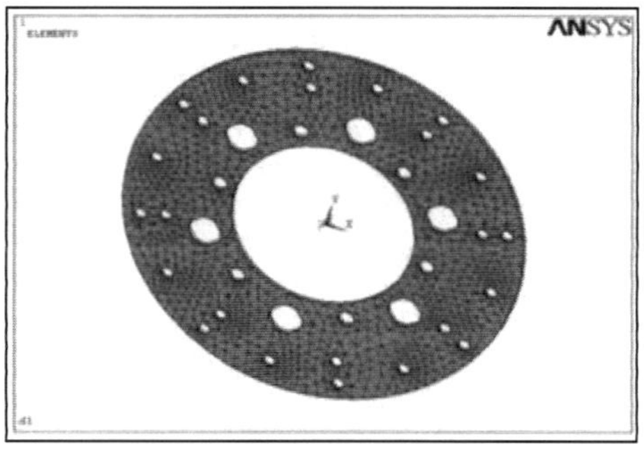

Figure. 2. 3-D meshing of a solid model

Type of Element : Solid90

Degrees of freedom : K_{xx}, K_{yy}, K_{zz}, C

Mesh size : 1.2

In thermal analysis, the functional is derived from governing differential equation with its boundary conditions. To get differential equation conversation of heat energy is applied to a control volume *dx, dy, dz* which has heat conduction rates q_x , q_y , q_z in *x, y, z* directions. Heat conduction rates in opposite direction q_{x+dx} , q_{y+dy} , q_{z+dz} can be given as [9],

$$q_{x+dx} = q_x + \left(\frac{\partial q_x}{\partial x}\right)dx$$
$$q_{y+dy} = q_y + \left(\frac{\partial q_y}{\partial y}\right)dy$$
$$q_{z+dz} = q_z + \left(\frac{\partial q_z}{\partial z}\right)dz \quad (1)$$

$$q_x = -K_c \times dy \times dz\left(\frac{\partial T}{\partial x}\right)$$
$$q_y = -K_c \times dz \times dx\left(\frac{\partial T}{\partial y}\right)$$
$$q_z = -K_c \times dx \times dy\left(\frac{\partial T}{\partial z}\right) \quad (2)$$

where,

K_c is conductivity and T is temperature.

Conservation of energy can be expressed as,

Rate of heat increment in the volume (E_{st}) = Rate of heat transfer across its surfaces + Rate of internal heat generation (E_g) (3)

where,

$$E_{st} = \rho c\left(\frac{\partial T}{\partial t}\right) dxdydz \quad (4)$$

$$E_g = \varphi dxdydz \quad (5)$$

where, φ is rate at which energy is generated per unit volume

Substitute equations (1), (2), (4), (5) in equation (3) gives the following,

$$-\left(\frac{\partial q_x}{\partial x}\right)dx \times \left(\frac{\partial q_y}{\partial y}\right)dy \times \left(\frac{\partial q_z}{\partial z}\right)dz + \psi dxdydz = \rho c\left(\frac{\partial T}{\partial t}\right)dxdydz \quad (6)$$

To determine temperature distribution along the disc, boundary conditions have to be substituted in equation (6).

Boundary Conditions

After the completion of finite element model, boundary conditions are applied for thermal as well as structural analysis. The thermal boundary conditions of the disc brake are adiabatic and applied at inner radius as well as outer radius of disc brake with initial temperature T=25°C along the radius of the disc.

Selection of the best material for the disc

The best material for the disc is selected based on the operating condition so that it can withstand at a constant hydraulic pressure P=1.05 MPa. In this study, results obtained for displacement, stresses and nodal temperature distribution are compared for different materials. Cast iron, stainless steel and carbon-carbon composites are taken under consideration. The analysis is carried out in ANSYS and by comparing the results, optimum material is selected. This study will also give the combination of material and vent which increases the life of the disc.

RESULTS AND DISCUSSION:

Design statement

The disc brake is designed for the motor cycle having the kerb weight of 143 kg. The motor cycle running at a speed of 60 km/hr, and it is brought to rest in 4 seconds on applying the brake. The radius of the rolling wheel is 480 mm and it is assumed that the kinetic energy of the rotating parts is 10% of the kinetic energy of the moving bike. The weight of the rider is assumed as 65 kg.

The optimum material for required application is obtained by comparing various parameters. To study the thermoelastic behavior of brake, a constant pressure of 1.05 Mpa is applied. The thermal boundary conditions are applied on the disc as mentioned. The temperature applied is as per the material obtained in the design calculations for the upper pad or on outer surface area of the disc for which most of the part is covered by the pad of the caliper. The material properties and operation conditions used for the validation are shown in Table 1 for the steady state condition of the 3 various materials of disc is as follows:

Disc brake calculations

From the above conditions,

m_1=143kg; m_2=65 kg;

$m=m_1+m_2$=143+65=208 kg

R= rolling radius of the wheel=480mm

Velocity of the moving vehicle; v_1=60 km/hr =16.66 m/s

Velocity of the vehicle after braking; v_2=0 km/hr =0 m/s

Step- 1:

E_v= kinetic energy of the motor cycle in Joules

E_r= kinetic energy in the rotating parts

$$E_v=0.5*m*(v_1^2-v_2^2) \quad (7)$$

$E_v=0.5*208*(16.66^2-0^2)$

E_v=28865.7824 J

$E_r=0.1*E_v$

E_r= 2886.57824 J

Step-2:

The total energy of the disc can be calculated by the following equation [2],

$$\text{Total energy}= E = E_v + E_r \quad (8)$$

E = 28865.7824 + 2886.57824 = 317252.36 J

Step-3:

$$E=T_b*\theta_b \quad (9)$$

θ_b= Angular displacement of the brake during braking period given by,

$$\theta_b= \omega*t-0.5*\alpha*t^2 \quad (10)$$

where, ω = Angular velocity in rad/sec.

$$\omega_1 = v_1/R = 16.66/0.48 = 34.7 \text{ rad/s} \quad (11)$$

α = Angular acceleration in rad/sec^2

$\alpha = (\omega_1 - \omega_2)/t = (34.7\text{-}0)/4 = 8.677$ rad/s^2 (12)

$\theta_b = (34.7*4) - (0.5*8.677*4^2) = 69.384$ rad

Table 1. Properties of different materials

Material properties	*Cast iron*	*Stainless steel*	*Carbon-carbon composites*
Thermal conductivity (K)	57	17.2	400
Density (ρ)	7200	7800	1800
Specific heat (c)	452	500	1420
Poisson's ratio (γ)	0.3	0.3	0.3
Thermal expansion (α)	11	16	0.31
Elastic modulus (E)	100	190	50.2
Coefficient of friction (μ)	0.1	0.1	0.1
Room temperature (T_1)	25	25	25
Temperature in disc (T_2)	221.8	202.88	87.6425
Hydraulic pressure (P)	1.05	1.05	1.05

Now from equation (9),

$T_b = E/\theta_b$ = Torque required to stop the wheel (13)

As, two brake pads are used

$E = E_t/2$ (14)

$T_b = 15876.18 / 69.384$

$T_b = 228.816$ N-m

$T_b = 228.816*10^3$ N-mm

Step-4:

Temperature and Energy dissipation calculations;

Let us assume 50% of the heat generated is dissipated to the surroundings during braking hence,

$E = E_1$ = Energy calculated in the design (15876.18 J)

m = mass of the disc brake assembly = 2.5 kg (Assumed)

Material 1: Cast Iron

$E = m*c_p*\Delta T$ (15)

$E_n = (E_{n-1}/2) + E_1$ (16)

For 1st braking,

$E_1 = m*c_p*\Delta T$

$\Delta T = E_1 / (m*c_p)$ (17)

$\Delta T = 15876.18 / (2.5*452) = 14.05$°C

For 2nd braking,

$E_2 = m*c_p*\Delta T$ (18)

$E_2 = (E_1/2) + E_1$ (19)

$E_2 = 23814.27$

$\Delta T = 21.07°C$

Hence, study is carried out for 8 brakes and found out that, for 8th braking, $\Delta T = 27.9896°C$

Initial temperature on the disc is T = 25°C

After 1st braking $T_1 = 25 + 14.05 = 39.05°C$

2nd braking $T_2 = 39.05 + 21.07 = 60.12°C$

3rd braking $T_3 = 84.71°C$

4th braking $T_4 = 111.05°C$

5th braking $T_5 = 138.27°C$

6th braking $T_6 = 165.93°C$

7th braking $T_7 = 193.81°C$

8th braking $T_8 = 221.08°C$

Thermal and stress analysis is carried out on the cast iron disc. The results such as nodal stresses and Von misses stress distribution are shown in figure 3 and figure 4. The rise in temperature and energy dissipation along the disc and average displacement are shown in figure 5 and figure 6.

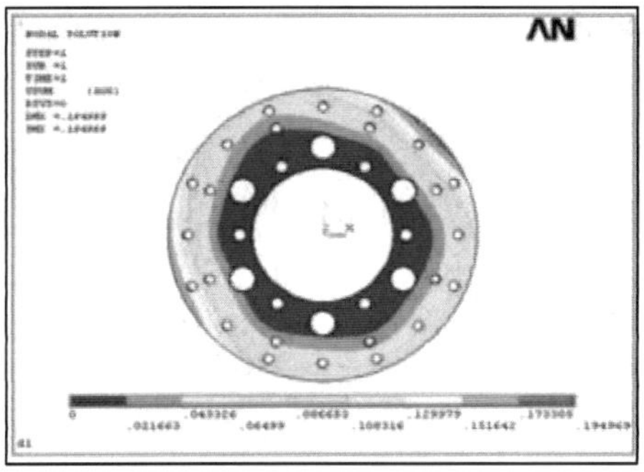

Figure. 3. Nodal stresses distribution

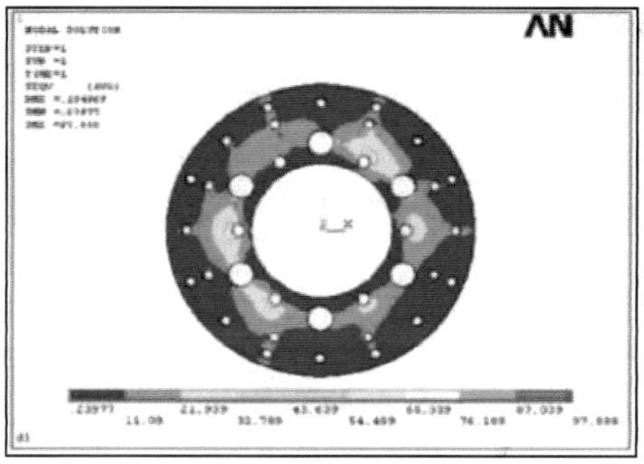

Figure. 4. Distribution of Von misses Stress

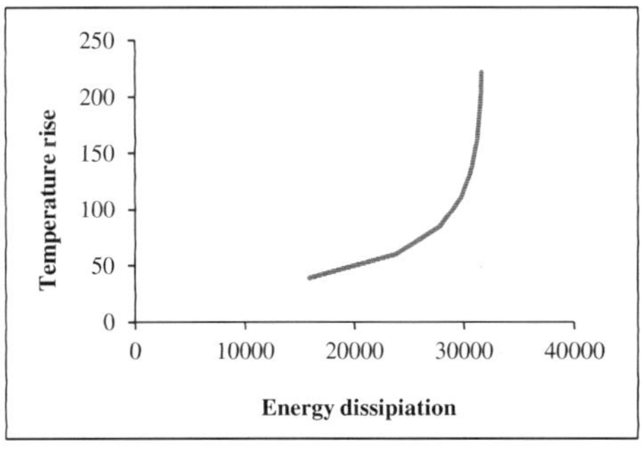

Fig. 5. Enegy dissipation in braking steps

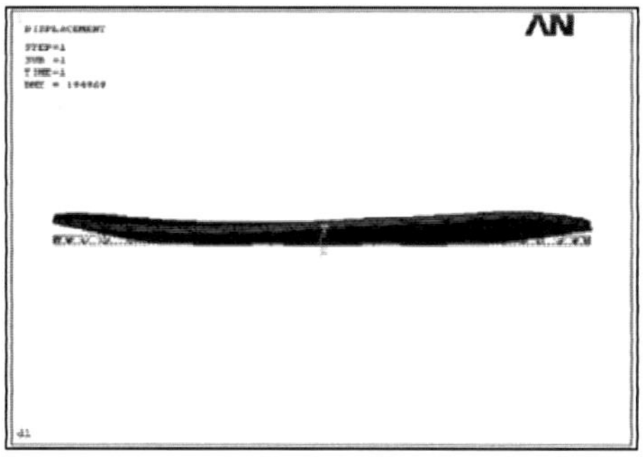

Fig. 6. Average displacement

Material 2: Stainless Steel

$E = m*cp*\Delta T$ (20)

$E_1 = 15876.18$ J

$\Delta T = 12.07°C$

$E_2 = 23814.25$ J

$\Delta T = 19.05°C$

Hence, study is carried out for 8 brakes and found out that, for 8th braking, $\Delta T = 25.30°C$

Initial temperature on the disc T=25ºC

After 1st braking T1 = 25 + 12.07 = 37.07ºC

2nd braking T2 = 37.07 + 19.05 = 56.12ºC

3rd braking T3 = 78.97ºC

4th braking T4 = 102.78ºC

5th braking T5 = 127.38ºC

6th braking T6 = 152.38ºC

7th braking T7 = 177.58ºC

8th braking T8 = 202.8ºC

Thermal and stress analysis is carried out on the stainless steel. The results such as nodal stresses and von misses stress distribution are shown in figure 7 and figure 8. The rise in temperature and energy dissipation along the disc and average displacement are shown in figure 9 and figure 10.

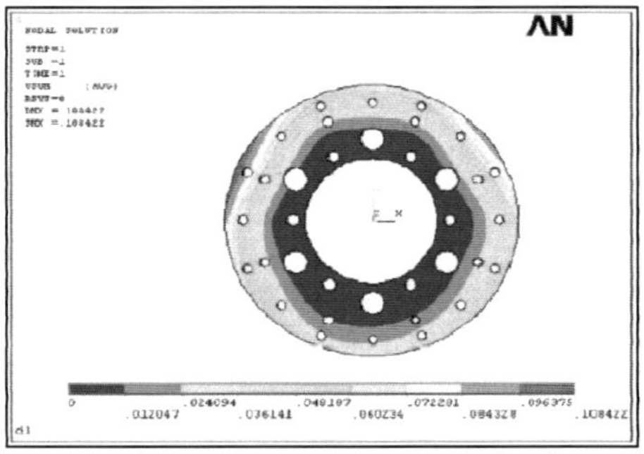

Figure. 7. Nodal stresses distribution

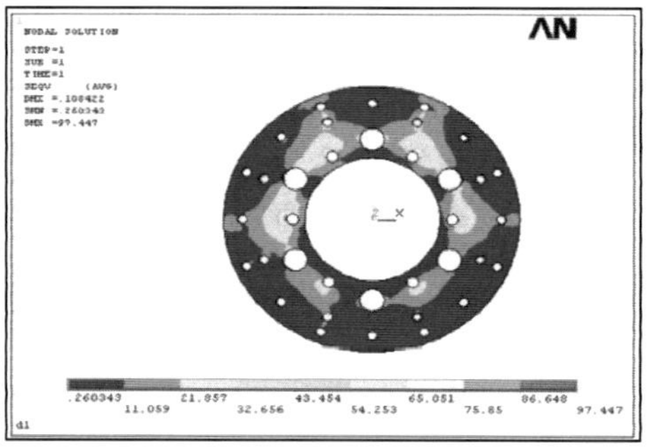

Figure. 8. Distribution of Von misses Stress

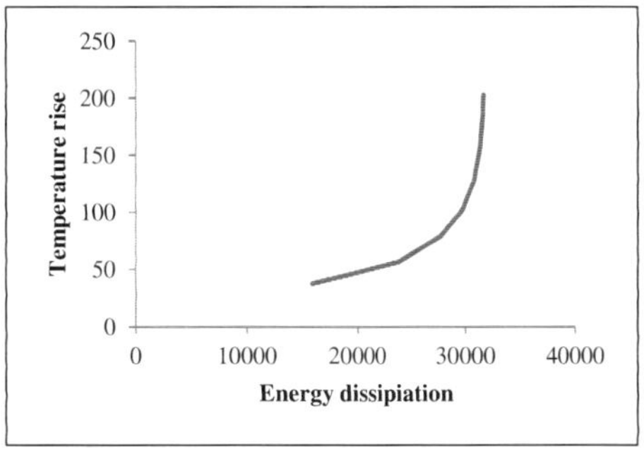

Figure. 9. Enegy dissipation in braking steps

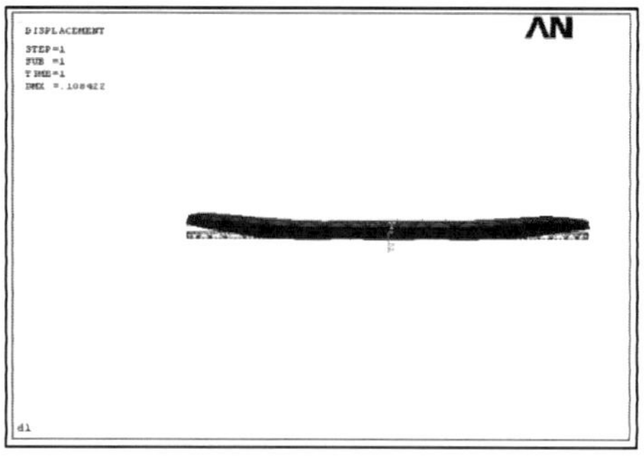

Figure. 10. Average displacement

Material 3: Carbon-carbon Composite

$E = m*cp*\Delta T$ (21)

$E_1 = 15876.18$ J and $\Delta T = 4.472°C$

$E_2 = 23814.29$ J

$\Delta T = 6.708°C$

Hence, study is carried out for 8 brakes and found out that, for 8th braking, $\Delta T = 8.909°C$

Initial temperature on the disc T = 25°C

After 1st braking T1 = 25 + 4.472 = 29.472°C

2nd braking T2 = 29.472 + 6.708 = 36.18°C

3rd braking T3 = 44.006°C

4th braking T4 = 52.391°C

5th braking T5 = 61.055°C

6th braking T6 = 69.8595°C

7th braking T7 = 78.7335°C

8th braking T8 = 87.6425°C

Thermal and stress analysis is carried out on the carbon-carbon composites. The results such as nodal stresses and von misses stress distribution in the disc are shown in figure 11 and figure 12. The rise in temperature and energy dissipation along the disc and average displacement are shown in figure 13 and figure 14. The results for cast iron, stainless steel and carbon-carbon composites are presented in table. 2.

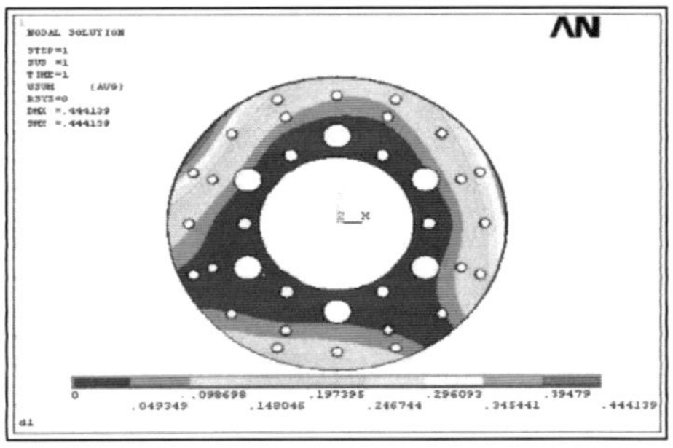

Figure. 11. Nodal stresses distribution

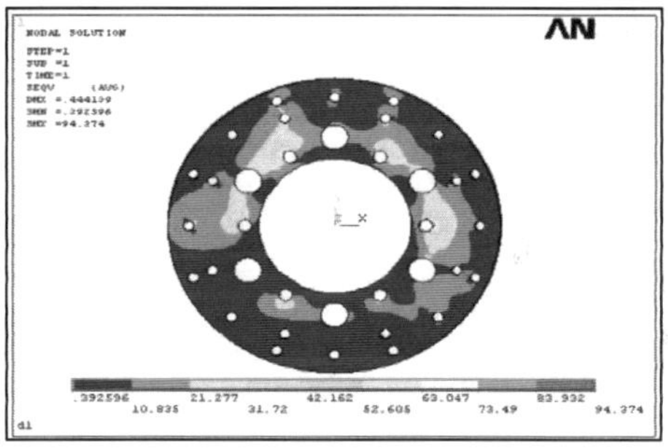

Figure. 12. Distribution of Von misses Stress

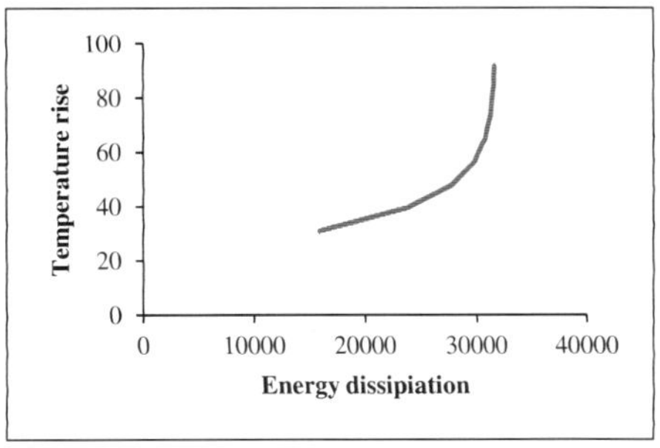

Figure. 13. Enegy dissipation in braking steps

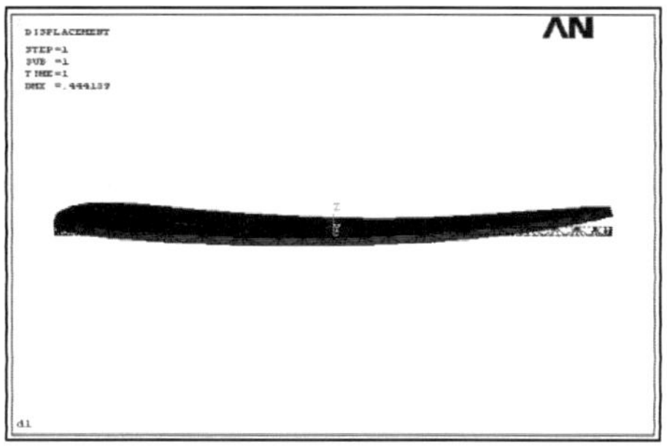

Figure. 14. Average displacement

Table 2. Comparison of Various Disc Parameters

Material parameters	*Cast iron*	*Stainless steel*	*Carbon-carbon Composites*
Diameter	240	240	240
Maximum temperature	221.8	202.8	91.55
Minimum temperature	25	25	25
Deflection	0.194	0.108	0.441
Von Misses Stresses	97.88	94.44	97.37
	0.239	0.26	0.392

CONCLUSION:

The steady state thermal analysis of the disc brakes is repeated in brake application that has been performed along with the theoretical design. Based on the results from ANSYS, the thermo elastic behaviors of the Carbon-Carbon composite, Stainless steel and Cast-iron disc brakes are investigated. In structural analysis deflection of disc, Von misses stresses and nodal stresses distribution is calculated. Same in case of thermal analysis heat flux distribution and nodal temperature distribution is calculated. All these parameters give the information that for a particular brake application which material is best suitable one. The present study can provide a useful design tool and improve the brake performance of disc brake system. Comparing the different results obtained from analysis, it is concluded that disc brake with 240 mm diameter and of stainless-steel material is the best possible combination for the present application.

Reference:

[1] Alibelhocine and Mostefa Bouchetara, "Simulation of fully coupled thermo mechanical analysis of disc brake rotor" WSEAS transactions on applied and theoretical mechanics, Volume 7 Issue 3, Page 169-181, 2012
[2] Bhandari V.B, "Textbook for design applications", Tata McGraw-Hill publications, 2010
[3] Faramarz Talati and Salman Jalalifar, "Analysis of Heat Conduction in a Disc Brake System", Heat and Mass Transfer, Volume 45, Issue 8, Page 1047-1059, 2009
[4] Guru Murthy Nathi, Charyulu, T.N., Goutham, K., Sathish Reddy, "Coupled structural/Thermal analysis of disc brake", International Journal of Research in Engineering and Technology, Volume 1, Issue 4, Page 539-553, 2012
[5] Masahira Kubota, Tsutomu Hamabe, Yasunori Nakazono, Masayuki Fukuda, & Kazuhiro Dio., "Development of a light weight brake disc rotor: a design approach for achieving an optimum thermal, vibration and weight balance", JASE Review, Volume 21, Issue 3, Page 349-355, 2000
[6] Maleque, M.A., Dyuti, S., Rahman, M.M., "Material Selection Method in Design of Automotive Brake disc", Proceedings of the World congress on Engineering, Vol. 3, 2010.
[7] Naveen kumar, S. and Kiran, M.B., "Redesign of Disc Brake Assembly with Lighter Material", International Journal of Engineering Research and Technology, Volume 1, Issue 7, Page 1-9, 2012
[8] Omar Maluf, Mauricio Angeloni, Marcelo Tadeu Milan, Dirceu Spinelli, Waldek Waladimir Bose Filho, "Development of materials for automotive Disc brakes", Minerva, Volume 4, Issue 2, Pages 149-158, 2007
[9] Saiprasit Koctniyom, "Thermal stress Analysis of Automotive Disc Brakes", Report submitted for Doctor of Philosophy, The University of Leeds, School of Mechanical Engineering, 2010

[10] Sung- Soo Kang, & Seong-Keun Cho (2012). "Thermal deformation and stress analysis of disc brakes by finite element method", Journal of Mechanical Sciences and Technology, Volume 26, Issue 7, Pages 2133-2137, 2012

[11] Sarip, S. Member, IACSIT, "Design Development of Light Weight Disc Brake for Regenerative Braking", International Journal of Applied Physics and Mathematics, Volume 3, Issue 1, Page 52-58, 2013

[12] Thilak, V.M.M, "Transient Analysis of Rotor Disc of Disc Brake using ANSYS" International Journal of management, IT and Engineering, Volume 2, Issue 8, Page 502-514, 2012

[13] ANSYS user manual

[14] CATIA V5R20 user manual

Light-weighting of single-sided motorcycle swing-arm through generative design

Adarsh Balasubramanian[1], G Shaktivel[2] and N Raghukiran[3]

[123]*Centre for automation, School of Mechanical Engineering, Vellore Institute of Technology, Chennai Campus, Tamil Nadu, India*

adarsh.balasubramanian2018@vitstudent.ac.in[1], sakthivel.g@vit.ac.in[2], raghukiran@vit.ac.in[3]

Abstract. *This work explores the prospects of Generative Design to generate the single-sided swing-arm of a homologated motorcycle to perform the process of light-weighting by selecting two aluminum alloys, namely Al 6061 (primary) and Al 7075 (reference), as materials and 5-axis milling as the manufacturing method. This work also cites and summarizes a couple of previous studies to establish the findings that Al 7075 is a superior material for the manufacturing of the single-sided swing-arm when compared to steel and further proceeds to expatiate as to why Al 6061 should be considered as opposed to Al 7075 for increased gains from the process of light-weighting with substantiations from the analysis of the single-sided swing-arm under the simulated loads.*

Key Words. Generative Design, Additive Manufacturing, Optimization, Automobile Engineering, 3D modeling.

1. INTRODUCTION

Motorcycles are one of the most affordable and most common modes of transport for many people around the world. Yet some motorcycles are homologations of purpose-built race motorcycles available to the public. These homologated motorcycles are loaded with the latest advancements in technology and sciences making them a road-legal version of their racing counterpart. One of the main characteristics of these motorcycles is that they are extremely light. Being extremely light, these motorcycles emit lesser pollution and have a lesser gross carbon footprint when compared to other traditional motorcycles as they consume lesser material and lesser fuel. This inference is substantiated [1] as it analyses the savings in energy and the reduction in emissions of vehicles using lightweight materials.

With automobile emissions contributing to global warming, automobiles must be as efficient as possible. The two main areas where major gains could be realized are the IC engine and material usage. With the maximum thermal efficiency achieved at about 50% [2] and further gains in the thermal efficiency of the

IC engines becoming increasingly expensive, manufacturers have resorted to the process of light-weighting to further optimize their vehicles. The process of light-weighting is axiomatic in the sense that manufacturers aim to achieve the lightest weight possible for each component while still maintaining and enhancing its intended purpose. Two major ways through which this is done are *Topology Optimization* and *Generative Design.* As there are numerous pieces of literature elucidating the principles and the application of *Topology Optimization*, this study focuses on the prospects of *Generative Design* [3,4]. If the process is estimated to be too expensive to be carried out for the entire automobile, it is implemented for the critical components as they offer the major gains. One such critical component is the swing-arm (as inferred from [5-7]), because it characterizes the behavior of the motorcycle as it regulates the interaction of the rear wheel with the road through a system of springs and dampers by connecting the rear wheel to the main chassis. There are two basic designs of the swing-arm that are in use today, namely single-sided swing-arm and double-sided swing-arm. Single-sided swing-arm is preferred over double-sided swing-arm motorcycles as it provides greater ease of access to the suspension system, chain assembly, brake assembly, and the wheel assembly, which helps with the maintenance of the motorcycle with minimal effort.

There are numerous types of single-sided swing-arms available in the market today most of which are manufactured through the standard process of welding multiple pieces together to create the swing-arm. This study focuses on the generation of the single-sided swing-arm as a single component through the process of *Generative Design* for maximum light-weighting and analysis of its performance against simulated forces.

Generative Design is an iterative design process that involves a program and a designer where the program, mainly originating from nature-mimicking procedures, generates a certain number of outputs that meet certain constraints while acknowledging user-defined criteria and limitations and the designer which can be either an algorithm or a human that fine-tunes the feasible region by changing minimal and maximal values of an interval in which a variable of the program meets the set of constraints, to reduce or augment the number of outputs to choose from.

2. MATERIALS SELECTION AND MODELLING

Almost all components of an automobile are made from alloys. This is because alloys offer a plethora of advantages over pure metal which includes increased strength and increased corrosion resistance. Aluminum alloys, especially, are used extensively as opposed to other alloys because they offer the fastest, safest, most environmentally friendly, and cost-effective way for increased performance and fuel economy while maintaining or enhancing safety and durability with reduced emissions.

2.1. MATERIAL SELECTION

Aluminum is 10% - 40% lighter than steel depending on the product, yet possesses significant malleability allowing engineers to design and manufacture shapes optimized for maximum performance. Automobiles manufactured from aluminum components are about 24% lighter than those made from steel and save about 0.7 gallons per 100 miles, which is a saving of about 15% in fuel consumption. These automobiles also have better acceleration, better braking, and better handling by virtue of the rigidity possessed by aluminum components which provides immediate and precise control to the driver. This study focuses on two aluminum alloys, namely Al 7075 and Al 6061. Al 6061 is considered as the primary alloy for the machining of the single-sided swing-arm and for analysis under simulated loads, while Al 7075 is considered as reference material because there is numerous literature exploring the performance of the single-sided swing-arm made from it and hence provides valuable reference points [4,5,8].

Al 7075 is an aluminum alloy with zinc as the primary alloying element. It has excellent mechanical properties and exhibits good ductility, high strength, toughness, and good fatigue resistance. However, it is more susceptible to embrittlement than many other aluminum alloys because of micro-segregation but has significantly better corrosion resistance. Al 6061 is a precipitation-hardened aluminum alloy, containing magnesium and silicon as its major alloying elements. It has very good corrosion resistance and very good weldability although reduced strength in the weld zone. It has medium fatigue strength and good cold formability. It is one of the most commonly used alloys of aluminum for general-purpose use.

Table 1. Mechanical Properties of Al 6061 & Al 7075

S. No	Properties	Al 6061	Al 7075
1	Density	2700 Kg/m^3	2810 Kg/m^3
2	Young's Modulus	68.9 GPa	71.7 GPa
3	Poisson's Ratio	0.33	0.33
4	Ultimate Tensile Strength	310 MPa	572 MPa
5	Tensile Yield Strength	276 MPa	503 MPa

Al 6061 is chosen for study because of its ease of fabrication compared to Al 7075. This is mainly due to Al 6061 alloy's lower hardness and tensile strength. The relatively lower hardness allows it to be machined more easily than Al 7075.

2.2. MODELLING AND MANUFACTURE OF SWING-ARM

MODELLING OF SWING ARM

An open-source CAD file of a homologated racing motorcycle is selected for this study. The essential components required for generative design studies such as frame, swing-arm pivot, linkage mount, axle blocks, wheel assembly, suspension linkage, mono-shock, footrest on the frame, engine block, and the chain assembly are isolated from the rest of the components for ease of work. These isolated components are termed *Derived components*. These derived components have to be classified as either preserved regions or as obstacle geometries.

The "*Preserve regions*" are those regions that are maintained and form the base from which additional geometries are created by the generative design algorithm i.e., these are regions that form a part of the final generated component. In this study, these are the swing-arm pivot, linkage mount, and axle block components. "*Obstacle geometries*" are those regions that are to be avoided and the algorithm has to build around these regions i.e., these are regions that do not form a part of the final generated component. In this study, these are the wheel assembly, suspension linkage, mono-shock, footrest on the frame, engine block, and the chain assembly components. The *Preserve regions* (green) and the *Obstacle geometries* (red) are illustrated in Figure 1.

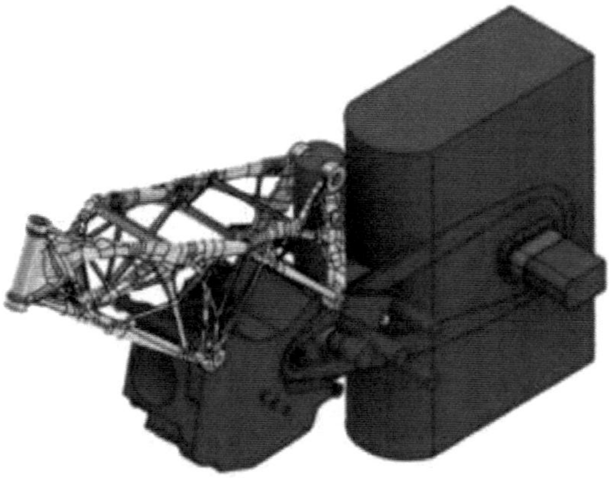

Figure 1: Illustration of Preserve regions and Obstacle geometries.

2.2.1 GENERATIVE DESIGN PARAMETERS

Once the preserve regions and the obstacle geometries are precisely defined, the study resolution is to be determined. The study resolution indicates the size of the mesh elements to be created. The coarser the mesh, the coarser the result is going to be. Because this study deals with a fairly complex part, the study resolution is set to the finest allowable limit. This ensures a fairly smooth output. Once the study resolution is defined, the materials for the study are selected. Aluminum alloys Al 6061 and Al 7075 are selected as study materials. These two aluminum alloys were selected over steel because other works have averred the superiority of Al 7075 over steel and the focus of this study is to assert that Al 6061 is a better material than Al 7071, for the manufacture of the swing-arm, on the account of gains achieved from light-weighting [4,5].

Characteristics such as deformation, stress developed, factor of safety, and weight are contrasted between materials Al 7075 and AISI 1020 steel [5]. It is observed from the literature that, despite having similar stress developed, the swing-arm made from Al 7075 deflects significantly less (about 71% lesser), has a higher factor of safety (about 18% more), and is significantly lighter (about 64% lighter) when compared to the swing-arm made of Steel. Hence the literature concludes that Al 7075 is an optimal material for the fabrication of the swing-arm.

A topology optimized swing-arm made from Al 7075 is compared to a standard swing-arm made from AISI 1010 steel [4]. It is observed from the literature that the swing-arm made from Al 7075, despite the weight reduction of about 44%, had significantly lower stress developed (up to 21% lesser), a notable increase in factor of safety, and significantly lower deflection. Hence, the literature concludes that the swing-arm made from Al 7075 is quite safe under static running conditions.

MANUFACTURE OF SWINGARM: -

Components of a motorcycle have to be stiff to support the mass and transfer loads with minimum losses. Losses usually occur when the component deforms under load. This is extremely undesirable and dangerous because it is extremely difficult to predict the behavior of a deformed component and might cause unpredictable failures. These components also have some amount of flex in them in certain directions. This is because the components have to sustain various vibrations generated while the vehicle is in motion. If enough flex is not provided, the components might fracture/snap under these immense vibrations. This is especially serious in the case of critical components such as swing-arms, suspension rods, etc. Hence for this study, the swing-arm of the motorcycle doesn't have to as rigid as possible and can have a limited amount of flex.

Once the materials are assigned, the study objectives are determined. The study objectives include the factor of safety (or safety factor) and the manufacturing method. The safety factor is a measure of the strength of the system that expresses how much stronger a system is than it needs to be for an intended set of loads. A target safety factor of 2 is set for this study. There are a plethora of manufacturing processes to choose from, but the two main processes under consideration are Additive Manufacturing and Milling. Since this is a fairly complex component, there would be increase preprocessing and postprocessing steps involved in the case of Additive Manufacturing. The volume of the component, because it needs to be created as a single component, would also prove as a deterrent for the additive manufacturing process. Hence this study focuses on the manufacturing of this part through milling. Due to the extreme complexity and the intricate nature of the part, the 5-axis milling is assigned as the manufacturing method.

3. DESIGN CALCULATIONS

Once the manufacturing method is defined, the constraints and loading conditions are defined. These are vital for the generation of the component because they help in the analysis of the component during the simulation and ensure that the component functions as intended.

3.1. WEIGHT AND LATERAL LOADING

The curb weight of the motorcycle is 250 kg and considering only one person (average weight of 75 Kg) is riding the motorcycle, the dead-weight of the motorcycle is 325 Kg. The weight distribution ranges from 58% - 65% on the rear axle in most motorcycles. Hence for this model, the weight distribution is taken to be 60% on the rear axle.

The net load on the rear axle of the motorcycle is
$L_S = 0.6*(M_T + M_P) = 0.6*(250+75) = 195$ Kg.

This load will be distributed on the single side of the swing-arm and will be acting at the damper at an angle of 88°. Thus, the vertical and horizontal components of the load are

Vertical load (L_{VS}) = $L_S \sin\theta = 195*9.81*\sin(88°) = 1911.8$ N
Horizontal load (L_{VH}) = $L_S \cos\theta = 195*9.81*\cos(88°) = 66.8$ N

The maximum acceleration on the motorcycle was found to be 5.5 m/s^2 and with a total mass (M_S) = 325Kg, the longitudinal force acting on swing-arm is,

$F_A = M_S \text{ x } accl^n = 325*5.5 = 1787.5$ N

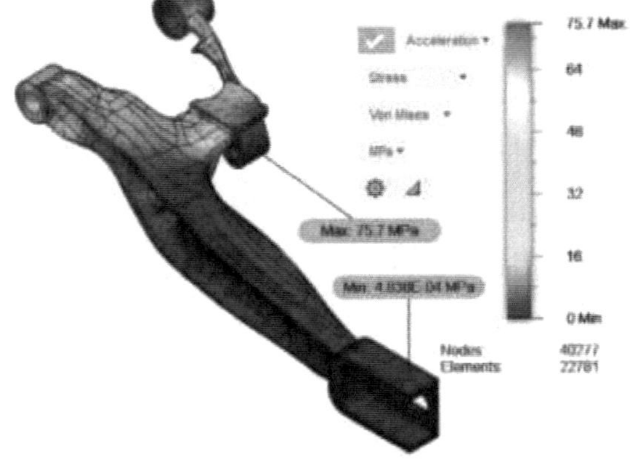

Figure 2: Von Mises Stress developed during acceleration.

The maximum deceleration on the motorcycle was found to be -6 m/s^2.
$F_B = M_S$ x deccl^n = 325*-6 = -1950 N (Negative sign indicates force acting on the opposite direction)

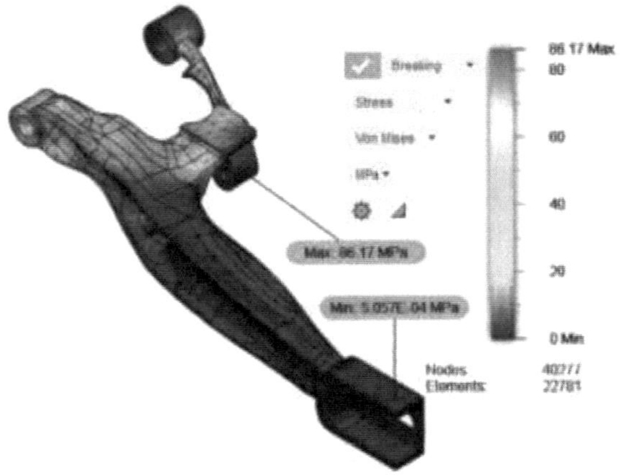

Figure 3: Von Mises Stress developed during deceleration.

3.2. CORNERING CONDITIONS

Cornering conditions are one of the most important criteria during the design of an automobile as different components are subjected to loads of different magnitudes acting in different directions. In the case of the swing-arm, high lateral forces act in an unbalanced state whose magnitude of variation depends on the angle of inclination and the speed of the motorcycle. It is assumed that 20% more load will be transferred to the inner side of the swing-arm while navigating a corner and hence the inner side of the beam has to have 70% of the total weight and the outer side of the beam will have the remaining 30%. A maximum lean angle of 40° is considered for cornering conditions. The forces, when divided into vertical and horizontal components, lead to a torsional and lateral imbalance in the middle part.

Thus, 70% of the total weight is
$F_{MAX} = 0.7*M_S*g = 0.7*325*9.81 = 2231.8$ N

And the remaining 30% is
$F_{MIN} = 0.3*M_S*g = 0.7*325*9.81 = 956.5$ N

Horizontal components (acting as lateral imbalance):

$F_{IH} = F_{MAX}\cos\theta = 2231.775 * \cos(40°) = 1709.6$ N
$F_{OH} = F_{MIN}\cos\theta = 956.4750 * \cos(40°) = 732.7$ N

Vertical components (acting as torsional imbalance):
$F_{IV} = F_{MAX}\sin\theta = 2231.775 * \sin(40°) = 1434.6$ N
$F_{OV} = F_{MIN}\sin\theta = 956.4750 * \sin(40°) = 614.8$ N

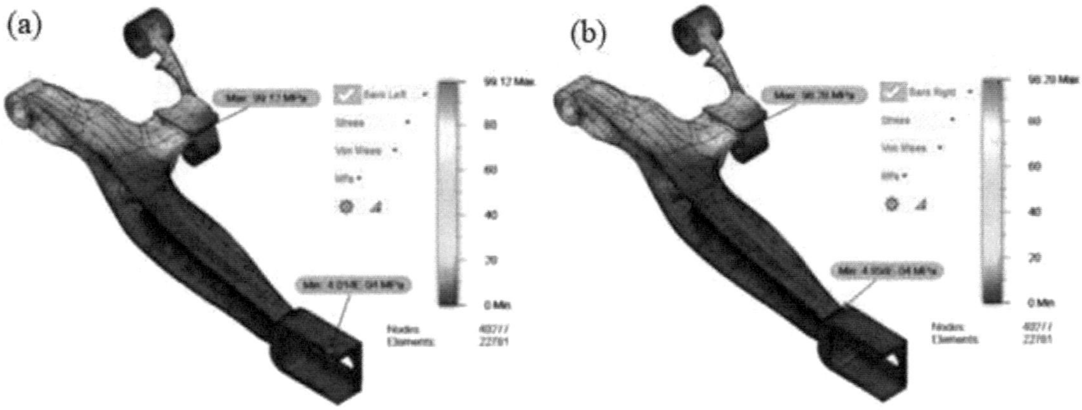

Figure 4: Von Mises Stress developed while navigating Left Turn (a) and Right Turn (b).

4. RESULTS AND DISCUSSIONS

4.1. SOLVING A GENERATIVE DESIGN STUDY

The core mechanism of Generative Design is a level set approach to topology synthesis. Level set methods, by their nature, use the surface area as their main acting stage. This means that changes in surface area to the starting shape (whether created by the solver or specified by the user) will alter both performance and outcomes. The solver inside generative design can explore a much larger design space than a traditional topology optimization system as it is currently based on linear-static FEA methods. By leveraging complex or exotic starting shapes and sizes, the design space can be expanded into new territories (all of which are based on the same problem but with varying starting shapes). The generative design study and analysis are carried out using AUTODESK Fusion 360. In this study, the design is fairly complex as the solver has to build around the prescribed obstacle regions. Owing to the complexity, the process of exploring all design facets is fairly time-consuming.

4.2. COMPARING OUTCOMES AND SELECTING APPROPRIATE GENERATED DESIGN

Once the outcomes are generated, the critical properties such as volume, mass, maximum permissible deflection, and the max permissible Von misses stress are compared between both the swing-arms. The critical properties are those properties that are of interest in the process of light-weighting.

Table 2. Properties Comparison of swing-arm components generated from Al 6061 and Al 7075

Component		
Material	Aluminum 6061	Aluminum 7075
Manufacturing method	5 - Axis Milling	5 - Axis Milling
Volume	1.06e+6 mm^3	1.88e+6 mm^3
Mass	2.9 kg	5.3 kg
Max permissible Von Mises Stress	137.5 MPa	72.5 MPa
Maximum permissible displacement	5.67 mm	2.04 mm
Guideline safety factor	2	2

The properties of both the swing-arm components generated from Al 7075 and Al 6061 are tabulated and the critical properties are highlighted.

It can be observed from the table that the component generated from Al 6061 is **45.7% lighter**, **43.5% smaller** than the component generated from Al 7075, and allows a max deformation up to 5.67mm. This is illustrated by Figure 5.

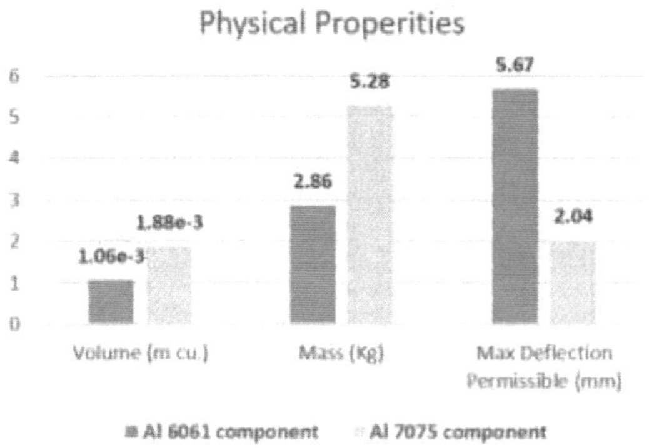

IOP Conf. Series: Materials Science and Engineering 1128 (2021) 012016 doi:10.1088/1757-899X/1128/1/012016

Figure 5: Comparison of physical properties of generated components.

It can also be observed from the table that, by virtue of having a greater permissible von mises stress limit, the component made from Al 6061 is stronger than the component made from Al 7075. This is illustrated as a graphical representation by Figure 6. The max permissible von mises stresses were derived from the volumetric analysis of the components made from Al 6061 and Al 7075. The increase in permissible von mises stress limit in the component generated from Al 6061 is attributed to its increased complexity and organic nature when compared to the component generated from Al 7075. This is because of the better formability possessed by the aluminum alloy Al 6061 as it has a lower hardness and a lower tensile strength compared to the aluminum alloy Al 7075.

The increased complexity and the enhanced organic nature of the component generated from Al 6061 ensure that it loads each of its support members along their strongest axis and allows the stress generated under loading to spread more effectively along its body, thus reducing the stress concentration developed at any particular point as opposed to the component generated from Al 7075. Hence, the component generated from Al 6061 can withstand higher loads before failure, making it stronger than the component generated from Al 7075.

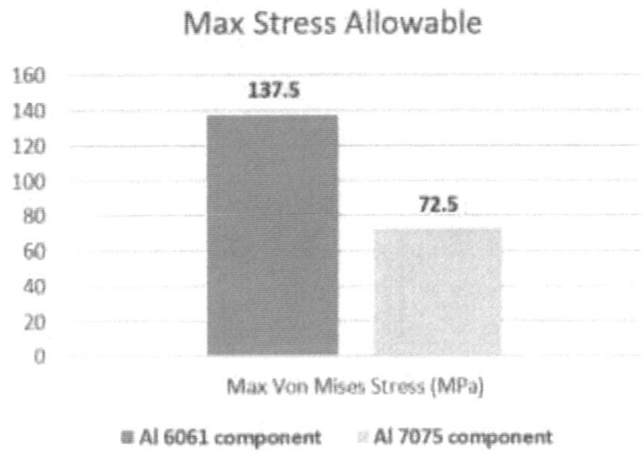

Figure 6: Comparison of Max permissible Von Mises Stress in the generated components.

Predicting on the above statements and observations from Table 2 and Figure 6, it is concluded that the component made from Al 6061 is around 89.7% stronger than the component made from Al 7075, thus making the aluminum alloy Al 6061 a superior material than the aluminum alloy Al 7075 for the manufacture of single-sided swing-arm.

4.3. COMPONENT GENERATION

Based on what is inferred from the table and graphs presented above, it is agreed that the swing-arm should be made of Al 6061. Since there are a few modifications to be done before the design could be machined, the Al 6061 outcome is exported as a component (Figure 7). These modifications include making slots for mounting the wheel assembly and a small change in the surface geometry of the linkage mount to allow for ease of movement under load.

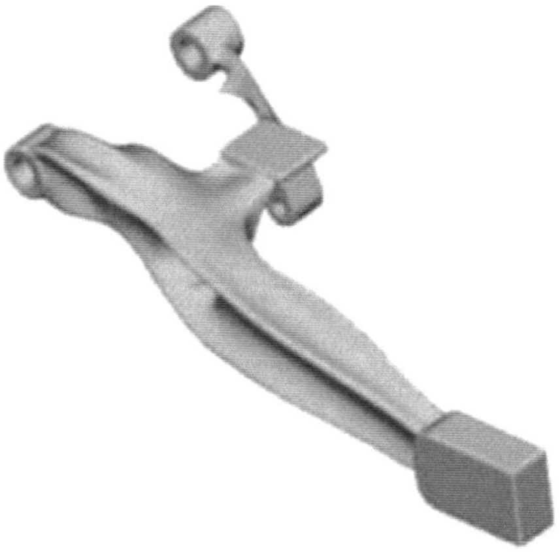

Figure 7: Component View of single-sided swing-arm.

The modifications to the component geometry and the meshing are done before the analysis. This is illustrated in Figure 8. The analysis is carried out on the single-sided swing-arm component made from Al 6061 based on the calculations presented under "DESIGN CALCULATIONS"

Figure 8: Modified and Meshed single-sided swing-arm component.

After the analysis, it is observed that the stresses developed under simulated loads were within the maximum permissible Von Mises Stress.

The Von Mises Stress developed under

i. Lateral Loading conditions are illustrated by Figure 2 for acceleration and by Figure 3 for deceleration.
ii. Cornering conditions are illustrated in Figure 4.

Figure 9 illustrates the observation that the maximum stress developed in the swing-arm component made from Al 6061 was 99.1 MPa under the simulated condition of navigating a left turn, as opposed to the maximum allowable stress of 137.5 MPa.

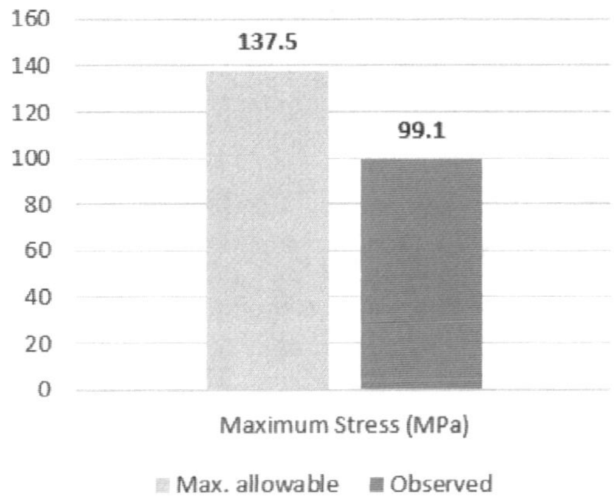

Figure 9: Maximum Stress developed.

4.4. SAFETY FACTOR AND DEFORMATION

Factor of safety (or) safety factor indicates how much stronger apart is that it needs to be. A safety factor of 1 indicates that any loads exceeding the simulated value will cause failure to the component subjected to those loads. Likewise, a safety factor of 2 indicates that the component can handle loads up to 2 times the simulated loads, and so on.

After the analysis, it is observed that the swing-arm reaches the lowest safety factor of 2.77 when it is being subjected to loads generated while navigating left turns. This minimum safety factor of 2.77 is still greater than the targeted safety factor of 2 set for the material Al 6061. The aforementioned statements are substantiated by the illustration given in Figure 10.

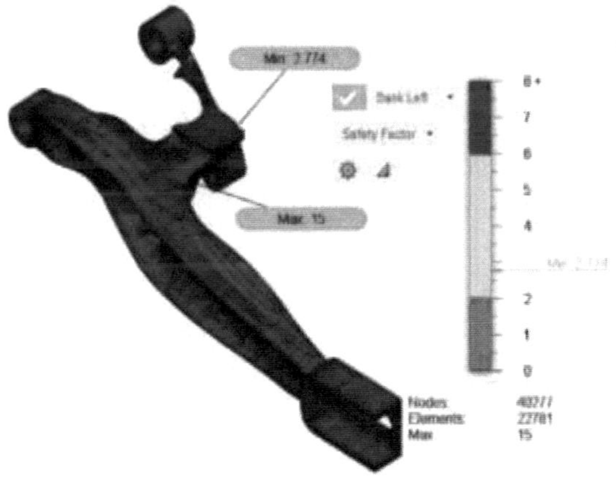

Figure 10: Minimum Safety Factor reached.

It is also observed that the swing-arm experiences a maximum deflection of 0.63 mm when it is being subjected to loads generated while navigating left turns. This maximum deflection of 0.63 mm is still lower than the allowable deflection of 5.67 mm beyond which the component would fail. Figure 8 substantiates the above statements by providing the illustration.

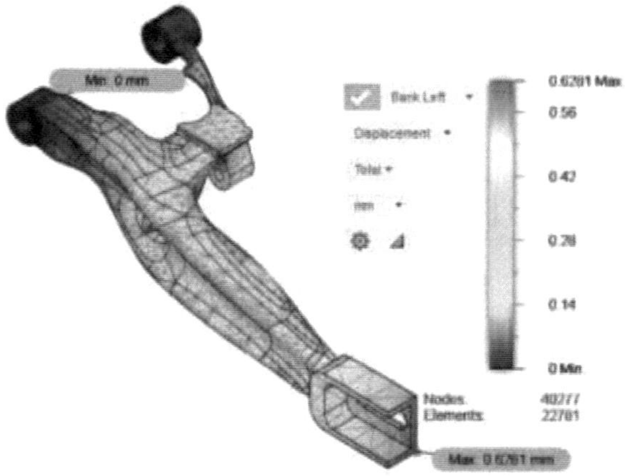

Figure 12: Maximum Deflection observed.

5. CONCLUSION

This work studied the behavior of the single-sided swing-arm generated from the simulated aluminum alloys Al 6061 and Al 7075. The Generative Design Study and the Simulation for this study were carried out in AUTODESK Fusion 360. The mechanical properties of both the swing-arms were compared and it was decided to proceed with the analysis on Al 6061. After the analysis was carried out with Al 6061, the following was observed. Maximum Von Mises stress developed was 99.1 MPa. The minimum safety factor reached by the swing-arm was 2.77. The maximum deflection experienced was 0.63. These values were observed under the conditions that simulate a bike navigating a left turn. The volume and the mass of the swing-arm were found to be around 1.06 mm^3 and 2.9 kg respectively.

The inference from the present work and the literature proves the superiority of Al 7075 over steels and considering the results from the analysis, it could be concluded that Al 6061 is a superior material than Al 7075 and steel for the fabrication of the single-sided swing-arm for intended gains from the process of light-weighting.

6. REFERENCES

[1] Wenlong S, Xiaokai C and Lu W 2016 Analysis of energy saving and emission reduction of vehicles using lightweight materials” *Energy Procedia* 889-893
[2] Cowell A 2018 Looking inside the heart of our championship winning car *Daimler Magazine*
[3] Patil K, Rajpurohit G, Magade R and Shirphule A 2019 Design and analysis of single-sided swing-arm for modified bike *Int. Res. J. of Eng. Technol* **06** 876-879
[4] Powar A, Joshi H, Khuley S, and Yesane D P 2016 Analysis and topological optimization of motorcycle swing-arm, *Int. J. of Curr Eng. Technol.* 270-274
[5] Satyanarayana K, Narendra M, Pavan Sai E, Sri Harsha, Babu G and Rao M 2020 Modelling and analysis of double sided monoshock swing-arm *JCT* **08** 510-519

[6] Schaller B 1996 Experimental validation of finite element analysis software applied to the design of a motorcycle swing-arm (Embry-Riddle Aeronautical University - Daytona Beach)
[7] Cossalter V 2006 Motorcycle dynamics, Second English Edition
[8] Abdullah S H, Ahmed M and Rahman W A 2018 Design of racing motorcycle swing-arm with shape optimisation *Int. J. Sci. Res. & Dev* **06** 179 - 183
[9] Chengzhi Han 2020 *J. Phys.: Conf. Ser.* **1676** 012085
[10] Rosenthal S, Maaß F, Kamaliev M *et al.* 2020 Lightweight in automotive components by forming technology. *Automot. Innov.* **3,** 195–209

Effects of Bracing System on Multistoryed Steel Building

Rahul Kumar Meena[1], G P Awadhiya [2], Abhishek Prakash Paswan[3] and Harshit Kumar Jayant[4]

[1,3 & 4]*Ph.D Scholar. Civil Engineering Department, Delhi Technological University, Delhi-110042*
[2]*Associate Prof. Civil Engineering Department, Delhi Technological University, Delhi-110042*

Abstract: In recent years, in regions of the highly seismic prone area, steel construction has changed from moment resistant frames to concentrically braced frames. To withstand earthquake forces, bracing part of the structural pattern by providing extra stiffness plays a crucial role in structural action. Concentric bracing in building frames is one of the most common horizontal load resistance patterns. For this reason, seven distinct models were created by changing the bracing pattern in the steel frame and analyzed for wind and seismic forces A 20-story steel frame structure is explored in this presented work. From this research, it can be inferred that the bracing factor will have a very significant impact on structural behavior under seismic loading. The most desirable bracing framework is the Backward bracing pattern. Horizontal displacement at the top floor is decreased by approximately 50 percent for Backward braced in frame structure compared to without bracing pattern.

Keywords: *Wind Load, Tall Building, Bracing Pattern, Staad Pro*

1. Introduction

The population is rising day by day, and the supply of land area is much lesser, so the only alternative is multi-story development. Due to earthquake and wind, a tall building is suspected of horizontal loads, and these forces cause storey drift. Various kinds of tall building structures reduce storey drift. Amol and Shelke (2016) analyzed high rise RCC twenty-story building using staad pro on different bracing patterns and concluded that chevron bracing is most effective in reducing bending moments for a wind load in columns. The bending moment can be reduced up to thirty-four percent in the elevation of the building and the plan, and it can reduce up to thirty-seven percent from without bracing pattern, and concluded that the use of a bracing pattern could effectively increase the performance of the building without adding considerable extra dead load into a structure. Sarita singla et al. (2012) examined the pattern of different shapes of building under wind loading. Saurabh and Komal (2018) analyzed the effect of seismic load on the x-type of the bracing pattern using staad pro and found that x-bracing is most efficient to resist story displacement and story drift reduction when bracing is provided on two parallel sides of the building.

Mohammed and Nazrul (2013) simulated the characteristics of RCC multi-story structure on various bracing patterns and concluded that displacement of RCC structure reduces after the use of bracing pattern. Cross bracing reduces the maximum horizontal displacement of the structure. Chadhar and Sharma (2015) considered the sixteen-story building with an alternate bracing arrangement and concluded that steel bracing decreases the shear force and bending moment and concluded that the doubled angle section effectively decreases the horizontal displacement. Amol Patil et al. (2018) examined the eccentric bracing pattern effect on a tall building, and they found that diagonal braced pattern has least horizontal displacement and reduces by fifteen percent concerning without bracing pattern also concluded that v braced pattern has less base shear and reduces by eleven percent as compare to without braced structure. Mishra et al. (2014) examined the reinforced concrete construction frames using various bracing patterns for seismic forces and concluded that steel bracing is an advantage in resisting seismic forces. The bracing pattern is an efficient member to manage the story drift in structure as opposed to bare frames. Shraddha J. Patil and R.S. Talikot (2015) considered wind load on tall elevation structure was examined and concluded that column reinforcement exceeds column reinforcement due to high bending moment compared to static building scenario. The coupling moment induces wind pressure that raises the stresses in the beam. Horizontal wind forces are strong on the building's upper floor, and it generates torsion and displacement. Mayank Walia et al. (2019) considered the effect of wind analysis of composite buildings with bracing patterns and founded that displacement of the twelve-story irregular building having x bracing as most suited to resist horizontal displacement.

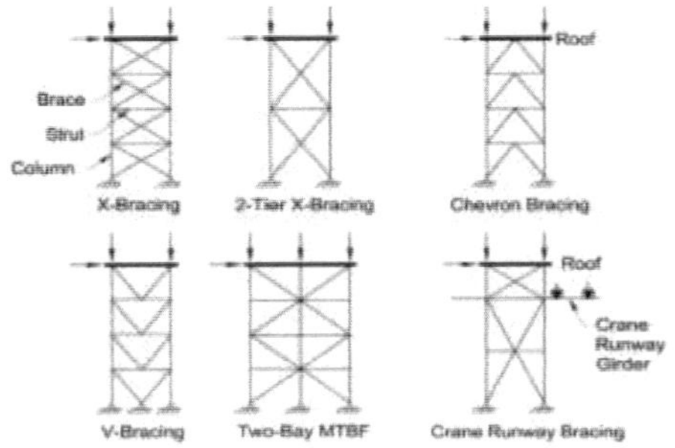

Figure 1 Concentrically Braced Frames

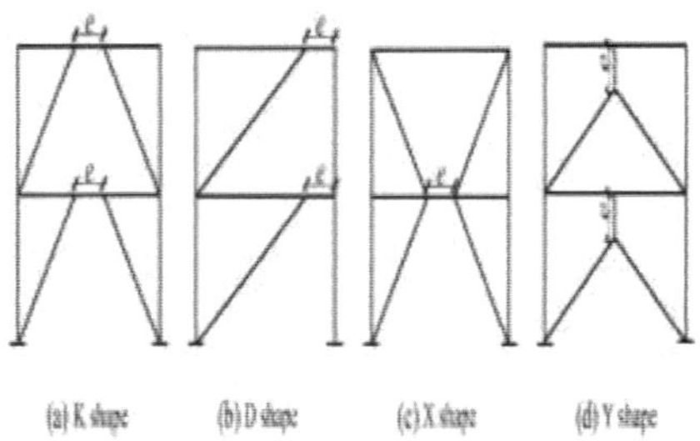

Figure 2 Eccentrically Braced Frames

2. Design Horizontal Forces

Unlike in the current design codes, the design horizontal force distribution in the proposed method is determined using a shear distribution factor, βi, which is extracted from and calibrated by detailed nonlinear time-history analyses of the EBF, (figure 2). When subjected to major earthquakes, this distribution of horizontal force accounts for EBF inelastic behavior and can be expressed as (Lee et al., 2004; Chao and Goel, 2005)

$$\beta_i = \frac{V_i}{V_n} = \left(\frac{\sum_i^n w_i h_i}{w_n h_n}\right)^{0.75T(-0.2)} \quad (1)$$

$$F_n = V\left(\frac{w_n h_n}{\sum_{j=1}^n w_j h_j}\right)^{0.75T(-0.2)} \quad (2)$$

$$F_i = (\beta_i - \beta_{i+1})F_n \quad (3)$$

When $i = n,\ \beta_{n+1} = 0$

Where β_i = shear distribution factor at the level, i

V_i, V_n = story shear forces at level *i* and at the top (*n*th) level, respectively

w_i, w_j = seismic weights at level *i* and *j*, respectively

h_i, h_j = heights of levels *i* and *j* from the ground, respectively

w_n = seismic weight of the structure at the top level

h_n = height of roof level from the ground

T= fundamental structure period obtained by code specified methods or elastic dynamic analysis

F_i, F_n = horizontal forces applied at level *i* and top-level *n*, respectively

V= Design shear force

3. Methodology

This paper analysis carried to find the displacement response of space steel frame under seismic and wind loadings in different bracing patterns. Also, determine the horizontal drift at each floor level for Concentric Braced Frames (CBFs) for fully restrained column base. Loading consideration is as per the Indian Standard codes

Dead load is determined by using considered components and their weight density. assumed live load as 4.0 kN/m^2. Wind load calculated as per IS 875 part 3 (2015),

$$P_Z = 0.6(V_Z{}^2) \tag{4}$$

$$V_z = V_b \times K_1 K_2 \times K_3 \times K_4 \tag{5}$$

Where P_Z is wind pressure design in N/m^2 at height, z and Vz is wind velocity design in m/s at height z, and Vz determined by simple wind velocity accounting V_b probability factor(k_1), terrain height and structure factor(k_2), topography factor(k_3) and Importance factor for the cyclonic region (k_4). The seismic load is measured using a static method equivalent to:

Base shear:

$$V_B = (A_h \times W) \tag{6}$$

Where, A_h = design horizontal seismic coefficient for a structure, W = seismic weight of the building

$$A_h = \left(\frac{ZIS_a}{2Rg}\right) \tag{7}$$

Z, in Table 2 of IS 1893:2016, is the zone factor (part 1). I is the important factor and R is the reduction factor of response; Sa/g is the average acceleration coefficient of response for rock and soil sites as defined in IS 1893:2016 (part 1). The values for the 5 percent damping of the structure are given. According to this equation, base shear distribution along the height is carried out

$$Q_i = V_B \times \frac{W_i h_i{}^2}{\sum_{j=1}^{n} W_j h_j{}^2} \tag{8}$$

Where,
Q_i = Design horizontal force at floor i, W_i = Seismic weight of floor at i, h_i = Height of floor i measured from the base, and n = Number of the story in the building at which the masses are located. Load taken into consideration are Seismic Loading, Zone factor – 0.24 ,Soil type – 2 (medium Soil),Importance Factor – 1.50 ,Response reduction – 4.00 ,Foundation Depth- 3.00 meter And Damping Ratio - 0.02 For Wind loading Basic Wind speed – 47.00 m/s ,Terrain category – 2 ,Class – B ,Probability Factor 'k1' = 1.07 , Dead load is 3.88 Kn/m^2 Live load is 4.0 Kn/m^2.

Load Combinations

1) 1.5 (DL+ IL)
2) 1.2 (DL+ IL $\pm$ EL)
3) 0.9 DL$\pm$ 1.5 EL
4) 1.2 (DL+ IL $\pm$ WL)
5) 0.9 (DL$\pm$ 1.5 WL)

4. Analysis of Models

The building used for this study was designed to the standards presented in the IS 800: 2007 and is 1893 (part 1): 2016 In STAAD Pro, a multi-story steel building is analyzed. The building's design depends on the least specifications, as mentioned in the Indian Standard Codes.

4.1 Steel Frames

The frame used for this study is a 20 (G+19) story, steel braced structures. The typical floor height is 3 m with a total of 60 m of the building. In the plan, the sides span 20 meters by 20 meters divided into 5-meter square bays as represented in figure 6

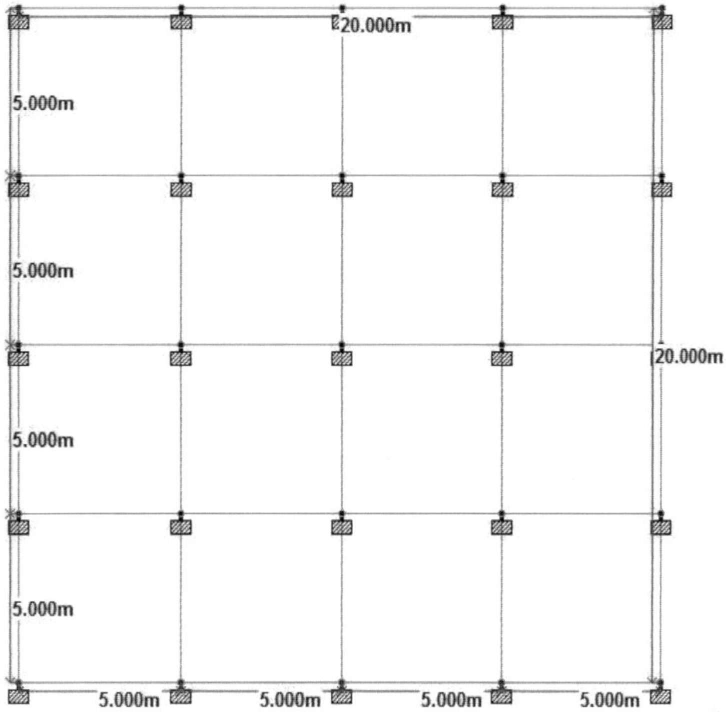

Figure 3 Structural Floor Plan of Steel Concentric Frame

Below mentioned models are considered for the analysis:

1. Steel frame without bracing
2. Steel frame with backward bracing
3. Steel Frame with forward bracing
4. Steel frame with X- bracing
5. Steel frame with inverted V- bracing
6. Steel frame with K- bracing
7. Steel frame with V- bracing

Model 1:- Steel frame without bracing

The steel section chosen for the column is ISMC 400 and for beam ISMB 600. The elevation view of the concentric frame is represented in figure 4. 4.The elevation view in 3D is represented in figure 5 and the storey displacement in x and z direction for without bracing is represented in figure 6.

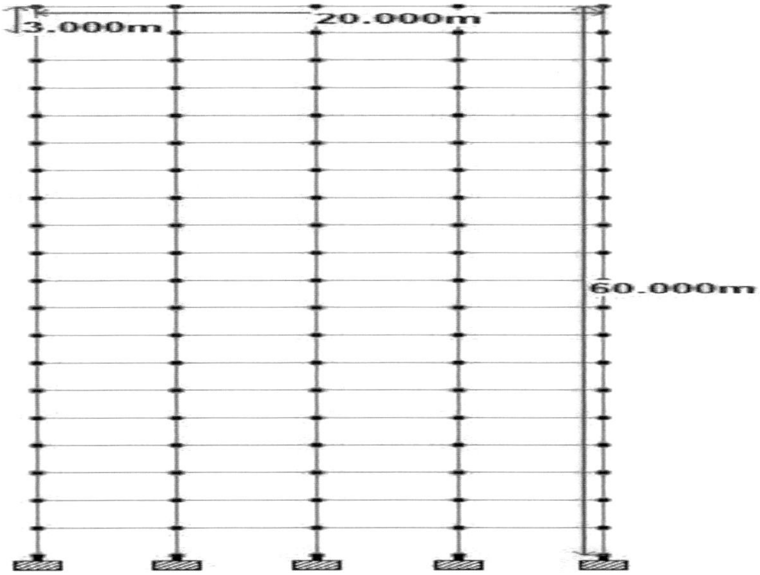

Figure 4 Elevation of the Steel frame without bracing

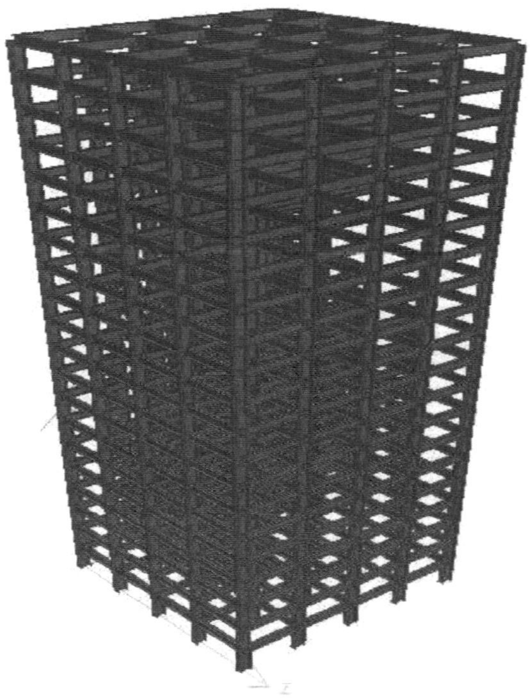

Figure 5 3D view of steel building without bracing

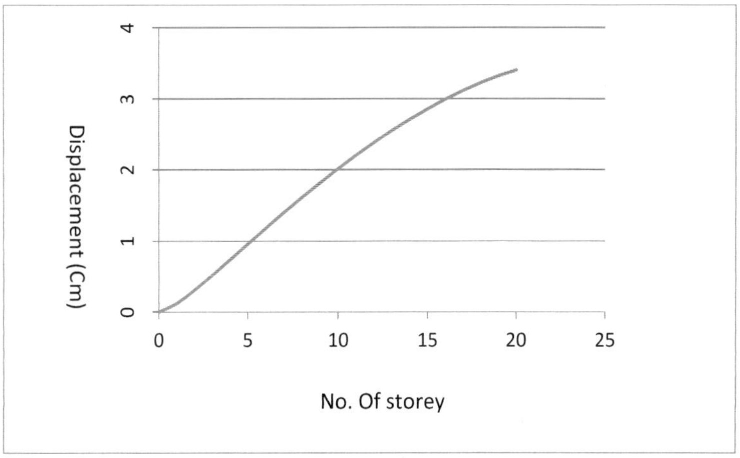

Figure 6 Storey displacement in X & Z - direction without bracing

Model 2: - Steel frame with backward bracing

All supports are fixed and the steel section used for beams is ISMB 600 and columns are ISMC 400 and for the bracing pattern it is ISA 200×200×12. The elevation view of the frame with backward bracing is represented in figure 7.the storey displacement in x and z direction for backward bracing is represented in figure 8.

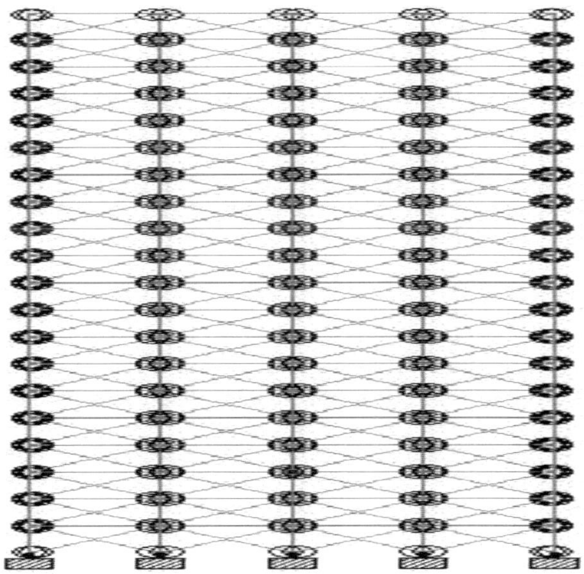

Figure 7 Elevation of Backward Bracing

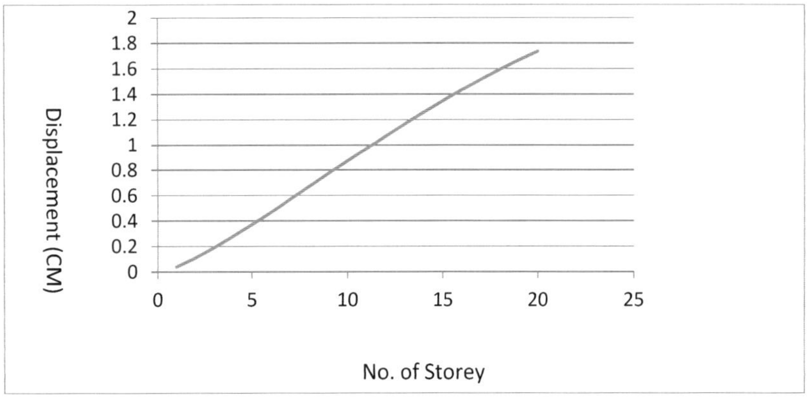

Figure 8 Storey displacement in X & Z-direction Backward Bracing

Model:- 3 Steel Frame with forward bracing

All supports are fixed and the steel section used for beams is ISMB 600 and columns are ISMC 400 and for the bracing pattern it is ISA 200×200×12. The elevation view of the frame with forward bracing is represented in figure 9. the storey displacement in x and z direction for forward bracing is represented in figure 10.

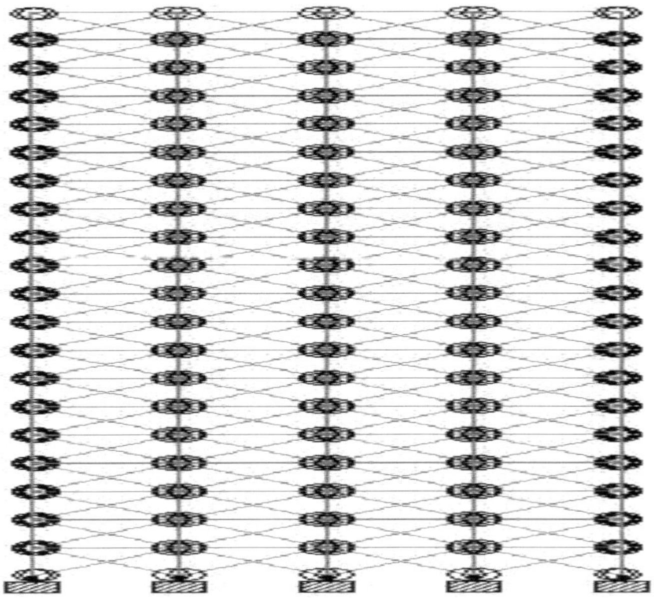

Figure 9 Elevation of Forward Bracing

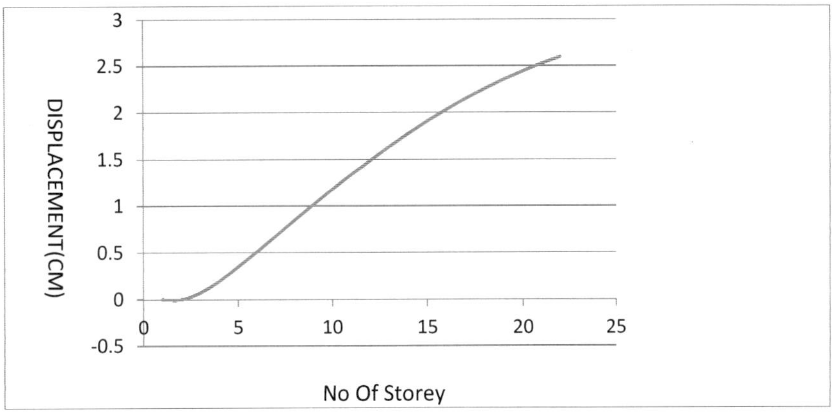

Figure 10 Storey displacement in X & Z-direction Forward Bracing

Model 4:- Steel frame with X- bracing

All supports are fixed and the steel section used for beams is ISMB 600 and columns are ISMC 400 and for the bracing pattern it is ISA 200×200×12. Elevation view of the frame with X- bracing as represented in figure 11. the storey displacement in x and z direction for x-bracing is represented in figure 12.

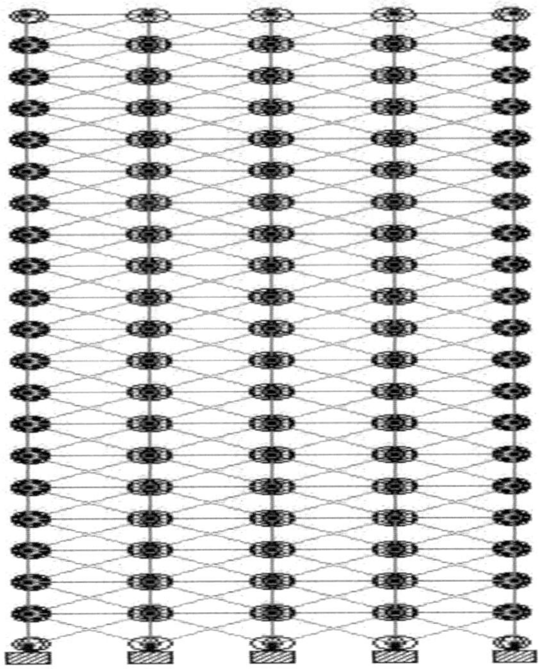

Figure 11 Elevation of the Steel frame with X- bracing

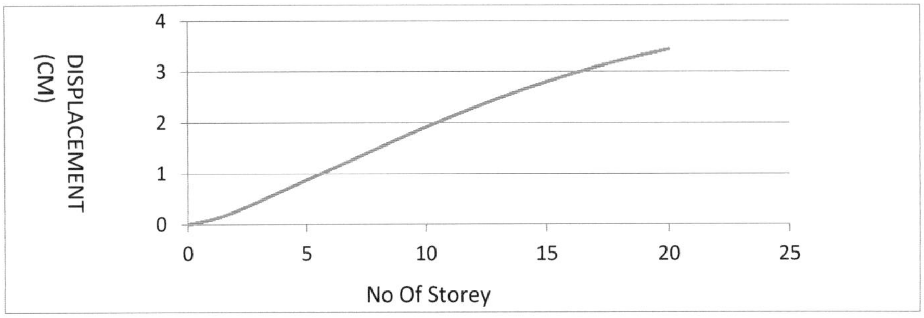

Figure 12 Storey displacement in X & Z - direction with X- bracing

Model 5:- Steel frame with inverted V- bracing

All supports are fixed and the steel section used for beams is ISMB 600 and columns are ISMC 400 and for the bracing pattern it is ISA 200×200×12. Elevation view of the frame with INVERTED V-bracing as represents in figure 13. the storey displacement in x and z direction for v-bracing is represented in figure 14.

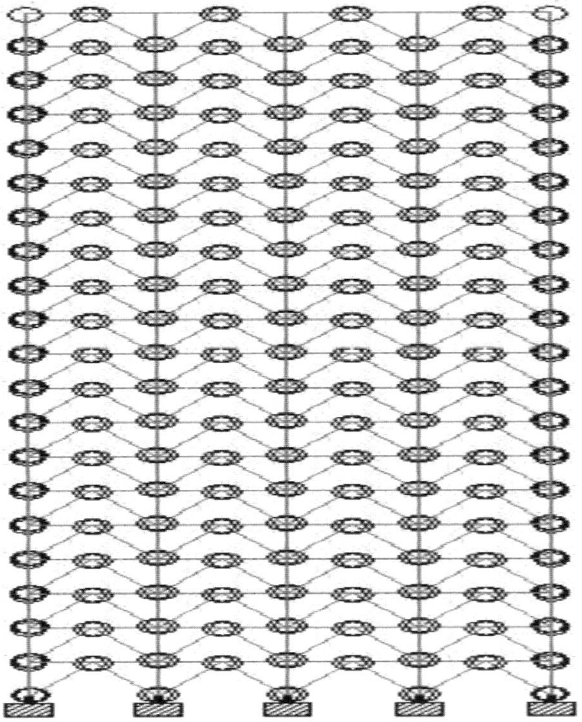

Figure 13 Elevation of the Steel frame with inverted V- bracing

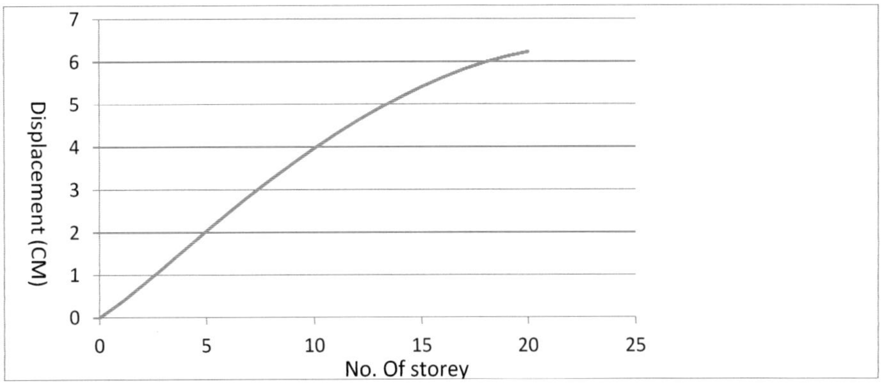

Figure 14 Storey displacement in X & Z- direction with inverted V- bracing

Model 6:- Steel frame with K- bracing

All supports are fixed and the steel section used for beams is ISMB 600 and columns are ISMC 400 and for the bracing pattern it is ISA 200×200×12. Elevation view of the frame with K- bracing as represented in figure 15. the storey displacement in x and z direction for k-bracing is represented in figure 16.

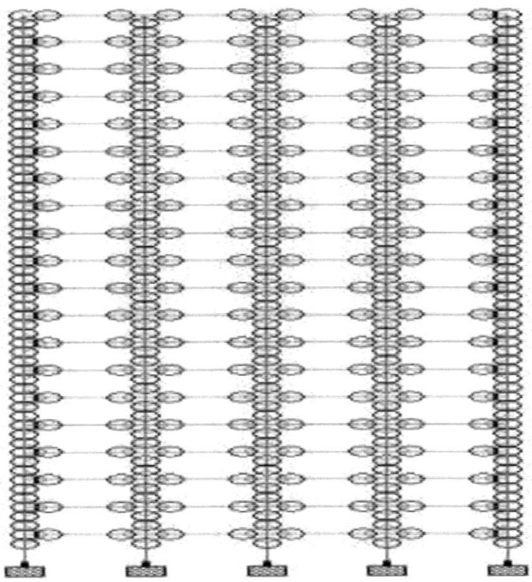

Figure 15 Elevation of the Steel frame with K- bracing

K-bracing is symmetric for steel building as represented in the figure 15 .

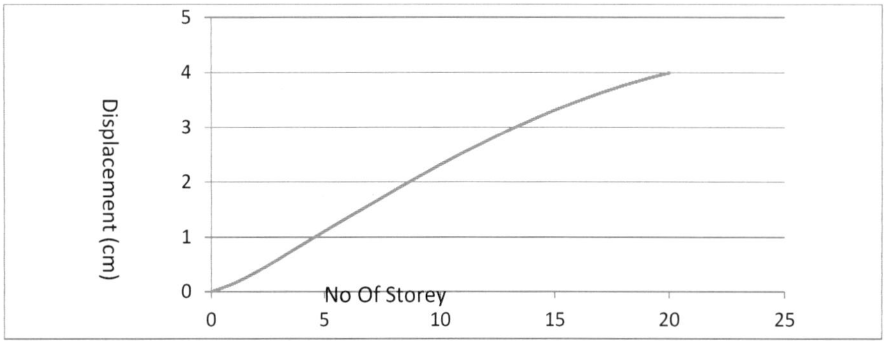

Figure 16 Storey displacement in X & Z - direction with K- bracing

Model 7:- Steel frame with V- bracing

All supports are fixed and the steel section used for beams is ISMB 600 and columns are ISMC 400 and for the bracing pattern it is ISA 200×200×12. Elevation view of the frame with V- bracing as represented in figure 17. the storey displacement in x and z direction for v-bracing is represented in figure 17.

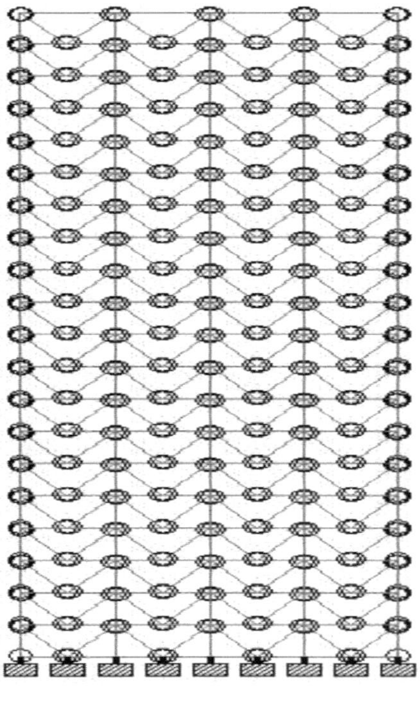

Figure 17 Elevation of the Steel frame with V- bracing

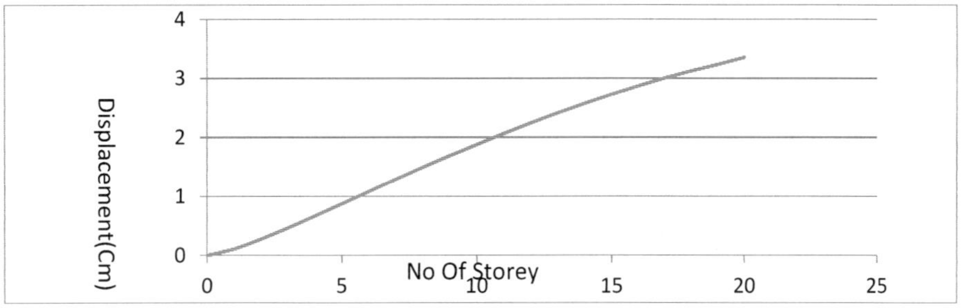

Figure 18 Storey displacement in X & Z - direction with V- bracing

We can also see in this case that top story displacement for inverted V-bracing is less as compared to without bracing but more as compared to X-bracing.

5. Results:

5.1 Displacements:

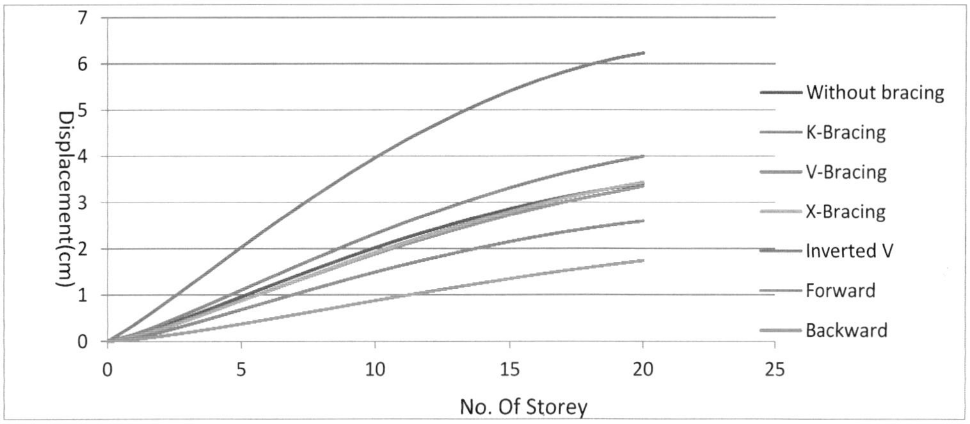

Figure 19 Comparison of displacements in X & Z-direction for different bracing pattern

Figure 19 illustrates that as the number of floors increases, displacement increases without bracing and with bracing. The maximum top story displacement for without bracing is maximum that is 6.2869 cm and the least for X-bracing which is 1.2491 mm.
Displacement on the first floor for without bracing is 4.802 mm and for X-bracing, it is 1.52 mm.

IOP Conf. Series: Materials Science and Engineering 1128 (2021) 012017 doi:10.1088/1757-899X/1128/1/012017

5.2 Storey Shear :

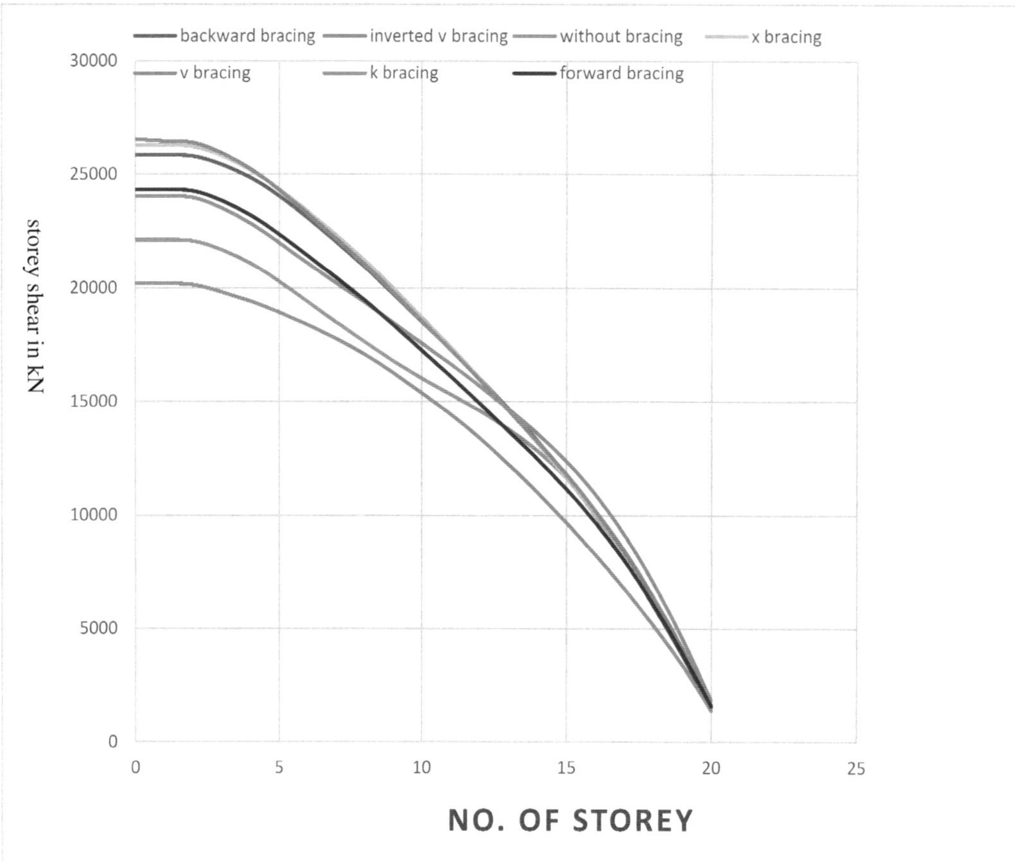

Figure 20 Comparison of Storey Shear for different bracing pattern

The maximum storey shear for a steel building is in v bracing and the least for inverted v bracing is shown in figure 20.

5.3 Shear Force:

Table 1 Maximum Shear Force

Models	Shear Force(kN)
Without bracing	37826.570
X-bracing	59496.387
Inverted-V bracing	45395.121
K-bracing	36305.188
V-bracing	42193.262
Backward Bracing	71376.016
Forward Bracing	46996.742

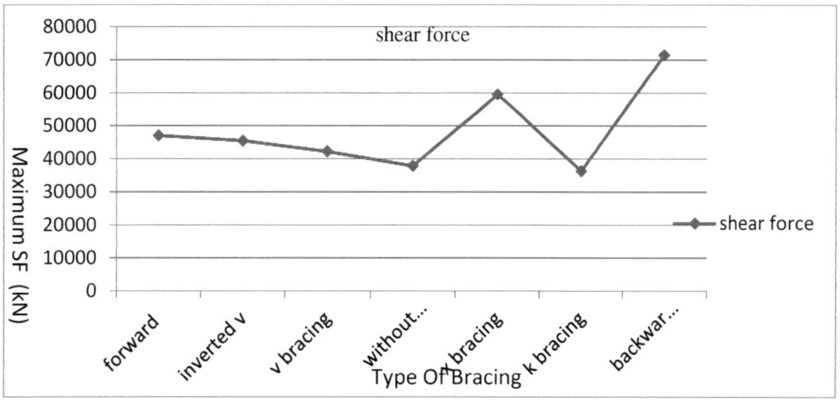

Figure 21 Comparison in Shear Force for different bracing patterns

The maximum shear force for a steel building without bracing and with bracing is represented in figure 21 and table 1 Maximum shear force found for backward bracing is 71376.016 kN and the least for without bracing is 37826.570 kN.
For Backward -bracing, the shear force is 71376.016 kN, and the difference between without bracing and backward bracing is less.

5.4 Bending Moments:

Table 2 Bending Moments

Models	Bending moments(kN-m)
Without bracing	7459.444
X-bracing	8551.276
Inverted-V bracing	6867.884
K-bracing	5784.231
V-bracing	7021.314
Backward bracing	10266.527
Forward bracing	10606.850

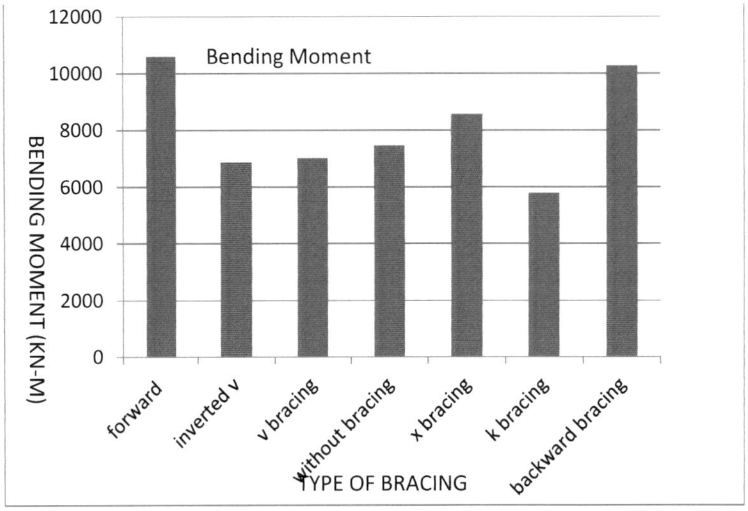

Figure 22 Comparison in Bending moments for different bracing patterns

Maximum bending moment M_y and M_z (whichever is greater) is represented in figure 22 and table 2 for steel building without bracing and with bracing. The maximum bending moment found for Forwarding bracing is 10606.850 kN-m and least for K-bracing which is 5784.231 kN-m.
For Forward bracing, the maximum bending moment is 10606.850 kN-m.

6. Conclusion:

A general analysis of structural structures for tall buildings has been presented in this article. A pattern-based broad classification has been proposed, unlike the height-based classifications in the past. Whenever Bracings are supported with a structure even if it is concentrated or eccentric, it gives more resistance to horizontal deflection and also beneficial in areas that are vulnerable to earthquakes. The following conclusions may be drawn based on the present research.:

- from all bracing patterns, the maximum shear force is provided by the inverted backward bracing pattern.
- In contrast to the other bracings, the forward bracing pattern produces maximum bending moment.
- The most appropriate bracing scheme is the method of backward bracing.
- For Backward braced in frame construction, horizontal displacement at the top floor is reduced by approximately 50 percent compared to without the bracing pattern.

Future research can be done for braces that are likely to develop substantial bending moments and shear forces in specific applications. The impact of bending moment and shear force on braced frame behaviour is unknown, and in order to simulate this, the empirical analysis may be carried out.

7. Acknowledgment:

The authors are greatly thankful to Prof. S. Anbukumar, Associate Professor G.P. Awadhiya and Dr RituRaj, Mr. Manvendra Verma, Mr. Deepak Singh(Ph.D. Scholar, Delhi Technological University), and all the Faculty and Student in the Civil Engineering Department, Delhi Technological University, Delhi without their support this paper, cannot be completed at the time. Mr. Deepak Singh's support is greatly appreciated.

8. Reference:

[1] Amol S. Rajas and N.L. Shelke (2016) *"Wind analysis of high rise building with different Bracing Pattern"*, International Journal of Advanced Research in Science, Engineering and Technology, ISSN: 2350-0328.

[2] Sarita singla, Taranjeet Kaur, *et al* (2012) *"Behavior of RCC tall building having Different Shapes Subjected to wind load"*, International Conference on Advances in Civil Engineering.

[3] Saurabh Kanungo and Komal Bedi (2018) *"Analysis of a tall structure with X-type Bracing considering seismic load using analysis tool staad pro"*, International Journal of Engineering Sciences and Research Technology, 7(12),366-373.

[4] Nauman Mohammed and Islam Nazrul (2013) *"Behavior of multistory RCC structure with Different type of bracing pattern"*, IJIRSET, Vol. 2 Issue 12, Dec 2013, ISSN: 2319-8753.

[5] Shachindra Kumar Chadhar and Abhay Sharma(2015) "*seismic behavior of RC building frame with Steel Bracing Pattern using various Arrangements*" International research Journal of Engineering and Technology Vol.02, Issue 5, Sep. 2015, PP 08-12.

[6] Amol Patil, Dhiraj Deshmukh, Viswajeet Duble, somesh Patil (2018) "*To Study of Efficient Eccentric Bracing Pattern in Tall Steel building*" International Journal of Advanced Research in Science and Engineering Vol.07, Special Issue NO 03, Feb 2018.

[7] Rishi Mishra, Abhay Sharma, Vivek Garg (2014) "*Analysis of RC building Frames for seismic forces using Different Types of Bracing Patterns*" International Journal of Engineering Research & Technology Vol.03, Issue 7, July 2014.

[8] Shardha J. Patil, R.S. Talikot (2015) "*effect of wind load on high rise structure*" International Journal of Recent Technology and Engineering Vol.04, Issue 3, July 2015.

[9] Mayank Walia, Nirbhay Thakur, Nitish Kumar Sharma (2019) "*wind analysis of composite building with a bracing pattern*" International Journal of Innovative and Exploring Engineering Vol.08, Issue 8, June 2019.

[10] IS 1893 (Part 1): 2016, *"Criteria for earthquake resistant design of structure"*.

[11] IS 875 (Part 3): 2015, *"Code of Practice for Design Loads (other than Earthquake) for building and structures"*.

[12] IS 875 (Part 2): 1987, *"Code of Practice for Design Loads (other than Earthquake) for building and structures"*.

[13] IS 800: 2007, "*Code of practice for General Construction in steel"*.

[14] IS 456: 2000, "Code *of Practice for Plain and Reinforced Concrete*"

[15] IS 875 (part 5):1987 "*Special load and Combination*"

[16] IS 875(part 1): 1987 "*Code of practice for Design Loads(other than Earthquake) for building and structure*"

Cohesive Zone Modeling for Predicting Interfacial Delamination in over mold Components

Sai Krishnan M[1], Jeyanthi .S*[2], Pradeep Kumar Mani[3], Hareesh K T[4],M.C Lenin babu[5]

[1,2]School of Mechanical and building science, VIT University, Chennai-600127,Tamilnadu,India.
[3,4]Valeo India Private Limited, CEE DEE YES IT Park, 2nd floor, Block-1, No.63, Rajiv Gandhi Salai, Navalur, Chennai- 600130, Tamilnadu, India. [5]Electric Vehicles Incubation and Testing division and School of Mechanical Engineering,VIT Chennai.

Corresponding author's e-mail address:Jeyanthi.subramanian@vit.ac.in

Abstract. Delamination is one among the significant issues in over molding and electronic packaging. Notably, because of material compatibility and also the weak adhesion-strength of interfaces makes it complex. Typically, delamination takes place at the bimaterial interfaces. In order to predict interface delamination, the cohesive zone technique may be applied as an attainable tool. However, the proper alternative of the cohesive parameters still forms an obstacle for the general usage. The objectives of the work, analytical & numerical modeling of delamination of interfaces between molding compound and metal frames and its subsequent validation using numerical double cantilever beam test and four-point bend tests. In this work, the double cantilever beam and four-point bend tests are used to obtain the strain energy release rate of mode I and mode II values, these values are used as parameters for the cohesive zone method.

1. Introduction

Overmolding is a method in which a single part is producing the usage of or more different materials in combination. Usually the first material, referred to as the substrate, is partially or absolutely covered by next materials (overmold substances) during manufacturing process. Within the overmolding process, the base layer part is molded first, after which the additional plastic layer is molded over and across the original part. The substrate is probably anything like a machined metal component, like threaded inserts, screws, or electrical connectors. In general, interfacial delamination is one of the common failure mechanism in electronic packages, consist of multilayered structures made of dissimilar materials. Interfacial delamination failure includes techniques: crack initiation and crack propagation, numerous experimental setups and theories have been constructed to take a look at these strategies. Amongst that research the cohesive zone approach in various finite element packages are becoming a popular tool for bimaterial interfacial delamination modeling. Cohesive zone modeling for constitute the delamination behaviour of electronic packaging was reported in various literature. In this work, to identify the interface behaviour of cohesive characterization of copper/PBT GF30 bimaterial, through experiment as well as numerical methods. Parameters that define a cohesive-zone model cannot be measured in conventional away [5], and a well labelled method for acquiring the parameters isn't available. Special works have prospered in replicating experimental behaviour [6], but a well-designated technique for counting mixed-mode cohesive zone parameters stays lacking. In this work, the resistance of delamination is characterized by fracture toughness (or) strain energy release rate. To measure the delamination fracture toughness, numerical double cantilever beam (DCB) test is used for Mode I (opening mode), and four-point bend test (FPB) is used for Mode II (shearing mode). In this paper, the mode mix dependency was not considered. Energy release rate values obtained from finite element method of double cantilever and four-point flexure tests are given to the parameters for cohesive zone method which will be explained in the below 3.1 and 3.3 sections.

2. Experiment setup

Test sample consists of the copper lead frame over mold with the PBT GF30 which is thermoplastic material. PBT means polybutylene terephthalate which is grouped under the thermoplastic material family, 30% of glass fiber added that is why it's represented as the PBT GF30, because of extra added glass fiber, the material behaves excellent mechanical strength and great dimensional accuracy in a wide temperature range. Copper led frame consists of total length 70 mm, and width of copper is 9mm and total plastic overmold a length of 55 mm, and also width of plastic material 11mm.

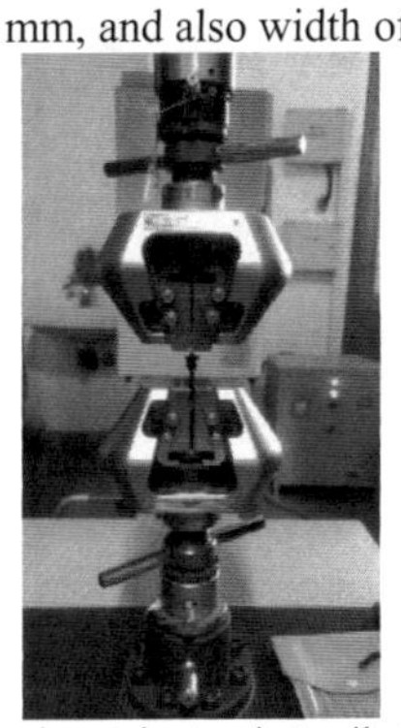

Figure 1 Experimental set up in tensile test machine

The sample is tested in the tensile test machine as shown in figure 1. The sample is vertically placed in the two dies. For bottom die, sufficient pressure is applied for holding purpose, feed rate is given as 2mm/min, load vs displacement data as recorded and shown in figure 2

2.1. Load Vs displacement

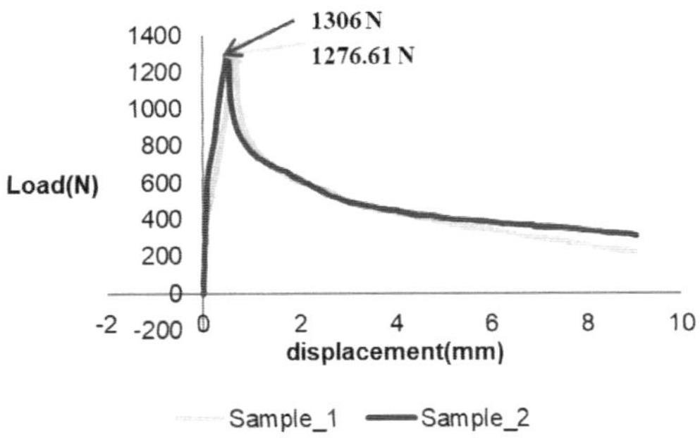

Figure 2 Load Vs Displacement data

From the load vs displacement data, for the sample 1 peak load or critical load observed is 1272.61N. At this critical load separation has started between metal and plastic. For the sample 2 critical load observed at 1306N.From the load displacement data, for the sample 1 and sample 2 delamination separation starts at the average of 1.27kN to 1.3kN. Up to the 1.3kN interface bonding behaves like a linearity, the point at which it loses its stiffness the curve follows the nonlinear behaviour. The following figure 3 shows after and before the test samples.

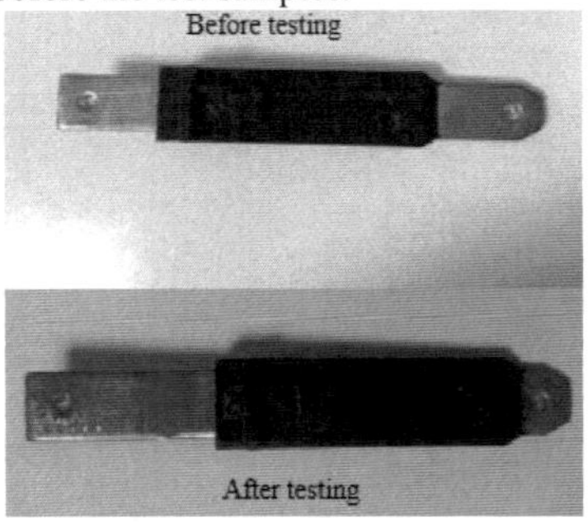

Figure 3 Sample after and before experiment

3. Energy release rate via VCCT

In this work, numerical and analytical bimaterial double-cantilever beam (DCB) and four-point bend(FPB) test have been performed to determine the strain energy release rate for near opening and shear loading conditions for PBTGF30&copper bimaterial interface ,by using the virtual crack closure technique (VCCT).Obtained energy release rate values are assigned as parameters for cohesive layer in between metal and plastic interface and then simulating load displacement data with cohesive-zone finite element modeling.

$$G = -\frac{d\Pi}{dA} \tag{1}$$

Where G is the energy release rate which is calculated from finite element method over the virtual crack closure technique (VCCT) first presented by rybicki and kanninen [9], and in a while enhanced for 3D-cases by krueger [10]. They showed a way for calculating the strain energy release rate in the finite element method. In this work mode mix dependency was not yet considered. VCCT calculates energy-release rate, with the assumption that the energy required to separate a surface is the same as the energy required to close the same surface [10], and also assumes that stress states near the crack tip do not change significantly when the crack grows by a small amount. For 2-D crack geometry with a low-order element mesh, the interfacial toughness is defined as

$$G_I = -\frac{1}{2\Delta a} Ry\Delta v \tag{2}$$

$$G_{II} = -\frac{1}{2\Delta a} Rx\Delta u \tag{3}$$

Where G_I and G_{II} are mode I and II energy-release rate, respectively. Δu and Δv are relative displacement between the top and bottom nodes of the crack face in local coordinates x and y, respectively, and Rx and Ry are reaction forces at the crack-tip node. Δa is crack extension, as shown in the following figure 4.

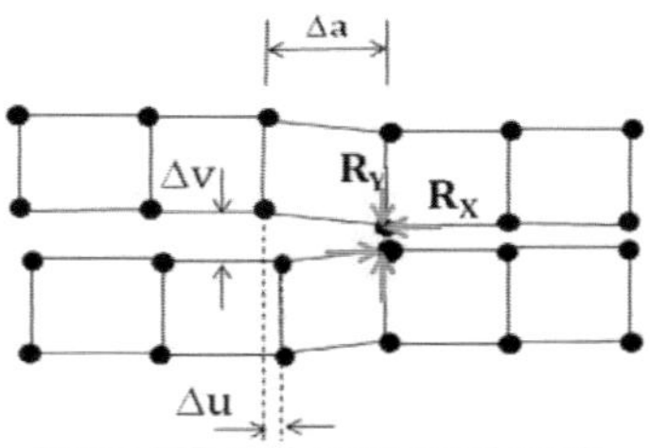

Figure 4 2D crack geometry schematic

3.1. Finite element procedure of double cantilever beam test

The test is used to calculate the initiation and propagation value of opening mode under cyclic as well as static loading situations. The dimensions of finite element model consist, length of 100 mm was considered for both copper and PBT GF30 material. Thickness of copper considered as the 1.5 mm and thickness of PBT GF30 considered as 2.6 mm. Width of 9 mm considered for both copper and PBT GF30, a predefined varying crack was assigned in between copper & PBT GF30 interface. Finite element model of DCB was shown in figure 5.

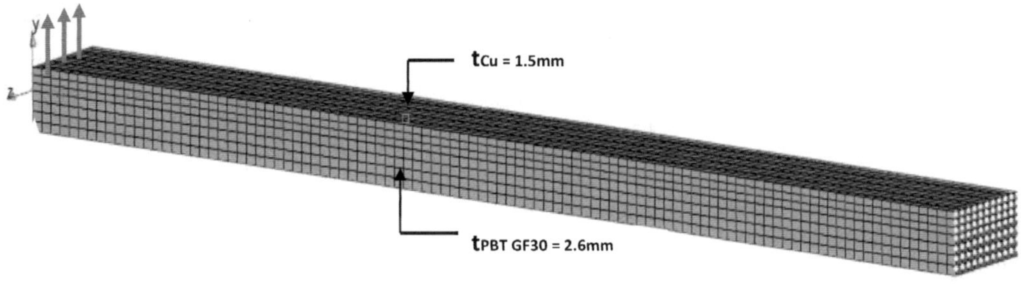

Figure 5 loading and boundary condition of copper and PBT GF30 double cantilever beam specimen

Material properties of copper and PBT GF30 are mentioned below table 1

Table 1 Material properties of bimaterial specimen

Sl. No.	Material	Young's modules(E)	Poisson's ratio(μ)
1	Copper	121GPa	0.33
2	PBT GF 30	9.7GPa	0.35

Proper boundary conditions and loading are essential for any finite element simulation. In this analysis at the right end of the cantilever beam all degrees of freedom constrained, and at the left end of the beam the average value of critical load i,e P_{crt} =1.29 kN is assigned ,for bottom portion of left end of beam all degrees of freedom is constrained. In this simulation we used ANSYS® apdl 18.2 version. For both copper and PBT GF30, structural solid element was used. Some assumptions taking into considered for analysis they are: For PBT GF30 material linear elastic property was considered because at room temperature PBT GF30 behave like elastic material and also delamination process considered to be quasi static process. The numerical simulation is performed for various crack lengths for 1 mm ≤ a ≤ 6 mm. The important consideration about the finite element analysis of DCB test, strain energy of mode I value is varied until the simulated load vs displacement graph closely

IOP Conf. Series: Materials Science and Engineering 1128 (2021) 012018 doi:10.1088/1757-899X/1128/1/012018

matches the experimental data [17]. Average value of strain energy up to crack length of 6 mm obtained by finite element method 7.56 N/mm.

3.2. Double cantilever beam analytical calculation

For a bimaterial interface with layers of distinct materials, a robust analytical answer for strain energy release rate isn't available. Some sources provide derivations from plate theory or stress intensity factor analysis that results in complicated equations [12], the compliance technique can be applied by using treating the thinner layer as a single cantilever beam [14]. A modified compliance technique may produce a great approximation for cracks with various thickness [13]. On this technique, the layers of the DCB specimen are treated as separate cantilever beams, resulting within the following equation from soboyejo et al. for mode I SERR [11].

$$G = \frac{6P^{2}a^{2}}{E_{PBT}\, b^{2}\, t_{PBT}^{3}}\left(1 + \frac{1}{\beta_T^{3}\beta_E}\right) \tag{4}$$

$$\beta_T = \frac{t_{Cu}}{t_{PBT}} \tag{5}$$

$$\beta_E = \frac{E_{Cu}}{E_{PBT}} \tag{6}$$

Where G is the strain energy release rate for opening mode loading, produced by the DCB loading configuration from applied load P and also a is the crack length and b is specimen width. β_E is the stiffness ratio between copper and PBT GF30, and β_T is the thickness ratio. Material property young's modules appears in Table.1 for both materials. Knowledge of crack length a is required to apply this formula, all obtained analytical strain energy release rates values are summarized and average value of strain energy up to 6mm delamination length obtained by analytical method = 7.68 N/mm, which is much less than 3% distinct from the average value computed using VCCT.

3.3. Four point bend test numerical modeling

The four-point bend (FPB) test is one of the popular test to find out strain energy release rate characterization because it gives strong delamination in the interface and does no longer rely on length of the crack. To calculate the strain energy release rate with the FPB test, the VCCT technique was applied using finite element method by using ANSYS®, a 3D model of the FPB test is prepared. For both copper and PBT 30GF 8-node structural solid element was assigned. The dimensions of finite element model consist of the, width of the copper and PBT GF30 considered as the 9mm and thickness of copper and PBT GF30 considered as the 1.5 mm and 2.6 mm respectively. The total length of the model is 100 mm. The crack length is 9 mm on both side of the symmetry line.

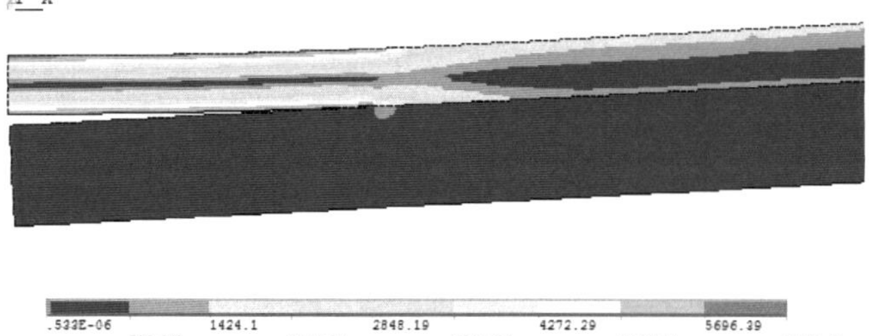

Figure 6 Finite element result of the four-point bend specimen with interfacial crack

The distance between inner supports is 74mm, and the outer rollers separated by 90 mm with force P_{crt}. To decrease the computation time, a one-half symmetry model is considered for simulation. Both copper and PBT 30GF are assumed to be linear elastic at room temperature. Material properties are assigned for this analysis are from Table 1. Nodes of copper at the boundary of symmetry have displacement was constrained in the direction of x. At the narrow support pin, y- directional displacement that is vertical displacement is constrained, the contours of near crack stress field are shown in figure 6.

Initial simulations do not show crack-interpenetrations and stresses are primarily tensile, so contact pairings are not used in the model. The FPB model shows similar stress contours to those at the DCB crack tip. The contours are asymmetric due to the dissimilar layers. The stresses vary with distance and angle from the crack tip. The maximum von mises stress is found to be below the copper yield stress, so the elastic assumption is accepted. Average value of mode II strain energy release rate value obtained from finite element analysis was $G_{II} = 12.96$ N/mm.

3.4. Four point bend test analytical calculation

For the bimaterial interface, strain energy release rate can be calculated by the following equations from Charalambides et al. [15]. I_{Cu} is the area moment of inertia of the copper, while I_C is the area moment of inertia of the overall composite beam. λ is a non-dimensional parameter that gives the stiffness ratio between copper and PBT GF30, FPB test was independent of arbitrary crack length [15], so for this study a crack length of 9mm as considered. Average strain energy release rate is obtained by calculating G_{II} during the steady-state delamination, $G_{II} = 13.1$ N/mm. In this work analytical and finite element method calculations of G_{II} for the FPB test assume a perfect symmetry within the specimen.

$$G_{II} = \frac{(1-\mu_{Cu}^2)P^2 L^2}{8b^2 E_{cu}}\left(\frac{1}{I_{cu}} - \frac{\lambda}{I_c}\right) \tag{7}$$

$$I_{Cu} = \frac{t_{Cu}^3}{12} \tag{8}$$

$$\lambda = \frac{(1-\mu_{PBT}^2)}{(1-\mu_{Cu}^2)} \tag{9}$$

$$I_c = \lambda\, I_{cu} + \frac{t_{PBT}^3}{12} + \frac{\lambda t_{Cu}\, t_{PBT}(t_{Cu}+t_{PBT})^2}{4(\lambda t_{Cu}+t_{PBT})} \tag{10}$$

4. Interface cohesive zone modelling

Interfacial delamination failure include techniques of crack initiation and crack propagation. Numerous experimental setup and theories have been constructed over the years. Amongst that research the cohesive zone approach as being carried out in various finite element packages is becoming a popular tool and also, for cohesive zone method does no longer require starter cracks. Accordingly, Cohesive zone modeling is perfectly suitable for interfacial cracking since such cracks generally tend to propagate alongside the interfaces of a structure, Parameters that describe a cohesive zone model cannot be determine straight away [5], and a well designated method for getting the parameters isn't available yet. So far special works have succeeded in repeating experimental behaviour [6], but a well-described technique for enumerating mixed-mode cohesive zone parameters stays lacking. In this finite element cohesive method. strain energy release rate values gained from the numerical double cantilever beam test and four-point bend test are used has the parameters for the interfacial cohesive elements. In a cohesive zone model, cohesive elements are placed along material interfaces. Science the model is symmetric about the x-axis. In this study one half the model is considered for the simulation, thickness of copper considered as the 0.75 mm and thickness of PBT GF30 taken as the 1.3 mm and length of copper considered as the 70 mm and length of PBT GF30 considered as the 55 mm replicating the original specimen model. The bottom portion of PBT GF30 nodes is constrained in all degrees of freedom, and nodes of copper elements are constrained in all degrees of freedom except in translation y, and also enforced displacement of 2mm/min given to the translation of positive y-direction. The cohesive finite element model shown in the figure 7

Figure 7 Cohesive zone finite element Symmetry Model

For both copper and PBT GF30 finite element models, 8-noded structural solid element was assigned. For mesh, aspect ratio 1 is maintained throughout the mesh geometry. Cohesive elements i.e inter205 elements are placed in between copper and PBT GF30 interface. From finite element model the load-displacement data releases the, critical forces and displacements for delamination as well as initial loading stiffness behaviour.

4.1. Correction factor

In actual sample there is total of 1135 mm^2 mating surface area between metal and plastic but in simulation 86% area only considered due element capability.

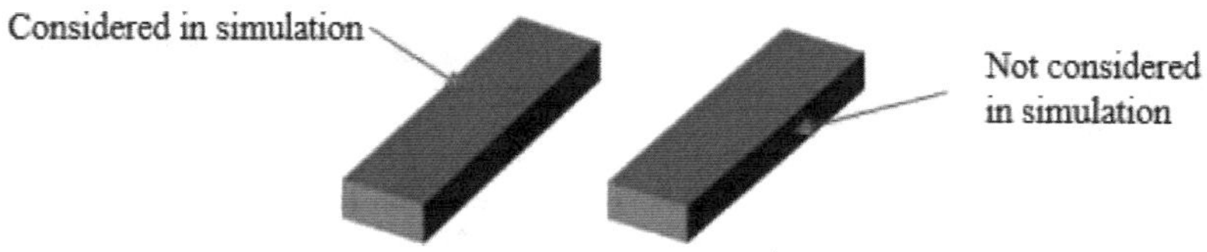

Figure 8 Considered Surface are in simulation

For this reason, we are getting a more deviation. For the actual consideration of total surface area, due to the linearity, width of the FEA model increased from 9mm to 10.5 mm to compensate the total surface area and then performed the same above-mentioned finite element analysis. After the correction factor has done, finite element analysis performed as mentioned above procedure and load displacement data has recorded, this time critical load observed at the 1158.3N which is less than the 10% deviation.

4.2. Experiment Vs Numerical results

From the experiment, average critical load identified from the range of 1.25kN to the 1.3kN. From the simulation, load displacement data critical load observed less than 10% deviation from the experiment which is 1158.3N.

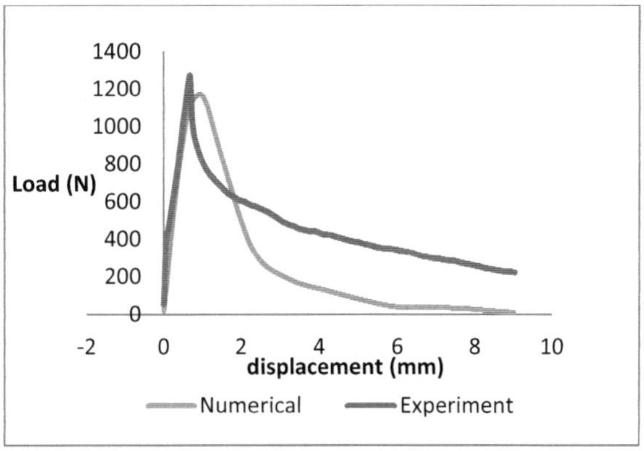

Figure 9 Experiment Vs Numerical results

4.3. Challenges involved in cohesive zone modeling

When cohesive zone method is used with finite element analysis, delamination path must be known priori in order to place the cohesive elements appropriately in a finite element mesh and also cohesive zone elements capable of creating in flat surfaces only. The cohesive zone method strategies are difficult for predicting crack growth inside a solid body under general loading conditions. The cohesive zone method strategies best suited for monotonic loading conditions.

5. Conclusion

An attempt was made for understanding the interface behaviour of copper/PBT GF30 materials. From this work we identified the copper/PBT GF30 cohesive characteristic behaviour, which follows the exponential cohesive behaviour. Based on experimental data, critical failure load identified for the Copper/PBT GF30 interface. Parameters required for cohesive zone method were identified, to do so numerical double cantilever beam and four-point bend test were performed. While performing the DCB and FPB test, critical load as considered from the experiment data. Based on fracture mechanics technique requires the calculation of fracture parameters, which can be extracted from finite element analysis results using the virtual crack closure technique. This study is useful for the better understanding of the interface behaviour of copper/PBT GF30 interface for future models.

References

[1] Naghipour P, Bartsch M, Voggenreiter H. Simulation and experimental validation of mixed mode delamination in multidirectional CF/PEEK laminates under fatigue loading. International journal of solids and structures. 2011 Mar 15;48(6):1070-81.

[2] Biel A, Alfredsson KS, Carlberger T. Adhesive tapes; cohesive laws for a soft layer. Procedia materials science. 2014 Jan 1;3:1389-93.

[3] Zhang B, Yang D, Ernst L. Cohesive zone modeling of mixed-mode delamination test. In2013 14th International Conference on Electronic Packaging Technology 2013 Aug 11 (pp. 559-565). IEEE.

[4] Zhang W, Yang D, Ernst L, Zhang B, Yang W, Cai M. Interface crack propagation between epoxy moulding compound and copper. In2016 17th International Conference on Thermal, Mechanical and Multi-Physics Simulation and Experiments in Microelectronics and Microsystems (EuroSimE) 2016 Apr 18 (pp. 1-8). IEEE.

[5] Mei H, Gowrishankar S, Liechti KM, Huang R. Initiation and propagation of interfacial delamination in integrated thin-film structures. In2010 12th IEEE Intersociety Conference on Thermal and Thermomechanical Phenomena in Electronic Systems 2010 Jun 2 (pp. 1-8). IEEE.

[6] Raghavan S, Schmadlak I, Leal G, Sitaraman SK. Framework to extract cohesive zone parameters using double cantilever beam and four-point bend fracture tests. In2014 15th International Conference on Thermal, Mechanical and Mulit-Physics Simulation and Experiments in Microelectronics and Microsystems (EuroSimE) 2014 Apr 7 (pp. 1-5). IEEE.

[7] Liu S, Mei Y, Wu TY. Bimaterial interfacial crack growth as a function of mode-mixity. IEEE Transactions on Components, Packaging, and Manufacturing Technology: Part A. 1995 Sep;18(3):618-26.

[8] Tran HT, Shirangi MH, Pang X, Volinsky AA. Temperature, moisture and mode-mixity effects

on copper leadframe/EMC interfacial fracture toughness. International Journal of Fracture. 2014 Jan;185(1):115-27.

[9] Rybicki EF, Kanninen MF. A finite element calculation of stress intensity factors by a modified crack closure integral. Engineering fracture mechanics. 1977 Jan 1;9(4):931-8.

[10] Krueger R. Virtual crack closure technique: history, approach, and applications. Appl. Mech. Rev. 2004 Mar 1;57(2):109-43.

[11] Soboyejo WO, Lu GY, Chengalva S, Zhang J, Kenner V. A modified mixed-mode bending specimen for the interfacial fracture testing of dissimilar materials. Fatigue & Fracture of Engineering Materials & Structures. 1999 Sep;22(9):799-810.

[12] Sundararaman V, Davidson BD. An unsymmetric double cantilever beam test for interfacial fracture toughness determination. International journal of solids and structures. 1997 Mar 1;34(7):799-817.

[13] Xiao F, Hui CY, Kramer EJ. Analysis of a mixed mode fracture specimen: the asymmetric double cantilever beam. Journal of Materials Science. 1993 Oct;28(20):5620-9.

[14] Cao HC, Evans AG. An experimental study of the fracture resistance of bimaterial interfaces. Mechanics of materials. 1989 Jun 1;7(4):295-304.

[15] Charalambides PG, Lund J, Evans AG, McMeeking RM. A test specimen for determining the fracture resistance of bimaterial interfaces.

[16] Auersperg J, Dudek R, Michel B. VCCT and integral concepts of bi-material interface fracture in low-k structures—Going to understand relation. In2010 12th Electronics Packaging Technology Conference 2010 Dec 8 (pp. 632-636). IEEE.

[17] Kwatra A, Samet D, Sitaraman SK. Effect of thermal aging on cohesive zone models to study copper leadframe/mold compound interfacial delamination. In2015 IEEE 65th Electronic Components and Technology Conference (ECTC) 2015 May 26 (pp. 1531-1537). IEEE.

[18] Shirangi MH, Wunderle B, Wittler O, Walter H, Michel B. Modeling cure shrinkage and viscoelasticity to enhance the numerical methods for predicting delamination in semiconductor packages. InEuroSimE 2009-10th International Conference on Thermal, Mechanical and Multi-Physics Simulation and Experiments in Microelectronics and Microsystems 2009 Apr 26 (pp. 1-8). IEEE.

IOP Conf. Series: Materials Science and Engineering 1128 (2021) 012019 doi:10.1088/1757-899X/1128/1/012019

Significance of Taguchi Design compared to Response Surface Methodology in predicting optimal machining conditions for PCM process

K A V Uday Kiran[1]*, N Vigneshwar[2], Darius Gnanaraj Solomon[1] and Joseph Jeyapaul Arulraj[3]
[1, 2] Department of Design and Automation, School of Mechanical Engineering, Vellore Institute of Technology, Vellore - 632014, INDIA
[3] Department of Economics and Statistics, University of Dodoma, Tanzania
Corresponding author Email Id: [1]ananta.udaykiran2019@vitstudent.ac.in

Abstract: Photochemical Machining (PCM) is a non-conventional machining process which produces stress free complex flat metal components. Optimization of process parameters for Photochemical machining of SS316L steel is done using Taguchi method and the results are compared with results obtained through Response surface methodology (RSM) available in literature. The objective of this work is to predict the optimal machining conditions of PCM to achieve minimum undercut for SS316L steel. The Factors affecting the response of the model are: Etching Time (Ti), Temperature (T) and Etchant Concentration (C) in PCM. This work shows that Taguchi method is faster compared to RSM in predicting the optimal conditions for minimizing defects in PCM process. Taguchi method requires only 9 experiments for finding optimal conditions whereas RSM requires 20 experiments saving about 55% of time and effort.

1. Introduction

Photochemical Machining (PCM) is a non-conventional machining process which uses the principle of metal erosion for manufacturing. This process is mostly used to fabricate sheet metal components, where the work piece is masked to differentiate the machining area with the surrounding area in order to create metal erosion using chemical etchants only in the selected area. The factors like Etching Time (Ti), Temperature (T) and Etchant Concentration (C) are significant in this process. Machining defects such as undercut arise due to change in the conditions of the factors. An experimental study has been conducted to predict minimal undercut of SS316 L steel with Photochemical Machining process (PCM) using Response Surface Methodology (RSM) [1]. The optimum conditions are found by conducting 20 experiments and by processing the data. In this work, a comparative study is done between Taguchi and RSM to find the optimal conditions to achieve minimal undercut of SS316L steel using PCM process. RSM is a statistical approach for the optimization of manufacturing processes introduced by George E. P. Box and K. B. Wilson in 1951[2] . It uses two design approaches such as Box-Behnken Design (BBD) and Central Composite Design (CCD) to find the optimal parameters affecting the response a machining process. Genichi Taguchi an engineer and statistician from Japan has introduced a method called 'Taguchi Designs' for the optimization problems in manufacturing industry during late 1980's. He stated the use of Orthogonal Arrays (OA), Linear Graphs and Loss functions [3-4] in identifying the optimal conditions of a process with a less number of experiments to be carried out. This process reduces the number of experiments to be carried out for the experimentation.In recent years most of the researchers stated that the Taguchi method is simple and efficient Design of Experiments. Optimal surface roughness of sintered PA 2200 prototypes is evaluated using Taguchi method [5], and this method showed advantages over other methods like RSM and Full Factorial Designs [FFD] in identifying the optimal conditions. Few of the works include optimizing polishing efficiency and torque evaluation in robotic polishing process [6], Optimization of parameters for a welding unit in manufacturing sector and carbon nano tubes stiffened speaker design [7,8], Performance evaluation of quality in PLA 3D printed parts having thin walls [9] and laser additive manufacturing in Inconel 625 [10]. The application of Taguchi design for the photochemical machining of aluminium is done in [11], where the optimal parameters are obtained with 9 experimental trials with three factor and three level interaction. The procedural steps to be followed in Taguchi approach is mentioned by Krishniah K, Shahabudeen P [3] as stated by Taguchi. This includes, finding the affecting factors and their levels in the response of the process and identifying a standard Orthogonal Array (OA) for assigning the order of the factors to be followed in experimentation using linear graphs. The optimal factors can be achieved by constructing an average response plot of the process. In this work, the optimal parameters in Photochemical Machining of SS316L steel is identified using Taguchi approach and the results are compared with the results through RSM available in literature [1]. The factors like Etching Time (Ti), Temperature (T) and Concentration (C) are considered as affecting parameters and the undercut of SS316L steel with PCM process is chosen as the response of the model. The optimal conditions are identified with only 9 experiments using Taguchi approach compared to 20 experiments required by RSM [1].

2. Design Methodology

The procedure of Taguchi design is followed as described by Krishniah K and Shahabudeen [3] . Genichi Taguchi has introduced few procedural steps to be followed during the experimentation. Initially the factors affecting the undercut (response) in the Photo Chemical Machining (PCM) is chosen at three different levels. The Orthogonal Array (OA) was chosen based on factors and levels to identify the total number of experiments to be conducted to predict the optimal response. Then, factors are assigned accordingly with the aid of linear graph which suggest the sequential order of levels to be used for the experiment. The Response graphs and ANOVA have been carried out to identify the optimal and significant factors in the experiment. The step by step analysis of experiment is illustrated in the following sections.

2.1. Factors and Levels

Photo Chemical Machining (PCM) is a machining process that utilizes chemical reaction for etching of the material. In this process, the etching time, etchant temperature and concentration are the significant parameters affecting the unmasked area exposed to the chemical. Any abrupt change in the process parameters may result in the material defect called 'Undercut'. So, the optimal conditions of factors should be chosen to minimize the effect on to the material. In this work these factors are considered at three different levels taken from [1] with notations of factors as Ti, T, C along with coded values -1, 0, 1 as shown in the table 1 below.

Table 1. Factors and their Levels [1]

Factor	Notation	Low (-1)	Medium (0)	High (1)
Time (min)	Ti	30	45	60
Temperature (^{o} C)	T	50	65	80
Concentration (gm /lit)	C	600	750	900

2.2. Identifying an Orthogonal Array (OA)

The Orthogonal Array (OA) defines the standard order of experimentation to be carried out. An OA is chosen by the number of experiments to be conducted on to the process selected. The formula mentioned below is used to identify the number of experiments to be used in Taguchi method.

$$N = 1 + N_f\,(L-1) \quad (1)$$

Where,
N = Number of total Experiments to be conducted
N_f = Number of input Factors
L = Number of Levels of Factors

As the experiment is a combination of three factors and three levels, a three level standard orthogonal array must be chosen, where the total number of experiments to be conducted as per the above equation are seven, based on the required number of experiments, the nearby OA $L_9\,(3^4)$ is chosen as per the guidelines given in [3].Now the factors are assigned to the chosen $L_9\,(3^4)$ OA by following the standard linear graph of L9 OA as shown in the below figure1.

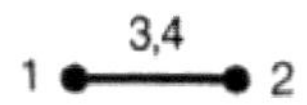

Figure 1. Linear Graph for L_9 OA

The assignment of the factors in OA is analyzed and is done accordingly by fixing a dummy factor at column 4. The results obtained with this assignment are similar with fixing the dummy factor at column 3 of the OA. The standard orthogonal array with the assigned factors is shown in the table2.

Table2. Assigned Factors $L_9\,(3^4)$ OA

Experiment No	Time (Ti)	Temperature (T)	Concentration (C)
1	-1	-1	-1
2	-1	0	0
3	-1	1	1
4	0	-1	0
5	0	0	1
6	0	1	-1
7	1	-1	1
8	1	0	-1
9	1	1	0

2.3. Response Analysis

The responses of the Taguchi model are predicted using the regression equation 2 shown below and the values obtained are tabulated in table 3.

$$U_d = 0.09139 + 0.06062\,(Ti) + 0.0583\,(T) + 0.0041\,(C) - 0.00985\,(Ti^2) + 0.02955\,(T^2) - 0.00545\,(C^2) + 0.024875\,(Ti \times T) - 0.000875\,(Ti \times C) + 0.001625\,(T \times C) \quad (2)$$

Table 3. Predicted Responses of the Taguchi Model

Experiment No	Time (Ti)	Temperature (T)	Concentration (C)	Predicted Response
1	-1	-1	-1	0.008
2	-1	0	0	0.021
3	-1	1	1	0.085
4	0	-1	0	0.063
5	0	0	1	0.090
6	0	1	-1	0.168
7	1	-1	1	0.085
8	1	0	-1	0.133
9	1	1	0	0.255

The response totals and the average response of the model are evaluated by summing up the responses of the factors at each level and dividing by the number of levels of the respective factor. The Response totals and the average response values are shown in the tables 4 and 5.

Table 4. Response Totals of the Model

Levels	Time (Ti)	Temperature (T)	Concentration (C)
-1	***0.1141***	***0.1553***	0.3096
0	0.3206	0.2443	0.3383
1	0.4727	0.5078	***0.2595***

Table 5. Average Response Values of the Model

Levels	Time (Ti)	Temperature (T)	Concentration (C)
Low (-1)	***0.0380***	***0.0517***	0.1032
Medium (0)	0.1068	0.0814	0.1127
High (1)	0.1575	0.1692	***0.0865***
Difference	0.2260	0.1980	0.0961
Ranking	1	2	3

The Response plot is plotted between the average responses of the factors and their levels as shown in figure 2 which illustrates the optimal conditions obtained from the experiment as Low Time (Ti_{-1}), Low Temperature (T_{-1}) and High Concentration (C_1).

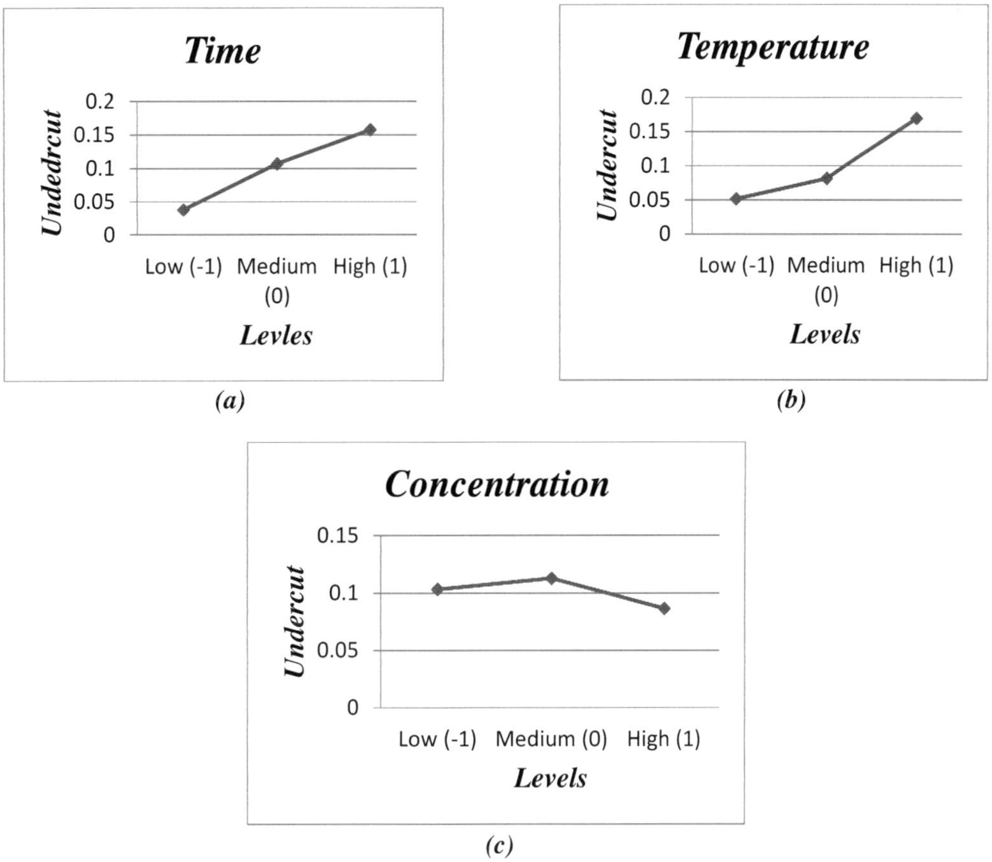

Figure 2. Average Response plot

2.4. ANOVA

The Analysis of Variance (ANOVA) is conducted using the correction factor and sum of squares of the factors. The Mean Square error of the factors is calculated by dividing the Sum of Squares by the Degree of Freedom (DOF) of each factor. The F-value andp-value obtained showed that, Etching time (Ti) and Temperature (T) aresignificant parameters affecting the response of the model. The table 6 below shows the ANOVA of the Model.

Table 6. ANOVA of the Model

Source of Error	Sum of Squares	DOF	Mean Square Error	F-Value	p-value
Etching Time (Ti)	0.0215	2	0.01075	26.875	***0.0121***
Temperature (T)	0.0224	2	0.0112	28	***0.0114***
Concentration (C)	0.001	2	0.0005	1.25	0.4028
Error	0.0012	3	0.0004		

Total	0.0463	8

The table 6 shows that the F-Value is greater than the standard F-Value i.e. 9.55 and the calculated P-values of the factors Etching Time (Ti) and the Temperature (T) are less than 0.05. It is observed that the factors Etching Time (Ti) and Temperature (T) are showing significant effect on undercut of the material with Photochemical Machining process.

3. Signal to Noise (S/N) Analysis

Signal to Noise (S/N) ratios are calculated to find the effects of noise factors on the responses of the model. The analysis is done using 'smaller the better' quality characteristic approach. The S/N response of the model is calculated by using the equation 3 mentioned below.

$$\eta = -10 \log\left[\frac{1}{n} \sum Y_i^2\right] \tag{3}$$

In S/N analysis, the maximum response is chosen as the optimal response irrespective of minimization or maximization condition. The responses of the model are calculated and tabulated in Table 7 and the Average Responses of the model are plotted as shown in figure 3. The tables 8 and 9 illustrate the total and average response values of the model.

Table 7. S/N responses of the model

Ti	T	C	R_a	η
-1	-1	-1	0.008	***41.72***
-1	0	0	0.021	33.59
-1	1	1	0.085	21.41
0	-1	0	0.063	24.06
0	0	1	0.090	20.91
0	1	-1	0.168	15.49
1	-1	1	0.085	21.46
1	0	-1	0.133	17.49
1	1	0	0.255	11.87

Table 8. S/N response totals of the model

Factor	Ti	T	C
Low (-1)	96.73	87.25	74.71
Medium (0)	60.47	72.00	69.54
High (1)	50.83	48.78	63.78

Table 9. Average response totals of the model

Factor	Ti	T	C
Low (-1)	32.24	29.08	24.90
Medium (0)	20.15	24.00	23.18
High (1)	16.94	16.26	21.26
Difference	4.860	11.17	19.53
Ranking	3	2	1

From the table 9 it is clear that the optimum noise factors are Low Time (Ti), Low Temperature (T) and Low Concentration (C) and the Optimum response in terms of S/N ratio is evaluated from the equation 4 below.

$$\eta_{Opt} = \bar{\eta} + (Ti - \bar{\eta}) + (T - \bar{\eta}) + (C - \bar{\eta}) \tag{4}$$

Where,
Ti, T, C = Response Totals of factors at respective levels

$$\bar{\eta} = \text{Grand Average Mean} = \frac{GrandTotalofallObservations}{TotalNoofObservations}$$

Therefore,

$$\eta_{Opt} = 40.01$$

The optimal factors obtained with S/N analysis are Low Time (Ti), Low Temperature (T) and Low Concentration (C) and the predicted response is 40.01, is close to the value 41.72 as reported in table 7. From this, we can conclude that the optimal conditions obtained through S/N analysis are satisfactory. The optimal conditions with raw data analysis reported in table 5 are Low Time (Ti), Low Temperature (T) and High Concentration (C). The predicted optimal factors found through RSM [1] satisfy all these conditions.

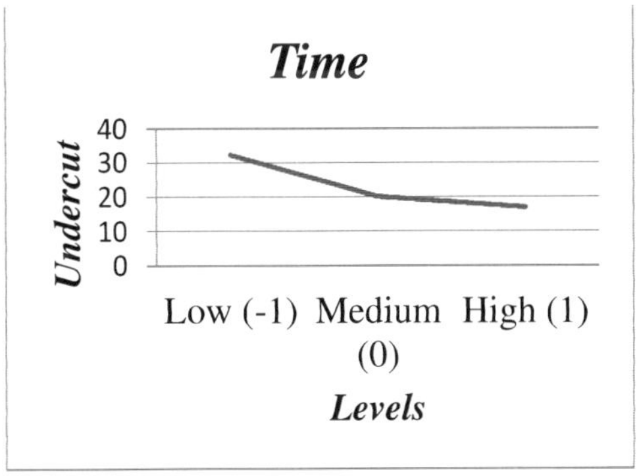

(a)

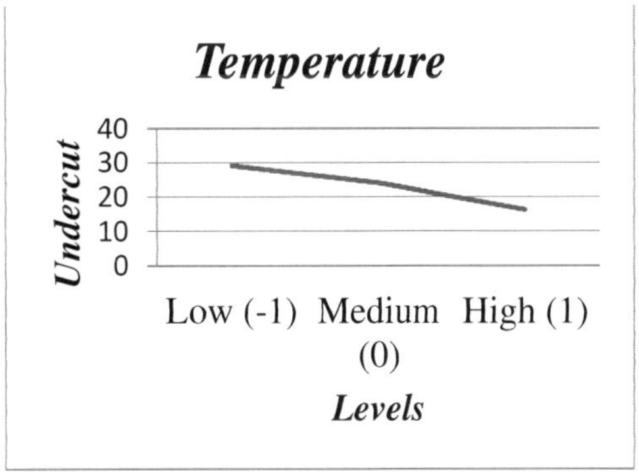

(b)

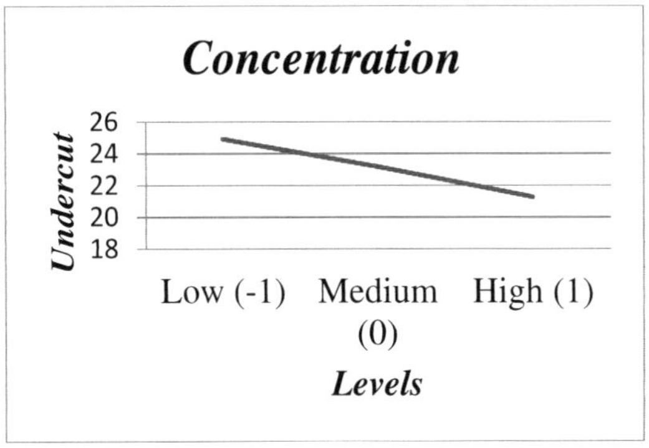

(c)

Figure 3. Average Response Plot of the S/N analysis

4. Results and Discussion

The optimal conditions obtained and the significant factors with the Taguchi experiments are compared with the results obtained through RSM available in literature. The number of experiments required by Taguchi method is equal to 9 which are considerably less compared to 20 experiments required by RSM approach. This shows that Taguchi is better approach than RSM in predicting the optimal machining conditions of Photo Chemical Machining (PCM) process. The results of Taguchi method through raw data analysis gives the following optimum conditions for achieving the minimum undercut are Low Etching Time (Ti), Low Temperature (T) and Low Concentration (C) levels. The S/N ratio analysis gives the following set of conditions such as Low Etching Time (Ti), Low Temperature (T) and High Concentration (C) levels. The analysis of variance (ANOVA) shows the Etching Time (Ti), Temperature (T) as the significant factors affecting the response of the model in both the cases. Since the concentration is not significant and also both high and low concentrations give the minimum values of undercut as per figure 4 (c), it can be concluded that the use of Taguchi method saves 55% of the effort and time required for finding the optimum conditions.

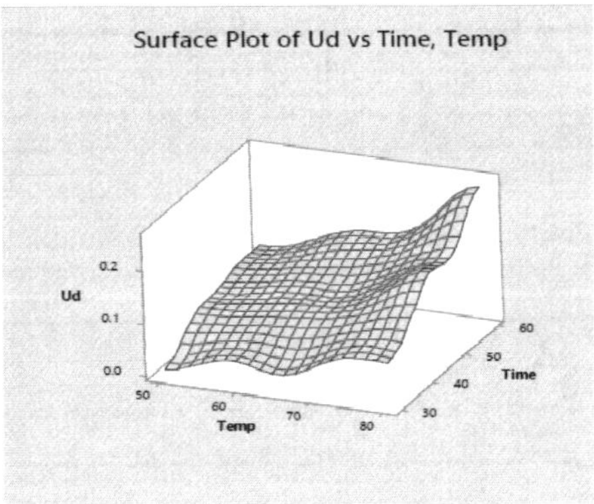

(a)

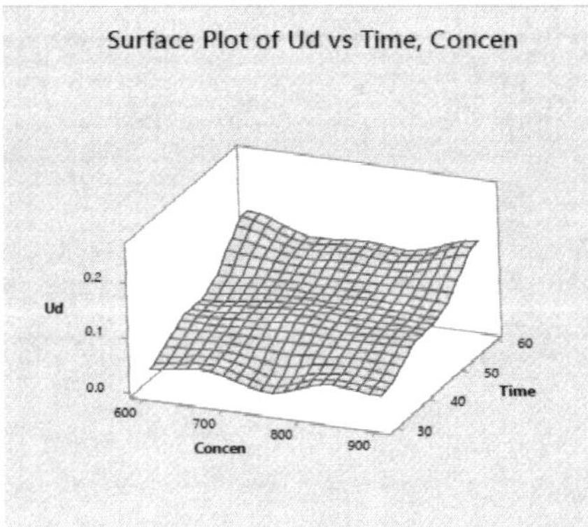

(b)

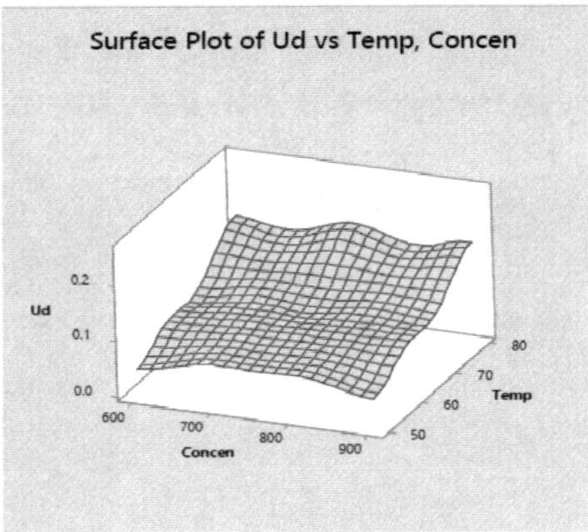

(c)

Figure 4. Surface Plots obtained using Response Surface Methodology (RSM) between (a) Etching Time and Temperature, (b) Etching Time and Concentration and (c) Temperature and Concentration [1]

5. Conclusion

In this work, the optimal etching conditions of Photo Chemical Machining (PCM) process on SS316L Steel is predicted using Taguchi method. The results are compared with the results obtained by Response Surface Methodology (RSM) available in literature.

- The etching conditions obtained by Taguchi method are similar to the results obtained through RSM and the number of experiments with TAGUCHI method is less than RSM which shows about 55% reduction.
- The Etching time (Ti) and Temperature (T) showed significant effect on the Undercut of the SS316L steel with PCM and the concentration of etchant is not significant.
- Optimal conditions for achieving optimum undercut are: Low etching time (30 min.), Low temperature (50°C) and low concentration (600 gm/lit) or high concentration (900 gm/lit).
- Taguchi method is better than RSM approach in predicting the optimal machining conditions of PCM to achieve minimal undercut.

References

[1] Bhasme A B and Kadam M S 2016 *IJMET* **7(2)** pp 25- 32.
[2] Montgomery D C 2013 *Design and Analysis of Experiments* (8thEdn John-Wiley and Sons Inc).
[3] Krishnaiah K and Shahabudeen P 2012 *Applied Design of Experiments and Taguchi Methods* Eastern Economy Edn (New Delhi: PHI Learning Pvt).
[4] Roy R K 2010 *A Primer on the Taguchi Method (2nd Edn* Society of Manufacturing Engineers).
[5] Mavoori N K, Vekatesh S and Hussain M M 2019 *Rapid. Prototyp. J.* **25/3** pp 454-461.
[6] Mohsin I, He K, Li Z, Zhang F and Du R 2020 *Appl. Sci.***10** 824.
[7] Zahraee S M, Chegeni A and Toghtamish A 2016 *JIEM* **9(2)** pp 487-504.
[8] Lai F M and Tu C W 2016 *Applied Acoustics.* **113** pp 81- 88.
[9] Aslani K, Chaidas D, Kechagias J, Kyratsis P and Salonitis K 2020 *J.Manuf. Mater.Process.* **4** 47.
[10] Yang B, Lai Y, Yue X, Wang D and Zhao Y 2020 *Scanning.*
[11] Mazarbhuiya R M and Rahang M 2019 *IOP Conf. Ser.: Mater. Sci. Eng.* **491** 012033.

Improving Sound Absorption coefficient and analysis of polyurethane foam reinforced with wood powder

M Gokul[1], T Gowtham[2], *S Jeyanthi[3], Lenin babu M C[4],Venkatachalam.G[5] Elammaran jayamani[6]
[1,2]PG Student, [3]Asso Professor,[5]Professor, School of Mechanical Engineering, Vellore Institute of Technology, Chennai campus, India,[4]Electric Vehicles-Incubation, Testing and Research Centre, Vellore Institute of Technology, Chennai, India
.[6]Sr lecturer, Swinburne University, Sarawak campus, Malaysia.
*jeyanthi.subramanian@vit.ac.in[3]

Abstract:
When it comes to acoustic insulation Polyurethane foam plays a very vital role for various sound absorption applications. The interest of this paper is to spot out the best combination of bio-based porous materials to obtain the best acoustic property. Here the porous materials used are polyurethane foam, pine wood with air gap. The sound absorption depends on the material properties such as Flow resistivity, porosity, tortuosity, thermal characteristic length and physical characteristic length, the different combinations of materials with air gap are being tested for sound absorption coefficient using impendence tube method using COMSOL. The results are being compared with numerical values obtained using impedance tube that helps us to present the best combination which provides the maximum sound absorption.

Keywords: Polyurethane foam, pine wood, impedance tube, COMSOL, sound absorption analysis.

1. Introduction:

One of the major problems in the industry is the accumulation of the wastes and the other major problem is the noise pollution being increased drastically which is a major concern for the human health. Thus, an environment friendly material and efficient enough with the sound absorption property has to be developed [1].

The fluid phase and the solid phase combine together to form a porous material [2]. A better absorbing porous material is obtained when the fluid phase percentage is higher than the solid phase. At lower frequencies better acoustic damping is obtained, thus it is obtained by increasing the water content and decreasing the cell diameter [3]. By adding a highly active catalyst an improved cellular structure would be obtained which will improve the sound absorption of the flexible foam.

Polyurethane foam is formed by the combination of di-isocyanate (NCO) and polyol (OH) [4]. The different of polyurethane foams can be obtained when they are mixed in different proportions. Flexible foam is obtained when the NCO group is higher than the OH group and when the OH group is more in ratio then a foam with increased sound absorption coefficient is obtained [5]. To increase the absorption of the incident pressure waves, natural fibres are added, which in term increases the voids in the foam [6].

Currently the main area of research is acoustics and experiments are being conducted widely around the globe. Wood powder [7-9] which is used as one of the porous materials is one the most easily and abundantly available. These are unburned by-products that can be obtained cheaply. This has man other advantages as they are renewable and abundant, don't cause damage to the health of living beings and are also nonabrasive. The most important of all is that wood has a very good sound absorbing property [10-14] which would enhance the sound absorption in this experiment.

Thus, the above experiment was carried out for various combinations of porous materials; i.e. wood powder and Polyurethane using COMSOL. The best combination was chosen and the sound absorption experiment was conducted for that combination. The same combination was used in COMSOL Multiphysics for the numerical prediction [15-16].

2. Experimental Setup:

2.1. Materials

In various fields Polyurethane foam is used as a generally for the noise control applications. Since the foam has effect on the human health the usage of the foam is limited. A new study has been made to improve the effective use of the foam. Thus, Wood Powder also known as Saw Dust has been used. Wood Powder helps in the improvement in the sound absorption effectively. Wood Powder contains of vast number of voids and is a natural porous material which helps in the investigation of sound absorption coefficient. The Wood Powder is mixed with the matrix and a thickness of 30 mm has been used. Bio-based Polyurethane foam is used for obtaining the composite matrix material needed for this experiment. Polyol and isocyanate methylene diphenyl diisocyanate together form the bio-based polyurethane foam.

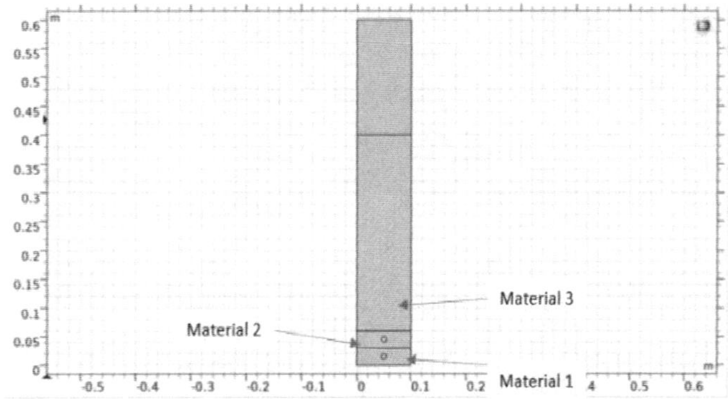

Fig. 1 Material Description in COMSOL

2.2. Measurements of Acoustic Properties

Sound absorption coefficient of the foam is measured using the ASTM E1050-12. The frequency ranges between 220 – 4500 Hz. Circular cross sections of diameter 33mm and 96mm are used form the sample. Samples of different combinations are being used for the test. Random signals are being generated from the source. The power amplifier is connected to a DAQ at one end and to the other end a speaker of 16ohm is connected. Two 1/2" MicrotechGefell microphone is connected to the M+P Vibpilot DAQ. The M+P Vibpilot DAQ spectrum analyser, time domain is extracted from the bandwidth of 6400Hz with a frequency resolution of 0.5Hz.

Fig. 2 Impendence tube setup

3. Numerical Model:

Here we will look into two formulations by which we can study and predict the acoustic behaviour. It is known as the displacement-pressure (u^s, P), which has 6 variables for 3D; 3 for the solid phase and the other 3 for the fluidic phase. The combined variable is used by the 4-space variable. When the poroelastic material is subjected to a single acoustic wave of angular frequency ω, then the formulation be as such for:

For solid phase:

$$N\nabla^2 u^s + (P - N)\,\nabla\nabla\, u^s + Q\nabla\,(\nabla\, u^f) + \omega^2\,(\overline{\rho_{11}} u^s + \bar{\rho}_{12}\, u^f) = 0 \quad (1)$$

For fluid phase

IOP Conf. Series: Materials Science and Engineering 1128 (2021) 012020 doi:10.1088/1757-899X/1128/1/012020

$$(\nabla u^s) - R\nabla(\nabla u^s) + \omega\,(\overline{\rho_{12}}u^s + \bar{\rho}_{22}u^f) = 0 \tag{2}$$

Here, u^s and u^f is used to represent the displacement vector in structured and saturated fluid.

Whereas,

$\bar{\rho}_{11}, \bar{\rho}_{12}, \bar{\rho}_{22}$ are obtained from the factor of mass coupling.

$$\bar{\rho}_{11} = \rho_{11} - j\frac{b}{\omega}, \tag{3}$$

$$\bar{\rho}_{12} = \rho_{12} - j\frac{b}{\omega} \tag{4}$$

$$\bar{\rho}_{22} = \rho_{22} - j\frac{b}{\omega} \tag{5}$$

Here,

b is the viscus coupling factor

Mass coupling factor is obtained by:

$$\rho_{12} = \rho_{1-}\ \rho_{22} \tag{6}$$

$$\rho_{12} = -\Phi\rho_0(\alpha\infty - 1) \tag{7}$$

$$\rho_{22} = -\Phi\rho_0 - \ \rho_{12} \tag{8}$$

Viscous coupling parameter:

$$b = \sigma\phi^2[\sqrt{1 + j\frac{.4\alpha_\infty^2\eta\rho air\omega Pr}{\sigma^2 z^2\phi^2}}] \tag{9}$$

Where,

ϕ is the porosity

For fluidic region dynamic tortuosity is given by:

$$\rho_{\text{eff}} = \frac{\alpha_\infty\rho\ air}{\phi}[1 + \frac{\sigma\phi}{j\omega\alpha_\infty\rho\ air}\sqrt{1 + j\frac{.4\alpha_\infty^2\eta\rho air\omega}{\sigma^2\Lambda^2\phi^2}}] \tag{10}$$

$$\mathrm{K}_{\text{eff}} = \frac{\gamma P_0 I/\phi}{\gamma - (\gamma-1)[1 + \frac{\sigma\phi}{j\omega\alpha_\infty\rho\ air}\sqrt{1+j\frac{.4\alpha_\infty^2\eta\rho air\omega}{\sigma^2\Lambda'^2\phi^2}}]} \tag{11}$$

Here,

Λ is viscous characteristic length

Λ' is thermal characteristic length

σ is the flow resistivity

$\rho\ air$ = 1.21 kg/m^3

γ = 1.14 is the specific heat ratio of air at standard temperature

P_o = 101320 N/m^2

η = 1.84 X 10^{-5} Ns/m^2 (for air)

Pr = 0.702 (for air)

Characteristics impedance:

$$(Z_p) = \sqrt{\rho\text{eff} * \text{Keff}} \tag{12}$$

Wave number:

IOP Conf. Series: Materials Science and Engineering 1128 (2021) 012020 doi:10.1088/1757-899X/1128/1/012020

$$(k) = \omega\sqrt{\rho eff / Keff} \tag{13}$$

Normalized surface impedance is obtained by:

$$Z_{sneq} = \frac{z_p}{\phi} * \cot(k*L) * Z_{air} \tag{14}$$

Where L is the thickness of the porous material

Compute sound reflection:

$$R = \frac{z\ sneq-1}{z\ sneq+1} \tag{15}$$

Absorption Coefficient:

$$(\alpha) = 1 - |R|^2 \tag{16}$$

3.1. Numerical prediction of COMSOL

The acoustical behavior of the samples is obtained by simulating them using the finite element approach. High physical parameters are necessary to simulate the porous materials since it a combination of fluid and solid phases. There are different models for determining the acoustics properties of the materials; and of those JCA model is being used in this paper. This model is used to predict the important factors such as the material porosity, airflow resistivity, tortuosity, thermal characteristic length and viscous characteristic length. This model is used to express the equations in equations (10) and (11).

3.2. Model Definition

In the system that is modeled, the incident sound frequencies hit the layer of the porous material at angle of 0 degrees and it has a wave vector *k*. In this, the model mentioned has a height of 600mm and consists of a width of 100mm. The height of each porous layer; i.e. Polyurethane and Wood is 30mm. The model of the incident field was made in such a way that a background pressure is applied to air domain. The boundary condition for the model is that an air pressure is applied to the boundary model. Sound hard boundary is applied to the top and bottom of the model so that there is no leakage of the pressure outside of the model. Another boundary called the poroelastic material is applied to the porous domain. In these five domain parameters are applied to porous materials. One of the advantages that COMSOL Multiphysics carries is that two or more phases can be coupled in this software. A free triangular mesh is used in this analysis. The mesh is of the size 0.21 times of the wavelength. Finally, the important parameter; i.e. sound absorption coefficient can be evaluated. The absorption results can be seen in the figure 4(a) and 4(b).

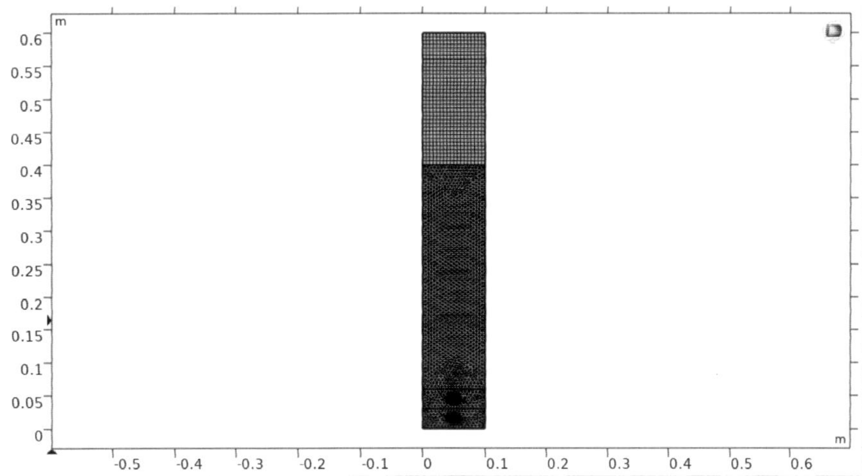

Fig. 3 Mesh of Impendence Tube

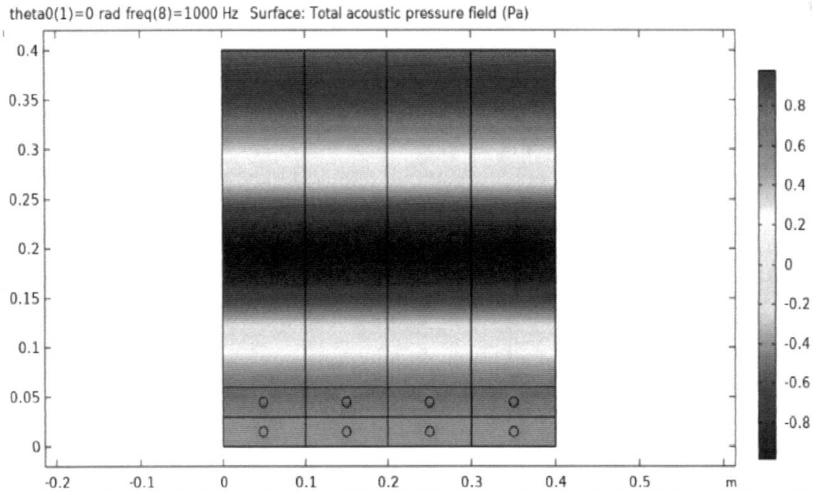

Fig. 4(a) Simulation results of Sound Absorption

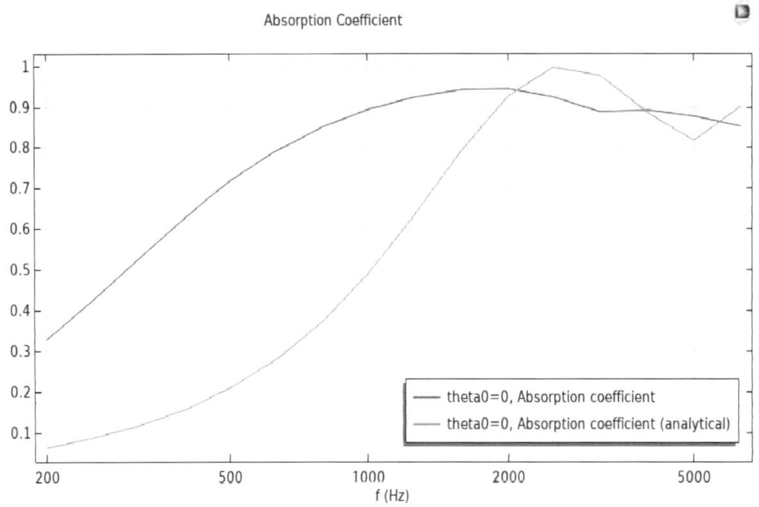

Fig. 4(b) Absorption Coefficient

4. Results and discussion:

Using the impedance tube, we out find out the sound absorption coefficient for the different combination of 30mm thick wood powder and the foam along with air and using COMSOL Multiphysics with the help of the JCA model the same is predicted.

4.1. Effect of wood on porous materials

Sound absorption coefficient results for different combinations of wood and foam is shown in figure 5. From the result it can be observed that with the presence of wood the sound absorption is enhanced and particularly for the lower frequency it is more effective and leads to added benefits. Without combining the porous materials, it was observed that the sound absorption was in average but when the combinations of the porous materials were used the lower frequencies were effectively absorbed. Meanwhile, the combinations were altered it acted as a resonator and it showed a good result of sound absorption in lower frequencies and the oscillating tends to happen at higher frequencies.

After plotting the results for the different combinations such as Air-Polyurethane-Wood, Polyurethane-Air-Wood, Polyurethane-Wood-Air, it was found out that Polyurethane-Wood-Air combination showed better sound absorption compared to the other two. Another set of combinations were also made by keeping the first material constant and reciprocating the last two materials and in such case also Wood-Polyurethane -Air combination showed better sound absorption. It can be seen that when air as a material was positioned before the porous materials it affected the sound absorption by reducing it but when the porous materials were positioned ahead of air as a material, we could see that the sound absorption was better. Thus, it can be clearly stated that air has a significant role in the sound absorption.

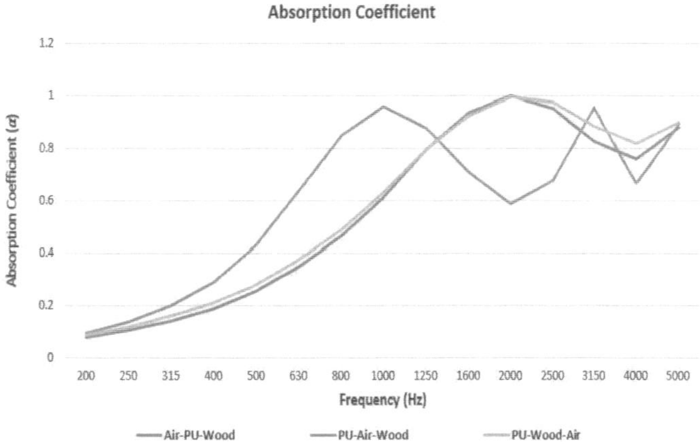

Fig. 5 Absorption Coefficient for Different combinations

4.2. Validation using Minitab

The predicted sound absorption for different combinations using the COMSOL Multiphysics are been validated using Minitab. The sound absorption is being is plotted with respect to the frequency as shown in figure 6. The demonstrated validation shows that the predicted values as the inputs make good agreement for the different combinations of the porous material and air medium. For the case of Polyurethane-Wood-Air it was found sound absorption was the maximum as per the experimentation though it does not go in agreement with the prediction made in Minitab. But it was also found that of the different combinations, Polyurethane-Wood-Air had the shown up with strong correlation values whereas Wood-Air- Polyurethane and Air -Polyurethane-Wood had an intermediate correlation values clearly stating that Polyurethane-Wood-Air is the most appropriate combination to go with for obtaining the maximum sound absorption as shown in figure 7. In other cases, though it might have a good predicted value the combination is not strong enough which might reduce the effectiveness of the absorption.

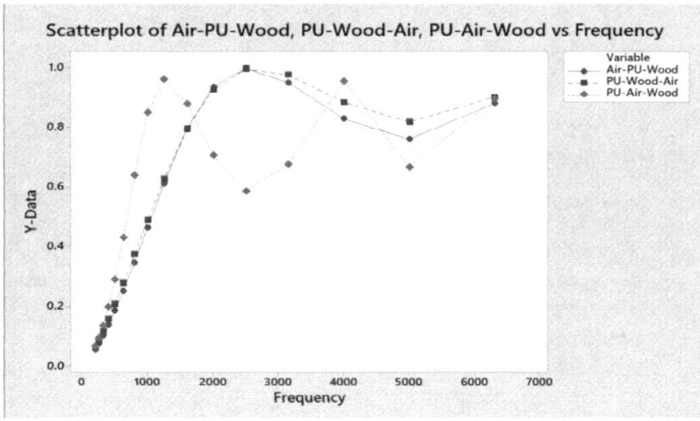

Fig. 6 Validation of Absorption Coefficient for Different combinations

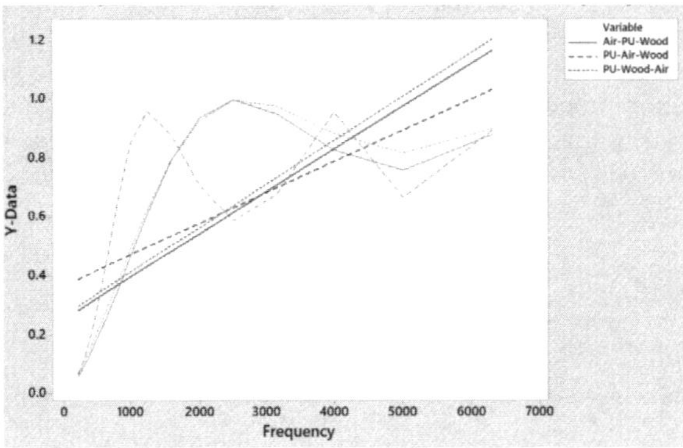

Fig. 7 Correlation factors for different combinations

$$\mu = \frac{\sum xy - \sum x \sum y}{\sqrt{\left(\Sigma_x^2 - \frac{(\sum x)^2}{n}\right)\left(\Sigma y^2 - \frac{(\sum y^2)}{n}\right)}}$$

Here μ is the correlation factor

$0.25 \leq \mu < 0.75$ it would be an intermediate correlation

$0.75 \leq \mu < 1$ it would be a strong correlation

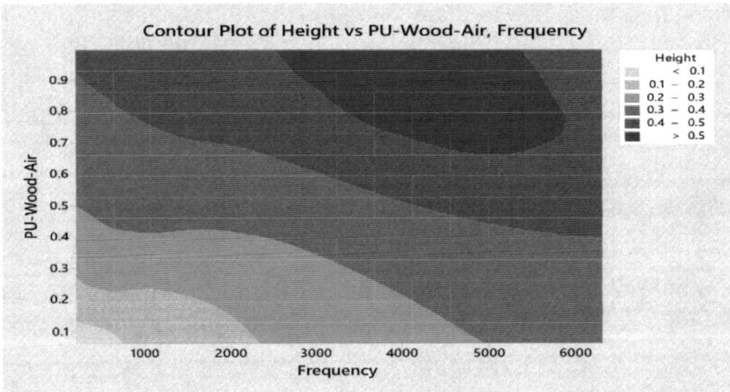

Fig. 8 Contour plots for Frequency vs Porous material with respect to the height

5. Conclusion

To conclude, this study was been focused to develop a new combination among the porous materials which is bio-based and which is eco-friendly as well. The samples were developed using the wood powder and polyurethane foams along with air medium when used for experimentation as well as for simulation. Different possible combinations were made by shuffling the arrangements of the fibres and air. The samples were then tested using the impedance tube. Using the COMSOL Multiphysics, sound absorption predicted by numerical method is being compared with that of the experimental results. The comparison results were in good but the prediction and it was to be in match with the correlation factors the combination that has been chosen is being said as a correct combination.

Furthermost, the results states that:

1. The inclusion of the wood powder as a porous material increases the sound absorption coefficient even at the lower frequencies.
2. As observed from the physical it can be seen that the flow resistivity increases due to the inclusion of the wood powder.
3. Due to the formation of the different combinations of the porous materials and analysing the results, based on the different applications the combinations can be altered and can be used.
4. Better sound absorption can be obtained by using a powder fibre material.

References

[1] Jeyanthi, S., Yuvaraj, L. & Mailan Chinnapandi, L.B., 2019. Experimental and finite element approach for finding sound absorption coefficient of bio-based foam. Journal of Vibroengineering, 21(6), pp.1761–1771. Available at: http://dx.doi.org/10.21595/jve.2019.20335.

[2] Ju Hyuk Park, Sei Hyun Yang, Hyeong Rae Lee, Cheng Bin Yu, Seong Yeol Pak, Chi Sung Oh, Yeon June Kang, Jae Ryoun Youn,Optimization of low frequency sound absorption by cell size control and multiscale poroacoustics modeling,Journal of Sound and Vibration,Volume 397,2017,Pages 17-30,ISSN 0022-460X, https://doi.org/10.1016/j.jsv.2017.03.004.

[3] Jae Gyoung Gwon, Seok Kyeong Kim, Jung Hyeun Kim,Sound absorption behavior of flexible polyurethane foams with distinct cellular structures,Materials & Design,Volume 89,2016,Pages 448-454,ISSN 0264-1275,
https://doi.org/10.1016/j.matdes.2015.10.017.

[4] Tanaka, Ryohei & Hirose, Shigeo & Hatakeyama, Hyoe. (2008). Preparation and characterization of polyurethane foams using a palm oil-based polyol. Bioresource technology. 99. 3810-6. 10.1016/j.biortech.2007.07.007.

[5] Ibrahim, M., & Melik, R. (2003). Optimized sound absorption of a rigid polyurethane foam. Archives Of Acoustics, *28*(4).
Retrieved from https://acoustics.ippt.pan.pl/index.php/aa/article/view/459

[6] Ekici, Bülent & Kentli, Aykut & Kucuk, Haluk. (2012). Improving Sound Absorption Property of Polyurethane Foams by Adding Tea-Leaf Fibers. Archives of Acoustics. 37. 10.2478/v10168-012-0052-1.

[7] S. Kassim, Muna. (2017). Experimental study of sound absorption properties of reinforced polyster by some natural materials,ResearchGate, Publication 314403335.

[8] Tiuc, Ancuta Elena & Nemes, Ovidiu & Vermesan, Horatiu & Toma, Adina. (2018). New sound absorbent composite materials based on sawdust and polyurethane foam. Composites Part B: Engineering. 165. 10.1016/j.compositesb.2018.11.103.

[9] Kang, CW., Oh, SW., Lee, TB. et al. Sound absorption capability and mechanical properties of a composite rice hull and sawdust board. J Wood Sci 58, 273–278 (2012). https://doi.org/10.1007/s10086-011-1243-5

[10] Smardzewski, Jerzy & Kamisiński, Tadeusz & Dziurka, Dorota & Mirski, Radosław & Majewski, Adam & Flach, Artur & Pilch, Adam. (2015). Sound absorption of wood-based materials. Holzforschung. 69. 431–439. 10.1515/hf-2014-0114,.

[11] Con Wassilieff, 1996 Sound Absorption of wood-based materials, Elsiver Applied Acoustics, Volume 48, Issue 4, Pages 339-356.

[12] Peter Zvolenský, Ľubomír Kašiar, Peter Volna, Dalibor Barta, 2017 Simulated Computation of the Acoustic Energy Transfer through the Structure of Porous Media in Application of Passenger Carriage Body, 10th International Scientific Conference Transbaltica : Transportation Science and Technology.

[13] Milan Červenka, Michal Bednařík, Tomasz G. Zieliński, 2018 Direct numerical simulation of sound absorption in porous media, Euronoise ISSN: 2226-5147.

[14] Saati, Ferina & Moheit, Lennart & Marburg, Steffen. (2017). Parameter variations for a porous layer simulation. The Journal of the Acoustical Society of America. 141. 4016-4016. 10.1121/1.4989232.

[15] Tomasz g. Zieli´n ski, 2015 Multiphysics Modelling of Sound Absorbing Fibrous Materials, COMSOL Conference– Grenoble.

[16] Ahmadinejad, Parvin & ahmadinejad, parvin & monazzam, MR. (2018). 3D Simulation of Impedance Tube: Effect of Mesh Size on Sound Absorption Characteristic. 18.

Numerical analysis of hat stiffened composite panels for pre and post buckling conditions

Ram Murthy V[1], Annamalai K*[2], M Elango[3]
[1]PG Student, [2*]Professor, [3], Assistant Professor (Selection Grade)
School of Mechanical Engineering, VIT Chennai, India.

rammurthy1997@gmail.com[1], kannamalai@vit.ac.in[2]
elango.m@vit.ac.in[3]

Abstract

This research aims in studying the pre-buckling and post-buckling behaviour of composite panels stiffened by hat stiffeners under compressive loading. Stiffened panels are used in aerospace, marine and also in bridge structures. They are used to make the plate girders stiff. By using composite materials, the strength can be increased, and the weight can be reduced. In Composite panels which are stiffened, local buckling and failure of panel happens because of the applied compressive loads and due to combination of compressive and shear loads. The Collapse of the stiffened composite structure starts from the intersection between the plate and the stiffeners due to stress effect caused by deformation of lamina skin and stiffeners then leads to the structural buckling of skin of the panel. This research uses nonlinear Finite element approach to investigate the collapsing behavior using ABAQUS software.

Keywords: Stiffeners, Finite Element Method, Composites, Laminated Panels, Buckling.

1. Introduction:

The need for strong, efficient and light weighted structures led all experts to use composite materials in a lot of areas. The purpose of designing the laminated composite materials is to achieve better performance, which are not possible with the traditional materials. This is because of their high stiffness and strength to weight ratio. Some of the applications are Aircrafts, Automobiles and Space vehicles. However, there are many difficulties, when it comes in choosing the suitable material for the structural applications. Composite panels are nowadays used in bridges for retrofitting of the structures as they possess high strength. In stiffened composite panels, the in-plane material possesses strong material properties when compared to the direction which is normal to the fibre orientation. They have very low resistance against Shear and compression because of very weak inter laminar interfaces. Because of the weak inter-laminar interfaces, cracks are formed which leads to failure. This is known as delamination and it will reduce the load carrying capability of the structure. Delamination in compressive loaded structures causes buckling. Loss of stiffness is the main cause of this failure. Manufacturing defects and dropping of tools are some of the causes of buckling. The Ply orientation also have impact on the strength and stiffness of the composite laminates. Many researches have been done on stiffened composite panels with different type of stiffeners.

Stiffeners are used to control local buckling. They stiffen the panels against out of plane deformations. Most of the researchers have worked on different type of stiffeners like T-shaped and J-shaped stiffeners. (Faruk Eladi,[1]) studied the post buckling durability of J-stiffened and hat stiffened boards by comparing the experimental results with the analytical results. He concluded that Hat stiffened panel is 25% more efficient than the J-stiffened panels. (L.Boni et al, [2]) concentrated on developing a proven method for modeling, by means of Finite Element method, the post-buckling behaviour of stiffened composite flat panels exposed to compression loads. The experimental values for model validation were taken during a test on two sets of CFRP flat stiffened panels.

The similarity between the numerical and experimental outcome shows the efficiency of the FEM method in anticipating the buckling behavior. (Shashi Kumar et al, [3]) used artificial neural networks for estimating the buckling load of laminated hat stiffened panels under compressive loading. It is found that artificial neural networks can be used in an efficient manner to identify the buckling load for different type of loading conditions. (Vishwanath Vashisht, [4]) investigated the stiffened panel structure which is made up of different materials under pressure loads. He found that the combination of HYBOR/Boron composite and Aluminium and steel is the suitable material for stiffening in re-entry vehicles in aerospace industry. (NitinKumar et al, [5]) studied the structural weight optimization of hat stiffened aircraft wing by varying the distance between the stiffeners and the number of stiffeners. He found that150 mm stringer gap is optimum. (Weibgraeber et al, [6]) presented about linear investigation of buckling behavior of an orthotropic plate under compressive load. He determined the absolute minimum bending stiffness which enables the reinforcement to act as near rigid support for arbitrary long plates. (Riming Tan et al,[7]) studied the effect of impact location on the composite panels with Single-L shaped stiffener due to compression. He concluded that the bending stiffness in the impact location affects the damage formation. In order to reduce the damage, the stiffness discontinuity in structural design should be avoided. (Yao Yao ye et al, [8]) investigated and analyzed about how sub-stiffeners can enhance the anti-buckling performance of the composite panel stiffened using T stiffeners. The optimal stacking ply sequence of the sub-stiffeners also increased the critical load value of T-stiffened panel by112.6%. (WeiSun,[7]) investigated the effect of failure of a composite panel with three T-shaped stiffeners due to low velocity impact when it is subjected to compressive buckling. When the impact energy is increased there is serious debonding of stiffener and panel. The increase of compressive load decreases the compressive stiffness of the panel as result of the damage propagation of the stiffener and buckling of panel. The buckling phenomenon of composite panels depends on the properties of the material used and also the type of fibre orientation of the composite panel. In this paper, CFC and Kevlar have been considered to analyze the clasp behaviour of the composite panel using non-linear FEM with ABAQUS software.

2. Methods:

The Linear buckling analysis is also known as eigen value buckling analysis, it predicts the buckling strength of a structure. However, eigen value buckling does not consider non linearities and imperfections. So, more precise post- buckling analysis needed to determine the load value at which the structure will become unstable.

2.1 Mathematical Model:

Considering Eigen value pre-buckling analysis, appropriate stiffness matrix is given by,

$$[K_0] = \int_{\Omega}[B_0]^T\,[C][B_0]d\Omega \qquad \text{..... (1)}$$

Where $[B_0]$ corresponds to strain-displacement matrix. The solution can be found from,

$$[K_0]\{q_0\} = \{f_0\} \qquad \text{... (2)}$$

Where $\{q_0\}$ = are pre-buckling response
$\{f_0\}$ = load vectors respectively.
The Geometric stiffness matrix [Kp] is given to be,

$$[K_\rho] = \int_{\Omega} [G][\sigma][G]\,d\Omega \qquad \text{... (3)}$$

Where $\{\sigma\}$ = stress vector, Hence, the linear buckling problem is given by,

$$([K_0] - [K_\rho]\{\chi\} = 0 \qquad \text{... (4)}$$

3. Geometry:

The Geometry of hat stiffened laminated composite panel can be seen from the figure 1. The dimensional parameters of the Composite panel have a thickness of 356mm and height 356 mm. There are three hat stiffeners and the spacing between the stiffeners is 100mm. The Ply configuration of the panel is [[45/-45/0/90/] s] s and it has thickness of 2 mm. The thickness of each ply is 0.125mm and the panel is made up of 16 layers.

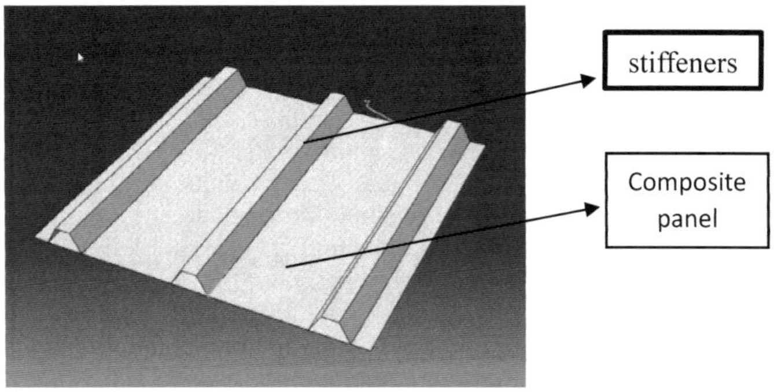

Figure. 1 Composite panel with hat stiffeners

4. Materials Properties:

Two materials, CFC/epoxy and Kevlar/epoxy were used for this buckling analysis of composite panel. The material properties are given below in Table 1.

Table.1 Material properties.

Symbol	Units	CFC	Kevlar
E_{11}		164.00	195.00
E_{22}= E_{23}	Gpa	12.80	14.60
G_{12}=G_{13}		4.50	7.50
G_{23}		2.50	5.00
$\vartheta_{12} = \vartheta_{13}$	None	0.32	0.30
ϑ_{12}		0.45	0.45
X_{1t}		2724.00	3100.00
X_{1c}		111.00	500.00
X_{2t}		50.00	150.00
X_{2c}		1690.00	1800.00
X_{2t}	Mpa	290.00	600.00
X_{2c}		290.00	600.00
S_{12}		120.00	250.00
S_{13}		137.00	320.00
S_{23}		90.00	200.00
ρ	Kg/m^3	1800	1400

IOP Conf. Series: Materials Science and Engineering 1128 (2021) 012021 doi:10.1088/1757-899X/1128/1/012021

5. Meshing:

The composite panel is discretized with S4R element for the analysis. The S4R element has the bending capability and can be used when the loading of the structure is asymmetric. The S4R element possesses a DOF of 3 per node. The size of the mesh is 5mm and have generated 12400 elements as shown in Figure2.

Figure. 2 Mesh

6. Boundary conditions:

The composite panel is completely fixed at the bottom edge and it is allowed to move in axial direction along the top edge. The Vertical edges are simply supported. The uniform axial load of 1 KN/m is applied axially in the Z-Direction of the panel.

Figure. 3 Boundary conditions

6.1 Pre-Buckling analysis:

The composite panel is made up of 16 layers and the material properties are assigned for each layer. The thickness of each layer of the composite panel and the stiffener is 0.125 mm. The total thickness of the composite panel and the stiffeners is 2 mm. considering the linear analysis the Eigen value step is used in place of general static step. Axial load of 1KN/m is applied along the edge of the composite panel. After the necessary conditions being assigned, the finite element model is tested for pre-buckling analysis. The buckling modes can be viewed in the visualization module. The buckling load can be acquired through the product of the Eigen value and load magnitude.

Buckling load = Eigen value x Axial load magnitude

6.2 Post Buckling analysis:

Under post buckling study, the non-linear F-S curve data for the corresponding FE model can be obtained by utilizing the modified Riks algorithm [9]. It is thereby intended to study the post buckling behavior, hence the governing conditions pertaining to pre buckling behavior, as discussed previously, is updated to obtain the load displacement criteria corresponding to post-buckling behavior. In the post-buckling analysis, the non-linear load vs displacement curve is found using the modified Riks algorithm. The study for pre-buckling is updated to conduct a nonlinear analysis of load deflection to predict Post-buckling behavior. For post buckling analysis, Eigen value step is replaced by Riks method. The Load is replaced by critical load and a new job is created which can be submitted for post buckling analysis.

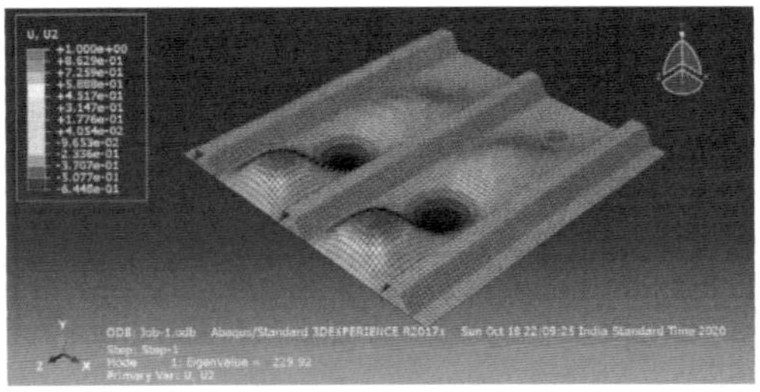

Mode1

Mode 2

Mode3

Mode 4

Mode 5

Figure.4 Buckling modes of (CFC/epoxy)

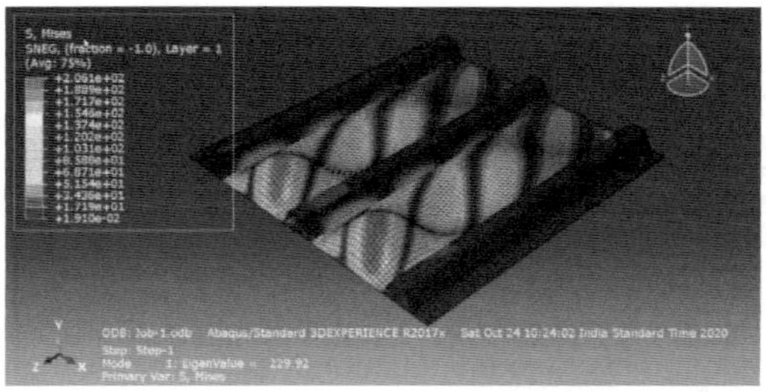

(a)

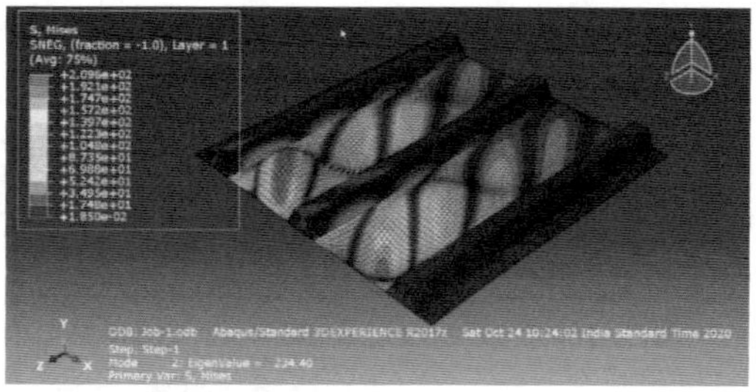

(b)

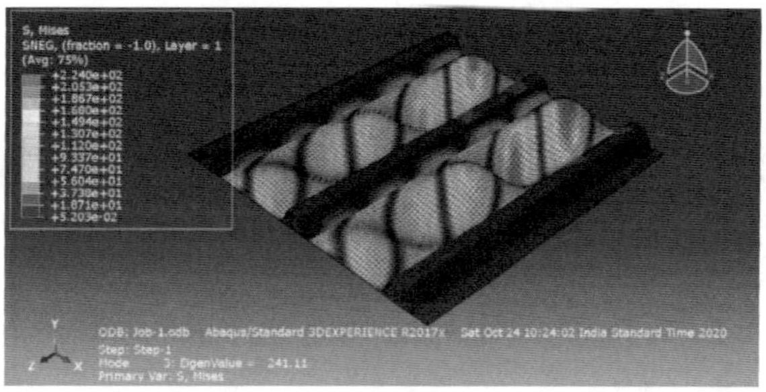

(c)

Figure.5 *Von Mises (CFC/epoxy):*

Mode1

Mode2

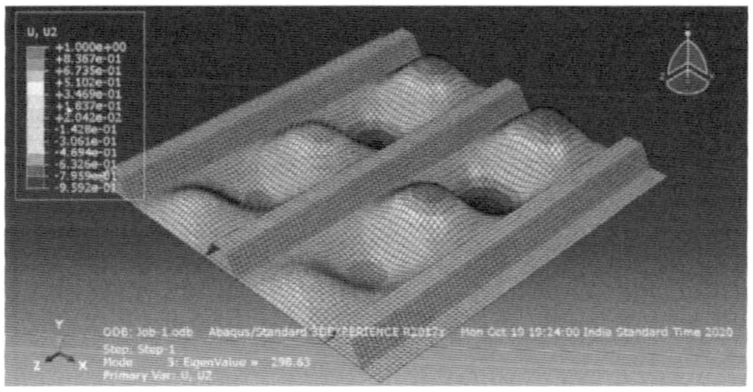

Mode3

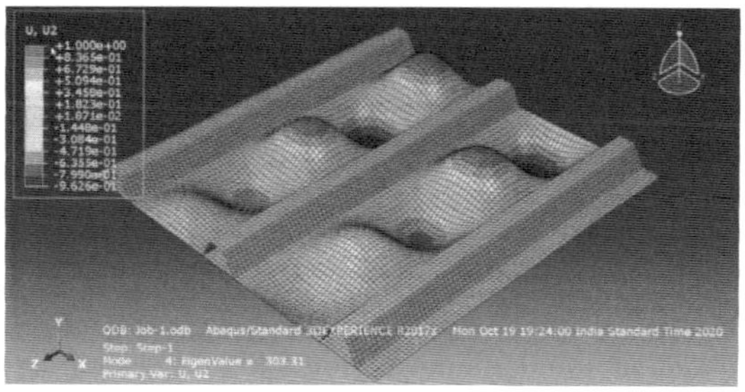

Mode 4

Mode 5

Figure.6 B*uckling modes* (Kevlar/epoxy)

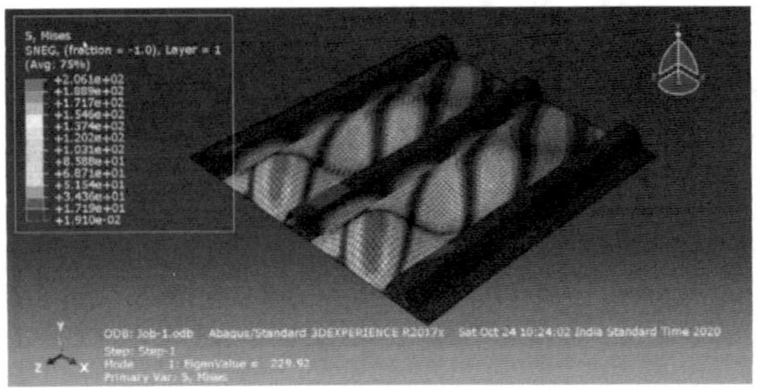

(a)

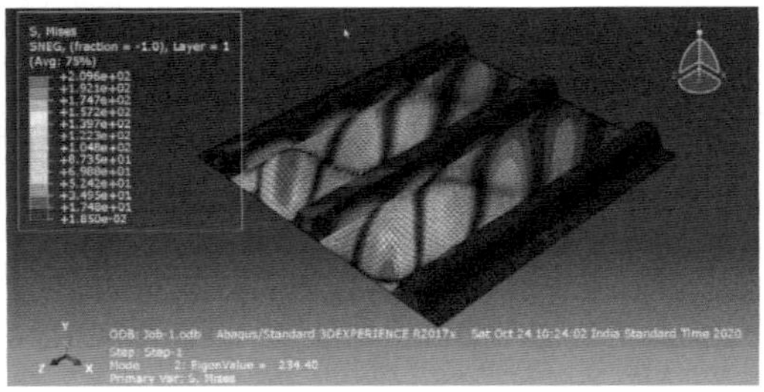

(b)

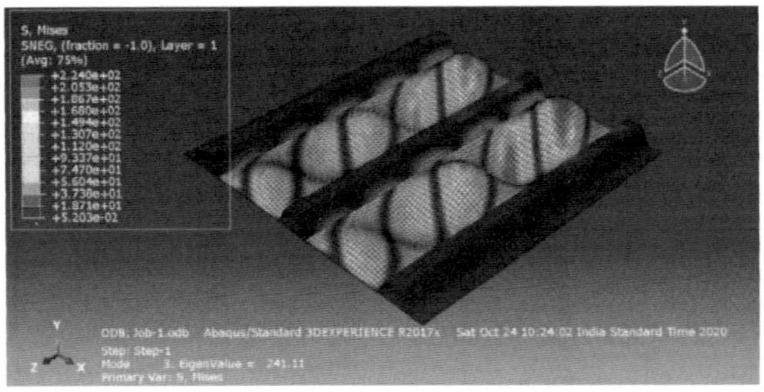

(c)

Figure.7 Von Mises (Kevlar/epoxy)

7. Results and discussion:

The buckling analysis of hat stiffened composite panel is done which is made up of CFC/epoxy and Kevlar/epoxy. The linear buckling analysis is done in order to identify the modes of the stiffened composite panel. The composite panel is loaded axially with load of 1KN/m and submitted for analysis. Five buckling modes are taken. Then the Eigen value buckling step is replaced by static Riks method to proceed to the post buckling analysis in ABAQUS. Now the model is submitted for post-buckling analysis which uses static riks method to solve the problem. The Critical load is applied axially in post buckling analysis.

Hashin's damage criteria is applied to identify the anisotropic damage in elastic-brittle materials. It is used to find the failure modes in fibre reinforced composite materials. The elastic properties of CFC/epoxy and Kevlar/epoxy are given in table.1 for hashin's damage criteria. In case of composite panel made of CFC/epoxy, the first buckling load value is229KN/m. The buckling load value is a product of Eigen value and applied load. The von mises contours are also taken. In case of composite panel made of Kevlar/epoxy the buckling load is 245 KN/m. Kevlar/epoxy possess more stress than CFC/epoxy. The buckling load is also higher in Kevlar/epoxy when compared to CFC/epoxy.

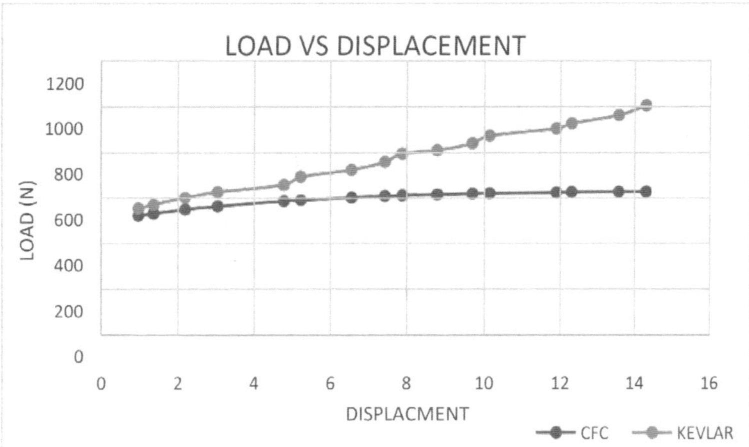

Figure.8 Load vs Displacement

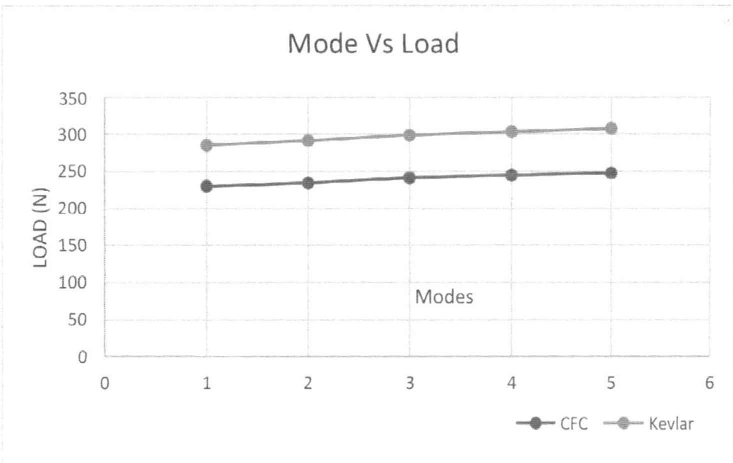

Figure.9 Load vs mode

Figure.8 shows the Load vs mode graph comparison between CFC/epoxy and Kevlar/epoxy. In post buckling analysis, the load at which the panel becomes unstable is identified. The curve of load vs displacement curve increases linearly and after the increase of load it get reduced. Composite panel with stiffeners made up of Kevlar/epoxy with stands maximum load when compared to composite panel with stiffeners made of CFC/epoxy. Figure.9 shows the Load vs Displacement curve comparison between CFC/epoxy and Kevlar/epoxy.

8. Conclusion:

The pre-buckling and post-buckling analysis of composite panels stiffened with hat stiffeners made up of CFC/epoxy and Kevlar/epoxy is done using the finite element method. The hat stiffened composite panel made up of Kevlar/epoxy has the maximum load capacity when compared with CFC/epoxy at the ply configuration of ([45/-45/0/90]s)s. The Kevlar/epoxy panel is stiffer when compared to CFC/epoxy hat stiffened panels.

References:

[1] Faruk Eladi (2010). Structural Efficiency and Post-buckling Strength of J- and Hat-stiffened Composite Panels, *Reinforced plastics and composites*, vol.29,No.10/2010.

[2] L. Boni, D. Fanteria , A. Lanciotti (2012). Post-buckling behaviour of flat stiffened composite panels: Experiments vs. analysis, *Composite Structures 94 (2012)3421–3433.*

[3] Shashi Kumar*, Rajesh Kumar, Sasankasekhar Mandal and Atul K. Rahul, The Prediction of Buckling Load of Laminated Composite Hat-Stiffened Panels Under Compressive Loading by Using of Neural Networks, *The open civil engineering journal,* Volume 12, pages468-480.

[4] K.S.VishwanathVashisht (2016). Experimental Investigation of the Aircraft Stiffened Panel Structure under Pressure Loads, *International Journal of Innovations in Engineering and Technology (IJIET),* Volume 7 Issue 4 , ISSN: 2319 –1058.

[5] K.C. NithinKumar , Subhash Chavadaki , Amir Shaikh , Durgeshwar Pratap Singh , ShwetankAvikal(2019).Weightoptimizationofahatstiffenedpanelofatypicaltransportaircraftcomposite wing. *Materials Today: Proceedings.* Volume 26, Part 2, 2020, Pages471-474.

[6] P.Weißgraeber,C.Mittelstedtb,W.Becker(2012).Bucklingofcompositepanels: Acriterionfor optimum stiffener design, *Aerospace Science and Technology*,Vol 16 Issue 1Pages10–18.

[7] Riming Tan, Jifeng Xu, Zhidong Guan , Wei Sun, Tian Ouyang, Shoucai Wang (2020). Experimental study on effect of impact locations on damage formation and compression behaviour of stiffened composite panels with L-shaped stiffener, *Thin–Walled Structures*, Volume 150, 106707.

[8] YaoyaoYe,WeidongZhu,JunxiaJiang,QiangXu,YinglinKe(2020).Designandoptimizationof composite sub-stiffened panels, *Composite Structures* Volume 240,112084

[9] Y.B. SudhirSastry,Pattabhi R. Budarapu , N. Madhavi , Y. Krishna(2015). Buckling analysis of thin wall stiffened composite panels, *Computational Materials Science* 96 (2015) 459–471.

Study on performance of various stent materials using explicit finite element analysis

Ajinkya R Mane[1], **Davidson Jebaseelan D**[2]

[1] PG student , [2] Professor, School of Mechanical Engineering,
Vellore Institute of Technology, Chennai, Tamilnadu, India. 600127

Email: maneajinkya108@gmail.com[1] , davish2k4@gmail.com[2] .

Abstract. In many countries, cardiovascular disease is the main cause of mortality. Stent is tiny wire mesh tube which is inserted in coronary artery to get rid of blockage. Stents are available in various types Bare metallic, Drug eluting and Bioresorbable. The use of bioresorbable stents reduces the threat of post-implantation effects like stent thrombosis and bleeding, it also dissolves in the body over the period of time once the healing is done while re-narrowing of the artery is more likely to occur in bare metal stents. Metals are important materials for stent due to mechanical properties and ease for Image processing. The objective of the work makes an effort to compare various materials used in stent using explicit analysis. For analysis, we used AISI 316L stainless steel, Nitinol and Mg alloy (WE43) which are standard materials for stent application. Explicit finite element study was done to study the performance of stent at varying pressures from 100 to 120mm of Hg, which is normal blood pressure range of a human body. This computational study found that Nitinol has stress values from 2.18MPa-2.62MPa and was less compared to other materials. Further study are planned to study the impact of coating that can enhance the biocompatibility.

Keywords- Stent; finite element analysis;

1. Introduction

Cardiovascular disease is a very saviour issue as number of deaths caused by this disease is greater than that of any other diseases worldwide. Cardiovascular diseases (CVDs) is a term which accounts coronary heart disease, cerebrovascular disease, peripheral arterial disease, rheumatic heart disease, congenital heart disease, deep vein thrombosis and pulmonary embolism. Main cause of Cardiovascular disease is related to blockage of arteries which supplies blood to the heart which is mainly due the deposition of the fats on the inner periphery of the arteries resulting in reduced cross section which is available for the blood flow. Deposition of the fat on the inner periphery of the arteries results in reduced blood flow and oxygen towards heart, leads to improper heart function. In order to open the arteries expandable metal/polymeric mesh is placed during the angioplasty surgery which is called as stent. Conceptually stent came in to the picture in 1964, this concept of stent was firstly introduced by Dotter and Judkins. Actual implementation of stent was done for the first time in 1977 for bypass graft surgery, followed by angioplasty of another patient after four months.[1] The perfect stent must have the option to be conveyed to the sore through any calcified fragments without getting disfigured from the conveyance framework.[2,3].

Different types of the stents are available like-Dual therapy stent (DTS), Bio-resorbable vascular stent (BVS), Bio-engineered stent, Drug eluting stent (DES) and Bare metal stent (BMS). Bare metal stent is the simple mesh of metal which is used to reopen the artery, healing of the artery takes place once the artery starts growing around the stent. Re-narrowing of the artery is more likely to take place in bare metal stents. [4-6] Main issue associated with the stents are- slow healing rate of artery, premature absorption of the stent. In case of bio-resorbable stents these issues lead to re-narrowing the arteries and ultimately misfunctioning of the heart. In order to overcome these issues changes in the stent design are made so that the healing of artery will speed up and chances of re-narrowing the artery will reduce.[7] Selection of proper material plays a major role in stent application. Metals, Ceramics and Polymers are available as stent material but Ceramics are strong not preferred due to lack of flexibility and plasticity. Moreover fracture toughness is also low for ceramics. While in Polymers high flexibility property alone is not sufficient. Polymers shows high flexibility but it is hard to acquire polymers with both high adaptability and strength at body temperature. Therefore, Metals are most important materials for stent due to mechanical properties and ease for Image processing. The improvement of appropriate materials for stents falls behind from the perspective of materials science[8]. This work aim is to compare the materials amongst Mg alloy,316L stainless steel and Nitinol under specific loading conditions.

2. Methodology

Methodology is followed as shown in figure 1. Finite element method is one of the numerical methods that is pretty flexible and might offer repeated solutions as compared to other in less time. This method is used to offer solutions to troubles which are ruled by way of partial differential equations. The methodology used for the analysis of the stent is the finite element modelling.[9] For simulating the results of stent geometry ANSYS v19.2 software is used.

Figure 1. Methodology

2.1. Modeling of stent:

With the help of SOLIDWORKS CAD software, geometrical model of stent is designed with specific dimensions. Figure 2 shows the overall view of stent. In stent the distance between struts also makes impact on its results, to avoid that total length of stent model is 7.01mm which is made with consideration of size of arteries in human body. Proper size of the stent depends on the size of artery it is to be positioned. Geometrical details of the stent are provided in table no.1. Following figures 2(a) & (b) shows the various views of stent.

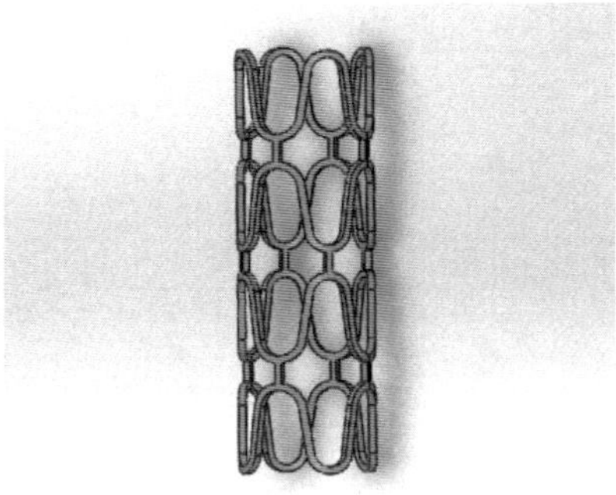

Figure 2(a). Front View of Stent

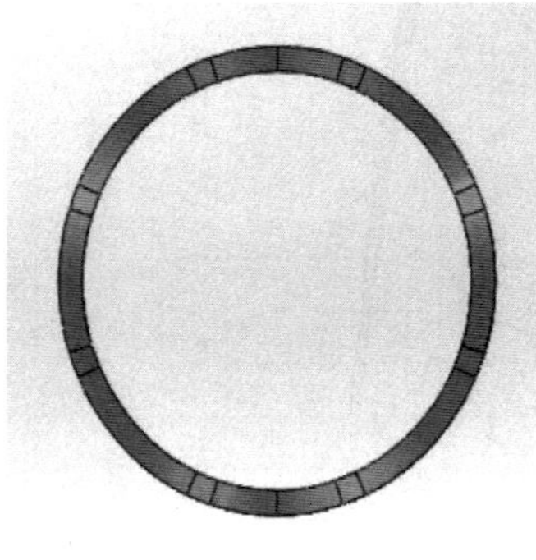

Figure 2(b). Top view of Stent

The reference values considered for geometrical design of the stent is given in Table 1.

Table 1. Geometric details of the stent.

S. No.	Parameters	Dimension
1.	Total length	7.01mm
2.	Inner diameter	2mm
3.	Outer diameter	2.24mm
4.	Thickness	0.12mm

2.2. Meshing:

Meshing is important part of simulation which helps to obtain accurate result from an FEA model where complex geometries are divided into simple elements. Typically the accuracy depends on the mesh size. Smaller mesh size gives more accuracy in results of the provided physical domain. Various mesh views of stent are show in Figure 3(a) & (b).

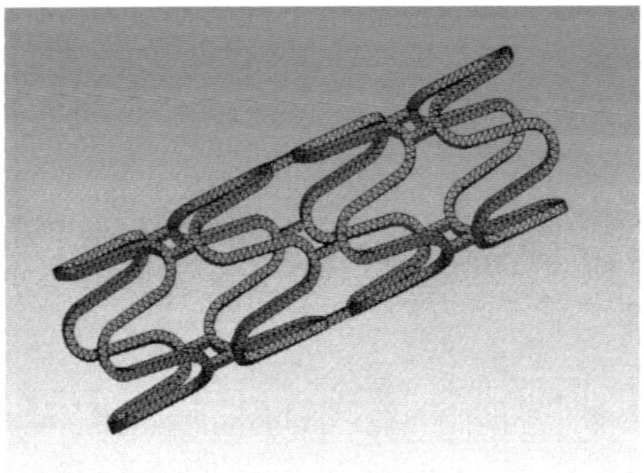

Figure 3(a). Complete mesh view of stent

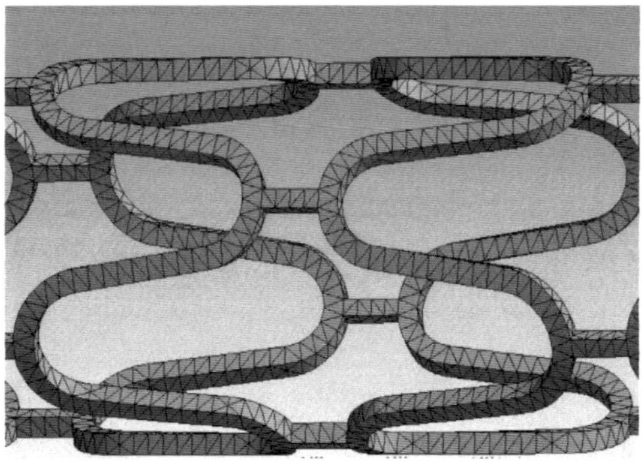

Figure 3(b). Detailed mesh view near the strut

2.3. Materials and their properties:

For this analysis we are taking 316L stainless steel, Mg alloy (WE43) and Nitinol. Austenitic stainless steel type 316L is most commonly used stent material with outstanding corrosion resistance. Mg alloy (WE43) is also known as biodegradable material for stent which consists of lowest density. It also shows high strength. While Nitinol is alloy of Ni and Ti which is used in stents due to its super-elasticity, shape memory effect and biased stiffness. Those material properties are shown Table 2 as follows.

Table 2. Material Properties.

	AISI 316L SS	Mg Alloy (WE43)	Nitinol
Density	7850 kgm^{-3}	1840 kgm^{-3}	6450 kgm^{-3}
Poisson's ratio	0.31	0.27	0.32
Young's Modulus	200000 MPa	44000 MPa	28000 MPa
Specific heat	0.76986 cal/(g)(K)	0.966 cal/(g)(K)	0.32 cal/(g)(K)

where m= meter, kg=kilogram, Pa = Pascal (N/m^2).

2.4. Solver setting:

For performing the simulation the commercial software ANSYS R19.2 is used. We know that blood pressure of humans vary from 80 to 120 mm of Hg. For calculation purpose we are considering the 100 to 120 mm of Hg pressure which is common for young adults and old peoples. Explicit Dynamics analysis system is used for specific time integration.

3. Results and conclusion

The stent model was subjected to FEA using three materials 316L Stainless steel, Mg alloy (WE43) and Nitinol with pressure ranging from 100mmHg to 120mmHg. The results of the stent analysis study are tabulated in Table 3.

Table 3. Result of stent finite element analysis.

Sr No	Material	Pressure (mmofHg)	Von mises stress (MPa)		Equivalent strain (m/m)		Resultant deformation (m)	
			Min	Max	Min	Max	Min	Max
1	AISI 316L Stainless Steel	100	0.044698	2.4641	2.7977e-7	1.3958e-5	0	1.2103e-7
		110	0.049496	2.7095	3.0462e-7	1.5349e-5	0	1.3310e-7
		120	0.053838	2.9556	3.3159e-7	1.6739e-5	0	1.4517e-7
2	Mg alloy (WE43)	100	0.10686	3.0533	3.2119e-6	8.3344e-5	0	6.1068e-7
		110	0.11745	3.3585	3.5269e-6	9.1773e-5	0	6.7175e-7
		120	0.12804	3.6635	3.8423e-6	0.0001001	0	7.3281e-7
3	Nitinol	100	0.077030	2.1871	3.1450e-6	9.8089e-5	0	7.0017e-7
		110	0.084824	2.4078	3.4638e-6	0.00010797	0	7.7076e-7
		120	0.092635	2.6296	3.7832e-6	0.00011787	0	8.4144e-7

The colour plots of equivalent stress and Equivalent strain are shown in figure 4 and figure 5 respectively.

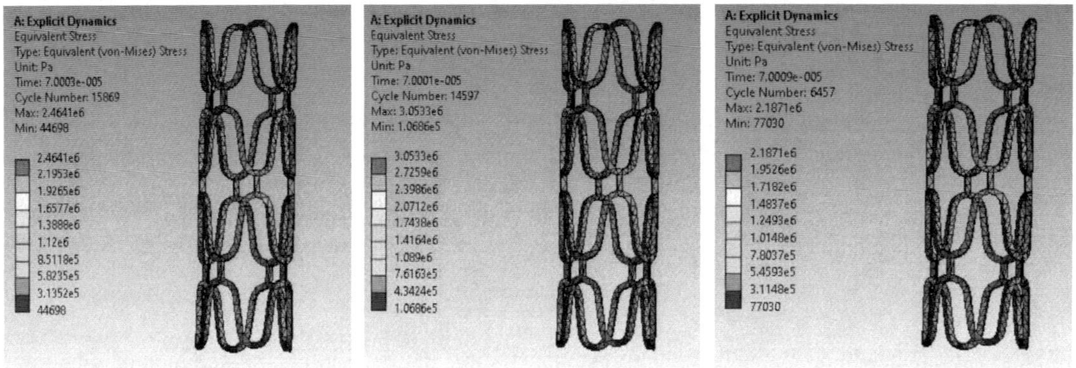

(i) (ii) (iii)

Figure 4(a). Equivalent stress at 100mmHg pressure for (i)316L SS, (ii)Mg alloy, (iii)Nitinol

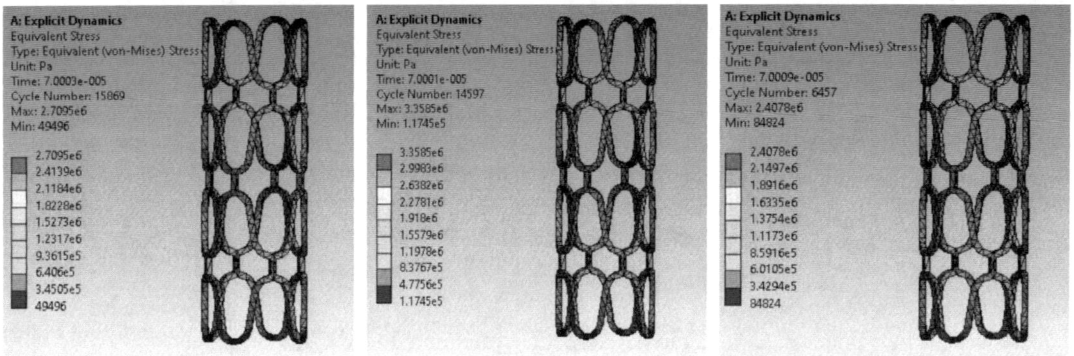

(i) (ii) (iii)

Figure 4(b). Equivalent stress at 110mmHg pressure for (i)316L SS, (ii)Mg alloy,(iii) Nitinol

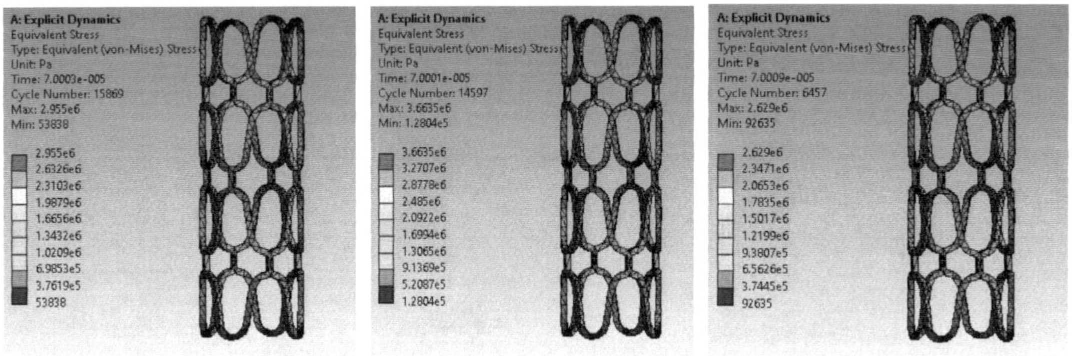

(i) (ii) (iii)

Figure 4(c). Equivalent stress at 120mmHg pressure for (i)316L SS,(ii) Mg alloy,(iii) Nitinol

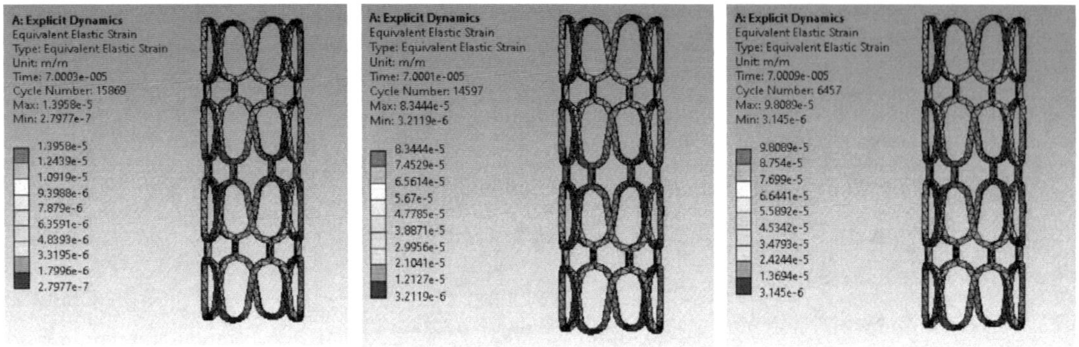

(i) (ii) (iii)

Figure 5(a). Equivalent elastic strain at 100mmHg pressure for (i)316L SS,(ii) Mg alloy, (iii) Nitinol

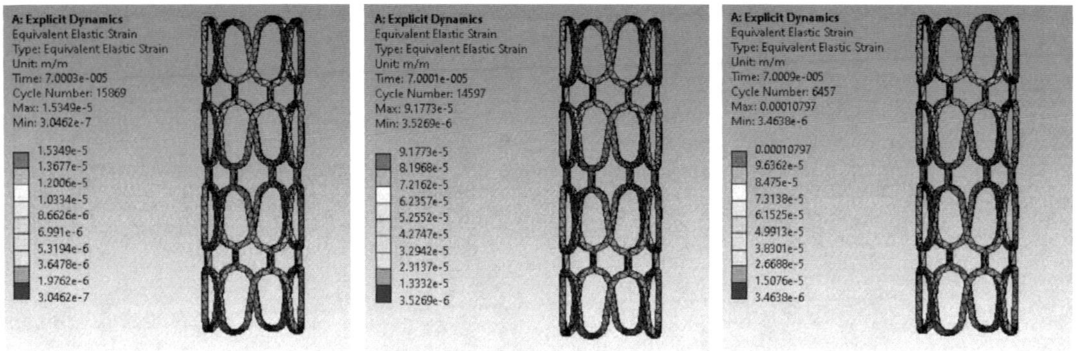

(i) (ii) (iii)

Figure 5(b). Equivalent elastic strain at 110mmHg pressure for (i)316L SS,(ii) Mg alloy, (iii) Nitinol

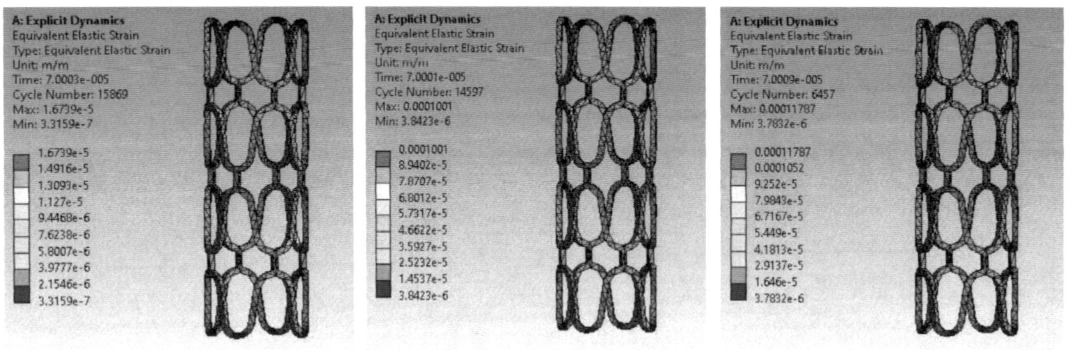

(i) (ii) (iii)

Figure 5(c). Equivalent elastic strain at 120mmHg pressure for (i)316L SS,(ii) Mg alloy, (iii) Nitinol

From the Table.3 it is clearly evident that Nitinol shows lowest equivalent stress values ranging from 2.18MPa-2.62MPa whereas equivalent stress for 316L stainless steel ranging from 2.46MPa-2.955MPa and for Mg alloy 3.05MPa- 3.66MPa which is highest of all. In case of equivalent elastic strain, Nitinol shows highest mean value which is 4.7477×10^{-3} whereas 316L SS shows lowest value. This computational study proves that Nitinol is best amongst all other materials in this analysis. Nitinol shows less stress values and high strain under the Pressure range for a stent model and remains stable under hostile environment in human body for small sized lesions. Thinner struts improve flexibility which signifies increase in inner diameter and decrease in potential injuries. Literature study shows that considering high-risk patients with small vessels, diabetes arteries shows impermissible high rate of restenosis. Mechanical properties of Nitinol ensure biomechanical characteristics to stents. Biased stiffness confirms stent recovery from the crimped position which is significantly more impervious to compression than to extension. With reference to the stress-strain curve, due to presence of the plateau it also shows good kink resistance. Not only for stent application but for orthodontic use it shows corrosion resistance as compare to stainless steel. But nickel release from nitinol indicates issue with biocompatibility. It can be improved with the help of coating. However coating processes are different for stent patterns.

4. Future Scopes

For the Biodegradable Drug eluting Stent effect of Nitinol stent with various coating materials to be studied which will be useful to overcome the current issues related to Peripheral artery Diseases.

5. References

[1] Javaid Iqbal, Julian Gunn, Patrick W. Serruys, (2013)Coronary stents: historical development, current status and future directions, *British Medical Bulletin*, Volume 106, Issue 1,Pages 193–211, https://doi.org/10.1093/bmb/ldt009

[2] Sangiorgi G, Melzi G, Agostoni P,et al (2007), A.Engineering aspects of stents design and their translation into clinical practice. Annali dell'Istituto Superioredi Sanita,Volume 43,Issue1,Pages89–100.

[3] Watson T,Webster, M.W, Ormiston J.A, Ruygrok P.N & Stewart J.T. (2017) Long and short of optimal stent design.Open Heart, Volume 4,Issue 2.https://doi.org/10.1136/openhrt-2017-000680

[4] Lim W-H, Seo W-W, Choe W. (2011) "Stent Coated With Antibody Against Vascular Endothelial-Cadherin Captures Endothelial Progenitor Cells, Accelerates Re-Endothelialization, and Reduces Neointimal Formation". Arteriosclerosis, Thrombosis, and Vascular Biology,Volume 31,Pages 2798-2805. https://doi.org/10.1161/ATVBAHA.111.226134

[5] Lüscher, T. F., Steffel, J., Eberli, F. R., Joner, M., Nakazawa, G., Tanner, F. C., & Virmani, R. (2007). Drug-eluting stent and coronary thrombosis: biological mechanisms and clinical implications. *Circulation*, Volume 115,Issue 8,Pages 1051-1058. https://doi.org/10.1161/CIRCULATIONAHA.106.675934

[6] American Heart Association ACCF/AHA/SCAI Guideline for Percutaneous Coronary Intervention, (2011).

[7] Dangas, G., Kuepper F. (2002) "Restenosis: Repeat Narrowing of a Coronary Artery."Circulation 220, Volume 105, Pages 2586-2587. https://doi.org/10.1161/01.CIR.0000019122.00032.D

[8] Hanawa, T. (2000) Materials for metallic stents. *J Artif Organs* ,Vol.12, Pp73–79. https://doi.org/10.1007/s10047-008-0456-x

[9] Paul J. Kolston (2000), *Finite element modelling: a new tool for the biologist* ,Volume 358, Pages 611-631. http://doi.org/10.1098/rsta.2000.0548

[10] Stoeckel D, Pelton A, Duerig T. (2004) Self-expanding nitinol stents: material and design considerations. European Radiology.;14(2):292-301. DOI: 10.1007/s00330-003-2022-5

[11] Maleckis, K., Anttila, E., Aylward, P., Poulson, W., Desyatova, A., MacTaggart, J., & Kamenskiy, A. (2018). Nitinol Stents in the Femoropopliteal Artery: A Mechanical Perspective on Material, Design, and Performance. *Annals of biomedical engineering*, Volume 46,Issue 5, Pages 684–704. https://doi.org/10.1007/s10439-018-1990-1

Tensile properties of hybrid kenaf/glass fiber reinforced epoxy composites with different stacking sequence using experimental and FEA simulation method

M N M Ansari[1,2], K S Vinoth[2], Afeefah Nordin[1], Alaseel Bassam[2], and A Topa[3,4]

[1] Institute of Power Engineering, Universiti Tenaga Nasional, 43000 Kajang, Selangor,Malaysia.
[2] Mechanical Engineering Department, Universiti Tenaga National (UNITEN), 43000 Kajang, Malaysia.
[3] Ocean and Ship Technology Group, Institute of Transportation Infrastructure, Universiti Teknologi PETRONAS, Perak, 32610, Malaysia.
[4] Department of Maritime Technology, Faculty of Ocean Engineering Technology and Informatics, Universiti Malaysia Terengganu, Kuala Terengganu, 21300, Malaysia.

Corresponding author: ansari@uniten.edu.my; ansari.uniten@gmail.com

Abstract. The conventional overhead transmission line insulators are usually made of glass and ceramics, but the materials are heavy and bulky. Therefore, kenaf fibre was chosen as partial replacement material to be reinforced with Glass/epoxy composite to reduce the weight and bulkiness of the composite structures. This study used Epoxy as matrix and kenaf/glass fibres to reinforce fillers to fabricate the composites in various stacking sequences. Finite Element Analysis (FEA) was carried out using LS-DYNA software to investigate the tensile properties of the hybrid composite. The results showed a significant improvement in tensile values of the modelled composite exhibited by the kenaf/glass fibre. The tensile strength of the hybrid composites was comparable with the Glass/Kenaf/Epoxy hybrid composites' experimental results. About 200% tensile strength improvement can be observed for both experimental and numerical for G/K/G/E sample than pure epoxy samples, thus making the new composite a potential substitute for insulator applications.

1. Introduction

The environmental issues, production cost and the decreasing of petroleum sources are the primary concern for the industrial sectors for the past three decades [1]. Extensive research has been deployed in renewability and sustainability to find a suitable material that can replace the existing conventional materials. This has resulted in an extensive growing interest in studying natural fibres where these natural fibres are used as a reinforcement in polymer matrix composites [2]. Researchers, scientist and polymer engineers are attracted towards natural fibre-based composites. Due to its environmental friendliness, lightweight, high durability, non-toxic and safer health and biodegradable properties. Natural Fibre is considered a potential substitute material that can be used in automotive, construction, furniture, and other mass-production industries [3]. Natural fibres are readily available fibres that can be extracted from animals, plants and mineral resources. Generally, plant-based natural fibres are suitable for natural reinforcement materials in polymer composites [4]. The most common plant-based natural fibres used in applications are bast and core fibres such as flax, jute, hemp, ramie, sisal and kenaf [5]. Among all the plant-based natural fibres, kenaf is considered cheap renewable material, offering a reasonable choice to commercialize composite materials to advance high-performance engineering products [6]. Kenaf fibre belongs to species of *Hibiscus Cannabinus* derived from plants of the genus *Hibiscus*. Kenaf plants can grow up to more than 3m in height and a base diameter of 3 to 4 cm. Kenaf is ready to be harvested in 4-5months and consists of 65% core and 35% bast fibres [7]. Kenaf fibre has superior tensile strength, flexural strength, and good impact strength and low density [8].

The tensile characterization of kenaf fibre reinforced polypropylene (PP) composite has been studied by Yuhazri et al. (2017). The author made 6 different samples through a hand lay-up method that consisted of one layer of kenaf fibre in a sample to 6 layers of kenaf fibre. The result showed that the sample consisted of 6 layers of kenaf fibre reinforced P.P. composites achieved 5.531kN and the lowest was performed by one layer of kenaf fibre reinforced P.P. composites (0.613kN) [9]. On a separate note, research was conducted by Kotresh et al. (2014) to study the tendon properties as a part of their study on tendons on the human body. Kenaf fibres were used with various weight percentages (10%, 20%, 30% and 40%) in polyester resin. The experiment resulted in 40 wt. % kenaf fibre reinforced polyester composites having the highest tensile strength, and it was proven that the addition of kenaf fibres led to an increase in tensile strength of the material [10]. Despite its advantages, kenaf fibre application alone will not achieve the required properties targeted for particular application because of the plant based natural fibres characteristics that are hydrophilic [11]. High moisture sensitivity of natural plant fibre can affect the durability of the resultant material [12]. Therefore, the research on hybridization between natural and synthetic fibres has increased as this class of composite material can resolve the drawbacks of single natural fibre reinforced composites. Hybrid composites are generally defined as a composite that consists of two or more types of fibres within a single matrix [13,14,15]. Recent studies have reported that several factors influence the performance of hybrid composites, including fibre orientation, fibre length, stacking sequence, configurations, the intermingling of the fibre and fibre-matrix bonding [16]. Glass fibre is a synthetic fibre commonly used in various applications. It is considered the cheapest among all synthetic fibres and imposed high temperature, non-corrosive, and possesses high tensile strength [17]. Several studies have been conducted to study the effect of hybridization on mechanical properties of composites. Zhafer et al. (2017) compared the flexural and impact strength between varying weight fraction utilized in kenaf/glass reinforced unsaturated polyester composite fabricated by hand lay-up method. The outcome revealed that flexural strength and impact increased with the increasing kenaf fibre and glass fibre loading until 20 wt.% kenaf/ 10wt.% glass fibre composite [18].

Meanwhile, another finding by Nurshairatun et al. (2017) on the flexural properties and moisture resistance ability of non-woven kenaf/glass fibre mat reinforced unsaturated polyester (UPE) composites revealed that the existence of glass fibre at the outermost layer of the hybrid composite such as UPE/GF/KF/GF showed highest flexural properties. However, the results of water absorption showed that samples with kenaf fibre at the outermost layer had the highest water absorption due to kenaf fibre's hydrophilic nature [19]. Since the experimental work involves purchasing, preparing, fabricating and testing the material, the experimental works' time can be reduced by the introduction of numerical simulation methods where the experimental work will be conducted by using Finite Element Analysis (FEA) software. The FEA software can be utilized to obtain solutions to a large class of engineering problems involving stress analysis, heat transfer, electromagnetism and fluid flow. A previous research utilized ABAQUS simulation analysis and compared the results with the composites prepared using hand lay-up technique. The percentage of error was calculated to determine the simulation analysis's accuracy to the experimental result [20]. Another work was done on tensile properties of short glass fibre/reinforced polyamide composites (PA66-GF) and short carbon fibre reinforced polyamide composites (PA66-CF). The experimental sample was prepared using an injection moulding process at different orientations relative to the preferential fibre direction (0°,

45°,90°). The numerical simulation was done by utilizing LS-DYNA software with 2D and 3D model designs. The results indicated that the 2D model design for both composites had better numerical-experimental correlation than the 3D model. The stress-strain curves from the 2D model for both composites retraced the experimental result [21].

In this paper, the tensile strength of hybrid kenaf/glass fibre reinforced epoxy composites with different stacking sequences was examined through FEA using LS-DYNA software and compared with data from the authors' previous experimental work [22]. For experimental analyses, ASTM D3039 was adopted for the specimen dimensions, and six different stacking sequences of composite layers were investigated viz. E, K/E, G/K/G/E, K/G/K/E, K/G/K/G/E and G/E as explained in our preliminary study [22].

2. Finite Element Modelling Details

2.1. Finite Element Discretization and Element Type

The geometry of the specimen with the appropriate dimensions was generated and meshed with LS-Prepost. Auto Mesher tool was used to discretize the geometry into shell elements with an average size of 0.5 mm, which yields to a total number 6480 element per composite layer. Both ends of the specimen have coarse mesh and were modelled as rigid bodies. These portions of the specimen were firmly gripped in the experimental work, and thus their deformation was negligible. Each composite layer has the same mesh scheme as it was duplicated from the first layer using Translate command with the Copy Elements option. The specimen having the stacking sequence of K/G/K/G is shown in Fig. 1. The default shell element type (Belytshko-Tsay formulation) was used with two integration points through-thickness and a recommended shear factor of 0.833. This element formulation was selected due to its computational efficiency. The layers' cohesiveness was modelled with tiebreak contact (TIEBREAK_SURFACE_TO_SURFACE), and hourglass control was adopted to prevent hourglass modes.

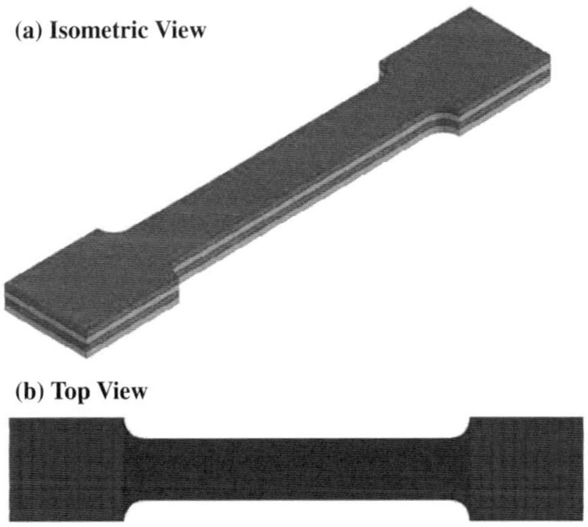

Figure 1. 3D model of the tensile specimen: (a) Isometric View showing the stacking of different layers: K/G/K/G specimen (b) Top view showing the meshing scheme.

2.2. Material Model

The assignment of material properties was implemented by selecting the appropriate material model. While LS-DYNA provides a massive library of material models, the enhanced composite damage material model (MAT054) is widely used in modelling composite structures due to its versatility and robustness [23-28]. The material properties can be defined along with three orthogonal directions, and brittle failure can be incorporated. When undamaged, the material behaves linearly whereas a nonlinearity option is available through damage parameters. The mechanical properties and the damage parameters of the composite layers are listed in Table 1. These properties are adopted from previous studies [29-32] and calibrated to obtain the experimental results' stress-strain curves [22]. The pure epoxy specimen was modelled with the piecewise linear plasticity material model (MAT024) with an Elastic modulus of 266 MPa, a tensile strength of 26.5 MPa, which was defined using MAT_ADD_EROSION. The nonlinear plastic behaviour was captured by adopting the experimental stress-strain curve into the material model. Similar approach of modelling plastics is reported in [33].

Table 1: Mechanical Properties of the Composite Laminate (MAT054).

Non-woven kenaf fibre reinforced with epoxy matrix					
	Value	Units		Value	Units
EA	0.9	GPa	GAB	0.335	GPa
EB	0.9	GPa	GBC	0.335	GPa
EC	0.072	GPa	GCA	0.335	GPa
PRBA	0.340	-	RO	1300	kg/m^3
PRCA	0.0193	-	XC	5.769	MPa
PRCB	0.0193	-	XT	25.00	MPa
EFS	0.135	-	YC	2.692	MPa
SC	1.154	MPa	YT	0.577	MPa
Non-woven glass fibre reinforced with epoxy matrix					
	Value	Units		Value	Units
EA	38.50	GPa	GAB	4.20	GPa
EB	38.50	GPa	GBC	4.20	GPa
EC	32.98	GPa	GCA	2.60	GPa
PRBA	0.0787	-	RO	1174	kg/m^3
PRCA	0.0787	-	XC	29.31	MPa
PRCB	0.4206	-	XT	90	MPa
EFS	0.0152	-	YC	85	MPa
SC	7.41	MPa	YT	27.68	MPa

Abbreviation: E.A.: E_a, Young's Modulus in a-direction, E.B.: E_b, Young's Modulus in b-direction, E.C.: E_c, Young's Modulus in c-direction, PRBA: v_{ba}, Poisson Ratio, ba, PRCA: v_{ca}, Poisson ratio, ca, PRCB: v_{cb}, Poisson ratio, cb, EFS: Effective failure strain, SC: Shear Strength in A.B. plane.

2.3. Boundary Conditions

One end of the specimen is fixed in all degrees of freedom via the constraints option in MAT_RIGID card definition, whereas the other end is allowed to translate only along the direction of force (x-axis). PRESCRIBED_MOTION_RIGID keyword is used to impose the displacement on the moving end at a constant rate of 5 mm/min. Due to the slow speed of the test, mass-scaling is adopted to expedite the simulation run time.

3. Results and Discussion

A comparison between the FEA simulation and experimental results' stress-strain curves is shown in Fig. 2 and Fig. 3. Comparison between the two figures showed that there were good agreements between experimental and numerical results. The numerical model was found to be very accurate in reproducing the mechanical behaviour of the samples. Indeed, the stress-strain curves obtained from the simulations retraced the experimental ones up to the fracture of the samples.

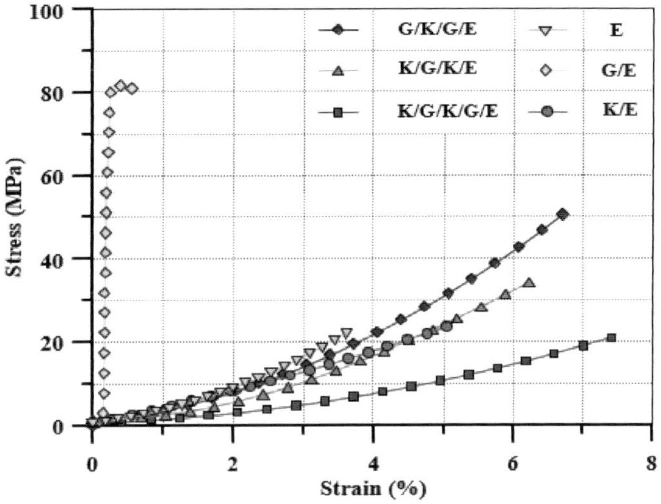

Figure 2. Simulation results of stress strain of E, K/E, G/K/G/E, K/G/K/E, K/G/K/G/E and G/E.

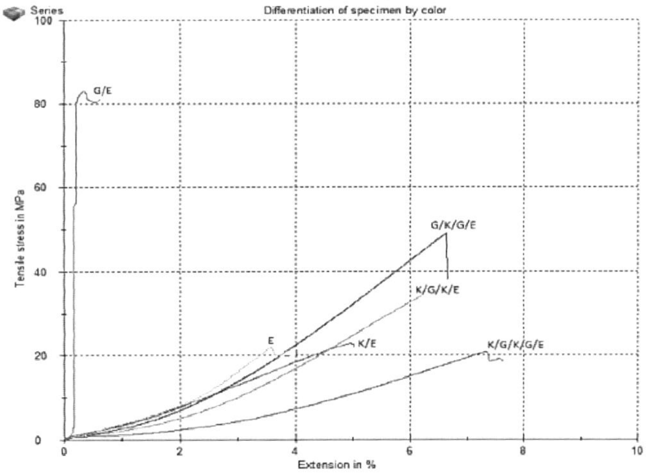

Figure 3. Experimental test results of stress-strain of E, K/E, G/K/G/E, K/G/K/E, K/G/K/G/E and G/E composite [22].

The G/K/G/E specimen had the maximum tensile strength among the hybrid Kenaf composites with a value reaching 49.0 MPa and the second-highest fracture strain of 6.6%. It was observed that the K/G/K/G/E hybrid composite had the highest value of fracture strain at 7.25%. Nevertheless, it was the weakest in terms of tensile strength while compared to the other specimen. Therefore, in this case, it can be seen that adding another layer of kenaf on top of G/K/G/E resulted in a considerable reduction of the tensile strength. This could be due to poor interfacial bonding between kenaf and glass fibre and the ineffectiveness of Kenaf fibre to transfer stress to the hybrid composite compared to specimens with glass fibre surfaces that finally resulted in such failure.

The K/G/K/E hybrid composite performance was slightly different, with a tensile strength of 33.2 MPa and fractured strain at 6.2%. Both K/G/K/E and K/G/K/G/E had lower tensile strength than G/K/G/E. It can be concluded that having glass fibre on the surface side of the composite is the better option regarding the high tensile property of glass fibre. It is worth noting that the tensile strength would be more significant when the excessive strength material is used as the surface, which is the primary load-bearing component in tensile strength calculation. This indicated that glass fibre on the surface was able to withstand the tensile stress. The kenaf core was bearing the pressure and distributed them consistently along with the hybrid composite.

Consequently, this shows that the stacking sequence plays a significant role in the performance of hybrid composites. Besides, a natural fibre hybrid composite's strength is highly dependent on the failure strain of the individual fibres [34]. Epoxy matrix has plastic behaviour and should display large plastic deformation, but the plastic zone was hardly observed in all hybrid samples. This could be due to the inefficiency of the surface quality of the composites and resulted in stiff samples. Plastic deformation is the permanent distortion occurring when a material is subjected to stresses that exceed its yield strength and finally cause it to elongate or bend before it eventually ruptures. The tensile strength (MPa) values and strain (%) for both experimental and simulation samples were tabulated in Table 2. The values were taken at four points along with the graph at 25%, 50%, 75% and 100%, i.e. when the sample breaks.

Table 2: Tensile strength (MPa) and strain (%) for experimental and simulation samples of kenaf/epoxy hybrid composites.

Sample ID	% increase in the stress-strain curve	Experimental		Simulation	
		Tensile Strength (MPa)	Strain (%)	Tensile Strength (MPa)	Strain (%)
K/E	25%	5	1.6	5	1.3
	50%	13	2.6	11	2.5
	75%	16	3.8	17	3.7
	100%	22	5.2	21	5.1
G/K/G/E	25%	5	1.6	7	1.7
	50%	15	3.3	18	3.6
	75%	31	4.8	27	4.7
	100%	49	6.5	50	6.6
K/G/K/E	25%	4	1.7	4	1.6
	50%	12	3.2	11	3.2
	75%	22	4.6	23	4.8
	100%	34	6.2	35	6.3
K/G/K/G/E	25%	2	1.8	3	1.8
	50%	6	3.7	7	3.7
	75%	13	5.5	13	5.6
	100%	21	7.4	20	7.4
G/E	25%	2	0.2	8	0.1
	50%	20	0.3	32	0.3
	75%	55	0.4	70	0.4

	100%	80	0.5	81	0.5
E	25%	4	0.8	4	0.8
	50%	8	1.7	7	1.6
	75%	12	2.5	15	2.7
	100%	21	3.5	22	3.5

*25%, 50%, 75% and 100% = percentage in each stress/strain curve

It can be seen that the values of experimental and simulation for tensile strength and strain were 95% accurate, and hence the model is said to be reliable for such analysis. Nonlinearity behaviour was observed in the hybrid composites and the G/E composites, which is due to the micro failure of the glass fibres. The G/E and E had the lowest strain and was very brittle compared to the samples with the addition of Kenaf fibre. As natural fibre is flexible, it contributes to the elongation of the hybrid composite before it finally ruptured. In contrast, the K/E composite demonstrated perfect linear response. The sample G/K/G/E had the highest tensile strength value than other kenaf hybrid composites which was 49 MPa. The simulation sample result showed a value of 50 MPa, which was very close to the actual value. The same trend was also observed for the strain value, which was 6.5% and 6.6% for experimental and simulation, respectively. The FEA also showed high accuracy at 50% stress/strain curve of the same sample, where the value of tensile strength was 15 MPa (experimental) and 18 MPa (simulation). In comparison, strain was 3.3% (experimental) and 3.6% (simulation). The overall comparison between experimental and simulation results for each graph's four points showed good agreement except for sample G/E at 50% stress/strain curve. The simulation result showed tensile stress's value to be 32 MPa, but the actual value was 20 MPa. There was some deviation of the simulation value compared to the experimental result.

4. Conclusion and Future Work

In the present study, the numerical simulations of the tensile test of the hybrid composite were successfully performed using the nonlinear finite element software LS-DYNA. The numerical simulations details were presented, and an orthotropic material model with damage was adopted for modeling the behaviour of the composite specimens. The material properties were adopted from previous studies, and the model was calibrated to obtain satisfactory results compared to the experimental data. The model was capable of reproducing the experimental stress-strain curve, and the brittle failure mode was captured. Micro damage of the material was observed through the nonlinearity of the stress-strain curve, and the influence of the stacking sequence was highlighted. About 200% improvement was observed in the hybrid composite in comparison with pure epoxy composite. Stacking sequences had proven that it could affect the properties of hybrid composites as portrayed by both experimental and numerical results. There were 2% of deviation errors between experimental and simulation findings, thus highlighting the reliability of the numerical model.

Author's contribution

M.N.M.Ansari[1,*] - Idea & concept generation, suggestion and advice on the simulation studies, suggestion on interpretation of the data, supervision of the project and manuscript editing. **K.S.Vinoth**[2]**:** first draft and manuscript writing, experimental data collection, analyses of data. **Noor Afeefah Nordin**[1] - project management, suggestion and advice on the analyses, manuscript editing and revision. **Alaseel Bassam** - simulation and interpretation of the data, manuscript editing. **Ameen Topa**[3] - simulation and analytical tools application, simulation results and interpretations, manuscript editing.

Acknowledgements

The authors would like to acknowledge UNITEN R&D Sdn.Bhd. (URND) of the Universiti Tenaga Nasional (UNITEN) for the TNB Seed Fund Program funded by the Tenaga Nasional Berhad, Malaysia (Project code: U-TS-RD-18-10) for the funding to carry out this research and Universiti Teknologi PETRONAS and Universiti Malaysia Terengganu for their support on the software simulation part of the project.

References

[1] Khalil HPSA, Suraya NL, Atiqah N, Jawaid M, Hassan A. Mechanical and thermal properties of chemical treated kenaf fibers reinforced polyester composites. Journal of Composite Materials 2012;47(26):3343-3350.
[2] Nirmal U, Lau STW, Hashim J. Interfacial Adhesion Characteristics of Kenaf Fibers Subjected to Different Polymer Matrices and Fiber Treatments. Journal of Composites 2014; Article ID 350737:12 pages | https://doi.org/10.1155/2014/350737.
[3] Paukszta D, Borysiak S. The influence of processing and the polymorphism of Lignocellulosic fillers on the structure and properties of composite materials: A Review. Materials 2013;6(7):2747-2767.
[4] Akil HM, Omar MF, Mazuki AAM, Safiee S, Ishak ZAM, Bakar A.A. Kenaf fiber reinforced composites: A review. Materials and Design 2011;32:4107-4121.
[5] Dehbari N, Moazeni N, Rahman WA. Effects of kenaf core on properties of poly (lactic acid) bio-composite. Polymer Composites 2014;35(6):1220-1227.
[6] Radzi MKFM, Muhamad N, Akhtar MN, Razak Z, Foudzi FM. The Effect of Kenaf Filler Reinforcement on the Mechanical and Physical Properties of Injection Moulded Polypropylene Composites. Sains Malaysiana 2018;47(2):367-376.
[7] Salleh Z, Taib YM, Hyie KM, Mihat M, Berhan MN, Ghani MAA. Fracture toughness investigation on long kenaf/woven glass hybrid composite due to water absorption effect. Procedia Engineering 2012;41:1667–1673.
[8] Suresh R, Reddy SBA, Ravi VS. Natural Fiber Reinforced Green Composites with Epoxy Polymer Matrix and its Mechanical Properties Analysis. International Journal on Recent Technologies in Mechanical and Electrical Engineering (IJRMEE) 2018;5(6):12-19.
[9] Yuhazri MY, Amirhafizan MH, Abdullah A, Yahaya SH, Lau STW. Potentially of utilizing non-woven kenaf fiber composite for car door map pocket. Journal of Advanced Manufacturing Technology 2017;11(2):129-138.
[10] Sardar K, Veeresh K, Gowda M. Characterization and Investigation of Tensile Test on Kenaf Fiber Reinforced Polyester Composite Material. International Journal of Recent Development in Engineering and Technology 2014;2(6): 104-112.
[11] Hidayah IN, Syuhada DN, Khalil HPSA, Ishak ZAM, Mariatti M. Enhanced performance of lightweight kenaf-based hierarchical composite laminates with embedded carbon nanotubes. Materials and Design 2019;171
[12] Rahul K, Kumar K, Karunakar CH. National Conference on Trends in Science, Engineering & Technology by Matrusri Engineering College & IJCRT 2018, ISSN: 2320-2882.
[13] Nam TH, Ogihara S, Tung NH, Kobayashi S. Effect of alkali treatment on interfacial and mechanical properties of coir fiber reinforced poly (butylene succinate) biodegradable composites. Composites Part B: Engineering 2011;42(6):1648-1656.
[14] Ashori A. Hybrid composites from waste materials. Journal of Polymers and the Environment 2010;18(1):65-70.
[15] Sivakumar D, Ng L.F., Chew RM, Bapokutty O. Investigation on failure strength of bolted joints woven fabric reinforced hybrid composite, International Review of Mechanical

Engineering 2017;11(2):138-143.
[16] Bajpai PK, Singh I. and Madaan J. Comparative studies of mechanical and morphological properties of polylactic acid and polypropylene based natural fiber composites. Journal of Reinforced Plastics and Composites 2012;31(24):1712–24. doi:10.1177/0731684412447992.
[17] Sivakumar DM, Kathiravan S, Ng L.F., Siti Hajar SMDF, Sivaraos S. Mechanical properties of plain woven kenaf/glass fiber reinforced polypropylene hybrid composites. Materials Testing 2019;61(11):1095-1100.
[18] Zhafer SF, Rozyanty AR, Musa SBSSL. Flexural Properties and Impact Strength of Kenaf-Glass Fiber Unsaturated Polyester Hybrid Composites. Advanced Materials Engineering and Technology. AIP Conference Proceedings 2017;1835,020019.
[19] Nurshairatun AR, Bakar MBA, Masri MN, Sulaiman MA, Mohamed M, Thirmizir MZM. Mechanical and Water Absorption Properties of Hybrid Kenaf/Glass Fiber Mat Reinforced Unsaturated Polyester Composites. Material Science Forum 2017;888:228-233.
[20] Nabila H, Hamat S, Arumugam L, Umaira A, Alibe IM, Golshan M, Majid DL. Glass/Epoxy woven composite laminate design based on nonlinear results. ARPN Journal of Engineering and Applied Sciences 2015;10(21):9992-9998.
[21] Molaro A, Lanzillo MS, Zimbardi F, Causa A, Villacci B. Numerical-Experimental Correlation of Mechanical tests on Fiber-Reinforced Polyamide Composites. 11th European LS-DYNA Conference 2017.
[22] Vinoth K, Ansari MNM, Begum S, Yahya Z, Atiqah A. Effect of Fiber Loading on Tensile Strength of Kenaf/Glass Fiber Epoxy Hybrid Composite for Insulator Application. Materials Today 2019:1-8.
[23] Charoenphan S, Bank LC, Plesha ME. Use of LS-DYNA Shell Element in the Analysis of Composite Plates with Unbalanced and Unsymmetric Lay-ups. In The 6th International LS-DYNA Users Conference-Simulation 2000;8:47-55.
[24] Her S-C, Liang Y-C. The finite element analysis of composite laminates and shell structures subjected to low velocity impact. Composite Structures 2004;66:277–285
[25] Shokrieh MM, Tozandehjani H, Omidi MJ. Effect of fiber orientation and cross section of composite tubes on their energy absorption ability in axial dynamic loading, Mechanics of Composite Materials 2009;45(6):567-576.
[26] Chandekar GS, Thatte BS, Kelkar A.D. On the behavior of fiberglass epoxy composites under low velocity impact loading. Advanced in Mechanical Engineering 2010;2:1-11
[27] Wade B, Feraboli P, Osborne M. Simulating laminated composites using LS-DYNA material model MAT54 part I: [0] and [90] ply single element investigation. FAA JAMS: 2012, Baltimore, MD. USA.
[28] Osborne M. Single-element characterization of the LS-DYNA MAT54 Material Model (Doctoral Dissertation). 2013.
[29] Chang FK, Chang KY. A progressive damage model for laminated composites containing stress concentrations. Journal of composite materials 1987;21(9):834-855.
[30] Chang FK, Scott RA, Springer GS. Failure strength of nonlinearly elastic composite laminates containing a pin loaded hole. Journal of Composite Materials 1984;18(5):464-477.
[31] Yousif BF, Shalwan A, Chin CW, Ming KC. Flexural properties of treated and untreated kenaf/epoxy composites. Materials and Design 2012;40:378–385.
[32] Alkhatib F, Mahdi E, Dean A. Crushing response of CFRP and KFRP composite corrugated tubes to quasi-static slipping axial loading: Experimental investigation and numerical simulation. Composite Structures 2020;246: 1-29.
[33] Dar UA, Mian HH, Abid M, Topa A, Sheikh MZ, Bilal M. Experimental and numerical investigation of compressive behavior of lattice structures manufactured through projection micro stereolithography. Materials Today Communications. 2020;25:101563.
[34] Saba N, Jawaid M, Alothman OY, Paridah MT. A Review on Dynamic mechanical analysis of natural fiber reinforced polymer composites. Construction and Building Materials

2016;106:149-159.

Investigation on the Effect of Charge Carriers in Chemical Shift Anisotropy of Si Single Crystal

E. Viswanathan[1,2,3*], K. Padma Prasad[2,4*] and S. Ganapathy[2]

[1]Department of Nuclear Physics, University of Madras, Guindy Campus, Chennai 600025

[2]NMR Facility, Physical Chemistry Department, National Chemical Laboratory, CSIR, Pune 411008

[3]Department of Physics, Sri Sankara Arts and Science College, Enathur, Kancheepuram 631561

[4]JSS College of Arts, Commerce and Science, Ooty Road, Mysore 570025

*email-id:vsuscient@gmail.com, kppsix@gmail.com

Abstract. The Chemical Shift Anisotropy (CSA) of single crystalline silicon (Si) was investigated using ^{29}Si static NMR for commercially procured *p*-type and *n*-type silicon. The observed CSA has been distributed between -80 *ppm* to -90 *ppm* for *n*-type and *p*-type Si. The NMR parameters such as Isotropic Chemical Shift (δ_{iso}), Chemical Shift Anisotropy tensor ($\Delta\delta$) and the asymmetry parameter (η) of the CSA were extracted by using DMFIT with the CSA-static model. The $\Delta\delta$ and η are found to be -1.73 *ppm* and 0.3 *ppm* respectively for the *n*-type and the asymmetry parameter of the CSA was found one order different from *p*-type material. The changes observed in CSA pattern and spin-lattice relaxation (T_1) time are ascribed to strength (mobility) of the shallow donor level and shallow acceptor level effects in the Fermi contact term.

1. Introduction

Silicon based materials are very effective in energy storage applications, photonics and photovoltaic cells with high efficiencies and high densities[1–8]. It has been reported recently that microstructure and electronic properties probes the detection of nitrogen impurities in phosphorous doped silicon[9]. The solid state NMR spectroscopy were sensitive at the subatomic scale as they influence dramatically the NMR parameter with respect to their chemical and structural nature[10]. Most of the parameters in NMR spectroscopy, are decided by the electron and nuclear interaction terms from the total Hamiltonian. For instance, the total Hamiltonian of any system is encompasses the individual Hamiltonian's of electron (H_e) from the group of interacting electron, nuclear (H_n) from nuclear interaction with external magnetic field or Zeeman energy plus the magnetic dipolar coupling among the nuclei, and electron-nuclear ($H_{e\text{-}n}$) from magnetic coupling between the nucleus and the electron spin respectively[11]. Any shift observed in NMR parameters primarily due to the electron probability from non-zero-probability, which is known to be contact Knight shift or Knight shift. The quantum computation is an complimentary technique to understand the spin manipulation at the atomic level. There are some reports which has explained the chemical shielding tensor parameters of the boron phosphide nanotubes by using DFT technique[12]. In order to engineer the spin dynamics in semiconductors, which is much warranted, the study of NMR parameters are essential[13,14]. With this motivation the authors investigated the spin dynamics of silicon single crystals using ^{29}Si NMR measurements.

2. Methods

The commercially available *n*-type and *p*-type silicon single crystals were utilized for ^{29}Si static NMR studies and ^{29}Si *spin-lattice* relaxation NMR studies. The experiments were carried out in broad-brand probe equipped with 7 mm zirconia rotor tuned at Larmor frequency of 79.495 MHz in a superconducting magnet of 9.39 T at room temperature, associated with Bruker -AVANCE DRX- 400 NMR spectrometer. The silicon possess spin ½ nuclei (with abundance 4.7%), showing large spin lattice relaxation (T_1) in crystalline form (T_1= 3 *h*). In order to study such a long T_1 materials, we have adopted a initial combo-pulse such as series (50 number of pulses) of $\pi/2$ (1.2 μs) pulses, with a fixed recovery time delay τ after the combo-pulses, and the actual $\pi/2$ (90°) pulse. In the case of spin-lattice relaxation time, the delay between the pulses were varies in order to probe the exponential behavior of magnetization. The measured chemical shifts were presented in *ppm*, which are referenced with respect to DSS (4, 4-Dimethyl-Silapentane-1-Sulfonic acid) taken as the standard reference. The spectral simulations made with DMFIT2006[15] and the measured saturation recovery of the delay time dependent change in magnetization were fitted with Topspin 2.1.

3. Research and Discussion

3.1. Spectral Studies

This present study focuses on comprehensive understanding of doping effect in semiconducting material, which is an extension work of our previous study on the effect of SHI irradiation in nitrogen doped 6H-SiC samples[16]. Figure 1 shows the ^{29}Si static NMR spectra of *n*-type silicon (a) and *p*-type silicon (b) recorded the Larmor frequency of 79.495 MHz. The solid-black line shows the experimental chemical shift anisotropy spectra and the solid red line shows the chemical shift anisotropy simulated using DMFIT[15]. The NMR parameters such as, isotropic chemical shift (δ_{iso}), chemical shift anisotropic tensor ($\Delta\delta$) and the asymmetry parameter (η) of the chemical shift anisotropy were extracted by using DMFIT[15] with chemical shift anisotropy static model,

$$\sigma_{iso} = 1/3\,(\sigma_{xx}^{PAF} + \sigma_{yy}^{PAF} + \sigma_{zz}^{PAF}) \quad (1)$$

$$\Delta\delta = (\sigma_{zz}^{PAF} - \sigma_{iso}) \quad (2)$$

$$\eta = (\sigma_{xx}^{PAF} - \sigma_{yy}^{PAF}) \quad (3)$$

The best spectral simulations were achieved with standard deviation between 1 to 2, for both *n*-type and *p*-type silicon static spectra, by allowing the parameters to reach the best convergence. The position of the peak is dictated by the electron density around the nucleus. The electron density makes the shielding of the external magnetic field to the nucleus. If the electron density is more near to nucleus which causes effect of external magnetic as low. Then the nucleus experienced low magnetic field and the peak appear further up field (to the left) in the spectrum. If the electron density is low, the peak of the nucleus appear further downfield (to the right) in the spectrum.

In this work, the experimentally measured chemical shift anisotropy is distributed between −80 *ppm* to −90 *ppm* for both *n*-type and *p*-type silicon. However from the DMFIT simulation[15], the actual isotropic chemical shifts were appeared at −83.73 *ppm* and −84.61 *ppm* for *n*-type (δ_{iso}(*n*-Si)) and *p*-type (δ_{iso}(*p*-Si)) respectively. The electron density around the nucleus in *n*-type Si is higher than the electron density in *p*-type due to the nitrogen doping introduce additional electronic shielding in *n*-Si. The observed chemical shift anisotropic tensor $\Delta\delta$ is found to be at −1.73 *ppm* and −3.38 *ppm* for *n*-

type (δ_{iso}(*n*-Si)) and *p*-type (δ_{iso}(*p*-Si)) respectively. On the other hand, the asymmetry parameter (η) found to be in 0.3 *ppm* and 0.082 *ppm* for *n*-type (δ_{iso}(*n*-Si)) and *p*-type (δ_{iso}(*p*-Si)) respectively.

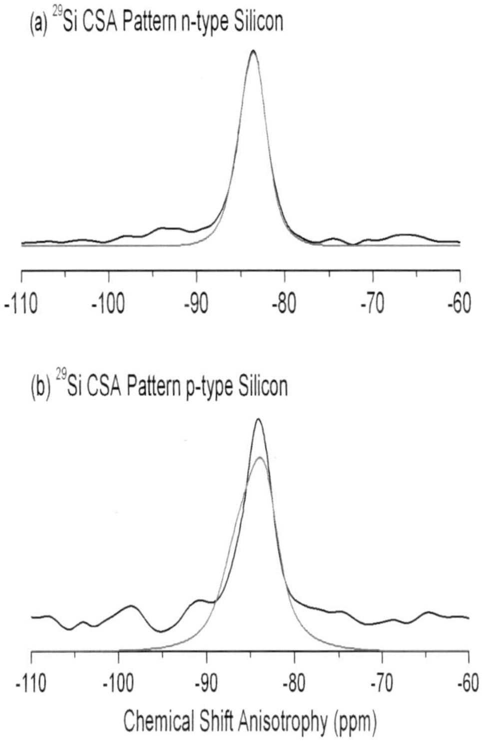

Figure 1. ^{29}Si chemical shift anisotropy pattern of (a) *n*-type and (b) *p*-type silicon.

Table 1. DMFIT simulation of ^{29}Si static NMR spectrum

Type of doping	Isotropic Chemical Shift (δ_{iso} in *ppm*)	Chemical Shift Anisotropic Tensor ($\Delta\delta$ in *ppm*)	Asymmetry Parameter (η in *ppm*)	Standard deviation
n-type silicon	-83.73	-1.73	0.3	1.8
p-type silicon	-84.61	-3.38	0.08	1.3

3.2. Spin-Lattice Relaxation Studies

Figure 2 shows the *spin-lattice* relaxation times (T_1) for both *n*-type and *p*-type samples under static condition using saturation recovery method. A combination of initial combo pulse with fixed delay between the pulses were adopted to magnetize saturation of ^{29}Si in longitudinal plane the for such a long T_1 samples. The net or average magnetization $\langle M_z(t)\rangle$ for each recovery delay time τ after the $\pi/2$ pulse were recorded. The spin lattice relaxation time were determined using Eq.(4) by analyzing acquired recovery of magnetization as a function of saturation recovery delay time. The driven conventional form of relaxation function can be justified with magnetization recovery using the following expression,

$$\langle M_z(t)\rangle = M_0(1 - \exp(-t/T_1)) \tag{4}$$

The gradual increase of ^{29}Si magnetization as a function of the saturation recovery delay time (τ) is shown in Figure 2.

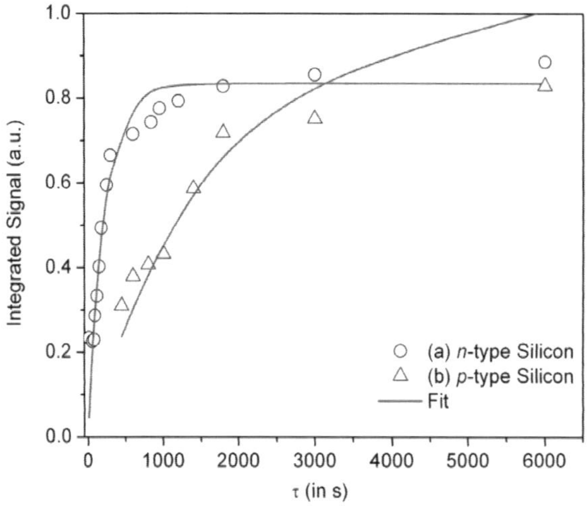

Figure 2. ^{29}Si Magnetization evolution as a function of the saturation recovery delay (τ) for (a) *n*-type silicon, (b) *p*-type silicon.

The mechanism of relaxation of ^{29}Si is mainly due to the interaction between nuclear spin and electrons, which is consistent with the mechanism explained by N.C. Halder[11]. It has been reported the relaxation time is a comprehensive tool to probe the effect of doping in solids as the conduction electrons readily interacts directly to the fixed nuclear site. The measured spin lattice relaxation times of silicon immediately led us to estimate the effect arising from the doping or impurity atom. The fluctuation of spin state of the conduction electrons or holes will always influence the NMR spin lattice relaxation in semiconductor[16]. The T_1 in semiconductor were mainly arises from the fluctuations in spin state of the conduction electrons (*n*-type) or holes (*p*-type) and has been studied in the present investigation.

In doped Si, the mobile electrons or holes that are the dominant effect in relaxation time corresponding as the electrons shielding or de-shielding nature due to nitrogen and phosphorous doping respectively. Compared to undoped silicon the value T_1 (3 h) is very much reduced, because of hyperfine coupling between the large electron density at the conduction band and the ^{29}Si nuclei. From the discussion, the general factors which affect the relaxation time T_1 in semiconductor are the conduction electrons (depending up on the doping concentration)[17], localized paramagnetic centers, and the mechanism of transport phenomena.

An early study by Stephen Hartman[18] reported that the relaxation of ^{29}Si site in silicon carbide is affected by the given carrier concentration and it has been reported that the relaxation rate ($1/T_1$) was affected eight times faster in *n*-type silicon than the *p*-type silicon. The similar such magnitude of eight times faster relaxation rate is observed in *n*-type silicon than in *p*-type silicon, from this present investigation of the spin-relaxation study of *p*-type and *n*-type silicon. This is because of the spin-lattice relation is mainly affected due to the charge carriers, in our case the relaxation is affected because of the effective mass is directly proportional to the resonance frequency.

As it has been known that the resonance frequency $\vartheta = eB_0/2\pi m^* c$ of any nuclei is mainly depend on the effective mass[19], hence the difference observed in our experimental are due to large effective mass of for *n*-type with respect to less in *p*-type. For instance, the effective mass of electron and holes are 1.08 (m_e*) and 0.56 (m_h* = 0.56) respectively and hence the effect is pronounced. The observed T_1 is 218 for *n*-type silicon. In the present work, the relaxation time T_1 for *p*-type silicon is found as 1740 s. It reveals that the hyperfine interaction between the hole and the ^{29}Si nuclei is weak compared to that between the electron and the nuclei. In order to understand further the fundamentals of conduction band electrons to electronic shielded nucleus, the theoretical calculations using Density Functional Theory (DFT) were employed. The current study clearly emphasis the effect of charge carriers in chemical shift anisotropy of solids, which brings the motivation for the futuristic study in more complex structure, such as doping in layered structures using in solid state NMR and DFT study.

4. Conclusions

The changes observed in CSA pattern and spin-lattice relaxation time are ascribed to strength (mobility) of the shallow donor level and shallow acceptor level affects in the Fermi Contact term. From the ^{29}Si static measurement, the change in NMR parameters like δ_{iso}(static), $\Delta\delta$ and η parameters in *n*-type and *p*-type Si are discussed. The spin-lattice relaxation time in *p*-type Si is lower than *n*-type Si because of the hyperfine interaction of electrons in Si. The effect of electron nuclear interaction in *n*-type and *p*-type Si, through ^{29}Si spin-lattice relaxation studies, are discussed.

References

[1] D.M. Chen, J.-Y. Tian, Z.-W. Wang, C.-S. Liu, M. Chen and M. Du, *Chem. Commun.* 2017, **53**, 10668 – 10671.
[2] L. Pavesi and D.J. Lockwood, Eds., *Silicon Photonics*, Springer, 2004.
[3] S.S. Lee and C.M. Park, *Electrochim. Acta*, 2018, **284**, 220 – 225.
[4]. P. Li, G. Zhao, X. Zheng, X. Xu, C. Yao, W. Sun and S.X. Dou, *Energy Storage Mater.* 2018, **15**, 422 – 446.
[5] X. Wang, C. Yang, X. Xiong, G. Chen, M. Huang, J.H. Wang, Y. Liu, M. Liu and K. Huang, *Energy Storage Mater.* 2019, **16**, 344 – 353.
[6] Y. Xu, Q. Li, H. Xue and H. Pang, *Coord. Chem. Rev.* 2018, **376**, 292 – 318.
[7] L. Pavesi and R. Turan, Eds., *Silicon Nanocrystals: Fundamentals, Synthesis and Applications*, Wiley, 2010.
[8] H. Zhou, X. Li, Y. Li, M. Zheng and H. Pang, *Nano-Micro Lett.* 2019, **11**, 40.
[9] Deng-Hao Ma, Wei-Jia Zhang, Rui-Ying Luo, Zhao-Yi Jiang, Qiang Ma, Xiao-Bo Ma, Zhi-Qiang Fan, Deng-Yuan Song, Lei Zhang, *Superlattice Microst* 2016, **93**, 269 - 279.
[10] R.F. Moran, D.M. Dawson, S.E. Ashbrook, *Int. Rev. Phys. Chem.* 2017, **36**, 39.
[11] N.C. Halder, *J. Chem. Phys.*, 1970, **52**, 5450.
[11] Mahdi Rezaei-Sameti, *Physica B* 2012, **407,** 22–26.
[12] V. Privman, I.D. Vagner, G. Kventsel, *Phys. Lett. A* **1998**, 239, 141.
[13] B.E. Kane, Nature **1998**, 393, 133.
[14] D. Massiot, F. Fayon, M. Capron, I. King, S. Le Calvé, B. Alonso, J.O. Durand, *Mag. Res. Chem.* 2002, **40**, 70-76.
[15] E. Viswanathan, D. Kanjilal, K. Sivaji, and S. Ganapathy, *J. Phys. Chem.* 2011, **115**, 7766 - 7772.
[16] J. Stephen Hartman, Arjun Narayanan, and YouXiang Wang *J. Am. Chem. Soc.* 1994, **116**, 4019–4027.
[17] Jutzi, U. Holtmann, D. Kanne, C. Krüger, R. Blam, R. Gleiter, I. Hyala-Kryspin; *Chem. Ber.* 1989, 122.
[18] Z.K. Keane, M.C. Godfrey, J.C.H. Chen, S. Fricke, O. Klochan, A.M. Burke, A.P. Micolich, H.E. Beere, D.A. Ritchie, K.V. Trunov, D. Reuter, A.D. Wieck, and A.R. Hamilton, *Nano Lett.* 2011, **11**, 3147–3150.
[19] David R. Lide, 84th Eds., *Handbook of Chemistry and Physics*, CRC Press, Inc. 2003-2004.

IOP Conf. Series: Materials Science and Engineering 1128 (2021) 012025 doi:10.1088/1757-899X/1128/1/012025

Effect of porosity and profile axial loading on elastic buckling and free vibration of functionally graded porous beam.

B P Harsha[1(a)], P Jeyaraj[1(b)] and Lenin Babu Mailan Chinnapandi [2,3]

[1] Department of Mechanical Engineering, NITK, Mangalore 575 025, India
[2] School of Mechanical Engineering, Vellore Institute of Technology, Chennai, India
[3] Electric Vehicles-Incubation, Testing and Research Centre, Vellore Institute of Technology, Chennai, India

[(a)] harsha.12patil06@gmail.com, [(b)] pjeyaemkm@gmail.com

Abstract. Elastic buckling and free vibration behaviour of porous functionally graded beam under profile axial mechanical loading is presented analytically. Symmetric and asymmetric porosity distribution with non linear material functional grading is presented. Buckling study is carried out first, then free vibration study under buckling loads is presented with profile loading. The effect of porosity, porosity pattern, slenderness ratio, boundary condition on critical buckling load and fundamental frequency is also studied. Ritz method with shear and normal deformable beam theory (SNDBT) is employed for formulation. It is observed that, with increase in porosity, buckling load decreases for both porosity distribution whereas, fundamental frequency increases for symmetric porosity distribution and is vice versa for asymmetric porosity distribution under all loading conditions.

1. Introduction

Functionally graded material (FGM), are inhomogeneous composite structures, whose properties are tailored in the desired directions based on the applications. Porosity induced structure with lower specific weight in combination with the FGMs have broad scope in defense, automobiles, aerospace and various other engineering fields. Buckling and free vibration study are both fundamental in structural design. Simsek et al. [1] studied static bending and buckling behaviour of functionally graded nano beams by nonlocal-Timoshenko theory analytically. Sunil et al. [2] experimentally studied vibration and buckling behaviour of synthetic foam core sandwich composite beam under compressive axial loading. Chen et al. [3] studied static bending and buckling of FG porous metal foam beam by shear deformable Timoshenko beam theory. Free vibration and buckling of FG-Carbon nano tubes reinforced composite cylindrical panel subject to thermal loading is presented by Vinod et al. [4] by finite element approach. Amol et al. [5] studied the effect of hybridization of glass and sisal fabric composite on thermal buckling characteristics experimentally. The hygro-thermal effect on the mechanical buckling of angle-ply and cross ply composite beams is studied by Mokhtar and Ashraf [6] analytically using Hyperbolic refined shear deformation theory. Nivesh et al. [7] studied vibration and buckling of FG-carbon nano-tube reinforced composite plate when subject to non uniform thermal loading by finite element approach. Vijay et al. [8] presented free vibration behaviour of FG-graphene reinforced composite plates, under non uniform edge loading analytically based on Reddy's theory. Mirmeysam and Bekir [9] investigated free vibration and static buckling of porous FG reinforced plates, based on first order theory using spectral Chebyshev approach. Yang et al. [10] studied the dynamic response of Euler Bernoulli beam having open edge cracks, under axial compressive load and moving concentrated transverse load. Simsek and Kocaturk [11] studied the dynamic output of FG simply supported beam analytically, under the application of moving concentrated harmonic load. Arunkumar et al. [12] studied the bending and free vibration behaviour of foam filled truss panel by numerical approach. Vibration response of porous FG beam is presented by Chen et al. [13] using Ritz approach. Fang et al. [14] studied the free vibration and thermal buckling analysis of rotating axial non-local, FG nano beam based on the integration of Eringen's non-local elasticity theory with Euler-Bernoulli beam theory. Armagan and Metin [15] studied the elastic buckling of beams under axially varying in-plane mechanical loads for different boundary conditions. They employed Ritz method coupled with SNDBT theory. Thuc et al. [16] presented free vibration behaviour of composite beams under axial mechanical loading using SNDBT theory. The effect of axial loads on buckling and its mode shapes for FG beam is investigated by Melaibari et al. [17] using differential qudarature method(DQM). Ashish et al. [18]studied the vibration and acoustic response of porous graphene foam, by using refined plate theory. Vijay et al. [19] presented the vibro acoustic behaviour of isotropic plate under steady state excitation, with non uniform edge loading analytically by employing Reddy's theory.

From the above literature survey, it can be said that the effect of profile axial mechanical loading on buckling and free vibration in the presence of porosity for a nonlinear FGM beams is a field still unexplored. Hence effort is made in presenting the behavior of different porosities under profile loading as a comparative study.

2. Methodology

A beam having length L, width b and height h, is taken under consideration with three boundary conditions [Pinned-Pinned(P-P), Fixed-Fixed (F-F) and Fixed-Pinned (F-P)]. Symmetric (Porosity pattern-I) and Asymmetric (Porosity Pattern-II) porosity distribution are studied. The reader is referred to Chen et al. [13] for details regarding material distribution. Four different profile loadings(N(x)) are studied as shown in figure 1. The external axial in-plane profile mechanical loading is given by equation (1). The critical buckling load is determined from equation (2) and natural frequencies are obtained from equation (3).

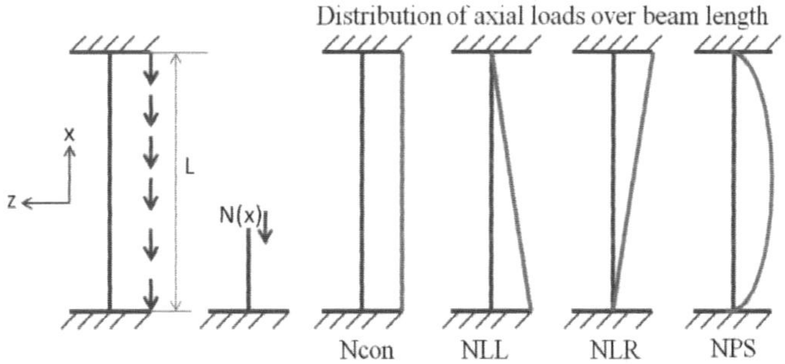

Figure 1. Fixed-Fixed beam with profile load description.

The formulation is based on Ritz method using SNDBT theory, the reader is referred to Armagan et al. [15] and Thuc et al. [16] for more details regarding derivation and profile loading.

$$N(x) = N_0\left[\alpha_1\left(x+\frac{L}{2}\right)^2 + \alpha_2\left(x+\frac{L}{2}\right) + \alpha_3\right] \tag{1}$$

$$([K] - N_0[M])\{U\} = 0 \tag{2}$$

$$([K-G] - \omega^2[M])\{U\} = 0 \tag{3}$$

Where, α_1, α_2 and α_3 are constants which depend on the type of profile loading, ω is fundamental natural frequency and N_0 is critical buckling load, $[M]$ mass matrix, $[K]$ stiffness matrix and [G] the geometric stiffness matrix.

3. Results

In this study, a beam of L=1m, b=h, is used and studied with two slenderness ratios (L/h) of 20 and 100. Critical buckling load and critical fundamental frequencies are calculated for different porosity distribution and porosity intensities(ζ) under various profile loading. The buckling load (P_0) and fundamental natural frequencies (ω_0) are expressed in non-dimensional form which are given in equation (4) and equation (5) respectively, the symbols E, I, ϑ, ρ meaning young's modulus, moment of inertia, poission's ratio and density respectively.

$$\overline{Pcr} = \frac{P_0 L^2}{EI} \tag{4}$$

$$\bar{\omega} = \omega_0 L \left(\frac{\rho(1-\vartheta^2)}{E}\right)^{1/2} \tag{5}$$

3.1. Mechanical Buckling Study

Effect of different profile loading, slenderness ratio (L/h) and porosity co-efficient(ζ), on the dimensionless buckling co-efficient ($\overline{Pcr}$), for a beam of L=1m and F-F boundary condition(BC), for two porosity patterns is presented in table 1. Results show that, as porosity increases, the stiffness of the structure decreases, hence a decreasing trend of $\overline{Pcr}$ is observed for all the investigated cases. With increase in slenderness ratio from 20 to 100, the critical buckling load decreases by a large margin for all loading cases, this is because of decrease in structural stiffness with increasing (L/h) ratio. It is observed that, NLR provides the highest critical buckling load where as NLL profile loading provides the lowest critical buckling load considering both PP-I (porosity pattern-I) and PP-II (porosity pattern-II). It can be said that, PP-I with symmetric porosity distribution has higher critical buckling load compared to PP-II, for all the loading cases studied.

Referring to figure 2 (a) and (b), a decreasing trend of $\overline{Pcr}$ for both PP-I and PP-II is observed with increase in porosity, with PP-II showing a higher slope. It can be observed that, $\overline{Pcr}$ for Ncon and NPS profile load is comparable for both the porosity distributions. Referring to figure 2 (c) and (d), it can be said that, as the boundary conditions change from F-F, F-P to P-P, the stiffness of the structure decreases, hence a proportionate decrease in $\overline{Pcr}$ is observed for both porosity distributions, under all the loading cases. The effect of (L/h) ratio on critical buckling load is studied in figure 2 (e) and (f) for F-F BC, it can be observed that, there is proportionate decrease in critical buckling load as (L/h) ratio increases from 20 to 100, for all the loading conditions.

Hence from the buckling study, it can be concluded that, symmetric porosity distribution provides better buckling strength compared to asymmetric porosity distribution.

Table 1. Effect of porosity intensity(ζ), porosity pattern, slenderness ratio(L/h) and type of profile loading on dimensionless buckling co-efficient of beams with PP-I and PP-II (L=1m, F-F BC).

Dimensionless Buckling Co-Efficient ($\overline{Pcr}$)									
		Loading Pattern							
L/h	ζ	Porosity Pattern-I (PP-I)				Porosity Pattern-II (PP-II)			
		Ncon	NLL	NLR	NPS	Ncon	NLL	NLR	NPS
	0.2	70.2940	52.9290	101.8571	68.5309	66.4236	50.0137	96.2537	64.7580
20	0.4	64.6961	48.7163	93.7353	63.0726	56.2689	42.3691	81.5326	54.8575
	0.6	59.0454	44.4645	85.5340	57.5625	44.9084	33.8176	65.0601	43.7813
	0.2	69.6640	52.4830	100.8105	67.8968	65.8206	49.5865	95.2524	64.1508
100	0.4	64.1360	48.3200	92.8041	62.5091	55.7700	42.0157	80.7040	54.3552
	0.6	58.5657	44.1254	84.7360	57.0802	44.5336	33.5521	64.4375	43.4040

3.2. Free Vibration Study

Effect of different profile loading, slenderness ratio (L/h), and porosity co-efficient(ζ), on the dimensionless fundamental frequency($\bar{\omega}$), for a beam of L=1m and F-F boundary condition(BC), for two porosity patterns is presented in table 2. Fundamental frequencies are extracted under the buckling load of 0.5 times the lowest critical buckling load considering all the loading scenario for L/h of 20 and 100 respectively. The lowest critical buckling load for (L/h) of 20 is 36,98,783N and 5,886N for slenderness ratio of 100. It can be observed that, with increase in porosity, $\bar{\omega}$ increases for PP-I and decreases for PP-II, under all loading conditions and slenderness ratios. As slenderness ratio increases from 20 to 100, $\bar{\omega}$ decreases for all loading cases. This can be explained as, with increase in length of structure, the stiffness of the structure decreases hence, fundamental frequency decreases.

Referring to figure 3 (a),(b),(c) and (d), it can be said that, $\bar{\omega}$ increases with increase in porosity for PP-I and it decreses for PP-II . It is also observed that, as buckling load increases from 0.25Pcr to 0.75Pcr, the fundamental frequencies shift towards higher frequency domain, due to increase in stiffness of the structure with increasing loading. This trend is same for all four loading cases. It is observed that, NLL profile loading offers higher fundamental frequencies whereas NLR loading offers the lowest of the four cases for both PP-I and PP-II FGM beams.

Referring to figure 4 (a) and (b), the effect of boundary condition has been presented here for PP-I and PP-II FGM beams respectively. As BC changes from F-F, F-P to P-P, the stiffness of the structure decreases, hence $\bar{\omega}$ decreases proportionately for all four loading cases. The decrease in $\bar{\omega}$, due to increase in slenderness ratio from 20 to 100, because of decrease in stiffness is observed in figure 4 (c) and (d) for PP-I and PP-II FGM beams respectively.

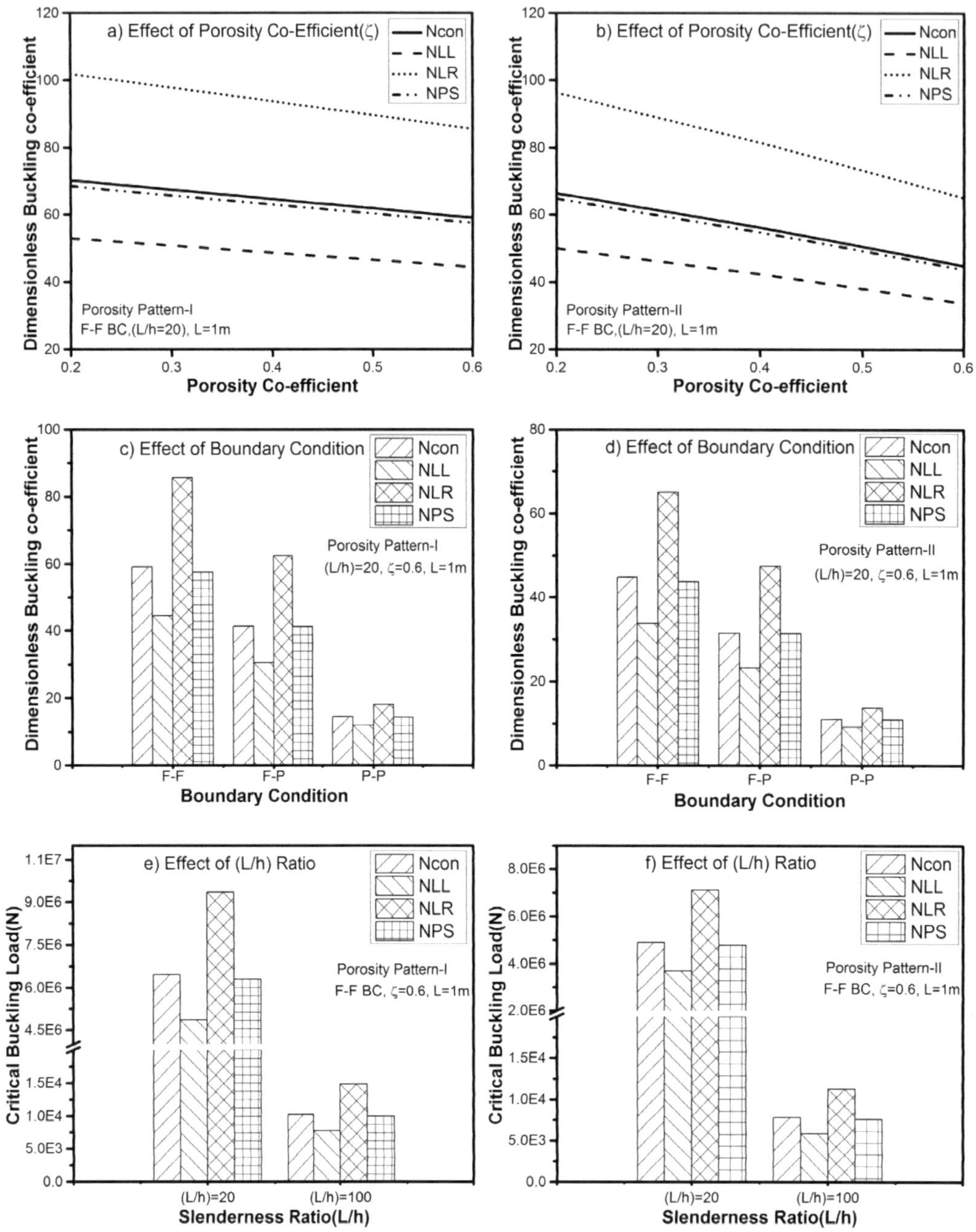

Figure 2. Effect of porosity coefficient (ζ), slenderness ratio (L/h) and boundary conditions (BC) on buckling characteristics of FGM-beams with PP-I and PP-II.

Table 2. Effect of porosity intensity(ζ), porosity pattern, slenderness ratio(L/h) and type of profile loading on dimensionless fundamental frequency of FGM-beams with PP-I and PP-II. (L=1m, F-F BC, buckling load ΔP=0.5Pcr, Pcr=36,98,783N for (L/h)=20 and Pcr=5,886N for (L/h)=100).

Dimensionless Fundamental Frequency ($\bar{\omega}$)									
		Loading Pattern							
L/h	ζ	Porosity Pattern-1(PP-I)				Porosity Pattern-2(PP-II)			
		Ncon	NLL	NLR	NPS	Ncon	NLL	NLR	NPS
20	0.2	0.33724	0.34731	0.32672	0.33708	0.33001	0.34032	0.31923	0.32985
	0.4	0.34010	0.35094	0.32875	0.33993	0.32278	0.33422	0.31073	0.32256
	0.6	0.34650	0.35834	0.33406	0.34629	0.31368	0.32678	0.29974	0.31338
100	0.2	0.06739	0.06941	0.06529	0.06736	0.06594	0.06801	0.06379	0.06591
	0.4	0.06798	0.07015	0.06571	0.06794	0.06450	0.06679	0.06209	0.06446
	0.6	0.06928	0.07165	0.06679	0.06924	0.06269	0.06531	0.05991	0.06263

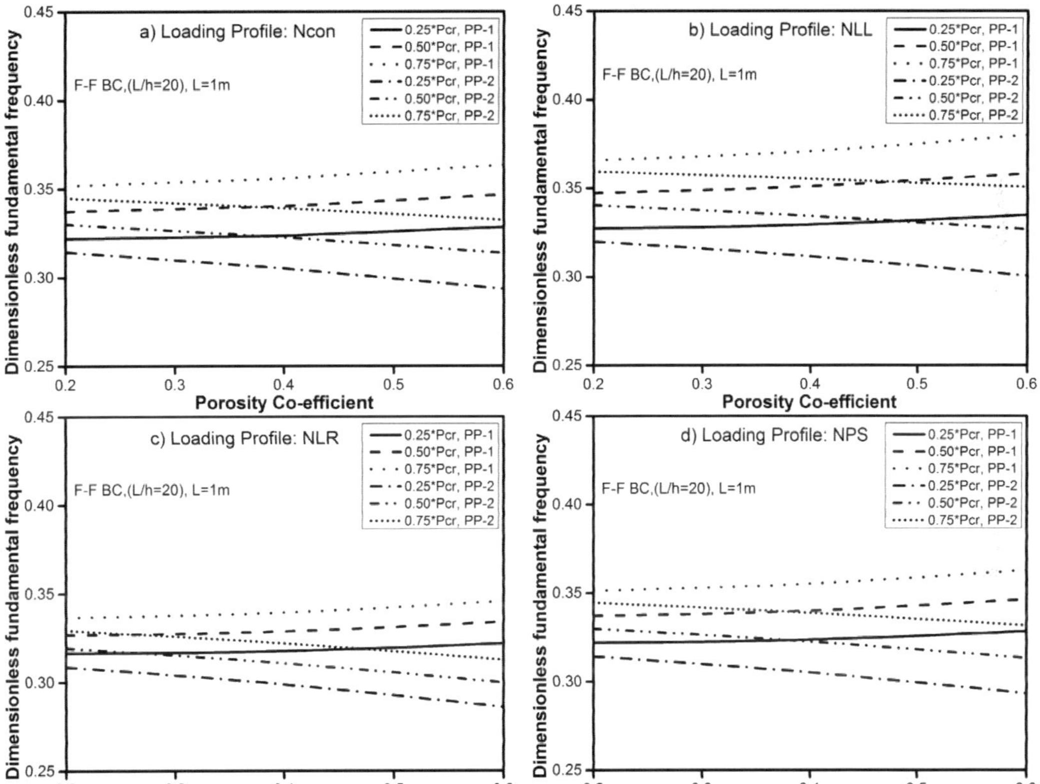

Figure 3. Effect of porosity coefficient(ζ) and porosity intensity on dimensionless fundamental frequency of FGM-beams with PP-I and PP-II under varying buckling load ΔP, for various profile loadings. (Pcr=36,98,783N).

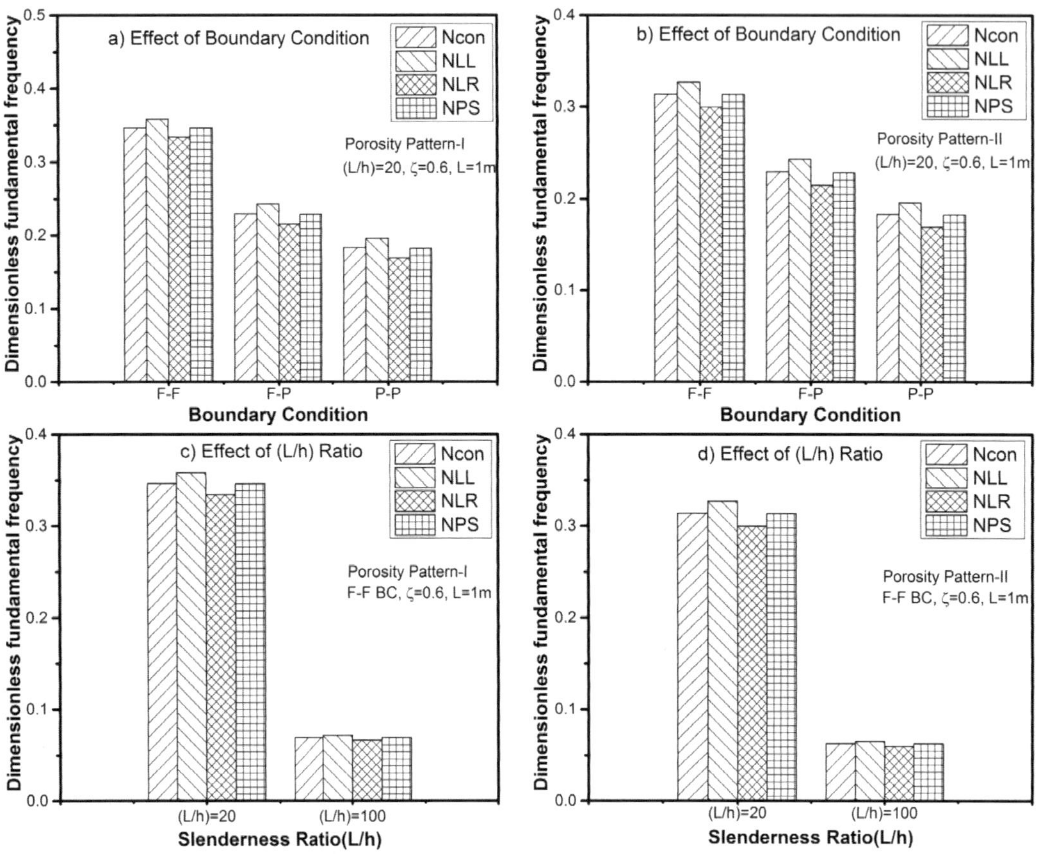

Figure 4. Effect of BC and slenderness ratio on dimensionless fundamental frequency of FGM-beams with PP-I and PP-II under buckling load ΔP=0.5Pcr, Pcr=36,98,783N.

4. Conclusions

Effect of profile in-plane axial mechanical loading, porosity pattern, porosity intensity, (L/h) ratio and BC on buckling and free vibration of FGM beam is presented in this study. Ritz method is used in conjunction of SNDBT theory to carry out the study in a comparative manner. The following fundamental conclusions can be derived from this study.

- Profile axial loading has significant effect on the buckling and vibration characteristics.
- Porosity effect decreases the critical buckling load.
- Porosity pattern has significant contribution on the free vibration characteristics of the P-FGM beam.

References

[1] Simsek M and Yurtcu H H 2013 Analytical solutions for bending and buckling of functionally graded nanobeams based on the nonlocal Timoshenko beam theory *Compos. Struct.* **97** 378–86

[2] Waddar S, Pitchaimani J, Doddamani M and Barbero E 2019 Buckling and vibration behaviour of syntactic foam core sandwich beam with natural fiber composite facings under axial compressive loads *Compos. Part B Eng.* **175** 107133

[3] Chen D, Yang J and Kitipornchai S 2015 Elastic buckling and static bending of shear deformable functionally graded porous beam *Compos. Struct.* **133** 54–61

[4] Bhagat V, Jeyaraj P and Murigendrappa S M 2018 Buckling and Free Vibration Behavior of a

Temperature Dependent FG-CNTRC Cylindrical Panel Under Thermal Load *Mater. Today Proc.* **5** 23682–91

[5] Gilorkar A, Murugan R and Pitchaimani J 2020 Thermal buckling of sisal and glass hybrid woven composites: Experimental investigation *Compos. Part C Open Access* **2** 100012

[6] Bouazza M and Zenkour A M 2020 Hygro-thermo-mechanical buckling of laminated beam using hyperbolic refined shear deformation theory *Compos. Struct.* **252** 0263–8223

[7] George N, Jeyaraj P and Murigendrappa S M 2017 Buckling and Free Vibration of Nonuniformly Heated Functionally Graded Carbon Nanotube Reinforced Polymer Composite Plate *Int. J. Struct. Stab. Dyn.* **17** 1750064

[8] Gunasekaran V, Pitchaimani J, Babu L and Chinnapandi M 2020 Analytical investigation on free vibration frequencies of polymer nano composite plate : Effect of graphene grading and non-uniform edge loading *Mater. Today Commun.* **24** 100910

[9] Mirmeysam R A and Bediz B 2020 Free vibration and buckling behavior of functionally graded porous plates reinforced by graphene platelets using spectral Chebyshev approach *Compos. Struct.* **253** 112765

[10] Yang J, Chen Y, Xiang Y and Jia X L 2008 Free and forced vibration of cracked inhomogeneous beams under an axial force and a moving load *J. Sound Vib.* **312** 166–81

[11] Simsek M and Kocatürk T 2009 Free and forced vibration of a functionally graded beam subjected to a concentrated moving harmonic load *Compos. Struct.* **90** 465–73

[12] Arunkumar M P, Pitchaimani J and Gangadharan K 2018 Bending and free vibration analysis of foam-filled truss core sandwich panel *Sandw. Struct. Mater.* **20** 617–38

[13] Chen D, Yang J and Kitipornchai S 2016 Free and forced vibrations of shear deformable functionally graded porous beams *Int. J. Mech. Sci.* **108–109** 14–22

[14] Fang J, Zheng S, Xiao J and Zhang X 2020 Vibration and thermal buckling analysis of rotating nonlocal functionally graded nanobeams in thermal environment *Aerosp. Sci. Technol.* **106** 106146

[15] Karamanli A and Aydogdu M 2019 Buckling of laminated composite and sandwich beams due to axially varying in-plane loads *Compos. Struct.* **210** 391–408

[16] Thuc P V, Huu-Tai T and Aydogdu M 2017 Free vibration of axially loaded composite beams using a four-unknown shear and normal deformation theory *Compos. Struct.* **178** 406–14

[17] Melaibari A, Abo-bakr R M, Mohamed S A and Eltaher M A 2020 Static stability of higher order functionally graded beam under variable axial load *Alexandria Eng. J.* **59** 1661–75

[18] Kumar A, Gunasekaran V, Chinnapandi M and Pitchaimani J 2020 Acoustic response behavior of porous 3D graphene foam plate *Appl. Acoust.* **169** 107431

[19] Gunasekaran V, Pitchaimani J and Chinnapandi L B M 2020 Vibro-acoustics response of an isotropic plate under non-uniform edge loading: An analytical investigation *Aerosp. Sci. Technol.* **105** 106052

Design of frontal longitudinal for enhancement in crashworthiness performance

Samer Fakhri Abdulqadir[1], Alaseel Bassam[2,3], M N M Ansari[4], and A Topa[5,6]

[1] Department of Mechanical Engineering, University of Anbar, Iraq.
[2] Mechanical Engineering Department, Universiti Tenaga National (UNITEN), 43000 Kajang, Malaysia.
[3] Department of Administration, State company of Phosphate, Anbar, Iraq.
[4] Institute of Power Engineering, Universiti Tenaga Nasional, 43000 Kajang, Selangor,Malaysia.
[5] Ocean and Ship Technology Group, Institute of Transportation Infrastructure, Universiti Teknologi PETRONAS, Perak, 32610, Malaysia.
[6] Department of Maritime Technology, Faculty of Ocean Engineering Technology and Informatics, Universiti Malaysia Terengganu, Kuala Terengganu, 21300, Malaysia.

Corresponding author: ansari@uniten.edu.my; ansari.uniten@gmail.com

Abstract. The increase in the number of vehicles due to increase in population leads to an increase in the number of accidents. To find a design that keeps occupants as safe as possible by enhancing the energy absorption capabilities and hence decreasing the number of fatalities is taking place. Non-linear finite element is used in this study to predict new design performance and to compare with other conventional S-shaped designs. It was shown that the proposed design increased the specific energy absorption (SEA) by 265% in comparison with the conventional S-Shaped frame design with octagonal geometries and increased in SEA 16% of the S-Shaped with foam filled-square cross-sectional. In this study, the specific energy absorption will take in consideration the energy absorber used is made of mild steel.

1. Introduction

Collisions cause thousands of people either killed or injured around the world every year [1], [2]. In Malaysia, the road network had been developed by 650% and the length of the road reached to 71,814 km from 1971 to 2005 due to the population growth and the number of vehicles [3]. According to (Royal Malaysian Police 2011), the number of fatalities due to accidents has increased by 1.8% (2009: 6745 to 6872 in 2010) [4]. the (World Health Organization 2009) has predicted that more than 1.2 million people die due to road accident each year and the number of injuries about 20 to 50 million. To enhance the crashworthiness of the structure, the proposed design must support an offset 40% collisions as well as the direct one. Australian New Car Assessment Program ANCAP (2006)[5] demonstrated that about 35% of all vehicle collisions were oblique impacts as shown in figure 1 below. Therefore, around 70% of all accidents happen in the front car whether in the direct or oblique collisions. So, it is important to concentrate on how to design frontal car structures that able to absorb the most impact energy during collisions in different directions.

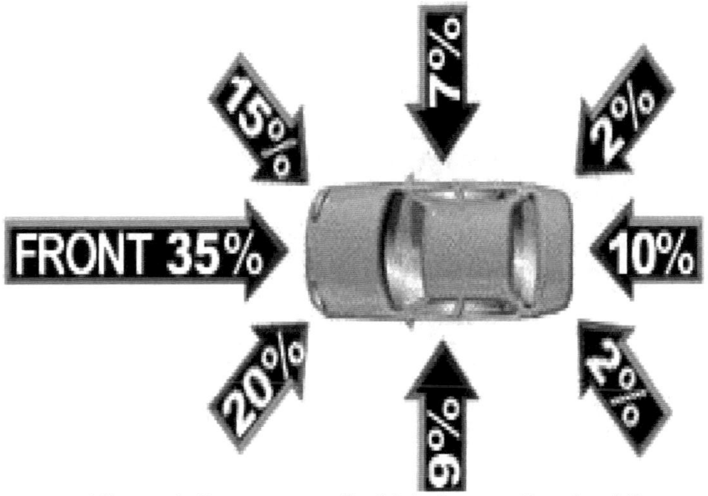

Figure 1. Percentages of vehicle impact direction [4].

Researchers and manufacturers paid their attention to occupants' safety, which became the most critical design objective in ground vehicles [6]. To save and protect vehicle's occupants, the design of structures must be able to absorb as much as possible energy caused by collision due to plastic deformation and keep car compartment away from severe deformation and decrease the intrusion of the vehicle. To carry out these conditions, the design should take into consideration that both longitudinal must work together during a collision, this enables the structure to absorb enough energy and keeps occupants safer. The decreasing of absorbed energy comes when one or both longitudinal is not loaded axially and hence the bending will happen. The proposed design tries to achieve two roles, the first one is getting higher absorbed energy and the second one is in oblique or 40 percent offset collision, the energy absorbed by the longitudinal should be almost as same as in a direct collision. Witteman (1993) stated that it is important to design a stiff longitudinal structure to be able to dissipate enough energy when loaded partially in overlap collision.

The structure should be strengthened to prevent or decrease the bending during collisions which will decrease in the amount of absorbed energy. The structure should be flexible when loaded axially to reduce excessive-high deceleration [3]. Many studies [3], [7], [8] have been implemented using different cross-section tubes subjected to direct and 40 per cent offset dynamic loading. These studies have found that the hexagonal tube was the best choice depending on multi-criteria decision making (MCDM) process. The parameters selected were included energy absorption (EA) and crash force efficiency (CFE). Samer [3] has studied the using of the hexagonal tube as a longitudinal absorber subjected to dynamic load with different directions. The study was compared with the conventional S-shaped used in the car. The study has concluded that using thin-walled hexagonal tube has superior performance than conventional one since it has higher energy absorption capability and lower acceleration within acceptable limits.

2. Finite element modeling

ABAQUS (non-linear finite element) has been employed to study the thin-walled hexagonal structure. The finite element was used to predict the behavior of the structure under dynamic loading. The hexagonal tube was modelled by using 4-node shell continuum (S4R) element along the thickness direction of the tube. When the aluminum foam is incorporated, the foam was modelled using 8-noded continuum elements with reduced integration techniques with incorporated hourglass control. The size of the mesh used was 5mm. The contact interaction between all parts was general contact algorithm. Coefficient of friction was incorporated between the foam and tube wall and it was set at 0.2 [9]–[11].

Plates were modelled as a rigid body. Moving plate (striker) was permitted to move axially along the structure while other translational and rotations degree of freedom were fixed. The impact velocity of the striker on the structure was set as 56km/h with a mass of 275kg (figure 3). The velocity was used according to the New Car Assessment Program (NCAP) by the National Highway Traffic Safety Administration (NHTSA). The study has taken a mass with 25 % of the compact car (1100) kg. Each longitudinal was assumed to absorb an equivalent kinetic of 275kg mass since the maximum absorbed energy by the two longitudinal is less than 50% [12].

3. Materials

3.1. Steel

A36 steel material was used in this study to simulate the structure. The A36 steel is a very common material, cheaper, most available and high energy absorption capability used in design as energy absorption. A36 was modelled as per Johnson-Cook constitutive isotropic model. It's more accurate to take in consideration the strain rate varies since this material is strain rate dependent [13], [14]. Material characterizations are combined as in Equation 1. [15].

$$\sigma_T = \left[A + B\left(\varepsilon_{eff}^p\right)^N\right]\left(1 + C\ln\frac{\dot{\varepsilon}_{eff}^p}{\dot{\varepsilon}_0}\right)\left[1 - \left(\frac{T - T_0}{T_{melt} - T_0}\right)^M\right] \quad (1)$$

Where σ_T is the dynamic flow stress.
The ε_{eff}^p is the effective plastic strain
$\dot{\varepsilon}_{eff}^p$ is the effective plastic strain rate.
$\dot{\varepsilon}_0^p$ is a reference strain rate,
T_{melt} is the melting temperature, and
A, B, N, M and C are material parameters.
whereas T_0 is the transition temperature and it is usually taken as the room temperature of 293-297 ° Kelvin [16], [17]. The parameters of Johnson-Cook are given in Table 1[18].

Table 1. Summary of parameters for A36 steel[19].

Parameter	Value	Description
A	146.7 MPa	Material parameter
B	896.9 MPa	Material parameter
N	0.320	Strain power coefficient
C	0.033	Material parameter
M	0.323	Temperature power coefficient
$\dot{\varepsilon}_0$	1.0 sec-1	Reference strain rate
r	7850 kg/m3	Density
Tm	1773 °K	Melting Temperature
Cp	486 J/kg-°K	Specific Heat

3.2. Aluminum foam

The CRUSHABLE FOAM and the CRUSHABLE FOAM HARDENING options are assigned in the ABAQUS/Explicit software package to consider the plastic aluminium foam behaviour. This option based on the foam model by Deshpande and Fleck [20]. The foam properties used in this study are listed in Table 2.

Table 2. Summary of parameters for aluminum foam model[9].

r_f(kg/m^3)	s_p (MPa)	a	a_2 (MPa)	b	g	e_D
534	12.56	2.12	1544	3.680	1.00	1.6206

4. Results and Dissection

4.1. Full overlap collision

In this type of collision, both absorber members are loaded axially. Figure 2 shows the structure of the longitudinal member. The SEA which is equal to the energy absorbed by the structure divided by both structure masses in addition to the foam mass. The structure is shown in Figure 3. The thickness used in this study was 2mm. The structure was fixed from one end and the free end subjected to the striker of 1100 kg which represents an average car with two occupants and luggage [20], with striker velocity of 54km/h. Foam-filled steel enhances the energy absorption and gives the folding more stability and hence more folded, which increases the energy absorption and the crash force efficiency (CFE).

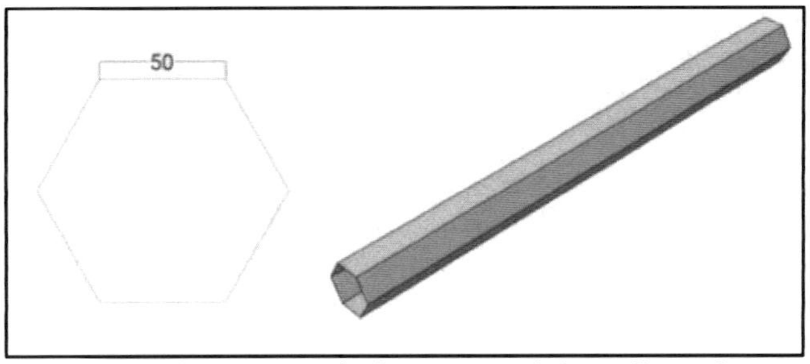

Figure 2. structure designs two and three dimensions.

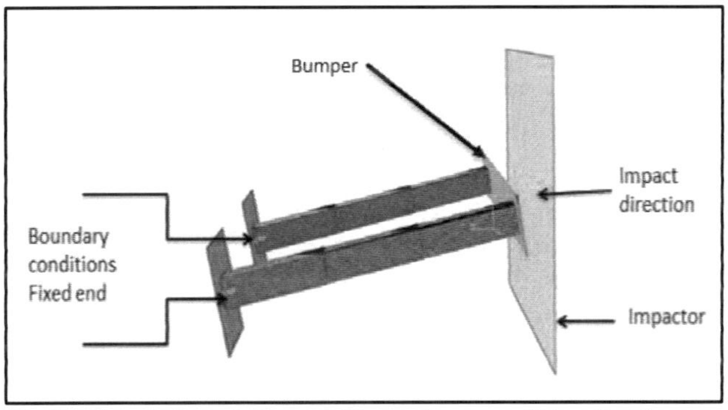

Figure 3. Finite element modeling and boundary condition for the proposed design.

The specific energy absorption (SEA) diagram is shown in Figure 4 which represents the function of displacement. The figure has been demonstrated that the highest specific energy absorption was 4.4 (kJ/kg). In this case, both members are loaded axially. From the Figure (4), the curve starts with almost no energy absorbed until reach 10 mm which represents the impactor hitting the bumper without absorber, when the impactor reaches the absorbers, the absorbed energy starts to increase with the displacement and the energy is directly proportional with the displacement. The curve is almost linear and reaches the maximum displacement no rebound has occurred.

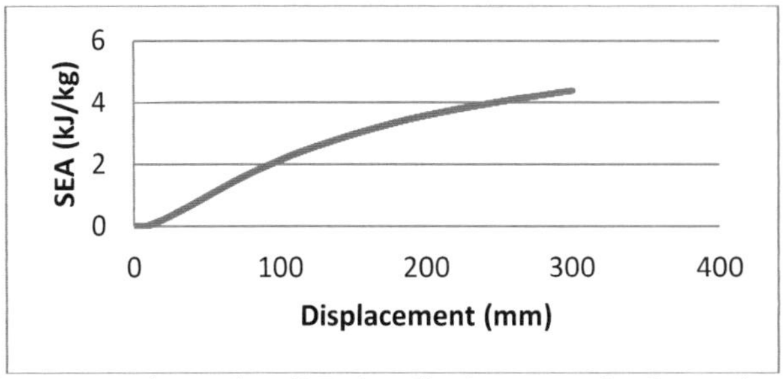

Figure 4. Specific Energy absorption (SEA) – displacement for foam aluminum filled steel.

Figure 5 represents the force-displacement curve which begins to rise with low steep until it reaches (25 kN) which represents the impactor hitting the bumper before reaching the absorber. When the impactor reaches the absorbers, the curve starts to rise rapidly with a very short time until it reaches the maximum value at (160kN), representing the force needed to form the first fold of each absorber. The absorber after forming the first fold starts to be less stiff, so the force needed to form more folds will be less. The force begins to decelerate gradually with more time, which is preferable to avoid high rapid deceleration less serious injuries.

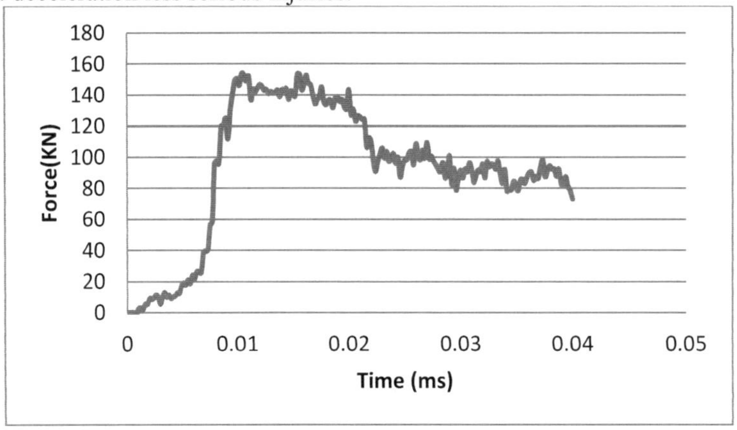

Figure 5. Force - displacement curve for the proposed design.

Figure 6 represents the deceleration-time curve. The deceleration behaves as a result of the force. The figure shows that the maximum deceleration is at high peak force when the first fold is formed. The maximum acceleration is 16 g, this magnitude is acceptable according to [21] and it's much lower than the permissible level. The basic levels of HIC for different occupant sizes are shown in Table 3 below::

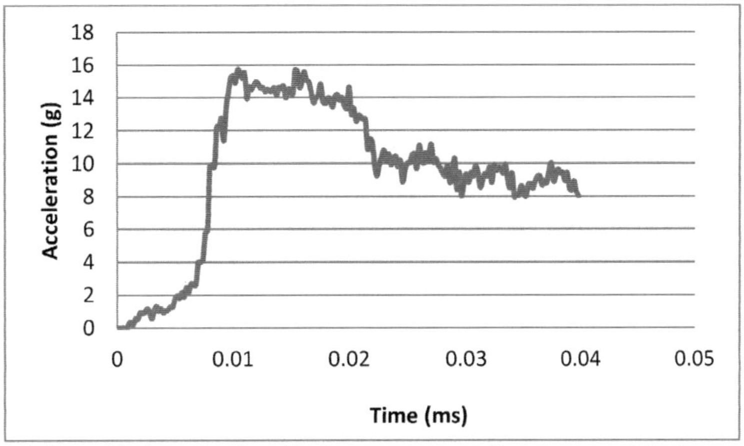

Figure 6. Acceleration-displacement curve for a proposed design.

Table 3. HIC level for different Occupant Sizes[21].

Dummy Type	Mid-Sized Male	Small Female	6 Year Old Child	3 Year Old Child	12 Month Old Infant
Existing/Proposed HIC Limit	1000	1000	1000	900	660

According to the table, the acceptable acceleration level is 60 g. The acceleration below this limit means that the structure is safer and recommended and vehicle occupants will be at a lower risk level. Equation 2 shows the parameters affecting HIC.

$$HIC = \left\{ \left[\frac{1}{t_2 - t_1} \int_{t_1}^{t_2} a(t)dt \right]^{2.5} (t_2 - t_1) \right\}_{max} \quad (2)$$

Where t_1 and t_2 refer to any two random times during the acceleration pulse [21] a is a resultant head acceleration [22].

4.2. Forty per cent (40%) offset collision

The purpose of using 40 per cent offset collision is to assess the proposed design when subjected to this type of loading. In this collision, one of the longitudinal members is loaded axially and the other one is unloaded. When the impactor starts to collide the structure, first it will reach the bumper. The bumper will absorb a small amount of energy comparing with the absorber. The impactor continues to move on towards the structure for 110 mm as in figure 7. When the impactor starts to contact the stiff absorber, the energy starts to increase rapidly with a short time, and it increases as displacement increased. The purpose of using 40 per cent offset collision is to assess the proposed design when subjected to this type of loading. In this collision, one of the longitudinal members is loaded axially and the other one is unloaded. When the impactor starts to collide the structure, first it will reach the bumper. The bumper will absorb a small amount of energy comparing with the absorber. The impactor continues to move on towards the structure for 110 mm as in figure 7. When the impactor starts to contact the stiff absorber, the energy starts to increase rapidly with a short time, and it increases as displacement increased.

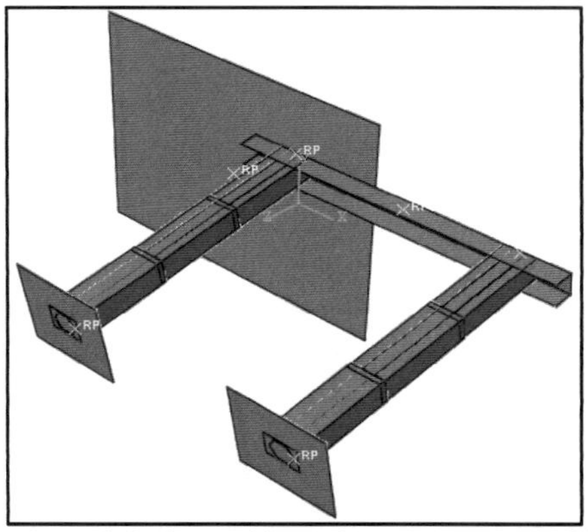

Figure 7. Proposed design subjected to 40 percent offset loading.

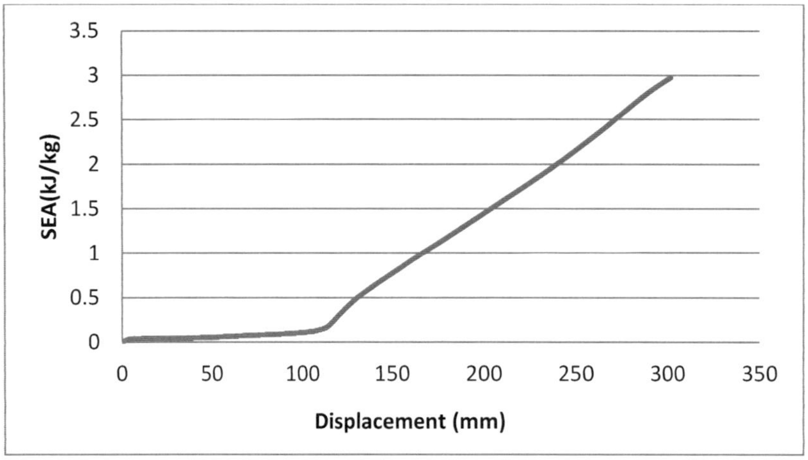

Figure 8. Specific Energy absorption (SEA) – displacement for foam aluminum filled steel in an offset collision (40%).

In the case of 40 percent offset crash, the rigid body has impacted the vehicle with a velocity of 54 km/h. In the 40 percent offset overlap. From figure 8 it can be said that the SEA of the structures is up to 3 (KJ/Kg). In an overlap collision, the vehicle started to rotate due to non-symmetrical loading, and this leads to non-symmetrical deformation mode, reactions, displacements and velocities of the two energy absorbers. Figure 9 represents the force-time curve for both left and right frontal longitudinal. The maximum force level of loaded side is nearly 152 KN after 10 ms, after that time the curve fluctuates below that level due to buckling until it reaches 68.6KN at time of 40 ms, while the unloaded side shows increasing slowly with low force level until the end of simulation and the force at that time is about 24 KN.

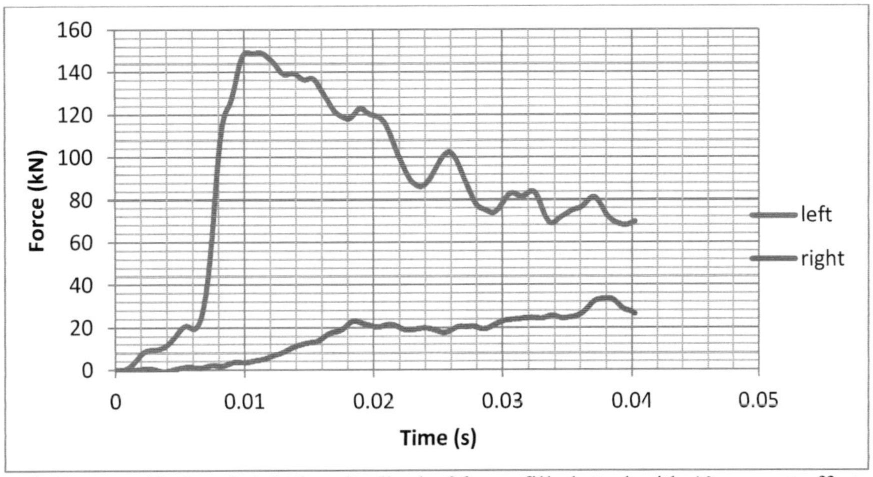

Figure 9. Forces of left and right longitudinal of foam-filled steel with 40 percent offset crash.

The deceleration curve of the 40 percent offset crash for both left and right absorber is plotted in Figure 10. The same difference can be seen in the force figure. The peak deceleration in the left absorber was 15.2 g at 10 ms, after that the deceleration fluctuates and decreases until it reaches 7.1 g at 40 ms, due to the buckling. The right absorber starts to increase due to increasing the force until it reaches the peak value of 2.5 g at 40 ms.

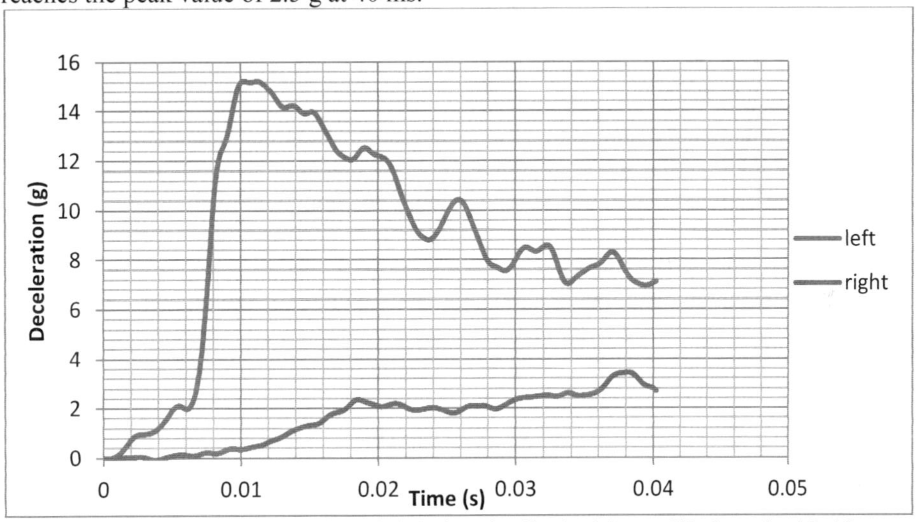

Figure 10. Deceleration of left and right longitudinal of foam-filled steel with 40 percent offset crash.

5. Structure design comparison

Ahmed Elmarakbi et al. [23] used the conventional design (S-shaped) for both aluminum and steel. The study included the different cross-section and different inner stiffening as shown in Figure 11 below. The maximum specific energy absorption that has got were in these types type 2, type 3, type 7 (selected design), type 8 and type 10 (1.377, 1.342, 1.549, 1.411 and 1.656 (KJ/Kg) when using octagonal with different inner stiffening. From the other hand, another study done by Zhang and Saigal [24] they designed S-Shaped structure with different square cross-sectional shapes.

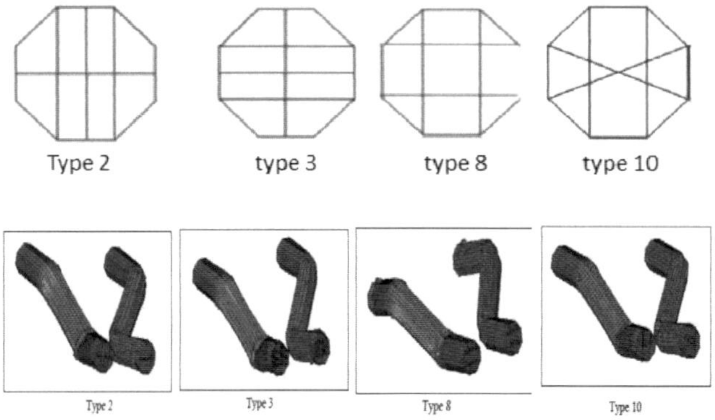

Figure 11. Two and three-dimension of the S-Shaped designs.

Table 4. SEA of different profiles with proposed design.

Profile	Name	SEA (KJ/Kg
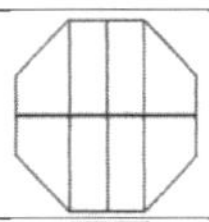	S-Shaped type 2	1.377
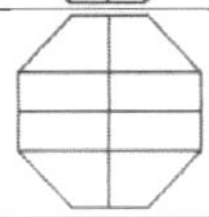	S-Shaped type 3	1.342
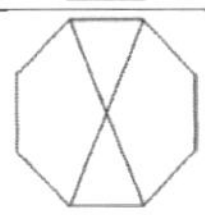	S-Shaped type 7 (selected design)	1.549
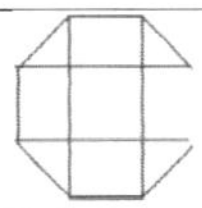	S-Shaped type 8	1.411
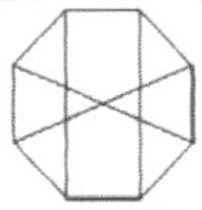	S-Shaped type 10	1.656
	Square S-Shaped	3.8
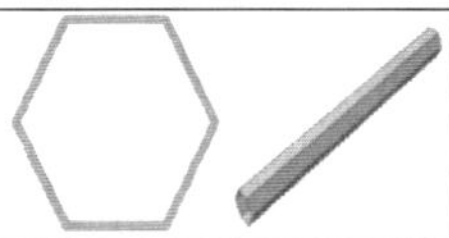	proposed design	4.4

They stated that the highest SEA is 1.375 KJ/Kg and when the foam incorporated with their design, the highest SEA is 3.8 kJ/kg. they concluded that the foam-filled structure increases the SEA. Figures 12 show the comparison of different S-shaped design and proposed design.

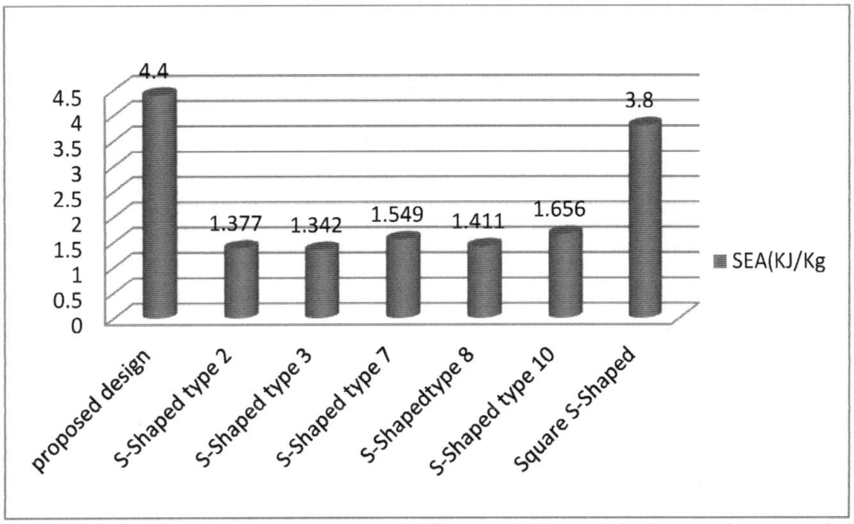

Figure 12. Specific energy absorption of different S-Shaped designs and proposed design.

6. Conclusion and Future Work

This research fulfilled two goals, at first a new design subjected to full overlap car collisions was evaluated and the second was the behavior assessment of 40 percent offset overlap impact on the proposed energy absorber. The proposed design is compared with other conventional S-shaped designs with different cross-sectional geometries used in the car. Numerical simulations were done by using ABAQUS. Results presented in this study were showed that the SEA of the frontal longitudinal members can be enhanced by using the proposed design. The results showed that the SEA of the proposed design was higher than other designs and the peak force and hence the acceleration is much lower than critical recommended by HIC. It can be concluded that the proposed design enhances the energy absorption and it increases the SEA by 265% of the S-Shaped designs with different octagonal geometries and increases SEA by 16% of the S-Shaped design when incorporated the foam filled-square cross-sectional. For 40 percent offset collision, the simulations showed good results the SEA is about 3(kJ/kg) which represented 68% of the full overlap and the deceleration is much lower than critical level.

Author's contribution

Samer Fakhri Abdulqadir[1]-Idea & concept generation**,** first draft and manuscript writing, simulation data collection, analyses of data,; **Alaseel Bassam**[2,3] **-** simulation and interpretation of the data, manuscript editing. **M N M Ansari**[4] suggestion and advise on the interpretation, presentation of the data and manuscript editing; **Ameen topa**[5,6] - simulation and analytical tools application, manuscript editing.

References

[1] A. A. Mohammed, K. Ambak, A. M. Mosa, and D. Syamsunur, "A Review of the Traffic Accidents and Related Practices Worldwide," Open Transp. J., vol. 13, no. 1, pp. 65–83, 2019, doi: 10.2174/1874447801913010065.

[2] Y. Maqbool, "Road safety and Road Accidents : An Insight Road safety and Road Accidents : An Insight," Int. J. Inf. Comput. Sci., vol. 6, no. 4, pp. 93–105, 2019.

[3] S. F. Abdulqadir, "Design a new energy absorber longitudinal member and compare with S-shaped design to enhance the energy absorption capability," Alexandria Eng. J., vol. 57, no. 4, pp. 3405–3418, 2018, doi: 10.1016/j.aej.2018.07.012.

[4] S. I. Ao, X. Huang, and O. Castillo, Lecture Notes in Electrical Engineering: Preface, vol. 70

LNEE, no. July. 2011.
[5] "CrashTest.com." [Online]. Available: http://www.crashtest.com/search_caf.php?src=mountains&uid=www5ea081acc95756.21598985&abp=1&country=MY&query=CarCrash&afdToken=3B1g-UGgKbOLnQTdwvTMZ2f6nL1Y3GkUmIqpon-b0D2e-_jslfg6AZ69rMFsG7uHbHI7OViuXl7Q_7QQycVxegqCwXIGXebTEbV64LeEUGw. [Accessed: 23-Apr-2020].
[6] M. M. Trivedi, T. Gandhi, and J. McCall, "Looking-in and looking-out of a vehicle: Computer-vision-based enhanced vehicle safety," IEEE Trans. Intell. Transp. Syst., vol. 8, no. 1, pp. 108–120, 2007, doi: 10.1109/TITS.2006.889442.
[7] F. Tarlochan, F. Samer, A. M. S. Hamouda, S. Ramesh, and K. Khalid, "Design of thin wall structures for energy absorption applications: Enhancement of crashworthiness due to axial and oblique impact forces," Thin-Walled Struct., vol. 71, pp. 7–17, 2013, doi: 10.1016/j.tws.2013.04.003.
[8] . S. F., "Improvement of Energy Absorption of Thin Walled Hexagonal Tube Made of Magnesium Alloy By Using Trigger Mechanisms," Int. J. Res. Eng. Technol., vol. 02, no. 10, pp. 173–180, 2013, doi: 10.15623/ijret.2013.0210025.
[9] Z. Ahmad and D. P. Thambiratnam, "Dynamic computer simulation and energy absorption of foam-filled conical tubes under axial impact loading," Comput. Struct., vol. 87, no. 3–4, pp. 186–197, 2009, doi: 10.1016/j.compstruc.2008.10.003.
[10] A. G. Olabi, E. Morris, M. S. J. Hashmi, and M. D. Gilchrist, "Optimised design of nested circular tube energy absorbers under lateral impact loading," Int. J. Mech. Sci., vol. 50, no. 1, pp. 104–116, 2008, doi: 10.1016/j.ijmecsci.2007.04.005.
[11] A. G. Olabi, E. Morris, M. S. J. Hashmi, and M. D. Gilchrist, "Optimised design of nested oblong tube energy absorbers under lateral impact loading," Int. J. Impact Eng., vol. 35, no. 1, pp. 10–26, 2008, doi: 10.1016/j.ijimpeng.2006.11.007.
[12] S. Alkhatib, "Sami Emad Alkhatib-201103668-Thesis," no. January, 2018.
[13] X. Kong, Q. Fang, J. Zhang, and Y. Zhang, "Numerical prediction of dynamic tensile failure in concrete by a corrected strain-rate dependent nonlocal material model," Int. J. Impact Eng., vol. 137, no. October 2019, 2020, doi: 10.1016/j.ijimpeng.2019.103445.
[14] Z. Liao, X. Yao, L. Zhang, M. Hossain, J. Wang, and S. Zang, "Temperature and strain rate dependent large tensile deformation and tensile failure behavior of transparent polyurethane at intermediate strain rates," Int. J. Impact Eng., vol. 129, no. January, pp. 152–167, 2019, doi: 10.1016/j.ijimpeng.2019.03.005.
[15] J. Dean, C. S. Dunleavy, P. M. Brown, and T. W. Clyne, "Energy absorption during projectile perforation of thin steel plates and the kinetic energy of ejected fragments," Int. J. Impact Eng., vol. 36, no. 10–11, pp. 1250–1258, 2009, doi: 10.1016/j.ijimpeng.2009.05.002.
[16] V. Kumar, L. Singh, and A. K. Tripathi, "Reliability analysis of safety-critical and control systems: A state-of-the-art review," IET Softw., vol. 12, no. 1, pp. 1–18, 2018, doi: 10.1049/iet-sen.2017.0053.
[17] A. Sedky and A. T. AlMotasem, "Fluctuation-Induced Excess Conductivity and Infrared Spectra in Y Doped BSCCO Superconductors," J. Mater. Appl., vol. 8, no. 1, pp. 41–49, 2019, doi: 10.32732/jma.2019.8.1.41.
[18] C. W. Isaac and O. Oluwole, "Numerical modelling of the effect of non-propagating crack in circular thin-walled tubes under dynamic axial crushing," Thin-Walled Struct., vol. 115, no. February, pp. 119–128, 2017, doi: 10.1016/j.tws.2017.02.012.
[19] C. Duan, H. Yu, Y. Cai, and Y. Li, "Finite element simulation and experiment of chip formation during high speed cutting of hardened steel," Appl. Mech. Mater., vol. 29–32, no. 5, pp. 1838–1843, 2010, doi: 10.4028/www.scientific.net/AMM.29-32.1838.
[20] V. S. Deshpande and N. A. Fleck, "Isotropic constitutive models for metallic foams," J. Mech. Phys. Solids, vol. 48, no. 6, pp. 1253–1283, 2000, doi: 10.1016/S0022-5096(99)00082-4.
[21] E. S. R. S. Michael Kleinberger, "Development of Improved Injury Criteria for the

Assessment," no. September, 1998.
[22] B. McHenry, "Head injury criterion and the ATB," ATB Users' Gr., no. February, pp. 5–8, 2004.
[23] A. Elmarakbi, Y. X. Long, and J. MacIntyre, "Crash analysis and energy absorption characteristics of S-shaped longitudinal members," Thin-Walled Struct., vol. 68, pp. 65–74, 2013, doi: 10.1016/j.tws.2013.02.008.
[24] C. Zhang and A. Saigal, "Crash behavior of a 3D S-shape space frame structure," J. Mater. Process. Technol., vol. 191, no. 1–3, pp. 256–259, 2007, doi: 10.1016/j.jmatprotec.2007.03.093.

IOP Conf. Series: Materials Science and Engineering 1128 (2021) 012027 doi:10.1088/1757-899X/1128/1/012027

Water absorption of kenaf/glass /epoxy hybrid composites for insulator core

M N M Ansari[1], Alaseel Bassam[2], A R M Nazim[2] and Noor Afeefah Nordin[1], Zainudin Yahya[2].

[1] Institute of Power Engineering, Universiti Tenaga Nasional, 43000 Kajang, Selangor,Malaysia.
[2] Mechanical Engineering Department, Universiti Tenaga National (UNITEN), 43000 Kajang, Malaysia.

Corresponding author: ansari@uniten.edu.my; ansari.uniten@gmail.com

Abstract. Kenaf fibres have gained substantial attracted considerable attention due to their economic feasibility and environmental acceptability. Natural fibre (kenaf) has started in mechanical and electrical applications by hybridising synthetic (glass) fibres. In recent years, the hybrid composite has been employed in various functions in industrial engineering. The present study deals with water absorption of glass/kenaf/epoxy hybrid composite materials used in high voltage polymeric insulator rods. Kenaf fibres have substituted two proportions of glass fibres (20 vol% IR20EP and 30 vol% IR30EP) with modified 60% of epoxy resins to fabricate the kenaf/glass/epoxy hybrid composites. The water absorption test was carried out at room temperature, and the water absorption resistance for composites was calculated regarding the rate of water absorption. Remarkable variations in the characteristics of water soaking up of the hybrid composite were obtained demonstrating that the water absorption influence on the insulator rod properties relies on the arrangement profiles and volume fraction of kenaf fibre of composite used. Based on the finding, a minor impact of water absorption on the glass fibre (non-hybrid) composite was noticed. Adding kenaf fibres to the composite fibreglass rod increases the composite's water absorption. Glass fibres around kenaf fibres have been shown to minimise water absorption. After 330 hours, all the composite specimens attained stability and ceased to soak up water.

1. Introduction

Currently, natural fibre is a compelling new form of reinforcement used in polymer composites [1]. Natural fibres have acquired pervasion over synthetic fibres due to their low cost, light-weight, and vitality as renewables[2]. In various utilizations such as automobile, fabric, texture, air cleaner and dielectric processing, researchers and manufactures have begun to favour natural fibres with good characteristics[3]. Kenaf, as one of these natural fibres, has economic significance and friendliness for the community. Kenaf is capable of consuming carbon dioxide in addition to the short growth period[4]. In high voltage HV-insulator industry, polymer insulation rods are a widespread replacement for porcelain[5]. Polymeric insulators are used in high voltage transmission lines in the range of 69-735 kV. The core of insulator is made using pultrusion machines with fibre-reinforced polymer FRP, and the fibres are axially aligned with the line of manufacturing. Although FRP provides advantages such as ease of installation, durability, and high mechanical strength, relative to porcelain insulating rods, a fracture during service leads to failing mechanically[6]. The water inside FRP will become nitric acid, which creates fractures. [7]. Fig.1 illustrates the composite insulator, which consists of three parts; the core (FRP), silicon rubber housing, and end-fitting[7]. The FRP insulator rods may be undefended under exceptional circumstances and allow humidity to enter into the composite. It can arise when the insulator's end-fitting is exposed against humidity [8]. Otherwise, the FRP rod is vulnerable from water penetration if the rubber housing coating thickness isn't adequate or is weakened due to corrosion[9]. Nevertheless, only if the acid content inside the FPR rod is high can a crack occur.

On the other hand, water runs around the FRP insulator chain, resulting in electrical breakdowns. The composite insulators' electrical properties have been strongly dependent on water absorption. The quantity of water absorbed must be measured to determine moisture absorption on the composite insulators' electrical properties. The humidity absorption of glass-reinforced-polymer composites used in polymeric insulator FRP rods has been investigated by Kumaso et al. [9]. It was noticed the high reliance on water absorption on the composite design. Akil et al.[10] have investigated the influence of water intake on hybrid jute/glass/polyester hybrid composites' mechanical characteristics. This study found that a cross-sectional configuration of kenaf fibre covered by glass fibre outer layer managed to decrease water absorption to 1000 h. Alaseel et al.[4] investigated the water absorption of glass/unsaturated polyester composites reinforced with kenaf fibre with three different volume fraction and three different fibre profile arrangement. This study discovered that water absorption significantly relies on the kenaf fibres arrangement profiles than its fibre content. Water absorption of pure glass fibres reinforced UPE composite has increased by ten times by hybridisation with kenaf fibres. Numerous laboratory processes can determine the resilience of hybrid composite materials to moisture absorbing [11]. In this study, the water absorption of kenaf/glass/epoxy hybrid composite were investigated for the insulator core application. The existing commercial glass/epoxy composite was used as a control sample for comparison in its properties.

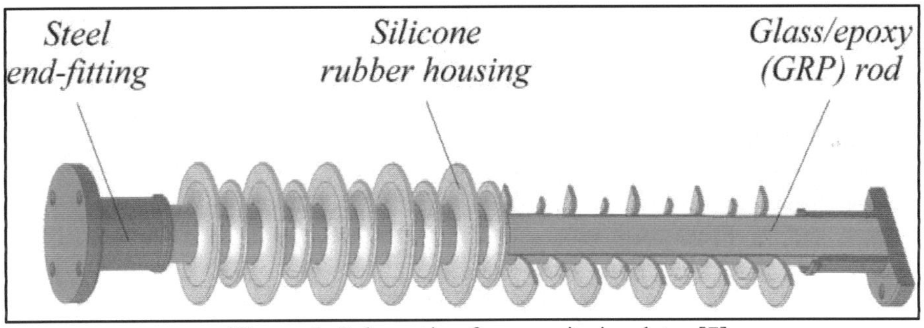

Figure 1. Schematic of composite insulator [7].

2. Experimental

2.1. Kenaf Fibre (KF)
Kenaf fibres (KF) possess good potential to be the alternative material in the production of fibre reinforced materials. In recent years, Kenaf fibre has shown promising characteristics for the use in natural fibre based applications as explored by the Malaysian Kenaf and Tobacco Council. Kenaf fibre has particular properties, like hardness, biodegradable, and high impact resistance, making it an attractive material for reinforced polymer composite. Kenaf fibre is made from the Kenaf (*Malvaceae* family) stem staple that takes less water to cultivate. It usually hits 3-4 metres in four to five months and can mould three times a year. Kenaf is economically inexpensive and readily widely abundant among other natural fibres. The density, specific gravity, diameter, tensile strength, Young's modulus,

and elongation to break of kenaf fibre are (1.19-1.4 g cm3, 1.04 g cm3, 17.7-100 µm, 295-930 MPa, 22-53 GPa, and 1.5-6.9 % respectively[12][12].

2.2. *Glass Fibre (GF)*

Glass fibres (GF) have been manufactured by melting of marbles of glass at approximately 2300 ^{0}F (1260 ^{0}C), and the blend was drawn with a simple cooling down with a coating device (usually platinum). A further drawing is employed to the fibre dimension. The fibres are then loaded onto a spool into a string. The fundamental forms of glass fibres are "E-glass" and "S-glass." Although E-glass has high resistivity and is known as (electric glass), S-glass has a so-called high tensile strength (structural glass). Glass fibre is typically isotropic. The density of glass fibre is 2.26 g.cm^3, specific gravity is 2.06 g/cm^3, Young's modulus is 72 GPa, tensile strength is 3450 MPa, the diameter is 3-20 µm, and elongation is 4.8%[13].

2.3. *Epoxy*

Epoxy is the most commonly used as a thermosetting resin in polymers. The epoxy resin is exceptional among all-polymer resins due to its distinct characteristics. Epoxy resin provides excellent thermal and mechanical properties, good electrical insulation properties, high corrosion, and chemical resistance. Epoxy is suitable for matrix materials because of its ability to work under varying conditions. Many researchers have used polymer composites using epoxy resin. Since epoxy is produced from the current insulators, several researchers have used epoxy resin to develop polymer composites. Epoxy (5192-1) is selected as the matrix material for the present study. Local resources supplied the epoxy resin and the hardener HY-5196. Table.1 shows the specification of the epoxy resin.

Table 1. Mechanical properties of epoxy [14].

Density (g/cm3)	Tensile strength (MPa)	Viscosity (MPa.s)	Hardness (ShoreD)	Compress strength (MPa)
1.15	85	600-800	88	300

2.4. *Fabrication of hybrid composite rod*

Pultrusion is generally a cost-effective technique to produce a low-cost reinforced fibre composite. It is a constant process for the manufacturing of a cross-sectional composite and offers an attractive performance-to-cost ratio in addition to easy processing. A solid round kenaf/glass/epoxy hybrid rod composites were fabricated at Innovative Pultrusion Sdn. Bhd industry, Seremban, Malaysia. The hybrid fibres soaked in epoxy resin were pulled in a PULTREX Px1000-8T automated pultrusion machine. Kenaf and glass fibres saturated with epoxy were pulled into the mold with 24.7 mm diameter designed to infeed area. The materials entered the heated die at temp 110 ºC with fixed speed around (1.7 x 10^{-2} m/s) and pulled out at temperature 160 ºC when it was completely dried. Fig.2 shows the two types of hybrid composites, (a) IR20EP vol.20% and (b) IR30EP vol.30% of kenaf fibres hybrid with glass fibres reinforced epoxy resin.

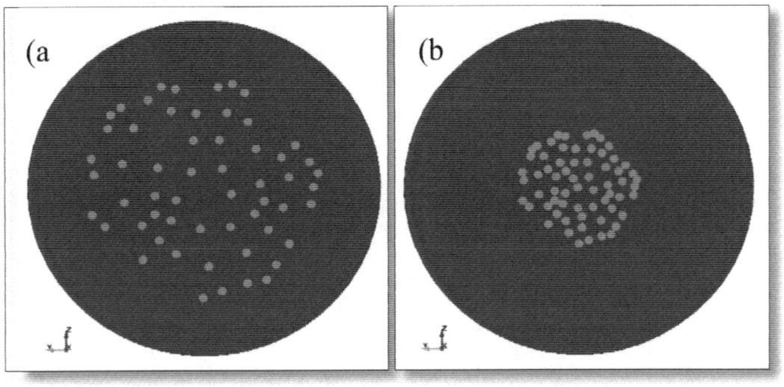

Figure 2. (a) IR20EP and (b) IR30EP hybrid composites.

2.5. Water Absorption Test

The water absorption test was carried out according to the procedures described in details in ASTM standard D5229/D5229M-92[15]. To eliminate the machining remains such as soil or grease, the specimens have been washed. The mass of the samples was initially measured then reported and the specimens submerged in water immediately. The weight difference was calculated using a 0.1 mg analytical scale. The specimens were immersed in water until they reached equilibrium state.

3. Results and Discussion

In general, the absorption of water in a hybrid composite depends on the form of fibre, the surface of the exposed and covered fibre, the matrix's density, temperature and volume fraction. The behaviour of water absorption considered as it rises over time until reaching the saturation level. The water absorption tests of control (epoxy), IR20EP, and IR30EP specimens are conducted under the same water absorption test conditions of kenaf/glass/Epoxy hybrid composites. The average reading of three specimens of each hybrid composite profile was recorded and compared with the control (epoxy) composite results. Fig.3 illustrated the water absorption behaviour in composite samples. The percentage of water gain and time are presented in the x-axis and y-axis respectively. The water absorption of hybrid composites needed more than 350 hours to reach saturation level. For the IR20EP composite, the percentage of water absorption raised steadily from zero to a saturated level at 0.84%. Simultaneously, the water absorption in the IR30EP composite elevated from zero to 0.25% on the first day, then decreased to 0.22% on the next day. Then it substantially rised up to 0.42% in the next three days. In the days (6, 7, 8, and 9), the water absorption percentage slightly increased. It increased significantly in day 10 (0.42% - 0.55%), then it stayed with small constant increased until the saturated level at 0.6% after 17 days. Based on this finding, it can be inferred that the control sample consists of glass/epoxy had the lowest water absorption which was 0.1 %. However, addition of 20 and 30 vol. % of kenaf fibre in the composites resulted in increment of water absorption to 0.84 and 0.6 % respectively. This was attributed to the cellulose structure of kenaf fibre which contains -OH groups that could form hydrogen bonding with water molecules thus increased the water absorption until it reached fibre saturation point.

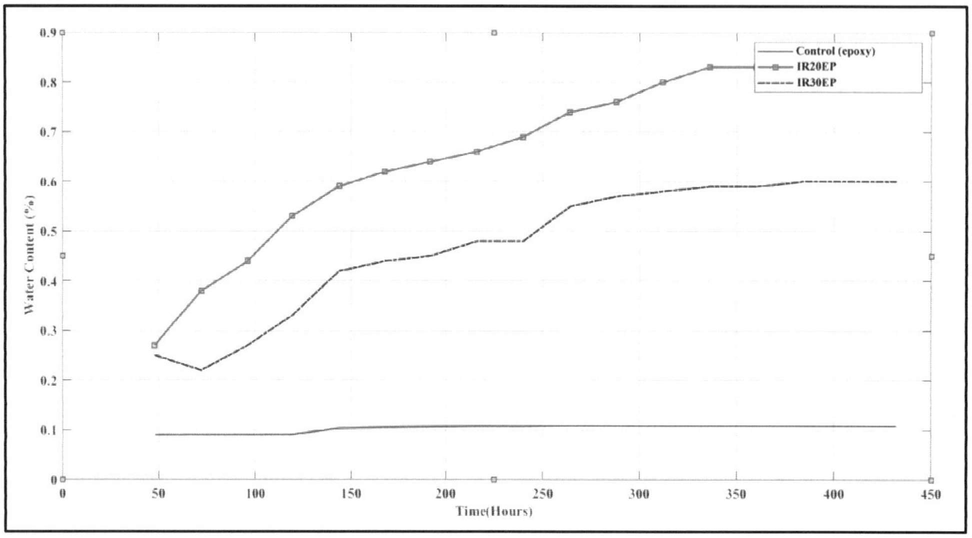

Figure 3. Water absorption of the hybrid and non-hybrid composites.

It was also observed that the specimen with 20 vol. % of kenaf fibre had higher water abrption compared to specimens with 30 vol. % of kenaf fibre. This is explained by the arrangements of fibre in the pultruded samples. Since the fibres were arranged loosely in the IR20EP composite, it had greater surface area with free -OH groups that could react with water molecules and finally absorbed more water. On the other hand, the fibres in specimen IR30EP were arranged tightly in the centre of the rod and had lesser exposed -OH groups thus resulted in lower water absorption. Therefore, it can be concluded that arrangement of fibres plays a very important role in determining its water absorption capacity.

4. Conclusions

Apart from the concern towards environmentally friendly material, the hybridization of natural fibres and synthetic fibres has subsequently reduced the manufacturing cost and its weight. However, the water absorption of the hybrid composites needs to be taken into consideration because it would affect the durability of the composites during its service as insulator core. The research emphasises on the effects of adding kenaf fibres in the water absorption of kenaf/glass/epoxy hybrid composite. The effect of kenaf fibre percentages (20 vol.% and 30 vol.%) and its fibre arrangement were studied. The following findings are drawn from this study:

- The control specimen (pure glass fibres reinforced epoxy) had a very minimal water absorption which was less than 0.11%.
- The incorporaton of kenaf fibres affects the composite's water absorption. Water absorption of kenaf/glass/epoxy hybrid composite had increased seven times more than pure glass/epoxy (non-hybrid) composite.
- The influence of the arrangement profile of kenaf fibre was more significant than its percentage. IR30EP hybrid composite had lesser water absorption due to its glass-material surroundings with kenaf fibre. Maximum absorption of water was achieved in IR20EP because the fibres were not completely enclosed with glass fibres and had more free hydroxyl groups.

- Overall, the percentage of water absorption of the epoxy hybrid composites was low and below 1% of the composite weight. Therefore, the kenaf fibres have an excellent possibility for hybridisation with glass fibres and binded with epoxy matrix.

Author's contribution

M.N.M.Ansari[1,*] - Idea & concept generation, suggestion and advice on the water absorption studies, suggestion on the interpretation of the data, supervision of the project and manuscript editing. **Alaseel Bassam**[2] – the first draft of the manuscript, analyses and interpretation of the data, **A.R.M.Nazim**[2] – analyses and interprets the results, manuscript editing, **Noor Afeefah Nordin**[1] - project management and co-supervision, suggestion and advice on the analyses, manuscript editing and revision. **Zainudin Yahya**[2] – co-supervision of the project.

Acknowledgements

The authors gratefully acknowledge Tenaga Nasional Berhad Malaysia's support through the Seed Fund Research Grant (Project No: U-TS-RD-18-10), the Institute of Power Engineering, the Department of Mechanical Engineering, and UNITEN R&D Sdn. Bhd., Universiti Tenaga Nasional to support the publishing of the results of this project.

References

[1] M. R. A. Karim, D. Tahir, E. U. Haq, A. Hussain, and M. S. Malik, "Natural fibres as promising environmental-friendly reinforcements for polymer composites," *Polym. Polym. Compos.*, 2020, doi: 10.1177/0967391120913723.

[2] R. Yahaya, S. M. Sapuan, M. Jawaid, Z. Leman, and E. S. Zainudin, "Effect of fibre orientations on the mechanical properties of kenaf–aramid hybrid composites for spall-liner application," *Def. Technol.*, vol. 12, no. 1, pp. 52–58, 2016, doi: 10.1016/j.dt.2015.08.005.

[3] M. K. F. Radzi *et al.*, "Influence of alkaline treatment and fiber loading on the physical and mechanical properties of kenaf/polypropylene composites for variety of applications," *Prog. Nat. Sci. Mater. Int.*, vol. 26, no. 6, pp. 657–664, 2016, doi: 10.1016/j.pnsc.2016.12.004.

[4] A. Bassam, M. N. M. Ansari, A. Atiqah, S. Begum, and A. R. M. Nazim, "Water Absorption Properties of Kenaf/Glass Reinforced Unsaturated Polyester Composites used in Insulator Rods," *Int. J. Eng. Adv. Technol.*, vol. 9, no. 2, pp. 4208–4212, 2019, doi: 10.35940/ijeat.b4931.129219.

[5] Y. Liu, X. Kong, B. Du, J. Li, and J. Sun, "Numerical Investigation on Collision of Pollution Particles on Outdoor Insulators," *IEEE Access*, vol. 7, pp. 56974–56985, 2019, doi: 10.1109/ACCESS.2019.2914075.

[6] M. Kumosa, L. Kumosa, and D. Armentrout, "Failure analyses of non-ceramic insulators: part II - the brittle fracture model and failure prevention," *IEEE Electr. Insul. Mag.*, vol. 21, no. 4, pp. 28–41, 2005, doi: 10.1109/mei.2005.1489987.

[7] K. Papailiou and I. N. Prassianakis, "Acoustic emission inspection and analysis of crimped metal-composite joints subjected to bending," no. February 2016.

[8] M. Kumosa, H. Shankara Narayan, Q. Qiu, and A. Bansal, "Brittle fracture of non-ceramic suspension insulators with epoxy cone end-fittings," *Compos. Sci. Technol.*, vol. 57, no. 7, pp. 739–751, 1997, doi: 10.1016/S0266-3538(97)00034-1.

[9] L. Kumosa, B. Benedikt, D. Armentrout, and M. Kumosa, "Moisture absorption properties of unidirectional glass/polymer composites used in composite (non-ceramic) insulators," *Compos. Part A Appl. Sci. Manuf.*, vol. 35, no. 9, pp. 1049–1063, 2004, doi: 10.1016/j.compositesa.2004.03.008.

[10] H. M. Akil, C. Santulli, F. Sarasini, J. Tirillò, and T. Valente, "Environmental effects on the mechanical behaviour of pultruded jute/glass fibre-reinforced polyester hybrid composites," *Compos. Sci. Technol.*, vol. 94, pp. 62–70, 2014, doi: 10.1016/j.compscitech.2014.01.017.

[11] N. L. Feng, S. D. Malingam, N. Razali, and S. Subramonian, "Alkali and Silane Treatments towards Exemplary Mechanical Properties of Kenaf and Pineapple Leaf Fibre-reinforced Composites," *J. Bionic Eng.*, vol. 17, no. 2, pp. 380–392, 2020, doi: 10.1007/s42235-020-0031-6.
[12] M. Nematollahi, M. Karevan, M. Fallah, and M. Farzin, "Experimental and Numerical Study of the Critical Length of Short Kenaf Fiber Reinforced Polypropylene Composites," *Fibers Polym.*, vol. 21, no. 4, pp. 821–828, 2020, doi: 10.1007/s12221-020-9600-x.
[13] M. Sheinbaum, L. Sheinbaum, O. Weizman, H. Dodiuk, and S. Kenig, "Toughening and enhancing mechanical and thermal properties of adhesives and glass-fiber reinforced epoxy composites by brominated epoxy," *Compos. Part B Eng.*, vol. 165, no. June 2018, pp. 604–612, 2019, doi: 10.1016/j.compositesb.2019.02.020.
[14] Y. Jiang and J. Wu, "Enhancement of interface strength of carbon fiber/epoxy resin composites filled with low-dimensional materials," *Compos. Interfaces*, vol. 00, no. 00, pp. 1–14, 2020, doi: 10.1080/09276440.2020.1766284.
[15] ASTM D5229-2004, "ASTM D 5229– 92 – Standard Test Method for Moisture Absorption Properties and Equilibrium Conditioning of Polymer Matrix Composite Materials," *Annu. B. ASTM Stand.*, vol. 92, no. Reapproved, pp. 1–13, 2010, doi: 10.1520/D5229.

Three-Dimensional (3D) printing and bioprinting for orthopaedic biomaterials – A short review

Sughanthy Siva A.P[1], M.N.M Ansari[2,*], Noor Afeefah Nordin[2], Ng Min Hwei[3]

[1] Department of Mechanical Engineering, Universiti Tenaga Nasional, 43000 Kajang, Selangor, Malaysia.

[2]Institute of Power Engineering, Universiti Tenaga Nasional, 43000 Kajang, Selangor, Malaysia.

[3] Tissue Engineering Centre, Universiti Kebangsaan Malaysia Medical Centre, Jalan Yaacob Latif, Bandar Tun Razak, Kuala Lumpur, 56000, Malaysia

Corresponding author: ansari@uniten.edu.my; ansari.uniten@gmail.com

Abstract. Tissue engineering is a non-conventional approach in creating artificial organs and tissues to heal from bone fracture or organs malfunction by regenerating the damaged tissues. This review paper focuses on bone scaffold fabrication methods using three-dimensional (3D) printing and bioprinting for tissue engineering applications. Various types of biomaterials and fabrication techniques of bone scaffolds aimed for tissue engineering were thoroughly discussed. Suitability of biomaterial and polymer for bone scaffolds is deliberated as part of the review paper. Several biomaterials, mechanical properties and porosity of scaffolds were reviewed to give an in depth understanding on 3D printing and tissue engineering field.

Keywords: *Bone Scaffold; Tissue Engineering; Three-Dimensional (3D) Printing; Bioprinting*

1. Introduction

Bone tissue generally comprises of 65-70% inorganic crystals, with hydroxyapatite (HA) as the main constituent. On other hand, due to unique hierarchical structural organization with multi-scales, bone is considered as a natural solid biocomposite which contributes to fracture, toughness, and high strength. Bone injuries and defects are among the significant clinical problems faced by the patients that needs to be treated before it could deteriorate one's health. These defects can be treated by various methods including autografts, allografts and alloplastic materials. Additional surgical site is needed for autografts which also carries the risks of postoperative infections or donor-site illness. Meanwhile, allografts normally exhibit the potential risk of disease transmission and immunologic rejection upon its treatment [1,2]. However, the advancement in bone tissue engineering could provide an alternative approach to repair the bone defects and expedite the healing process. Bone tissue engineering conceivably gives elective opportunities to bone substitutes rather than allograft. This is on the grounds that the artificial bones would give a structure for the cells to attach, differentiate, proliferate, and form new bone tissues. Moreover, these artificial bones would go about as cell transporter, factors of development, or other signals of biomolecular and most importantly, an ideal material is crucial to be chosen for the tissue engineering in order to get a notable healing process. Table 1 summarizes the composition and mechanical properties of corticol and cancellous bones.

Table 1. Bone composition and mechanical properties [4].

Bone Type	Composition (wt %)			Mechanical Properties	
	Protein	Calcium Phosphate (CaP)	Water	Strength (MPa)	Elasticity (GPa)
Cortical	28	60	12	70-200	3-30
Cancellous	26	54	20	0.1-30	0.02-0.5

Bones generally possess compressive strength which is high ~170 MPa, tensile strength which is low ~104-121 MPa and shear strength which is very low ~51.6MPa [16]. High compressive strength is well known to be an essential aspect in order to withstand any loads/stresses subjected to the bones. Bone repair can be executed in three potential methodologies, (1) substitute of cells that supply a necessary capacity; (2) conveying the substances that induce tissue, for example, development factors to be focused on the spot; and (3) a three-dimensional (3D) scaffold that develop cells. Nonetheless, the initial two methodologies are more suitable for modest and all around contained damages, while the 3D scaffold is appropriate for creating a bigger tissue blocks with predesigned shapes, customized to the patient's type of injury [3,4]. Therefore, this paper discusses the basic concepts of tissue engineering, the suitable biomaterials and method of printing for fabrication of bone scaffolds.

2. Tissue Engineering

The term "*Tissue Engineering*" was introduced by the National Science Foundation (NSF) in the year 1988 (USA) as "the application of principles and methods of engineering and life sciences toward fundamental understanding of structure-function relationships in normal and mammalian tissues and the development of biological substitutes to restore, maintain or improve tissue function". The three major interactive elements fundamental for tissue engineering to succeed are cells, signaling factors, and scaffolds. Tissue engineering is a field of multidisciplinary whereby it facilitates the advancements in chemistry, physics, clinical sciences and life, and engineer in solving the basic therapeutic issues problem, for example loss of organ failure or tissue. The broad areas of tissue engineering applications are orthopaedics, skin development, cartilage regeneration and reconstruction of neurons and organs [5, 6].

Optogenetic is an emergent technology among the most recent research fields, such as three-dimensional (3-D) bioprinting, nanotechnology, and stem cells. For the TE systems to be fruitful, a mixture of scaffolds, growth factors, and cells should be the character of the utilized material. In order to ensure the success of the method, the scaffolds average pore size should be 300-400 μm range in order to allow osteons to grow into the scaffold. Fig.1 shows the chronological development of the tissue engineering scaffolds [7].

Tissue engineering and regenerative medication (TERM) has an extraordinary part as it gives a significant idea in planning more effective unions with the point of upgrading osteoconduction, osteoinduction, osteogenesis and osteointegration in bone deformities [8].

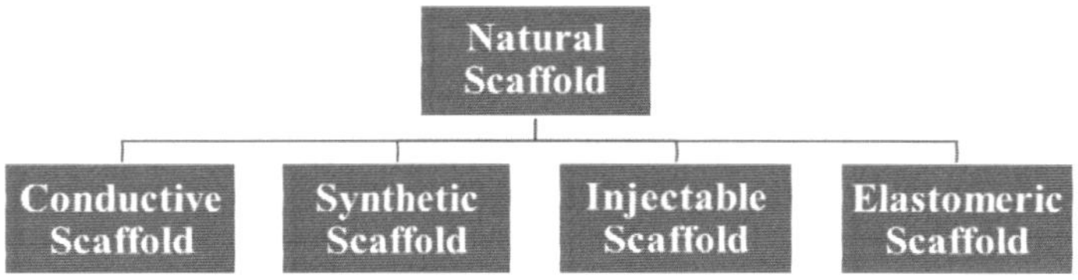

Figure 1. Scaffolds in Tissue Engineering.

The investigation of novel ideas in designing bone scaffolds and other complex biomedical models, for example, prostheses and implants can be cultivated by the use of computational modelling. This model will utilize patient-specific anatomical information and can be emerged by Additive Manufacturing (AM) procedures to ease and expedite the treatment process [9].

3. Biomaterials

Biomaterials from a health care perspective can be defined as "materials that possess some novel properties that makes them appropriate to come into immediate contact with the living tissue without eliciting an adverse immune rejection reaction." [10]. Biomaterials can give the necessary mechanical and chemical prompts, just as other regulatory flagging significant for cell self-restoration or obligation to a separation heredity [11]. For a better biocompatibility properties, biomaterials ought to be non-antigenetic, non-cytotoxic, high cell suitability properties that can induce cell movement and proliferation, just like the requirements for an extracellular matrix needed for tissue growth. Depending on the tissue to regenerate, there are now many options for biomaterials in orthopaedic [12].

Osteogenic biomaterials are an option in contrast to conventional bone repair strategies, prompting quicker healing of the elaborate joint. Biomaterials utilized for artificial bones are partitioned into two kinds; natural materials and synthetic, for example, PLA, polycaprolactone (PCL), collagen (COL) and gelatin (GEL) [13]. Calcium orthophosphates, a sort of inorganic biomaterials is right now standing out enough to be noticed for its expected clinical applications in numerous territories such as in bone substitutions and other interior obsessions. A past work revealed that HA shows the ideal properties such as nontoxic, biocompatible, bioactive, non-immunogenic, non-inflammatory, osteoconductive and possesses good osteointegration properties [14].

In applications of orthopaedic, the usage of biomaterial is usually intended to replace the injured bone or reestablish the structural integrity. Fig.2 shows some of the important factors to be considered in designing orthopaedic materials. The classical biomaterials can be additionally arranged into: (i) metals and alloys (ii) non-metals such as polymers, ceramics, porcelain, glass, and composites [15]. In a modern context of orthopaedic biomaterials are sub-categorized into two significant classifications namely, nanophase biomaterials and classical biomaterials as shown in Fig.3.

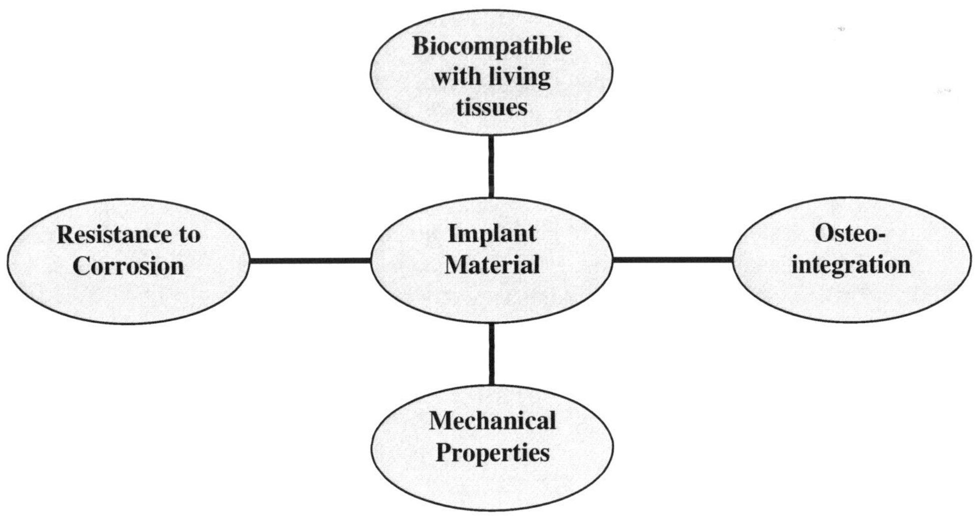

Figure 2. Important factors for orthopaedic biomaterial design.

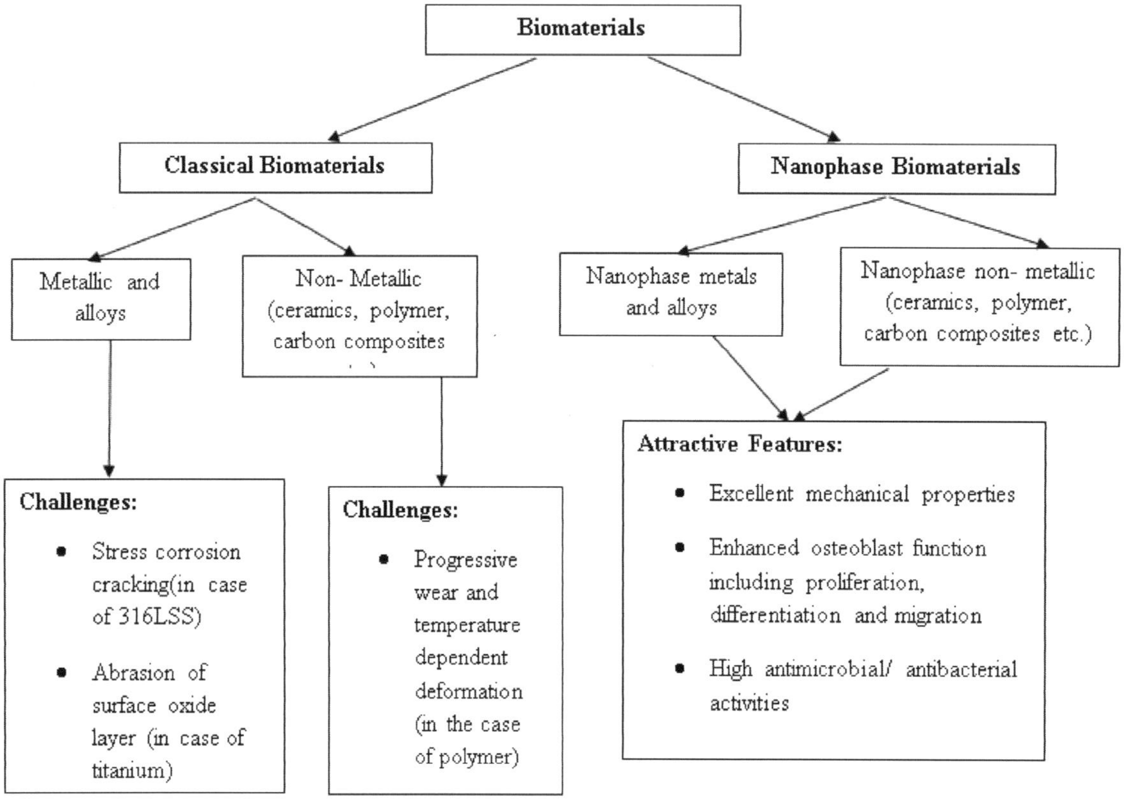

Figure 3. Categories of orthopaedic biomaterials: classical and nanophase biomaterials [15].

4. Three-Dimensional (3D) Printing

The 3D printing is an emerging technology that can be used to construct complex structures inexpensively for modelling, prototyping, or production through deposition or solidification. This technology was developed by the Massachusetts Institute of Technology (MIT) during 1990s' and it was utilized for prototype fabrication and design validation purpose. The concept of 3D printing is the deposition or addition of materials in molten form by one layer over the other in 2-dimensional area in succession through an automated layer-by-layer process to construct a 3D structure [16, 17]. The 3D printed object structure is customized by a computer-aided design (CAD) model which is connected to a 3D printer. The CAD models designate 3D objects in a series of cross-sectional layers, and permits 3D printers to reproduce models through an additive process. By converting computed tomography (CT) or magnetic resonance imaging (MRI) of clinical defects into a CAD software, the specific CAD models of the patients can be developed. A slicing software is then used to section CAD models into layer by layer information using G-code, which encodes the 3D CAD models in a 3D printer controllable format. The G-code will then modify and optimize the required parameters such as print head speed, nozzle temperature, layer height and pressure [19, 20]. The schematic diagram of a 3D printer is illustrated in Fig.4.

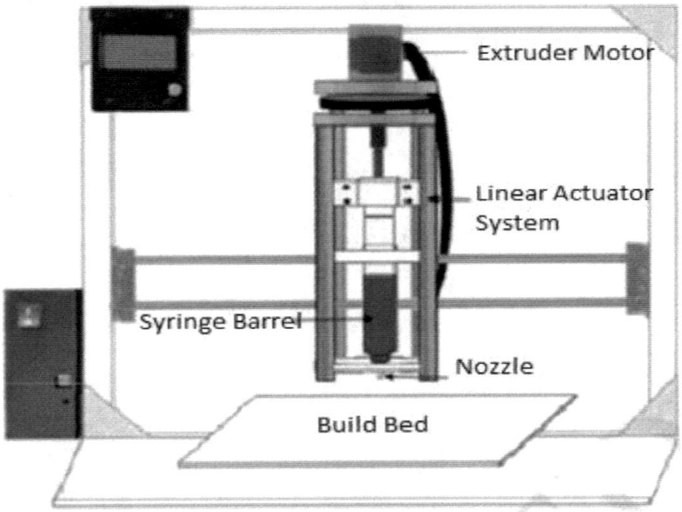

Figure 4. Schematic diagram of a 3D printer [18].

In 3D printing, the architectures of scaffolds at every angle including its porosity, pore shape and interconnectivity can be discretely controlled to achieve patient-specific demands besides having high structural complexity and flexibility. Once incorporated in the patient's body, the 3D printed scaffolds would provide a growth-directing structure that facilitates cell migration and proliferation to regenerate the bone tissue. In addition, the fabrication of scaffolds using 3D printing is a quick process and could decrease experimental errors [21]. A previous study by Fahimipour *et al.* (2017) constructed avascular endothelial growth factor (VEGF)-loaded with gelatine/alginate/ Beta-TCP composite scaffold by method of 3D printing with usage of a computer-assisted design (CAD) model. The scaffolds compressive modulus, 98 ± 11 MPa, was seen to be in the cancellous bone range, with suggestions of their potential applications for craniofacial tissue engineering [22].

Hassanajili *et al.* (2019) applied the 3D printing approach to develop bone scaffolds using polylactic acid, polycaprolactone and hydroxyapatite (PLA/PCL/HA) composites. A dissolvable 3D printed negative mold was used to caste the composite suspensions to achieve simultaneous macro- and micro-porous composites using freeze drying/particle leaching method. They found that the scaffold with PLA/PCL 70/30 w/w and 35wt% of HA composition showed better properties in terms of its mechanical properties, microstructure, pore size and porosity [23]. With the existence of various 3D printing processes that are currently evolving, designers need to fully understand the merits and demerits of these concepts and associated materials in order to make a better decision and choice in selecting the 3D printing process to optimize the bone scaffolds for tissue engineering purposes [24].

5. Bioprinting

Bioprinting is one of the progressing and inventive innovation of this century that has got developing intrigue around the world such as in reformed drug industry, tissue engineering, and regenerative prescriptions areas, and so on. It suggests the usage technology of three-dimensional (3D) printing to print distinctive biomaterials with joined suitable cells. The triumph of bioprinting technology depends upon the profitability of 3D bioprinters and availability of the bioink (involved polymer arrangement and feasible cells) and its primary, physio-chemical, mechanical, thermal, and biological features [25]. The contrast between "*3D printing*" and "*3D Bioprinting*" must be differentiated clearly as both of these terms are invariably used by the scientific community. Both of these processes construct a 3D structure layer-by-layer from a 3D CAD model. In any case, 3D Bioprinting includes the utilization of cell-laden bioinks and other living cells to develop a structural tissue model whereas 3D printing technologies utilizes only materials in molten form without any biologics [26, 27]. The technology of bioprinting itself has been the focal point of research for almost 15 years and will keep on growing to cater the tissue engineering needs for quite a long time to come [28]. The overall process of bioprinting is summarized in Fig.5 while Fig.6 shows the three different techniques of bioprinting.

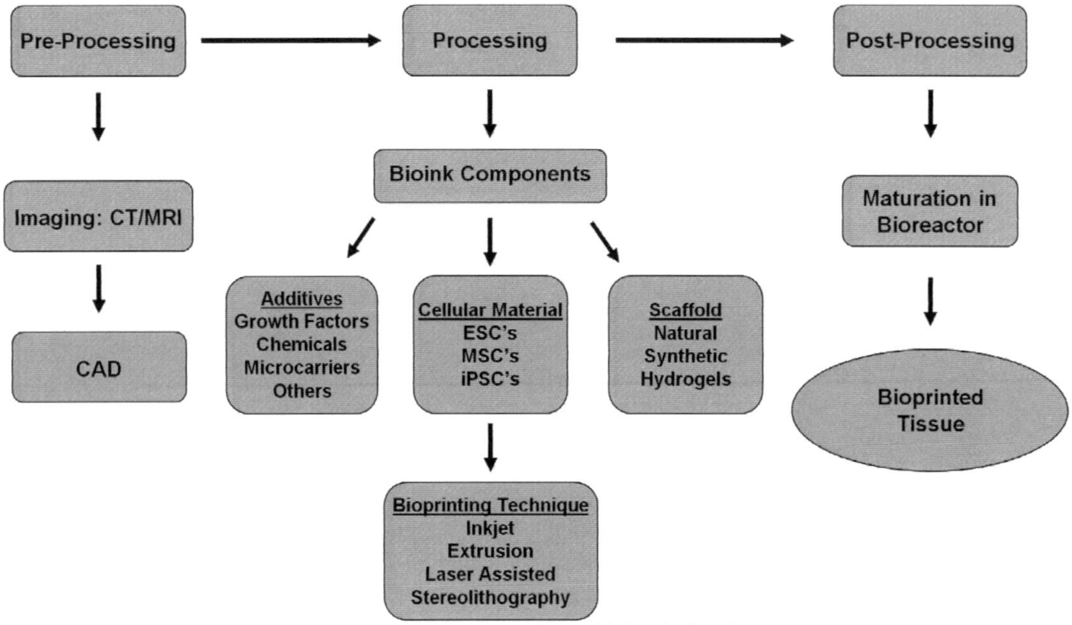

Figure 5. The overview of bioprinting [27].

The three major modalities of bioprinting technologies are categorised into extrusion-based bioprinting (EBB), droplet-based bioprinting (DBB) and laser-based bioprinting (LBB), which have been utilized for vascular or vascularized tissue fabrication [29].

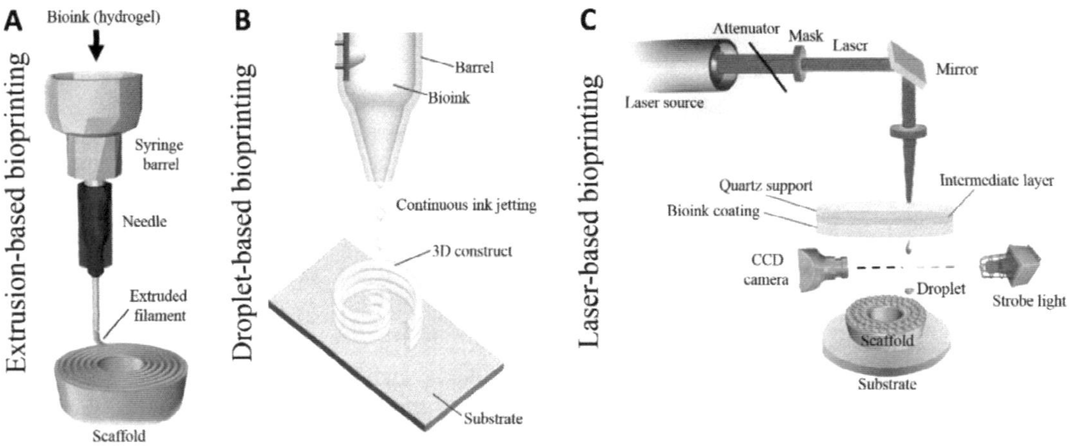

Figure 6. Mechanisms of bioprinting techniques: A) extrusion-based bioprinting, B) droplet-based bioprinting, C) laser-based bioprinting [30].

Demirtas *et al.* (2017) studied the effect of blending of cells of chitosan solution and it's composite with nanostructured bone-like hydroxyapatite (HA) using 3D bioprinter. MC3T3-E1 pre-osteoblast cell laden chitosan and chitosan-HA hydrogels, utilize extruder-based bioprinter and were characterized by comparing these hydrogels to alginate and alginate-HA hydrogels. In addition, the cell viability, proliferation and osteogenic differentiation had improved in the presence of bone-like nanostructured HA in alginate and chitosan hydrogels and it was also been proved by the authors [31].

At first, exosomes were considered as futile cellular metabolic wastes, but then it had been recognized that they have numerous significant cellular capacities. One strand of the exosomal miRNAs contains miRNA-induced silencing complex (miRISC) which are normally grouped into Argonaute protein structure and interatomic fusion with target mRNA transcripts which leads to the restraint of comparing gene expression. Some biomolecules found in exosomes are usually treated as biomarkers for diagnosis of any disease, prognosis, and damage conditions, since their levels or substance might change its conditions in response to few illnesses or wounds [32, 33]. Wei *et al.* (2019) investigated BMP2/macrophage-derived exosomes to improve the bio-functionality for titanium implants and to modify titanium nanotube implants to favour osteogenesis. As an emerging functional material for bone tissue engineering, the exosome-integrated titanium nanotube may promote and serve the tissue regeneration [34].

6. Summary and future perspective

As it has been discussed in this review, orthopaedics is one of the areas in the tissue engineering field. There are different strategies/methods and wide selection of materials that can be utilized in fabricating scaffold. Essentially, scaffolds are utilized as a site in connecting the cells, proliferating, separating and migrating by considering the protein synthesis and developing factors. Scaffold plays an important part in giving mechanical support and conveying cells or particles that are inductive to the influenced site. In addition, scaffolds offer assistance in controlling the tissues that are recently shaped in their organizing and functioning. Each strategies or procedures have their parameters that ought to be considered such as the materials utilized, temperature, pressure applied and time retained that will result in a prodigious scaffold. On the off chance that there's any changes in those process parameters, it may result in a scaffold with some microstructural defects. Other than that, the chemical composition for scaffold fabrication, surface roughness, material-cell interaction and degradation play an important part in the characteristics of the scaffold surface. To overcome this, a vast knowledge and reliable data from the distinctive field is required such as biomedical designing and chemical engineering field. This could be taken into thought in the future for the improvement of the tissue engineering field and its applications.

Acknowledgment

The authors would like to thank the Department of Mechanical Engineering, iRMC and UNITEN R&D Sdn. Bhd., Universiti Tenaga Nasional for the support to publish the finding of this project. Also, the facilities and support provided by Tissue Engineering Centre, HUKM, Cheras, Malaysia are greatly acknowledged.

Author's contribution

S.A.P.Sughanthy[1] – first draft of the manuscript, data collection, analyses and review of the data, **M.N.M.Ansari**[2,*] - Idea & concept generation, suggestion and advice on the review manuscript, supervision of the project and manuscript editing, **Noor Afeefah Nordin**[2] - project management and co-supervision, suggestion and advice on the analyses, manuscript editing and revision, **Ng Min Hwei**[3] – suggestion and review manuscript editing, co-supervision of the project.

References

[1] Li, X., Zhang, S., Zhang, X., Xie, S., Zhao, G., & Zhang, L. (2017). Biocompatibility and physicochemical characteristics of poly (ε-caprolactone)/poly (lactide-co-glycolide)/nano-

hydroxyapatite composite scaffolds for bone tissue engineering. *Materials & Design*, *114*, 149-160.
[2] Zhang, D., Wu, X., Chen, J., & Lin, K. (2018). The development of collagen based composite scaffolds for bone regeneration. *Bioactive materials*, *3*(1), 129-138.
[3] Khojasteh, A., Fahimipour, F., Eslaminejad, M. B., Jafarian, M., Jahangir, S., Bastami, F., ... & Tayebi, L. (2016). Development of PLGA-coated β-TCP scaffolds containing VEGF for bone tissue engineering. *Materials Science and Engineering: C*, *69*, 780-788.
[4] Michael, F. M., Khalid, M., Walvekar, R., Ratnam, C. T., Ramarad, S., Siddiqui, H., & Hoque, M. E. (2016). Effect of nanofillers on the physico-mechanical properties of load bearing bone implants. *Materials Science and Engineering: C*, *67*, 792-806.
[5] Hasnain, M. S., Ahmad, S. A., Chaudhary, N., Hoda, M. N., & Nayak, A. K. (2019). Biodegradable polymer matrix nanocomposites for bone tissue engineering. In *Applications of Nanocomposite Materials in Orthopedics* (pp. 1-37). Woodhead Publishing.
[6] Mondal, S., & Pal, U. (2019). 3D hydroxyapatite scaffold for bone regeneration and local drug delivery applications. *Journal of Drug Delivery Science and Technology*, 101131.
[7] Zarrintaj, P., Bakhshandeh, B., Saeb, M. R., Sefat, F., Rezaeian, I., Ganjali, M. R., ... & Mozafari, M. (2018). Oligoaniline-based conductive biomaterials for tissue engineering. *Acta biomaterialia*, *72*, 16-34.
[8] Oryan, A., Kamali, A., Moshiri, A., Baharvand, H., & Daemi, H. (2018). Chemical crosslinking of biopolymeric scaffolds: Current knowledge and future directions of crosslinked engineered bone [27] scaffolds. *International journal of biological macromolecules*, *107*, 678-688.
[9] Madrid, A. P. M., Vrech, S. M., Sanchez, M. A., & Rodriguez, A. P. (2019). Advances in additive manufacturing for bone tissue engineering scaffolds. *Materials Science and Engineering: C*.
[10] Triplett, R. G., & Budinskaya, O. (2017). New frontiers in biomaterials. *Oral and Maxillofacial Surgery Clinics*, *29*(1), 105-115.
[11] Barata, D., Dias, P., Wieringa, P., van Blitterswijk, C., & Habibovic, P. (2017). Cell-instructive high-resolution micropatterned polylactic acid surfaces. *Biofabrication*, *9*(3), 035004.
[12] Cheng, K. Y., Chang, C. H., Yang, Y. W., Liao, G. C., Liu, C. T., & Wu, J. S. (2017). Enhancement of cell growth on honeycomb-structured polylactide surface using atmospheric-pressure plasma jet modification. *Applied Surface Science*, *394*, 534-542.
[13] He, X., Fan, X., Feng, W., Chen, Y., Guo, T., Wang, F., ... & Tang, K. (2018). Incorporation of microfibrillated cellulose into collagen-hydroxyapatite scaffold for bone tissue engineering. *International journal of biological macromolecules*, *115*, 385-392.
[14] Akindoyo, J. O., Beg, M. D., Ghazali, S., Heim, H. P., & Feldmann, M. (2017). Effects of surface modification on dispersion, mechanical, thermal and dynamic mechanical properties of injection molded PLA-hydroxyapatite composites. *Composites Part A: Applied Science and Manufacturing*, *103*, 96-105.
[15] Kumar, S., Nehra, M., Kedia, D., Dilbaghi, N., Tankeshwar, K., & Kim, K. H. (2020). Nanotechnology-based biomaterials for orthopaedic applications: Recent advances and future prospects. *Materials Science and Engineering: C*, *106*, 110154.
[16] Do, A. V., Smith, R., Acri, T. M., Geary, S. M., & Salem, A. K. (2018). 3D printing technologies for 3D scaffold engineering. In *Functional 3D Tissue Engineering Scaffolds* (pp. 203-234).
[17] Cuiffo, M. A., Snyder, J., Elliott, A. M., Romero, N., Kannan, S., & Halada, G. P. (2017). Impact of the fused deposition (FDM) printing process on polylactic acid (PLA) chemistry and structure. *Applied Sciences*, *7*(6), 579.
[18] Koski, C., Onuike, B., Bandyopadhyay, A., & Bose, S. (2018). Starch-hydroxyapatite composite bone scaffold fabrication utilizing a slurry extrusion-based solid freeform fabricator. *Additive Manufacturing*, *24*, 47-59.

[19] Turnbull, G., Clarke, J., Picard, F., Riches, P., Jia, L., Han, F., ... & Shu, W. (2018). 3D bioactive composite scaffolds for bone tissue engineering. *Bioactive materials*, *3*(3), 278-314.

[20] Koons, G. L., & Mikos, A. G. (2019). Progress in three-dimensional printing with growth factors. *Journal of controlled release*, *295*, 50-59.

[21] Ma, H., Feng, C., Chang, J., & Wu, C. (2018). 3D-printed bioceramic scaffolds: From bone tissue engineering to tumor therapy. *Acta biomaterialia*, *79*, 37-59.

[22] Fahimipour, F., Rasoulianboroujeni, M., Dashtimoghadam, E., Khoshroo, K., Tahriri, M., Bastami, F., ... & Tayebi, L. (2017). 3D printed TCP-based scaffold incorporating VEGF-loaded PLGA microspheres for craniofacial tissue engineering. *Dental Materials*, *33*(11), 1205-1216.

[23] Hassanajili, S., Karami-Pour, A., Oryan, A., & Talaei-Khozani, T. (2019). Preparation and characterization of PLA/PCL/HA composite scaffolds using indirect 3D printing for bone tissue engineering. *Materials Science and Engineering: C*, *104*, 109960.

[24] Zhang, L., Yang, G., Johnson, B. N., & Jia, X. (2019). Three-dimensional (3D) printed scaffold and material selection for bone repair. *Acta biomaterialia*, *84*, 16-33.

[25] Aljohani, W., Ullah, M. W., Zhang, X., & Yang, G. (2018). Bioprinting and its applications in tissue engineering and regenerative medicine. *International journal of biological macromolecules*, *107*, 261-275.

[26] Vijayavenkataraman, S., Yan, W. C., Lu, W. F., Wang, C. H., & Fuh, J. Y. H. (2018). 3D bioprinting of tissues and organs for regenerative medicine. *Advanced drug delivery reviews*, *132*, 296-332

[27] Bishop, E. S., Mostafa, S., Pakvasa, M., Luu, H. H., Lee, M. J., Wolf, J. M., ... & Reid, R. R. (2017). 3-D bioprinting technologies in tissue engineering and regenerative medicine: Current and future trends. *Genes & diseases*, *4*(4), 185-195.

[28] Blaeser, A., Campos, D. F. D., & Fischer, H. (2017). 3D bioprinting of cell-laden hydrogels for advanced tissue engineering. *Current Opinion in Biomedical Engineering*, *2*, 58-66.

[29] Lepowsky, E., Muradoglu, M., & Tasoglu, S. (2018). Towards preserving post-printing cell viability and improving the resolution: Past, present, and future of 3D bioprinting theory. *Bioprinting*, *11*, e00034.

[30] Datta, P., Barui, A., Wu, Y., Ozbolat, V., Moncal, K. K., & Ozbolat, I. T. (2018). Essential steps in bioprinting: From pre-to post-bioprinting. *Biotechnology advances*, *36*(5), 1481-1504.

[31] Demirtaş, T. T., Irmak, G., & Gümüşderelioğlu, M. (2017). A bioprintable form of chitosan hydrogel for bone tissue engineering. *Biofabrication*, *9*(3), 035003.

[32] Jing, H., He, X., & Zheng, J. (2018). Exosomes and regenerative medicine: state of the art and perspectives. *Translational Research*, *196*, 1-16.

[33] Scimeca, J. C., & Verron, E. (2017). The multiple therapeutic applications of miRNAs for bone regenerative medicine. *Drug discovery today*, *22*(7), 1084-1091.

[34] Wei, F., Li, M., Crawford, R., Zhou, Y., & Xiao, Y. (2019). Exosome-integrated titanium oxide nanotubes for targeted bone regeneration. *Acta biomaterialia*, *86*, 480-492.

Experimental Investigations on the Drilling of Titanium Metal Matrix Composite

Arun Rajesh M[1, a], Bala Kumar S S [1, b], Ashok R[1, c] &
Vijay Sekar K S [2, d]

[1] Final Year Student of BE, Department of Mechanical Engineering, SSN College of Engineering,

Kalavakkam - 603 110, Tamil Nadu, India

[2] Associate Professor, Department of Mechanical Engineering, SSN College of Engineering,

Kalavakkam - 603 110, Tamil Nadu, India

[a] arun16020@mech.ssn.edu.in, [b] bala16026@mech.ssn.edu.in, [c] ashok16021@mech.ssn.edu.in,

[d] vijaysekarks@ssn.edu.in

ABSTRACT

Titanium Metal Matrix Composites have high specific strength and offer high stiffness compared with Nickel alloys and steel. While maintaining equivalent strength, it can reduce 50% weight relative to other super alloys. In this work, preparation of Silicon Carbide reinforced Titanium Metal Matrix Composite by powder metallurgy and drilling of the composite to evaluate the drilling forces are discussed and the effect of drilling process parameters on thrust force, surface finish and chip formation is studied.

Key-words: Titanium Metal Matrix Composite, Si-C particles reinforced Titanium MMC, Powder Metallurgy, Machinability, Drilling, chip Morphology.

INTRODUCTION

A Metal Matrix Composite (MMC) is a material with minimum two constituent parts, one being a metal while the other may be of any other metal or material, such as a ceramic or organic compound. The two distinctive constituents of a composite material are the matrix and the reinforcement. The matrix is a continuous monolithic material in which the reinforcement material is embedded. Structural applications demand light weight high strength materials such as titanium, magnesium, or aluminium. These materials are commonly used as matrix in MMC, as they provide a compliant support for the reinforcement. Cobalt and cobalt-nickel alloy matrices are used for high temperature applications. The reinforcement materials are either continuous or discontinuous and are used to enhance physical properties such as friction coefficient, conductivity, wear resistance. A variety of material is used as reinforcing material in composites like aluminium oxide, titanium oxide, silicon carbide, titanium carbide, boron carbide, titanium boride, zirconium boride, aluminium nitride, silicon nitride, etc. In the current industrial scenario, the need for light weight and high-performance material has risen. Thus, MMCs have emerged as the strongest contenders against their monolithic alloys. MMCs are used in aerospace, marine and automotive industries. Titanium MMC is very relevant in Industries such as aerospace sector where high strength materials coupled with low weight is required. Many researchers have worked on MMCs by powder metallurgy (PM) routes to improve the mechanical, thermal, and electrical properties of monolithic alloys. Titanium MMC has potential for high strength, corrosion protection, and excellent heat resistance [1].

In drilling of the Titanium Composite, thrust force and surface roughness are largely influenced by the feed rate and the processing technique [2]. Powder Metallurgy processes are commonly used for Titanium Metal Matrix Composite components used for automotive, aerospace and motorcycle industries [1].

In this work, Titanium Metal Matrix Composite reinforced with Silicon Carbide particles is prepared and the drilling on the MMC is conducted. In addition, the drilled surface and chips obtained during drilling are examined using Scanning Electron Microscope (SEM).

METHODS

[Following paragraphs explain in brief the 3 adopted methods in sequence: 1.Fabrication of MMC 2.Experimentation]

1. Fabrication of MMC:

The Metal Matrix Composite is fabricated through Powder Metallurgy route. *Figure 1 and Figure 2* shows the images of the Titanium and Si-C powders respectively. The Titanium and Si-C powder are taken in the ratio *85:15* by volume [3]. The powder mixture is ball milled for 1hour at 300 rpm with tungsten carbide balls [4]. *Table 1 and Table 2* shows the specifications of Titanium and Si-C powders.

Table 1 Specifications of Ti powder

Purity	High purity (99%)
Size	40-50 microns
Density	4.506 g/cm^3
Melting Point	1668°C

Table 2 Specifications of Si-C powder

Purity	Highly pure (99%)
Mesh size	220 mesh
Density	3.21 g/cm^3
Melting point	2730°C

IOP Conf. Series: Materials Science and Engineering 1128 (2021) 012029 doi:10.1088/1757-899X/1128/1/012029

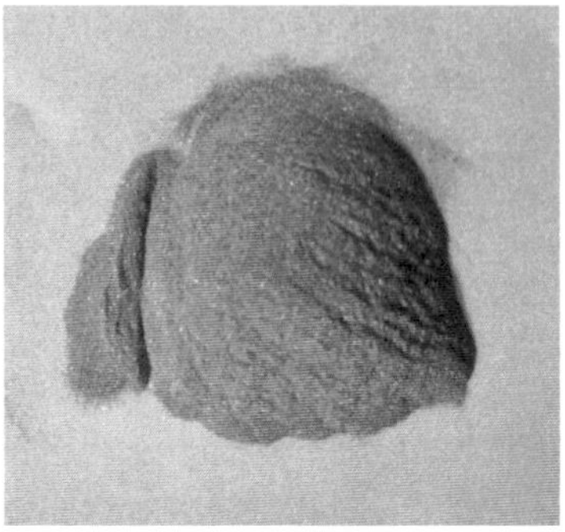

Fig 1 Titanium Powder *Fig 2 Silicon Carbide Powder*

Few (*3-4*) drops of polyvinyl alcohol is used as a binder. *Figure 3* and *Figure 4* shows the powder before and after ball milling respectively. The powder is compacted at *160 bar* using *10 tonne* hydraulic press. The dimension of the die is $32\times32\ mm^2$ with *8 mm* thickness (*Figure 5*). The compacted powders (*Figure 6*) are loaded on a tubular furnace under controlled atmosphere (Argon) and sintered at a temperature of *1300 degree Celsius* for a holding time of 1 hour [5].

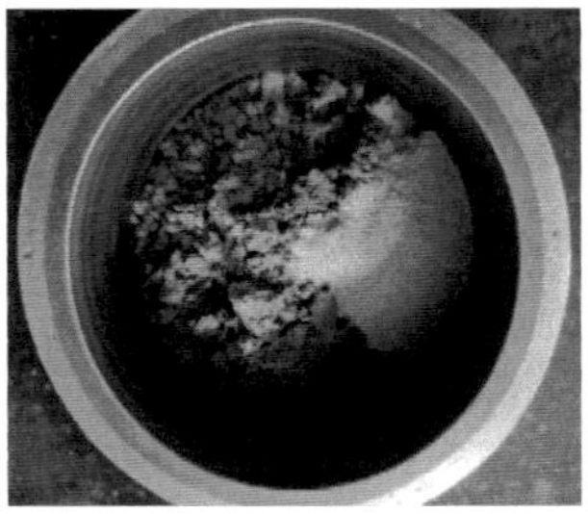

Fig 3 Before Ball milling *Fig 4 After Ball milling*

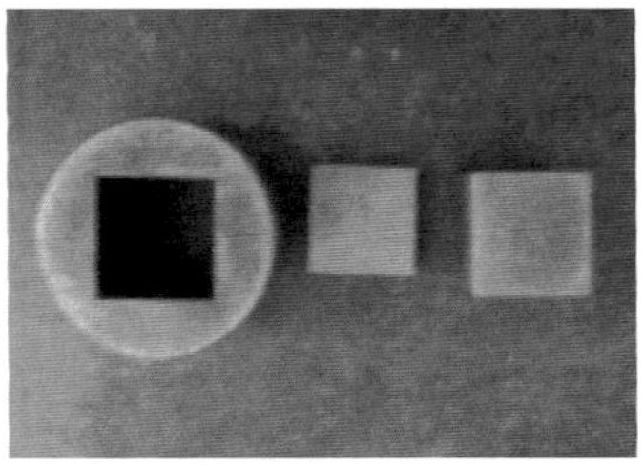

Fig 5 Die 32×32 mm^2 Fig 6 Compacted powder

The heating and cooling rate are maintained at *5 degree Celsius* per minute [5]. *Table 3* shows the heating and cooling temperature and time taken. The sample is kept at *1300 °C* for a holding time of 1 hour. *Figure 7* shows the samples after being sintered in the furnace.

Table 3

Heating (°C)	Time (Hours)	Cooling (°C)	Time (Hours)
30 - 600	3	1300 - 400	4.5
600 - 1300	3	400 – 30	6

Fig 7 composite samples after sintering

2. Experimentation:

The Titanium MMC samples prepared are drilled using solid carbide tool (*ϕ10mm*). At different drilling speeds, 7 holes are drilled (*Figure 10*). The thrust force and the torque for drilling of the MMC composite are measured using dynamometer (*Figure 10)* and multicomponent force indicator (*Figure 11). Figure 8* shows the drilling machine and *Figure 9* shows the drill bit (*dia 10 mm*) on the wise.

IOP Conf. Series: Materials Science and Engineering 1128 (2021) 012029 doi:10.1088/1757-899X/1128/1/012029

Fig 8 Drilling Machine

Fig 9 drill bit & workpiece on wise

The drilling is carried out on the four samples. Figure 10 shows the blind holes with the numbers.

Fig 10 Drilled Samples

RESULTS AND DISCUSSION

(The abridged form of results is being given here.)

The data collected during experimentation are given in the *Table 4* and the *figure 11* is plotted between Cutting speed versus Thrust and Cutting speed versus Torque. It can be seen in *figure 12* that the change in torque with cutting speed is relatively less. The variation of thrust force with respect to the cutting speed is shown in *figure 11.*

Table 4 Thrust Force and Torque for different spindle speed

S.NO	Speed (RPM)	Thrust Force (N)	Torque (Kg-m)
1	515	960	0.2
2	720	1270	0.1
3	103	1030	0.1
4	1152	1090	0.1
5	164	950	0.1
6	257	960	0.2
7	73	1730	0.3

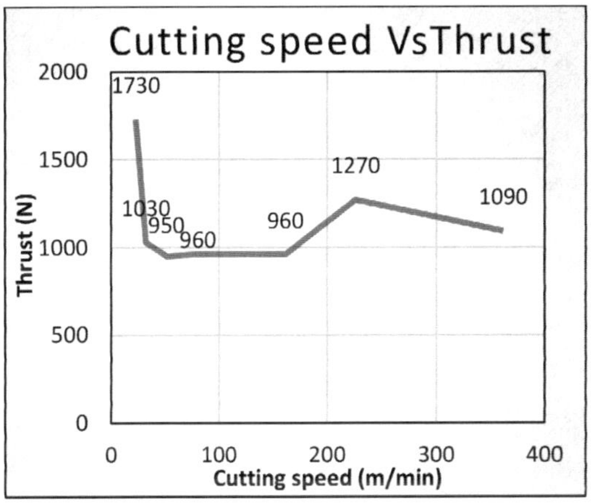

Fig 11: Cutting Speed Vs Thrust

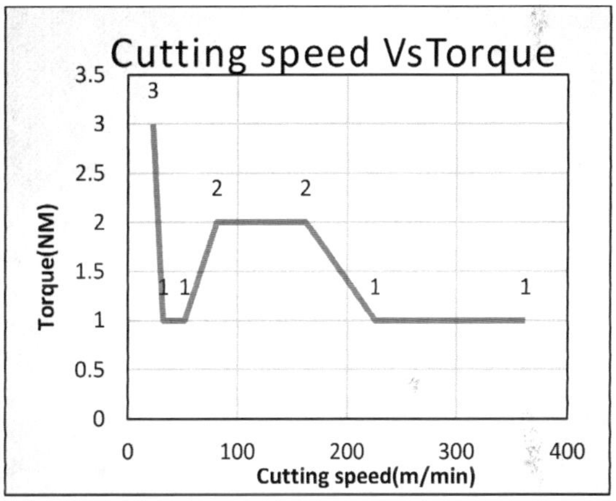

Fig 12 Torque vs cutting speed

The SEM Macrograph of the drilled specimen and SEM images of the chips formed during drilling are shown in figures. The SEM images of drilled surface of drilled workpieces and the chips formed are taken at two different speeds (103rpm and 720rpm). The chips formed are powdery and this concludes the poor ductility property of the material. This can be explained on the basis of the powder metallurgy technique that is used for the fabrication. The burrs formed during drilling are of negligible size.

Fig 11 Hole Surface drilled at 720RPM

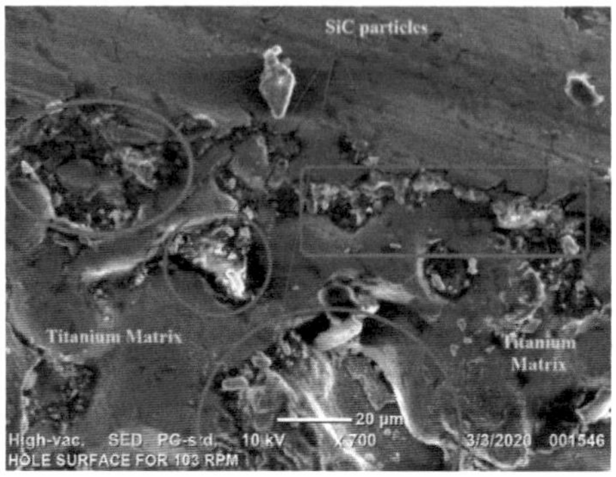

Fig 12 Hole Surface drilled at 103RPM

Fig 13 Sample drilled at 720 RPM

From the above shown SEM images, it can be seen that the surface layer is highly sheared due to the drilling process. Titanium and Si-C particles are severely fragmented during drilling operation. Silicon Carbide particles present in the Titanium matrix are shattered on the surface of the workpiece (*Figure 11 & 12*). Titanium is a hard material, in addition to which the abrasive nature of the Si-C particle makes the MMC even harder for machining. Silicon carbide particles are unevenly dispersed in the Titanium Matrix. More over the presence of pores can be noticed from SEM image. Due to porous nature of the material, application of more thrust force will result in the failure. *Figure 13 shows* the sample drilled at 720 rpm.

Fig 14 SEM micrographs of chips produced during drilling at 103 rpm

Fig 15 SEM micrographs of chips produced during drilling at 720rpm

Figure 14 and *Figure 15* shows the SEM images of the powdery chips formed during drilling.

CONCLUSIONS

The following are the conclusions drawn from the experimentation,

1. Reductions of thrust force improve the surface finish and reduce the power consumption in drilling. From 100 to 500 rpm the thrust force is least. At this speeds the tool wear will be less and the power consumed will also be less relatively.

2. Surprisingly no solid chip is formed during machining instead powders were formed which is because of the powder metallurgy preparation process.

3. The chips formed are discontinuous and segmented in nature. They are powdery and this concludes the poor ductility property of the material.

4. Due to presence of Silicon Carbide in the Titanium matrix, it makes it hard to drill the workpiece and abrasive action of Si-C particle causes considerable wear of the Tool.

5. Due to the increased porosity nature of the workpiece, the workpiece gets damaged on applying more force on it. All MMC that are fabricated through Powder metallurgy route shows porosity.

6. Burr formation is negligible when Composite prepared by powder metallurgy.

REFERENCES

[1] Kondoh K. Titanium metal matrix composites by powder metallurgy (PM) routes. InTitanium Powder Metallurgy 2015 Jan 1 (pp. 277-297). Butterworth-Heinemann.
[2] Ramkumar T, Selvakumar M, Mohanraj M, Chandrasekar P. Experimental investigation and analysis of drilling parameters of metal matrix (Ti/TiB) composites. Journal of the Brazilian Society of Mechanical Sciences and Engineering. 2019 Jan 1;41(1):8.
[3] Poletti C, Balog M, Schubert T, Liedtke V, Edtmaier C. Production of titanium matrix composites reinforced with SiC particles. Composites Science and Technology. 2008 Jul 1;68(9):2171-7.
[4] P.S. Kulkarni and S. Ratnaparkhi Fabrication of Ti-CNT composites through powder metallurgy. International Engineering Research Journal Page No 1724-1726
[5] Bolzoni L, Ruiz-Navas EM, Gordo E Understanding the properties of low-cost iron-containing powder metallurgy titanium alloys. Materials & Design. 2016 Nov 15;110:317-23.

Mechanical properties of Tungsten Tri-oxide (WO_3) reinforced poly (lactic-acid) (PLA) nanocomposites

M.L.M. Shaath[1], M.N.M Ansari[2,*], Noor Afeefah Nordin[2], M.S.H.Al-Furjan[3]

[1] Department of Mechanical Engineering, Universiti Tenaga Nasional, 43000 Kajang, Selangor, Malaysia.

[2]Institute of Power Engineering, Universiti Tenaga Nasional, 43000 Kajang, Selangor, Malaysia.

[3]Institute: School of Materials Science and Engineering, State Key Laboratory of Silicon Materials, Zhejiang University, Hangzhou 310027, China.

Correspnding author's email: ansari@uniten.edu.my ; ansari.uniten@gmail.com

Abstract. In this paper, Tungsten trioxide reinforced poly (lactic acid) (PLA) blend nanocomposites were prepared using twin screw extrusion and injection moulding with different nanofillers concentrations (0.5wt%, 1wt%, 3wt%, and 5wt%) together mixed into a PLA matrix. Mechanical properties such as tensile and flexural properties were investigated. The tensile and flexural results indicate a noticeable enhancement in the mechanical strength of the PLA material. The WO_3/PLA nanocomposites were compared with neat PLA to identify the attributes and loading effects on the mechanical behaviour of the nanocomposites. A high content of tungsten led to a brittle, fragile, and inelastic composite, while a lower content resulted in a high elastic modulus and stiffness of the composite. The composites reinforced with 3wt% of WO_3 has improved its tensile strength and flexural strength by 30.77% and 41.38% respectively. The extrusion stability and composite molten state were compromised in the presence of high concentration of tungsten while a lower percentage eased processing and injection.

1. Introduction.

Polylactic acid (PLA) is an aliphatic polyester synthetically made from lactic acid with the help of microorganisms. It has been widely used in the bio-medical and pharmaceutical fields for several decades due to its biocompatibility and biodegradability in contact with mammalian bodies[1]. Lactic acid (LA) can be frequently found naturally in animals and plants as a by-product of metabolism; therefore, PLA and its degradation products are non-toxic and harmless to the environment. Lactic acid is produced in an industrial scale by using sugar containing agricultural bi-products such as starch, sugarcane, rice husk and corn. The PLA is unable to degrade by itself but after the hydrolysis process, it becomes biodegradable with the presence of microorganisms when it is placed in an industrial composting facility due to its high temperatures(60°C) and humidity [2]–[5].

PLA like other synthetic biopolymers is criticized for the need of high processing energy usually from fossil fuels which is needed to turn the raw materials into PLA pellets used in industries. Meanwhile, 1 kg of PLA consumes less energy than 1 kg of equivalent polymers from petrochemical feedstock. Fig.1 shows the fossil fuel energy consumed by various polymers.

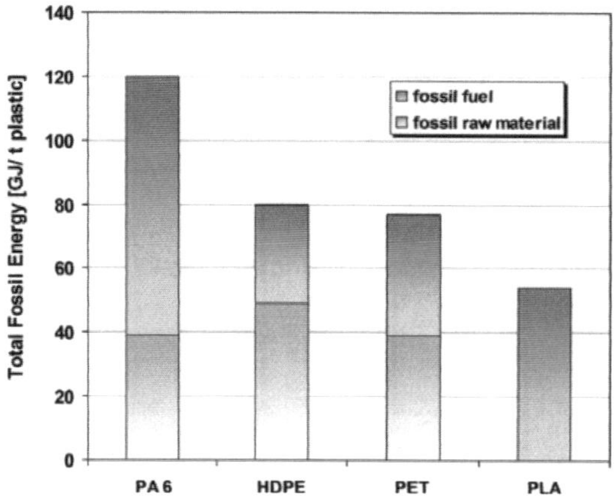

Figure 1. Fossil energy consumed per kg of material produced [8].

One of the biggest technical challenges to widespread acceptance of bio-origin polymers is difficulties in achieving barrier and mechanical properties[6]. Researchers have been developing and patenting new composites of PLA, since the discovery of this green/renewable polyester from 1980s. In 2010, PLA was the second most important bioplastic of the world with regard to consumption volume[6]. It has been intensively used after the revolution of 3D printing where it is used to print everything from models to scaffolds used in surgeries. In the recent years, governments have started encouraging industries to use this polymer for packaging due to the alarming dangers of petroleum derived plastics to the environment. Since then, PLA is being widely used as a plastic packaging from foods and beverages to cosmetics and electronic gadgets as shown in Fig.2 [1], [4], [7]–[10] . PLA has the potential to become the most used plastic in Malaysia for packaging purposes. When the government imposed new tax in 2017 whereby the customers are required to pay 20c per plastic bag in order to preserve the environment, PLA plastic bags have become an alternative option to be used in business sectors. [11]. Apart from that, it is also being used in food and beverages packaging such as yogurt, vegetable wrappings and plastic cups. However, PLA does not meet the requirements of some products which needs high mechanical strength and thermal stability; hence, it was necessary to reinforce it with some additives in order to enhance the properties and to replace the non-biodegradable plastics derived from petroleum base. The government of Malaysia is aiming to ban the single-use plastic by the year 2030 thus an alternative material needs to be developed [12]. In recent years, transition metal oxides gained lots of attention by industries and researchers for their unique properties. Among numerous transition metals, Tungsten oxide was of a special interest because of its distinctive characteristics which have led to numerous applications and prospective for further developments. Such applications include fire proofing fibers, optical devices, photo catalyst, gas censors and electro chromatic windows (smart windows). It is also being used as a color pigment due to its rich yellow that can be found in the formulation of paints and ceramics.

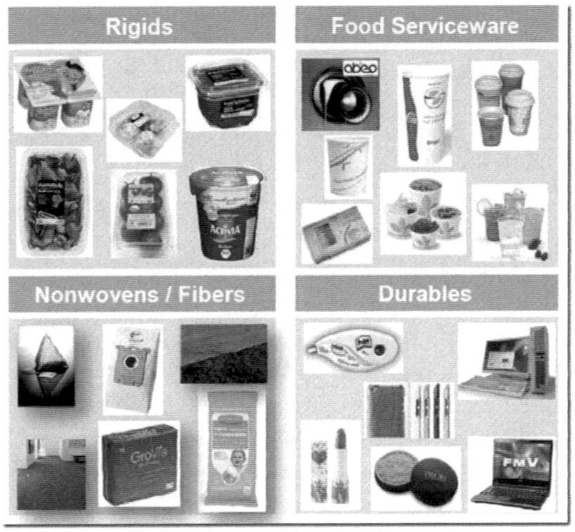

Figure 2. Common uses of Polylactic acid [7].

2. Experimental

2.1. Materials

2.1.1. Poly (Lactic acid) - PLA
PLA used in this research are solid pellets suitable for extrusion and injection moulding process. This material was manufactured by NatureWorks grade Ingeo 2003D with 1.24g/cm^3 and other standard characteristics reported in our previous articles. These materials were supplied by Innovative Pultrusion Sdn.Bhd., Malaysia.

2.1.2. Tungsten Tri-oxide (WO_3)
WO_3 physical and technical properties are very impressive with a melting point of 1473° C, boiling point of 1700° C and a density of 7.16 g/cm^3. These significant properties makes it a potential reinforcement nanoparticle to enhance the properties of PLA [13], [14].

2.2. Methods

To test the tensile stress-strain, the samples were prepared as per the ASTM D638 standards where the measurements are as follows, dumbbell shape type I model with a length of 180mm from end to end, 135mm from neck to neck, and an overall width of 20mm with a thickness of 6.5mm in average, while the gauge length of 50mm. All the samples' measurements were carefully checked and numbered one by one, before testing or obtaining the data. The melting temperatures were chosen as per the material maker's recommendation with an increase of 20° C because of the nanoparticle addition, which requires extra heat to infuse the particles into the polymer matrix and to attain a slender and smooth melting of the substances. Table 1 shows the material composition used in this study, and Table 2 shows the barrel temperatures chosen during the injection process.

Table 1. Material composition and sample formulation.

Sample ID	Material Composition
WPLA0	0wt% WO_3 100wt% PLA
WPLA0.5	0.5wt% WO_3 99.5wt% PLA
WPLA1	1wt% WO_3 99wt% PLA
WPLA3	3wt% WO_3 97wt% PLA
WPLA5	5wt% WO_3 95wt% PLA

2.2.1. *Material Preparation*

Tungsten trioxide nano powder was weighed using a scale of high accuracy up to 0.0001grams, and a weight percentage of (0.5wt%,1wt%, 3wt% and 5wt%) was added to PLA pellets samples, after coating the pellets with a coupling agent to ease the mixing process and to increase surface adhesion. Then the material was shaken for 15 minutes in a glass container in order to ensure a homogenous dispersion of tungsten in the PLA raw material. The same process was repeated for other samples.

2.2.2. *Twin Screw Extrusion*

After the plastic material was mixed with tungsten nanoparticles, a twin-screw extruder was used at SIRIM Malaysia to fabricate the material under a temperature starting from 160° C to 190° C as in table 1 in the heater bands train and a pressure of 55-60bar. The molten material was then pulled using a conveyor belt and rolled manually into rolls which was later crushed in order to construct the tensile and flexural samples using injection moulding. Table 2 shows the temperature profile during the extrusion of compounds.

2.2.3. *Injection Moulding*

The extruded material wires were crushed and then used in an injection moulding machine (HAITAN MA600) at UNITEN labs. After trial and error, the machine was operated under a pressure of 50bars with temperatures of 162° C, 168° C, 173° C and 180° C in the heater band train, and an ASTM-D 638 standard mould was used to fabricate the specimens with dimensions of 180mm in length, 6.5±0.2mm in width and 3.1±0.2mm in thickness. The gauge length was measured to be 50mm and the limits are marked using a permanent marker.

Table 2. Distance in the barrel to feeding zone with temperature (^{0}C).

Distance in the barrel to feeding zone	Temperature [°C]
0 to 90 mm	160
90 to 180mm	175
180 to 270mm	180
270 to 360	185
360 to 450mm	187
450 to 540mm	190
From the Die	190

During injection moulding process, the melting and moulding temperatures were set at 210°C and 40°C respectively for PLA and PLA-WO_3 composites based on the manufacturer's data sheet.

2.2.4. Tensile Test

The variation of the mechanical properties of pure PLA and the reinforced PLA was determined. The tensile test was conducted using ZWICK ROEL universal tensile machine. Fig.3 shows the tensile set-up used for each specimen in the normal room temperature. The universal tensile testing machine was used to obtain the tensile properties data. Using a computerized system, the data sheet showing each specimen's test results can be analysed. Five specimens were tested in each sample designated in the Table 1. and the average value of those tests results were analysed. The specimen stability between the machine chalks is critical to the accuracy of the tensile analysis result, hence, the specimen under testing is cautiously grasped amid the two load chalks as shown in Fig.3.

Figure 3. Tensile test set-up (ASTM-D638).

2.2.5. Flexural Test

The universal testing machine (ZWICK ROEL) was used to determine the flexural strength and to study the effects of the nanoparticle employed in the polymer reinforcement. As per the ASTM-D 790 standards, a 3-point flexural test was followed to determine the flexural load versus strain pattern. A computer software was used to plot the graph and analyse the data. The readings are tabulated and given in Table 3. The flexural test set-up as per the ASTMD790 standard is shown in Fig.4.

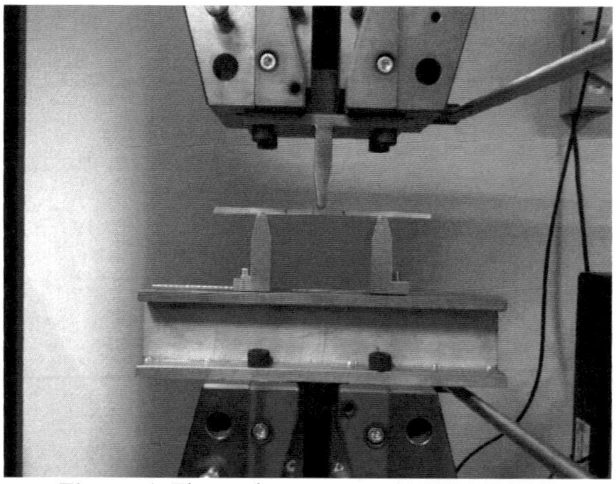

Figure 4. Flexural test set-up (ASTM-D790)

3. Results and Discussion

The samples endured the analyses under diverse settings and diverse printing parameters. With regard to the Tensile strength, it is the fundamental property to be concerned about but another aspect to be taken into account is the ductility which can substantially affected by the dimension and strength of the end product. With this in consideration, the materials were assessed for competent tensile properties. The resulting values are compared and shown in Fig.5. The outcomes of the tensile assessments indicated that all the composite samples ranging from 0.5wt%~5wt% of WO_3 reinforced PLA (WPLA0.5, 1, 3 and 5) shows greater tensile strength than the neat PLA sample with a maximum strength of 61.1MPa (WPLA3) while the neat PLA sample (WPLA0) shows the lowest tensile result of 48.0 MPa. This is attributed to the change in the reinforcement of WO_3 which has made significant improvement in the tensile strength irrespective of the loading level. Similar results were found from many previous research studies [15]–[17]. Zhang *et al.* (2020) reported that magnesium reinforcement in PLA showed an increase in the flexural strength compared to those neat PLA membranes [18].

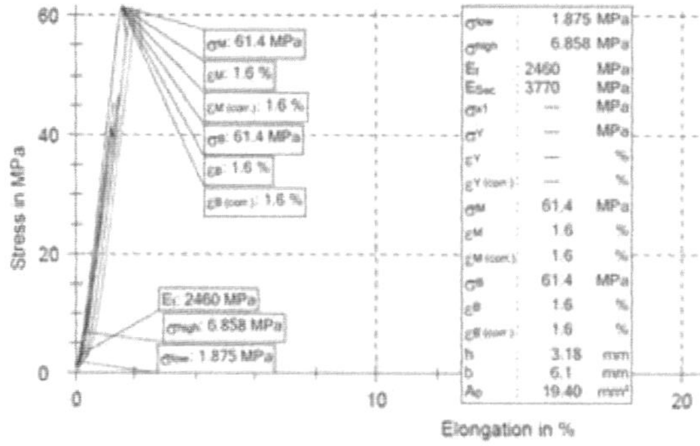

Figure 5. Tensile stress-strain graph of neat PLA and PLA/WO_3 composites.

Flexural test, unlike the tensile doesn't show much information regarding the material properties but it's a compilation of attributes of the other properties enhanced with the nanoparticles reinforcement in PLA matrix. The results of flexural strength for all samples are summarized in Table 3. It can clearly be seen from Fig.6 that the flexural strength of the composites increased proportionally to the percentage of filler loading until 3wt%. The neat PLA (WPLA0) had the lowest flexural strength which was 85 MPa with an elongation of 5.6% while the highest flexural strength obtained from this study was exhibited by sample WPLA3 (145.0 MPa) with 4.0 % elongation. It can be observed that adding greater amount of WO_3 to 5wt% resulted in decrease in the flexural strength. This could be due to agglomeration of the nanoparticles that resulted in poor integration between PLA and WO_3, and subsequently caused the composite to be unable to withstand more load. The results obtained give us a proof of the enhancement in mechanical properties after being reinforced with the WO_3 nanoparticles. Similar studies were done by other researchers on reinforcing nanomaterials in PLA for improvement in their mechanical strength. Ajay Kumar *et al.* (2020) investigated the effect of printing speeds on the tensile strength of carbon fibre reinforced PLA, and it was found that the carbon fiber reinforced PLA is of superior mechanical properties compared to neat PLA [19].

Table 3. Flexural properties of WO_3/PLA composites.

Sample ID	Flexural Strength (MPa)	Flexural Strain (%)	Flexural Modulus (GPa)
WPLA0	85.0	5.6	1.52
WPLA0.5	130.0	5.2	2.50
WPLA1	140.0	4.0	3.50
WPLA3	145.0	3.2	4.53
WPLA5	102.0	2.9	3.52

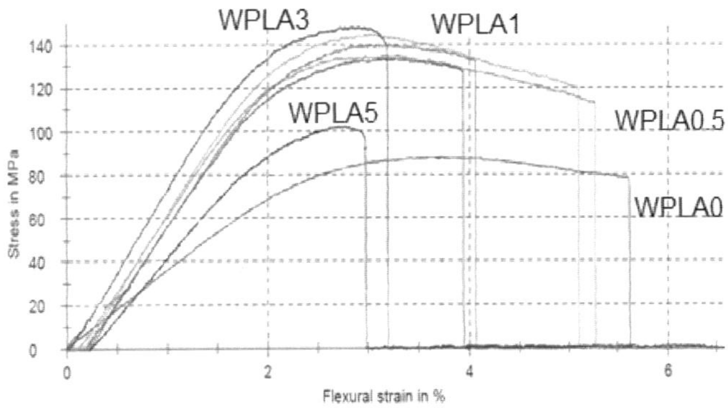

Figure 6. Flexural stress-strain graph of neat PLA and PLA/WO_3 composites.

4. Conclusions

Although PLA is considered a biodegradable plastic, its setbacks could be tailored by reinforcing the polymer, using different fillers to introduce new desired characteristics. This study was conducted to identify the effects of reinforcing the neat PLA with a nanoparticle material and to verify the enhancement in mechanical properties of the renewable biodegradable material PLA that can be a potential replacement for petroleum-based plastics such as polyolefins.

- This concept of reinforcing renewable bioplastics such as PLA with nano-fillers like WO_3 has acquired attractive mechanical properties required for various applications.
- Based on the study conducted, the applied force at breaking load ratio is greater for the PLA-WO_3 with 3wt% of filler loading.
- The results obtained showed that the reinforced composites had greater mechanical properties compared to the neat PLA.
- The findings also indicated that there was good compatibility between PLA and WO_3 which resulted in notable improvement for mechanical properties of the composites.

Author's contribution

M.L.M. Shaath[1]- first draft of the paper, conduct experiment, data collection and analyses of the results, M.N.M **Ansari**[2,*] - Concept development, conduct experiment, advise on the manuscript writing and editing, **Noor Afeefah Nordin**[2]- project management, manuscript editing and revision. **M.S.H.Al-Furjan**[3]- manuscript editing and revision.

Acknowledgement

Authors would like to thank the Department of Mechanical Engineering, Universiti Tenaga Nasional, for providing the facilities to conduct this research. Also the support from Research Management Center (iRMC) and UNITEN R&D Sdn. Bhd., Universiti Tenaga Nasional to publish the findings of this project are greatly acknowledged.

References

[1] M. Dash, F. Chiellini, R. M. Ottenbrite, and E. Chiellini, "Chitosan - A versatile semi-synthetic polymer in biomedical applications," *Prog. Polym. Sci.*, vol. 36, no. 8, pp. 981–1014, 2011, doi: 10.1016/j.progpolymsci.2011.02.001.

[2] S. A. P. Sughanthy, M. N. M. Ansari, and A. Atiqah, "Dynamic mechanical analysis of polyethylene terephthalate/hydroxyapatite biocomposites for tissue engineering applications," *J. Mater. Res. Technol.*, vol. 9, no. 2, pp. 2350–2356, 2020, doi: 10.1016/j.jmrt.2019.12.066.

[3] R. Hagen, "PLA (Polylactic Acid)," *Ref. Modul. Mater. Sci. Mater. Eng.*, no. April 2015, pp. 1–7, 2016, doi: 10.1016/b978-0-12-803581-8.01530-7.

[4] L. T. Sin, A. R. Rahmat, and W. A. W. A. Rahman, *Overview of Poly(lactic Acid).* Elsevier, 2013.

[5] L. T. Sin, A. R. Rahmat, and W. A. W. A. Rahman, *Synthesis and Production of Poly(lactic Acid).* 2013.

[6] K. Madhavan Nampoothiri, N. R. Nair, and R. P. John, "An overview of the recent developments in polylactide (PLA) research," *Bioresour. Technol.*, vol. 101, no. 22, pp. 8493–8501, 2010, doi: 10.1016/j.biortech.2010.05.092.

[7] L. T. Sin, A. R. Rahmat, and W. A. W. A. Rahman, "Applications of Poly(lactic Acid)," *Polylactic Acid*, pp. 301–327, 2013, doi: 10.1016/b978-1-4377-4459-0.00008-1.

[8] R. Chandra, H. Takeuchi, and T. Hasegawa, "Methane production from lignocellulosic agricultural crop wastes: A review in context to second generation of biofuel production," *Renewable and Sustainable Energy Reviews*, vol. 16, no. 3. Pergamon, pp. 1462–1476, Apr. 01, 2012, doi: 10.1016/j.rser.2011.11.035.

[9] L. T. Sin, A. R. Rahmat, and W. A. W. A. Rahman, *Mechanical Properties of Poly(lactic Acid)*, no. 2008. 2013.

[10] L. T. Sin, A. R. Rahmat, and W. A. W. A. Rahman, *Overview of Poly(lactic Acid).* 2013.

[11] R. A. Ilyas *et al.*, *Properties and Characterization of PLA, PHA, and Other Types of Biopolymer Composites.* Elsevier Inc., 2020.

[12] M. A. Salim, Z. H. Termiti, and A. M. Saad, *Mechanical Properties on ABS/PLA Materials for*

Geospatial Imaging Printed Product using 3D Printer Technology. Elsevier Ltd., 2019.
[13] C. G. Granqvist, *Switchable glazing technology for eco-efficient construction*. 2013.
[14] J. Wang, A. Han, S. Jaenicke, and G. K. Chuah, *Advances in Sorbents and Photocatalytic Materials for Water Remediation*. Elsevier B.V., 2013.
[15] S. A. Raj, E. Muthukumaran, and K. Jayakrishna, "A Case Study of 3D Printed PLA and Its Mechanical Properties," *Mater. Today Proc.*, vol. 5, no. 5, pp. 11219–11226, 2018, doi: 10.1016/j.matpr.2018.01.146.
[16] V. Sharma, S. Kapoor, M. Goyal, and P. Jindal, "Enhancement of the mechanical properties of graphene based acrylonitrile butadiene styrene (ABS) nanocomposites," *Mater. Today Proc.*, vol. 28, pp. 1744–1747, 2020, doi: 10.1016/j.matpr.2020.05.155.
[17] S. K. Dhinesh, P. S. Arun, K. K. L. Senthil, and A. Megalingam, "Study on flexural and tensile behavior of PLA, ABS and PLA-ABS materials," *Mater. Today Proc.*, 2020, doi: 10.1016/j.matpr.2020.03.546.
[18] Hao Yang Zhang, Heng Bo Jiang, Ji-Eun Kim, ShuXin Zhang, Kwang-Mahn Kim, Jae-Sung Kwon, Bioresorbable magnesium-reinforced PLA membrane for guided bone/tissue regeneration, *Journal of the Mechanical Behavior of Biomedical Materials*,Volume 112,2020.
[19] M. Ajay Kumar, M. S. Khan and S. B. Mishra, Effect of fused deposition machine parameters on tensile strength of printed carbon fiber reinforced PLA thermoplastics, *Materials Today*: Volume 27, Part 2, 2020, Pages 1505-1510

Design and Simulation of Henna-based Composite Substrate for UHF-RFID Application on Dielectric Properties

Nadzeefah Zamil[1], M N M Ansari[1], Y E Jalil[2], Noor Afeefah Nordin[1], Zainudin Yahya[1] and A A Azlan[3]
[1] Institute of Power Engineering, Universiti Tenaga Nasional, Selangor, Malaysia
[2] Department of Electrical and Electronics Engineering, Universiti Tenaga Nasional, Selangor, Malaysia.
[3] Politeknik Sultan Salahuddin Abdul Aziz Shah, Selangor, Malaysia
nadzeefah96, afeefahnordin1, 9m2agc (@gmail.com), ansari, yantierana (@uniten.edu.my)

Abstract. This paper presented the effect of dielectric properties of henna-based composite towards the performance of the tag antenna for UHF-RFID application. The dielectric properties focused in this work were dielectric constant and loss tangent. Henna-based composite was developed with different amount of henna loading as the substrate for RFID passive tag. The design of antenna was optimized using CST Microwave software to achieve resonance frequency at 915 MHz. The output parameters from this work were reflection coefficient (dB), bandwidth (MHz), gain (dB), directivity (dBi), antenna efficiency (%) and voltage standing wave ratio (VSWR). In conclusion, dielectric properties of the substrate affect all the output parameters significantly except for directivity value.

1. Introduction

Radio-Frequency Identification (RFID) gain its attention across the world since 1990's [1]. It is a system that can be used for item identifications, tracking activities and logistic purposes due to its flexibility, ready-ability and longer read range compared to barcode. RFID system is generally composed of three main components which are computer host, reader and tag. RFID tags can be divided into three categories based on type of their power supply which are active tag, semi-active tag and passive tag. As the reader send the electromagnetic wave to the passive tag, the antenna will inductively couple with IC chip, instruct the IC chip to reply its ID and the ID will be sent back to the reader via backscattering signal and finally, the data will be displayed in the computer host. The passive tag has become the most friendly-user because of its low cost, low maintenance and small in size since no battery is needed. Meanwhile, the performance of passive RFID tag depends on the properties of its antenna and substrate.

In RF application, the most popular commercial substrates are Rogers and FR-4. Rogers has a higher dielectric constant and remain stable although with higher frequencies and lower dissipation factor compared to FR-4. However, due to its expensive price, FR-4 is more affordable and commonly chosen for mass production. FR-4 is made from laminated epoxy with woven fiberglass sandwiched by copper sheet on both sides and has dielectric constant of 4.3. This rigid value of dielectric constant for FR-4 drives the researchers to look forward for flexibility by applying plant loading in the polymer during fabrication of substrate. The value of dielectric properties can be altered depending on the amount of plant loading since the plant contains higher weightage of carbon element.

The dielectric properties of the composite are depending on the polarization of the material [2] and amount of plant fiber loading [3]. This is due to interfacial and orientation polarization between filler (plant fiber) and matrix (polymers). As known, plant has permanent-polar molecule from the hydroxyl group which contributes to high hydrophilic characteristics while most of the polymers are non-polar [4]. Due to these circumstances, this will lead to interfacial and orientation polarization in the composite. The interfacial polarization occurred due to the difference of conductivity between two materials; the higher the differences of conductivity, the higher the interfacial polarization [5][6]. At the same time, orientation polarization occurred because of the existence of polar molecules in plant structure which is lignin, contributing to poor adhesion between filler and matrix [7]. Error! Reference source not found. shows the example of composite with different type of plants.

Table 1. Dielectric constant value (e') for different type of composites.

Composites	Weightage (%)	Frequency (Hz)	Dielectric Constant (e')
Jute/Bamboo/Polyester	15:15:70	3.0 – 6.0	3.50 – 5.50 [2]
Sisal/Polyester	50:50	0.7 -4.2	3.25 – 3.55 [5]
Jute/Polypropylene	55.9:44.1	3.0 – 6.3	3. 10 – 7.10 [4]
Kenaf/Epoxy	14:86	12.4G – 16.4G	0.50 – 2.50 [6]

One of the methods to enhance adhesion between plant and polymer is chemical treatment such as sodium hydroxide (NaOH) or potassium permanganate ($KMnO_4$) [8][9]. The chemical treatment will help to remove the lignin structure from the plant and increase the roughness of the surface and will simultaneously reduce the hydrophilic nature of plant [10]. Reducing of OH^- in fibers will decrease the interfacial and orientation polarization occurred between fillers and composites. Thus, this work will discover the potential of henna-based composite for UHF-RFID application at 915 MHz with various value of dielectric constant and loss tangent.

2. Methodology

2.1. Preparation of Henna-based Composite Substrate

Henna particles were oven-dried, sieved and treated with sodium hydroxide solution for 12 hours with a ratio of 1:20. The size of henna particles was kept constant not exceeding 150 µm. Then, the treated henna particles were dried in the oven for 24 hours after had been washed by tap water for 3 times. Next, henna particles was mixed with epoxy polymer with different proportion before mixing with hardener for curing process. Finally, the mixture was poured in silicon mould for solidification process.

2.2. Dielectric Testing

The dielectric testing towards henna-based composite substrates was conducted using Agilent 85070 coaxial probe method based on ASTMB150 standard. They were placed between the fixtures which composed of a few holes and screwed up to tighten it. Then, the frequency was set between 860 – 960 MHz specifically for ultra-high frequency. As the probe placed within the hole, the dielectric constant and loss tangent were measured and exported as raw data.

2.3. Simulation Work

CST Microwave Software is the software to run the simulation especially for RF application to observe the behaviour of the antenna. The result of dielectric testing was implemented in CST Microwave software for simulation stage and the antenna design was optimized to maintain the resonance frequency at 915 MHz. Figure 1 and Table II show the design of the antenna using pure copper deposited on the henna-based composite. In order to determine the effect of dielectric properties towards the performance of tag is approaching accuracy, only m dimension was adjusted during optimization process and the resonance frequency was maintained at 915 ± 0.5 MHz (can refer Table III). This antenna design composed of capacitive loading, meandered line, T-matching loop, 18 Ω IC chip and a lumped element which is 0.8 pF capacitor. Each substrate is specified with an ID number based on dielectric constant value followed by loss tangent value. For example, the substrate that has 4.3 dielectric constant and 0.025 loss tangent is specified as 4325.

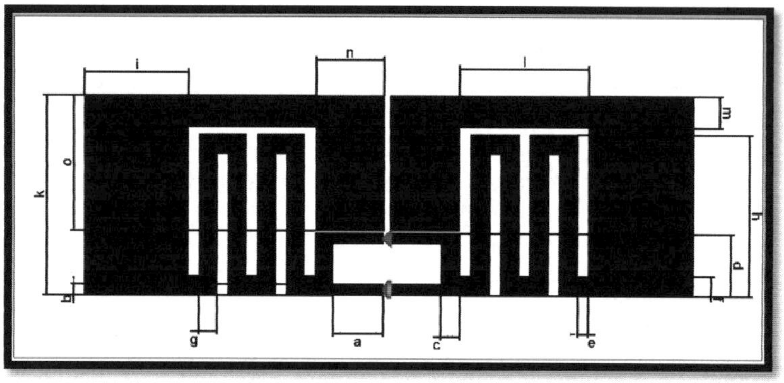

Figure 1. Design of antenna.

Table 2. Specification of antenna design.

Name	Dimension (mm)	Name	Dimension (mm)
a	8.0	**h**	25.0
b	1.7	**i**	17.0
c	3.0	**k**	31.0
d	9.6	**l**	20.6
e	1.7	**m**	4.9
f	3.0	**n**	11.0
g	3.0	**o**	21.0

3. Result and Discussions

In RFID application, the properties of the antenna really exert an impact towards the performance of tag. Nevertheless, the dielectric properties of the substrate also contribute to the tag performance although it is just a medium where the antenna, IC chip and lumped element are deposited. This is because the higher permittivity value of the substrate helps to reduce the dissipation of the RF wave. Hence, the main parameters of the dielectric properties that will be considered in this research are dielectric constant and loss tangent.

3.1. Effect on Reflection Coefficient and Bandwidth

In this context, the reflection coefficient or return loss describes the ratio of the complex amplitude of wave reflected from the tag antenna to the complex amplitude of wave transmitted from the reader where the reflection coefficient is considered accepted when it can go down below than -10 dB. Same goes to bandwidth, the bandwidth can be calculated by the difference between higher frequency and lower frequency where reflection coefficient is at -10 dB or VSWR is 2 which its value tells the amount of data can be transmitted. Figure 2 shows result of reflection coefficient with various value of dielectric constant and loss tangent. As the value of dielectric constant increases, the value of reflection coefficient decreases which is good for performance of the tag. Higher dielectric constant and lower loss tangent values will shift the resonance frequency to the left. Hence, the m dimension needs to be increased so that the resonance frequency at 915 MHz can be achieved. Lowest reflection coefficient was experienced by 4335 and followed by 3825. Meanwhile, the calculation of the bandwidth had been calculated and tabulated in Table III. By comparing various value loss tangent, it shows that bigger bandwidth could be obtained by applying higher loss tangent. This can be observed by comparing 4345 with 4325 and 4335.

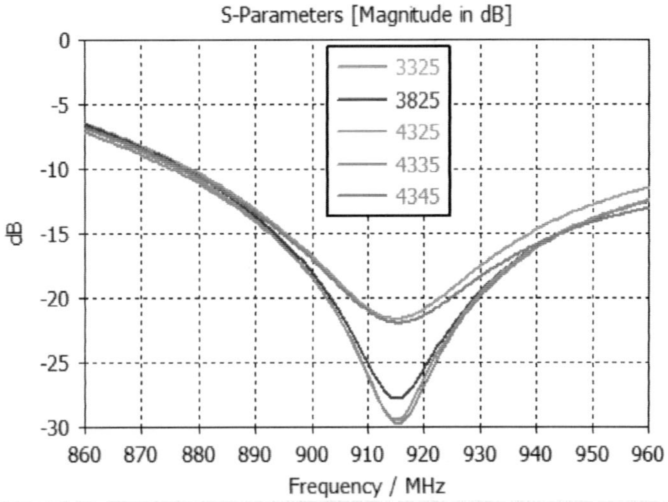

Figure 2. Result of Reflection Coefficient.

3.2. Effect on Gain, Directivity and Antenna Efficiency

Antenna directivity describes the degree of the RF wave transmitted where antenna gain refers to the intensity of the RF wave in particular degree. The product of directivity and antenna efficiency can be expressed as the value of gain. In other word, the value of gain depends on the value of directivity and antenna efficiency. Figure 3 and Figure 4 show the radiation pattern of gain and directivity for each substrate respectively with omnidirectional pattern since it is dipole antenna. Based on Table III, it is clear that higher gain can be obtained by applying lower loss tangent substrate. The directivity value keeps increases as the value of dielectric constant and loss tangent also increase but not very significantly. This is proven that dipole antenna has omnidirectional properties of radiation compared to directive antenna such as Yagi. However, the highest gain and antenna efficiency was achieved by 3825 and followed by 3325.

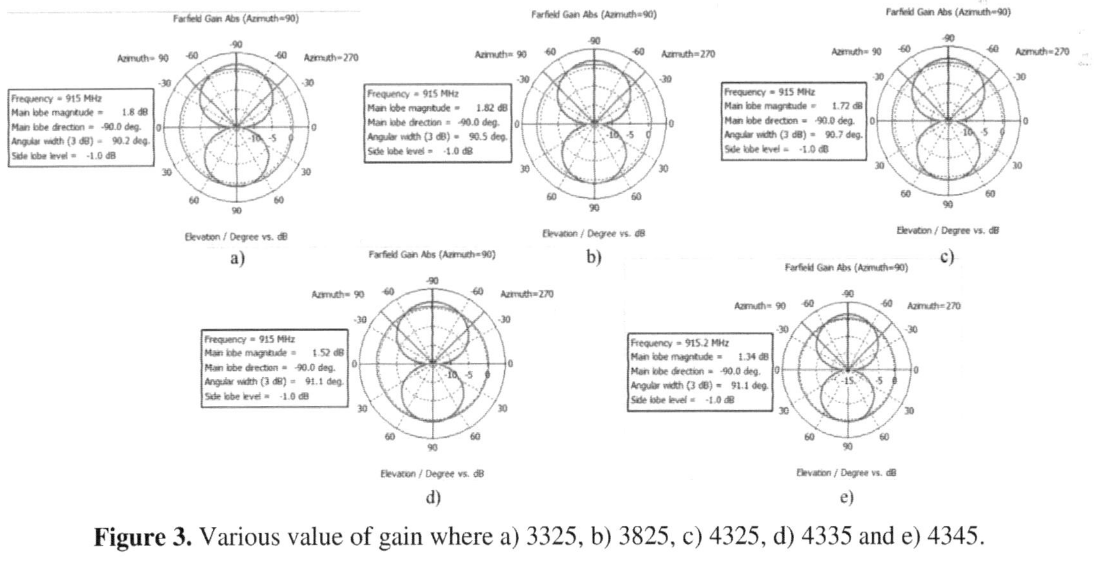

Figure 3. Various value of gain where a) 3325, b) 3825, c) 4325, d) 4335 and e) 4345.

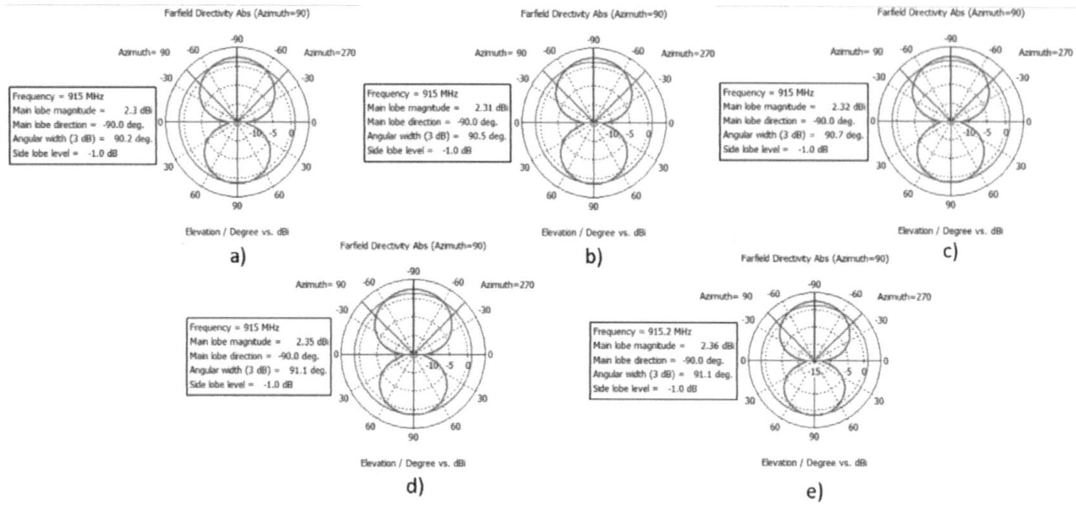

Figure 4. Various value of directivity where a) 3325, b) 3825, c) 4325, d) 4335 and e) 4345.

3.3. Effect on VSWR

VSWR stands for Voltage Standing Wave Ratio describes the efficiency of the power received from the reader to the antenna. As the power received decreases, the antenna might have not enough power to transmit back. The ideal value of VSWR with 0% reflected power is 1:1. Hence, the best VSWR that can be achieved in this research is 1.07 experienced by 4325 and 4335. Table I showed the summarization of simulation result.

Table 3. Summarization of simulation result.

ID Number	Dielectric Constant	Loss Tangent	m (mm)	Reflection Coefficient (dB)	Bandwidth (MHz)	Gain (dB)	Directivity (dBi)	Antenna Efficiency (%)	VSWR
3325	3.3	0.025	1.80	-21.63	102.27	1.80	2.30	78.26	1.18
3825	3.8	0.025	2.95	-27.79	203.44	**1.82**	2.31	**78.79**	1.09
4325	4.3	0.025	4.90	-29.46	175.97	1.72	2.32	74.14	**1.07**
4335	4.3	0.035	3.50	**-29.68**	195.75	1.52	2.35	64.68	**1.07**
4345	4.3	0.045	2.67	-22.02	**205.09**	1.33	**2.36**	56.36	1.17

4. Conclusion

The amount of henna loading can give different value of dielectric properties which contributes to flexibility of substrate in UHF-RFID application. Based on these simulation result, (1) lowest reflection coefficient experienced by 4334, (2) bigger size of bandwidth can be achieved by implementing higher loss tangent, (3) highest gain and antenna efficiency was achieved by 3825 (4) directivity has changed but not so significant due to properties of dipole antenna which radiated omnidirectional wave and (5) the best VSWR was obtained by 4325 and 4335. In conclusion, higher dielectric constant contributes to stability of overall performance as the value of reflection coefficient significantly decreased.

Author's contribution

Nadzeefah Zamil[1]: Idea & concept generation, first draft and manuscript writing, data collection and simulation work, analyses of data and interpretation. **M N M Ansari[1,*]** - Idea & concept generation, suggestion and advice on the simulation studies, suggestion on interpretation of the data, supervision of the project and manuscript editing. **Y E Jalil[2]** - suggestion and advice on the simulation studies, simulation and analytical tools application, interpretation of the data, co-supervision of the project and manuscript editing. **Noor Afeefah Nordin[1]** - project management, suggestion and advice on the analyses, manuscript editing and revision. **Zainudin Yahya** – co-supervision of the project. **A A Azlan[3]** - Idea on the design of the antenna, simulation and analytical tools application, co-supervision of the project and manuscript editing.

Acknowledgement

The authors would like to thank Innovative Research and Management Centre (iRMC), Universiti Tenaga Nasional for BOLD grant project number RJO10517844-105 for the funding, Department of Mechanical and Electrical Engineering, Universiti Tenaga Nasional and Antenna Research Centre, UiTM Shah Alam for providing the facilities to complete this research.

References

[1] C. C. Lee, S. Der Chen, C. T. Li, C. L. Cheng, and Y. M. Lai, "Security enhancement on an RFID ownership transfer protocol based on cloud," *Futur. Gener. Comput. Syst.*, vol. 93, pp. 266–277, 2019, doi: 10.1016/j.future.2018.10.040.

[2] E. Jayamani, S. Hamdan, M. R. Rahman, and M. K. Bin Bakri, "Comparative study of dielectric properties of hybrid natural fiber composites," *Procedia Eng.*, vol. 97, no. December, pp. 536–544, 2014, doi: 10.1016/j.proeng.2014.12.280.

[3] M. A. A. Mohsin, L. Iannucci, and E. S. Greenhalgh, "Fibre-volume-fraction measurement of carbon fibre reinforced thermoplastic composites using thermogravimetric analysis," *Heliyon*, vol. 5, no. 1, p. e01132, 2019, doi: 10.1016/j.heliyon.2019.e01132.

[4] G. George, K. Joseph, E. R. Nagarajan, E. Tomlal Jose, and K. C. George, "Dielectric behaviour of PP/jute yarn commingled composites: Effect of fibre content, chemical treatments, temperature and moisture," *Compos. Part A Appl. Sci. Manuf.*, vol. 47, no. 1, pp. 12–21, 2013, doi: 10.1016/j.compositesa.2012.11.009.

[5] P. A. Sreekumar, J. M. Saiter, K. Joseph, G. Unnikrishnan, and S. Thomas, "Electrical properties of short sisal fiber reinforced polyester composites fabricated by resin transfer molding," *Compos. Part A Appl. Sci. Manuf.*, vol. 43, no. 3, pp. 507–511, 2012, doi: 10.1016/j.compositesa.2011.11.018.

[6] M. J. M. Ridzuan, L. S. S. Liew, M. S. A. Majid, E. M. Cheng, A. Khasri, and A. Z. A. Firdaus, "Dielectric properties of kenaf filled epoxy composites," *IOP Conf. Ser. Mater. Sci. Eng.*, vol. 670, no. 1, 2019, doi: 10.1088/1757-899X/670/1/012047.

[7] A. Saidah, Y. Yudinata, and S. E. Susilowati, "Design of composite material of rice straw fiber reinforced epoxy for automotive bumper," *3rd Int. Conf. Comput. Eng. Des. ICCED 2017*, vol. 2018-March, pp. 1–4, 2018, doi: 10.1109/CED.2017.8308105.

[8] J. E. R. Dhas, P. Pradeep, D. Gladson, R. Dinesh, and M. Aran, "Comparison of mechanical properties for alkali treated and untreated palm/glass sandwiched fiber reinforced polymer composite," *Proc. Int. Conf. Recent Adv. Aerosp. Eng. ICRAAE 2017*, pp. 1–5, 2017, doi: 10.1109/ICRAAE.2017.8297216.

[9] Z. Xia, J. Y. Xu, and X. G. Li, "Effect of alkaline treatment of straw fiber on mechanical properties of cement-bonded straw fiber board," *2010 Int. Conf. Mech. Autom. Control Eng. MACE2010*, no. 2008, pp. 3006–3008, 2010, doi: 10.1109/MACE.2010.5536208.

[10] M. A. Kassim, M. N. A. M. Taib, M. A. Jamaludin, N. Zakaria, and K. Nordin, "Mechanical properties of treated Kenaf (Hibiscus cannabinus) polyester composite at different alkaline concentration," *BEIAC 2013 - 2013 IEEE Bus. Eng. Ind. Appl. Colloq.*, pp. 421–424, 2013,

doi: 10.1109/BEIAC.2013.6560161.

Numerical Investigations into the Effect of Confinement on the Stability of an Oscillating Planar Liquid Jet

Ashish Arote*, Mukund Bade, Jyotirmay Banerjee
Department of Mechanical Engineering, SV National Institute of Technology, Surat, India-395007
E-mail: ashisharote4@gmail.com

Abstract. The liquid jet under certain conditions when perturbed will lead to instability. The understanding about an effect of confinement on these jets is still obscure. Hence, numerical investigations are reported in the present study for two phase spatially oscillating planar jet in a quiescent air under the confined conditions. Simulations are performed by solving the *Navier-Stokes* equations using *volume of fluid* (VOF) method to track the air-water interface. In the present study an oscillating jet is subjected to various *confinement ratios* (CR) to understand the jet behaviour under such conditions. It is demonstrated that the amplitude of oscillation increases as the confinement ratio decreases. This behaviour is a result of increased intensity of vortex rotation at each jet peaks as the confinement ratio increases. Moreover, it is also demonstrated that the vortex structure interact with the side-walls and disturb the incoming flow causing non-linear growth of oscillations leading to primary breakup in these jets.

1. Introduction

Oscillating jet is a spatially perturbed periodic jet which is produced using a *fluidic oscillator*. These jets undergo an oscillation instability in which its oscillation amplitude increases as the jet propagates into the domain. These jets are found to be of significant importance in the applications involving mixing, atomization and dispersion. The typical applications involve cooling of electronic circuits [1][2], flow control strategies [3], mixing and separation of chemical components [4].

Initial study of an instability associated with planar jet was performed by Squire [5] and, Hagerty and Shea [6]. They identified the as symmetric and asymmetric modes of instability in their study. The limitation of these studies was the assumption of uniform sheet thickness which is unable to predict jet breakup. Therefore, Clark and Dombrowski [7] performed a second order non-linear analysis to predict the instability in these jets. The results of these studies were based on temporal analysis which assumes that the perturbations grow only with time. Thus, spatio-temporal analysis was needed to understand the instability associated with these jets. This issue was resolved by Asare [8], Li [9], Chuech [10] and, Sirignano and Mehring [11] in their work and highlighted the effect of non-linearity in the growth of induced perturbations. Thus, it was demonstrated that the sinuous instability of jet is well predicted in the near-field whereas as the aero-dynamic effect increases the jet takes a saw-tooth shape due to non-linear growth.

The effect of confinement on the inviscid planar jets was studied by Juniper [12] and demonstrated that the jet are most unstable at confinement of $h = 1$. The similar observation was made by Juniper [13] for circular jets. Rees and Juniper [14] demonstrated that for viscous flows also the flow destabilizes when the confinement is increased. Healey [15] showed that semi-confinement (confining only one side of the faster stream) can also lead to instability in the flow. Wang *et al.* [16] studied the instability of confined non-Newtonian liquid jet in an annular gas layer and demonstrated that an increase in thickness of gas layer destabilizes the jet. Mosavati *et al.* [17] studied the jet instability and vortex structures in a confined cavity where the jet undergoes self oscillations. Hao *et al.* [18] have reported the numerical results of the confined impinging jet mixers, where the flow structures in the shear layer are studied to understand the mixing characteristics of the jets.

The fundamental understanding about the effect of confinement on the stability of oscillating jets is vital in improving the efficacy of industrial applications. Thus, the numerical study can provide better insights into the cause of instability and the breakup mechanisms in these jets. Moreover, to the best of authors' knowledge there are no numerical studies in the previous literature that investigate influence of confinement on the stability of these jets.

Structure of the present paper is as follows. Section 2 discusses the governing equations and their solution methodology. The results are discussed for the different confinement ratios in Section 4. The present study is summarized in Section 5.

2. Numerical Methodology

The numerical methodology adapts the one fluid model where both phases of the fluid share one set of governing equations [19]. Both phases are assumed to in a continuum and are incompressible. The interface between both the phases is considered to be sharp and under the influence of forces such as surface tension and gravity. The present study is carried out using an in-house high resolution solver [23].

2.1. Governing Equations

The fluid flow for the present problem is governed by *Navier-Stokes* equations. Non-dimensional integral form of these equations is given as,

$$\iint\limits_{CS} u_j^*.d\vec{A} = 0 \tag{1}$$

$$\begin{aligned}\iiint\limits_{C\forall} \frac{\partial u_i^*}{\partial t^*} d\forall + \iint\limits_{CS} (u_i^* u_j^*).d\vec{A} = -\frac{1}{\rho^*} \iiint\limits_{C\forall} \frac{\partial p^*}{\partial x_i^*} d\forall + \frac{1}{\rho^* \mathrm{Re}} \iint\limits_{CS} \left(\mu^* \overrightarrow{\nabla^*} u_i^*\right).d\vec{A} \\ + \frac{1}{\rho^* We} \iiint\limits_{C\forall} F_{ST_i} d\forall - \frac{1}{Fr^2} \iiint\limits_{C\forall} d\forall\end{aligned} \tag{2}$$

such that the variable with the superscript * is non-dimensional. Term u_i and u_j represent the velocity vector components wherein $j = 1, 2, 3$ and $i = 1, 2, 3$ denote the repeating and non-repeating indices respectively. Variables p, ρ and μ denote the pressure, density and viscosity of the fluid with a computational cell having a control surface (CS) area $\overrightarrow{dA}$ which encloses a cell volume of $d\forall$ inside a control volume ($C\forall$). Moreover, $Re = \rho_l U h_0/\mu_l$, $Fr = U/(gh_0)$ and $We = \rho_l U^2 h_0/\sigma$ denote the Reynolds number, Froude number and Weber number respectively; such that U, g and σ are the characteristic velocity (unity), acceleration due to gravity and surface tension coefficient, respectively. On the right hand side of Equation 2 the term F_{ST} represents the surface tension force whereas, the last term on the right denotes the force due to gravity. The mixed cell properties such as density and viscosity are obtained using the fluid volume fraction (ϕ) as

$$\begin{aligned} \rho^* &= \phi + (1-\phi)\rho_g/\rho_l \\ \mu^* &= \phi + (1-\phi)\mu_g/\mu_l \end{aligned} \tag{3}$$

where the variables with subscript g and l represent the mixture cell properties of gas and liquid phase respectively. Equation 1 and Equation 2 is scaled with the help of half jet width as presented in [23].

The equations governing the flow field are discretized using the *finite volume methodology* with the flow properties situated at the cell centroid. Temporal terms are discretized using *Runge-Kutta* (RK2) method [20] which is explicit and second order accurate. Advection term is descritized with the help of *total variation diminishing* (TVD) scheme [21] respectively.

Navier-Stokes equations are solved using the *semi-explicit* formulation. The solution for pressure Poisson equation is obtained with the hep of *bi-conjugate gradient stabilized* (Bi-CGSTAB) method [22] and the residue of 1×10^{-6} is maintained for the present computation. *Courant number* limit is set as 0.1 using the average time stepping of 10^{-4}. The present computations are carried out in parallel using *OpenMP*. The details of compiler directives followed by the code is discussed in detail previously by [23].

2.2. Surface Tension Modelling

Surface tension force F_{ST} in Equation 2 is discretized using *continuum surface force* (CSF) method [24]. In the study, spurious velocities at an interface are minimized with the help of balanced CSF method [25]. An interface curvature is evaluated using the *height function* (HF) methodology [26].

2.3. Interface Tracking

An interface between two phases is tracked with the help of *volume of fluid* (VOF) method [27]. In the present study, the advection equation for scalar volume fraction (ϕ) is solved by second order accurate operator split method, where the fluid is advected consecutively in x and y directions during a single time step. Fluxes at the cell interface are evaluated using the case-wise geometric PLIC [28] [29] method. Therefore, interface reconstruction is performed on the basis of *Rudman algorithm* [28] that identifies the fluid orientation cases in each computational cell using the interface normals. This methodology is adopted due to its computational efficiency.

3. Simulation Setup and Validations

The domain considered for the present study and the boundary conditions used in it is demonstrated in Fig. 1. Here, an oscillating planar water jet having a width of $2h_0$ enters the quiescent surrounding of air at a velocity of V_{inlet}. An inlet section is positioned at centre of the top face and the jet develops into the negative $y - direction$. An inlet velocity vector of the jet consist of the two velocity components, out of which, one provides the jet the required inlet velocity (V_{inlet}) whereas, the other provides the periodic perturbations to it. The u^* and v^* velocity components for this inlet vector are given as,

$$\begin{aligned} u^* &= A_0 \sin(2\pi S \times t^*) \\ v^* &= V_{inlet} \end{aligned} \tag{4}$$

where, S represents the *Strouhal number* that accounts for the perturbation frequency (f) and an initial amplitude of oscillation A_0 such that, $S = fU/h_0$. The vertical side walls are treated as with the free slip boundary condition whereas; the bottom face of the domain acts as a typical pressure outlet. The values for physical properties of air and water used for the study with the non-dimensional numbers are provided in the previous study by the authors [23]. The flow conditions used for the present study are provided in Tab. 1.

Table 1. Flow parameters used for the present study.

Fr	We	Re	V_{in} (m/s)	w_0 (m)	h_0 (m)	S	A_0
5.08	54.72	4426.96	2.05	0.075	0.00395	0.495	0.2531

The *confinement ratio* (CR) for the present study is defined as the ratio of domain width to the jet width. The jet width in the computational domain is maintained as unity and the domain width is varied as $50h_0$, $70h_0$, $90h_0$ and 130 to change the confinement ratio.

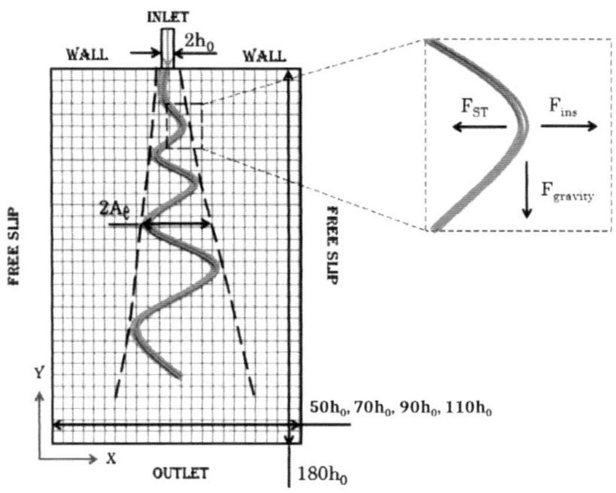

Figure 1. Details of the computational domain and flow conditions.

3.1. Grid Refinement Study

The grid refinement studies based on *Richardson's extrapolation* [30] were previously carried out in [23]. It is demonstrated that when the spatial refinement in all three directions is performed using three grid levels, the *Grid Convergence Index* (GCI) between medium and the finest grid was within the convergence limits. Therefore, for the present study, grid resolution of 250×700, 350×700, 450×700 and 550×700 are selected for the confinement ratio of $CR_1 = 50$, $CR_2 = 70$, $CR_3 = 90$ and $CR_4 = 110$ respectively.

3.2. Validations of the Numerical Model

Validations for the present numerical model with the experimental results of Asare et al. [8] were previously discussed by Arote et al. in [23]. These validations are carried out for the growth envelope (A_e/A_0) that evolve due to the instability associated with the oscillations of these jets and were found to be in agreement with the experimental results.

4. Results and Discussion

The present section provides the discussion on the effect of confinement ratio on the growth of oscillation amplitude. The growth envelope (A_e/A_0) is tracked as demonstrated in Fig. 1. The bulges at the oscillation peaks are neglected to avoid any ambiguity in the measurement of an amplitude.

Fig. 2 demonstrates the development of oscillating jet for different CR values. It is observed that at the near field development of the jet is independent of the confinement effects. However,

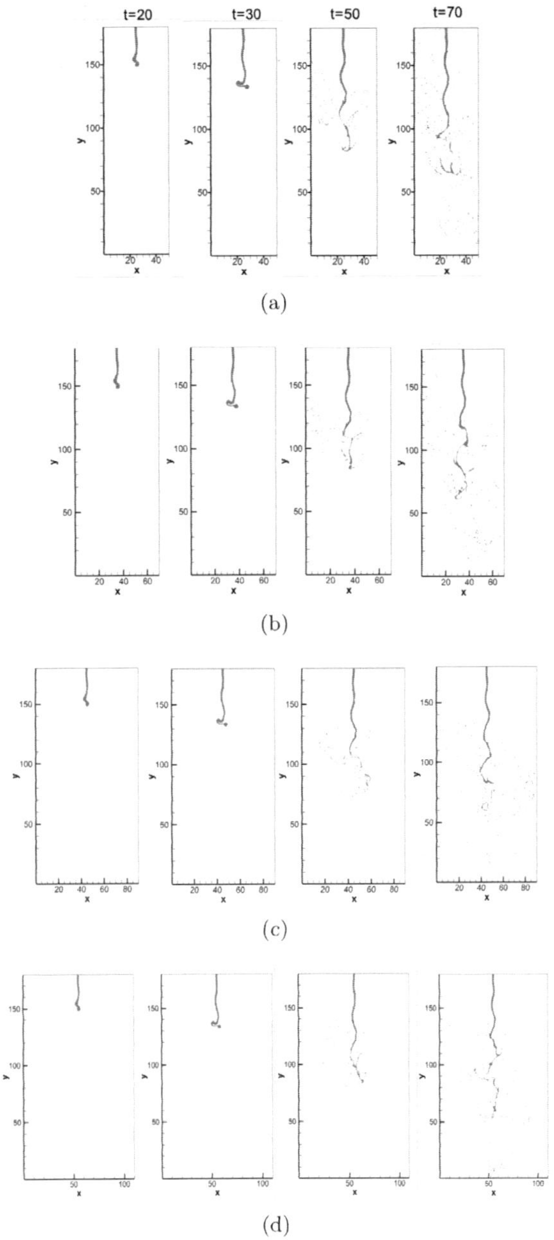

Figure 2. Development of oscillating jets for a) $CR_1 = 50$, b) $CR_2 = 70$, c) $CR_3 = 90$ and d) $CR_4 = 110$.

at the downstream, as the aerodynamic interaction between the gas-liquid phases increase the confinement is having significant impact on the development of jet. Here, the oscillation amplitude tends to increase at the downstream due to the low pressure region created at every oscillation peak [23]. Moreover, it is observed from Fig. 2 that as the confinement ratio increases the gas interactions become dominant and oscillations grow non-linearly which results in primary breakup of the jet.

The comparison of growth rates for oscillation amplitude for various confinement ratios is depicted in Fig. 3. It is observed that the instability associated with the amplitude of oscillation increases with increase in the confinement ratio. A reason behind the destabilization of oscillating jets can be understood through the analysis of enstrophy (ω^2) values at the typical oscillation peak where ω is the vorticity in flow. Fig. 4 demonstrates the comparison of enstrophy values at a oscillation peak between different CR values. It is observed that the intensity of vortex rotation at the oscilation peaks increases as the confinement ratio increase. As this rotation intensity increases the pressure at the eye of the vortex reduces and leads to the jet being drawn towards it which promotes instability [23].

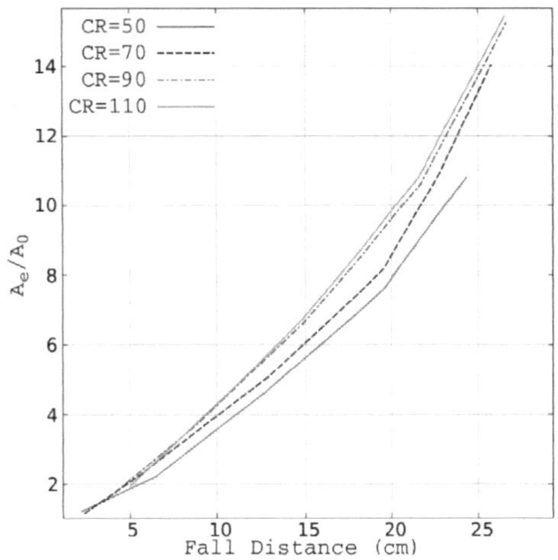

Figure 3. Comparison of the instability growth rates in terms of the growth envelope A_e/A_0 between various confinement ratios.

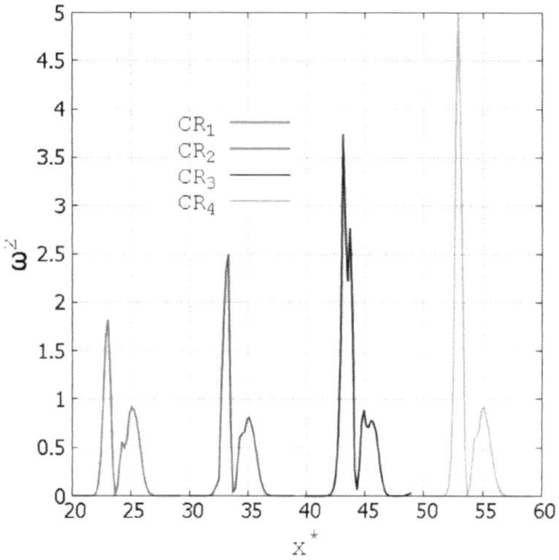

Figure 4. Comparison of enstrophy values along x^* at $y^* = 152$ for various confinement ratios.

Fig. 5 demonstrates the vorticity contours for various CR values. It is observed that the vortices interact with the side-walls and are directed towards the incoming jet oscillations. These vortices affect the development of these oscillations more when the CR values are less. Moreover, the momentum transfer into the surrounding fluid is observed to increase as the CR value increases.

5. Conclusions

The effect of confinement on the growth envelop of the oscillation amplitude is studied. The development oscillating jets is studied. The reason behind the associated instability in the oscillating jet is also identified in the present study. It is demonstrated that the jet development in the near filed is not affected by the confinement. Moreover, at the downstream, the primary atomization is accelerated if the CR values increase. Furthermore, it demonstrated that an increase in the confinement ratio destabilizes the oscillating jet; which is caused by the increased intensity of vortex rotation at the jet peaks. Stability is also affected by the vortex structures disturbing the incoming flow of jet.

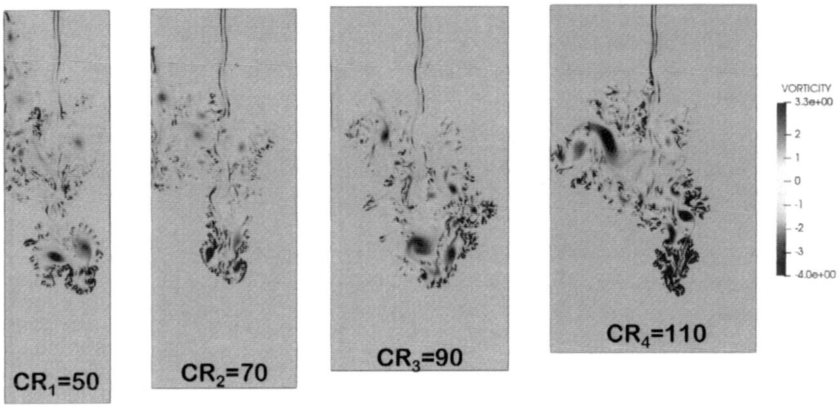

Figure 5. Vorticity plot for various confinement ratios at $t^* = 70$.

Acknowledgement

Authors would like to acknowledge the financial assistance provided by the Science and Engineering Research Board, India (Grant No. MTR/2019/000941).

References

[1] Mon G. Pulsatile Impinging Cooling and Systems Using Fluidic System for Elecronic Modules Oscillators. US Patent; 5,190,099, 1993.
[2] Sutton TG. Fluidic Feedback-Controlled Liquid Cooling Module. US Patent. 5,815,370, 1998.
[3] Raghu S. Fluidic oscillators for flow control. Exp Fluids. 2013;54(2).
[4] Tesar V. Oscillator micromixer. Chem Eng J. 2009;155(3):78999.
[5] Squire HB. Investigation of the instability of a moving liquid film. Br J Appl Phys. 1953;4(6):1679.
[6] Shea WW. Hagerty J. F. 1955 A study of the instability of plane fluid sheets. J. Appl. Mech.;22:509-14.
[7] Clark CJ, Dombrowski N. Aerodynamic Instability and Disintegration of Inviscid Liquid Sheets. Proc R Soc A Math Phys Eng Sci. 1972;329(1579):46778.
[8] Asare HR, Takahashi RK, Hoffman MA. Liquid Sheet Jet Experiments: Comparison With Linear Theory. J Fluids Eng. 1981;103(4):595.
[9] Li X. On the instability of plane liquid sheets in two gas streams of unequal velocities. Acta Mech. 1994;106(34):13756.
[10] Chuech SG. Spatial instability of a viscous liquid sheet. Int J Numer Methods Fluids. 2006;50(12):146174.
[11] Sirignano WA, Mehring C. Review of theory of distortion and disintegration of liquid streams. Prog Energy Combust Sci. 2000;26(4):60955.
[12] Juniper MP. The effect of confinement on the stability of two-dimensional shear flows. J Fluid Mech. 2006;565:17195.
[13] Juniper MP. The effect of confinement on the stability of non-swirling round jet/wake flows. J Fluid Mech. 2008;605:22752.
[14] Rees SJ, Juniper MP. The effect of confinement on the stability of viscous planar jets and wakes. J Fluid Mech. 2010;656:30936.
[15] Healey JJ. Destabilizing effects of confinement on homogeneous mixing layers. J Fluid Mech. 2009;623:24171.
[16] Wang XT, Ning Z, L M. Temporal instability analysis of a confined non-Newtonian liquid jet with heat and mass transfer. Eur J Mech B/Fluids. 2020;84:3506. Available from: https://doi.org/10.1016/j.euromechflu.2020.07.005
[17] Mosavati M, Barron RM, Balachandar R. Characteristics of self-oscillating jets in a confined cavity. Phys Fluids. 2020;32(11). Available from: https://doi.org/10.1063/5.0023833
[18] Hao Y, Seo JH, Hu Y, Mao HQ, Mittal R. Flow physics and mixing quality in a confined impinging jet mixer. AIP Adv [Internet]. 2020;10(4):110. Available from: https://doi.org/10.1063/5.0002125
[19] Tryggvason G, Scardovelli R, Zaleski S. Direct Numerical Simulations of GasLiquid Multiphase Flows. Cambridge: Cambridge University Press; 2011.

[20] Sigal Gottlieb, David Ketcheson, Chi-Wang S. Strong Stability Preserving Runge-Kutta and Multistep Time Discretizations. I^{st} Edition. World Scientific; 2011.
[21] Harten A. High Resolution Schemes for Hyperbolic Conservation Laws. J Comput Phys. 1983;49(3):35793.
[22] 1. Vorst HA van der. Bi-CGSTAB: A Fast and Smoothly Converging Variant of Bi-CG for the Solution of Non-Symmetric Linear Systems. SIAM J Sci Stat Comput. 1992;13(2):63144.
[23] Arote A, Bade M, Banerjee J. Numerical investigations on stability of the spatially oscillating planar two-phase liquid jet in a quiescent atmosphere. Phys Fluids. 2019 Nov 1;31(11):112103.
[24] 1. Brackbill JU, Kothe DB. ScienceDirect - Journal of Computational Physics : A continuum method for modeling surface tension*1. J Comput Phys. 1992;335354.
[25] Francois MM, Cummins SJ, Dendy ED, Kothe DB, Sicilian JM, Williams MW. A balanced-force algorithm for continuous and sharp interfacial surface tension models within a volume tracking framework. Vol. 213, Journal of Computational Physics. 2006. p. 14173.
[26] Lopez J, Zanzi C, Gomez P, Zamora R, Faura F, Hernandez J. An improved height function technique for computing interface curvature from volume fractions. Comput Methods Appl Mech Eng [Internet]. 2009;198(3336):255564.
[27] C. W., Hirt; B. D. N. Volume of Fluid (VOF) Method for the Dynamics of Free Boundaries. Int J Comput Phys. 1979;39(1):20125.
[28] Murray Rudaman. Volume-Tracking Methods for Interfacial Flow Calculations. Int J Numer Methods Fluids. 1997;24(7):67191.
[29] Saincher S, Banerjee J. A redistribution-based volume-preserving PLIC-VOF technique. Numer Heat Transf Part B Fundam. 2015;67(4):33862.
[30] Roache PJ. Perspective: A method for uniform reporting of grid refinement studies. Journal of Fluid Engineering. 1994; 116(3):40513.

IOP Conf. Series: Materials Science and Engineering 1128 (2021) 012033 doi:10.1088/1757-899X/1128/1/012033

Design and analysis of energy-efficient solar panel cooling system using computational fluid dynamics

Sarthak Vaishnav, Vivek Kapadiya, R. Harish*, R. Mohan
School of Mechanical Engineering, Vellore Institute of Technology, Chennai, Tamil Nadu 600127, India
*Corresponding author E-mail address: harish.r@vit.ac.in

Abstract

This paper highlights the design of an effective liquid cooling system that utilizes the heat generated from the solar panel as a cooling medium to maintain the optimal desired temperature of the solar panel. The coolant for this finned cooling system is selected based on its vaporizing temperature range and thermal cycle characteristics. For analysis purposes, a CAD model is generated in Solidworks CAD package, and further meshing and numerical simulations are performed using Ansys Fluent software. Flow and heat transfer characteristics of the cooling system are investigated by plotting stream functions, velocity, and temperature distributions inside the system. Probing of variations in temperature, pressure and turbulent kinetic energy along vertical as well as longitudinal direction is graphically analysed. Thus, elucidating characteristic of parametric conditions under observation. This study's results can be the potential background for designing an efficient solar panel cooling system with superior thermal performance.

1. Introduction

Even though, in the presence of surplus sunlight, high-temperature regions fail to exploit their maximum potential due to high thermal generations in the solar panel.. The temperature has an adverse effect on panel output. Advanced cooling technology uses external thermal absorbent to reduce panel temperature. This research paper has proposed a model to use low vaporization point liquid chemical as a heat absorbent to reduce solar panel surface temperature. Without any external source energy and by using material properties to our advantage, we accomplished the same. The paper includes a simulation study conducted using Ansys fluent. Peng et al. [1] carried out an experiment in two different phases. In the first phase, they analyzed the relation between the effect of solar radiation, power output, and efficiency of the solar panel. In the second phase, they conducted a cooling experiment and concluded that it increases solar output efficiency. Agarwal [2] main idea behind the study is to review the literature on photovoltaic cooling techniques with the help of phase change materials (PCM), including PV-thermal systems and building integrated photovoltaic systems. The objective is to identify important research areas to verify the reliable performance and commercial viability of the technology. Various parameters need to be evaluated and optimized for the geographical location of interest for getting the best output, a hike in electrical efficiency as high as 5% is observed with PV-PCM integrated systems. Moharram et al. [3] objective of his research was to minimize the amount of water and electrical energy required for cooling solar panels, especially in hot arid regions, e.g., desert areas of Egypt. A cooling system was developed based on a water spraying mechanism for PV panels. With the help of the mathematical model, the cooling of the solar panel is determined. A cooling model is proposed to determine how long it takes to cool down the PV panels to their normal operating temperature. Bhaskar et al. [4] Objective of the present work is to design a Water-cooling system for the solar cell to optimize electrical efficiency and to extract heat energy. A hybrid system that simultaneously generates both electrical energy and heat energy is considered. This hybrid system contains photovoltaic cells attached to an absorber plate with fins attached to its reverse side. A simulation model is prepared, and performance curves are obtained for a single pass, single duct solar collector with fins. Waqas [5] discussed the electrical and thermal management of PV systems integrated with Phase Changing Materials. The purpose of this article is to provide current know-how of PV-PCM technology, simultaneously stating research gaps and challenges faced. They provided a detailed elaboration of different aspects such as PCM selection, system development, performance evaluation, simulation, and heat transfer enhancement, along with cost estimation. Rajvikram M. et al. [6] proposed solution focuses on the efficiency of PV panels with Phase Changing Materials and aluminum sheet as TCE. This experiment was performed under direct sun rays for three months. They performed an experiment using two 5 W panels, and results were compared with naturally ventilated panels without PCM and aluminum. Li et al. [7] Research demonstrate a high operating temperature of crystalline silicon-based photovoltaic (PV) module would eventually lead to a decrease in efficiency and lifetime. They attached PCM at the back of the PV module in PV-PCM system format in order to absorb excess heat. Kang et al. [8] heat transfer performance of silicon oil immersion cooling densely packed solar cells were studied and analyzed, with and without fin structure, using experiment and simulation methods. The results showed that solar cells temperature declined and distributed fairly as silicon oil inlet flow rate increases; however, the temperature of the solar cells improved linearly. Mehrotra et al. [9] In this research paper author says that If we submerged PV cells in water, that is beneficial to maintain its surface temperature. It's also provided better efficiency at very high or extreme temperatures. The author calculated the electrical parameter and the cooling factor. This calculation elaborates upon the cooling factor that also favors electrical efficiency enhancement. Here temperature variation is from 304-312K. Abdulgafar et al. [10] Photovoltaic cell cooling is studied in distilled water immersion conditions of a polycrystalline panel at different depths. With the increase in depth, an evident increase in efficiency was observed. Thakur et al. [11] In this research paper author proposed the utilization of heat generated by the panel with a heat pump, and it can be used in various fields. He supported his conclusion through mathematical relationships. Haller et al. [12] In this paper, the author proposed two methods to increase the efficiency of a solar panel system. First, by liquid immersion, and secondly, he proposed a design that utilizes an air blower to curb temperature surge. Singh and Kumar [13] a shell and heat exchanger tube are considered where; hot water is flowing through a tube, and cold water runs outside but still in the shell. CFD technique resolves the heat exchanger in individual elements to find the temperature gradients, pressure distribution, and velocity vectors. Egon et al.[14] this research paper answers all quires related to CFD analysis along with information about boundary conditions, numerical grid, and initial condition information. Chen et al. [15] give a detailed 3-D computational fluid dynamics analysis of gas particles that are falling on the solar receiver panel using the Euler-langrage method and also give momentum analysis of these particles. Cho et al.[16]used CFD and DEM (discrete element method) to analyze the flow of gas-solid particle collision model. There are also numerous experimental [16-20] and analytical models [21-25] developed to estimate the flow and heat transfer characteristics in enclosures. Sumedha et al.[26]mentioned that because of using the non-renewable resources people have noticed the bad impact of it on Earth pollution, Climate Change so people are forced to use renewable resources like solar panel. He has mentioned drawback of the solar cells like it will not work in the night time , cleaning problems etc and mentioned innovative ways and technologies to solve this problem. Arifin et al.[27] mentioned the bad impact of using conventional resources. When solar cell is placed where sun intensity is too high its efficiency decrease with the increase in temperature. Author has proposed passive cooling system where we use heat sink and fins to increase the efficiency of the solar panel. Author has carried out experiment with different number

IOP Conf. Series: Materials Science and Engineering 1128 (2021) 012033 doi:10.1088/1757-899X/1128/1/012033

of fins with heat sink and finds out that if anybody use 15 number of fins with heat sinks it will where temperature is 10 degree Celsius efficiency will increase by 2.74%.

Our concept idea aim is to iterate and perform CFD analysis for a system that can be a potential candidate for an efficient cooling system by using material's characteristic properties to our advantage. While comparing other materials, carbon disulphide's boiling point is not too high that may turn it into vapour immediately nor too low that may resist it's vapour generation at experimentation conditions. From the experimental data we have collected carbon disulphide show optimum chemical properties for this experiment. Even though it being a bit reactive material when exposed to hot surfaces, it's reactivity can be undermined by mixing it with water and that may increase the boiling point/ vapourizing point of the liquid.

Molar mass	76.13 g·mol−1
Appearance:	Colourless liquid
Impure Appearance	light-yellow
Melting point	-111.61 °C
Boiling point	46.24 °C (115.23 °F; 319.39 K)
Refractive index (nD)	1.627
82.4 kPa (40 °C)	82.4 kPa (40 °C)

Table 1: Properties of carbon disulphide:

Thus, performing a virtual simulation and obtaining expected results for the same. Though, no physical experimentation was performed using carbon disulphide, a guideline to follow for handling and using carbon disulphide is as follows.

Handling – Keep it away from any flammable source, wear a body suit before operating hence avoid breathing vapours. Use personal equipment like gloves and shoves to cover whole body.

Solar panels are made of material that are highly corrosion resistant. Modules of the solar panels are sealed with the help of the vacuum between their sheet and materials. This sealing will prevent any corrosion due to carbon disulphide until and unless there is crack in your panel. Our system is a solar panel enclosed and immersed it in carbon disulfide liquid with the top of enclosing being transparent for sunlight to enter. The transparent surface has augmented surface area by incorporating a step patterned surface, facilitating heat transfer rate enhancement. Due to low vaporization temperature, the liquid absorbs heat and starts evaporating in the range of 45-50-degree Celsius. In the upper section of the casing, due to increased surface area, the heat transfer rate accelerates, and vapors start condensing. This liquid confluence and adds up to the liquid source at the bottom of the case, adhering to the slope of the enclosure. For this process, we analyze heat flow, vapor velocity, pressure, and heat transfer with the assist of ANSYS fluent software.

IOP Conf. Series: Materials Science and Engineering 1128 (2021) 012033 doi:10.1088/1757-899X/1128/1/012033

2. Methodology

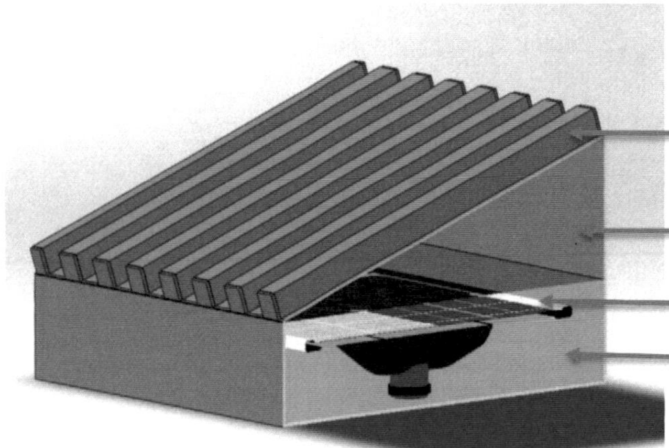

Figure1: Cross-section of the enclosure

The solar panel is enclosed and fixated to the bottom of an aluminum casing (thermal conductivity 205 W/mK) of net volume 0.03m^3 of base length and width being 0.577m and 0.354m respectively. Covered by transparent glass (thermal conductivity 0.8 W/mK) for inflow of sunlight, with enhanced surface area; dimensioned - 0.366m, 0.019m and 0.024m for length, breadth and height respectively of the fins. immersed in carbon-disulfide, which has a boiling point of 46 degrees Celsius. Thermal energy is harnessed by the liquid, and when the temperature approaches boiling point, the liquid starts vaporizing. These vapors move upwards, clings, and amalgamate on the wall of the case. These droplets slide over the slant surface and flow back into the sink. This cycle repeats until the temperature falls below 35 degrees Celsius. The system utilizes heat energy generated by solar panels to cool panels and thereby increasing efficiency.

The system is designed to automatically regulate solar panel surface temperature by using low vaporization temperature liquid and augmented surface area. The lower section of the system is filled with Carbon disulfide with a panel immersed in it.

2.1.1 Calculation

As shown in the figure in the c1 are material will go through conduction:

$$R = \left(\frac{1}{ka}\right) \tag{1}$$

R=resistance

K=thermal conductivity of material
A=surface area

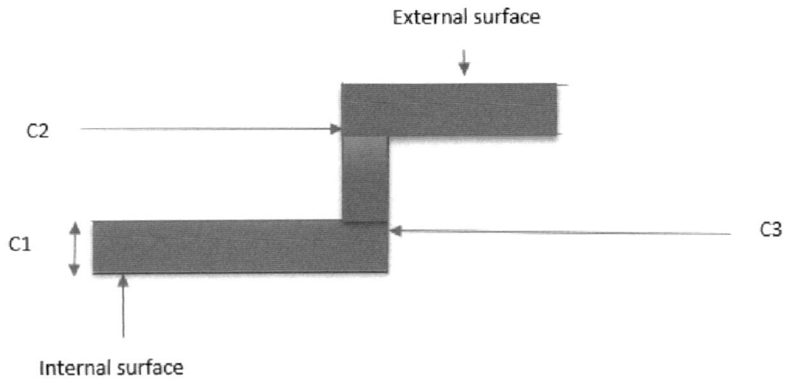

Figure 2: dimensional diagram of an outer surface of the case

In the C2 region, material will go through convection: (Air and glass surface convection)

Case 1:
Laminar-turbulent constant wall temperature

$$Nu_x = 0.037 \cdot (Re_x^{0.8} - 871) \cdot Pr^{1/3} \tag{2}$$

Re=Reynold number
Nu-Nusselt number
Critical Reynold number $=5*10^5 < Pr < 60$

Case 2:
High-speed flow, constant wall temperature:
Q=hA [Tw-Tat]

And our final C3 region will go under conduction: (vapour glass surface contact)

$$R = \left(\frac{1}{ha}\right) \tag{3}$$

Using C1, C2, C3, we will find
Total heat transfer/Area by C1, C2, C3

Amount of heat required to be transferred to maintain the temperature at 40 degree Celsius at an ambient temperature of 30 degree Celsius (at steady state). In the final step, we will equate both to get the amount of surface area required at a steady state.

Grashof Number - $G = \frac{\beta g(T_0 - T\infty)L_c^3}{v^3}$

$G = (1.15 * 9.81 * |35 - 50| * ((108.01958)^3)) / ((0.298^2))$

G = 2401778.936

= 2.40 x 10^6

Abbreviations	Nomenclature
R	Resistance (K·m^2/W)
K	Thermal conductivity of material(W/m·K)
A	Surface area(m^2)
Re	Reynold number
Nu	Nusselt number
Pr	Prandtl number
Tw	Wall temperature(° C)
h	Heat transfer coefficient (W/m^2k)
Tat	Temperature of surrounding
C1	Region for heat flow
C2	Region for heat flow
C3	Region for heat flow
Re	Reynolds number
Q	Heat transfer rate(W)
G	Grashof Number
g	Gravitational acceleration (m/s^2)
β	Coefficient of volumetric expansion
Ts	Temperature at the surface (°C)
T∞	Temperature far from surface (°C)
L	Characteristic length (m)
v	Kinematic viscosity of fluid (m^2/s)

3. Results and Discussions:

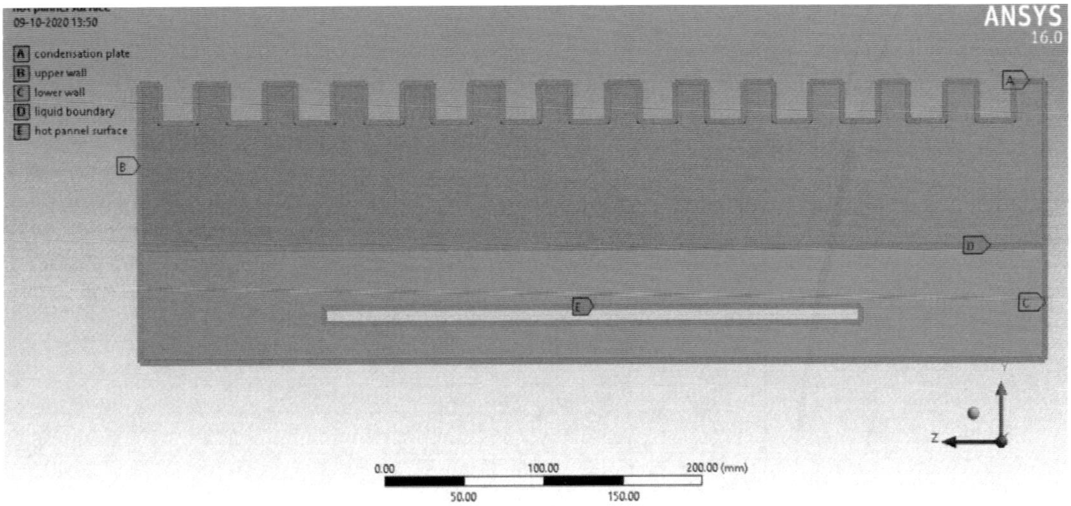

Figure 3-CFD analysis boundary nomenclatures, mesh, and results

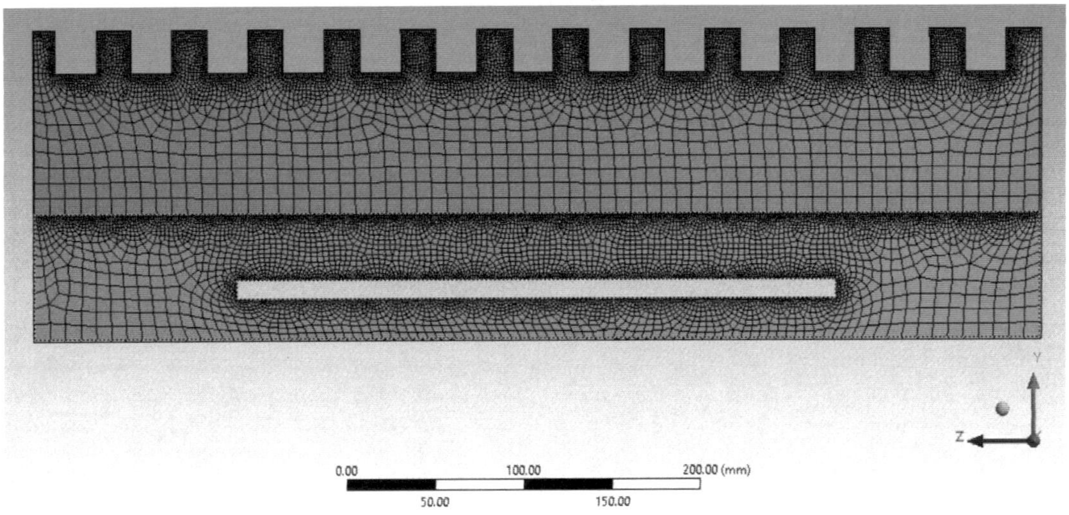

Figure 4: Mesh design of solar PV cell cooling system model

Figure 3 indicates the schematic diagram of the solar panel cooling system and meshed geometry is shown in Fig. 4. Figure 5 shows the velocity distribution of carbon disulphide inside the solar panel enclosure and multi recirculating convective cells are seen evolving inside the solar panel cooling system. Figure 6 and 7 indicates the turbulent kinetic energy and energy dissipation rate inside the solar panel system. Since excited vapours tend to move upwards due to high kinetic potential and exert force on each other by random collision, high turbulent kinetic energy is observed in the gas filled section of the chamber compared to lower liquid filled section. Moreover, due to smaller volume availability to vibrate and collide freely near the fins due to walled boundaries, more turbulence is observed in the middle section of gas volume then in upper section adjoin to fin's upper section. Similarly, due to the very same reason higher dissipation is observed on the lower fin section then upper fin section. Figure 8 shows the temperature distribution due to convective heat transfer and carbon disulphide near the hot panel surface absorbs heat and diffuses the heat inside the solar panel system.

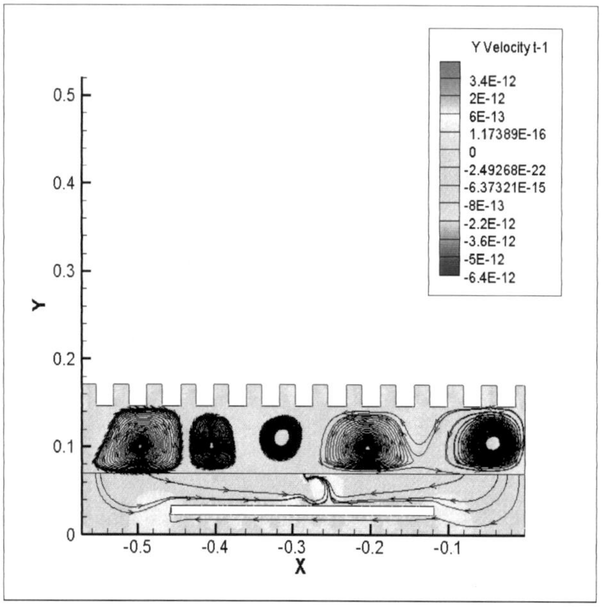

Figure 5: Y directional velocity flow of carbon disulfide

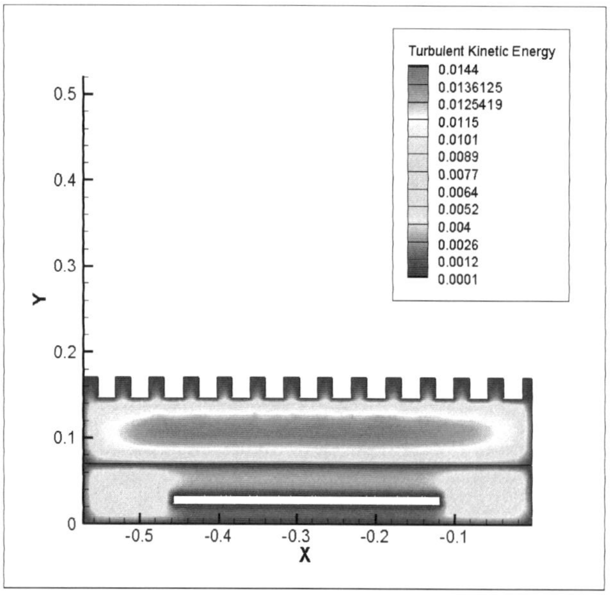

Figure 6: Turbulent kinematic energy flow in the enclosure

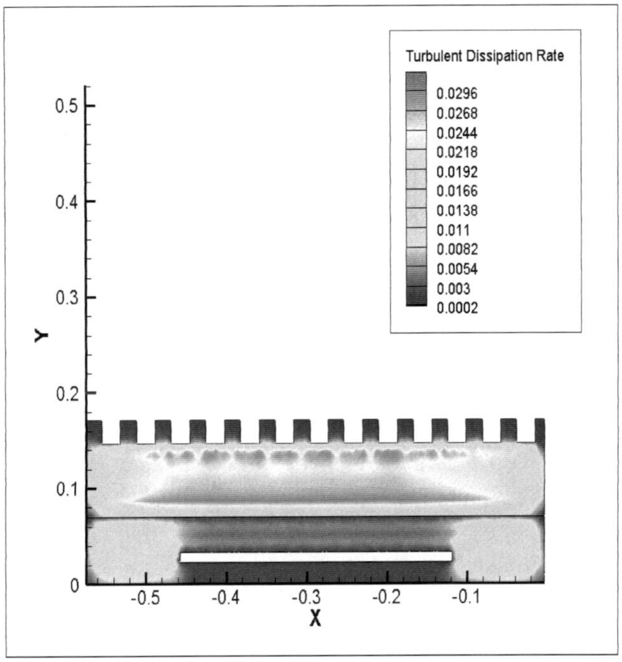

Figure 7: Turbulent Dissipation rate in the enclosure

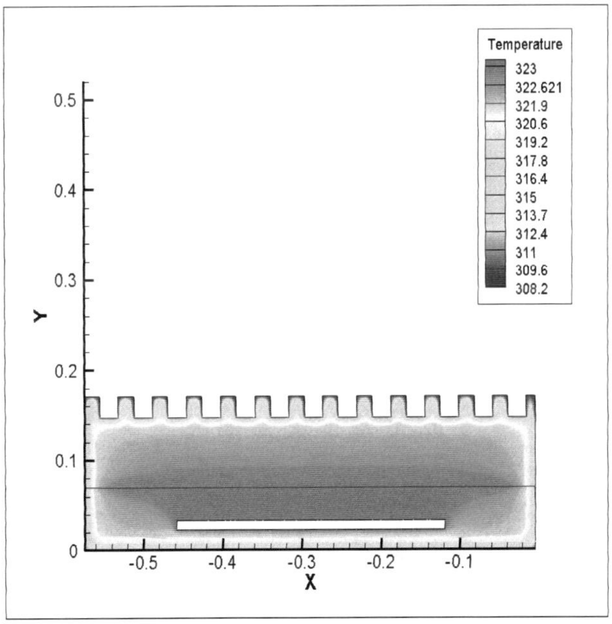

Figure 8: Temperature variation of the enclosure

Figure 9 and 10 indicates the temperature and pressure variation inside the solar panel along the longitudinal direction for different inclination angles of solar panel. It is seen that the temperature distribution inside the enclosure decreases with increase in the inclination angles of solar panel.

However, an opposite trend is observed in the pressure distribution inside the enclosure. Figure 11 and 12 represents the pressure and turbulent kinetic energy variation for different solar panel elevations. Table shown in Figure 13 is tabulated by probing variations in temperature, pressure and kinetic turbulent energy along vertical as well as longitudinal axis at specific intervals.

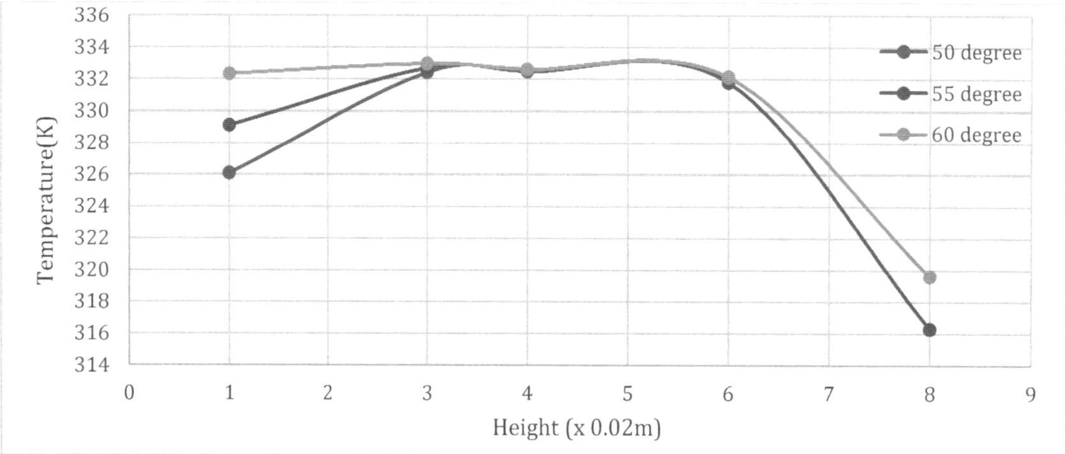

Figure 9: Temperature variation along the longitudinal direction

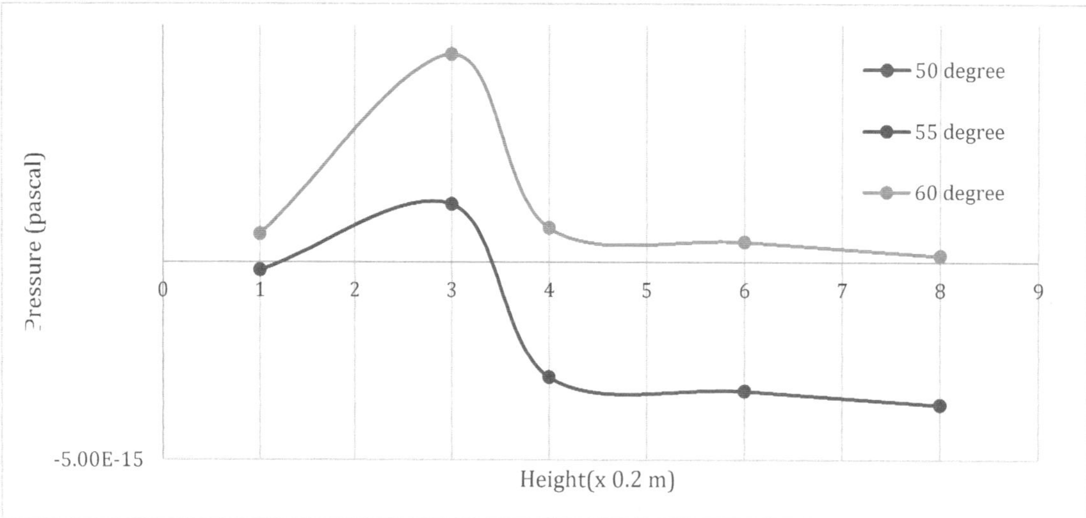

Figure 10: Pressure variation along the longitudinal direction

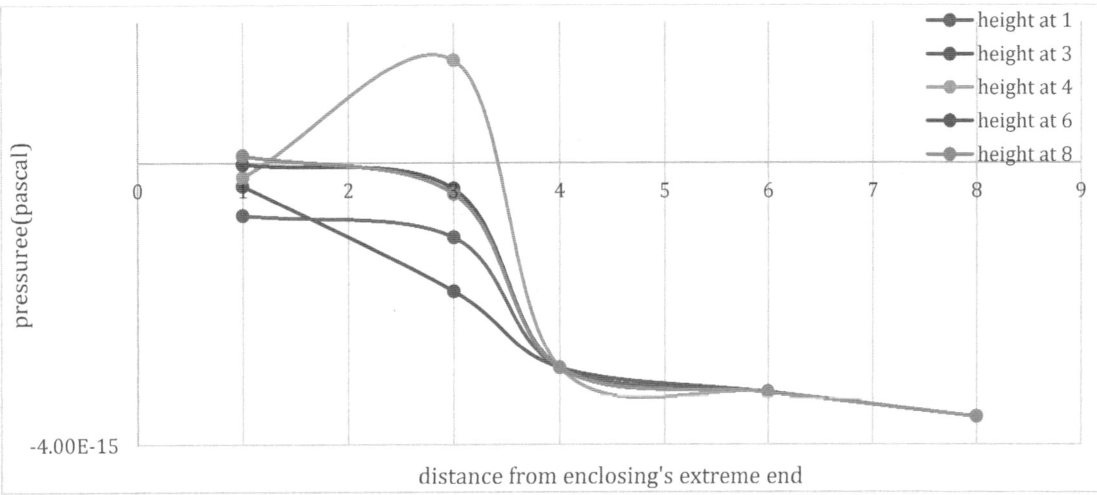

Figure 11: Pressure variation inside solar panel

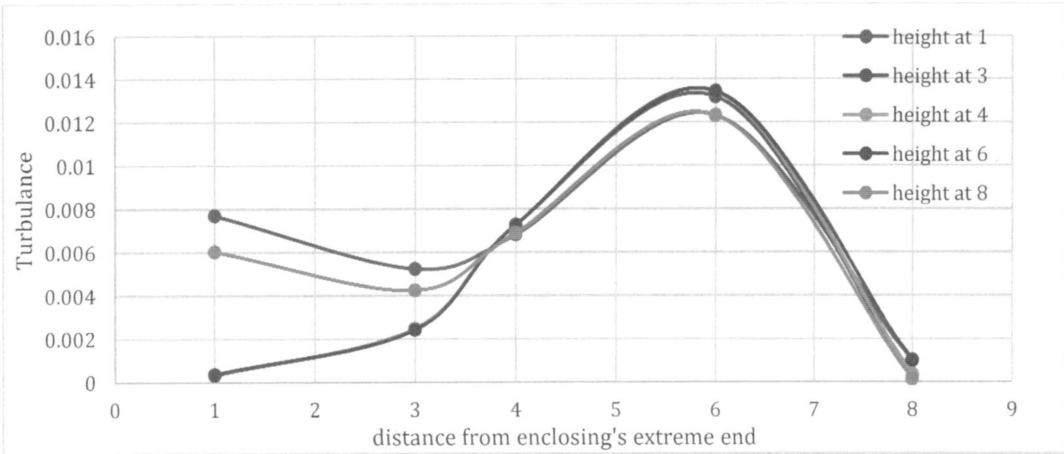

Figure 12: Turbulence kinetic energy variation inside solar panel

X (m)	Y (x0.1m)	Temperature (K)	Pressure (P)	Turbulent K E (J/Kg)
0.1	1	331.27	-7.46E-16	0.007691876
0.1	3	332.38	-1.06E-15	0.005234927
0.1	4	332.41	-2.91E-15	0.006809396
0.1	6	331.67	-3.26E-15	0.012323185
0.1	8	321.96	-3.60E-15	0.000978513
0.2	1	332.06	-3.23E-16	0.000320973
0.2	3	332.99	-1.82E-15	0.002436296
0.2	4	332.49	-2.91E-15	0.007264425
0.2	6	331.8	-3.26E-15	0.013181454
0.2	8	315.32	-3.60E-15	0.000266505
0.3	1	332.12	-2.03E-16	0.000317751
0.3	3	332.99	1.46E-15	0.002485905
0.3	4	332.49	-2.91E-15	0.007275655
0.3	6	331.81	-3.25E-15	0.013414463
0.3	8	316.29	-3.60E-15	0.000355239
0.4	1	332.06	-1.55E-17	0.000358472
0.4	3	332.99	-3.63E-16	0.002419203
0.4	4	332.49	-2.91E-15	0.007262155
0.4	6	331.81	-3.25E-15	0.013435755
0.4	8	321.86	-3.60E-15	0.001005924
0.5	1	331.17	1.07E-16	0.00603221
0.5	3	332.49	-4.37E-16	0.004244912
0.5	4	332.47	-2.91E-15	0.006890195
0.5	6	331.73	-3.26E-15	0.012328816
0.5	8	313.04	-3.60E-15	0.000121543

Table 2 : Tabulation of Temperature, Pressure and T.K.E. with respect to X-Y plane

4. Conclusion:

Using carbon-disulfide as a phase changing heat absorption material, panel heat could be absorbed, and the efficiency of the system could be increased. The numerical simulation is performed using finite volume method to analyse the heat and flow characteristics inside the solar panel with carbon-disulfide as phase change material. Based on the graphical interpretation of data, we can conclude that pressure, kinetic turbulence energy, and the ambient temperature inside the enclosure increase with the increase in panel temperature and is maximum above the panel and minimum along the sides of the casing. In the vertical direction, the temperature and pressure are maximum at the liquid-gas interface and minimum on the condensing boundary. Contrasting to it, the turbulent kinetic energy is maximum in the air-vapor phase region. The results from the present study will be useful in designing an effective solar panel cooling system.

5. References:

1. Peng Z, Herfatmanesh MR, Liu Y. Cooled solar PV panels for output energy efficiency optimization. Energy Conversion and Management. 2017 Oct 15; 150:949-55.
2. Chandel SS, Agarwal T. Review of cooling techniques using phase change materials for enhancing the efficiency of photovoltaic power systems. Renewable and Sustainable Energy Reviews. 2017 Jun 1; 73:1342-51.
3. Moharram KA, Abd-Elhady MS, Kandil HA, El-Sherif H. Enhancing the performance of photovoltaic panels by water cooling. Ain Shams Engineering Journal. 2013 Dec 1;4(4):869-77.
4.Gardas BB, Tendolkar MV. Design of a cooling system for the photovoltaic panel for increasing its electrical efficiency. International J Mechanical Prod Engineering. 2012; 1:63-7.5.
5.Waqas A, Ji J, Xu L, Ali M, Alvi J. Thermal and electrical management of photovoltaic panels using phase change materials–A review. Renewable and Sustainable Energy Reviews. 2018 Sep 1; 92:254-71.
6. Rajvikram M, Leoponraj S, Ramkumar S, Akshaya H, Dheeraj A. Experimental investigation on the abasement of operating temperature in the solar photovoltaic panel using PCM and aluminum. Solar Energy. 2019 Aug 1; 188:327-38.
7. Li Z, Ma T, Zhao J, Song A, Cheng Y. Experimental study and performance analysis on solar photovoltaic panel integrated with phase change material. Energy. 2019 Jul 1; 178:471-86.
8. Kang X, Wang YP, Xin GC, Shi XS. Experiment and simulation study on silicon oil immersion cooling densely packed solar cells under high concentration ratio. Int. J. Energy Power Eng. 2016;5(3):90-6.
9. Mehrotra S, Rawat P, Debbarma M, Sudhakar K. Performance of a solar panel with water immersion cooling technique. International Journal of Science, Environment, and Technology. 2014 Jun;3(3):1161-72.
10. Abdulgafar SA, Omar OS, Yousif KM. Improving the efficiency of the polycrystalline solar panel via the water immersion method. International Journal of Innovative Research in Science, Engineering, and Technology. 2014 Jan;3(1):96-101.
11. Haller M, Frank E. On the potential of using heat from solar thermal collectors for heat pump evaporators. In Proceedings of the ISES Solar World Congress 2011 2011 Aug.
12. Singh G, Kumar H. Computational Fluid Dynamics Analysis of Shell and Tube Heat Exchanger. Journal of Civil Engineering and Environmental Technology. 2014 Aug;1(3).
13.Thabet S, Thabit TH. Computational Fluid Dynamics: Science of the Future. International Journal of Research and Engineering. 2018;5(6):430-3.
14. Chen H, Chen Y, Hsieh HT, Siegel N. CFD modeling of gas-particle flow within a solid particle solar receiver. International Solar Energy Conference 2006 Jan 1 (Vol. 47454, pp. 37-48.
15. Bellan S, Matsubara K, Cho HS, Gokon N, Kodama T. A CFD-DEM study of hydrodynamics with heat transfer in a gas-solid fluidized bed reactor for solar thermal applications. International Journal of Heat and Mass Transfer. 2018 Jan 1; 116:377-92.
16. Harish. R, Venkatasubbaiah. K, Mathematical modeling and computation of fire induced turbulent flow in partial enclosures, Applied Mathematical Modelling, vol.37, pp: 9732-9746, 2013.
17. Harish. R, Venkatasubbaiah. K, Transport phenomena of turbulent fire spread through compartment connected to vertical shaft in tall building, Fire Safety Journal, vol.61, pp: 160-174, 2013.
18. Harish. R, Venkatasubbaiah. K, Effects of buoyancy induced roof ventilation systems for smoke removal in tunnel fires, Tunnelling and Underground Space Technology, vol.42, pp: 195-205, 2014.
19. Harish. R, Venkatasubbaiah. K, Numerical investigation of instability patterns and nonlinear buoyant exchange flow between enclosures by variable density approach, Computers & Fluids, vol.96, pp: 276-287, 2014.

20. Harish. R, Venkatasubbaiah. K, Large Eddy Simulation of thermal plume behavior in horizontally partitioned dual enclosure, Building Simulation, vol.8(2), pp: 137-148, 2015.
21. Harish. R, Venkatasubbaiah. K, Numerical study of water spray interaction with fire plume in dual chambers connected to tall shaft, Fire Safety Journal, vol. 74, pp: 1-10, 2015.
22. Harish. R, Venkatasubbaiah. K, Non-Boussinesq approach for turbulent buoyant flows in enclosure with horizontal vent and forced inlet port, Applied Mathematical Modelling vol. 40, pp: 927-941, 2016.
23. Harish. R, Venkatasubbaiah. K, Numerical simulation of turbulent plume spread in ceiling vented enclosure, European Journal of Mechanics-B/Fluids, vol.42,pp:142-158, 2013.
24. Harish. R,Effect of heat source aspect ratio on turbulent thermal stratification in a naturally ventilated enclosure, Building and Environment,143,pp:473-486, 2018.
25. Harish. R, Buoyancy driven turbulent plume induced by protruding heat source in vented enclosure, International Journal of Mechanical Sciences,148,pp:209-222, 2018.
26. Sumedha R.G. Weliwaththage, Udara S.P.R. Arachchige, Solar Energy Technology,University of Sri Jayewardenepura, Sri Lanka,vol 1:JRTE;2020.
27. Arifin Z, Suyitno S, Tjahjana DD, Juwana WE, Putra MR, Prabowo AR. The Effect of Heat Sink Properties on Solar Cell Cooling Systems. Applied Sciences. 2020 Jan;10(21):7919.

IOP Conf. Series: Materials Science and Engineering 1128 (2021) 012034 doi:10.1088/1757-899X/1128/1/012034

Computation of a Compartment Fire with Smagorinsky Sub-Grid Scale Model

Bhisham K Dhurandher[1], R Kumar[2], A K Dhiman[3] and Santhosh K Dubba[4]
[1]School of Mechanical Engineering, VIT Chennai, Tamil nadu, India
[2]Department of Mechanical & Industrial Engineering, IIT Roorkee, India
[3]Department of Chemical Engineering, IIT Roorkee, India
[4]Department of Mechanical Engineering, Aditya Institute of Technology & Management, India

Email: bhishamkumar.d@vit.ac.in

Abstract. A fire tests using plywood crib as fire source was done in a cubical compartment to investigate the fire development and temperature distribution. The dimension of the experimental compartment was 4 m high, 4 m wide and 4 m length with a door opening of 2 m × 1 m wide located at one of the compartment wall. Several parameters were measured which includes fuel mass loss rate, compartment gas temperature, centerline doorway temperature, heat flux and wall surface temperature. Three-dimensional numerical simulations were performed using large eddy simulation with different values of Smagorinsky sub-grid scale constant. In order to accurately investigate the behavior of fire within the compartment by computational fluid dynamics (CFD), it is vital to study these Smagorinsky sub-grid scale constant. Numerically predicted results with various Smagorinsky sub-grid scale constant was compared with the physical results. It was found that the varying the Smagorinsky constant affect the numerical simulation results.

1. Introduction

Better knowledge of the improvement in compartment fire is vital for fire analysts and firefighters for foreseeing and assessing temperature, heating of adjacent structures and smoke creation [1]. The experimental data and the numerical simulation are two principal techniques used to explore the progress of compartment fires. No, doubt, progressively fire safety assurance concerns to utilize fire modeling software, including two-zone models and CFD models. As computational technologies are more accessible now at lower cost, CFD software's are progressively being used for engineering applications [2]. Nonetheless, due less comprehensive information about material properties and incomplete demonstrates on the pyrolysis and fire behavior at different scenarios, there is still a requirement for development of numerical re-enactment strategies. Several turbulence models are present in literature for simulating the turbulent mixing phenomenon in various fire scenarios, such as Direct Numerical Simulation (DNS) model [3], Large Eddy Simulation (LES) model [4,5] and RANS [6]. The Sub-grid scale coefficient, (C_s) depends on flow and studied with the range from 0.1 to 0.25 by different flow applications as studied by different researchers [7,8]. While the simulation of indoor airflow under different conditions (i.e, Natural, forced and mixed convection) the value of 0.16 and 0.2 were found to be better agreement with experimental results [9,10]. The sub-grid scale turbulent Prandtl number (P_r) is derived by empirical correlation from the range from 0.2 to 0.9.

The physical results of compartment fire were studied and found that the effect of C_s, P_r and gird size (Δx) for weak buoyant plume there is no significant effect with the variation of model coefficient C_s and P_r. But for the case of the strong, buoyant plume, the (C_s) has an significant effect on the predicted temperature and heat flux [11]. The effect of Prandtl number was studied for enclosure fire and concluded that the (P_r) has no significant effect on Heat release rate (HRR), temperature and CO yield as compared to other coefficient, whereas, from S_c=0.1 higher value of CO yield was predicted with relative difference of 59 % than that of default value S_c= 0.5 [12].

It has been found that the effect of (C_s), on the simulation of compartment fire needs more investigations. Therefore, in the present study, the numerical simulations were performed using FDS code for different SMG coefficients for two different mess sizes, i.e. 0.1 m and 0.14 m. In FDS a default value of C_s, P_r, S_c are 0.2, 0.5 and 0.5 respectively [13].

2. Experimental Setup & Numerical Model

2.1. Compartment Configuration

The experimental compartment with dimension (4 m × 4 m × 4 m) was constructed to provide data from fire experiments. At center of the front wall of the compartment there was a door opening of dimension 1 m × 2 m. The experimental compartment walls were made up of normal bricks and plaster of 3 cm thickness on both sides of the wall. The ceiling was made up of Reinforced Cement Concrete (RCC). The schematic sketch of experimental fire compartment is presented in Figure 1. The ply board cribs were used as fuel for the experiment. The cribs were burned on insulated platform at the center, beneath which a digital load cell was provided to record the mass loss rate of the burning crib during experiment [14]

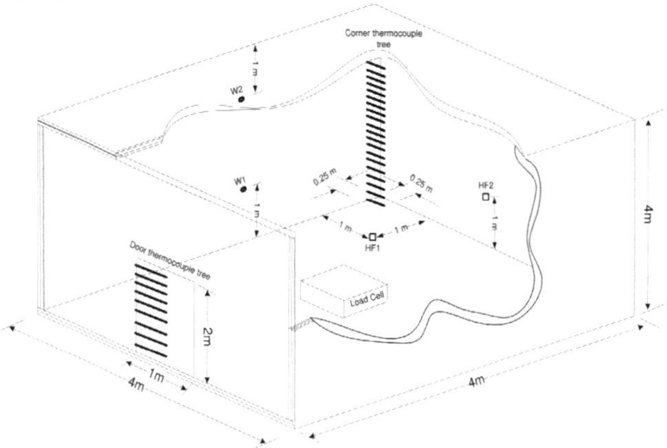

Figure 1. Schematic view of experimental compartment

2.2. Instrumentations

Several instrumentations were placed to measure various parameters such as, MLR, gas temperature and door gas velocity. The heat flux at different locations was measured using SBG01 heat flux guage of rating 0-20 kW/m^2. These gauges were installed in the experimental room to record the heat flux at floor and walls, as shown in Figure 1. HF1 and HF2 of working range 0-20 kW/m^2 were placed on the floor facing vertically upward and on back side of wall. Data logger was used to collect the readings of heat flux gauges with sampling rate of 5 sample per second. All heat flux gauges installed were having field view of 180^0, emissivity > 0.95 and had error margin of ± 6%.

For measuring the gas temperature profile of compartment, one thermocouple array was placed at one of the back corner of the compartment. A total of 20 K-type thermocouples of 1 mm diameter were located with continuous vertical spacing of 200 mm throughout the tree. The K type thermocouple had the error of ± 1.5 ^{0}C. The bottom most thermocouple was located 50 mm above the floor and the top most thermocouple was located at 150 mm below the ceiling. An array of thermocouple array was installed to record the door temperature. A total of 9 thermocouples were installed stating from 150 mm height with a pitch of 150 mm upto 900 mm and thereafter a pitch of 300 mm up to 1800 mm height. All output readings heat flux gauges and thermocouples were collected using a data acquisition.

2.3. Description of FDS Model

The present transient numerical simulation was conducted with the FDS code Version 5.5.3 developed by NIST [15]. The dimension of the computational domain which encloses the compartment is 4.5 m ×

5.0 m × 4.5 m having a opening of dimension 1.0 m × 2.0 m located at the middle of front wall. At the center of the compartment a fire source was placed with a height of 0.2 m above the floor. To investigate the gas movement the computational domain has been extended 1 m from the doorway and the boundary conditions was set as open. The temperature sensors and heat flux gauges are defined in present simulation exactly at the same location as located in the experimental compartment. The detailed description of the location of instruments is as explained in Section 2.2. In the present study, the ambient temperature is was 20 ^{0}C and the relative humidity was 70 %. The default velocity and thermal wall boundary conditions as imposed in FDS, have been used for present numerical investigation.

2.4. Grid Sensitivity Analysis

As mentioned in the previous section, CFD tool FDS is used for present simulation. The size of the grid is an important parameter in CFD dictating its numerical accuracy. The size of the eddies which can be solved will mainly depend on the refinement of the numerical grid. Froude number was used to derive the characteristic diameter of the plumes, (D^*) and the gird size (Δx) was used to solve the flow field in buoyant plumes [16].

$$D^* = \left(\frac{\dot{Q}}{\rho_\infty c_p T_\infty \sqrt{g}} \right)^{2/5} \quad (1)$$

Finer grid size may additionally provide extra correct solution; however, refinement of the accuracy of computation will be small after the variety of grid factors is expanded past a sure level. At the equal time, computational expense gets more prominent. In the present numerical simulation, five different gird size (Δx) has been chosen which ranges from 0.05 m to 0.14 m. It was found that for grid size of 0.05, 0.08 and 0.10 no significant deviation was found. The average percentage change for upper hot gas layer with reference to grid size 0.10 m for 0.08 and 0.05 m was 2.5 % and 2.1 % respectively. So for the present study the grid size of 0.10 m was adopted. The grid sensitivity analysis is illustrated in Figure 2. As the size of the cell increases, the computation time also increases which refines the predicted results. The initial time step for calculation is set to be 3 s. All calculations were carried out on a Dell-PC with a 3.60 GHz dual processor and 12 GB of RAM. It took approximately 12 h for a typical run.

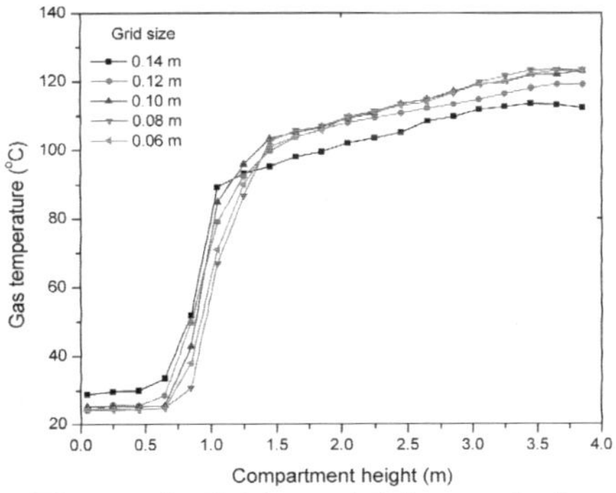

Figure 2. Grid sensitivity analysis

3. Results and Discussions

The fire experiment was conducted using cribs made up of plywood as fire source. The compartment gas temperature, wall temperature and heat flux are measured at various locations and are validated with FDS code with varying SGS model coefficient are shown in Figure 3. To study the effect of turbulence model constant two grid size one of fine resolution of filter width 0.1m and coarse resolution of filter width of 0.14m have been taken. Figure 3 represents the comparison of experimental data with numerically predicted results by varying SGS constants in fine as well as coarse resolution. In Figure 3(a) the numerical results were obtained for different Smagorinsky sub-grid constant C_s=0.13, 0.15, 0.17, 0.2, 0.23 and compared with the measured values. In Figure 3(b) the Prandtl number (P_r) has been varied for value 0.1 to 0.9 and compared with the measured values, similarly the effect of turbulent Schmidt number (S_c) has been represented in Figure 3(c). With the default value of C_s, P_r and S_c (0.2, 0.5, 0.5) the result found to be in good agreement with experimental value. On further inspection it can be noticed that the effect of variation of C_s and S_c has less significant effect on the numerical results. Whereas, the numerical result obtained with the lesser value of Prandtl number, P_r=0.1 has significant effect on the predicted results. From Fig. 3(a,b,c) the thermal discontinuity height was predicted 0.95 m from the floor and when measured with the corner thermocouple is was found to be 1.05 m from the floor. However, no significant effect was seen when determined with different values of all SMG coefficients.

Temperature distributions at doorway centerline are illustrated in Figure 4. With default values for C_s, P_r and S_c the predicted results follow the similar trend to those of experimental values and found to be in good agreement. The percentage deviation between the predicted and experimental results are around 14 % for the cold layer temperature below the neutral plane and approximately 13 % for hot gas layer temperature above neutral plane. However, when experimental results were compared with different value of Prandtl number it was found to be under predicted with the percentage difference of about 36 % for P_r = 0.1 at the hot gas layer zone. On the other hand, no significant difference was found with the variation of Schmidt number, as shown in Figure 4(c). When the experimental results of doorway temperature profile were simulated with coarse grid the predicted centerline doorway temperature profile was found under predicted with deviation of 12 % with experimental values for default CFD turbulence model constants. However no significance change was noticed with the variation of C_s. With P_r= 0.1 the largest deviation was found about 32 % and a noticeable variation can be seen with different values of P_r as shown in Figure 4(b). The deviation between numerical and experimental temperature profiles may be due to the fact that the temperature sensor absorbs the radiation coming from the heated surface boundaries such as wall, ceiling, and floor, or due to radiative heat transfer from hot gas layer to the cold layer which is present below the thermal discontinuity height and from the flames. Figure 5 presents the variation of heat flux on the floor of the compartment with different values of SGS constants. The location of the heat flux sensor HF1 is shown in Figure 1. The predicted peak heat flux prediction for HF1 was about 8 % when compared with measured heat flux. The variation of C_s and P_r has no significant effect on the predicted heat flux for both the heat flux sensors and found good agreement with the experimental results. Whereas with P_r = 0.1 the discrepancy in measurements is about 18 % for HF1. The simulation result was also found in good agreement with the experimental values when the grid size was coarse

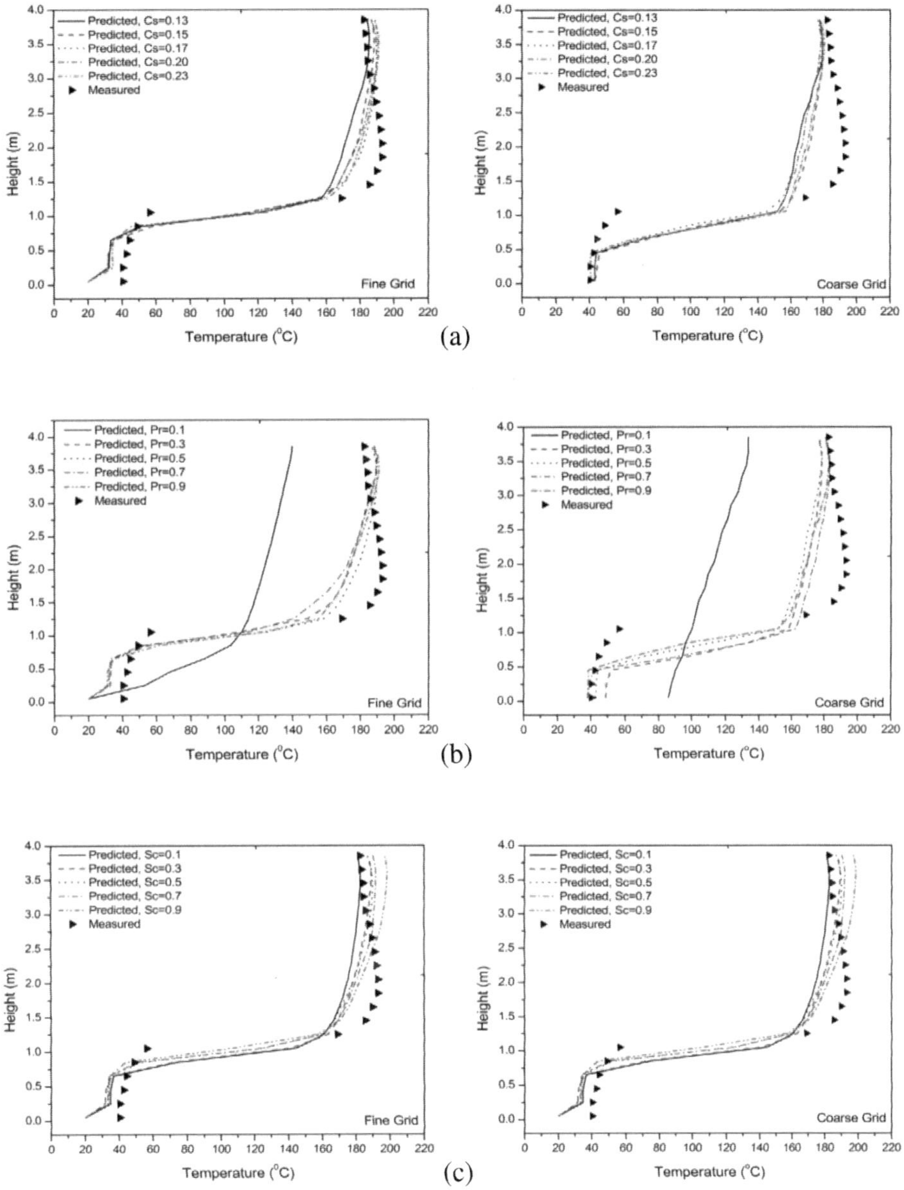

Figure 3. Comparison of compartment gas temperature profile. (a) Variation of C_s for fine and coarse resolution. (b) Variation of P_r for fine and coarse resolution. (c) Variation of S_c for fine and coarse resolution.

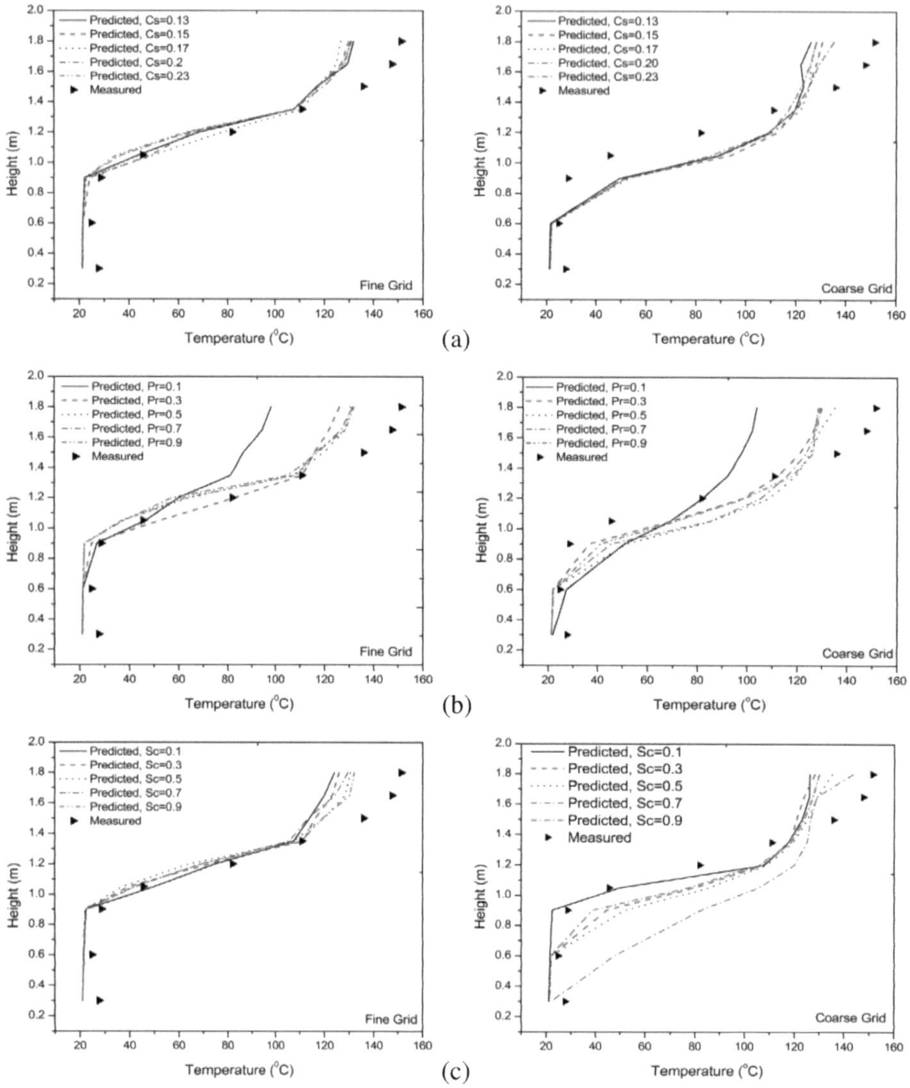

Figure 4. Comparison of door temperature profile. (a) Variation of C_s for fine and coarse resolution. (b) Variation of P_r for fine and coarse resolution. (c) Variation of S_c for fine and coarse resolution.

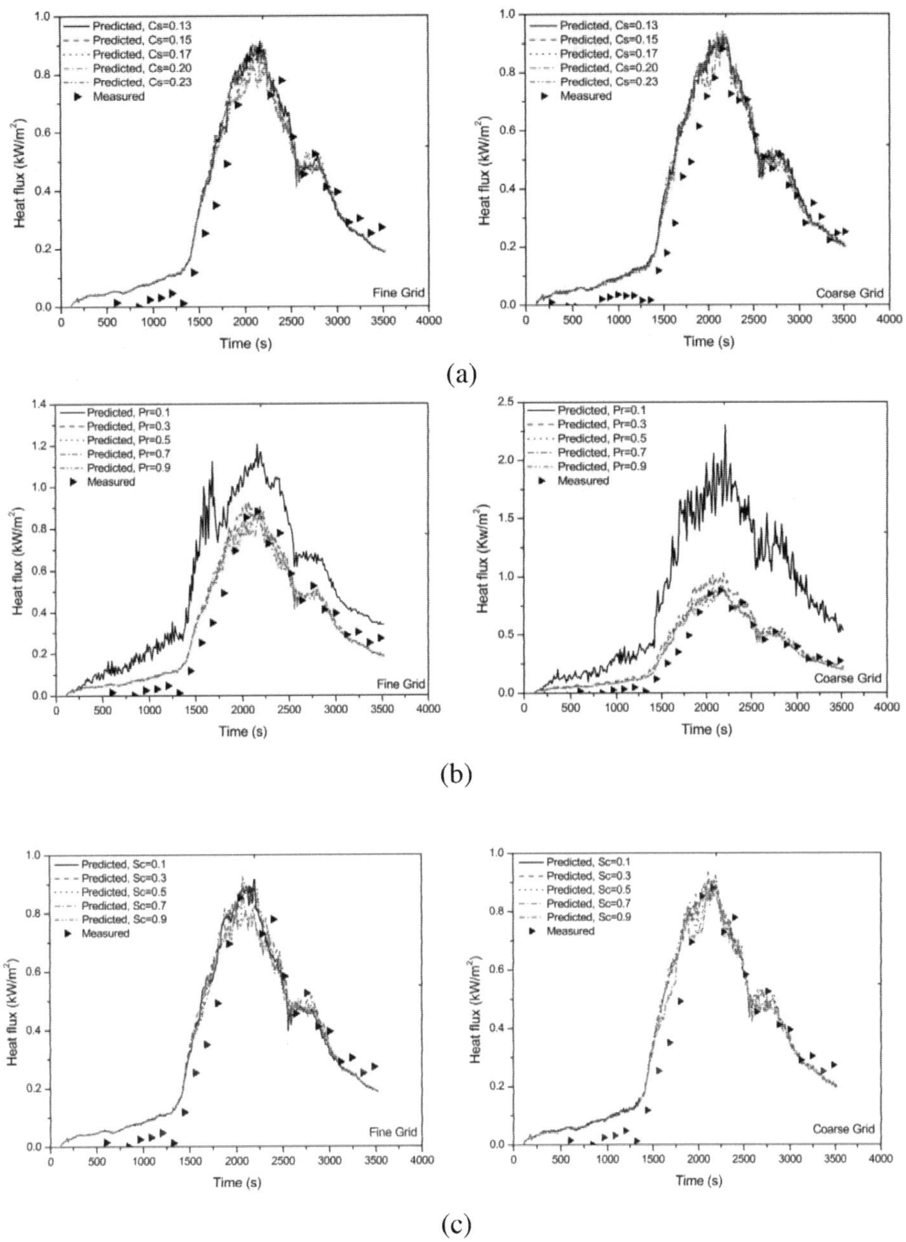

Figure 5. Comparison of heat flux on the compartment floor.(a) Variation of C_s for fine and coarse resolution. (b) Variation of P_r for fine and coarse resolution. (c) Variation of S_c for fine and coarse resolution.

4. Conclusion

The present paper illustrates the comparison between experimental fire test data with numerical simulation in an enclosure under natural ventilation conditions. Twenty sets of simulation by large eddy simulation were carried out to investigate the effect of turbulence model constant on the temperature profile and heat flux in compartment fires. And also to investigate the effect of turbulence model constants on grid resolution, simulation were performed for fine grid as well as coarse grid. The following observations were made:

- When simulations were performed with Δx of 0.1 m the simulation results has good agreement with the experimental values than that of coarse grid with filter width 0.14 m.
- The Smagorinsky sub-grid constant C_s on fine grid shows less significant effect on the predicted results as compared to coarse grid.
- Lower value of turbulent Prandtl number has large deviation with both fine and coarse grids as compared with higher turbulent Prandtl number. Whereas the effect of turbulent Schmidt number on both grid had less significant effect.
- The buoyant plume in present experiment is not very large therefore the predicted results show less dependency on the grid resolution. The effect of Smagorinsky sub-grid constant C_s has less influence due to weak buoyant plumes been as choice should be made for Prandtl number.

Nomenclature

$\dot{Q}$	Heat release rate, KW
T_∞	Ambient Temperature, k
g	Gravitational acceleration constant, m/s^2
Δx	Grid size, m

Greek Symbols

c_p	Specific heat capacity, kJ/kg-k
ρ_∞	Air density, kg/m^3

Non-dimensional Numbers

C_s	Smagorinsky Constant
P_r	Turbulent Prandtl number
S_c	Turbulent Schidmt number
D^*	Characteristic fire length

References

[1] Dhurandher B K, Kumar R, Dhiman A K 2017 An experimental study on crib fires in a closed compartment *Thermal Science* **21** 1431-1441

[2] Moinuddin K A M, Thomas I R 2007 Factors Affecting Grid-independent Results for Compartment Fire Modelling In: 6th Australasian Fluid Mechanics Conference Crown Plaza, Gold Coast, Australia

[3] William E M, McGrattan K B, Baum H R 1996 Numerical simulation of combustion in fire plumes *In: Proceedings of the 26th Symposium (International) on Combustion, Pittsburgh* 1523.

[4] McGrattan K B, Baum H R and Rehm R G 1998 Large eddy simulation of smoke movement *Fire Safety Journal* **30** 161–178.

[5] Zhang W, Hamer A, Klassen M, Carpenter D and Roby R 2002 Turbulence statistics in a fire room model by large eddy simulation *Fire Safety Journal* **37** 721–752.

[6] Wen J X and Huang L Y 2001 CFD modeling of confined jet fires under ventilation controlled conditions *Fire Safety Journal* **34** 1–24.

[7] Yaga M, Endo H, Yamamoto T, Aoki H and Miura T 2002 Modeling of eddy characteristic time in LES for calculating turbulent diffusion flame *Int. J. Heat Mass Transfer* **45** 2343–2349.

[8] Murakami S 1998 Overview of turbulence models applied in CWE-1997 *J. Wind Eng. Ind. Aerodyn* **74** 1–24.

[9] Zhang W and Chen Q Y 2000 Large eddy simulation of indoor airflow with a filtered dynamic subgrid scale model *Int. J. Heat Mass Transfer* **43** 3219–3231.

[10] Emmerich S J and McGrattan K B 1998 Application of a large eddy simulation model to study room airflow *ASHRAE Transactions* **104** 1-9.

[11] Zou G W, Tan H P, Chow W K, Gao Y 2007 Effects of varying Smagorinsky constant on simulating post-flashover fires *International Journal of Computational Fluid Dynamics* **21** 107–119.

[12] Yang D, Hu L H, Jiang Y Q, Huo R, Zhu S, Zhao X Y 2010 Comparison of FDS prediction by different combustion models with measured data for enclosure fires *Fire safety Journal* **45** 298-313.

[13] Wei Z, Jinjin Lv, Zhaozhou Li and Tianshui L 2013 A study of bifurcation flow of fire smoke in tunnel with longitudinal ventilation *International Journal of Heat and Mass Transfer* **67** 829-835.

[14] Dhurandher B K, Kumar R, Dhiman A K, Sharma P K 2017 An investigation on compartment fires with plywood cribs *Fire and Material* **41** 209-224

[15] McGrattan K B, Hostikka S, Floyd J E, Baum H R, Rehm R G 2007 Fire dynamics simulator (Version5), technical reference guide. NIST Special Publication 1018–5, National Institute of Standards and Technology, Gaithersburg, Maryland, October 2007

[16] Merci B, Maele K V 2008 Numerical simulations of full-scale enclosure fires in a small compartment with natural roof ventilation *Fire Safety Journal* **43** 495–511

Numerical Investigation on Pylon and Flush Wall Injection in Cold Coaxial Jets

J. Sarathkumar Sebastin[a], S. Jeyakumar[a*], K. Karthik[a], R. Sivakumar[b]

a Aeronautical Engineering, Kalasalingam Academy of Research and Education, Krishnankoil, India – 626126.
b School of Mechanical Engineering, VIT University, Chennai, India.

*corresponding author: sjeyakumar1974@gmail.com

Abstract. The dual combustion ramjet (DCR) engine is considered one of the promising engines for the hypersonic missile propulsion system. This paper reveals the non-reacting flow characteristics of the DCR engine with the influence of pylon and wall injections in coaxial jets using numerical investigation. The supersonic combustor of the DCR engine is modeled and analysed using the commercial CFD software ANSYS 18.0. The three-dimensional compressible Reynolds-averaged Navier-Stokes (RANS) equations coupled with the SST $k - \omega$ turbulence model have been used to analyse the coaxial mixing characteristics of the jets. The numerical study is validated with the experimental data and agrees with it for further investigation. The pylon and wall injectors are located symmetrically at the gas generator's nozzle exit, and the air is used as the injectant to simulate gaseous fuel. The pylon and wall injection results are compared with the actual DCR engine. In a typical DCR engine, the shock waves generated from the gas generator nozzle enhance the mixing of the coaxial jets with minimum total pressure loss. The pylon injection induces more shock interactions along the flow direction within the supersonic combustor leading to higher total pressure loss than the wall injection. The pylon injection provides the spatial distribution of fuels compared to the wall injection in the coaxial supersonic flow field.

1. Introduction

The scramjet engines are the propulsion systems for the hypersonic airbreathing vehicle[1–3]. The scramjet engine's key features are fuel injection, air-fuel mixing, and combustion, particularly for heavy hydrocarbon fuels that need a relatively long time for chemical reactions to continue. The well-known propulsion concepts pursued in scramjet applications are the Dual-mode Combustion Ramjet (DMCR) and Dual Combustion Ramjet (DCR). In DMCR, subsonic and supersonic combustion modes occur within a solitary combustor itself; however, the DCR consists of a gas generator and the supersonic combustor[4]. Billig[5] and Waltrup[6] proposed the DCR engine concept for hypersonic missile applications in the 1980s. The DCR engine combines the advantages of both ramjet and scramjet combustion modes of operation. The DCR engine schematic illustration is shown in Figure 1.

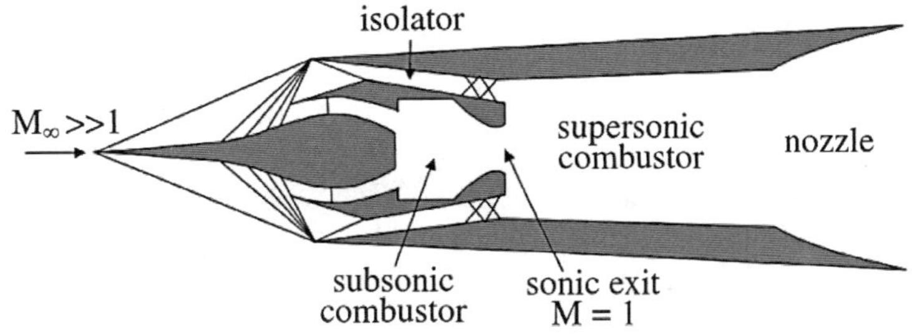

Figure 1: DCR engine with shock structures[2]

The DCR engine operates in the ramjet mode with a limited volume (typically one-fourth) of the high-speed incoming air, which is allowed through a gas generator where the fuel is combined and ignited. The gas generator's significance is to supply the supersonic combustor with partly reacted and preheated fuel-rich streams for complete combustion than direct injection into the supersonic flow[7]. The combustion flame from the gas generator serves as a pilot flame for the supersonic combustion. A significant volume of remaining air (nearly three-fourths) flows through the annular duct between the gas generator and the outer casing of the DCR engine into the supersonic combustor at supersonic velocity. This annular duct serves as an insulator between the inlet and the supersonic combustor to avoid combustion-induced oscillations from inlet flow. The gas generator nozzle's rim helps create a recirculation zone for the coaxial jets that forms the low-speed high-temperature region to enhance the flame holding and stabilisation phenomenon. Hence, no separate flame holding device is necessary to sustain the combustion as a direct scramjet engine.

Tan et al.[8] have shown that the quasi-one-dimensional approach can be applied to evaluate various modes in the entire dual-mode scramjet combustor operation. Stockbridge[9], Waltrup, and Billig[10] studied the DCR combustor and the inlet interactions and formulated a semi-empirical relationship for the shock/boundary-layer interactions. The combustion characteristics of the DCR engine are numerically investigated by Zang et al. [11]. The authors revealed that stable combustion could be observed at the nozzle rim, and the mixing layer regimes are formed by increasing the fuel equivalence ratios. The numerical review on the analysis of splitter plate thickness[12] reveals that an increase in plate thickness (gas generator nozzle rim) improves the mixing of coaxial jets, enhancing complete combustion. Choi and Yang[13] numerically study the coupling process between the compressibility effects and the turbulent combustion in a dual combustion ramjet engine. Experimental and numerical studies by Tan et al.[14] revealed that the DCR engine achieves appreciable performance under operating conditions at Mach 4 and Mach 6.

Kurian and Habeeb[15] demonstrated that a petal nozzle could be used at the gas generator's exit instead of the conventional nozzle, which provides better mixing and combustion efficiency under the same operating conditions. However, these nozzles cause a higher loss of stagnation pressure than standard nozzles. Jeyakumar et al. [16] demonstrated that coaxial jets' mixing efficiency could be enhanced using the tandem cavities at a marginal rise in stagnation pressure loss. The authors analysed the influence of cavities in the axisymmetric supersonic field [17–22]. Byun et al.[23] developed a new correlation method to determine the pre-combustion shock train length and the combustion performance of the DCR engine.

The effect of transverse injection in the coaxial supersonic flow field has not been explored systematically. The present study investigates the characteristics of the coaxial jets in a modified DCR engine by incorporating transverse injection under the non-reacting flow field. The fuel injection schemes include flush wall and pylon injections adopted at the supersonic combustor inlet. The study details the modified DCR engine's performance parameters in terms of wall static pressure distribution, mixing, and stagnation pressure loss, and compared with the conventional DCR engine.

2. Numerical Modelling

A three-dimensional compressible Reynolds Averaged Navier Stokes (RANS) equation with a density-based double precision solver is used to resolve the flow field of the DCR engine. The RANS equation provides accurate results even with the coarse meshes and resolves the steady flow more efficiently than other models [24]. The transport equation model called Shear Stress Transport (SST) k- ω model with traditional constants have been used to solve the turbulent flow field[25]. The SST k-ω turbulence model yields accurate results in predicting mixing layers and jet flows[26]. The fluid is considered an ideal gas and the mass-weighted-mixing-law is used to compute the thermal conductivity and viscosity. For spatial discretisation, a second-order upwind system (SOU) is used along with the Advection Upstream Splitting Process (AUSM) flux splitting to provide an effective solution with reliable convergence characteristics[27]. The Courant-Friedrichs-Levy (CFL) number is chosen as 0.5 to ensure computational stability[28]. The governing equations, i.e., mass, momentum, and energy equations written for the total enthalpy are expressed as:

Continuity equation

$$\frac{\partial \rho}{\partial t}+\frac{\partial}{\partial x_i}(\rho u_i)=0 \tag{1}$$

Momentum equation

$$\frac{\partial}{\partial t}(\rho u_i)+\frac{\partial}{\partial x_i}(\rho u_i u_j)+\frac{\partial P}{\partial x_i}=\frac{\partial}{\partial x_i}(\tau_{ij}) \tag{2}$$

Energy equation

$$\frac{\partial}{\partial t}(\rho H)+\frac{\partial}{\partial x_i}(\rho H u_i)=-\frac{\partial}{\partial x_i}(\tau_{ij}u_i)+\frac{\partial q_i}{\partial x_i} \tag{3}$$

The turbulence kinetic energy, k, and the specific dissipation rate, ω is obtained from the following transport equations:

$$\frac{\partial}{\partial x_i}(\rho k u_i)=\frac{\partial}{\partial x_j}\left(\Gamma_k\frac{\partial k}{\partial x_j}\right)+G_k-Y_k+S_k \tag{4}$$

and

$$\frac{\partial}{\partial x_j}(\rho \omega u_j)=\frac{\partial}{\partial x_j}\left(\Gamma_\omega\frac{\partial \omega}{\partial x_j}\right)+G_\omega-Y_\omega+D_\omega+S_\omega \tag{5}$$

The terms, G_k represents the production of turbulent kinetic energy and G_ω as the generation of ω, Γ_k and Γ_ω represent the effective diffusivity of k and ω respectively, Y_k and Y_ω represent the dissipation of k and ω due to turbulence, D_ω represents the cross-diffusion terms and S_k and S_ω as the user-defined source terms.

3. Numerical setup

3.1. Computational domain

Figure 2(a) shows the DCR engine test facility's schematic layout, consisting of a gas generator, isolator, and supersonic combustor. The symmetric portion of the supersonic combustor is chosen for the study to reduce the computational expense (Figure 2b). One-fourth of the incoming air is allowed to pass through the gas generator. The flow issued from the gas generator nozzle is in choking condition. The remaining air flows at a Mach number of 2.0, through the annular passage located between the outer casing and the gas generator, which acts as the isolator. The coaxial jets enter into the supersonic combustor, and the flow details are listed in Table 1.

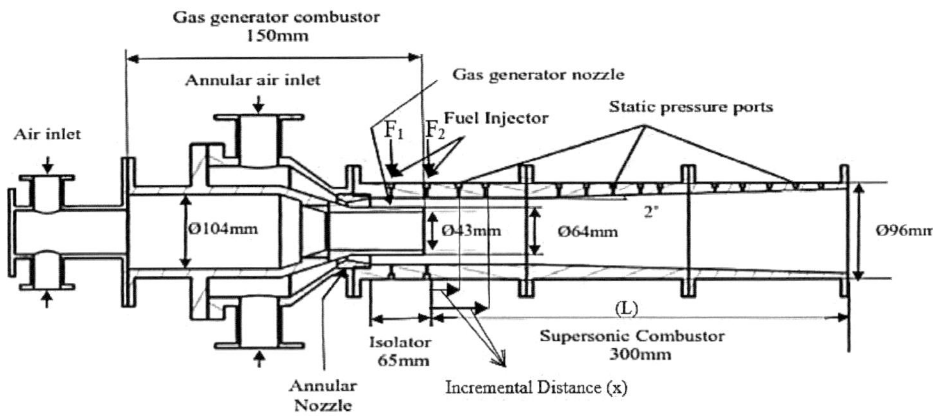

Figure 2(a). Experimental test facility of DCR Engine

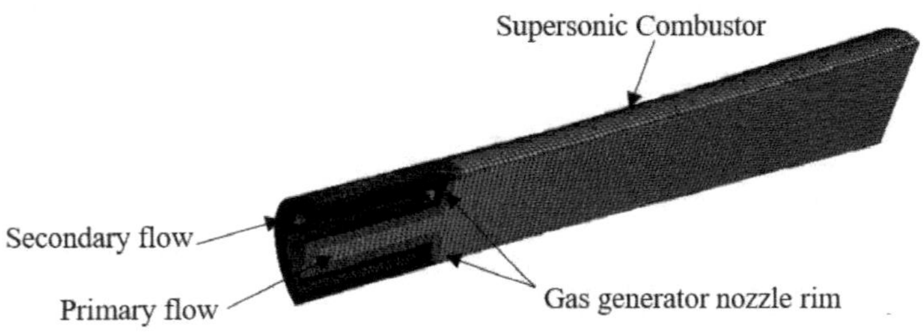

Figure 2(b). Computational domain of the DCR engine with cut shell mesh

Table 1: Operating conditions of DCR engine

Parameters	Gas generator	Annular flow
Stagnation pressure (MPa)	0.1856	0.8
Stagnation temperature (K)	305	305
Mass flow rate (kg/s)	0.25	0.75
Mach number	1.0	2.0

In the supersonic combustor, a secondary fuel is injected normal to the flow direction by flush wall injection and Pylon injection. In the case of pylon injection, the fuel is injected directly to the rim of the gas generator nozzle to study the coaxial jets' flow characteristics. The air is injected to mimic gaseous fuel in the non-reacting flow study. The injection pressure ratio (= injection pressure, P_j /annular airflow pressure, P_{oa}) is 1.875 is tested, and the results are compared with the typical DCR engine.

3.2. Boundary Condition

The DCR experimental setup's symmetric outline is described in the experimental section. The numerical investigations are performed from the gas generator nozzle to the supersonic combustor exit to reduce the computational time. The imposed boundary conditions for the DCR engine domain are the same as that of the experimental data listed in Table 1. The mass flow inlet condition is adopted for air and fuel inlet to the DCR engine, and the pressure outlet condition is considered for the combustor exit. The wall is applied with no-slip and adiabatic boundary conditions.

3.3. Grid generation

A 3D axisymmetric unstructured grid is generated from the gas generator nozzle to the supersonic combustor exit. Three mesh grid structures of different sizes are incorporated here, namely coarse (0.5 million), medium (0.65 million), and fine (0.75 million). The y+ value is maintained at less than 1.0 in all the simulations, where the corresponding first-row cell height is 0.001 mm. In this task, a mesh independence test is also performed to ensure and evaluate the CFD simulation features. The grid independence study reveals that the static pressure values have a variance of less than 1 percent for all the mesh sizes. However, no further error analysis is needed to illustrate grid convergence (Figure 3). Furthermore, to analyse the flow fields in the DCR combustor, the present CFD technique with medium cell size is preferred to save computational time and cost.

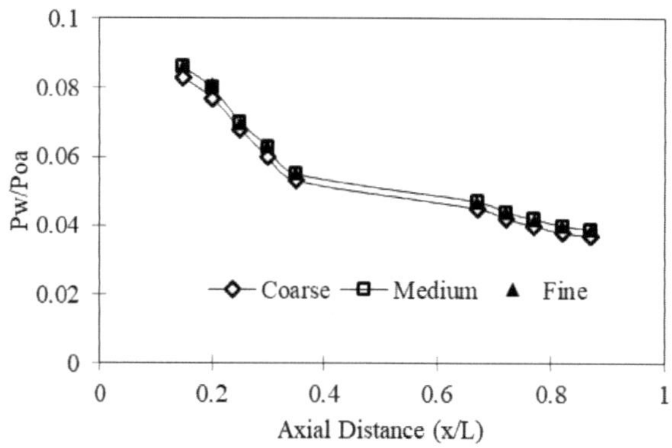

Figure 3. Wall pressure values along combustor for different grid sizes

4. Results and Discussions

4.1. Validation of the numerical model

The numerical simulation results of the actual DCR engine wall pressure values are validated with the authors' experimental data. The experimental conditions and the computational details are reported in the preceding sections. Figure 4 illustrates the computed wall static pressure distributions with experimental data. The results indicated that the numerical results complied with the experiments' outcome (Figure 4). The small discrepancies observed in the results may be due to the model limitations, axisymmetric assumptions, and machining uncertainty.

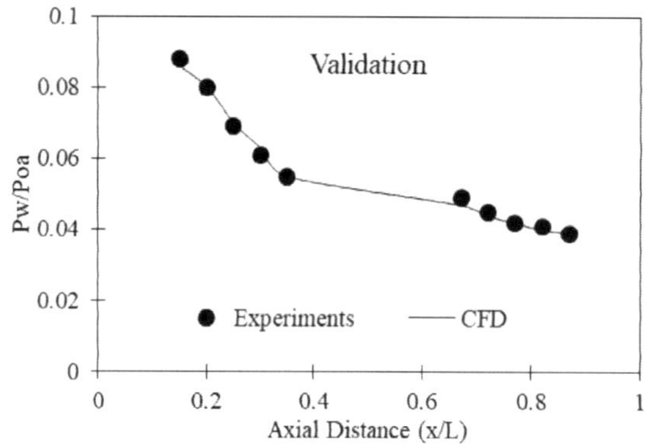

Figure 4. Comparison of numerical wall static pressures with the experimental results

4.2. Wall static pressure distribution

The coaxial jets enter the supersonic combustor at different operating conditions, which drag over the combustor's internal surface. The static pressure distribution on the combustor wall indicates the intensity of drag generated by the flow field. The static-pressure distribution on the inner wall of the DCR engine with and without injection is shown in Figure 5.

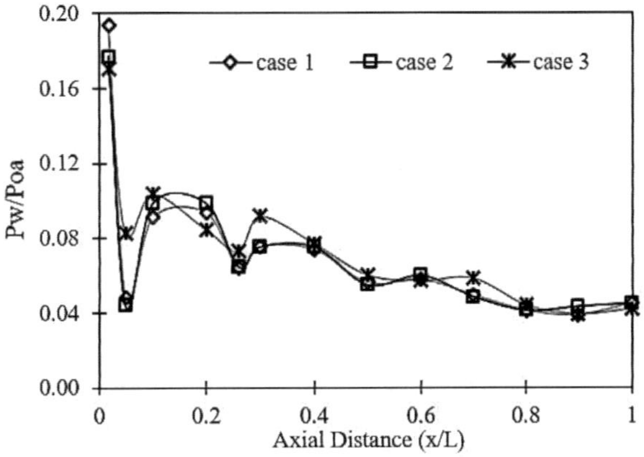

Figure 5: Wall pressures distribution in the supersonic combustor for various injection and no-injection cases

Here, the ratio of wall static pressure (Pw) of the supersonic combustor to the annular stagnation pressure (Poa) at different axial locations of the DCR engine is represented. The Mach number contours for cases 1, 2, and 3 are presented in Figure 6. The Mach number contours from the numerical simulations are taken for 2D planes in the injection location in the combustor's flow directions.

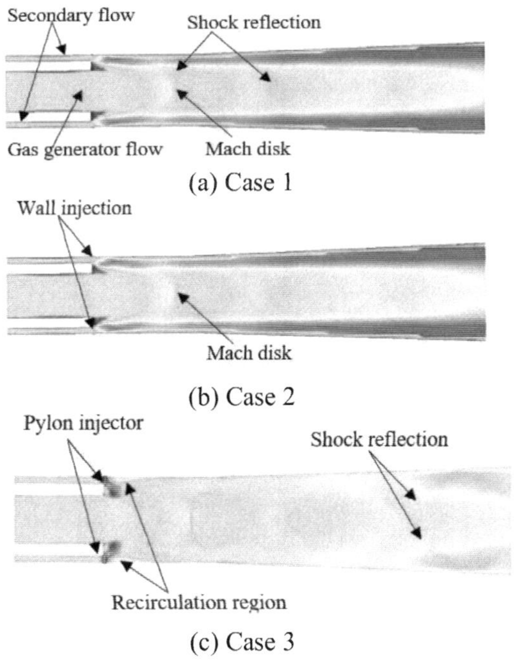

Figure 6. Mach number contour for various injection cases of the DCR engine

The plot shows peak pressure is noted for all the injection and no injection cases at the supersonic combustor inlet. This is due to the oblique shock waves generated from the isolator exit, which interacts with the wall boundary layer creates a compressive zone. For no-injection and wall injection, static pressure profiles follow a similar pattern with a marginal variation in the pressure values compared to the pylon injection. The static pressure profile's wavy nature is due to the reflected shocks from the combustor wall and the shock to shock interactions. The Mach number of contour plots show that the shock interactions vanish downstream of the flow for the no-injection and wall injection cases. In contrast, for pylon injection, a shock to shock interactions are seen throughout the combustor with diminishing strength.

4.3. *Mixing of coaxial jets*

In Figure 7, planes created at various axial locations of the supersonic combustor envisage the coaxial jets' mixing mechanism with the fuel injection schemes. The streamline plots show that the active counter-rotating vortices are present at the gas generator nozzle rim to enhance mixing and flame holding. The vortex pattern is not altered with the flush wall injection, and the fuel is carried away along with the stream in the flow direction without spatial distribution. However, in the case of pylon injection, the fuel is injected directly to the nozzle rim creates an active vortex in the normal direction of the flow that increases the fuel distribution spatially and enhances the mixing along the flow direction.

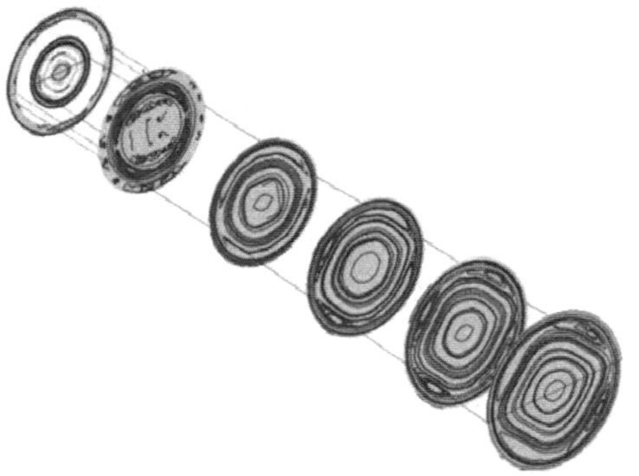

(a) **(b)**

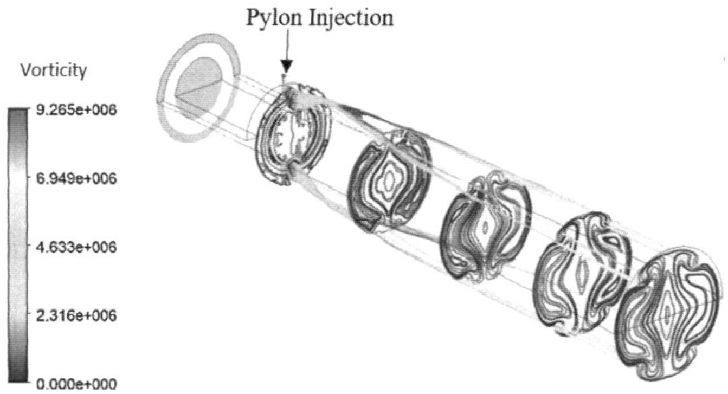

(c)

Figure 7 Vortex formation at the various axial locations of combustor along with the fuel streamlines (a) case 1, (b) case 2, (c) Case 3

4.4. *Total pressure loss*

The Total pressure loss (TPL) of the supersonic combustor for various injection cases are calculated as,

$$TPL = 1 - \frac{P_{oe}}{P_{oi}} \tag{6}$$

$$P_{oi} = \frac{P_{og} \times A_g + P_{oa} \times A_a}{A_g + A_a} \tag{7}$$

The difference in area-averaged stagnation pressures at the inlet (P_{0i}) to the outlet (P_{0e}) of the supersonic combustor to the inlet stagnation pressure is termed as stagnation pressure loss. In the equations, A- denotes the areas, and the subscripts 'g' and 'a' refers to the gas generator and annular flow conditions, respectively. The stagnation pressure loss for injection and no injection cases are depicted in Figure 8. The total pressure loss for the actual DCR engine across the supersonic combustor is 6.88%. The three-dimensional shock waves emanated from the gas generator nozzle rim cause shock and shock to boundary layer interactions resulting in total pressure loss across the combustor. Figure 7 shows that in both the wall injection and no-injection cases, the coaxial jets' interactions are significantly less compared to the pylon injection. In pylon injection, active vortices in the flow's radial directions generate an additional total pressure loss compared with the other cases.

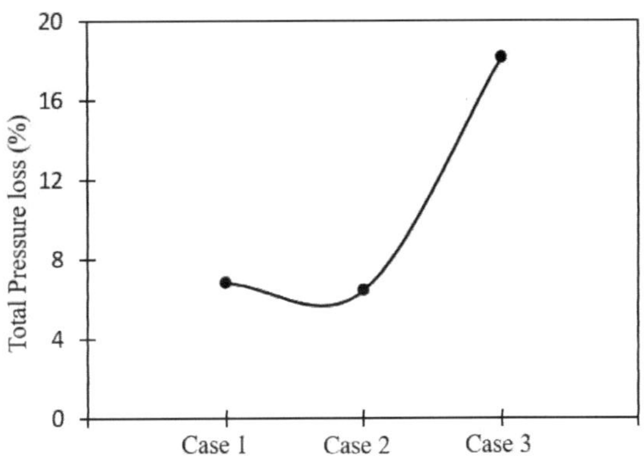

Figure 8. Total pressure loss for with and without injection cases of DCR engine

5. Conclusions

The non-reacting flow characteristics of coaxial jets with flush wall injection and pylon injection are numerically investigated. A three-dimensional RANS equation with the SST k-ω turbulence model is used for the flow analysis. The numerical results of the wall pressures are validated with the experimental data. The major conclusions of the study are:

- The supersonic combustor's wall static pressure profile reveals that the shock to boundary layer interaction at the combustor's inlet shows peak pressure values for all the cases. The intensity of the shock interactions with the centre flow has vanished downstream of the gas generator nozzle in the flush wall and no-injection. Simultaneously, the pylon injection indicates the shock interactions with reducing intensity in the combustor.
- The mixing of the coaxial jets is enhanced by oblique shock waves formed from the gas generator nozzle rim for no injection and flush wall injection schemes. However, the fuel injection from the wall is not spatially distributed throughout the combustor. Pylon injection provides an enhancement in mixing coaxial jets with the fuel stream. The improvement in mixing is evident from the series of shock waves formed from the injection location towards the combustor's downstream.

- Pylon injection induces more stagnation pressure loss than the no injection and wall injection cases of the supersonic combustor. The pressure loss is mainly due to the shock train's strength emerges from the gas generator nozzle and the injection strategies. Further investigations on reacting flow studies would reveal the injection schemes' performance in the DCR engine.

Nomenclature:

Ma	Mach number
Po	total pressure
P	Static pressure
ρ	density
u	velocity
ṁ	mass flow rate
k	turbulence kinetic energy
ω	specific dissipation rate
Gk	production of turbulent kinetic energy
Gω	generation of ω
Γk and Γω	effective diffusivity of k and ω
Yk and Yω	dissipation of k and ω
Dω	cross-diffusion terms

Subscripts

i	inlet
e	exit
g	gas generator
a	annular

Acknowledgement

The research work funding is supported under the Grant No.s: EMR/2016/004160 and ARDB/01/10411833.

References

[1] R.S. Fry, A Century of Ramjet Propulsion Technology Evolution, J. Propuls. Power. 20 (2004) 27–58. https://doi.org/10.2514/1.9178.
[2] F.S. Biling, Research on supersonic combustion, J. Propuls. Power. 9 (1993) 499–514. https://doi.org/10.2514/6.1992-1.
[3] P.J. Waltrup, M.E. White, F. Zarlingo, E.S. Gravlin, History of U.S. Navy Ramjet, Scramjet, and Mixed-Cycle Propulsion Development, J. Propuls. Power. 18 (2002) 14–27.
[4] R.S. Fry, The u.s. navy's contribution to airbreathing missile propulsion technology, AIAA 2011-2540, 2011.
[5] F.S. Billig, P.J. Waltrup, R.D. Stockbridge, Integral-rocket dual-combustion ramjets: A new propulsion concept, J. Spacecr. Rockets. 17 (1980) 416–424. https://doi.org/10.2514/3.57760.
[6] P.J. Waltrup, Liquid-fueled supersonic combustion ramjets - A research perspective, J. Propuls. Power. 3 (1987) 515–524. https://doi.org/10.2514/3.23019.
[7] M.B. Colket, L.J. Spadaccini, U. Technologies, E. Hartford, Scramjet Fuels Autoignition Study, J. Propuls. Power. 17 (2001) 315–323.

[8] L. Tian, L. Chen, Q. Chen, F. Li, X. Chang, Quasi-one-dimensional multimodes analysis for dual-mode scramjet, J. Propuls. Power. 30 (2014) 1559–1567. https://doi.org/10.2514/1.B35177.
[9] R.D. Stockbridge, Experimental investigation of shock wave/boundary-layer interactions in an annular duct, J. Propuls. Power. 5 (1989) 346–352. https://doi.org/10.2514/3.23159.
[10] P.J. Waltrup, F.S. Billig, Prediction of Precombustion Wall Pressure Distribution in Scramjet Engines., J. Spacecr. Rockets. 10 (1973) 620–622. https://doi.org/10.2514/3.27782.
[11] L. Zhang, H.-G. Sung, V. Yang, Flow and Flame Dynamics in a Hydrocarbon-fueled Dual-Combustion Ramjet Engine, AIAA 2020-0648. (2020). https://doi.org/10.2514/6.2020-0648.
[12] L. Zhang, J.Y. Choi, V. Yang, Supersonic Combustion and Flame Stabilization of Coflow Ethylene and Air with Splitter Plate, J. Propuls. Power. 31 (2015) 1242–1255. https://doi.org/10.2514/1.B35740.
[13] J.Y. Choi, V. Yang, High resolution numerical study on the coaxial supersonic turbulent flame structures, in: AIAA 2014-3744, 2014. https://doi.org/10.2514/6.2014-3745.
[14] J. Tan, J. Wu, Z. Wang, Experimental and numerical investigations on flow fields and performance of dual combustion ramjet, Proc. Inst. Mech. Eng. Part G J. Aerosp. Eng. 228 (2014) 920–929. https://doi.org/10.1177/0954410013482258.
[15] J. Kurian, K.. Habeeb, Cavities for Thermal and Momentum Mixing of Dual Supersonic Streams, AIAA 2002-4277. (2002).
[16] S. Jeyakumar, P. Balachandran, S. Indira, Experimental Investigations on Supersonic Stream Past Axisymmetric Cavities, J. Propuls. Power. 22 (2006) 1141–1144. https://doi.org/10.2514/1.21024.
[17] S. Jeyakumar, S.M. Assis, K. Jayaraman, Effect of Axisymmetric Aft Wall Angle Cavity in Supersonic Flow Field, Int. J. Turbo Jet Engines. 35 (2018) 29–34. https://doi.org/10.1515/tjj-2016-0027.
[18] S. Jeyakumar, K. Jayaraman, Effect of finite width cavity in axisymmetric supersonic flow field, Proc. Inst. Mech. Eng. Part G J. Aerosp. Eng. 232 (2018). https://doi.org/10.1177/0954410016674036.
[19] S.M. Assis, S. Jeyakumar, K. Jayaraman, The Effect of Transverse Injection Upstream of an Axisymmetric Aft Wall Angled Cavity in a Supersonic Flow Field, in: J. Phys. Conf. Ser., 2019. https://doi.org/10.1088/1742-6596/1276/1/012019.
[20] S. Jeyakumar, K.A. Yuvaraj, K. Jayaraman, F. Cardona, M.T.H. Sultan, Effect of cavity fore wall modifications in supersonic flow, in: IOP Conf. Ser. Mater. Sci. Eng., 2016. https://doi.org/10.1088/1757-899X/152/1/012002.
[21] S. Jeyakumar, S.M. Assis, K. Jayaraman, Experimental study on the characteristics of axisymmetric cavity actuated supersonic flow, Proc. Inst. Mech. Eng. Part G J. Aerosp. Eng. 231 (2017). https://doi.org/10.1177/0954410016667149.
[22] S. Jeyakumar, V. Venkateshwaran, N. Surjith, A.K. Raja, G.S. Samy, Experimental investigations on aft ramp cavities with fore wall modifications in scramjet combustors, 2017. https://doi.org/10.1007/978-81-322-2743-4_114.
[23] J.-R. Byun, C. Park, O.J. Kwon, J. Ahn, Experimental Study of Combustor–Isolator Interactions in a Dual-Combustion Ramjet, J. Propuls. Power. 31 (2015) 592–603. https://doi.org/10.2514/1.B35391.
[24] Zingg, D.W. and Godin, P., International Journal of Computational Fluid Dynamics A perspective on turbulence models for aerodynamic flows, International Journal of Computational Fluid Dynamics, 23 (2009) 4, 327–335.
[25] H. Wang, Z. Wang, M. Sun, N. Qin, Large eddy simulation of a hydrogen-fueled scramjet combustor with dual cavity, Acta Astronaut. 108 (2015) 119–128. https://doi.org/10.1016/j.actaastro.2014.12.008.
[26] S. Aravind, R. Kumar, Supersonic combustion of hydrogen using an improved strut injection scheme, Int. J. Hydrogen Energy. 44 (2019) 6257–6270. https://doi.org/10.1016/j.ijhydene.2019.01.064.

[27] X. Li, W. Liu, Y. Pan, L. Yang, B. An, J. Zhu, Characterisation of kerosene distribution around the ignition cavity in a scramjet combustor, Acta Astronaut. 134 (2017) 11–16. https://doi.org/10.1016/j.actaastro.2017.01.037.
[28] G. Choubey, K.M. Pandey, Effect of different strut + wall injection techniques on the performance of two-strut scramjet combustor, Int. J. Hydrogen Energy. 42 (2017) 13259–13275. https://doi.org/10.1016/j.ijhydene.2017.04.024.

Thermodynamic Assessment of a Gas Turbine System Combined with A Solid Oxide Fuel Cell Powered by Synthetic Gas from Municipal Solid Waste

Venkitesh K.S[1], Joseph Daniel[2], Manavalla Sreekanth[3]

[1]Research Scholar, School of Mechanical Engineering, VIT University, Chennai.

[2,3]Associate Professors (Sr.), School of Mechanical Engineering, VIT University, Chennai.

Corresponding author: venkiteshks@gmail.com

ABSTRACT

In this work, a detailed thermodynamic analysis is performed on a solid oxide fuel cell-gas turbine combined system. The fuel input to the fuel cell is the synthetic gas produced from the gasification of municipal solid waste. The efficiency and output power of the system are calculated and the system is analyzed for its performance by variation of parameters which includes compression ratio, fuel utilization factor and current density. The work is done based on the Municipal solid waste generated in Indian cities. Destruction of exergy is calculated from exergy analysis. System is modelled using flow sheeting program Cycle Tempo®. Components contributing to major exergy destruction are combustor and fuel cell. Energy efficiency of 44.5%, exergy efficiency of 39.2% and a power output of 170 kW is obtained in this system. This CHP system is very much acceptable for restaurants, shopping complex and other small-scale applications where power requirement is less.

Keywords

Hybrid system

SOFC

Municipal Solid waste (MSW)

Syngas

Exergy

1. Introduction

The idea of using a solid oxide fuel cell integrated with a gas turbine system is well recognized for many years. The fuel cell power generation system consisting of a gas turbine was initially analyzed by ide et al. [1]. The burning of fuel in the combustor contributes more to the irreversibility of the power generation system which leads to heavy efficiency losses. Research and development activities in the area of solid oxide fuel cell (SOFC) for large power generation system has made notable developments since early 1990s. The fuel cell can be made topping cycle or bottoming cycle in a hybrid system. A pressurized SOFC system which is integrated with a gas turbine bottoming cycle was analyzed by Harvey and Richter [2-5] and various optimum cycle parameters were obtained. The calculations were done by Aspen Plus simulation software. A heat engine such as gas turbine or steam turbine is combined with a non-heat engine such as fuel cell to form a hybrid energy system. The gas turbine requires high inlet temperature which is satisfied by the solid oxide fuel cell. The combustion chamber can be added to the system if required. A hybrid energy system with direct thermal contact involves the same working fluid passing through both the fuel cell and the gas turbine bottom cycle [6-7]. Different working fluids pass through the fuel cell and the gas turbine in the case of an indirect thermal contact system and it involves heat exchangers for heat transfer.

Solid oxide fuel cells fall under the category of high temperature fuel cells which are capable of producing high quality energy and heat. Various possibilities of combining solid oxide fuel cell with different types of turbines and micro gas turbines for electricity generation have been studied by lot of researchers [8]. Based on the operating pressure of the direct thermal contact solid oxide fuel cell, they are divided in to two types; atmospheric SOFC and pressurized SOFC. The atmospheric variant works at atmospheric pressure as the name indicates and is placed downstream of the gas turbine where the exhaust from the turbine is fed to the fuel cell. The pressurized variant of SOFC operates under a particular pressure and is placed between compressor and the turbine [9]. This increases power output but with a challenge in the design and control of the system. The high working pressure necessitates the secure sealing of the casing. Hybrid system preferably use atmospheric Solid oxide fuel cell to overcome this problem. Most of researchers are working on hybrid system with SOFC under pressure and the atmospheric variant attracted less attention which is evident from recent studies.

The solid oxide fuel cell offers fuel flexibility and almost any type of fuel can be used as input. The literature contains a huge number of research works which uses natural gas as fuel for the solid oxide fuel cell. The analysis of hybrid systems which uses synthetic gas as fuel input to SOFC is very rarely found in the literatures. The main objective of this work is to present a hybrid SOFC gas turbine system which is powered by syngas derived from municipal solid waste. The system is analyzed thermodynamically for its performance and destruction of exergy at various components are determined. A hybrid system consisting of a solid oxide fuel cell and a gas turbine was analyzed thermodynamically by Haseli et al. [10]. It was found that the hybrid system has 27.8% higher efficiency than a traditional gas turbine power plant. Exergy analysis of a hybrid gas turbine SOFC system was performed by Amati et al. [11]. The fuel is partly processed in the reformer and steam required for the reforming reaction is supplied by heat recovery steam generator. Exergetic performance of the hybrid gas turbine system integrating a solid oxide fuel cell was performed by Dincer et al. [12]. The integration of solid oxide fuel cell to a gas turbine power plant of 10 MW capacity operating at an efficiency of 30% was proposed by Cheddie et al. [13]. A Thermoeconomic analysis of SOFC-Gas turbine hybrid system of 500kW capacity fed by liquid fuels was conducted by Santin et al. [14]. Methanol and kerosene were used as fuels with four different fuel processing strategies. It was concluded that methanol fueled hybrid system perform well both thermodynamically and economically.

An ambient pressure SOFC-gas turbine hybrid system powered by coal derived syngas was modelled and analyzed to determine an optimum integration framework for fuel cell, gas turbine and other components by Zhao et al. [15]. Various biofuels such as methanol, ethanol, methane, hydrogen and ammonia are tested with a hybrid system and the performance was analyzed exergetically by Patel et al. [16]. It was found that the fuel cell efficiency depends on utilization of fuel per pass but the system efficiency is dependent of the nature of fuel and its composition. The efficiency of system with methane fuel is found to be higher. A thermoeconomic analysis of a hybrid SOFC steam injected gas turbine was done by Mazzucco et al. [17]. The system was integrated with a gasification unit fed by woodchips. Various configurations of the hybrid systems were studied and energy efficiency of 53% and exergy efficiency of 43% were obtained for the optimized plant. Syngas derived from the gasification of heavy fuel oil was used as fuel in a hybrid SOFC- gas turbine-steam turbine system and it was analyzed thermoeconomically and cogeneration opportunities were studied by Reyhani et al. [18]. Different configurations of hybrid systems were proposed and analyzed and a detailed thermodynamic model was simulated by Saisirirat et al. [19].
A comprehensive review of various feasible configurations of the hybrid gas turbine SOFC systems were presented by Buonomano et al. [20]. An optimal configuration of hybrid SOFC gas turbine system was modelled and analyzed thermoeconomically by Pirkandi et al. [21]. Hybrid systems which integrates solid oxide fuel cell with gas turbine have attracted interest of many researchers in the past years and extensive research studies have been conducted in this respect [22-27]. Energy, exergy and environmental analyses are performed on an externally fired gas turbine system which is integrated with a solid oxide fuel cell and a biomass gasifier unit was presented by Roy et al. [28]. The maximum efficiencies based on first and second laws were computed as 49.47% and 44.2% respectively. Exergy destruction of various components were determined and the gasifier unit contributes the maximum towards exergy destruction.
Combined cycles integrated with solid oxide fuel cell were compared thermodynamically by Biert et al. [29]. A combined powerplant operating with SOFC and gas turbine that uses ammonia as fuel was proposed by Ezzat et al. [30]. The system was analyzed based on first and second laws of thermodynamics and additional power is generated in a Rankine cycle utilizing the waste heat from the gas turbine exhaust. The performance of an advanced hybrid power generation system consisting of solid oxide fuel cell, gas turbine and organic Rankine cycle was analyzed by Roy et al. [31]. Energy efficiency of 49.29% and exergy efficiency of 44.57% were obtained for the system.

A review of the latest work shows that investigators have not focused much on hybrid system powered by syngas from solid waste. The system considered here in the present work is a hybrid system consisting of solid oxide fuel cell and gas turbine. The SOFC considered is a pressured one. The system is analyzed based on first and second laws of thermodynamics and efficiencies have been determined. The irreversibility associated with various system components which contributes to destruction of exergy are also determined and the assessment of performance has been done by varying parameters.

2. System description

The layout of the system proposed in this work is show in figure. 1. It consists of an air compressor, fuel compressor, air preheater, fuel preheater, solid oxide fuel cell, after burner and a gas turbine coupled with generator. The fuel used in this system is synthetic gas derived from gasification of municipal solid waste which has a heating value of 15000 kJ/kg and the air used in this system consists of 21% oxygen and 79% nitrogen. The preheated air stream enters the cathode and fuel enters the anode section of the fuel cell. respectively. Syngas undergoes internal reforming reaction at the anode producing pure hydrogen. This hydrogen reacts with oxygen from air and this process produces considerable electric power and efficiency of the hybrid system is increased.

Nomenclature

P	Pressure (bar)	**Subscripts/Superscripts**	
T	Temperature (^{0}C)	C	Compressor
h	Specific Enthalpy (kJ/kg)	Act	Actual
W	Work in kJ	s	isentropic
$\dot{m}$	Mass flow rate (kg/s)	SOFC	Solid oxide fuel cell
LHV	Lower heating value (kJ/kg)	AC	Alternating current
E	Reversible Cell voltage (volts)	DC	Direct current
$\bar{R}$	Universal gas constant (8.314 J/mol-K)	inv	Inverter
F	Faraday's constant (96485 C/mol)	AB	Afterburner
n_e	no: of electrons	GT	Gas turbine
N	Number of cells	gen	Generator
I	Current in ampere	i	Inlet
A	Area in m^2	o	Outlet
U_f	Fuel utilization factor	D	Destruction
Q	Heat transfer rate (J)	phy	Physical
$\dot{Ex}$	Exergy rate	che/ch	Chemical
y	Mole fraction	sys	System
e	Specific exergy	FC	Fuel cell
n	Molar flow rate	ac	Air compressor
s	Specific entropy (kJ/kg K)		
Greek Symbols			
η	Efficiency		
γ	Ratio of specific heats		
ϵ	Effectiveness		

The unreacted gases exiting from the fuel cells are further burned in the afterburner by reacting with each other. The output gases from the afterburner enters the gas turbine and produces work by expansion process. Two heat exchangers are used in this system to preheat air and fuel streams before entry in to the SOFC. The system is modelled in such a way that the fuel cell is forming the topping cycle and gas turbine unit forms the bottoming cycle. This is because, fuel cell bottoming cycle is more suitable for molten carbonate fuel cell applications.

3. Assumptions

- All gases within the hybrid system follow ideal gas behaviour.
- The potential and kinetic energy effects have been ignored.
- All the components in the hybrid system are adiabatic.
- Steady flow is assumed throughout the system.
- The temperature and pressure of exit gases from the anode and cathode sections of the fuel cell are same as that of the operating temperature and pressure of the fuel cell.
- The pressure and temperature distribution within the fuel cell have been neglected.

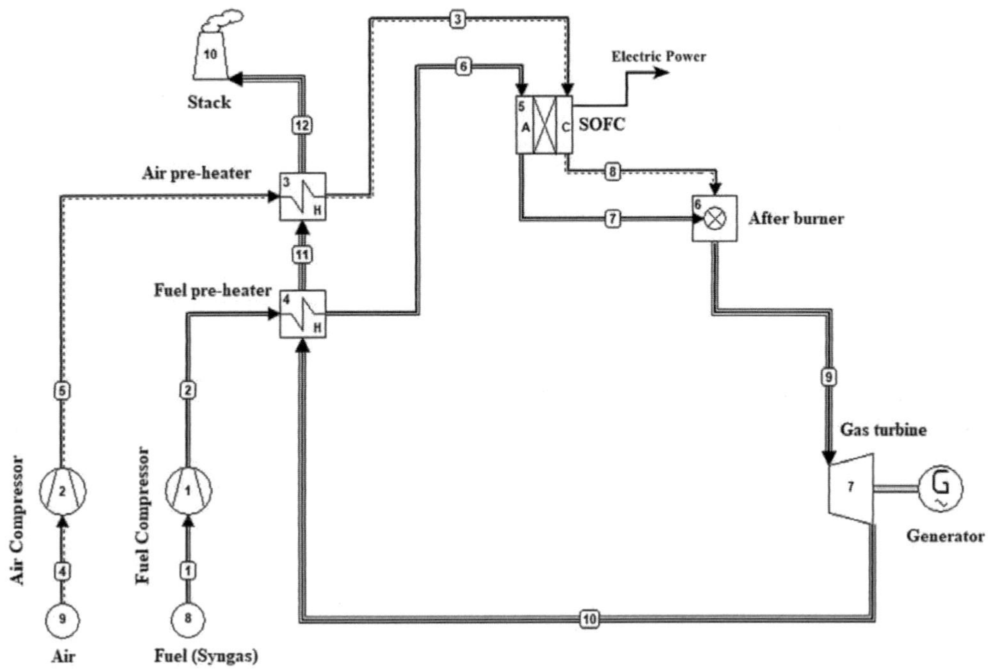

Figure. 1 Schematic layout of Hybrid-SOFC gas turbine system

4. Energy analysis

First law of thermodynamics is applied to each component in the cycle and system is modelled using relevant governing equations.

4.1. Air and fuel compressors

The governing equations for the compressors are given below. The input parameters are the compressor pressure ratio and isentropic efficiency with which the outlet temperature and work done by the compressor is calculated.

$$\frac{T_{5s}}{T_4} = \left(\frac{P_5}{P_4}\right)^{\frac{\gamma-1}{\gamma}} \quad (1)$$

$$\eta_C = \frac{W_{C-s}}{W_{C-act}} = \frac{T_{5s} - T_4}{T_5 - T_4} \quad (2)$$

$$\dot{W}_C = \dot{m}_1 \left(h_5 - h_4\right) \quad (3)$$

Fuel compressor model also have a similar set of equations.

4.2. Air and fuel heat exchangers

The hot exhaust gases from the gas turbine is utilized for preheating air and fuel streams. The governing equations are as follows

$$\varepsilon = \frac{T_6 - T_2}{T_{10} - T_2} \quad (4)$$

$$\dot{m}_2 \left(T_6 - T_2\right) = \dot{m}_{10} \left(T_{10} - T_{11}\right) \quad (5)$$

the air preheater can also be modelled in a similar way.

4.3. Solid oxide fuel cell

The fuel which is supplied to the SOFC undergoes direct internal reforming reaction producing hydrogen. The heat released by the electrochemical reactions are utilized for the endothermic reforming reactions. The SOFC is direct internal reforming type. The following reactions occur at the fuel cell

Steam reforming reaction

$$CH_4 + H_2O \rightarrow CO + 3H_2 \quad (6)$$

Water-Gas shifting reaction

$$CO + H_2O \rightarrow CO_2 + H_2 \quad (7)$$

The hydrogen thus produced reacts participates in the electrochemical reaction according to the following equation

$$H_2 + 0.5O_2 \rightarrow H_2O \quad (8)$$

The Nernst equation gives the reversible cell voltage of the fuel cell as follows

$$\text{E=E}^0 + \frac{\overline{R} \times T}{n_e \times F} \left(\frac{P_{H_2} \times \left(P_{O_2}\right)^{0.5}}{P_{H_2O}} \right) \quad (9)$$

Where E^0 is the ideal cell voltage under standard conditions (298 K and 1 bar)

The actual voltage of the SOFC is given by

$$V_{cell} = E - \Delta V_{loss} \quad (10)$$

The following equations gives the total current and the electric power developed by the fuel cell

$$I_{cell} = i \times A_{cell} \quad (11)$$

$$I_{Total} = N \times I_{cell} \quad (12)$$

$$W_{(DC-SOFC)} = V_{cell} \times I_{Total} \quad (13)$$

$$W_{(AC-SOFC)} = W_{(DC-SOFC)} \times \eta_{Inv-SOFC} \quad (14)$$

The mass and energy balance equation of the SOFC is given below

$$\dot{m}_3 + \dot{m}_6 = \dot{m}_7 + \dot{m}_8 = \dot{m}_3 + \dot{m}_6 \times U_f + \dot{m}_6 \left(1 - U_f\right) \quad (15)$$

$$\dot{m}_3 \times h_3 + \dot{m}_6 \times U_f \times LHV_{fuel} + \dot{m}_6 \times (1 - U_f) \times h_{fuel.Inlet} - W_{DC-SOFC} - \dot{m}_7 \times h_7 - \dot{m}_8 \times h_8 = 0 \quad (16)$$

Where $m_6 = m_{fuel}$

4.4. Afterburner

The electrochemical reaction inside the fuel cell uses only a part of the supplied air and fuel and the remaining unreacted fuel and gases further react each other in the afterburner. This reaction is exothermic and raises the temperature of the gases at the exit of the afterburner which is required by the gas turbine.

The mass and energy balance equation can be written as given below

$$\dot{m}_7 + \dot{m}_8 = \dot{m}_9 \quad (17)$$

$$\dot{m}_7\, h_7 + \dot{m}_8\, h_8 = \dot{m}_9\, h_9 + Q_{loss} \quad (18)$$

$$Q_{loss.AB} = \dot{m}_6 \times (1 - U_f) \times (1 - \eta_{AB}) \times LHV \quad (19)$$

4.5. Gas Turbine

The hot exit gases from the afterburner is expanded in the gas turbine producing mechanical work. A part of the energy is utilized for running the compressors and rest is used for electricity generation. From the knowledge of isentropic efficiency of the turbine, the exit temperature can be calculated.

$$\frac{T_{9s}}{T_{10}} = \left(\frac{P_9}{P_{10}}\right)^{\left(\frac{\gamma-1}{\gamma}\right)} \quad (20)$$

$$\eta_{GT} = \frac{W_{GT-Act}}{W_{GT-S}} \quad (21)$$

$$W_{GT} = \dot{m}_9 \times (h_9 - h_{10}) \quad (22)$$

4.6. Overall hybrid system

The energy efficiency of the entire hybrid system can be calculated as follows

$$\eta_{tot} = \frac{W_{net}}{m_{fuel} \times LHV} \quad (23)$$

$$W_{gen} = \eta_{gen} \times W_{GT} \quad (24)$$

5. Exergy equations

The exergy analysis of the system follows the second law of thermodynamics. This procedure evaluates various irreversibilities associated with the system components. Thermodynamic parameters such as pressure, temperature, enthalpy and entropy are determined at various points of the system and the conservation equations with respect to mass, energy and exergy are written for the evaluation of exergy loss and exergy destruction. Control volume concept is applied to each part of the system and exergy balance equation is written for that section

$$\sum\left(1 - \frac{T_0}{T}\right)Q_k + \sum \dot{Ex}_i = \sum \dot{Ex}_W + \sum \dot{Ex}_O + \dot{Ex}_D \quad (25)$$

IOP Conf. Series: Materials Science and Engineering 1128 (2021) 012036 doi:10.1088/1757-899X/1128/1/012036

Where Ex_i and Ex_O represents inlet and exit exergy rates of a control volume Ex_D represents the exergy destruction. The kinetic and potential exergy fluctuation rates are neglected in the analysis. The input and output exergy rates of the system are obtained by adding up physical and chemical exergies.

$$\dot{Ex} = \dot{Ex}_{phy} + \dot{Ex}_{che} \tag{26}$$

$$\dot{Ex}_{phy} = \dot{m}\left[\left(h - h_o\right) - T_0\left(s - s_0\right)\right] \tag{27}$$

$$\dot{Ex}_{ch} = \dot{n}\left[\sum\left(y_i \times e_i^{ch}\right) + \overline{R} \times T_0 \sum_i y_i \ln y_i\right] \tag{28}$$

The following equation is used to calculate the exergy efficiency

$$\eta_{ex.sys} = \frac{W_{net}}{\sum\left(Ex_{in}\right)_{sys}} \tag{29}$$

Parameters	Value
Ambient pressure	1 bar
Ambient Temperature	25^0C
Isentropic efficiency of Air Compressor	0.81
Isentropic efficiency of Fuel compressor	0.82
LHV of fuel (Syngas)	5 MJ/kg
Fuel utilization factor for SOFC	0.85
Cell Area	83.4 m^2
Current density	3000 A/m^2
Cell Resistance	5×10^{-5} ohm m^2
Afterburner efficiency	0.98
DC-AC inversion efficiency	0.89
Pressure loss in the SOFC	0.04
Pressure loss in the afterburner	0.05
SOFC stack temperature	850^0C
Gas turbine isentropic efficiency	0.81
Generator efficiency	0.95

Table. 1. Input parameters for the proposed Hybrid SOFC Gas turbine system

The equations for calculating destruction of exergy at various components of the system are given in table. 2. Table. 3 shows the studies from previous literatures in the similar domain

Component	Exergy destruction rate
Air compressor	$Ex_4 - Ex_5 + W_{ac}$
Fuel compressor	$Ex_1 - Ex_2 + W_{FC}$
Preheater	$Ex_5 - Ex_3 - (Ex_{12} - Ex_{11})$
SOFC	$Ex_3 + Ex_6 - Ex_7 - Ex_8$
Afterburner	$Ex_7 + Ex_8 - (Ex_9) - Ex_Q$
Gas turbine	$Ex_{10} - Ex_9 + W_{GT}$

Table. 2. Exergy destruction for various components of the plant

Sl. No.	Reference	Power plant capacity	Fuel Used	Fuel cell type	Fuel Cell parameters	Gas turbine parameters	Parameters considered	Maximum Exergy destruction
Unit →		MW			P (bar) & T(^{0}C)	P (bar) & T(^{0}C)		
1	Bellomare, 2013	34	MSW	SOFC	2.5/780	2.5 bar	Gasification temperature, SOFC temperature	Gasifier, combustor, SOFC, recuperator
2	Chen, 2015	10.4	Coal	SOFC	3.5/800	15/950	SOFC Temperature	Gasifier, CO_2 compressor, SOFC
3	Roy, 2017	36.1	Wood chips	SOFC	1/800	6 bar	Current density, cell pressure, cell temperature	Gasifier, vapour generator, SOFC.
4	Thattai, 2017	173	Coal + Biomass	SOFC	900^{0}C	8.78/919	Different retrofitting schemes considered	Gasifier
5	Nandwana, 2019	0.8-1	Coal + Cow Manure	SOFC	10/900	9.95/1200	Fuel composition	Gasifier

Table. 3. Studies from previous literature

6. Results and discussions

The system parameters such as compression ratio, fuel utilization factor and SOFC current density are varied to assess the performance of the plant.

6.1. Effect of compression ratio on energy and exergy efficiencies

Figure. 2 show the variation of efficiency values with compression ratio. The compression ratio was varied from 2 to 7 and the energy efficiency is found to decrease starting from a maximum value because, at higher compression ratios, the compressor requires more work input and thus the efficiency decreases on increasing compression ratios. Exergy efficiency is also following the similar trend.

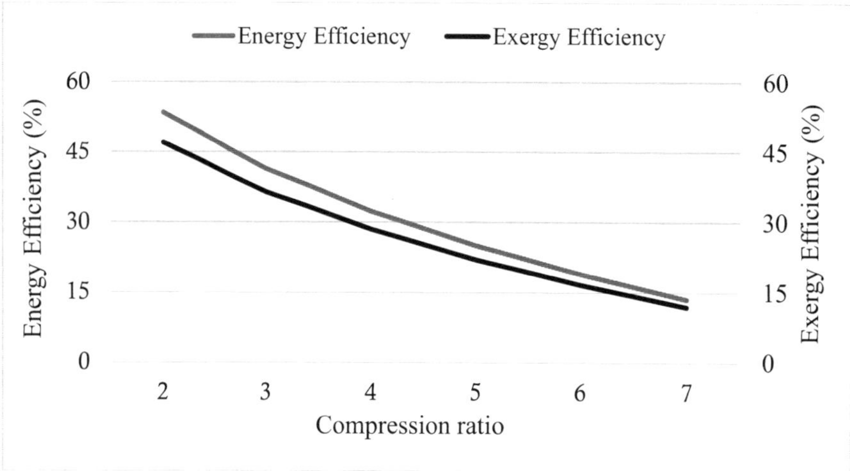

Figure. 2. Variation of energy & exergy efficiencies with compression ratio

6.2. Effect of fuel utilization factor

Figure. 3 and figure 4 shows the effect of fuel utilization factor on SOFC power and the efficiency. Fuel utilization factor represents the amount of fuel used up within fuel cell. It is varied from 0.7 to 0.86 and the SOFC power output is found to increase. The efficiency is also having a linear variation with reference to the fuel utilization factor. Exergy destruction rate have an increasing trend with compression ratio as shown in figure.5. Whenever the compression ratio is increased, the compressor requires more work input and some exergy loss occurs with the work transfer.

6.3. Exergy destruction of plant components

Figure 6 shows the destruction of exergy of individual components of the plant. Maximum destruction of exergy occurs at the fuel cell followed by the afterburner. This is because of the electrochemical reaction taking place inside the fuel cell which results in increased irreversibility. The combustion process at the afterburner also contributes to exergy destruction.

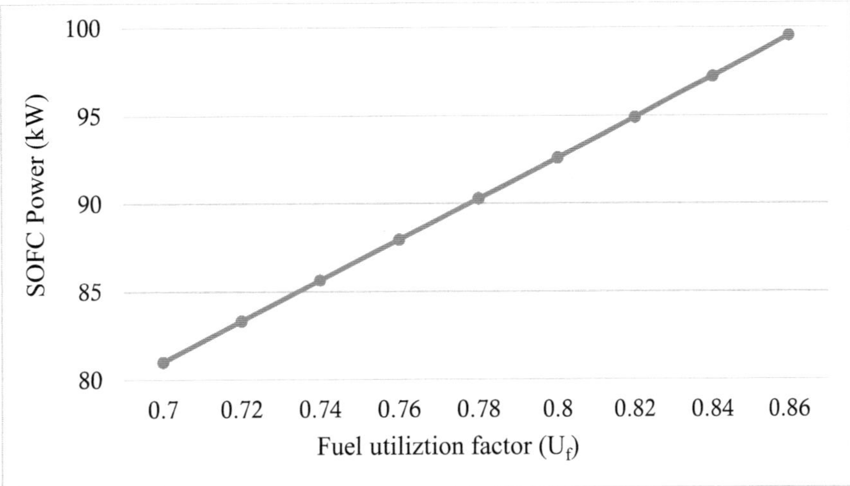

Figure. 3. Variation of SOFC power with fuel utilization factor

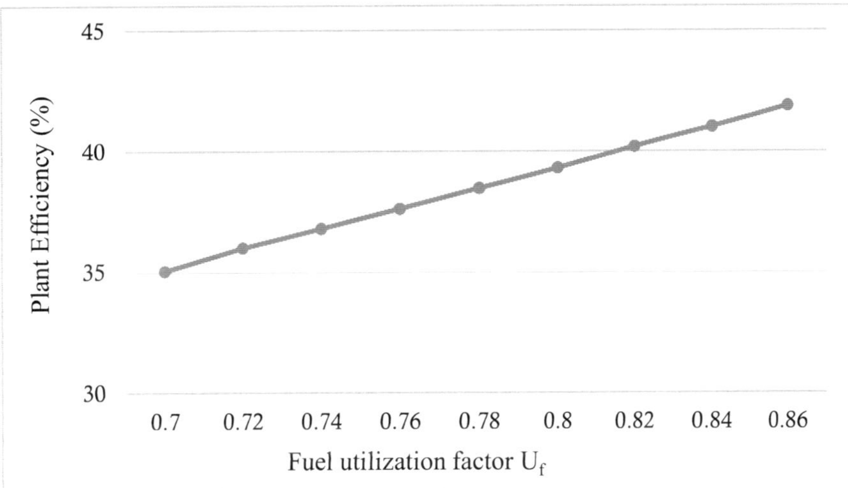

Figure. 4. Variation of efficiency with fuel utilization factor

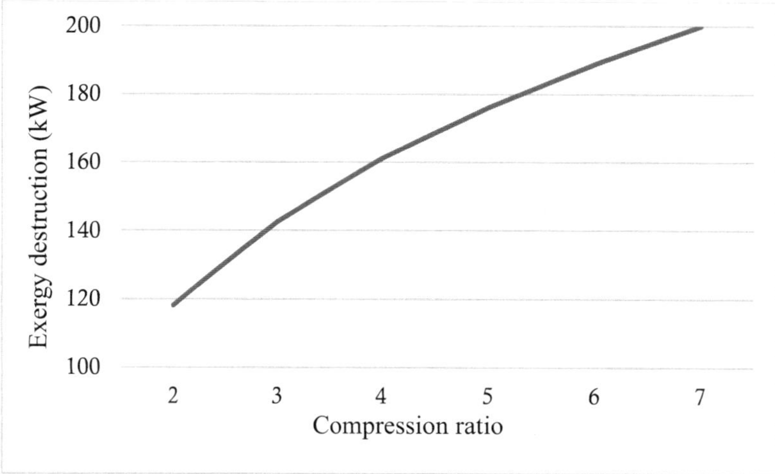

Figure. 5. Variation of exergy destruction with compression ratio

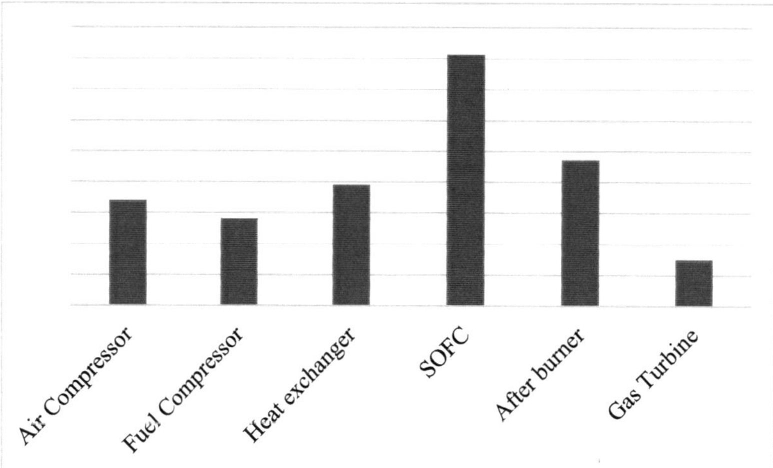

Figure. 6. Exergy destruction of various plant components

6.4. Effect of current density

The effect of current density on energy and exergy efficiencies are shown in figure 7 and figure 8 respectively. the cell pressure is 3 bar and the stack temperature is 1123K. The energy efficiency is having a decreasing trend with increasing current density. As the current density increases, the SOFC power output decreases and thus the efficiency. The exergy efficiency also follows a similar trend as that of energy efficiency. The exergy destruction increases with current density as shown in figure 9. This is due to the fact that, at higher current densities, there is a decrease in cell voltage and the requirement for the average number of cells is high which leads to low power output from the fuel cell.

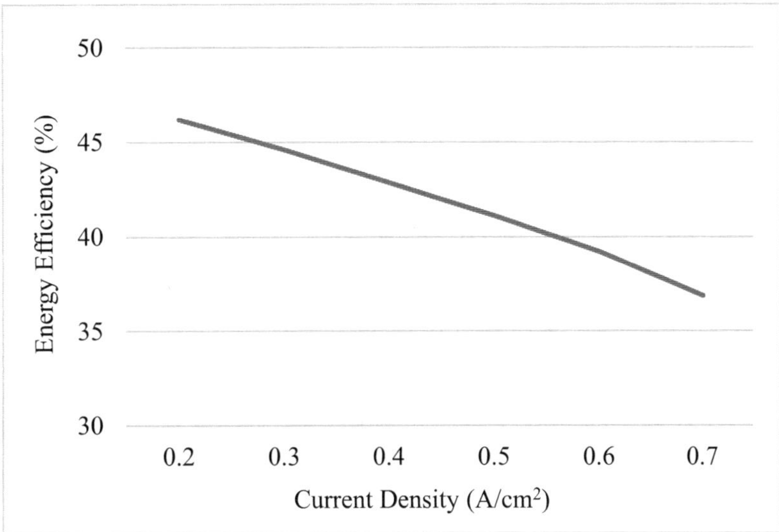

Figure. 7. Variation of energy efficiency with current density

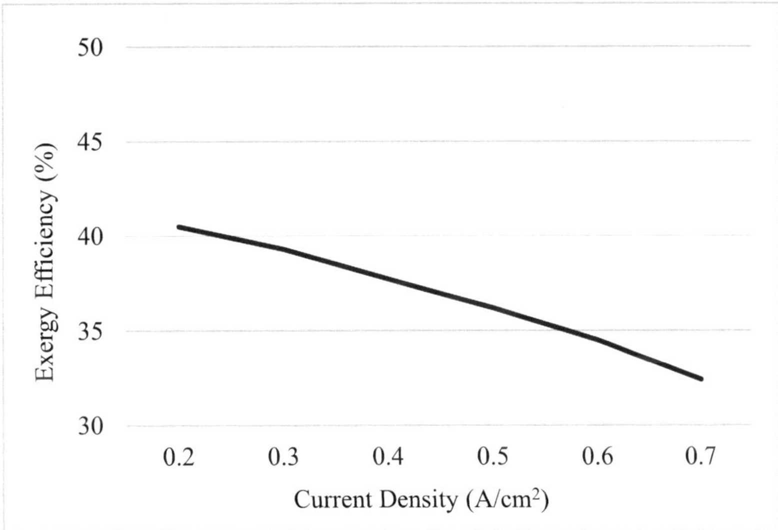

Figure. 8. Variation of exergy efficiency with current density

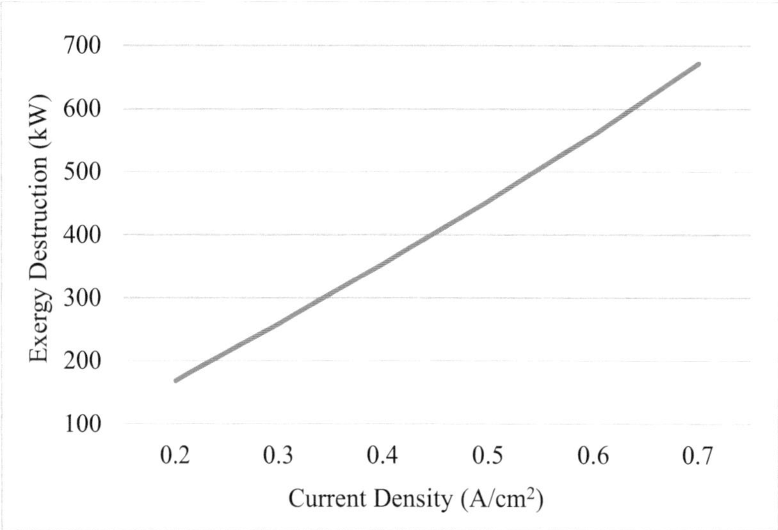

Figure. 9. Variation of exergy destruction with current density

7. Conclusion

A hybrid solid oxide fuel cell-gas turbine system was modeled and analyzed following the principles of first and second law of thermodynamics. The energy and exergy efficiencies of 44.5% and 39.2% respectively and a power output of 170kW was obtained. Exergy destruction at various components of the cycle was evaluated. Maximum exergy destruction occurs at the fuel cell due to the highest irreversibility resulting from the electrochemical and reforming reactions. It is followed by the afterburner due to the combustion reactions. The system can be considered as a CHP (combined heat and power) to satisfy the power requirements of domestic applications such as shopping centres, restaurants etc.

References:

[1]. Ide H, Yoshida T, Ueda H and Horiuchi N 1989 Natural gas reformed fuel cell power generation systems-a comparison of three system efficiencies *In Proceedings of the 24th Intersociety Energy Conversion Engineering Conference* **3** 1517-1522 Washington, DC.

[2]. Harvey S P, Richter H J 1994 A detailed study of a gas turbine cycle with an integrated internal reforming solid oxide fuel cell *In Proceedings of 29th Intersociety energy conversion engineering conference* **2** 961-973 Monterey, CA.

[3]. Harvey S P, Richter H J 1993 Improved gas turbine power plant efficiency by use of recycled exhaust gases and fuel cell technology. *In Proceedings of the ASME Winter Annual Meeting* **30** 199-207 New Orleans, LA.

[4]. Harvey S P, Richter H J 1994 Gas turbine cycles with solid oxide fuel cells. Part II: A detailed study of a gas turbine cycle with an integrated internal reforming solid oxide fuel cell. *Transactions of the ASME. J Energy Res Technol* **116** (4) 312-8.

[5]. Harvey S P, Richter H J 1994 Gas turbine cycles with solid oxide fuel cells. Part I: Improved gas turbine power plant efficiency by use of recycled exhaust gases and fuel cell technology. *J Energy Res Technol* **116** (4) 305-11. Transactions of the ASME.

[6]. Zhang X, Chan S H, Li G, Ho H K, Li J and Feng Z 2010. A review of integration strategies for solid oxide fuel cells. *Journal of Power Sources* **195**, 685-702

[7]. Brouwer J 2006 Hybrid Gas Turbine Fuel Cell Systems National Fuel Cell Research Center University of California.

[8]. Buonomano A, Calise F, Dentice M, Palombo A and Vicidomini M 2015 Hybrid solid oxide fuel cell-gas turbine systems for combined heat and power. *Journal of Applied Energy* **156** 32-85.

[9]. R Bove, S Ubertini 2008 Modeling solid oxide fuel cells: Methods, Procedures and Techniques **vol 1** (Springer)

[10]. Y Haseli, I Dincer and G F Naterer 2008 Thermodynamic modeling of a gas turbine cycle combined with a solid oxide fuel cell *International journal of hydrogen energy* **33** 5811-5822.

[11]. Valentina Amati, Enrico Sciubba and Claudia Toro 2008 Exergy analysis of a solid oxide fuel cell-gas turbine hybrid power plant *Proceedings of ASME International Mechanical Engineering Congress and Exposition* Boston, Massachusetts, USA.

[12]. Ibrahim Dincer, Marc A Rosen and Calin Zamfirescu 2009 Exergetic performance analysis of a gas turbine cycle integrated with solid oxide fuel cells *Journal of Energy Resources Technology* **131** 032001-9.

[13]. Denver F Cheddie 2010 Integration of a solid oxide fuel cell into a 10 MW gas turbine power plant *Energies* **3** 754-769.

[14]. Marco Santin, Alberto Traverso, Loredana Magistri and Aristide Massardo 2010 Thermoeconomic analysis of SOFC-GT hybrid systems fed by liquid fuels *Energy* **35** 1077-83.

[15]. Yingru Zhao, Jhuma Sadhukhan, Andrea Lanzini, Nigel Brandon and Nilay Shah 2011 Optimal integration strategies for a syngas fueled SOFC and gas turbine hybrid *Journal of Power Sources* **196** 9516-9527.

[16]. H C Patel, T Woudstra and P V Aravind 2012 Thermodynamic Analysis of Solid Oxide Fuel Cell Gas Turbine Systems Operating with Various Biofuels *Fuel Cells* **12** (6) 1115-1128.

[17]. Andrea Mazzucco, Masoud Rokni 2014 Thermo-economic analysis of a solid oxide fuel cell and steam injected gas turbine plant integrated with woodchips gasification *Energy* **76** (1) 114-129.

[18]. Hamed Akbarpour Reyhani, Mousa Meratizaman, Armin Ebrahimi, Omid Pourali and Majid Amidpour 2016 Thermodynamic and economic optimization of SOFC-GT and its cogeneration opportunities using generated syngas from heavy fuel oil gasification *Energy* **107** (15) 141-164.

[19]. Saisirirat P 2015 The solid oxide fuel cell (SOFC) and gas turbine (GT) hybrid system numerical model *Journal of Energy Procedia* **79** 845-850.

[20]. Buonomano A, Calise F, Dentice M, Palombo A and Vicidomini M 2015 Hybrid solid oxide fuel cell-gas turbine systems for combined heat and power. *Journal of Applied Energy* **156** 32-85.

[21]. Pirkandi Jamasb, Mahmoodi Mostafa and Ommian Mohammad 2017 An optimal configuration for a solid oxide fuel cell-gas turbine (SOFC-GT) hybrid system based on thermo-economic modelling. *J Cleaner Prod* **144** 375-86.

[22]. Rossi Iacopo, Traverso Alberto and Tucker David 2019 SOFC/gas turbine hybrid system: a simplified framework for dynamic simulation. *Appl Energy* **238** 1543-50.

[23]. Roshandel Ramin, Golzar Farzin and Astaneh Majid 2018 Technical, economic and environmental optimization of combined heat and power systems based on solid oxide fuel cell for a greenhouse case study *Energy Convers Manage* **164** 144-56.

[24]. Pirkandi Jamasb, Mahmoodi Mostafa and Ommian Mohammad 2017 Thermo-economic performance analysis of a gas turbine generator equipped with a pressurized and an atmospheric solid oxide fuel cell. *Energy Convers Manage* **136** 249-61.

[25]. Huang Yu, Turan Ali 2019 Fuel sensitivity and parametric optimization of SOFC-GT hybrid system operational characteristics *Thermal Sci Eng Progr.* 100407.

[26]. Eisavi Beneta, Chitsaz Ata, Hosseinpour Javad and Ranjbar Faramarz 2018 Thermo-environmental and economic comparison of three different arrangements of solid oxide fuel cell-gas turbine (SOFC-GT) hybrid systems *Energy Convers Manage* **168** 343-56.

[27]. Chan S H, Ho H K, Tian Y 2002 Modelling of simple hybrid solid oxide fuel cell and gas turbine power plant. *J Power Sources* **109** (1):111-20.

[28]. Dibyendu Roy Samiran Samanta and Sudip Ghosh 2019 Techno-economic and environmental analyses of a biomass-based system employing solid oxide fuel cell, externally fired gas turbine and organic Rankine cycle *Journal of Cleaner Production* **225** 36-57

[29]. L van Biert, T Woudstra, M Godjevac, K Visser and P.V. Aravind 2018 A thermodynamic comparison of solid oxide fuel cell-combined cycles *Journal of Power Sources* **397** 382-396.

[30]. M F Ezzat, I Dincer Energy and Exergy analyses of a novel ammonia combined power plant operating with gas turbine and solid oxide fuel cell systems *Energy* **194** 116750.

[31]. Dibyendu Roy, Samiran Samanta and Sudip Ghosh 2020 Performance assessment of a biomass fuelled advanced hybrid power generation *Renewable Energy* **162** 639-661.

Study of heat transfer enhancement through the non-circular ducts using three dimensional numerical investigations and their comparisons

Shambhu Kumar Mahato[1],Subhas Chandra Rana[2],Rabindra Nath Barman[3]

[1,2,3] Department of Mechanical Engineering, National Institute of Technology Durgapur, West Bengal, 713209,India.

E-mail: sambhu.16memahato@gmail.com

Abstract. The present study numerically surveyed fluid characteristics such as friction factor and heat transfer through the different geometry of twisted ducts. A three-dimensional analysis is performed for the Reynolds number of 100-50000 for the air that acts as a working fluid and a uniform wall temperature boundary case is studied numerically throughout the simulation of ducts. Numerical results of this present study are validated against the experimental results. Heat transfer improvements are made over the hexagonally twisted channels, straight square duct and clockwise-counterclockwise (CW-CCW) twisted square channels. Twist ratios of $7 \cdot 5$, $11 \cdot 5$ and $16 \cdot 5$ are considered for the current simulations. Laminar flow to turbulent flow regimes over the ducts are briefly presented and discussed. Results show enhancement in heat transfer through the CW-CCW , hexagonal and straight square ducts(SSD) for the laminar as well as for turbulent regimes are figured out and perfectly described.It is seen that in laminar regime, twisted CW-CCW square duct performed well than twisted square and hexagonal duct. A comparison of friction coefficients are also established for both the channels. Less friction factor is observed for twisted CW-CCW duct compared to other duct.

Keywords: Twisted CW-CCW square duct, Twist ratio, Pitch, Straight square duct, Hexagonal duct, Periodic flow

1. Introduction

Due to the industrial importance of the compact heat exchanger, many of the research publications are available in this area.So the twisted channels employed in this study specially applicable for the designing of industrial heat exchanger device. Heat transfer enhancement techniques mainly find applications in the design of more compact heat exchangers. Various industries, in particular refrigeration, automotive and chemical industries, are using this technology. The use of twisted ducts, fins, twisted strips, and ribs increases the heat transfer rates. These additives create a vortex that mixing the fluid in the direction of cross-stream which increases the heat transfer rate. Twisted channels used the passive heat transfer augmentation technique which does not require an external power supply.

Flow through an elliptic tube were analyzed up to a large twist ratio (TR=21, 53,106) by using the finite difference method (FDM) in the laminar flow regime. They determined the axial and circumferential velocities, resistance coefficients and the streamline patterns[1]. Friction factor and Nusselt numbers of twisted square ducts were experimentally and numerically investigated and found that Nusselt number obtained through the twisted ducts are higher than the straight square duct for Reynolds number of 1000-100000 and Prandtl number of $0 \cdot 7$-2. It was also identified laminar to transition point along the stream direction of the duct by the researcher [2, 3].

The effect of V-shaped twisted strip insertion on heat transfer, friction factor and thermal performance factor properties in a circular tube for three twisted ratios (y = 2, 4.4 and 6) and three different groups of depth and width ratios. The obtained results show that the average number of Nusselt and average friction factor in the tube with V-type twisted tape (VTT) increases with lower torsional ratios (y) and width ratios (WR) and increased depth ratios (DR)[4, 5].

Flow and heat transfer characteristics in a twisted elliptic tube (TET)were defined and the effects of the aspect ratio and the number of rotations in the TET were analyzed comparatively using three-dimensional (3-D) numerical simulation[6]. Experimental investigation of the fluid properties such as friction factor and heat transfer of square duct equipped with twisted tapes with different twisting conditions were analysed properly. It is found that at a minimum twist ratio, higher heat transfer enhancement occurred [7].

The influence of spiral screw tape coupled with ribbed tubulator on thermodynamic properties inside a circular tube for Reynolds number 6000 to 20000 and rib pitch ratio 1 to 3 were reported experimentally. It is revealed that by reducing the rib pitch ratio the heat transfer and friction factor increase. They also presented the correlations for the nozzle number and friction factor [8].

The characteristics of heat transfer and friction factor of the twisted oval tube were numerically studied by a realized k-epsilon method with varying the aspect ratios and twisted pitch length. They revealed that the Nusselt number and friction factor increases with the increasing aspect ratio as well as pitch length [9]. Laminar flow of water through an air cooled vertical channel is investigated experimentally with Reynolds number 110-1500 and twist ratio of 1.62-5.49. Correlations were also developed for heat transfer coefficients by [10].They found that the effectiveness of heat transfer increases with the growth of twisted-tape pitch at slow liquid flows after going beyond the critical speed.

Heat transfer enhancement using nanofluids in equilateral triangular duct (Al_2O_3/water) and also the effect of nanoparticle shape and size and its volume fraction (1%,2%,3%) for the 100-500 Reynolds number range was thoroughly studied by [11] .They find that the platelet's nanoparticle form shows the biggest improvement in heat transfer. The average Nusselt number rises up to 25% at the maximum Reynolds number when using platelet nanoparticle shape compared to pure water. Heat transfer is also increases by increasing the fraction of the nanoparticle volume.

He et al.[12] simulated the characteristics of fluid flow and heat transfer of CuO-water nanofluids with volume concentration in the range of 1-4% using single and double phase models at Reynolds number range of 300-36,000 inside the tube.They found that two-phase model gives the closest results in reality compared to single-phase model and at higher concentration of nanofluid i.e at 4%,performance efficiency is more.

The heat transfer and friction factor characteristics of a twisted hexagonal model were compared with the twisted square duct. It was reported that the twisted hexagonal model enhanced more heat transfer than the twisted square duct for the certain Reynolds number due to an increase in the number of sides of the hexagonal duct [13]. An experiment on convection inside a novel alternating elliptical axis tube and heat transfer and flow resistance correlations are made for Reynolds number of 500-50000. It is depicted that heat transfer enhanced greatly with less flow resistance [14].

The characteristics of fluid flow and heat transfer of helically finned tube with the assumptions of steady, three dimensional and incompressible fluid flows of continuity, momentum, and energy.

They also classified the effect of geometry on the calculation of local pressure coefficient, vorticity of fluid and the Nusselt number of the helically finned tube stated by [15]. Characteristics of friction factor and heat transfer of fluid flow through the twisted square duct (TSD) where nano-fluid used as a working fluid. It was found that nano-fluids used in TSD are greater friction factor and Nusselt number compared to the straight square duct [16].

kurnia et al.[17] were numerically investigated the flow behavior of air and the related heat transfer in a helical tube with a twisted tape insert subjected to constant wall temperature. Based on conservation equations of mass, momentum, and energy, a three-dimensional computational model is created and validated against the proven empirical correlations for the Reynolds number range of 100-2000 with different twist ratios. Lower twist ratios have been found to increase heat more than higher twist ratios.

A numerical study of the characteristics of heat transfer and fluid flow through swirling channels of different cross sections by varying the Reynolds numbers from 50 to 2000 and comparing these results with the results of a flat pipe and obtained that a twisted pipe enhances heat transfer more than a flat pipe [18].

Augmentation of heat transfer in heat exchanger by using twisted tape in inclined and horizontal tube is investigated numerically by [19] using twist ratios 4 and 6 for Reynolds number of 1056 to 2002 in FLUENT software.It showed that twisted tape enhanced the coefficient of heat transfer more than the smooth tube.They also presented the correlations of Nusselt number and friction factor.

It is clarified from literature survey that the most of the researchers have used twist ratios in between 2 to 20 for the design purpose of the industrial compact heat exchanger [2, 3].Based on this survey, this present study uses the twist ratios value of 7.5, 11.5 and 16.5.There are several investigations have done to improve the coefficient of heat transfer and characteristics of friction factor by the non-circular ducts. But the results of the heat transfer coefficients and properties of friction factors by the construction of twisted CW-CCW square ducts are not yet made by any researchers and the comparison with twisted hexagonal ducts and straight square ducts is not available either. So the objectives of this research work is to (a) determine the properties of the friction factor and the heat transfer coefficient through the twisted hexagonal and CW-CCW square duct with twist ratios $16 \cdot 5$ and $11 \cdot 5$ and (b) compare the results of friction factor and Nusselt number of twisted CW-CCW square duct with twisted hexagonal and straight square duct.

2. Phraseology

General terms related to twisted pipe are expressed in Figures (2) and (1):

Pitch (P):For tube length, the distance between two points in a row, since the direction of the cross section is directly aligned. The cross section rotates by 360° along the distance of one pitch.

Twist ratio (Tr):Relationship between the pitch (p) and the hydraulic diameter (d_h = 4A / p) of the duct is called the twist ratio.

3. Geometrical configurations

The inside diameter used for the twisted square section is $18 \cdot 5$ mm and the sides used for the hexagonal twisted duct are $18 \cdot 5$ mm. The thickness (δ)used for both the models are 2 mm.Tube rotates spirally along the channel axis and the axis extends through the center of the section. Twist ratios used are16·5,11 · 5, and 7 · 5. The outer cylinder inner shell is a straight circle (d_i). The outer diameter (d_o) for both models are 31 mm and 37 mm depicted in Figure (3). The tube rotates spirally along the channel axis and the axis extends through the center of the section. The twist ratios used are $7 \cdot 5$, $11 \cdot 5$ and 16·5.

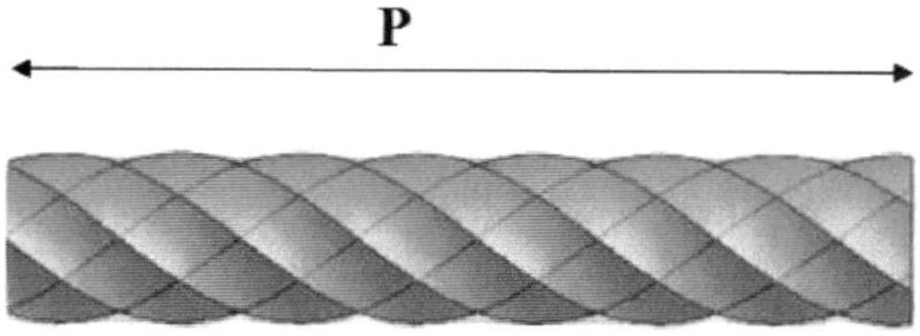

Figure 1. (Hex duct model)

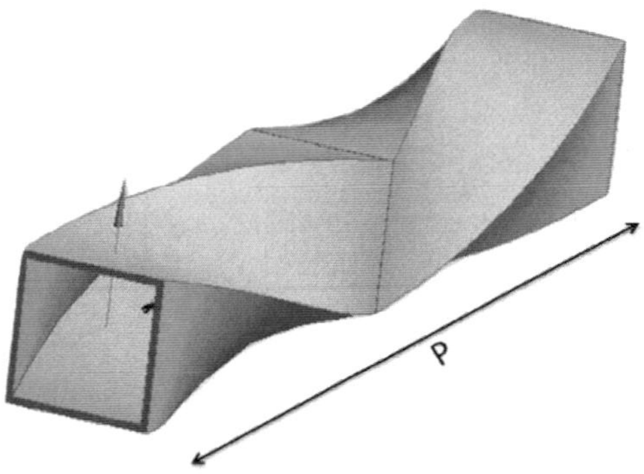

Figure 2. CW-CCW model

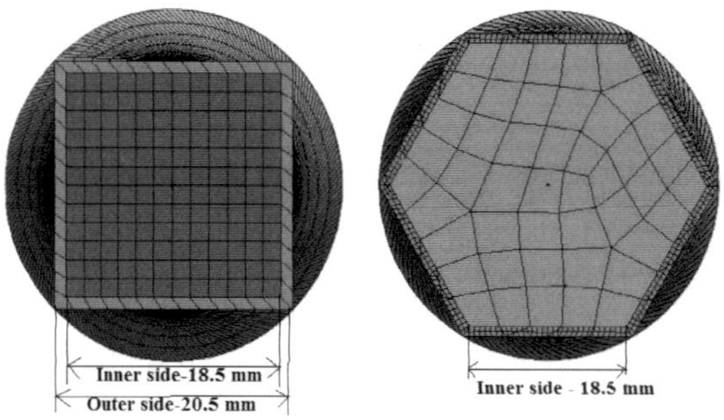

Figure 3. Twisting model (Tr-16.5)

4. Analytical calculations

In this present study, the concentric tubes were used in heat exchangers and cooled with reflux to provide bigger temperature gradient in the tube and in the housing surface. The flow is incompressible, steady, completely hydrostatic and thermally developed from time to time. Flow is deemed to be periodic when flow characteristics are repeated along the length (L) of the channel. The following are controlled equations for such flow are as

$$Continuity : \nabla \cdot \nu = 0 \tag{1}$$

$$Energy : \rho\nu \cdot \nabla\nu = -\nabla p + \nabla^2\nu \quad (2)$$

$$Momentum : \rho\nu \cdot \nabla T = \nabla^2 T \quad (3)$$

$\rho\nu \cdot \nabla$ T is the convective acceleration term that cannot be readily solved analytically due to its non-linear nature for the helically twisted tube. Thus, an approximate solution can be found by using the finite discretization method.

4.1. Boundary conditions

The twisted wall was maintained at a constant temperature at all times and the outer wall of the cylinder was considered thermal insulation. A no-slip boundary conditions were applied at the walls of the duct. A constant success rate ($\dot{m}$=0·35 kg/s) is used inside the outer cylinder. But with a constant flow rate at the entrance to each repeated model, the periodic boundary condition for the upper channel was considered. In general, the degree of dependence is 10^{-6}, 10^{-3}, and 10^{-5}, is considered for energy,continuity and velocity and the parameters that used for the simulation of ducts are represented in Table(3).

5. Grid independency test

The unstructured mesh was generally employed for both twisted hexagonal duct as well as for twisted CW-CCW square duct for the simulation due to their high complexity of the structures. Grid independence test was carried out for both the cases. Nusselt numbers are calculated at different unstructured grids at a constant Reynolds number and twist ratio of 600 and $16 \cdot 5$ for both structures. It is found that a grid that consists of 265745 elements of face size $1 \cdot 25{*}10^{-3}$ mm shows better results for the CW-CCW square duct which is seen from Table (1). Similarly, the same procedures were utilized for twisted hexagonal duct as shown in Table (2) and found that at grid elements of 86453 with a face size of $5 \cdot 6 * 10^{-3}$ mm are considered for the ducts shown in Figures (4) and (5).

Table 1. Constants used for twisted CW-CCW duct.

Element	Nusselt number	Face size*10^{-3}(mm)
140569	10.47	1.55
162432	10.510	1.48
175428	10.545	1.41
205797	10.539	1.35
236327	10.532	1.30
265745	10.521	1.25
290273	10.519	1.22

6. Methodology

Ansys' proficient software combined three-dimensional simulations for the inconsistent, stable stage flow across the model. As the pipe geometry is repeated over the length of the channel, it is stated that periodic flow is used. The numerical simulation may be restricted to a single repetition unit of total duct for periodic flows. Therefore, length of the model (L) of P/2 is sufficient for twisted hexagonal duct and CW-CCW duct. According to numerical results, the

Table 2. Constants used for hexagon duct.

Element	Nusselt number	Face size*10^{-3}(mm)
85475	6.71	6.05
83229	6.87	5.7
84843	6.93	5.5
86453	6.86	5.6
93842	6.85	5.4

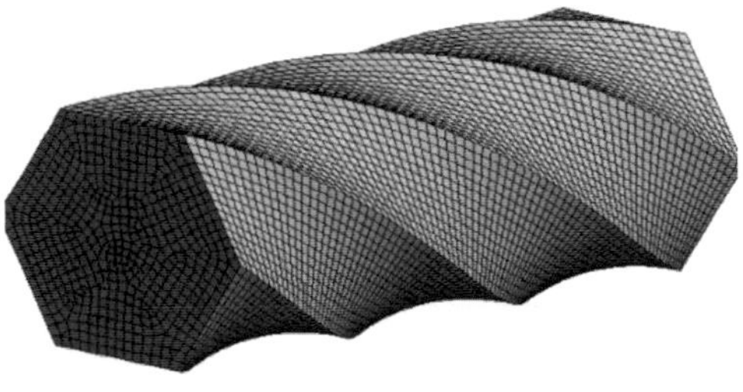

Figure 4. Mesh for Hexagonal duct (Tr-16.5)

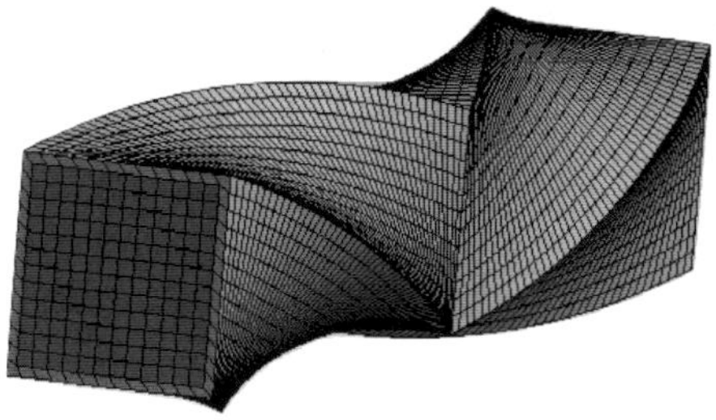

Figure 5. Mesh for CW-CCW duct (Tr-16.5)

range of laminar regime is considered within the Reynolds range of 100-3000. From the software processing module, we get the amount needed to calculate. The quantities used in the calculation:

$q^{'}$ = heat flux at wall
T_w = mean wall temperature
T_w = shear stress at wall
T_b = mean bulk temperature

The thermal strength of hot liquid (water) and cold liquid (air) is calculated taking into account the equations below (4) and (5)

$$Q_h = \dot{m}_h C_{ph}(T_{hi} - T_{ho}) \tag{4}$$

$$Q_c = \dot{m}_c C_{pc}(T_{ci} - T_{ho}) \tag{5}$$

Numerical models are isolated to prevent heat loss in the surrounding area. The average heat output($\bar{Q}$) can be calculated by Eq.(6)

$$\bar{Q} = (Q_h + Q_c) \div 2 = UA_p(\triangle T_{LMTD}) \quad (6)$$

With known values, it is possible to calculate the heat transfer area (AP) of the heat exchanger, $\triangle T_{LMTD}$(average temperature difference) and the total heat transfer coefficient (U). To calculate the total heat transfer coefficient (u) in the equation, the coefficient of heat transfer for sides of the flow, the hot and the cold sides expressed in Eq.(7)

$$1/(U_p) = 1/(hA_p)_c + 1/(hA_p)_h + R_w \quad (7)$$

Because the thickness of the THD (twisted hexagonal channel) is 2 mm. Therefore, the wall resistance of the channel = Thickness of the wall / thermal conductivity of steel = $0 \cdot 002/50 \cdot 2$ = $(3 \cdot 98)*(10^{-5})$ m^2k/w. Due to the small wall size and the high temperature flow of the square channel, the strength of the wall can be ignored. The coefficient of heat for the thermal transfer of the length of the twisted channel is calculated according to the formula:

$$h = q^{'}/(T_w - T_b) \quad (8)$$

Nusselt values can be calculated by Eq.(9)

$$Nu = (h * d_h)/k \quad (9)$$

In various places along the entire length of the twisted channel, the stress on the shear wall (t_w), heat flux ($q^{'}$), and volumetric temperature (T_b) are calculated using respective software. The average friction coefficient obtained due to the shear stress of the wall as

$$\bar{f} = 2 * (\tau_w/\rho V_m^2) \quad (10)$$

Table 3. Constants in Ansys fluent.

Models	Boundary conditions	Discretization
K/ϵ model - RNG	Condition - Mass flow specified	Turbulent Kinetic energy - Second order upwind
Viscous model - k - ϵ	Periodic type - Translational	Pressure/Velocity coupling - Second order upwind
Space - 3D	Outlet - Periodic	Energy - Second order upwind
Formulation - Implicit	Inlet - Periodic	Momentum - Second order upwind
Solver - Segregated	Wall - Constant temperature	Pressure - Second order upwind

7. Code Validation

The same geometric parameters such as inlet and outlet diameters, twist ratios, operating conditions, and the same boundary conditions of [2] are used to determine the standard for comparison for a twisted hexagonal duct, twisted clockwise-counter-clockwise square channel.Experimental verification of the heat transfer and friction coefficient characteristics of a twisted hexagonal duct and a clockwise-counterclockwise twisted square duct using air as a working fluid and numerical values obtained by [2] for the twist ratios of 16.5 shown in figures (6)and (7). Figure (6) shows that Nusselt number obtained in present study is validated against the Nusselt number of [2]. It can be observed that the results found from experimentally and

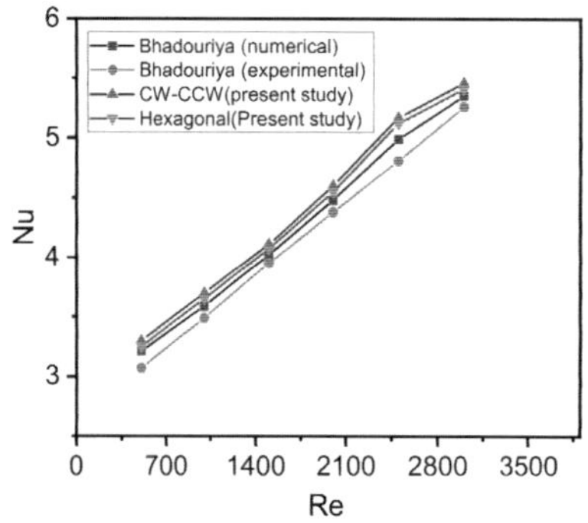

Figure 6. Nusselt number validation (Tr-16.5)

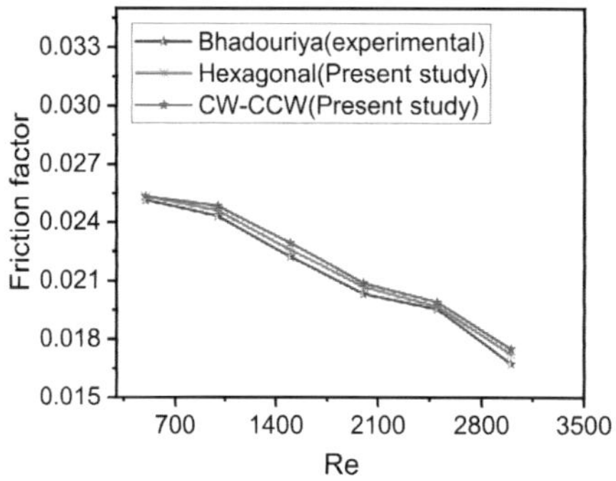

Figure 7. Friction factor validation (Tr-16.5)

numerically in [2] are closely agreed with the results of present study; average deviation which is observed for the Nusselt number is 1.33% for the twisted hexagonal duct and 1.52% for the twisted CW-CCW square duct compared to [2].Figure (7) depicts that average deviation for friction factor is found to be 1.13 % for twisted hexagonal duct and 1.42% for twisted CW-CCW square duct validated against the experimental data of [2].

8. Results and discussions

After checking the numerical methodology, modeling is performed for twisting factors $7 \cdot 5$, $11 \cdot .5$ and $16 \cdot 5$ for Re 100-50000. Nusselt values plotted against the Reynolds number shown in Figures (8),(9),(10),(11),(12),(13),(14),(15) for a twisting hex duct(THD) and a twisting square CW-CCW duct and the results are compared with the corresponding values of SSD(straight square duct) use air as the functioning fluid. The result shows that Nu values were higher in the twisted square duct CW-CCW using air as the working fluid than in the THD and SSD for a specific value of the twist ratios in the laminar flow regime. Comparison of results of twisted CW-CCW square duct, hexagonal duct with the experimental results of Bhadauriya et al.[2] at twist ratio 11.5 is shown in Table (4). It is seen that twisted CW-CCW duct has 55% higher Nusselt number than Bhadauriya et al.[2]in laminar flow regime i.e up to Re of 3000 at the same twist ratio. But in turbulent regime i.e Re value more than 3000, Nusselt number of Bhadauriya et al.[2] is high.

Numerical results for both laminar and turbulent systems are depicted for ratios between

Table 4. Comparison of results at Tr-11.5

Result type	Re	Tr	Twisted(CW-CCW)duct	Hex duct	Bhadauriya et al.
Nusselt number	1000	11.5	13.02	8.1	5.4
	2000	11.5	15.13	9.3	6.56
	3000	11.5	16.14	11.62	7.46
	10000	11.5	32.78	27.95	43.45
	30000	11.5	71.83	64.19	79.51
	50000	11.5	106.95	97.44	114.1

$16 \cdot 5$, $11 \cdot .5$ and $7 \cdot 5$. Figures(8) to (14) is a comparison between the Nusselt number and a friction factor with a twisted hexagonal duct and straight square duct. Figure (8) shows that the Nusselt number maximum for twisted squared duct CW-CCW compares to the twisted hexagonal duct and straight square duct for a twist ratio of $16 \cdot 5$ in the laminar flow system. But in the turbulent flow regime, it is seen from Figure (9) that the Nusselt values are more in straight square duct than the other CW-CCW and hexagonal twisted duct. Similarly, Figures (10) to (13) represents the Nusselt values for twist ratios of $11 \cdot .5$ and $7 \cdot 5$ in the laminar as well as in a turbulent flow regime. It is depicted that in the laminar regime, CW-CCW twisted duct dominant more than the other ducts but in turbulent regime, straight duct bears the high Nusselt number. Figures(14) and (15) indicated the comparison between the twisted hexagonal duct and the twisted CW-CCW square duct at different twist ratios such as $7 \cdot 5$, $11 \cdot .5$ and $16 \cdot 5$ in laminar and turbulent regime. It is found that in the laminar regime, the CW-CCW duct with a twist ratio of $7 \cdot 5$ has a higher Nusselt number. Likewise, in the turbulent regime, the straight square duct has higher Nusselt values than the twisted CW-CCW square duct and twisted hexagonal duct.

9. Average Coefficient of Friction

For different Reynolds numbers and twist ratios, the friction coefficient of the medium (f) of the fluid (air) flowing through the THD and the square twisted channel CW-CCW is calculated by the formula Eq.(10). The results of CW-CCW and THD are compared with the square twisted channel (SSD). The friction factor values in the CW-CCW twisted channel are very small as THD and SSD using air as the working fluid shown in Figure (15). It also shows that the friction factor first increases when the Reynolds number increases. When some Reynolds number is reached, the rate of decrease of the friction factor slows down. It is referred to the value of Reynolds number, which causes a abrupt changes in the slope of the coefficient of friction, as the critical Reynolds number. As shown in Figure (16), the critical Reynolds number was found between 2700-3000 for both structures, heralding the end of the laminar system and the beginning of the transition to the turbulent flow system.

10. Conclusion

The numerical results shown above concluded that the Nusselt number increases as the twist ratios decrease. In laminar mode, the twisted CW-CCW square channel has a higher Nusselt number than all the other ducts used for the design of heat exchangers such as twisted hexagonal channel, straight square channel and twisted square duct for the same twist ratios under the same operating and boundary conditions and constant pumping power . Percentage increase in Nusselt number of twisted CW-CCW square duct is nearly about 55% more than the results of Bhadouariya et al. [2] and 37.7% more than twisted hexagonal duct respectively. But in turbulent mode, it is observed that twisted CW-CCW square duct is not performed well compared to other duct used in this research. The friction factor can also be inferred that the

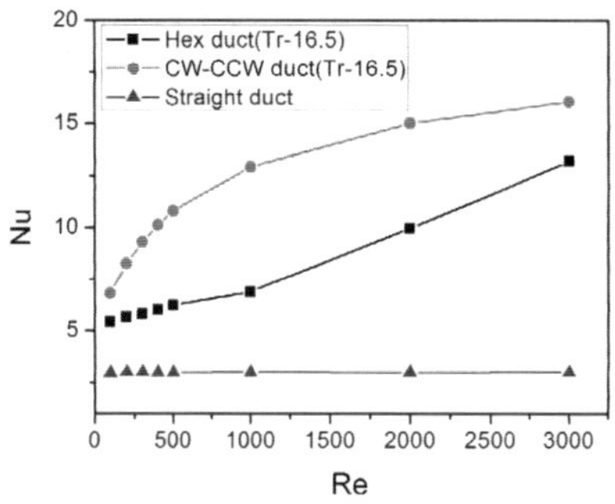

Figure 8. Laminar flow regime(Tr-16.5)

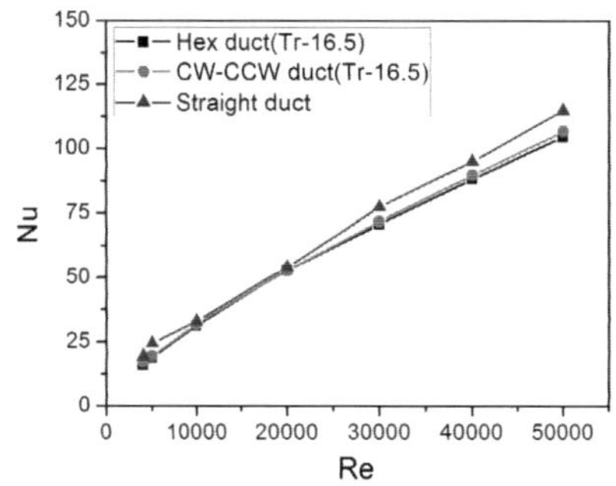

Figure 9. Turbulent flow regime(Tr-16.5)

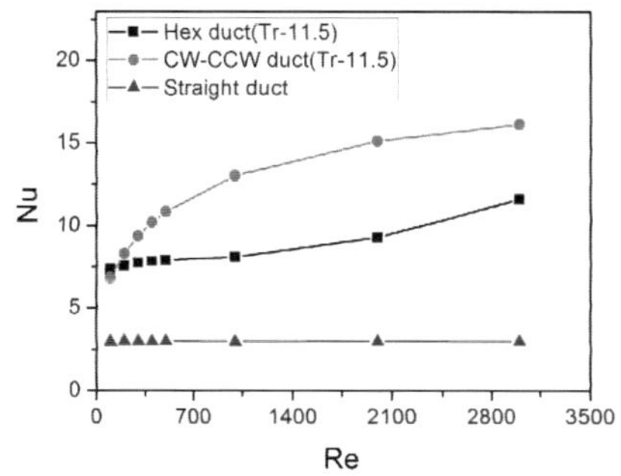

Figure 10. Laminar flow regime(Tr-11.5)

friction factor of the CW-CCW twisted square channel is smaller than the twisted hexagonal channel by the same values of the twist ratios.In this present work, the result shows that in the laminar regime, it is referred to use the twisted CW-CCW square duct than the other ducts due to the clockwise and counterclockwise design of the duct which generates higher secondary flow inside the duct resulting in high heat transfer. In future, the use of nanoparticle in the place of normal working fluid such as air may increase the heat transfer.

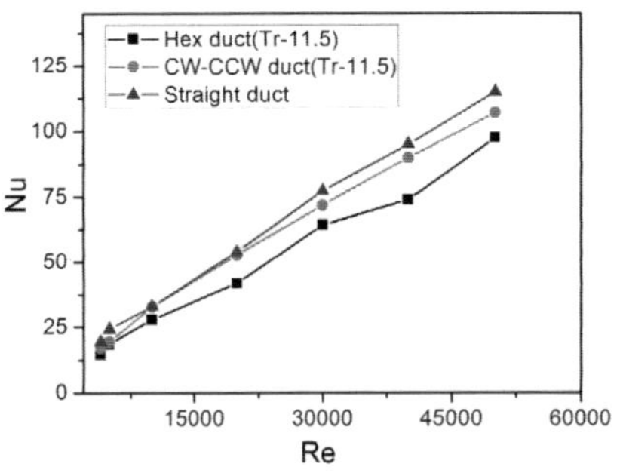

Figure 11. Turbulent flow regime(Tr-11.5)

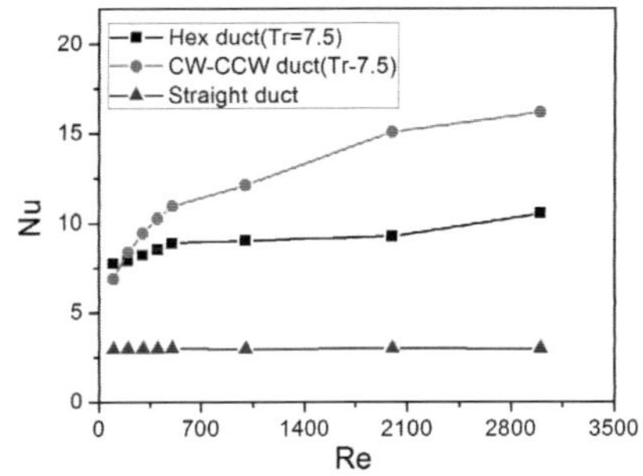

Figure 12. Laminar flow regime(Tr-7.5)

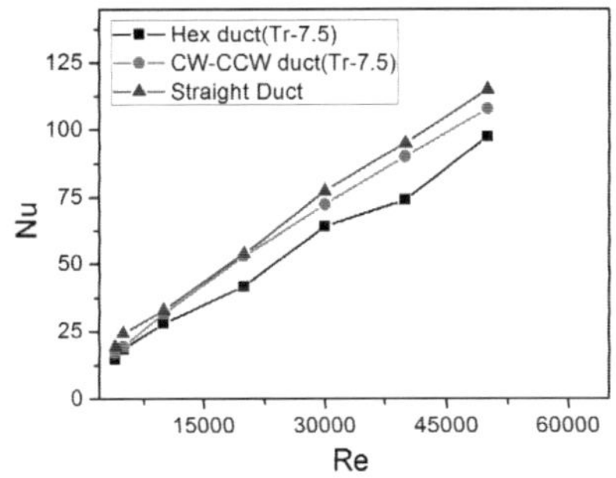

Figure 13. Turbulent flow regime(Tr-7.5)

11. Acknowledgement

We recognize the support of the Mechanical Engineering Department, NIT Durgapur for their valuable advise on the research work and also to provide computational support through out study.

12. Nomenclature

A : cross-sectional area of twisted duct, m^2
Tr : Duct twist ratio,dimensionless(P/d_h)
h : Coefficient of heat transfer, W/m^2 K

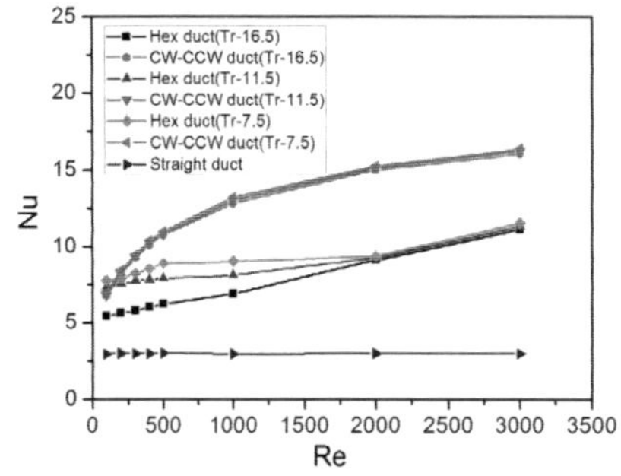

Figure 14. All ducts comparison at different twist ratios (Laminar)

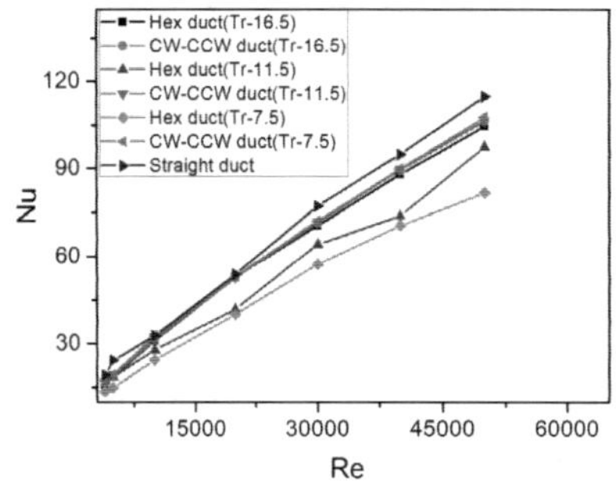

Figure 15. All ducts comparison at different twist ratios (Turbulent)

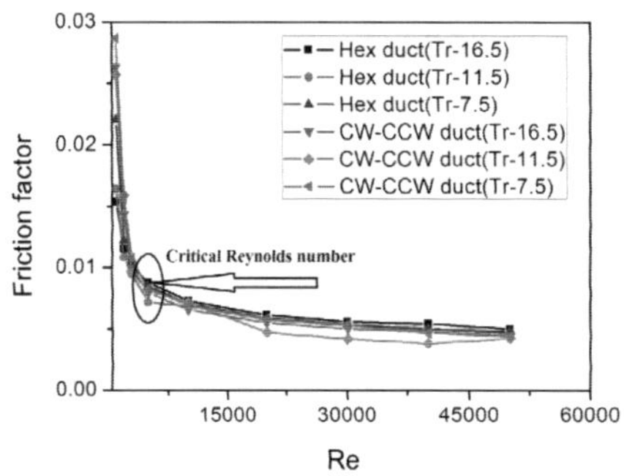

Figure 16. Friction factor comparison at several twist ratios

Q_c : heat duty of cold fluid, W
q' : heat flux of twisted wall, W/m^2
CW : clockwise twist
CCW : counter-clockwise twist
T : twist wall temperature, K
d_h : Hydraulic diameter, m
C_p : fluid specific heat, J/Kg K
Q_h : heat duty of hot fluid, W

P : pitch of the twisted channel, m
P_r : Prandtl number, dimensionless
$\dot{m}_c$: Mass flow rate of cold fluid, Kg/sec
T_{hi} : inlet temperature of hot fluid, K
T_{ho} : outlet temperature of hot fluid, K
T_{ci} : inlet temperature of cold fluid, K
Re : Reynolds number,(ρ v_md/μ)
T_{co} : outlet temperature of cold fluid, K
V : velocity of fluid flow, m/sec
d_i : internal diameter of cylinder, m
k : conductivity of fluid, W/m K
d_o : external diameter of cylinder, m
Nu : Nusselt number, dimensionless
SSD : straight square duct, m
L : periodic length, m
TSD : twisted square duct, m
$\bar{f}$: average friction factor
CW-CCW : clockwise-Counterclockwise
THD : twisted hexagonal duct
$\dot{m}_h$: mass flow rate of hot fluid, Kg/sec

Subscripts

m : mean
o : outlet
i : inlet
w : wall
b : fluid's bulk temperature
Greek symbols

ρ : fluid's density, kg/m^3
τ : shear stress of local wall, N/m^2
∇ : vector calculus operator
μ : dynamic viscosity, N-sec/m^2

References

[1] Chang K S, Choi J S and Kim J S 1988 *KSME Journal* **2** 44–51
[2] Bhadouriya R, Agrawal A and Prabhu S 2015 *International Journal of Heat and Mass Transfer* **82** 143–158
[3] Bhadouriya R, Agrawal A and Prabhu S 2015 *International Journal of Thermal Sciences* **94** 96–109
[4] Murugesan P, Mayilsamy K, Suresh S and Srinivasan P 2011 *International Communications in Heat and Mass Transfer* **38** 329–334
[5] Salman S D, Kadhum A A H, Takriff M S and Mohamad A B 2013 *The scientific world journal* **2013**
[6] Kim H R, Kim S, Kim M, Park S H, Min J K and Ha M Y 2016 *Journal of Mechanical Science and Technology* **30** 719–732
[7] Patil S and Vijaybabu P 2012 *Heat and Mass Transfer* **48** 1803–1811
[8] Nanan K, Pimsarn M, Thianpong C and Eiamsa-ard S 2014 *Journal of Mechanical Science and Technology* **28** 4771–4779
[9] Tan X h, Zhu D s, Zhou G y and Yang L 2013 *International Journal of Heat and Mass Transfer* **65** 244–253
[10] Klaczak A 2000 *Heat and Mass Transfer* **36** 195–199
[11] Ekiciler R, Çetinkaya M S A and Arslan K 2020 *Heat Transfer Research* **51**

[12] He W, Toghraie D, Lotfipour A, Pourfattah F, Karimipour A and Afrand M 2020 *International Communications in Heat and Mass Transfer* **110** 104440
[13] Mahato S K, Rana S C, Barman R N and Goswami S 2018 *CHEMICAL ENGINEERING* **71**
[14] Meng J A, Liang X G, Chen Z J and Li Z X 2005 *Experimental Thermal and Fluid Science* **29** 457–465
[15] Kim S, Kim M, Park Y G, Min J K and Ha M Y 2017 *Journal of Mechanical Science and Technology* **31** 3271–3284
[16] Mahato S K, Rana S C, Barman R N and Goswami S 2019 *Journal of Mechanical Science and Technology* **33** 5507–5514
[17] Kurnia J C, Chaedir B A and Sasmito A P 2020 *International Journal of Heat and Mass Transfer* **150** 119309
[18] Cheng J, Qian Z, Wang Q, Fei C and Huang W 2019 *Heat and Mass Transfer* **55** 823–844
[19] Mahdi A S, Al-Musawi S T M, Kadhim Z K, Hussain H M and Habeeb L J

3-E Analysis of a 3×500 MW coal fired thermal power plant using alternative coals species

Vishnu Rajeev[1], M Sreekanth[1,2]*, M Feroskhan[1], and Joseph Daniel[1]

[1]School of Mechanical Engineering, Vellore Institute of Technology Chennai,Vandalur-Kelambakkam Road, Chennai-600127, India
[2]Electric Vehicles Incubation, Testing and Research Centre, Vellore Institute of Technology Chennai,Vandalur-Kelambakkam Road, Chennai-600127, India

*Corresponding Author: manavalla.sreekanth@vit.ac.in

Abstract: This article reports a study conducted to understand the influence of different coal species on the performance of a thermal power plant. The aim is to understand the difficulties in replacing an existing coal with another available species. The methodology of 3-E (Energy, Exergy and Environment) analysis has been implemented on a 3×500 MW coal fired thermal power plant located in South India. The power plant is modelled using flow sheeting software called Cycle-Tempo. Key operating parameters like temperatures, mass flow rates were compared with the operating parameters and were found to be in good agreement. Maximum exergy loss for the design coal was found to occur in the combustor (36.5%), and the least exergy loss occurred in the de-aerator (0.13%). Optimization of the power plant was carried out with the aim of maximizing the efficiency, and the energy and exergy efficiencies obtained for the power plant were 34.35% and 33.64% respectively. Simulations were carried out with six different coal species, namely (i) Raniganj, (ii) Giridih, (iii) East Bokaro, (iv) West Bokaro, (v) Wardha valley, and (vi) Korba coals and various parameters that influence the performance were examined. An environment analysis was also carried out to study the emissions caused by the different species. East Bokaro coal turned out to be the best coal from efficiency viewpoint whereas Korba coal has the lowest amount of CO_2 produced per MW power generated.

Key Words: Indian Coal, Thermal Power Plant, Exergy Analysis, Environment Analysis, Modelling

1. Introduction

Thermal power plants cater to a large portion of world's energy requirements. Further, single large units of 2000 MW size is usually met by steam turbine based power plants. As the steam turbine power plants are limited by the laws of thermodynamics, the maximum thermal efficiency they can deliver using an optimized configuration is around 60% [1]. A basic steam turbine power plant consists of a boiler, turbine, condenser and a pump. Additional components like re-heaters and feed water heaters increase the thermal efficiency of the plant, mostly by increasing the heat added at the higher temperature [2]. A steady 2.3 % rise of power generation through coal is pointed out by Reddy et al. [3] and this is expected to trend till 2035. They have predicted a production of 15 trillion kWh energy from coal alone. In the back drop of diminishing coal reserves, a need to increase the efficiency of power plants arises. Although energy analysis is an inherent requirement during design, exergy analysis is what determines the usefulness of the available energy. Any modification in the operating parameters of the power plants, retro-fitting of new components etc. must be based on the exergy analysis rather than the energy analysis.

In two separate studies conducted [4, 5], boiler was identified as the device contributing to maximum exergy destruction. They also concluded that the efficiency increases with an increase in temperature and steam pressure, whereas the efficiency decreases with an increase in condenser pressure. Another study by Li and Liu [6] also concluded that the boiler is the source of the largest exergy destruction in their analysis of a 300 MW coal fired thermal power plant. Since boiler comprises of combustor and heat transfer surfaces, both these devices must be addressed separately in order to improve the performance. According to a study conducted by Elhelw et al. [7], turbines follow the boiler in their contribution to exergy loss. Exergy analysis can help in identifying the magnitude and the sources of such thermodynamic inefficiencies [8]. Li et al. [9] found that thermal exergy accounts for a large fraction of the total exergy losses. Erdem et al. [10] found that increasing the boiler output pressure leads to a rise in thermal efficiency. In their article, Kler et al. [11] stated that "Reducing the waste gas temperature can lead to a reduction in temperature difference in convective heat exchangers and an increase in the area of heating surfaces". Even when the temperature difference is very small at one end of the heat exchanger, there will still be appreciable irreversibility due to heat transfer over a finite temperature difference at other points in the heat exchanger [12]. Adibhatla and Kaushik [13]suggested a method of sliding pressure operation wherein the steam generator pressure can be varied, which could result in reduced exergy destruction. Apart from this, Vakilabadi et al. [14] suggested the regular addition of heat and water recovery systems to increase overall performance and exergy efficiency. Chauhan and Khanam [15] stated that combustion process contributes to the lion's share of exergy destruction. Gupta and Kaushik [16] concluded that condensers are the source of least exergy loss, but maximum energy loss. Oko and Njoku [17] found out that rising ambient temperature results in the drop of exergy efficiency and power output. Fu et al. [18] found that the decrease in isentropic efficiency of a component leads to an increase in exergy destruction. Periodic cleaning and maintenance of the condenser pipes can help in increasing the efficiency of both the condenser, as well as the plant [19]. Bidabadi et al. [20] suggested that using a counter flow arrangement in the boiler, not only does the boiler efficiency increase, but the amount of pollutants produced decreases.

India relies heavily on coal as an indigenous source of energy due to the vast reserves in various parts of the country [21]. However, most Indian coal species are of low quality due to high ash content [22]. Several studies on Indian coals have been studies under different operational and configurational conditions [22-24]. It is also important to have coal flexibility in operating a power plant to tackle problems of coal mine exhaustion and also issues related to logistics. Fuel flexibility involves issues of compatibility in composition, grindability, emission and economic factors. However, the thermal performance of the power plant mostly depends on the composition and operational parameters. So, a detailed investigation of thermal performance, especially from the viewpoint of 2nd law is required. To the best of our knowledge, such a study was not carried out on Indian coals. Hence, this work aims to study the performance of an existing power plant while using six different Indian coal species. Focus is on Energy, Exergy and Environment aspects of the performance. *Even though the overall power output and CO_2 emissions can be determined by the fuel composition and heating value, detailed component wise analysis needs to be carried out to identify and quantify the exergy destroyed. Exergy destruction can point out the process/components which need improved design for better utility. This cannot be carried out by energy analysis alone. Hence this study is necessary to evaluate the energy, exergy and environmental aspects of a thermal power plant and its ability to handle different coals.*

2. Methodology

All calculations related to energy and exergy are based on the standard concepts of 1st and 2nd laws of thermodynamics. The following notations and equations have been adopted from Cengel and Boles [2]. *These equations have been derived from first principles and hence are accurate.*

2.1. Energy balance in a steady flow system

$$\sum_{in} \dot{E} = \sum_{out} \dot{E} \tag{1}$$

For multiple streams,

$$\dot{Q} - \dot{W} = \Sigma_{out} \dot{m}\left(h + \frac{V^2}{2} + gz\right) - \Sigma_{in} \dot{m}\left(h + \frac{V^2}{2} + gz\right) \tag{2}$$

For a single stream,

$$\dot{Q} - \dot{W} = \dot{m}\left[h_2 - h_1 + \frac{V_2^2 - V_1^2}{2} + g(z_2 - z_1)\right] \tag{3}$$

2.2. Exergy balance in a steady flow system

$$\dot{X}_{in} - \dot{X}_{out} - \dot{X}_{destroyed} = 0 \tag{4}$$

$$\dot{X}_{heat} = \left(1 - \frac{T_0}{T}\right)\dot{Q} \tag{5}$$

$$\dot{X}_{work} = \dot{W}_{useful} \tag{6}$$

$$\dot{X}_{mass} = \dot{m}\psi \tag{7}$$

$$\psi = (h - h_0) - T_0(s - s_0) + \frac{V^2}{2} + gz \tag{8}$$

$$\dot{X}_{destroyed} = T_0\dot{S}_{gen} \tag{9}$$

However, the changes in kinetic and potential energies of flowing streams have been ignored in this study.

Reversible work, or the maximum possible work, in a combustion process that does not involve any work, is equal to the exergy destroyed, and can be calculated as follows:

$$\dot{W}_{rev} = \dot{X}_{destroyed} = \sum N_r\left(\bar{h}_f^0 + \bar{h} - \bar{h}^0 + T_0\bar{s}\right)_r - \sum N_p\left(\bar{h}_f^0 + \bar{h} - \bar{h}^0 + T_0\bar{s}\right)_p \tag{10}$$

$$\eta_I = \frac{W_{useful}}{Q_{input}} \tag{11}$$

$$\eta_{II} = \frac{W_{useful}}{W_{reversible}} = \frac{\eta_I}{\eta_{I reversible}} = \frac{X_{recovered}}{X_{supplied}} \tag{12}$$

The coal fired thermal power plant studied in this article consists of three turbines, a steam generator apparatus, five feed water heaters, a de-aerator and a condenser. Modelling and simulations have been carried out using Cycle-Tempo. The exergy analysis gives detailed insights on the amount of exergy lost in each apparatus. The inlet pressure values of the feed water heaters and the de-aerator were optimized in order to increase the overall efficiency of the plant. Coal samples from six different coal fields situated across India were used as fuel in the simulation, and the efficiencies were compared. Various parameters were studied and compared for each coal sample, and a detailed environmental analysis was carried out in order to come up with the best coal species.

3. Results and Discussion

Modelling of the 500 MW power plant was carried out in Cycle-Tempo. The fuel used was the design coal having a moderate carbon content of 32% and high ash content of 41%. This coal is similar to the Korba mine coal, India. Standard air was used as the source of air. The composition of fuel and air are mentioned in the Tables 1 and 2 given below. A pressure of 1 atm and a temperature of 30°C was defined as the environment conditions, which are required for carrying out the exergy analysis. The inlet pressure values of all the low-pressure feed water heaters and the de-aerator were optimized, to maximize the efficiency. *The deaerator is located between the low pressure and high pressure feed water heaters and it receives steam from the intermediate pressure turbine. It performs two functions viz. removing dissolved air from the feed water, which reduces the quality of water, and preheat the feed water heater. It is an important component of a thermal power plant using steam as the working fluid.*

Figure 1 shows the detailed schematic of the power plant and also the results, mostly in terms of pressure, temperature, enthalpy and mass flow rates at salient points. The following assumptions have been made:

i. All turbines operate at an isentropic efficiency of 95%.
ii. All pumps operate at an efficiency of 92%.
iii. Cooling water used in the condenser enters at 25°C and leaves at 30°C.
iv. Relative humidity of air is 50%.
v. The flue gas exits the stack at 350°C.
vi. The de-aerator is modelled as an open feed water heater, while the remaining feed water heaters are of non-contact type.
vii. Any water and steam leakages in the system are ignored.

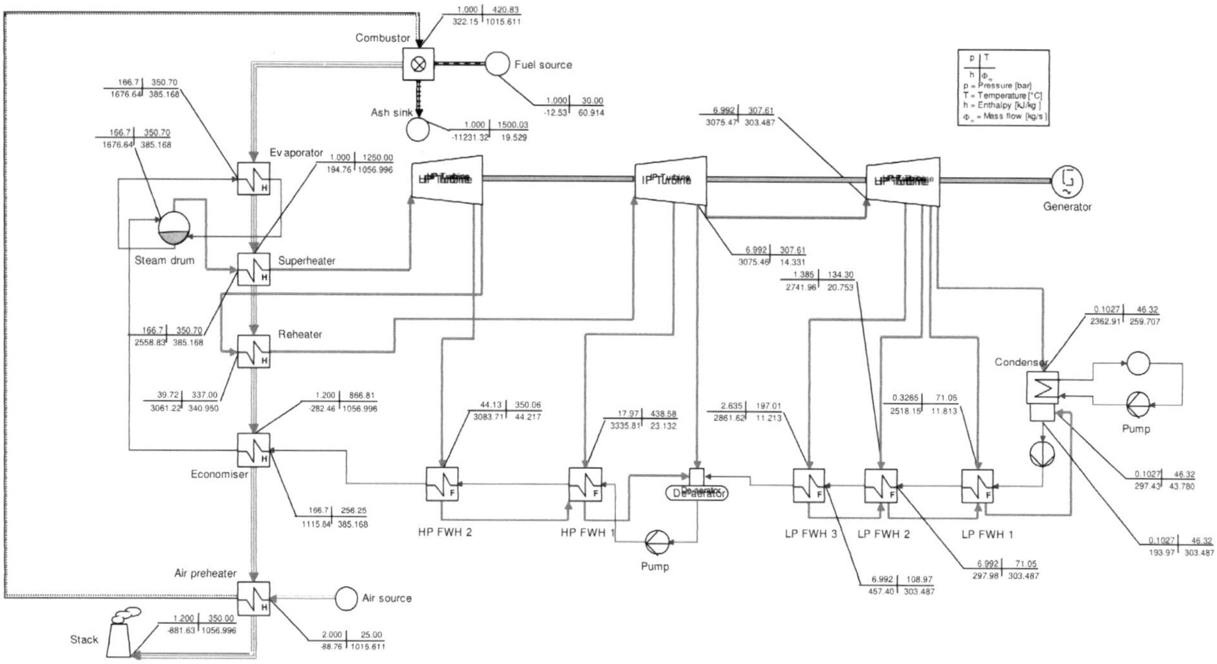

Figure 1. Model and Simulation result of the power plant

Table 1. Composition of the design coal used

Component	C	H_2	H_2O	N_2	O_2	S(s)	SiO_2
Mass %	32.05	3.3	14.0	0.83	8.54	0.28	41.0

Table 2. Composition of ambient air used

Component	Ar	CO_2	H_2O	N_2	O_2

Mole %	0.9 2	0. 3	1.0 1	77. 29	20. 75

3.1. Validation

In the comparative study that was carried out, the temperature and mass flow rates at 10 different salient points in the power plant scheme were compared and tabulated in Table 3 below. The power output was fixed at 500 MW. From the Table 3, it can be seen that except in the Low Pressure feed water heater inlet temperature, all other temperatures and mass flow rates are around an error of 5% from the actual design conditions. The discrepancies between the design and predicted parameters could be due to the differences in the choice of pinch and approach points, turbine and pump efficiencies and also due to the leakage of steam and water in the actual power plant, which have been ignored in the simulation. Thus we were convinced that the Cycle-Tempo model is accurate enough to carry out a parametric study. Further, various parameters like inlet air temperature, cooling water temperature gain, condenser pressure and power output were varied to monitor how these changes affected the energy and exergy efficiency. The results have been displayed in the form of graphs below. These figures are meant to check if the model predicts some established results or not. Figure 2 shows the influence of inlet air temperature on thermal efficiency. It shows an increasing trend because of the increase in the average temperature of heat addition to the working fluid. Figure 3 shows the influence of condenser cooling water temp rise on efficiency. Here, we can see a slight increase in efficiency because the amount of pump work has reduced due to the reduction of cooling water required. Figure 4 shows the influence of condenser pressure on efficiency and an established result of higher efficiency at lower pressures is evident. Figure 5 shows the variation of efficiency with power output. Here, there is no variation in efficiency. This is because efficiency depends only on operating parameters and not on the output power. The mass flow rate adjusts itself to the change in the output and hence, efficiency remains constant. However, in reality, higher mass flowrates result in larger pressure drops which will result in larger pump power consumption. *This emphasizes the importance of auxiliary power consumption and pressure drops, which need to be accurately estimated for precise analysis.* All the arguments presented above are valid for both energy and exergy efficiencies.

Table 3. Variation in temperatures and mass flow rates

Sl no	State point	Mass flow rate, m (kg/s)		Temperature, T (°C)		Error %	
		Actual	Cycle Tempo	Actual	Cycle Tempo	Mass flow rate	Temperature
1	Boiler outlet	404.72	385.16	537	537	4.83	0
2	Reheater inlet	361.45	340.95	337	337	5.67	0
3	LP turbine inlet	296.32	303.48	309.9	307.61	2.41	0.73
4	Condenser inlet	259.68	259.70	46.6	46.32	0.009	0.6
5	Condenser outlet	320.27	303.48	46.4	46.32	5.24	0.17
6	LP Heater 1 inlet	11.01	11.81	72.4	71.05	7.23	1.86
7	LP Heater 2 inlet	19.98	20.75	147.7	134.3	3.82	9.07
8	LP Heater 3	11.31	11.21	208.4	197.01	0.86	5.46

	inlet						
9	HP Heater 1 inlet	24.28	23.13	444.5	438.58	4.73	1.33
10	HP Heater 2 inlet	41.56	44.21	334.8	350.06	6.37	4.55

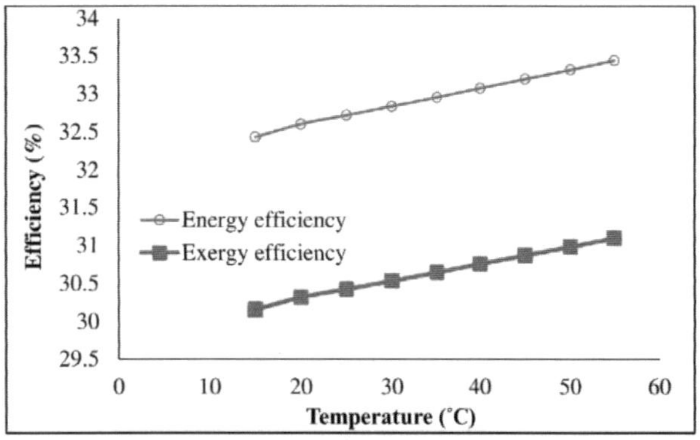

Figure 2. Variation of efficiency with inlet air temperature

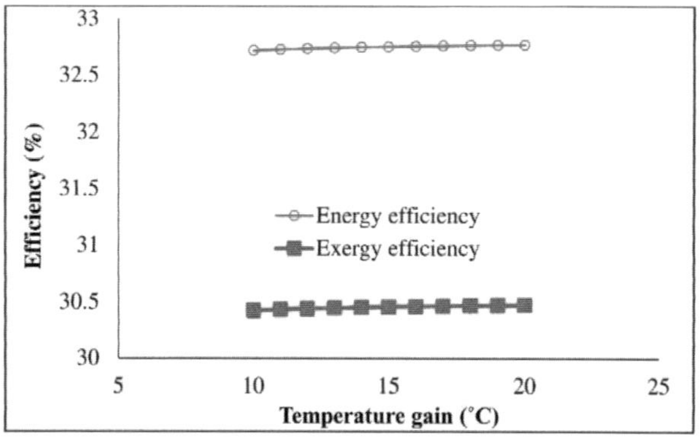

Figure 3. Variation of efficiency with gain in cooling water temperature

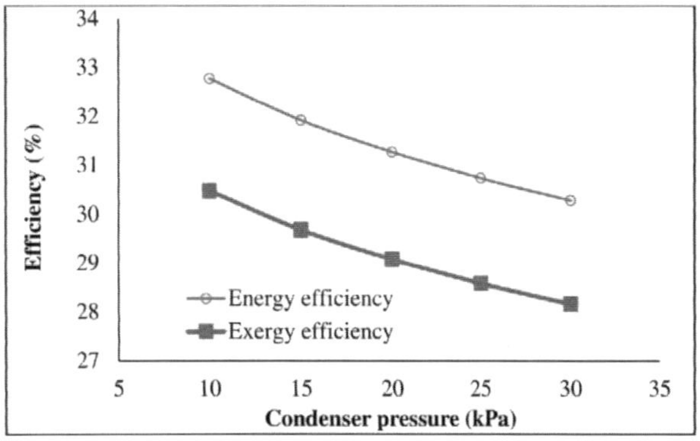

Figure 4. Variation of efficiency with condenser pressure

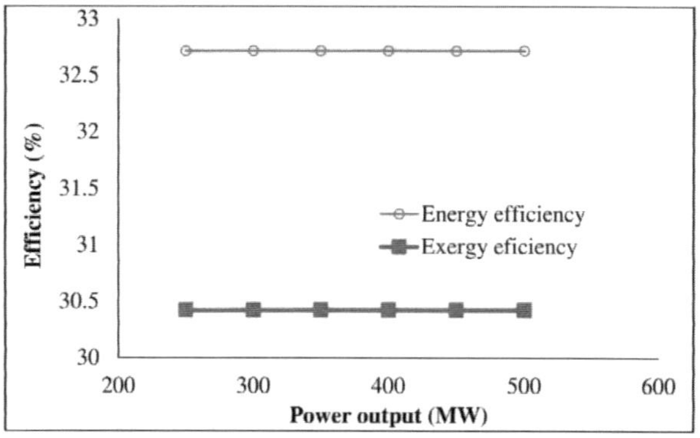

Figure 5. Variation of efficiency with power output

The exergy analysis was carried out using a dead state environment defined by Baehr [25]. A pressure of 1 atm and a temperature of 30°C was used as the environment condition for the exergy analysis. Exergy losses at each component of the power plant was determined through this exergy analysis. The results have been represented in the Table 4 and Sankey diagram given in figure 6.

Table 4. Exergy loss at each component

Sl no.	Apparatus	Exergy loss (%)
1	Turbines	2.06
2	Condenser	2.83
3	Stack	6.58
4	Combustor	36.5
5	Steam generator	15.61

	(excluding combustor)	
6	Feed water heaters	0.7
7	De-aerator	0.13
8	Pumps and pipes	1.95

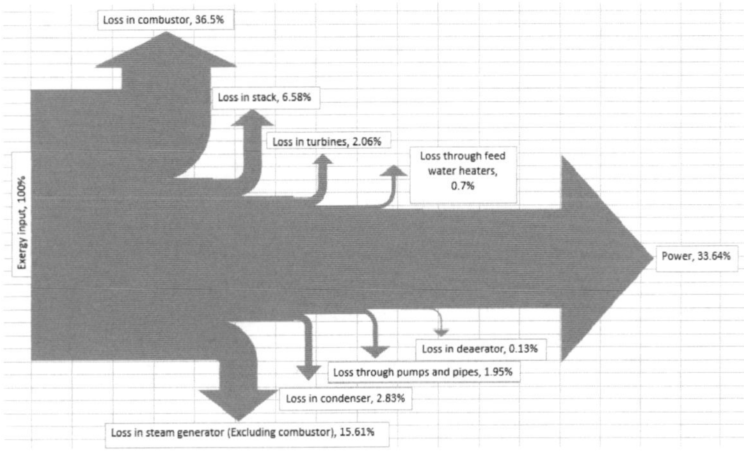

Figure 6. Sankey diagram representing exergy losses

The combustor was found to be the biggest source of exergy destruction, in which 36.5% of the total available exergy was lost. The steam generating apparatus including the combustor accounts for 52.11% of the total exergy loss. The least exergy loss took place in the de-aerator. The total exergy lost was found to be 66.36%. The remaining fraction of exergy is the second law efficiency or the exergy efficiency of the plant, which is 33.64%.

3.2. Simulation using different coal species

Simulations were carried out for six different coals taken from six coal fields located at various states across India. Their composition was obtained from Chandra and Chandra (2004). The composition of each coal sample is shown in Table 5 given below. In order to compare the results, the combustor outlet temperature and evaporator outlet temperature were taken as 1150°C and 1000°C respectively, for all the simulations that were carried out.

Raniganj and East Bokaro coals have high carbon content and low ash content while highest ash is seen in Korba coal. It can be inferred that the mass flow rate of flue gas and efficiencies depend on the fraction of carbon present in the fuel. Mass flow rate of the flue gas decreases with increase in the amount of carbon present in the fuel. This is because lesser carbon implies more fuel flow rate to produce the same amount of power, and hence more air and flue gas. *Moreover, lesser carbon in fuel implies that other components like moisture, oxygen and nitrogen would be more, all of them being heavier than carbon.* On the other hand, energy and exergy efficiencies were found to increase slightly with increase in the fraction of carbon present in the fuel. The reason for such a trend is the increase in boiler efficiency with increase in carbon content in the fuel. Hence exergy loss in the combustor is reduced.

Table 5. Composition of coal samples, mass flow rate of flue gas generated and efficiencies

Coal field	Elements (%)	Mass flow	Efficiency, %

	C	H	S	N	O	Ash	Moisture	rate of flue gas (kg/s)	Energy	Exergy
Raniganj, West Bengal (RW)	60.2	4.2	0.3	1.8	7.1	22.9	3.5	1834.92	28.64	26.63
Giridih, Jharkhand (GJ)	50.8	3.8	0.5	1.3	5.9	34	3.7	1840.34	28.39	26.4
East Bokaro, Jharkhand (EJ)	61	4.1	0.49	1.53	7.2	21	4.68	1834.61	28.69	26.67
West Bokaro, Jharkhand (WJ)	52.3	3.3	0.41	1.23	4.9	36	1.86	1844.74	28.35	26.36
Wardha Valley, Maharashtra (WM)	46.4	2.9	0.41	1.16	9.3	32	7.83	1845.29	28.42	26.43
Korba, Chhattisgarh (KC)	42.9	2.81	0.4	1.06	8.5	38	6.3	1848.05	28.29	26.3

3.3. Effect of combustor inlet air temperature

It can be observed from Figure.7 (a) and (b) that with increase in combustor inlet air temperature, there is an increase in both efficiencies. The reason for this trend is that a rise in air temperature results in the decrease of the amount of heat supplied in the form of fuel. Also, the total exergy at the inlet (thermo-mechanical as well as chemical exergy) is decreased. Further, there is a reduction in exergy loss at the air-preheater due to reduced temperature difference needed for heat transfer. Also, it can be seen that the Raniganj and East Bokaro coals stand apart at a higher efficiency compared to other coals, which could be due to their superior composition, i.e. carbon content.

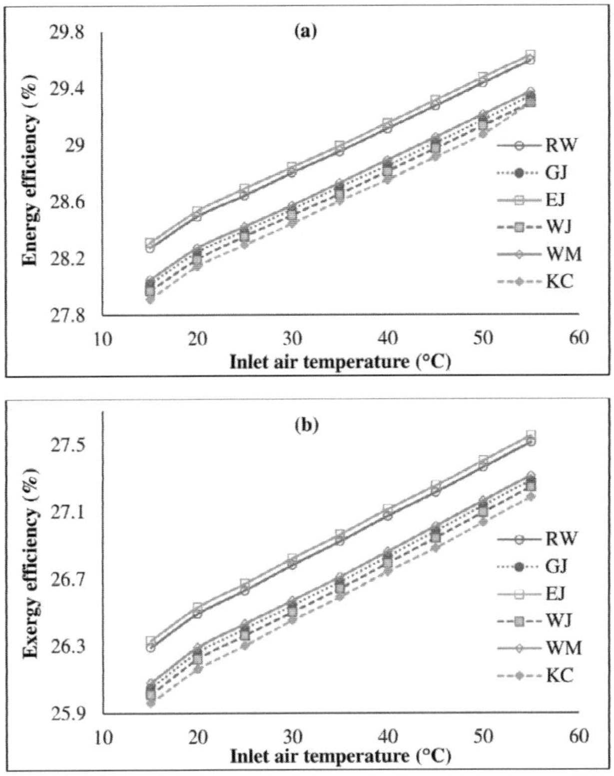

Figure 7. Variation of efficiency with inlet air temperature

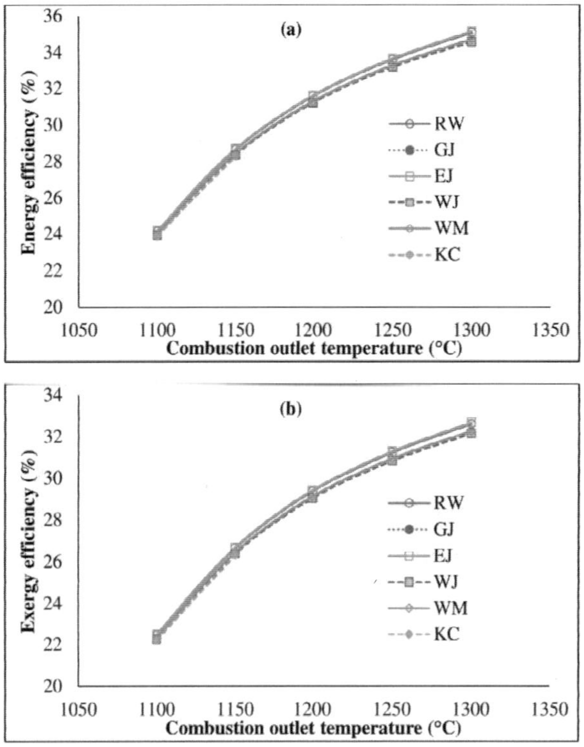

Figure 8. Variation of efficiency with combustor outlet temperature

3.3.1. Effect of combustor outlet temperature

It can be observed from Figure.8. (a) and (b) that efficiencies increase steeply with increase in combustor outlet temperature. This can be attributed to the fact that an increase in flue gas temperature at the combustor outlet leads to more heat transfer at higher temperature. This is in accordance to the 2nd law of thermodynamics. However, when the steam temperature remains the same, higher flue gas temperature will result in larger irreversibility, resulting in additional exergy destruction. Moreover, there is no appreciable difference between the various coal species as far as this result is concerned.

3.3.2. Effect of inlet steam temperature

Figure 9 shows the influence of steam turbine inlet temperature on efficiencies. A range of 500 to 700 °C has been chosen so as to cover the power plants with higher operating parameters too. However, the pressure has been kept constant, keeping the metallurgical limit in mind. With increase in steam temperature to the high-pressure turbine, an increase in efficiency was noticed. This is because, the degree of superheat is raised when steam temperature is raised, keeping the pressure constant. This influences both the power output as well as the heat input. However, the increase in the former will offset the increase in the latter resulting in an overall improvement of the power output, and hence efficiency. This is where, the choice of inlet pressure and temperature play a key role. Once again, there seems to be no influence of coal species on this result.

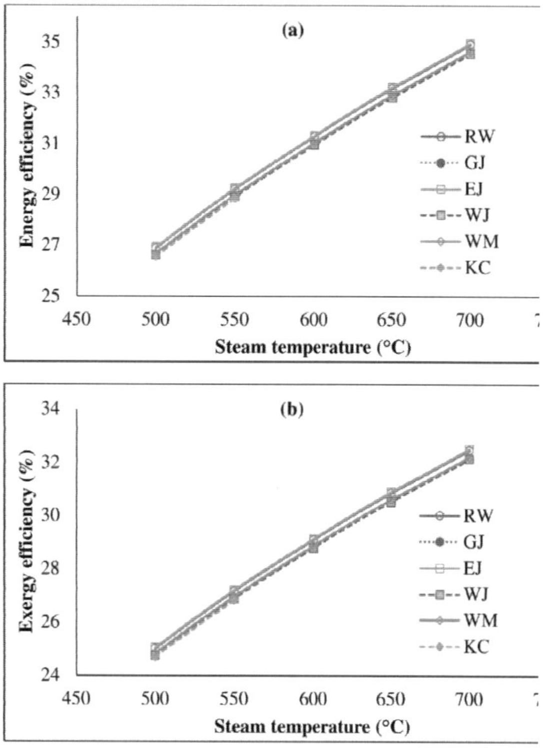

Figure 9. Variation of efficiency with inlet steam temperature

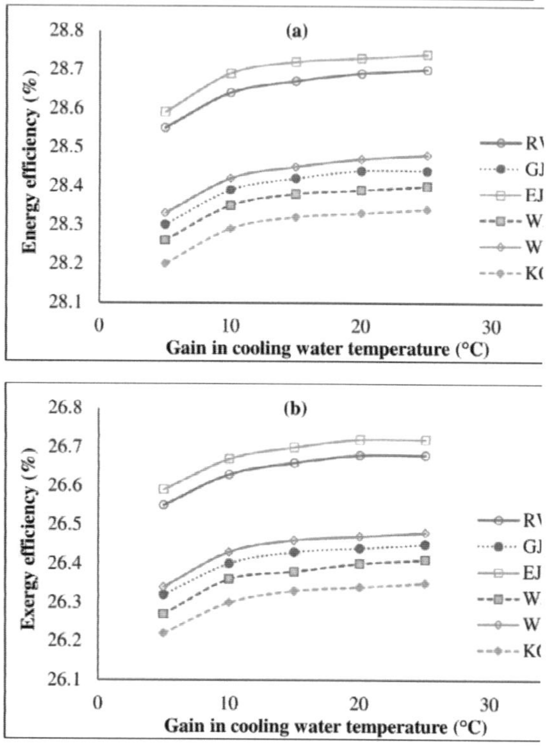

Figure 10. Variation of efficiency with gain in cooling water temperature

3.3.3. Effect of gain in cooling water temperature across condenser

Cooling water temperature rise has been varied from 5 to 25°C. This is only of academic interest as a rise of 25°C will be detrimental to the marine life living in the water body that supplies the cooling water. Also, such a high temperature rise puts additional load on the cooling tower. Moreover, the rise in cooling water temperature to 25°C implies a rise in the condenser pressure and hence temperature. This will reduce the power produced. However, an increase in efficiency is observed with increase in gain in cooling water temperature. This trend is seen in Figure 10. This could be because there is lesser exergy loss in condenser with increase in gain in cooling water temperature, as a result of reduction in the temperature difference between the exhaust cooling water and the incoming turbine exhaust system [24]. Also, the amount of cooling water needed by the condenser decreases with increase in the temperature gain. This reduces the pumping power required, resulting in a slight improvement in efficiency. In this case, the Raniganj and East Bokaro coals fare better than others, which could be attributed to lower flue gas flow rates and hence lower blower power consumption.

3.3.4. Effect of condenser pressure

The effect of condenser pressure is visualized in Figure.11(a) and (b). The chosen range of 6 to 14 kPa is typical of Indian power plants. A decrease in efficiency was observed with increase in condenser pressure. This can be attributed to the increase in mean temperature of heat rejection in the condenser as its pressure is increased. This observation is in accordance with the laws of thermodynamics. However, the amount of rise in efficiency may not be close enough to reality as the power needed to maintain vacuum is not considered. Hence, magnitudes of efficiency found in the figure 11 are ambitious while the real figures will be slightly lower. East Bokaro and Raniganj coals show a slightly better performance compared to other species. This is in no way related to the condenser pressure and can be attributed to their generally superior composition.

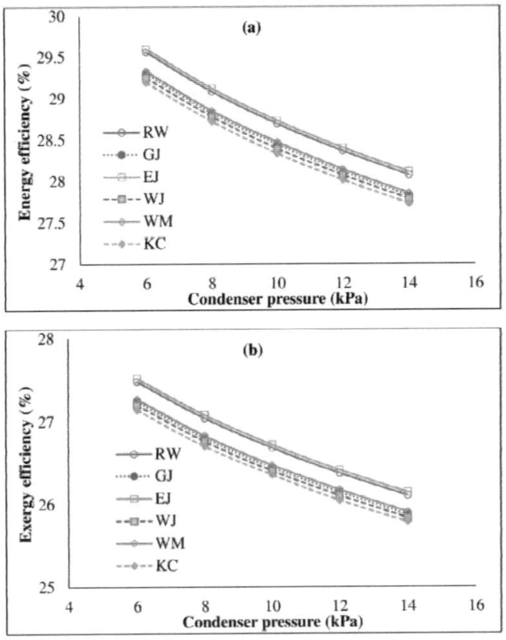

Figure 11. Variation of efficiency with condenser pressure

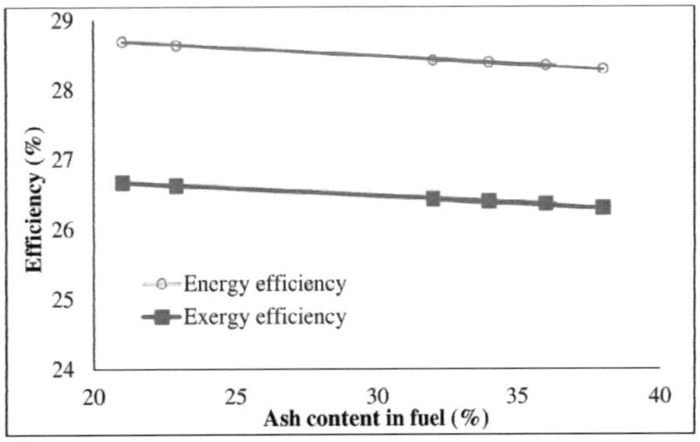

Figure 12. Variation of efficiency with ash content in fuel

3.3.5. Effect of ash content in fuel

Out of the fuel samples considered, the coal from Korba coal field situated in Chhattisgarh had the highest fraction of ash present in it, while the coal from East Bokaro coal field located in Jharkhand had the lowest. The efficiency values obtained when each sample was used as fuel is showed in Table 5. The coal from East Bokaro coal field delivered the highest efficiency, whereas the coal from Korba coal field delivered the lowest. It can be observed from Figure.12 that efficiency reduces with increase in ash content in the fuel. This is because, higher ash coals need to be fed at a larger rate to produce the same amount of power. Therefore, the fuel and air consumption increases thereby increasing the power consumed by the fuel and air supply systems. Although not modelled in this work, more ash results in thick deposits on the heat transfer surfaces resulting in poorer heat transfer rates. Also, ash leaves the combustor as molten slag at high temperature, thereby causing a direct heat loss.

3.3.6. *Effect of ambient temperature*

Figure 13 shows the variation of efficiency as a function of ambient temperature. Ambient temperature influences the temperature of combustion air as well as the heat rejected at lower temperature of the cycle. The range of 25 to 50°C is chosen for this study since the lower limit is a common temperature is several parts of the world while the higher limit is often found in the desert regions of the world. It is observed that efficiency slightly decreases with increase in ambient air temperature. This could be due to the increase of temperature of heat rejection of the cycle. It is in accordance with the laws of thermodynamics. Also, lower ambient temperature implies higher exergies of all streams. Hence, power plants in colder regions operate at a higher efficiency than those in hot regions. Here too, the Raniganj and East Bokaro coals demonstrate slightly better efficiencies which could be due to their superior quality.

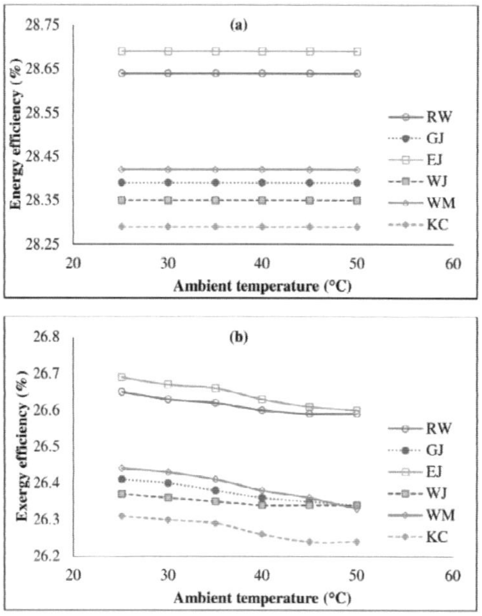

Figure 13. Variation of efficiency with ambient temperature

3.4. *Environment analysis*

Environment analysis was carried out by analyzing the composition of the exhaust gas which enters the stack. The composition of the exhaust gas along with their respective mass fractions are mentioned the Table 6 given below.

Table 6. Composition and mass fraction of exhaust gases

Coal field	Mass fraction of gases in emission						CO_2 per MW (kg/M)
	H_2O	N_2	O_2	Ar	CO_2	SO_2	
Raniganj	0.0236628	0.7893	0.0826	0.0133	0.0909	0.0002372	0.33358
Giridih	0.020661	0.7318	0.1633	0.0124	0.0714	0.000439	0.2628
East Bokaro	0.021962	0.7284	0.1519	0.0123	0.085	0.000438	0.31188
West Bokaro	0.018281	0.7325	0.1631	0.0124	0.0735	0.000219	0.27117

Wardha Valley	0.019181	0.7314	0.1718	0.0123	0.0651	0.000219	0.24025
Korba	0.018281	0.7329	0.1757	0.0125	0.0604	0.0002199	0.22324

The emission of CO_2 gas is the least when coal from Korba coal field is used as fuel. Hence from an emission point of view, the use of this fuel would cause the least harm to the environment.

4. Conclusions

Efficiency was found to increase with increase in parameters like inlet air temperature to the combustor, overall power output, gain in cooling water temperature in the condenser, combustor outlet temperature and steam inlet temperature at the high pressure turbine. Efficiency decreased with increase in condenser pressure, ash content in the fuel and ambient temperature. Exergy analysis shows that highest exergy loss takes place in the combustor (~37%), which is followed by the steam generator (~16%). Since, boiler is a combination of combustor and steam generator, this result is in general agreement with literature. The least exergy loss took place in the de-aerator (~0.13%). The overall exergy efficiency of the power plant was found to be 33.64%. Fuel with higher carbon content and low ash content will give higher efficiency. Among the fuel species studied, coal from East Bokaro coal mine located in Jharkhand produced the highest efficiency.

5. Nomenclature

Table 7. Nomenclature

Symbol	Name
$\dot{E}$	Rate of energy transfer (power), kW
g	Acceleration due to gravity=9.81 m/s2
h	Enthalpy, kJ/kg
h_0	Enthalpy at dead state, kJ/kg
$\dot{m}$	Mass flow rate, kg/s
$\dot{Q}$	Rate of heat transfer, kW
s	Entropy, kJ/kg-K
s_0	Entropy at the dead state, kJ/kg-K
T	Temperature, K
T_0	Temperature at the dead state, K
V	Velocity, m/s
$\dot{W}$	Rate of work (power), kW
$\dot{X}$	Rate of Exergy Transfer, kW
z	Height, m
η	Efficiency
η_I	First law efficiency or Thermal efficiency

η_{II}	Second law efficiency

References:

[1] Sreekanth, M., and Daniel, J. (2015) 'Implementing results of exergy analysis in steam turbine power plants', *International Journal of Applied Engineering Research*, Vol. 10, No. 12, pp. 32113-32123.
[2] Yunus A Cengel and Michael A Boles, 2010, "Thermodynamics: An Engineering Approach", 6th Edition, Tata Mc Graw-Hill Companies.
[3] Reddy, V. S., Kaushik, S. C., and Panwar, N. L. (2013) ‚Review on power generation scenario of India', *Renewable and sustainable energy reviews*, Vol. 18, pp. 43-48.
[4] Regulagadda, P., Dincer, I., and Naterer, G. F. (2010) 'Exergy analysis of a thermal power plant with measured boiler and turbine losses', *Applied Thermal Engineering*, Vol. 30, No. 8-9, pp. 970-976.
[5] Kumar, R. (2017) 'A critical review on energy, exergy, exergoeconomic and economic (4-E) analysis of thermal power plants', *Engineering Science and Technology, an International Journal*, Vol. 20, No. 1, pp. 283-292.
[6] Li, Y., and Liu, L. (2012) 'Exergy analysis of 300MW coal-fired power plant', *Energy Procedia*, Vol. 17, pp. 926-932.
[7] Elhelw, M., Al Dahma, K. S., and el Hamid Attia, A. (2019) 'Utilizing exergy analysis in studying the performance of steam power plant at two different operation mode' *Applied Thermal Engineering*, Vol. 150, pp. 285-293.
[8] Dong, Z., Li, D., Wang, Z., and Sun, M. (2018) 'A review on exergy analysis of aerospace power systems', *Acta Astronautica*, Vol. 152, pp. 486-495.
[9] Li, W., Liu, M., Yan, J., Wang, J., and Chong, D. (2018) 'Exergy analysis of centered water nozzle steam–water injector', *Experimental Thermal and Fluid Science*, Vol. 94, pp. 77-88.
[10] Erdem, H. H., Akkaya, A. V., Cetin, B., Dagdas, A., Sevilgen, S. H., Sahin, B., and Atas, S. (2009) 'Comparative energetic and exergetic performance analyses for coal-fired thermal power plants in Turkey', *International Journal of Thermal Sciences*, Vol. 48, No. (11), pp. 2179-2186.
[11] Kler, A. M., Potanina, Y. M., and Marinchenko, A. Y. (2020) 'Co-optimization of thermal power plant flowchart, thermodynamic cycle parameters, and design parameters of components', *Energy*, Vol. 193, pp. 1166-1179.
[12] Kaushik, S. C., Reddy, V. S., and Tyagi, S. K. (2011) ‚Energy and exergy analyses of thermal power plants: A review' *Renewable and Sustainable energy reviews*, Vol. 15, No. 4, pp. 1857-1872.
[13] Adibhatla, S., and Kaushik, S. C. (2014) ‚Energy and exergy analysis of a super critical thermal power plant at various load conditions under constant and pure sliding pressure operation', *Applied Thermal Engineering*, Vol. 73, No. 1, pp. 51-65.
[14] Vakilabadi, M. A., Bidi, M., and Najafi, A. F. (2018) 'Energy, exergy analysis and optimization of solar thermal power plant with adding heat and water recovery system', *Energy Conversion and Management*, Vol. 171, pp. 1639-1650.
[15] Chauhan, S. S., and Khanam, S. (2019) ‚Enhancement of efficiency for steam cycle of thermal power plants using process integration', *Energy*, Vol. 173, pp. 364-373.
[16] Gupta, M. K., and Kaushik, S. C. (2010) ‚Exergy analysis and investigation for various feed water heaters of direct steam generation solar–thermal power plant', *Renewable Energy*, Vol. 35, No. 6, pp. 1228-1235.
[17] Oko, C. O. C., and Njoku, I. H. (2017) ‚Performance analysis of an integrated gas-steam-

and organic fluid-cycle thermal power plant' *Energy*, Vol. 122, pp. 431-443.

[18] Fu, P., Wang, N., Wang, L., Morosuk, T., Yang, Y., and Tsatsaronis, G. (2016) 'Performance degradation diagnosis of thermal power plants: A method based on advanced exergy analysis', *Energy Conversion and Management*, Vol. 130, pp. 219-229.

[19] Bolatturk, A., Coskun, A., and Geredelioglu, C. (2015) 'Thermodynamic and exergoeconomic analysis of Çayırhan thermal power plant', *Energy Conversion and Management*, Vol. 101, pp. 371-378.

[20] Bidabadi, M., Vakilabadi, M. A., Poorfar, A. K., Monteiro, E., Rouboa, A., and Rahbari, A. (2016) 'Mathematical modeling of premixed counterflow combustion of organic dust cloud', *Renewable Energy*, Vol. *92*, pp. 376-384.

[21] Suresh MV, Reddy KS, and Kolar AK. (2010) '3-E analysis of advanced power plants based on high ash coal', International Journal of Energy Research. Vol. 34, No. 8, pp. 716-735.

[22] Suresh MV, Reddy KS, and Kolar AK. (2010) '4-E (energy, exergy, environment, and economic) analysis of solar thermal aided coal-fired power plants', *Energy for Sustainable Development*. Vol. 14, No. 4, pp. 267-279.

[23] Suresh MV, Reddy KS, and Kolar AK. (2011) 'ANN-GA based optimization of a high ash coal-fired supercritical power plant', *Applied Energy*. Vol. 88, No. 12, pp. 4867-4873.

[24] Kalimuthu S, Karmakar S and Kolar AK (2017) '3-E analysis of a Pressurized Pulverized Combined Cycle (PPCC) power plant using high ash Indian coal' *Energy*, Vol. 128, pp. 634-648.

[25] Baehr HD, Kabelac S. Thermodynamik. Berlin, Heidelberg, New York: Springer-Verlag; 1966.

IOP Conf. Series: Materials Science and Engineering 1128 (2021) 012039 doi:10.1088/1757-899X/1128/1/012039

Effect of Shape Imperfections on Collapse Moment of Strain Hardened Pipe Bends under in-plane Opening Bending Moment

M. Sai Santosh Pavan Kumar and T. Christo Michael

School of Mechanical Engineering, VIT Chennai, Chennai, TamilNadu, India-600127.

Corresponding author: christo.michael@vit.ac.in

Abstract. The effect of ovality and variable wall thickness on the collapse load of pipe bends subjected to in-plane opening moment is investigated using Finite Element (FE) analysis for an elastic plastic material. The collapse load was estimated from the moment-rotation curves using Twice-Elastic-Slope method. The presence of ovality reduced the collapse load by 13.1% for a model having bend characteristic 0.1, thus making its effect critical. The presence of thinning had no significant effect on the collapse load. Based on the FE results, a mathematical equation is proposed to estimate the collapse load of strain-hardened pipe bends for a range of bend characteristic from 0.1 to 0.6, considering the effect of ovality.

Keywords: elastic-plastic; ovality; pipe bends; collapse load; thinning.

Nomenclature	
C_0	Percentage of ovality
C_{th}	Percentage of thickening
C_t	Percentage of thinning
L	Straight pipe length (mm)
D	Pipe Outside diameter (mm)
D_{max}	Pipe Maximum outside diameter (mm)
D_{min}	Pipe Minimum outside diameter (mm)
t	Pipe Nominal thickness (mm)
t_{max}	Pipe Maximum thickness (mm)
t_{min}	Pipe Minimum thickness (mm)
R	Bend radius (mm)
r	Mean radius of pipe (mm)
h	Bend characteristic, $=Rt/r^2$
M	In-plane bending moment
M_o	Straight pipe Collapse in-plane moment
M_0^S	Straight pipe Limit in-plane moment, $=4rt^2\sigma_0$
E	Young's modulus
v	Poisson's ratio
σ_o	Yield stress

1. Introduction

The most pivotal component in any piping system are the pipe bends. Pipe bends are the most economical means of changing the direction of flowing fluid within a piping network. They are pliant (than straight pipes) and lead to lowered reaction forces and moment due to their elastic deformations. But if the load exceeds the elastic limit, plastic deformation occurs, and the load bearing capability of the pipe is lost significantly. This being the real scenario, analysis must be performed taking into consideration the effects of strain hardening.

The study of pipe bends comprises of analytical [1-3], numerical [4-30] and experimental [31-36]. The three main types of loading considered in the above studies include: in-plane opening moment [35], in-plane closing moment [17] and out-of-plane moment [12,14]. Most of the earlier studies were for pipe bends in static conditions but later, studies were performed to include the effects of pressure [2,6,11], fatigue [31], shakedown [21] and cracks [9]. A recent study [24], the analysis of 180° pipe bend to investigate the effect on collapse load, by considering the effect of geometric nonlinearity and the effect of ovality. The study indicated a decrease in the collapse load for pipes with thin wall and small bend radius and minor deviations for pipes with thick walls and large bend radius, in comparison to 90° pipe bends. Additionally, another recent work performed by Sherif. S. Sorour [25-26], wherein the entire process of rotary pipe bending [RPB] was performed using Finite Element Method (FEM) to account for the actual change in cross section rather than assuming the cross section to be elliptic. An experimental and FE analysis performed by Chattopadhyay [20,32] on effect of crack initiation load and its growth on collapse load, concluded to reduce the collapse load. Sumesh et al. [27] performed Finite element Analysis on structurally distorted cracked pipe bends under in-plane closing bending to estimate limit moments. J. Li et al. performed FEA to compute the limit loads under combined bending and torsion [28], with circumferential through wall crack subjected to torsion [29] and compared cracked and un-cracked wall considering geometric non-linearities subjected to torsion [30].

The early researches assumed pipe bends to have an ideal cross section i.e. circular. Kim et al. performed FE analysis of pipe bends, considering an Elastic Perfectly Plastic (EPP) material, imposed with internal pressure and in-plane loading with material non-linearity only [7], and considering geometric non-linearity [8]. Kim et al. [13] also proposed Twice Elastic Slope (TES) to estimate collapse moment. A mathematical equation to predict collapse load of bends imposed with internal pressure and in-plane moments was offered by Chattopadhyay et al. [19] for an Elastic Plastic (EP) material using the techniques of FE Analysis. Younan et al. [16] performed FEA for pipe bends exacted to out-of-plane moment, comparing EPP and EP material. These were some of the earliest researches considering the effects of strain hardened material although they were for ideal pipe bends.

However, the pipe bend's cross section changes, from circular to oval with variations in wall thickness, during the pipe bending process. Thus, the influence of shape imperfections viz. ovality and thinning became crucial for the FE analysis. The ASME standards allow 8% for ovality and thinning in pipe bends. However, the analysis conducted incorporated up to 20% for ovality and thinning.

An equation was proposed, on performing FE analysis, to evaluate collapse load of pipe bend comprising shape imperfections exacted to internal pressure and in-plane moments, by Christo Michael et al. [4-5] Another analysis on the influence of shape imperfections on the collapse load of bends imposed with in-plane opening bending with and without pressure was performed by Buckshumiyan et al. [22] The above studies were for pipe bends including the influence of shape imperfections although it was for an EPP material.

In practical applications and experiments, the material is subjected to the effects of strain hardening and hence must be considered in FEA. The collapse moment of pipe bends considering the material to be EPP is very conservative and thus gives a scope to investigate the effects of both strain hardening and shape imperfections on collapse moment. This gives scope for this paper to explore the effects of shape imperfections namely ovality and thinning/thickening on collapse load of strain hardened pipe bends imposed with in-plane opening bending load.

2. Geometry Modelling and Material

2.1. Geometry

The geometry in analysis is a 90° pipe bend having two attached straight pipes at the ends. The straight pipe's length is five times the mean pipe diameter i.e., $L = 5 \times D$. The pipe's mean radius is 50 mm. The thickness considered for the analysis are 2.5 mm, 5 mm and 10 mm and the bend radius considered for the analysis are 100 mm and 150 mm. From the above geometric dimensions, the bend characteristic ($h=Rt/r^2$) for the pipe bend ranges from 0.1 to 0.6. The degree of ovality is defined as the ratio of difference between the maximum and minimum diameter to the nominal diameter. The bend is assumed to have shape imperfections with an assumed shape of an ellipse [4].

$$C_O = \frac{D_{\max} - D_{\min}}{D} \times 100 \quad (1)$$

where nominal outside diameter: $D = \frac{D_{\max} + D_{\min}}{2}$

The thinning occurs at the extrados. It is defined as the ratio of difference between the nominal thickness and minimum thickness to the nominal thickness [4].

$$C_t = \frac{t - t_{\min}}{t} \times 100 \quad (2)$$

The thickening occurs at the intrados. It is defined as the ratio of the difference between the maximum thickness and nominal thickness to the nominal thickness [4].

$$C_{th} = \frac{t_{\max} - t}{t} \times 100 \quad (3)$$

The percentage rise in thickness in the intrados is equal to the percentage reduction in the thickness in the extrados.

The straight pipes are considered to have a circular cross section with the same thickness.

2.2. Material

The material is surmised as isotropic and elastic-plastic (strain-hardening) (Table 1)

The material used for the analysis is SS304, having properties as:

Young's modulus = 193 GPa

Poisson's Ratio = 0.2642

Yield stress = 271.929 MPa

Table 1. Plastic behaviour

Yield stress (MPa)	Plastic strain
271.929	0
345.91	0.00473
378.867	0.01264
403.619	0.02836
424.165	0.0491
527.587	0.105

3. Finite Element Modelling

FEA was performed using the ABAQUS software version 6.14 [37]. The geometry was also created using the same package.

3.1. Element and Mesh

A structured hexahedral mesh was adapted with the element type as C3D20R, a 20-node brick consisting of quadratic elements. The quantity of elements across thickness and length of the pipe are three and fifteen, respectively. The number of elements across the bend region and the cross section were varied for the mesh study. From the study it was concluded that 10 elements were required for cross section and bend region. The final mesh count of the model is 3000 elements and will be adopted for further analysis.

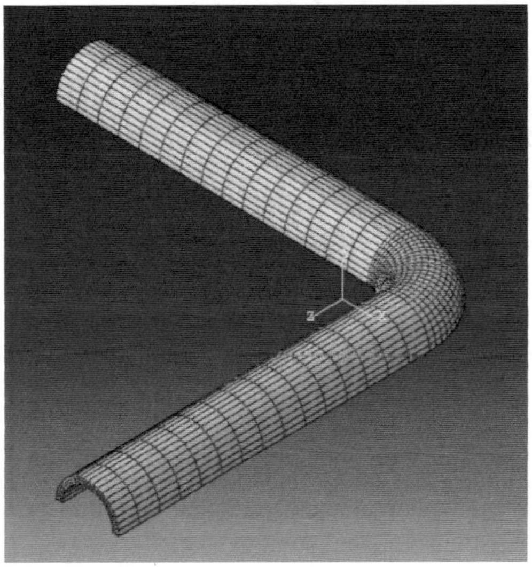

Figure 1. Structured Mesh

3.1.1. *Mesh study.* A mesh study was executed and ideal mesh density at bend region and the cross section were identified to perform the analysis. The study was performed for h = 0.2 and h = 0.4. The number of elements varied from 1656 to 4176. The mesh study was conducted having the response as collapse moment. From Figure 2(b) for the model h = 0.4 the percentage decrease in the value from change of number of elements from 3000 to 3696 is 0.008% which is negligible. Similarly, from Figure 2(a) for the model h = 0.2 the percentage decrease is 0.032%.

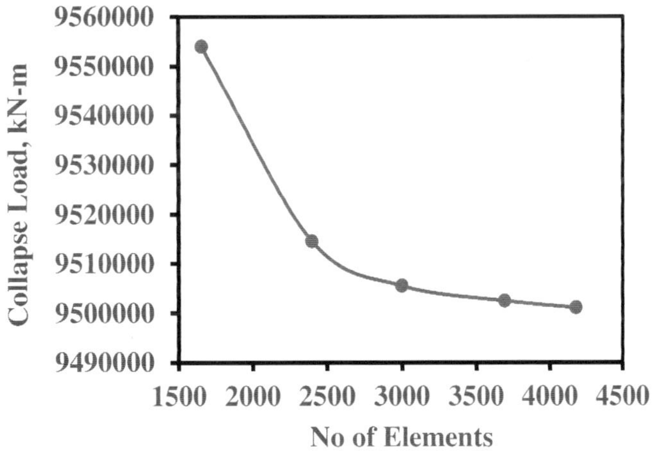

(a) h = 0.2

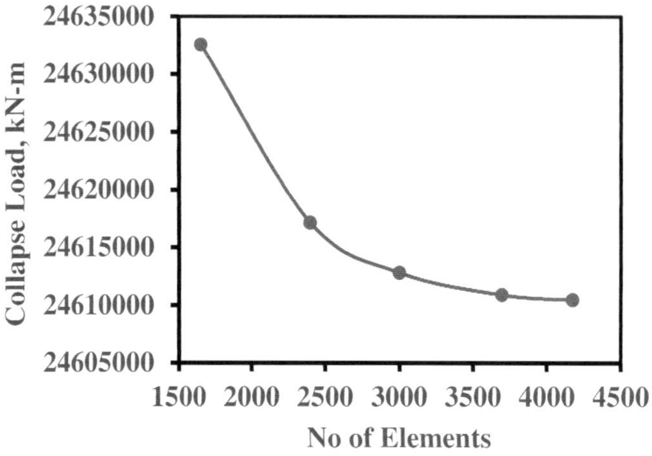

(b) h = 0.4

Figure 2. Mesh study

3.2. Boundary conditions and Loading

The analysis has been simplified by considering a symmetric model. The fixed and rotation boundary conditions are specified using the Multi Point Constraint (MPC) as shown in Figure 3. The MPC is used to connect all the surface nodes to a single reference point so that any boundary condition applied to this reference point will be applied to all the surface nodes it is linked to.

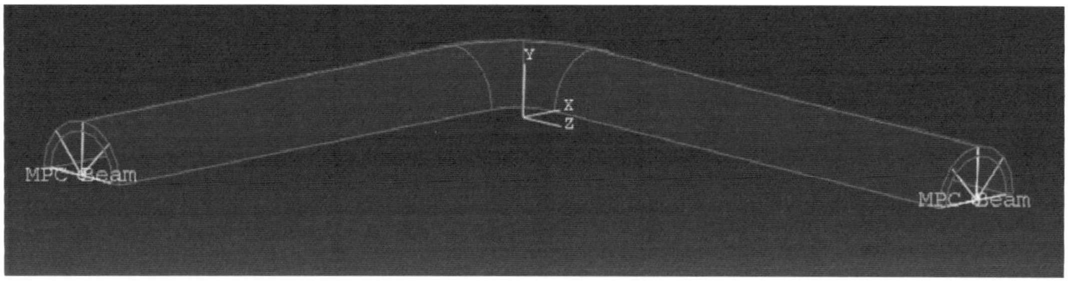

Figure 3. MPC constraint

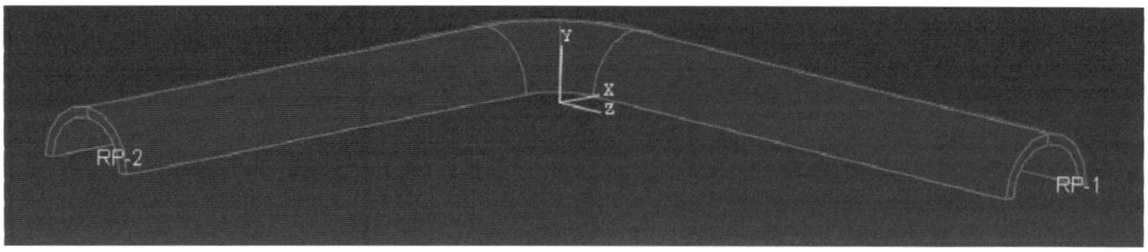

Figure 4. Reference Points

The reference point 1 shown in Figure 4 is given a fixed boundary condition. The reference point 2 is given a rotation boundary condition along the y vector (anti-clockwise rotation) and constraining other degrees of freedom. The final boundary condition is symmetry about the XZ plane. The Static RIKS method with non-linear geometry option is employed to solve this model, since it is used for predicting the unstable, geometrical non-linear collapse of a structure.

3.3. Post-Processing

The post processing was done using the same software. The moment and the rotation were extracted from the reference point 2 (Figure 4). The collapse moment is evaluated using the TES method, the point where the straight line having two times the slope of initial elastic region, in moment rotation curve, intersects the curve. The graph is plotted in Excel and TES load was obtained.

3.4. Python Script

The setup of the case from material, mesh, Boundary conditions, constraints and Post-Processing remains the same for all the 150 cases analyzed. The geometry varies for each of the above case. Hence to simplify the task a Python script was generated which upon executing creates the input files for all the cases. Then a separate code was used to run the INP files and generate the ODB files. Finally, the data was tabulated in Excel.

4. Verification

The FE model and the analysis was verified using the literature of Goodall [2]. This model considers pipe bend geometry having an ideal cross section. The material used was SS304 considering the effects of strain hardening. The percentage deviation in the collapse moment for both models is less than 1% as visualized from Figure 5. Thus, the code and the FEA procedure adopted is verified and can be used for the underlying analysis.

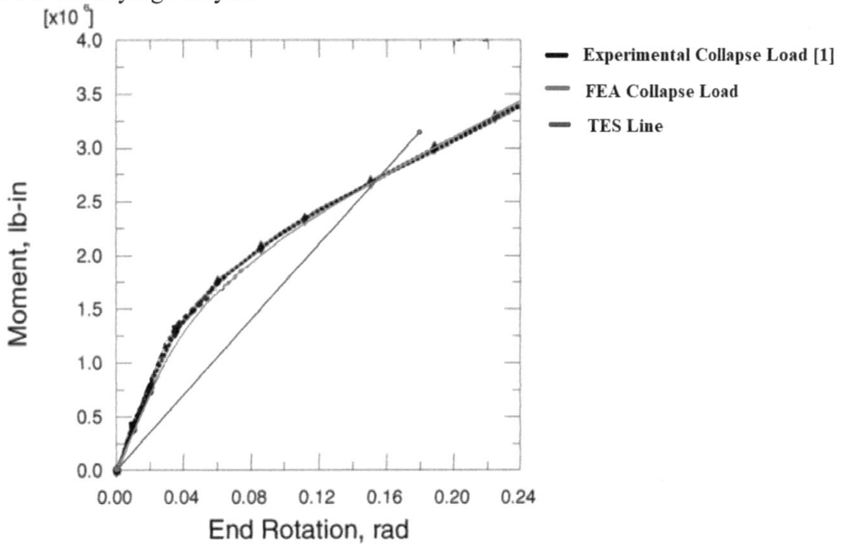

Figure 5. Verification of FEA procedure

5. Comparison of Strain hardened material and EPP material

The effects of strain hardened material and an EPP material on the collapse moment has been investigated. From Figure 6(a) and (b), it is known that the collapse load for a strain hardened material or Elastic Plastic material is higher than that for an EPP. There is a 19.88% difference between the collapse loads for a model with h=0.2 and 19.56% for the model with h=0.4. The pipe bends are not designed for the maximum collapse load but are designed for a load higher than minimum required by incorporating a safety factor. Hence designing the pipe bends by considering the material to be EPP makes the design very conservative which may be unnecessary.

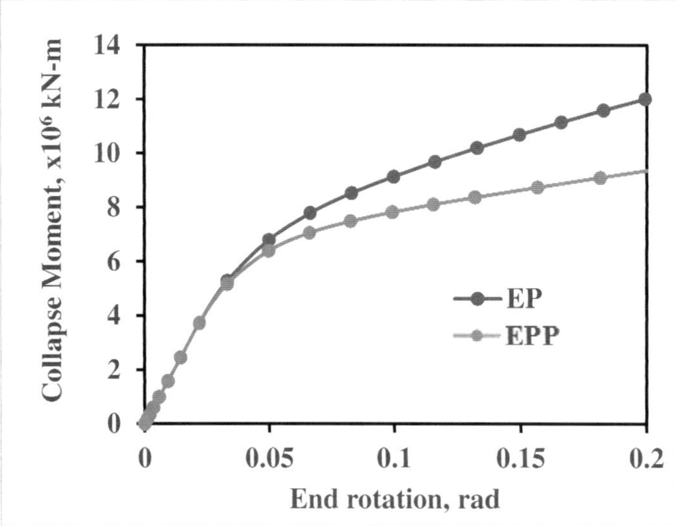

(a) h = 0.2

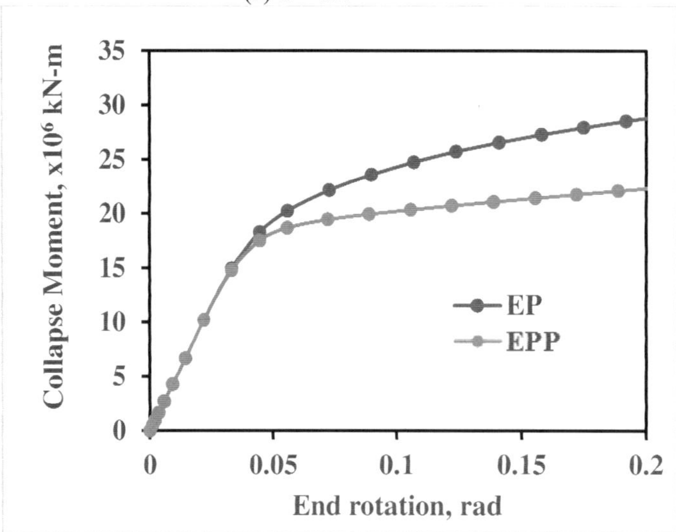

(b) h = 0.4

Figure 6. Comparison of EPP and EP material

6. Result and Discussion

The pipe bends usually are imposed with in-plane opening and closing and out-of-plane bending loads due to various dead and live loads namely wind load, seismic load, weight of the system and other loads acting on it. The pipe bends are further weakened due to geometric imperfections. Hence the effect of shape imperfections on collapse load is very crucial.

6.1. Effect of ovality on collapse moment

It is visualized from Figure 7; the collapse load reduces with rise in percent ovality. For the model having $h = 0.6$ the collapse load decreases form 26.62 kN-m to 24.99 kN-m with increase in percentage ovality from 0% to 20%.

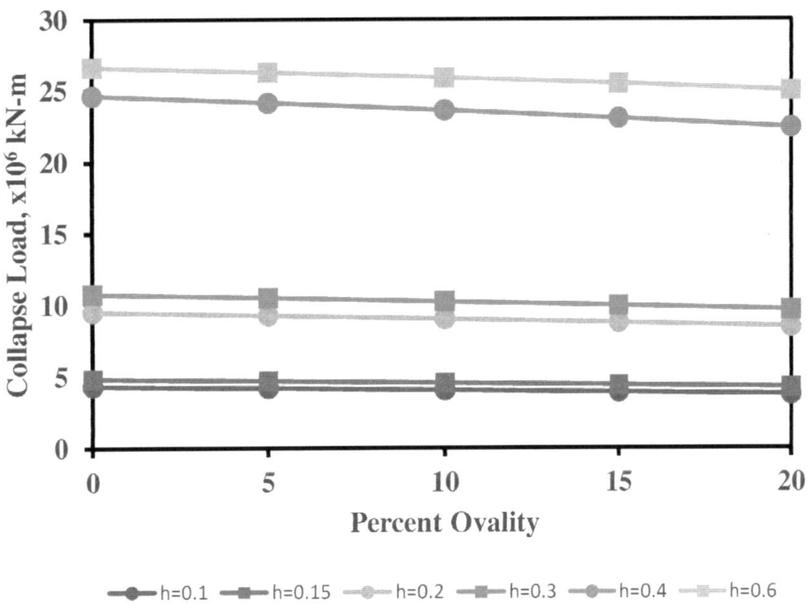

Figure 7. Collapse load vs Percent Ovality for various bend characteristic

To truly evaluate the effect of ovality, percentage difference in collapse moment must be investigated. The percentage difference in ovality is defined as the ratio of difference in the collapse moment of a model not considering the effect of shape imperfections (reference model) to the model with consideration of shape imperfections to the reference model. From Table 2, for the model having bend characteristic 0.1 the collapse load decreases by 13.09% with increase in percent ovality to 20%. This implies that the percentage of ovality in a model must be carefully interpreted for design of pipe bends.

Table 2. Effect of ovality on Collapse moment

	COLLAPSE_LOAD					
	BEND CHARATERISTIC					
OVALITY%	0.1	0.15	0.2	0.3	0.4	0.6
0	0	0	0	0	0	0
5	2.771789	2.819055	2.527388	2.396501	1.978124	1.266013
10	6.057354	5.810865	5.190525	4.893265	4.112683	2.753184
15	9.528685	8.875936	7.970253	7.469431	6.447588	4.313444
20	13.09255	11.96861	10.84699	10.17345	8.921972	6.11378

6.2. The effect of thickness and bend radius on collapse load

From Figure 8, we can visualize that the trends for collapse moment with change in thickness for various bend radii. The collapse load reduces with decrease in thickness. For model having bend radius 100 mm the collapse load decreases from 22.4 kN-m to 3.7 kN-m with decrease in thickness from 10 mm to 2.5 mm for percentage ovality of 20%. Also, the collapse moment decreases with decrease in bend radius. For the model having thickness 5 mm, the collapse load decreases from 9.7 kN-m to 8.5 kN-m with decrease in bend radius from 150 mm to 100 mm for percentage ovality of 20%.

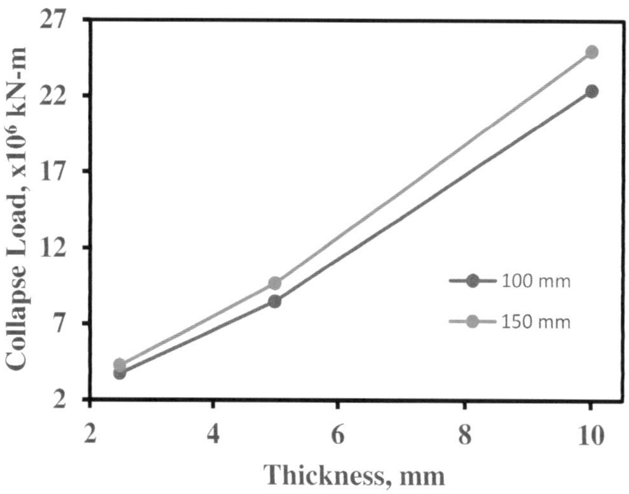

Figure 8. Collapse load vs Thickness for various bend radii

From Figure 9, we can understand the interaction of thickness of pipe, bend radii and percent ovality on percent collapse moment. With increase in thickness of the pipe, percent deviation in collapse moment reduces. Also, with increase in bend radius, percent deviation in collapse moment reduces. Thus, it is very crucial to design thin pipe bends subjected to ovality due to pipe bending process.

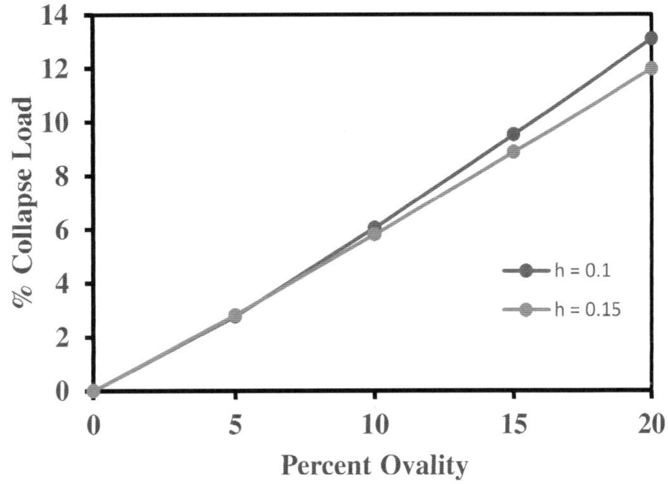

(a) t = 2.5 mm

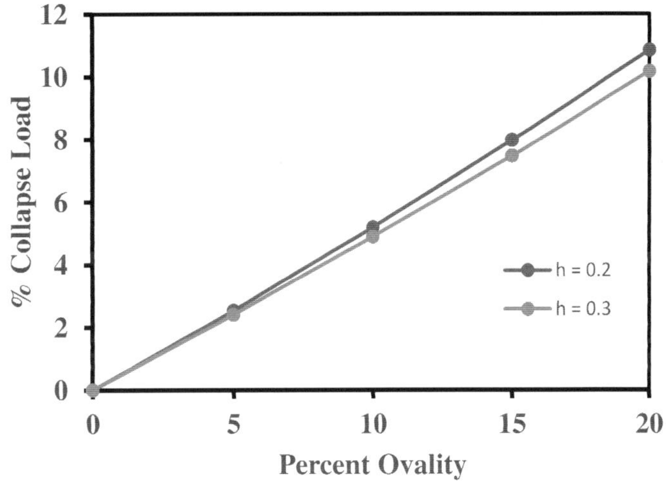

(b) t = 5 mm

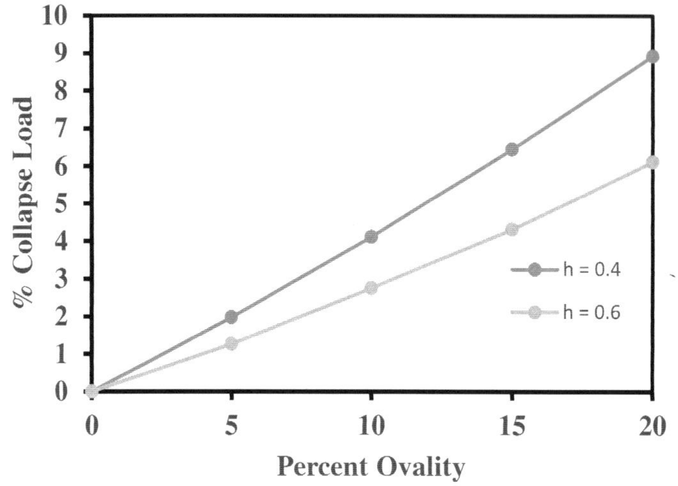

(c) t = 10 mm

Figure 9. Interaction of thickness, bend radius and percent ovality on percent difference of collapse load

6.3. *Effect of thinning on collapse moment*

From the Figure 10, it is observed that thinning has no effect on the collapse load. The percentage difference between the values differs by less than 2% which is insignificant for all the models considered. Thus, we can neglect the effect of thinning for the determination of collapse moment.

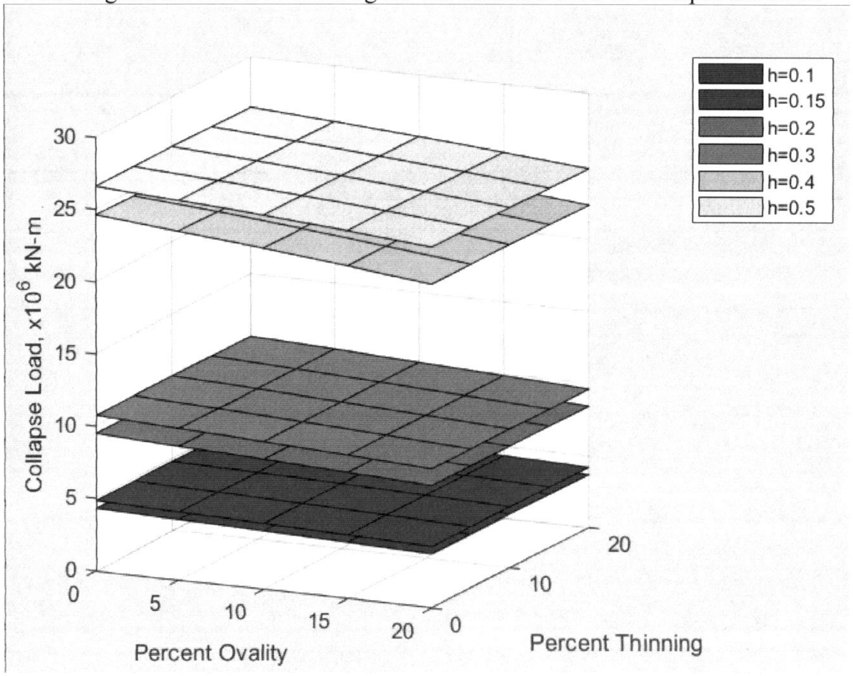

Figure 10. Effect of thinning on Collapse moment

6.4. Collapse load equation considering the effects of ovality

A mathematical expression is presented to determine the collapse moment of pipe bends including ovality for a strain hardened pipe bends using Data Fit [38] for a range of h = 0.1 to 0.6.

$$\frac{M_O}{M_O^{S^{0.35}}} = \frac{\left(\frac{R}{100r(72.8+1.46C_O)}\right)^{0.24}}{\left(\frac{r}{t}-3.04\right)^{0.36}} \tag{4}$$

The average deviation is 1.3% and the Coefficient of Multiple Determination (R^2) is 0.998. Thus, from Figure 11, the above equation is compatible with the FEA results for $h \in [0.1,0.6]$, with the maximum difference of 5.65%.

The previous equations proposed to estimate collapse load either included the effects of strain hardening only [8,19] or shape imperfections with EPP material [4,5]. However, the current study proposes an equation to estimate collapse load with both shape imperfections and strain hardening effects included.

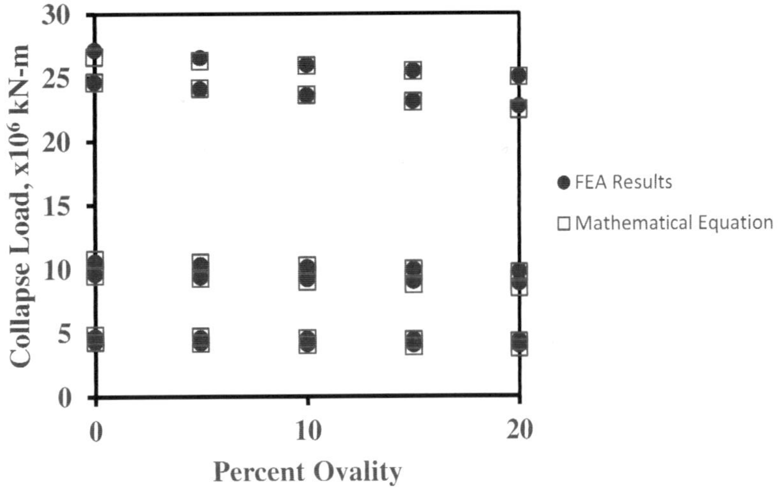

Figure 11. Comparison of proposed mathematical equation with FEA results

7. Conclusion

The following conclusions are deduced from the nonlinear analyses on pipe bends with shape imperfections under in-plane opening moment for a strain hardened material:

- The effect of thinning on collapse load is negligible and its effects are not considered for further study.
- The collapse load of pipe bends is significantly influenced by ovality. As the percentage of ovality increases the collapse moment reduces, hence considered during the design of pipe bends.
- It is also noted that the collapse moment increases with increase in bend radius and thickness.
- A mathematical expression is presented to determine the collapse moment of strain hardened pipe bends including the effect of ovality in the range of bend characteristics between 0.1 and 0.6.

References

[1] Calladine CR 1974 Limit analysis of curved tubes *J. Mech. Eng. Sci.* **16** 85–87

[2] Goodall IW 1978 Lower Bound Limit Analysis of Curved Tubes Loaded by Combined Internal Pressure and In-plane Bending Moment Research Division Report RD/B/N4360,

Central Electricity Generating Board, UK
[3] Spence J and Findlay GE 1973 Limit load for elbows under in-plane bending *Proc. 2nd. Int. Conf. on* Pressure Vessel Technology **1–28** 393–399
[4] Christo Michael T, Veerappan AR and Shanmugam S 2012 Effect of ovality and variable wall thickness on collapse loads in pipe bends subjected to in-plane bending closing moment *Eng. Fract. Mech.* **79** 138–148
[5] Christo Michael T, Veerappan AR and Shanmugam S 2013 Effect of internal pressure and shape imperfections on plastic loads of pipe bends under in-plane closing moment *Eng. Fract. Mech.* **105** 1–15
[6] Christo Michael T, Veerappan AR and Shanmugam S 2012 Comparison of plastic limit and collapse loads in pipe bends with shape imperfections under in-plane bending and an internal pressure *Int. J. Pres. Ves. Pip.* **99-100** 23-33
[7] Kim YJ and Oh CS 2006 Limit loads for pipe bends under combined pressure and in-plane bending based on finite element limit analysis *Int. J. Pres. Ves. Pip.* **83** 148–153
[8] Kim YJ and Oh CS 2006 Closed-form plastic collapse loads of pipe bends under combined pressure and in-plane bending *Eng. Fract. Mech.* **73** 1437–54
[9] Kim YJ, Kim YI and Song TK 2007 Finite element plastic loads for circumferential cracked pipe bends under in-plane bending *Eng. Fract. Mech.* **74** 643–668
[10] Kim YJ and Oh CS 2007 Effects of attached straight pipes on finite element limit analysis for pipe bends *Int. J. Pres. Ves. Pip.* **84** 177–184
[11] Kim YJ, Lee KH, Oh CS, Yoo B and Park CY 2007 Effect of bend angle on plastic loads of pipe bends under internal pressure and in-plane bending *Int. J. Mech. Sci.* **49** 1413–24
[12] Lee KH, Kim YJ and Park CY 2008 Plastic loads of pipe bends under combined pressure and out-of-plane bending *Int. J. Fract.* **49** 31–45
[13] Kim YJ, Kim NH, Oh CY and Oh CS 2008 A method to estimate plastic loads for elbows with non-uniform thicknesses *Fatigue. Fract. Eng. M.* **31** 822-837
[14] Mourad HM and Younan MYA 2001 Nonlinear Analysis of Pipe Bends Subjected to Out-of-Plane Moment Loading and Internal Pressure *J. Press. Vessel Technol.* **123** 253-8
[15] Mourad HM and Younan MYA 2002 Limit-Load Analysis of Pipe Bends Under Out-of-Plane Moment Loading and Internal Pressure *J. Press. Vessel Technol.* **124** 32-7
[16] Mourad HM and Younan MYA 2000 The Effect of Modeling Parameters on the Predicted Limit Loads for Pipe Bends Subjected to Out-of-Plane Moment Loading and Internal Pressure *J. Press. Vessel Technol.* **122** 450-6
[17] Shalaby MA and Younan MYA 1998 Limit Loads for Pipe Elbows with Internal Pressure under In-Plane Closing Bending Moments *J. Press. Vessel Technol.* **120** 35-42
[18] Shalaby MA and Younan MYA 1998 Nonlinear analysis and plastic deformation of pipe elbows subjected to in-plane bending *Int. J. Pres. Ves. Pip.* **75** 603–611
[19] Chattopadhyay J, Nathani DK, Dutta BK and Kushwaha HS 2000 Closed-Form Collapse Moment Equations of Elbows Under Combined Internal Pressure and In-Plane Bending Moment *Int. J. Pres. Ves. Pip.* **122** 431-6
[20] Chattopadhyay J and Tomar AKS 2006 New plastic collapse moment equations of defect-free and throughwall circumferentially cracked elbows subjected to combined internal pressure and in-plane bending moment *Eng. Fract. Mech.* **73** 829–854
[21] Balakrishnan S, Veerappan AR and Shanmugam S 2019 Determination of plastic, shakedown and elastic limit loads of 90°pressurized pipe bends with shape imperfections *Int. J. Pres. Ves. Pip.* **175** 103925
[22] A. Buckshumiyan, Veerappan AR and Shanmugam S 2014 Determination of collapse loads in pipe bends with ovality and variable wall thickness under internal pressure and in-plane opening moment *Int. J. Pres. Ves. Pip.* **123-124** 1-9
[23] Yahiaoui K, Moreton DN and Moffat DG 2002 Evaluation of limit load data for cracked elbows under opening bending and comparisons with existing solutions

Int. J. Pres. Ves. Pip. **79** 27–36

[24] Li J, Zhou CY and Zhu JG 2020 Limit loads for 180° pipe bends under in-plane bending moment considering geometric nonlinearity *Int. J. Pres. Ves. Pip.* **183** 1-13

[25] Sorour SS, Shazly M and Megahed MM 2019 Limit load analysis of thin-walled as-fabricated pipe bends with low ovality under in-plane moment loading and internal pressure *Thin-Walled Struct.* **144** 106336

[26] Sorour SS, Shazly M and Megahed MM 2019 Limit load analysis of thick-walled as-fabricated pipe bends under in-plane moment loading and internal pressure *Int. J. Pres. Ves. Pip.* **174** 1–12

[27] Sumesh S, Veerappan AR and Shanmugam S 2016 Assessment of plastic loads of critical throughwall circumferentially cracked pipe bends with structural distortions under in-plane bending *Thin-Walled Struct.* **104** 144-151

[28] Li J, Zhou CY, Cui P and He XH 2015 Plastic limit loads for pipe bends under combined bending and torsion moment *Int. J. Mech. Sci.* **92** 133–145

[29] Li J, Zhou CY, Miao XT, Chang L and He XH 2015 Plastic limit loads for pipe bends with circumferential through-wall crack under torsion moment *Int. J. Mech. Sci.* **100** 283–297

[30] Li J, Zhou CY, Chang L, Miao XT and He XH 2017 Limit loads for un-cracked and circumferential through-wall cracked pipe bends under torsion moment considering geometric nonlinearity *Thin-Walled Struct.* **116** 37–52

[31] Takahashi K, Tsunoi S, Hara T, Ueno T, Mikami A, Takada H, Ando K, Shiratori 2010 Experimental study of low-cycle fatigue of pipe elbows with local wall thinning and life estimation using finite element analysis *Int. J. Pres. Ves. Pip.* **87** 211-9

[32] Chattopadhyay J, Pavankumar TV, Dutta BK and Kushwaha HS 2005 Fracture experiments on through wall cracked elbows under in-plane bending moment: Test results and theoretical/numerical analyses *Eng. Fract. Mech.* **72** 1461–97

[33] Hilsenkopf P, Boneh B and Sollogoub P 1998 Experimental study of behavior and functional capability of ferric steel elbows and austenitic stainless steel thin-walled elbows *Int. J. Pres. Ves. Pip.* **33** 111–128

[34] Griffiths JE 1979 The effect of cracks on the limit load of elbows under in-plane bending: Experimental study *Int. J. Mech. Sci.* **21** 119–130

[35] Yahiaoui K, Moffat DG and Moreton DN 2000 Piping elbows with cracks. Part II. Global finite element and experimental plastic loads under opening bending *J. Strain Anal.* **35** 47–57

[36] Miyazaki K, Kanno S, Ishiwata M, Hasegawa K, Ahn SH and Ando K 1999 Fracture behavior of carbon steel pipe with local wall thinning subjected to bending load *Nucl. Eng. Des.* **191** 195–204.

[37] ABAQUS Version 6.14. User's manual Simulia RI USA 2018

[38] Data fit Version 9.0. Oakdale Engineering Oakdale PA USA 2009

Design and Vibration Analysis of Pump Assembly for Two Types of Cushy Mount

CH Mounika[1], **K Annamalai***[2], **S Vinoth kumar**[3], **Ponraj A**[4]
[1]M.Tech (CAD/CAM), School of Mechanical Engineering, VIT University, Chennai-600127, India.
*[2] Professor, School of Mechanical Engineering, VIT University, Chennai-600127, India.
[3] Mechanical Engineer, Chennai 600127, India.
[4] Mechanical Engineer, Chennai-600127, India.

Email:
[1] cmounika58@gmail.com
[2] kannamalai@vit.ac.in
3 bosrevs@gmail.com
4 ponraj.alagesan@gmail.com

Abstract. This paper is mainly focused on vibration characteristics of pump motor assembly and selecting suitable mount for the assembly using FEA and experimental approach. Pump motor assembly is connected to base structure by mount. Modal and FRF Analysis are performed to study the mode patterns, natural frequencies and vibration response of the entire structure. Experimental study helps to evaluate the vibration amplitude values of the structure. FFT analyser is used to measure the RMS acceleration of the entire assembly, BK (Bruel & kjaer) Connect software is used to extract the data from FFT analyser. Comparison study reveals and evaluates the vibration characteristics of two mounts and the best mount is suggested.The first section in your paper.
Keywords*:* Modal and FRF Analysis, FFT analyser, BK Connect.

1. Introduction:

Pump-Motor assembly is also known as power unit in injection molding machine. Pump-Motor assembly is placed on the cushy mounts. The type of pump used in this project is “Servo Pump”. They precisely convert electric energy into the hydraulic energy. While the machine is running, the produced vibration is transferred to the cushy mounts. Here Fast Fourier Transform (FFT) Analyzer is used for recording and predicting the data of vibration during experimentation. The first cushy mount is replaced with a second mount and again the analysis is done.

Podugu R et al [1], modal analysis of a centrifugal pump was performed to determine the natural frequencies using finite element approach. The first 10 natural frequencies were compared with centrifugal pump operating speed as per HIS (Hydraulic institute standards). Results showed that first natural frequency of pump 63.25Hz was found matching with pump operating speed of 62.25Hz which is having very less margin of 1.2%. As per HIS, 10% of variation has to be maintained between natural frequencies and pump operating speed. The model was further modified by adding stiffness to the pump pedestal so that natural frequency was increased to 74.31Hz which was above the limit as per HIS clause. Scheffer C et al [2], this study is an overview of vibration-based condition monitoring techniques for pumps. Vibration study and condition monitoring were used to identify faults in pumps. The main advantage of condition monitoring was to schedule the maintenance in advance. Specific technical problems can be identified and rectified with the implementation of vibration-based techniques which increases reliability and efficiency. Sinha JK [3], modal analysis was performed on the pump assembly to determine the root cause failure of the bearing. The obtained natural frequencies at 67.96Hz was matching with the pump components was found to be the main reason for failure. Analysing complete dynamic characteristics of pump will be beneficial for improving the performance compared to conventional condition monitoring. Tenali N et al [4], this paper states that vibration study and fault diagnosis are efficient techniques to find the defects in rotating shaft. Defects like unbalance, dry friction, misalignment and the surface cracks are the main reasons for vibration. The life of the rotating member can be identified by its peak value. The type of peaks in the spectrum indicates the type of fault. Results concludes that time domain and frequency domain techniques are used to determine the faults in the shaft, frequency domain technique can identify the location of fault. Ganesh et al [5], this study is to diagnosis and resolve vibration problem of vertical centrifugal pump using experimental and numerical approach. FFT analyser, impact hammer was used to carry out experimental vibration and modal analysis. Vertical centrifugal pump along with base frame were modelled using CATIA. Modal and harmonic analysis were performed in ANSYS to evolute natural frequencies and vibration amplitude respectively. Results showed that an error of 5.6% obtained by comparing numerical and experimental values. Peak velocity of 12.3mm/s was measured at coupling cover from experimental vibration analysis. Experimental modal analysis results showed that coupling cover natural frequency of 59.3Hz was close to operating speed. Modification of coupling cover was carried and performed numerical and harmonic analysis and found natural frequency was increased to 137.5Hz and vibration amplitude was reduced to 0.7mm/s.

2. Methodology:

In this research work the methodology as shown in figure 1 is adopted, the vibration analysis is performed on pump motor assembly. Modal analysis is performed on pump motor assembly for two types of cushy mount. Frequency Response Function (FRF) analysis is performed by exciting the pump assembly with 1**G** acceleration as an input at all the mounting points and the frequency range of 0-100Hz. Here the experimentation is done using FFT analyser and the results of two mounts are compared then the best mount is suggested.

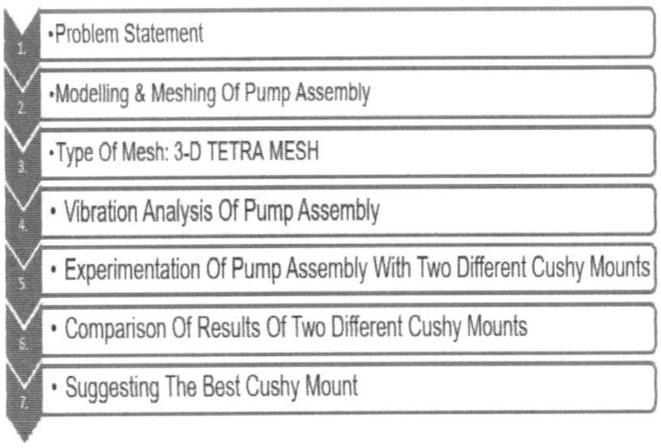

Figure 1. Methodology

3.Geometry:

3D model of pump and motor assembly is modelled using Solidworks and it is shown in figure 2.

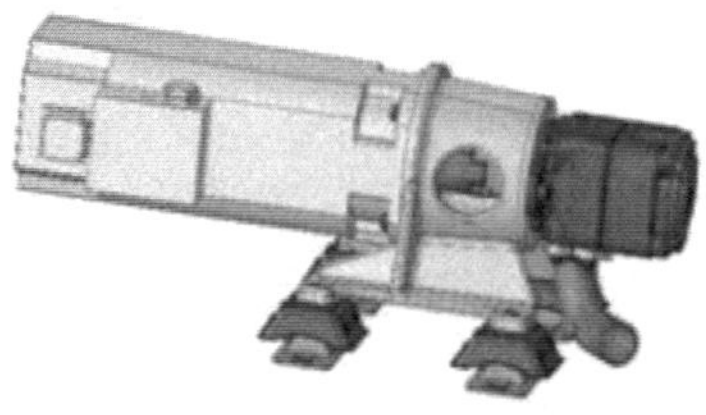

Figure 2. Pump-Motor Assembly

4.Meshing:

Finite element meshing of the pump motor assembly is done using Hypermesh. Here the type of mesh used is 3D tetra mesh with an element size of ten. After meshing, it is need to ensure that the mesh of each component should not intersect and penetrate with another components. And also, element quality (tetcollapse) of the mesh generated checked. Here the weld and bolt connections are used for connecting the different components of the geometry. Finite element meshed model is shown in figure 3 with total number of Elements: 717848 (Tetra Mesh).

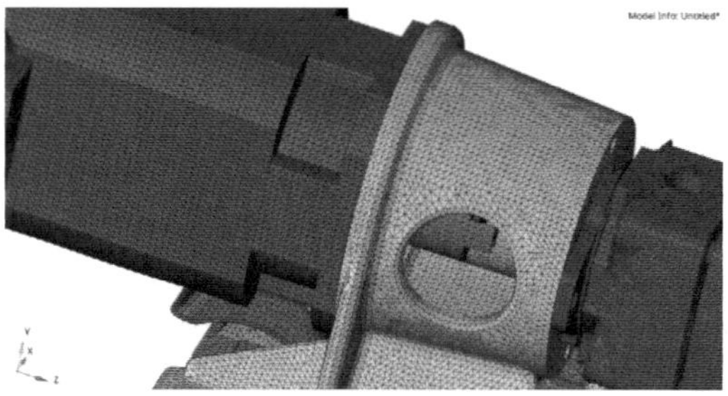

Figure 3. Enlarged View of Meshing

Table 1. Material Properties

Material Properties				
	Material	**Density (kg/m^3)**	**Poisson's Ratio**	**Young's modulus (GPa)**
Pump	FE410	7860	0.285	210
Motor	FE410	7860	0.285	210
Coupling	SS304	8000	0.29	193
Safety Manifold	SG400	7200	0.26	175
Suction Pipe	SG400	7200	0.26	175
Flange	SG400	7200	0.26	175

IOP Conf. Series: Materials Science and Engineering 1128 (2021) 012040 doi:10.1088/1757-899X/1128/1/012040

5.Boundary Conditions:

The following boundary conditions are used for the vibration analysis
Eight mounting points are restricted in all DOF.
Frequency range: 0-100 Hz.
Excited Acceleration: 1g (For FRF).
Figure 4 shows the pump-motor assembly with applied boundary conditions.

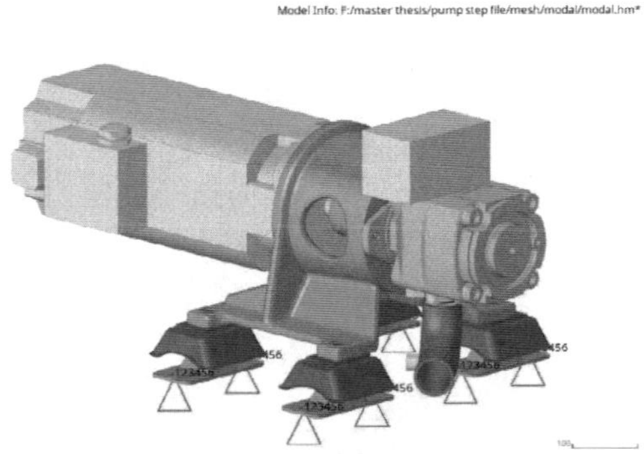

Figure 4. Pump Assembly Showing Constraints

6.Results:

6.1. Modal Analysis Results:

The results obtained through modal analysis are shown in figures 5 to 8 for respectively mode 1 to mode 4. The corresponding frequencies for mode 1 to mode 4 are given table 2.

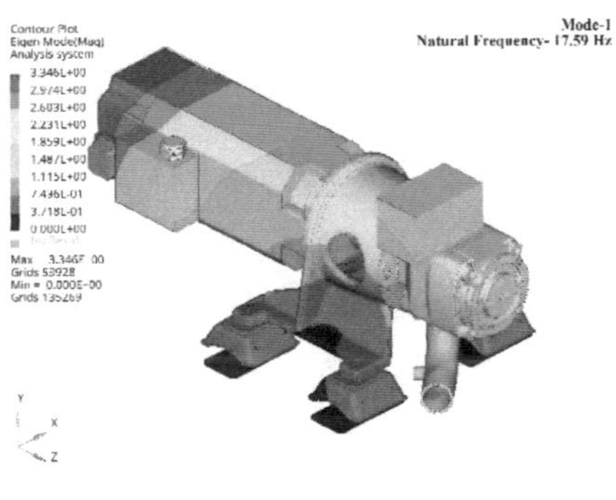

Figure 5. Mode 1 shows the bending mode along X-axis at 17.59 Hz.

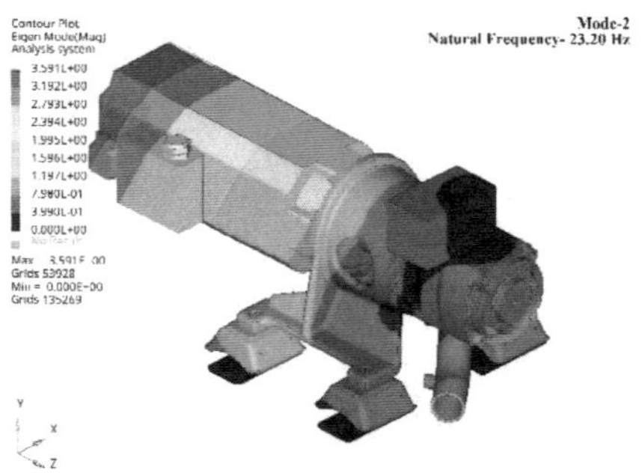

Figure 6. Mode 2 shows the lateral mode along Z-axis at 23.20 Hz.

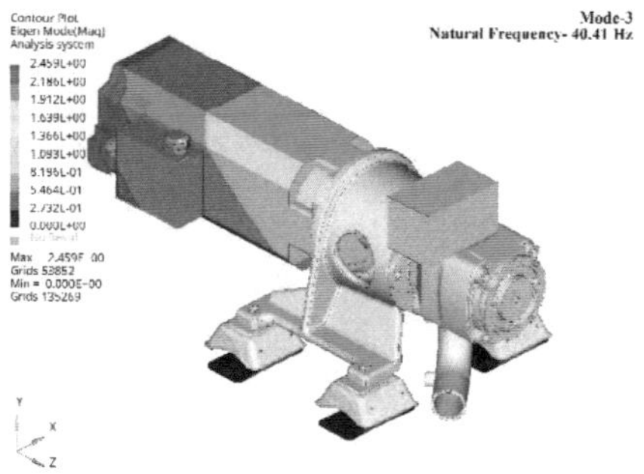

Figure 7. Mode 3 shows the vertical mode along Y-axis at 40.41 Hz.

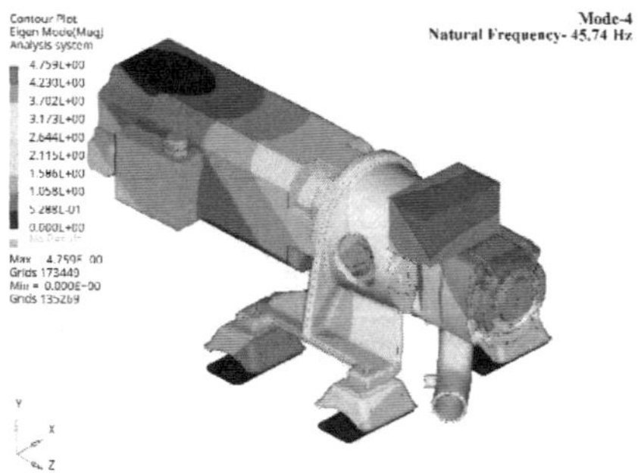

Figure.8. Mode 4 shows the twisting mode along X & Y-axis at 45.74 Hz.

Table 2. Natural frequency for mode 1 to mode 4

Mode	Frequency(Hz)	Type of Mode
Mode 1	17.59	Bending mode along X-axis
Mode 2	23.20	Lateral mode along Z-axis
Mode 3	40.41	Vertical mode along Y-axis
Mode 4	45.74	Twisting mode along X & Y-axis

6.2. FRF Analysis Results:

Operating Frequency of Pump-Motor Assembly – 41.66Hz
The Frequency response analysis is done and the critical points are identified shown in Figure.8.
In X-Direction the peak values are observed at point 1 and 4 with frequency 42Hz. (Figure.9)
In Y-Direction the peak values are observed at point 4 and 1 with frequency 60Hz. (Figure.10)
In Z-Direction the peak values are observed at point 4, 1, 3 and 2 with frequency 61Hz. (Figure.11)

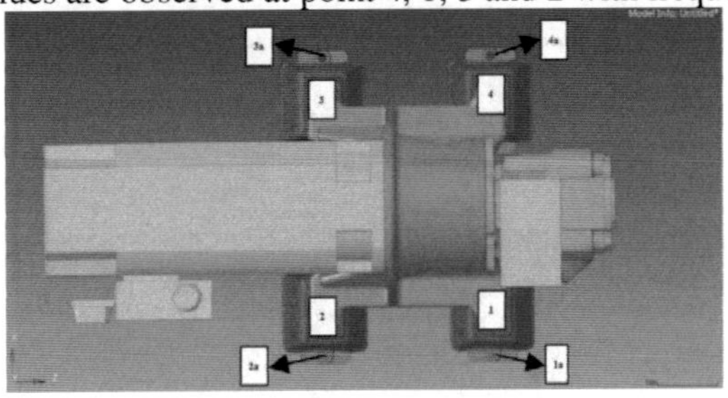

Figure 9. Critical points identified from FRF Analysis Results

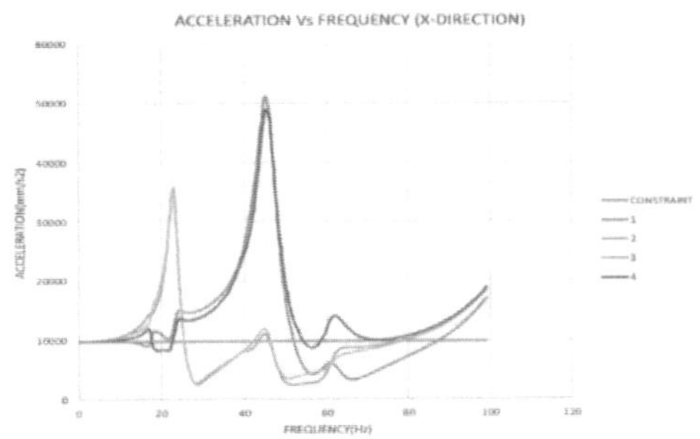

Figure 10. Acceleration Vs Frequency (X-Direction)

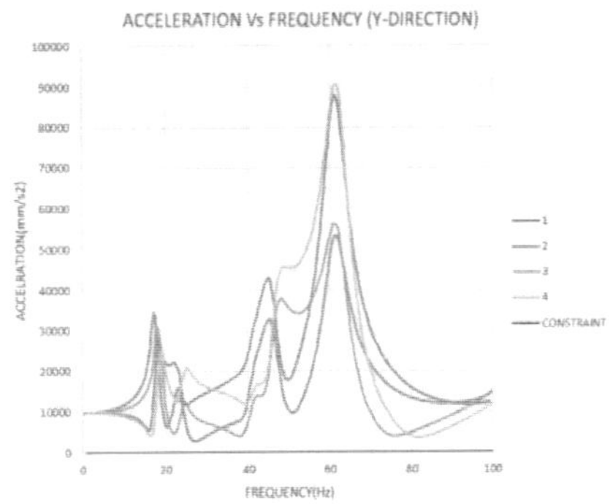

IOP Conf. Series: Materials Science and Engineering 1128 (2021) 012040 doi:10.1088/1757-899X/1128/1/012040

Figure 11. Acceleration Vs Frequency (Y-Direction)

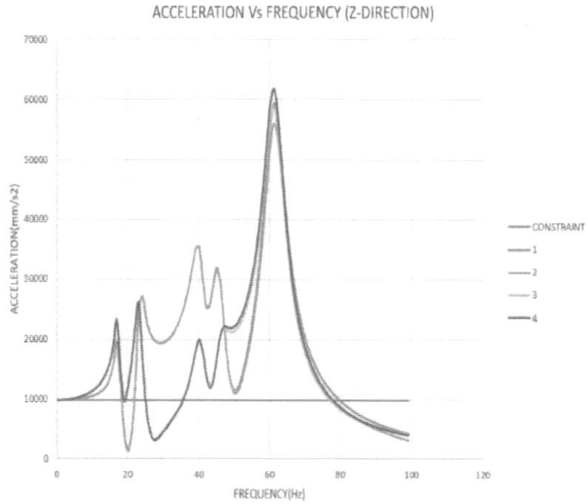

Figure 12. Acceleration Vs Frequency (Z-Direction)

7. Experimental Set Up:

Experimentation is done and the values are recorded at the critical points attained from FRF analysis. Here the FFT analyser and BK connect software are used for recording and analysing the experimental data.

Figure 13. Experimental Setup

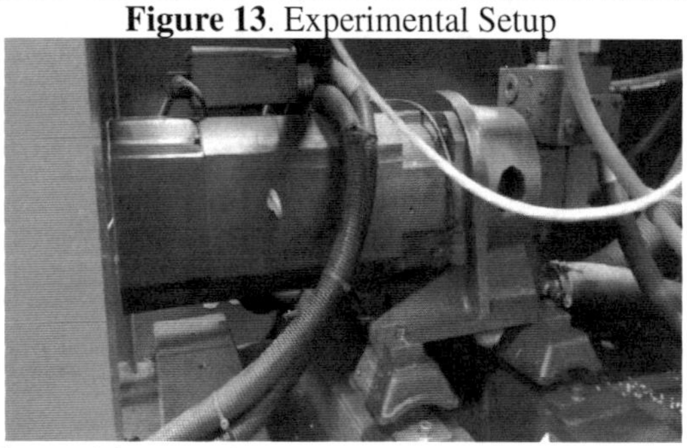

Figure 14. Pump-Motor Assembly with First Mount

IOP Conf. Series: Materials Science and Engineering 1128 (2021) 012040 doi:10.1088/1757-899X/1128/1/012040

Figure 15. Data Acquisition System

Figure 16. Measuring Point

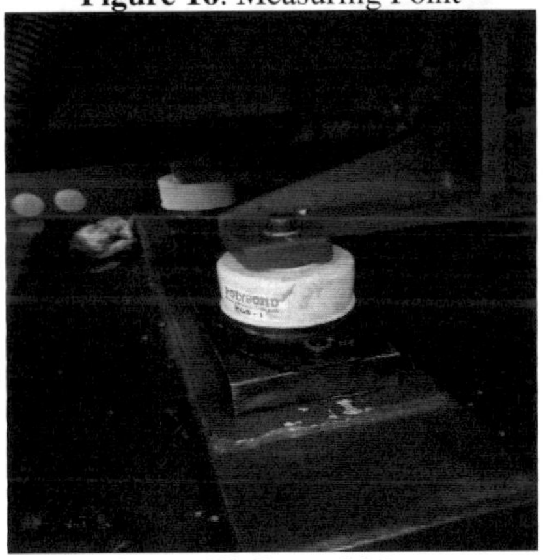

Figure 17. Second Mount

7.1 Experimental Results:

First Mount Results:

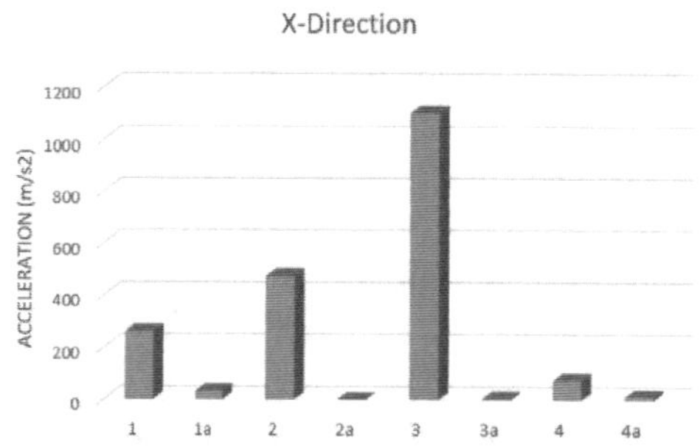

Figure 18. Acceleration Bar Graph in X-Direction

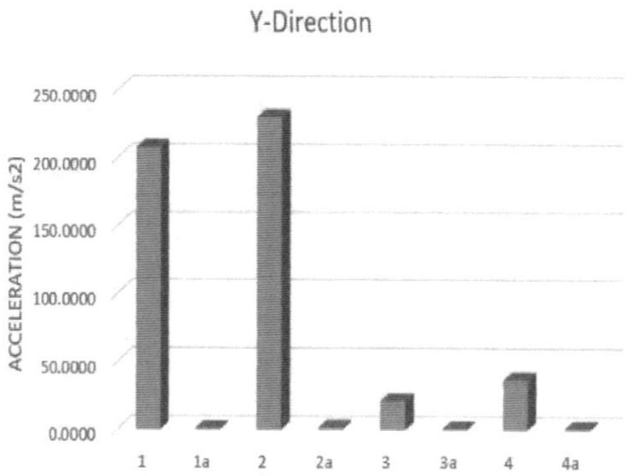

Figure 19. Acceleration Bar Graph in Y-Direction

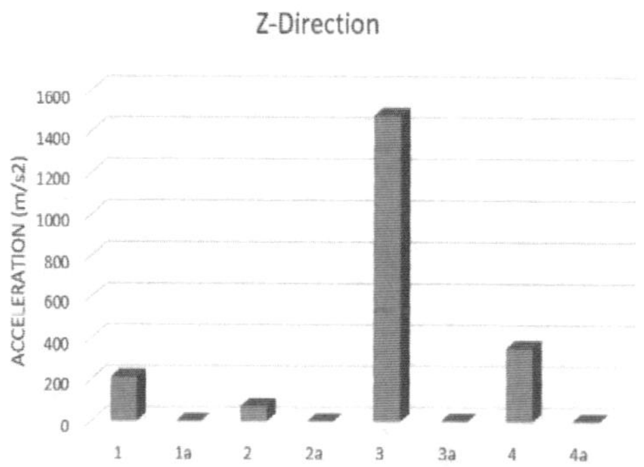

Figure 20. Acceleration Bar Graph in Z-Direction

Second Mount Results:

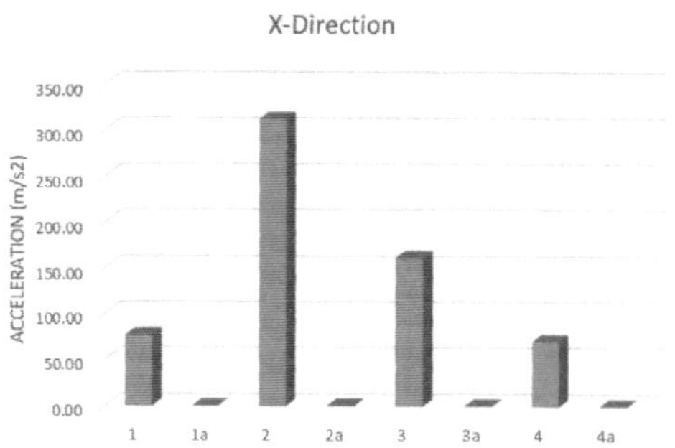

Figure 21. Acceleration Bar Graph in X-Direction for Second Mount

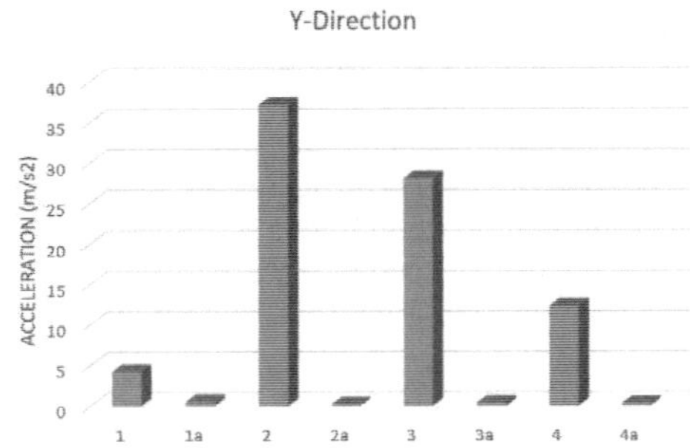

Figure 22. Acceleration Bar Graph in Y-Direction for Second Mount

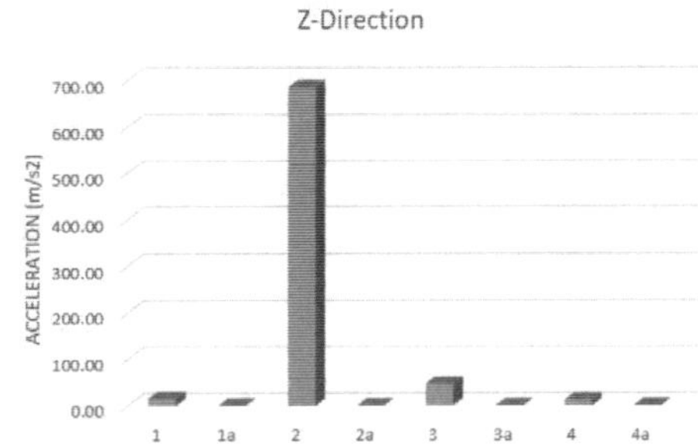

Figure 23. Acceleration Bar Graph in Z-Direction for Second Mount

Table 3. Comparison of First and Second Mount Results at Machine Base

Point	Acceleration(m/s2)					
	X-Direction		Y-Direction		Z-Direction	
	First Mount	Second Mount	First Mount	Second Mount	First Mount	Second Mount
1	260.74	76.98	207.52	4.25	212.3	15.78
2	473.97	312.09	229.66	37.17	72.14	68.46
3	1101.81	161.79	28.1	21.24	1477.7	48.18
4	76.28	70.86	37.01	12.32	354.38	11.97

Point	Acceleration(m/s^2)					
	X-Direction		Y-Direction		Z-Direction	
	First Mount	Second Mount	First Mount	Second Mount	First Mount	Second Mount
1a	32.48	0.45	1.26	0.53	1.43	0.07
2a	2.13	0.84	1.76	0.2	1.55	0.08
3a	5.02	0.48	1.06	0.39	2.54	0.07
4a	13.17	0.26	1.36	0.31	2.1	0.1

8. Conclusion:

From Table. 2. Comparison of first and second mount results at cushy mount points, the acceleration values of first mount are higher than second mount in all three directions. From Table. 3. Comparison of first and second mount results at machine base points, the acceleration values of first mount are higher than second mount in all three directions.

Considering the results of first and second mount at both the cushy mount and machine base points, the second mount is better than the first mount.

References

[1] Podugu R, Kumar JS, Ramanamurthy B, Kumar N. *A modal approach for vibration analysis and condition monitoring of a centrifugal pump. International Journal of Engineering Science and Technology* (IJEST). 2011;10.

[2] Scheffer C. *Pump condition monitoring through vibration analysis*. InPumps: Maintenance, Design, and Reliability Conference 2008 Apr 2 (pp. 1135-1146).

[3] Sinha JK, Rao AR. *Vibration based diagnosis of a centrifugal pump. Structural Health Monitoring*. 2006 Dec;5(4):325-32.

[4] Tenali N, Babu PR and Kumar KC. *Vibrational Analysis in Condition Monitoring and faults Diagnosis of Rotating Shaft-Over View. International Journal of Advanced Engineering Research and Science*.;4(1).

[5] Ganesh, Potnuru, and S. Rama Krishna. "*Diagnosis and Resolution of Vibration Issues in Vertical Centrifugal Pump*." Journal of Failure Analysis and Prevention (2020): 1-10.

[6] Patidar S and Soni PK. *An overview on vibration analysis techniques for the diagnosis of rolling element bearing faults*. International Journal of Engineering Trends and Technology (IJETT). 2013 May;4(5):1804-9.

[7] Farokhzad S, Bakhtyari N and Ahmadi H. *Vibration signals analysis and condition monitoring of centrifugal pump*. Technical Journal of Engineering and Applied Sciences. 2013; 1081:1081.

[8] Onari MM, and Boyadjis PA. *Solving structural vibration problems using operating deflection shape and finite element analysis*. In Proceedings of the Twenty-fifth International Pump Users Symposium 2009 pp. 85-102.

Design and Vibration Analysis of Injection Moulding Machine Base Structure

CH Kesava Manikanta Kumar[1], **K Annamalai***[2], **S Vinoth kumar**[3], **Ponraj A**[4]
[1] PG Student, School of Mechanical Engineering, VIT University, Chennai- 600127, India.
[2] Professor, School of Mechanical Engineering, VIT University, Chennai-600127, India.
[3] Mechanical Engineer, Chennai 600127, India.
[4] Mechanical Engineer, Chennai-600127, India.

Email:
[1] kesava4802@gmail.com
*[2] kannamalai@vit.ac.in
[3] bosrevs@gmail.com
[4] ponraj.alagesan@gmail.com

Abstract. In this study the dynamic behaviour and vibration parameters of Injection Moulding Machine Base Structure was performed by analysing the structure numerically as well as experimentally. Modal analysis and FRF analysis performed numerically to find the mode pattern, natural frequencies and critical points. Vibration parameters are measured by performing experimentation using FFT analyser and the data was taken by using BK (Bruel & kjaer) Connect software. The Fast Fourier Transform (FFT) analyzer is used to transform the continuous time domain data to continuous frequency domain data. The experimental results show that structure is vibrating high at clamping unit location along x-direction. Vibration level can be controlled at that particular location by increasing the stiffness of the structure.
Keywords: Modal and FRF Analysis, FFT analyser, BK Connect.

1. Introduction

Injection moulding is the cyclic process of producing plastic parts. Injection moulding machine consists of various units for performing several functions and all these units are mounted over the base frame. Base mainly consists of three frames. They are top frame, bottom frame and connecting vertical frame. All the frames are C-channel sections. Connecting vertical frame transfers weight and produced vibration to the bottom frame which is connected to the anti-vibration mount. During experimentation, Fast Fourier Transform (FFT) Analyzer is used for recording the data of vibration at differentconditions.

Mathes et al [1] this paper describes about vibration parameters of two-pole induction motor. The natural and forced vibration of motor are evaluated by numerical calculations using finite element approach and validated using experimental modal analysis. Forced vibrations are calculated when the motor is operated by convertor. By using FFT analyser frequency spectrum was analysed. Torres-Martinez R et al [2], This study includes static and dynamic analysis of high-speed machining Al-Cu alloy lathe bed. Vibration characteristics like deflection, natural frequencies and vibration amplitudes were calculated by performing simulations on FEM model. A cast iron bed was considered as a parametric model used in conventional speed machining. Deflection and natural frequencies of the machine structure was determined as per ISO starndards. This study concludes that Al-Cu alloy structure was more feasible than cast iron lathe bed. Jiao M et al [3], Evaluated dynamic and static characteristics of a large machine bed by using finite element analysis. Structural topology optimization was performed for the structure. Results showed that 8.58% of mass was reduced 7.41% of stiffness was improved. Static and dynamic properties are improved for the phase of design. Yang H et al [4], The author analysed CK61125 CNC lathe bed using FEA approach. Performed static and dynamic analysis on the lath bed and found maximum stress, strain and deformation as 100.98Mpa, 0.615 and 0.1455mm respectively. The natural frequencies of the lathe bed were obtained and results shows that 271.63 Hz, 290.41Hz and 305.88 Hz were first three resonant frequencies. Results shows that vibration characteristics of lathe bed gives strong theoretical parameters for the machine tool. Hong CC et al [5], the turning-milling CNC machine's primary, secondary shafts and machine bed were analysed. They selected the maximum displacement and natural frequencies values as the basic data to design the CNC machine in safety condition for avoiding resonance. The natural frequencies, linear dynamic stresses and displacements of CNC machine are obtained by using the SOLIDWORKS simulation module.

2. Methodology:

In this study vibration analysis is performed both numerically and experimentally. Modal analysis is done and the behavior of the mode shapes are identified. Frequency Response Function analysis is performed by exciting the machine base with 1G acceleration and a frequency range of 0-100Hz. Experimentation is performed and the data is acquired using FFT analyser. By observing the results,the idea for reduction of vibration is suggested.

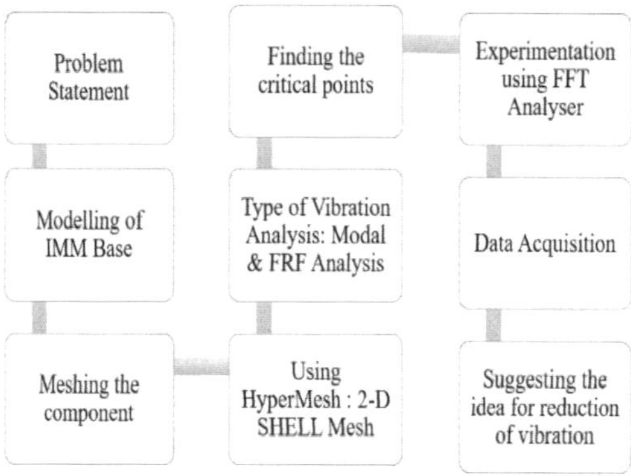

Figure 1. Methodology

3. Geometry:

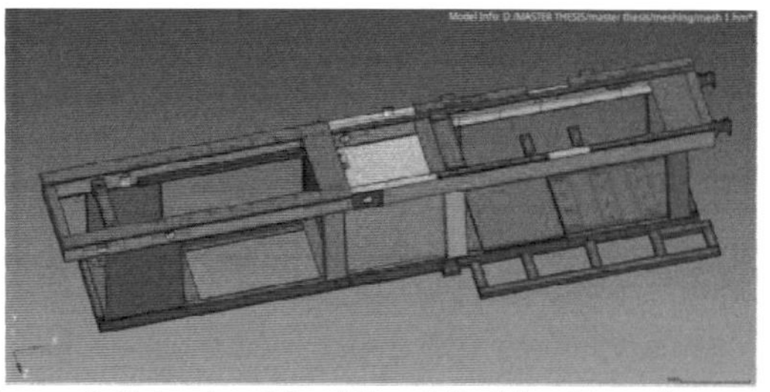

Figure 2. Injection Moulding Machine Base Structure

IOP Conf. Series: Materials Science and Engineering 1128 (2021) 012041 doi:10.1088/1757-899X/1128/1/012041

4. Meshing:

The meshing of the geometry is done in Hyper mesh. Here the type of mesh used is Shell mesh with an element size of fifteen. After meshing, we need to ensure that the mesh of each component should not intersect and penetrate with another components. We need to check the quality index of the mesh generated. Here the weld connections are used for connecting the different components of the geometry.

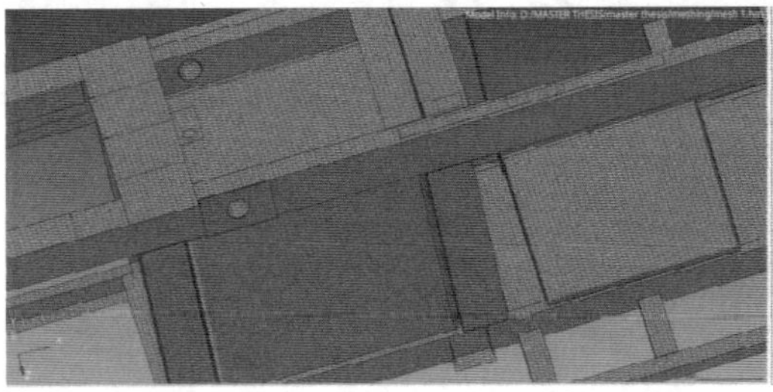

Figure 3. Enlarged View of Mesh

Number of Elements: 86797
Material Properties:
All the properties of the material are assigned to each and every component along with their thickness values.

- Material: FE410
- Density: 7860 kg/m^3
- Poisson's Ratio: 0.285
- Young's modulus: 210 GPa
- Weight of Injection unit: 1Tonne
- Weight of Clamping unit: 3Tonne

5. Boundary Conditions:

- Eight mounting points are restricted in all DOF.
- Frequency range: 0-200 Hz.
- Excited Acceleration: 1G (For FRF).
- The injection and clamping unit masses are laid through their CG points on the base.

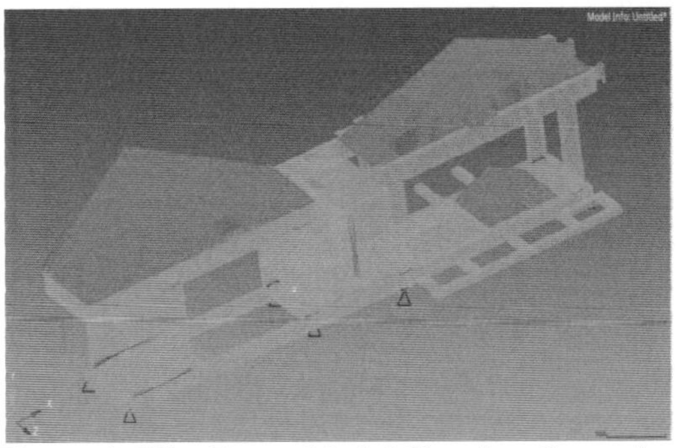

Figure 4. Wire Frame Modal Showing Constraints

6. Results:

6.1 Modal Analysis Results:

Modal analysis is done and the natural frequencies are obtained and from this modal analysis the behaviour is identified from its mode shapes.

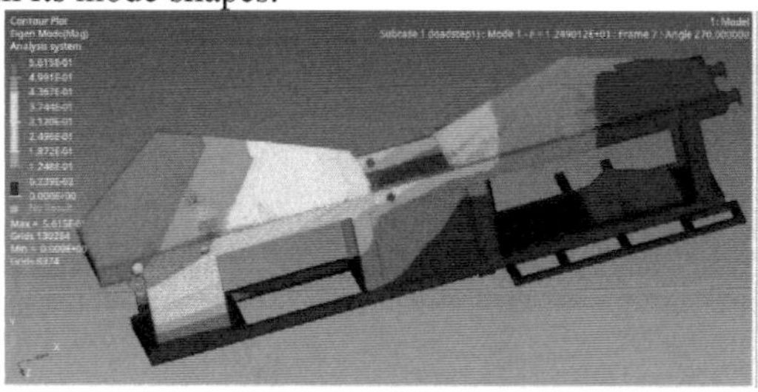

Figure 5. Mode No.1

From Figure.5, the type of mode is lateral mode along X-axis at frequency 12.49 Hz.

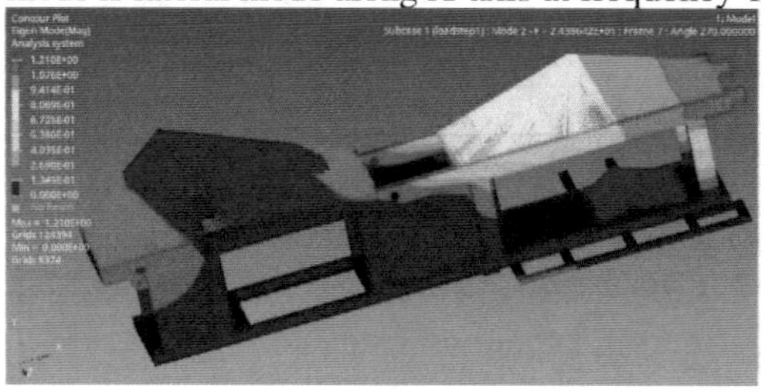

Figure 6. Mode No.2

From Figure.6, the type of mode is mixed mode along X & Y-axis at frequency 24.38 Hz.

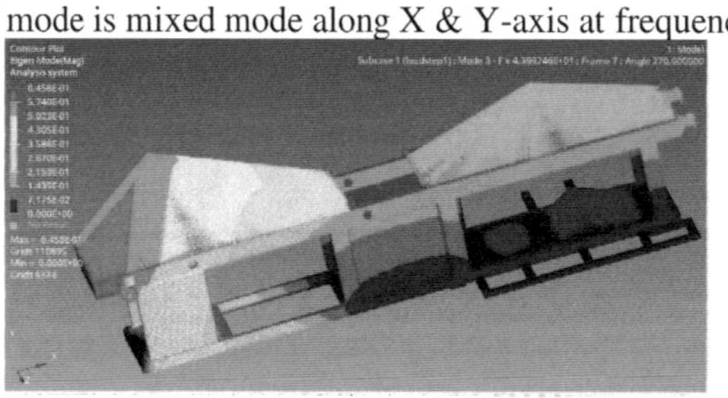

Figure 7. Mode No.3

From Figure.7, the type of mode is bending mode along Z-axis at frequency 43.99 Hz.

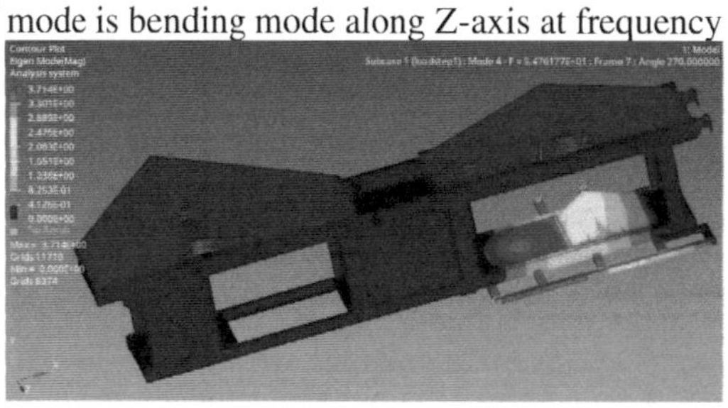

Figure 8. Mode No.4

From Figure.8, the type of mode is vertical mode along Y-axis at frequency 54.76 Hz.

IOP Conf. Series: Materials Science and Engineering 1128 (2021) 012041 doi:10.1088/1757-899X/1128/1/012041

Table. 1 Types of Modes

Mode	*Frequency(Hz)*	*Type of Mode*
Mode 1	12.49	Lateral mode along X-axis
Mode 2	24.38	Mixed mode along X & Y-axis
Mode 3	43.99	Bending mode along Z-axis
Mode 4	54.76	Vertical mode along Y-axis

Operating Frequency – 41.66Hz

6.2 Frequency Response Analysis (FRF):

The Frequency response analysis is done and the critical points are identified shown in Figure 9.
In X-Direction the peak values are observed at point 3 and 10 with frequency 44Hz. (Figure.10)
In Y-Direction the peak values are observed at point 3 and 19 with frequency 44Hz. (Figure.11)
In Z-Direction the peak values are observed at point 15 with frequency 24Hz and 4, 19 with frequency 12Hz (Figure.12)

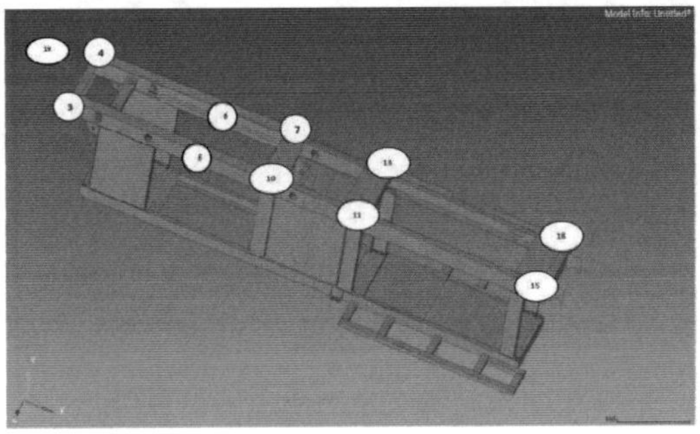

Figure 9. Critical Points of FRF Analysis

Figure. 9, shows the critical points of the model in FRF analysis The responses of the structure are captured at these critical points by exciting the structure with 1G acceleration.

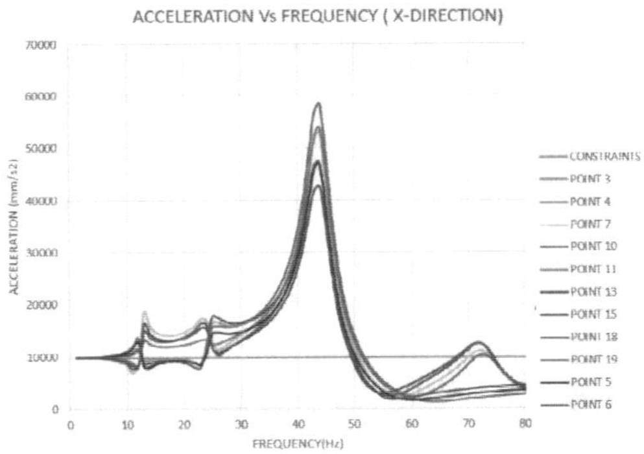

Figure. 10. Acceleration Vs Frequency (X-Direction)

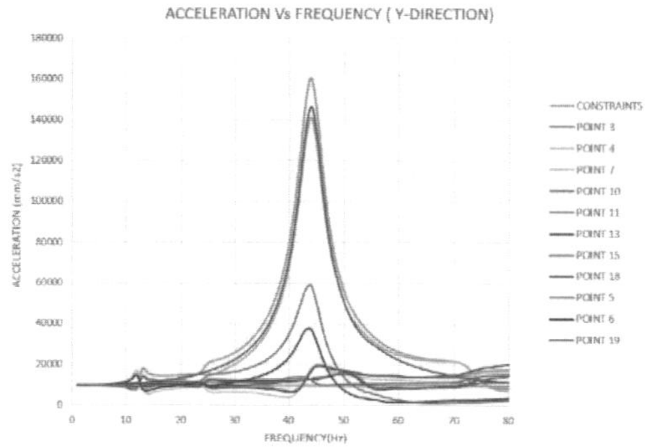

Fig. 11. Acceleration Vs Frequency (Y-Direction)

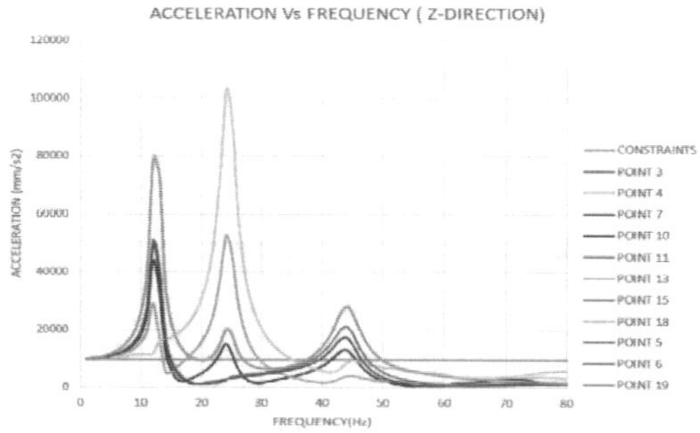

Figure. 12 Acceleration Vs Frequency (Z-Direction)

7. Experimentation Set Up:

The Experimentation is done and the values are recorded at the critical points attained from FRF analysis. Here the FFT analyser and BK connect software are used for recording and analysing the experimental data.

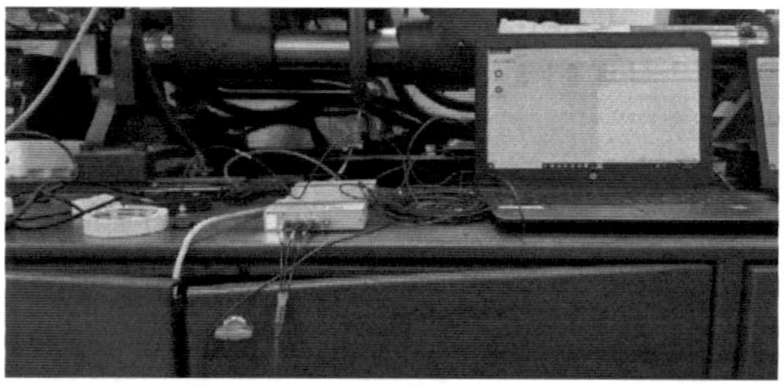

Figure 13. Experimental Setup

Figure 14. Injection Moulding Machine with Experimental Setup

IOP Conf. Series: Materials Science and Engineering 1128 (2021) 012041 doi:10.1088/1757-899X/1128/1/012041

Figure 15. Data Acquisition System

Figure 16. Measuring Point

7.1 Experimental Results:

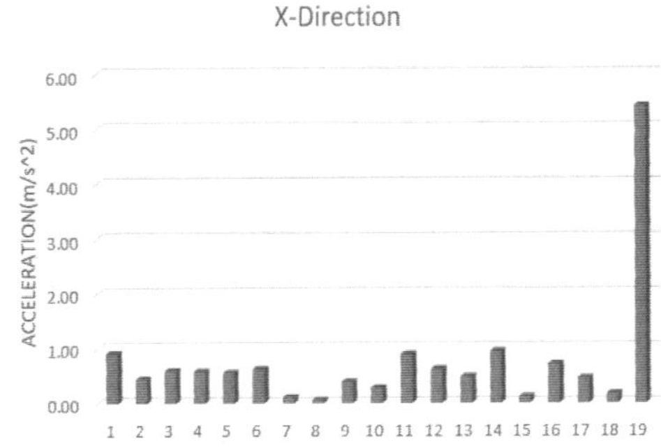

Figure 17. Acceleration Graph in X-Direction

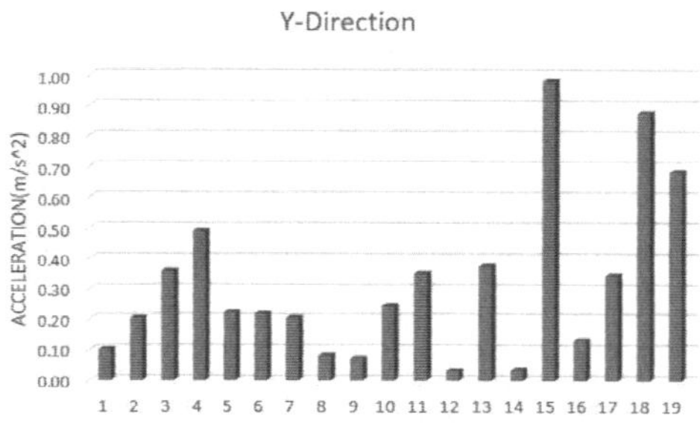

Figure 18. Acceleration Graph in Y-Direction

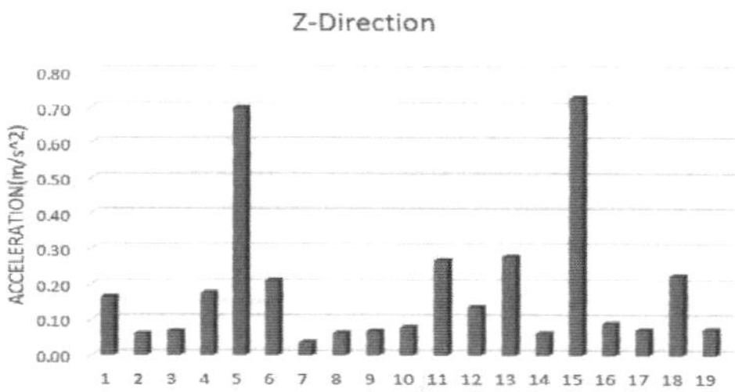

Figure19. Acceleration Graph in Z-Direction

Table 2. Experimental Results of All Points

Point	Acceleration(m/s^2)		
	X-DIRECTION	Y-DIRECTION	Z-DIRECTION
1	0.91	0.10	0.16
2	0.44	0.20	0.06
3	0.59	0.36	0.07
4	0.58	0.49	0.18
5	0.56	0.22	0.70
6	0.62	0.22	0.21
7	0.11	0.21	0.04
8	0.06	0.08	0.06
9	0.39	0.07	0.07
10	0.20	0.25	0.08
11	0.89	0.35	0.27
12	0.62	0.03	0.14
13	0.49	0.38	0.28
14	0.95	0.04	0.06
15	0.12	0.98	0.73
16	0.71	0.13	0.09
17	0.46	0.35	0.07
18	0.18	0.88	0.22
19	5.42	0.68	0.07

8. Conclusion:

From the experimental results, the acceleration values in X-Direction are observed high at point 19 i.e., 5.42 m/s^2. In Y-Direction, the highest acceleration value is observed at point 15 i.e., 0.98 m/s^2. In Z-Direction, the highest acceleration value is observed at point 15 i.e., 0.73 m/s^2. The acceleration values in both Y and Z-Directions are less than 1 m/s^2, but when compared to these directions the acceleration value in X-Direction is very high. So, the maximum vibration is producing at the point 19. The vibration amplitude at that particular location in the structure can be reduced by introducing ribs which serves as support and increases stiffness of the structure.

References

[1] Mathes S, Werner U, Bauer C, *Numerical and experimental vibration analysis of a two-pole induction motor mounted on an elastic machine test bed.*
[2] Torres-Martinez R, Urriolagoitia-Calderón G, Urriolagoitia-Sosa G and Espinoza-Bustos R, *Stress and vibration analysis of a lathe bed made of aluminum-copper alloy for high speed machining. In Applied Mechanics and Materials 2009* (Vol. 15, pp. 81-88). Trans Tech Publications Ltd.
[3] Jiao M, Guo XH, Wan DD 1 *Finite element analysis and lightweight research on the bed of a large machine tool based on HyperWorks. In Applied Mechanics and Materials 2012* (Vol. 121, pp. 3294-3298). Trans Tech Publications Ltd.
[4] Yang H, Zhao R, Li W, Yang C, Zhen L, *Static and dynamic characteristics modeling for CK61125 CNC lathe bed basing on FEM. Procedia engineering.* (2017 Jan 1;174:489-96.)
[5] Hong CC, Chang CL, Lin CY *Dynamic Structu`ral Analysis of Great Five-axis Turning-Milling Complex CNC Machine.* Global Journal of Research in Engineering. 2017 Jul 31.
[6] Zhang YW, Zhang WM Analysis of *Dynamics Characters of Bed Structure of CNC Machine Tool on FEM Method. In Applied Mechanics and Materials 2012* (Vol. 141, pp. 208-211). Trans Tech Publications Ltd
[7] Swami BM, Kumar KS, Ramakrishna CH. *Design and Structural Analysis Of CNC Vertical Milling Machine Bed.* International Journal of Advanced Engineering Technology. (2012;3(4):97-100).
[8] Li X, Zhao Z. *Dynamic analysis on and optimized design of the BED structure of CNC machine. In2011 Chinese Control and Decision Conference* (CCDC) 2011 May 23 (pp. 3998-4002). IEEE.
[9] Selvakumar A, Ganesan K, Mohanram PV *Dynamic analysis on fabricated mineral cast lathe bed. Proceedings of the Institution of Mechanical Engineers*, Part B: Journal of Engineering Manufacture. (2013 Feb;227(2):261-6.

Study of thermal behavior of Aluminium foams in batteries of electric vehicles

S. Harish[1], S. Jeyanthi[2*], M. Abinash[3], R. Suresh Kumar[4], Sahil Agarwal[5] G. Venkatachalam[6]

School of Mechanical Engineering, Vellore Institute of Technology, Chennai, Tamil Nadu 600127, India
Email address:jeyanthi.subramanian@vit.ac.in1

Abstract : The temperature of batteries in electric vehicles increases with time due to internal chemical reactions. This temperature rise is undesirable because it promotes corrosion and affects the battery life and thus a coolant can be used to prevent the temperature rise. The use of lightweight material and thermal management properties have a huge impact when designing a battery, especially, when using Aluminium foams it gives an exceptional advantage ranging from lightness, ductility to excellent absorption and conductivity making it a great deal for the future-oriented development of electric vehicles and market acceptance. The current research work tries to prove that Al foam can be used as a coolant in batteries of electric vehicles because of its porosity and density. The battery and Al foam are modelled in SOLIDWORKS and simulations are conducted in CFD and transient thermal to observe the heat transfer, flow characteristics and cooling effect on foam surface. The pores in Al foam are placed in three different patterns and two sizes (3mm and 5 mm) while modelling and hence six distinct models of Al foam are taken into study. The results from the simulations are compared and the Al foam that reduces maximum temperature at the outlet is chosen as the best coolant.

Keywords: Al Foam; Battery; Electric Vehicle; CFD; Thermal management

1. Introduction

The development in using metallic foams have become an enticing research field in scientific view point, industrial and various other daily-life applications. In this study it was focused on minimizing the required weight, improved efficiency and various other parameters that play a vital role in determining thermal characteristics. Metal foams are those that basically have low density while at the same time they exhibit high strength porous medium. One such novel material that possessed the required characteristics is the open cell Al foam. This open cell Al foam is identified by pore density which further gives us an extraordinary advantage like lightness, ductility, excellent absorption and conductivity. Al foams are manufactured through various methods such as direct foaming of Al alloy melting and direct metal powder processing. Based on the above properties, these Al foam is suitable for various application in vehicle design of electric vehicles like lightweight constructions, impact energy absorptions, acoustic damping and thermal insulations.

One of the vital fields where Al foams plays an integral role is the electric vehicle car design. Usually in electric vehicles they prove to be of greater impact in global warming as they satisfy with the reduction level of carbon dioxide. These metal foams when used as cores for the sandwich panels, provide an unexceptional stiffness strength in conjunction with the low weight, also their energy absorption doesn't allow them for larger deformation under constant stress. This Al foam sandwich technology acts as a vibrational damper by its design also helps in reduction of cost while improving its performance.

Many researchers were working in the area of metal foams and tried to study their behavior. Bhattacharya et al. [1] was the one who started the study on open cell foam's analytical characteristics like permeability, thermal conductivity along with their pore density. Some authors attempted to use PCM along with metal foam for better results. Lafdi et al. [2] had studied about the PCM in a highly conductive metallic foam along with which thermal conductivity of composites with different porosity and pore size. It had also been earlier proved that in the TMS design, the battery is well maintained at a quasi-state temperature as the latent heat of PCM helps in storing and releasing large quantities of heat. The other authors who worked on PCM are Morteza Alipanah [3] and Z.G. Qu [4] .The former presented about the formation of foam PCM(Phase Change Material) composite by saturation of paraffin into pores of copper foam and had observed that the thermal conductivity of PCM had depicted immense results by adding the foam structure whereas, the latter investigated on comparison in reduction of temperature using Al foam with different porosity and PCM such as Octadecane and Gallium. Further, Alvarez–Hernandez [5] and Dukhan [6] had presented speculative results for the pressure drop of airflow through samples of open cell aluminum foam for distinguished porosities and pore densities. Also, with respect to the Darcy-Forcheimer equation, there had been a greater decrease in the pressure drop for the higher porosity foam. Mohammadian et al. [7] had investigated the lithium-ion battery module through various simulations which talks about the effects of Al foam inside a pin fin heat sink. Numerous studies were conducted over past decades regarding sandwich structures consisting of metal foam core and composite material skin [8, 9]. These Al alloy were used for both withstanding static loads and impact energy and most importantly they have been used as a core material for sandwich structures and thin walled anti-buckling systems [10, 11]. The consistency of temperature among various cells was found to be critical to the battery pack's safety as large temperature differences were obtained which could cause a decrease in the overall battery pack's capacity [12].

Many authors tried to analyze the foam characteristics by placing them in a sandwich pattern. Lip Huat Saw [13] had studied about Al foam being used as a coolant medium and its impact on thermal management of Li-ion battery by which CFD analysis were carried out to check results with and without Al foam. Boomsma et al. [14] had discussed Al metal foam's forced convection in its various compressed parts and had been proved that they establish a good heat transfer and increased efficiency compared to the other conventional heat exchangers. Quasi-static and dynamic compression were the two tests that were conducted to determine the physical strength and mechanical properties of $AlSi_{10}$ - Epoxy hybrid foams. Locating the granules of polymer coated Al foam between the Al face sheets were the idea behind implementing the sandwich technology [15].

This study revolves on less weight of the battery system and also its cooling effect down on the battery with the help of metal foam. The pores in Al foam enhances the heat transfer characteristics and promotes cooling and hence, six different Al foam models with varying pore sizes and pore pattern is analyzed. Numerical simulations are carried out on battery and six different foam models in CFD and transient thermal platforms of ANSYS. The results of heat flux and temperatures are tabulated and compared in order to determine the pore size and pore pattern suitable for efficient cooling.

NOMENCLATURE

Al ALUMINIUM

CFD COMPUTATIONAL FLUID DYNAMICS

ρ DENSITY [kg/m^3]

k THERMAL CONDUCTIVITY [W/m-K]

μ VISCOSITY [kg/m-s]

C_p SPECIFIC HEAT AT CONSTANT PRESESURE [J/kg-K]

Table1: Thermophysical properties of multiphase materials

MATERIALS	ρ[kg/m^3]	k[W/m-K]	μ [MPa]	C_p[J/kg-k]
Al Foam	499.6	9.899	7.4	934.698509
Copper	8978	387.6	-	381
e-material(battery)	2092	18.2	-	678
t-material(terminal)	8978	387.6	-	381

2. Methodology

2.1. Geometry description

An open cell Al foam 3d model with dimensions of 106mm x 49mm is considered for the study along with a battery with copper as its material. The numerical simulation of battery (Figure 1) is done to observe its temperature contours and a main perspective of decreasing the temperature inside it. Therefore , simulations are done for six Al foam models which consist of 3 distinct patterns namely square, triangle and an irregular pattern and 2 different pore sizes - 3mm and 5 mm as shown in (Figure 2). There is an inlet, outlet, pores and two adiabatic walls (top and bottom) present in Al foam. The temperature from the battery outlet is fed to foam inlet, by heat spreader. The heat transfer rate and flow behavior in the foam is studied using ANSYS. This is done by combining two different platforms in ANSYS- Fluent Flow (CFD) and Transient Thermal. CFD is performed to determine the air flow in the pores while transient thermal is used for comparing the temperature and heat flux of various patterns.

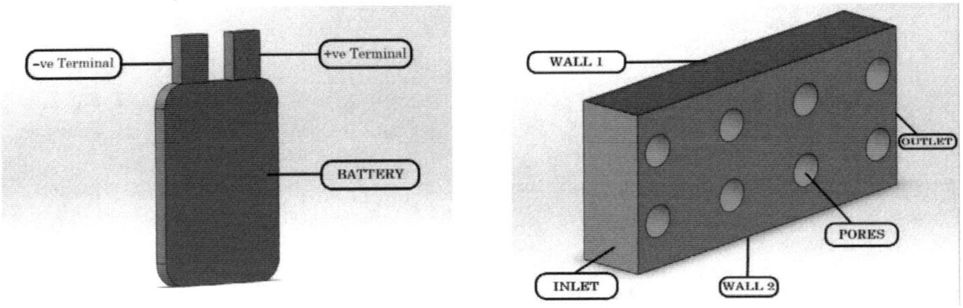

Figure 1: Schematic diagram of battery and Al foam

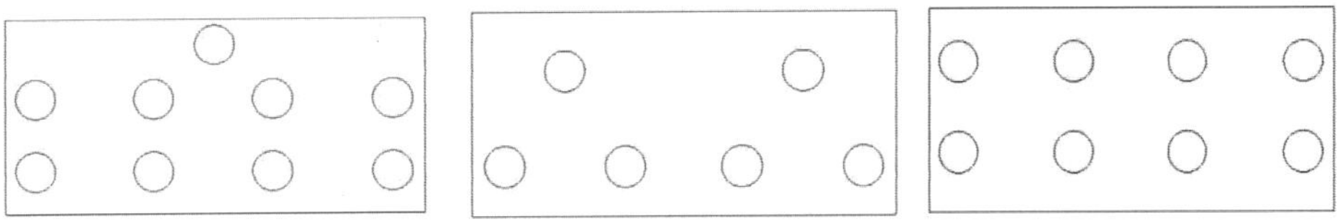

Figure 2: Schematic diagram of pores pattern on Al foam surface

2.2. Meshing

Fine meshing is done near the pores for 3 different patterns of various pore sizes by maintaining a proper span angle center and a resolution of 2.

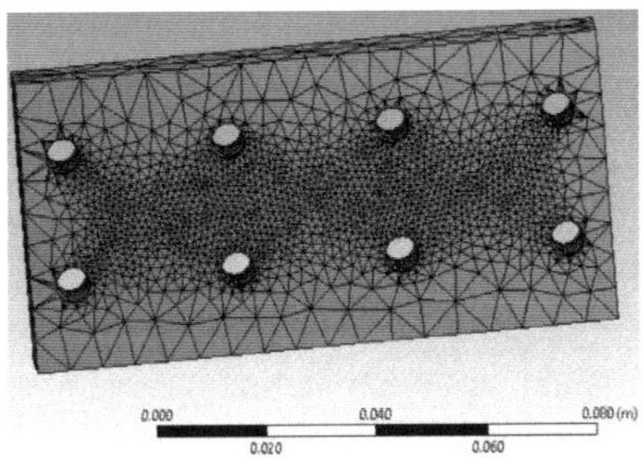

Figure 3: meshing of 3mm and 5mm pores for square pattern

2.3. Boundary conditions:

The battery model is enabled by adjusting its equivalent circuit model with an initial charge of 0.5Ah, reference charge of 10.5Ah and C-rate to be set to 0.3 under electric parameter conditions. Battery node, positive node and negative node are selected under the options active zone, positive zone and negative zone respectively. Four materials are taken into consideration namely, Cu, Al, e-material and t-material, where e-material and t-material are used in the battery and terminal respectively. Under convection, the heat transfer coefficient is set to 5 while the free stream temperature to 700k. The battery is then initialized using standard initialization method and the simulation results of the battery are observed.

The analysis of 3d Al foam of different patterns is carried out by combining Fluent Flow with Transient Thermal using ANSYS software. Initially, Fluent Flow is analyzed about the pressure based steady state to compute the inlet air flow. The energy equation is turned on and the viscous model is selected to laminar. The parameters of momentum, pressure, body forces, energy, liquid fraction and density are set to 0.7, 0.3, 1, 1, 0.9 and 1 respectively. Thermal conditions are set by adding temperature to the inlet and outlet of the Al foam also considering the walls to be adiabatic as shown in Table 2. The pores are chosen to be of the velocity- inlet with respect to the absolute frame to indicate the air flow in Al foam. The initialization is adjusted to standard and computed from inlet.

The transient thermal part is now combined to determine the temperature and heat flux differences in various patterns. In this analysis the estimated battery temperature is set at the inlet and an ambient temperature (T-α) at the outlet of the Al foam as shown in Table 3. Notable variations in heat flux and temperature for different patterns and pore sizes are observed.

Table 2: Boundary conditions of Al foam in CFD

NAME	TYPE	MAGNITUDE
Inlet Air	Velocity	1m/s
Inlet	Temperature	316k
Outlet	Temperature	295k
Wall 1	Temperature	Adiabatic
Wall 2	Temperature	Adiabatic

Table 3: Boundary conditions of Al foam in Transient Thermal

MATERIALS	Temperature1	T-α
Al Foam	313K	298K

3. Results and discussions

The temperature inside the battery increases from terminals to bottom which can be seen in Figure 4. This is because of the chemical reactions that takes place between the ions present in the battery.

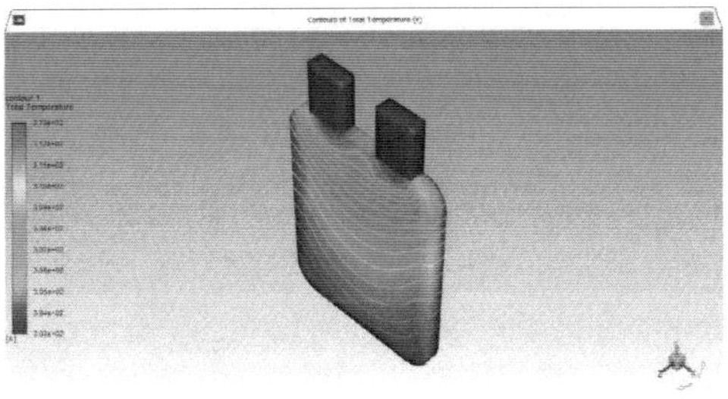

Figure 4: Temperature contours of a battery

The pores in Al foam are distributed along the foam surface in 3 different patterns namely square, triangle and irregular patterns. Two distinct pore sizes 3mm and 5mm are considered for this study (Table 4). The heat flux values are compared in order to find out the pattern and pore size in which cooling rate is higher.

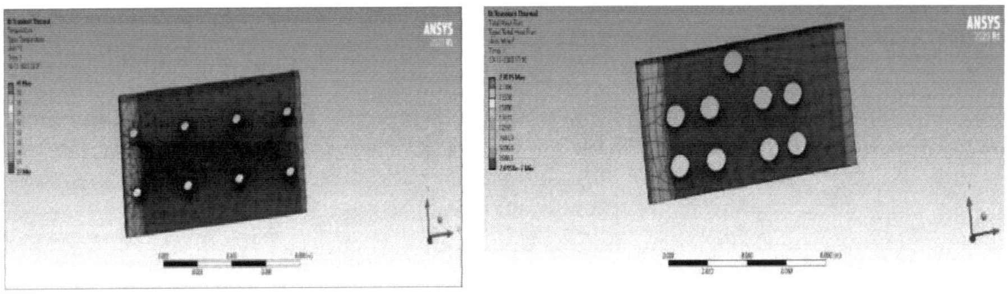

Figure 5: Temperature and heat flux contours of square and irregular pattern

Table 4: Heat flux values of different sized and pattern pores

Pore distribution pattern	Pore size(mm)	Inlet value(W/m^2)	Outlet value(W/m^2)
Square	3	24618	2735.4
Square	5	23668	2629.8
Triangle	3	23859	2651.1
Triangle	5	23861	2651.2
Irregular	3	23952	2661.3
Irregular	5	23835	2648.3

Heat flux is the amount of heat transfer at a particular area and its final value at the outlet explains the cooling action of Al foam. The heat flux vector usually points towards the region where there is lesser temperature and the temperature differences induce the heat flux. The flux value at outlet should be minimum as it denotes less heat transfer through the surface of the foam and more heat loss through the pores, eventually lowering the temperature at outlet compared to inlet and thus acts as coolant. The above contours represents the heat flux rate in six different models of Al foam at same boundary conditions. It can be noticed that among square and irregular patterns, 5mm sized pore has lesser heat flux value at the outlet than 3mm whereas in triangular pattern 3mm sized pore has lesser heat flux value and thus more cooling capacity. A separate comparison of 5mm square, 3mm triangle and 5mm irregular pattern shows that 5mm square pattern (foam 2) has the best cooling capacity out of all six foams used for study. Thus Al foam can be along with a battery as a coolant.

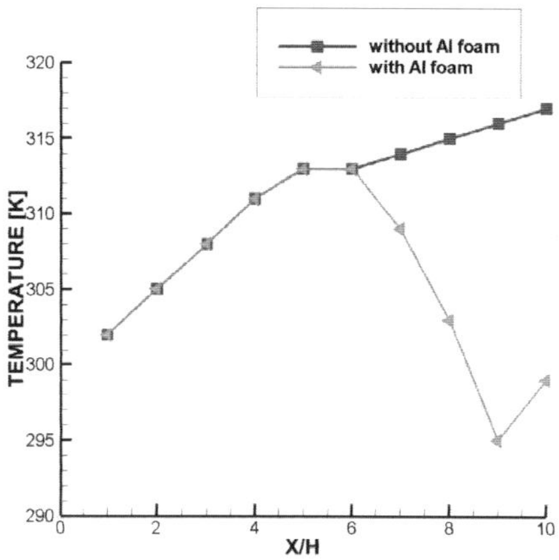

Figure 6: Comparison plot of temperature (with and without Al foam)

It can be observed that the temperature shoots up inside the battery and a temperature fall is seen in Al foam due to its porous nature and high energy absorption rate. These properties of metal foam can be utilized for cooling applications and hence it can be used along with battery. It is also noted that when battery is used with Al foam, the temperature at outlet is decreased unlike the condition where foam is not used (Figure 6).

4. Conclusions

Numerical simulations of 3D models of battery and metal foam are carried out to study the heat dissipation and flow characteristics on their surfaces. The reason behind choosing Al foam for the study is due to its high strength, high thermal performance, porosity and low density, which is a key factor for the designing battery. The battery temperature increases, which is reduced by connecting it with Al foam. Efficiency of pattern and pore size of Al foam are the main features that decide the cooling rate. The foam that has the least heat flux value at the outlet indicates that more heat is lost through the pores and enhanced the cooling process. The simulations are carried out in CFD and transient thermal platforms of ANSYS to get a deeper understanding of the heat flow process. The results from numerical investigation of six different models of Al foams indicate that maximum cooling occurs in the foam that has the pore size of 5mm and that follow square pattern. There is about 45 % reduction in temperature of the battery when used along with Al foam. From this, the pouch cell (battery) using Al foam can be implemented in electric vehicles for lightweight, enhanced heat transfer, increased cooling rate and long driving range.

References

[1] A Bhattacharya, VV Calmidi, RL Mahajan. 2002. Thermophysical properties of high porosity metal foams. Int. J. Heat Mass Transfer., 45, 1017–1031.
[2] Lafdi K, Mesalhy O, Shaikh S. 2007.Experimental study on the influence of foam porosity and pore size on the melting of phase change materials. Journal Of Applied Physics., 102, 83549.
[3] J. Baumeister, J. Weise, E. Hirtz, K. Höhne, J. Hohe. 2014. Applications of aluminium hybrid foam sandwiches in battery housings for electric vehicles. Mat.-wiss. u. Werkstofftech 45, No. 12.
[4] Z.G. Qu, W.Q. Li, W.Q. Tao. 2014. Numerical model of the passive thermal management system for
high-power lithium ion battery by using porous metal foam saturated with phase change material. International Journal of Hydrogen Energy Volume 39, Issue 8, Pages 3904-3913.
[5] AR Alvarez-Hernandez. 2005. Combined Flow and Heat Transfer Characterization of Open Cell Aluminum Foams. M.Sc. Thesis University of Puerto Rico.
[6] N Dukhan.2006. Correlations for the pressure drop for flow through metal foam. Exp. Fluids.,
41,665–672.
[7] Mohammadian S, Rassoulinejad-Mousavi S, Zhang Y. 2015. Thermal management improvement of an air-cooled high-power lithium-ion battery by embedding metal foam. J Pow Sour ., 296,305-313.
[8] Rizov V, Shipsha A, Zenkert D. 2005.Indentation study of foam core sandwich composite panels. Compos Struct., 69,95–102.
[9] Mohan K, Yip TH, Idapalapati S, Chen Z. 2011. Impact response of aluminum foam core sandwich structures. Mater Sci Eng A., 529, 94–101.
[10] Banhart J .2005. Aluminium foams for lighter vehicles. Int J Veh Des .,37,114–125.
[11] Idris MI, Vodenitcharova T, Hoffman M .2009. Mechanical behavior and energy absorption of closed-cell aluminium foam panels in uniaxial compression. Mater Sci Eng A ., 517,37–45.j.msea.2009.03.067
[12] R Kizilel, A Lateef, R Sabbah, MM Farid, JR Selman, S Al-Hallaj. 2008.J. Power Sources
183,370-375.
[13] Lip Huat Saw, Yonghuang Ye, Ming Chian Yew, Wen Tong Chong, Ming Kun Yew, Tan Ching Ng. 2017. Computational fluid dynamics simulation on open cell aluminium foams for Li-ion battery cooling system. Applied Energy Volume 204, Pages 1489-1499.
[14] K Boomsma, D Poulikakos, F Zwick. 2003. Metal foams as compact high-performance heat
exchangers, Mech. Mater., 35 ,1161–1176.
[15] J. Baumeister, J. Weise, E. Hirtz, K. Höhne, J. Hohe. 2014. Applications of aluminium hybrid foam sandwiches in battery housings for electric vehicles. Mat.-wiss. u. Werkstofftech 45, No. 12.

Buckling and Vibration Behavior of Composite Beam due to Axially Varying In-plane Loads

Sangram Anil Kanade[1], Lenin Babu Mailan Chinnapandi[1,2], P Jeyaraj[3] and Jeyanthi Subramanian[1]

[1]School of Mechanical Engineering, Vellore Institute of Technology, Chennai, India

[2]Electric Vehicles-Incubation, Testing and Research Centre, Vellore Institute of Technology, Chennai, India

[3]Department of Mechanical Engineering, NIT Surathkal, Mangalore

Corresponding author's e-mail address: lenin.babu@vit.ac.in

Abstract:
In this work buckling and vibration behavior of composite beam is analyzed based on the shear and normal deformable beam theory (SNDBT) by using Ritz method. The effects of aspect ratios, boundary conditions, fiber angles on the buckling and vibration subjected to non-uniform axial loads studied in detail. The obtained numerical results in terms of dimensionless fundamental frequencies and dimensionless first critical buckling loads are compared with the results from previous research for convergence studies.

Keywords: shear and normal deformable beam theory; Ritz method; Dimensionless critical buckling load

1. Introduction:

Composite laminates have been used increasingly in a variety of industrial areas due to their high stiffness and strength-to weight ratios, long fatigue life, resistance to electrochemical corrosion, and other superior material properties of composites. A true understanding of their structural behaviour is required, such as the deflections, buckling loads and modal characteristics, the through thickness distributions of stresses and strains, the large deflection behaviour and, of extreme importance for obtaining strong, reliable multi-layered structures, the failure characteristics[1]. Therefore, the researchers have developed various theories to present the bending, free vibration and buckling behaviours of LCBs. A review of these theories can be found in [2–4].

In some real applications, the beam structure is subjected to non-uniform in-plane loads. For example, the load exerted on the aircraft wings, load on the stiffened plate in the ship structures, or on the slabs of a multi-storey building by the adjoining structures usually are non-uniform in-plane loads, Panda and Ramachandra [6]. So, the behavior of the structures under the action of non-uniform in-plane compressive loading and shear loading is important in aircraft, civil and ship-building industries. Kang and Leissa [7] developed exact solutions to study buckling stability of rectangular plates under linearly varying in-plane loading. Panda and Ramachandra [6] investigated the effect of non-uniform in-plane loads on buckling stability of rectangular higher order shear deformation plates. Kim et al. [8] studied coupled stability analysis of thin-walled composite beams with closed cross-section subjected to various varing forces. Farajpour et al. [6].

The laminated plate theories are essential to provide accurate analysis of laminated composite plates, and a variety of laminated plate theories have been developed and reported in a large amount of literatures. It should be noted that the higher order beam theories (HBTs) which assume a higher-order variation of the axial displacement through the height of the beam, predict the structural behavior of the LCBs more accurate than classical beam theory (CBT) and the first-order shear deformable beam theories (FSDTs). The CBT neglects the shear deformation effect, underestimates the deflections and overestimates the natural frequencies and critical buckling loads. Moreover, the FSDTs require a shear correction factor which depends on the BC, loading condition and geometrical properties. Eventually, the use of HBTs has been increasing due to their advantages Over CBT and FSDTs.

According to literature and author's knowledge no researchers have attempted to study the buckling loads and their mode-shapes of composite laminated beam under varying axial load by using unified beam theories. The present study is intended to fill this gap in the literature. Six distributions of axial load are proposed, which are constant load, Linear Load-zero from left side, Linear Load-zero from right side, Parabolic Load-zero from left side, Parabolic Load-zero from right side, and Symmetric Parabolic Load. Unified higher order beam theories are proposed to consider all slenderness ratios of the beam ranging from thin, moderated thick, to thick cases. The differential quadrature method (DQM) is exploited to discretize the space and convert the governing differential equations into a set of algebraic equations that will be solved as an eigenvalue problem[10].

Applied new shape functions using the Ritz method for the thermomechanical buckling analysis of LCBs. It is noteworthy that the elastic buckling may also occur under the non-uniform variable in plane loading conditions. Some reported works can be found for the elastic buckling of isotropic columns with non-uniform loading conditions [11–14]. This work aims to study the buckling behavior of isotropic, laminated composite and sandwich beams based on a SNDBT [15-18] by using Ritz method. In this paper objective is to check effects and influence of BCs, aspect ratios, fiber angles on the buckling response of isotropic, laminated composite beams subjected to axially variable in-plane loads.

The following article is organized as follows: Section 2 focusing on problem formulation, which includes kinematics assumptions of displacement filed, constitutive equations of composite laminated beams, axial load functions, and equilibrium equations. Section 3 presents the solution strategy and discretization method of the beam using Ritz method. Section 4 is devoted to validation and parametric studies. These results preset effects of beam theories, type of loading, Aspect ratio, on critical buckling loads and corresponding mode-shapes. Main remarks and conclusion points are highlighted and summarized in Section

2. Kinematics, stress and strain relations :

The displacement field of the SNDBT to be used within this study is given with the following equations [15-18]:

$$U(x,z) = u(x) - z\frac{dw_b(x)}{dx} - \frac{4z^2}{3h^2}\frac{dw_s(x)}{dx} = u(x) - zw_b(x) - f(z)w_s(x) \quad \text{Eq. 1a}$$

$$W(x,z) = w_b(x) + w_s(x) + \left(1 - \frac{4z^2}{h^2}\right) w_z(x) = w_b(x) + w_s(x) + g(z)w_z(x) \quad \text{Eq. 1b}$$

where u = axial displacement,
w_b = bending components of the transverse displacement
w_s = shear components of the transverse displacement
w_z = an unknown which is defined to obtain the normal deformation effect.
When g(z) is set to zero, the present SNDBT is reduced to a HBT. With the linearity assumption for the deformations, the kinematic relations associated with the displacement field can be given by:

$$\varepsilon_{xx} = \frac{dU}{dx} = u - zw_b - f(z)w_s \quad \text{Eq. 2a}$$

$$\varepsilon_{zz} = \frac{dW}{dz} = g(z) \quad \text{Eq. 2b}$$

$$\gamma_{xz} = \frac{dW}{dx} + \frac{dU}{dz} = g(z)(w_s - w_z) \quad \text{Eq. 2c}$$

For each orthotropic lamina, the stress-strain relation is given by:

$$\begin{Bmatrix} \sigma_{xx} \\ \sigma_{zz} \\ \sigma_{xz} \end{Bmatrix}^k = \begin{bmatrix} Q_{11} & Q_{13} & 0 \\ Q_{13} & Q_{33} & 0 \\ 0 & 0 & Q_{55} \end{bmatrix}^k \begin{Bmatrix} \varepsilon_{xx} \\ \varepsilon_{zz} \\ \gamma_{xz} \end{Bmatrix} \quad \text{Eq. 3}$$

where (xx, zz, xz) and (xx, zz, xz) are the components of the stress and strain tensors, respectively. Q_{ij}'s are reduced transformed stiffness and can be given by:

$$Q_{11} = C_{11}cos^4\theta + 2(C_{12} + 2C_{66})cos^2\theta sin^2\theta + C_{66}sin^4\theta$$

$$Q_{13} = C_{23}sin^2\theta + C_{13}cod^2\theta$$

$$Q_{33} = C_{33}$$

$$\boldsymbol{Q_{55} = C_{55}sin^2\theta + C_{44}cos^2\theta} \quad \text{Eq. 4}$$

Where,

$$C_{11} = \frac{E_1}{\Delta};\ C_{12} = \frac{E_1\vartheta_{21}}{\Delta};\ C_{22} = \frac{E_2(1-\vartheta_{13}\vartheta_{31})}{\Delta};\ C_{13} = \frac{E_1\ \vartheta_{31})}{\Delta};;$$

$$C_{23} = \frac{E_2\vartheta_{12}\vartheta_{31})}{\Delta};\ C_{33} = \frac{E_3(1-\vartheta_{21}\vartheta_{21})}{\Delta};\ C_{66} = G_{12};\ C_{44} = G_{13};\ C_{55} = G_{23};$$

$\Delta = 1 - \vartheta_{12}\vartheta_{21} - \vartheta_{13}\vartheta_{31}$

E1, E2, G12, G13, G23, 12 and 21 are the six independent material properties.

Problem	Structure	Material properties
LCB	[15,19]	E1/E2 = Open; E3 = E2; G12 = G13 = 0.5E2; G23 = 0.2 E2 $\vartheta_{12} = \vartheta_{13} = \vartheta_{23} = 0.25$

3. Laminated composite beam:

A laminated composite beam (LCB) shown in Fig. 1 has the top and bottom layer thickness (h1) and core thickness (h2). The equations given above for laminated composite beam is used in the formulation of LCB. The main difference between laminated composite beam and laminated sandwich beam formulations are the material properties and the thickness of the layers. In the laminated composite beam formulation, all of the layers have the same thickness and material properties, but the core and face layers have different material properties and different thicknesses in the LCB formulation. It is noteworthy that the Zig-Zag theories where the discontinuity of the displacement at each layer interface is captured, can provide more accurate results than the classical higher order shear and normal deformable beam theories, especially for the sandwich or layered composite laminates.

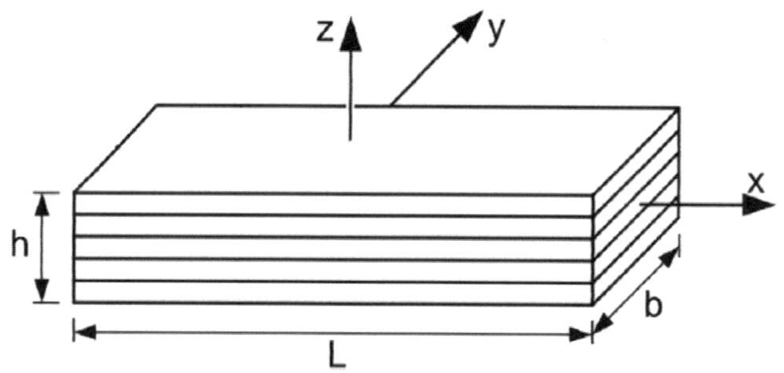

Figure 1:- Geometry of a laminated composite beam

Variational formulation

The potential energy of a shear and normal deformable beam can be written as:

$$U = \frac{1}{2}\int_v(\sigma_{xx}\varepsilon_{xx} + \sigma_{zz}\varepsilon_{zz} + \sigma_{xz}\gamma_{xz})dV \quad \text{Eq. 5}$$

where V is the volume of the beam. The potential energy of the external axially variable in-plane load shown in Fig. 2 is given by:

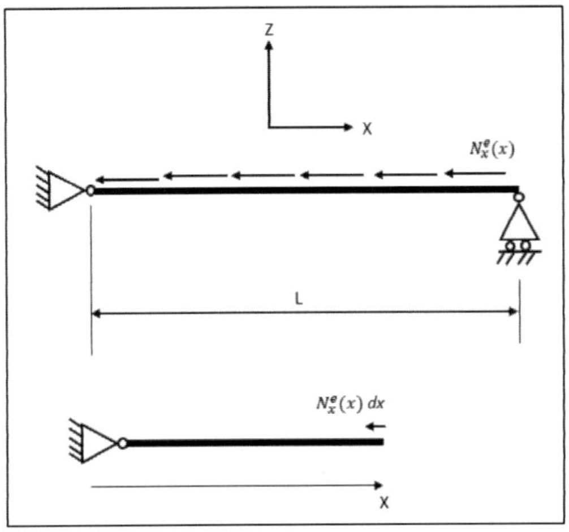

Figure 2:- Simply supported beam with a distributed in plane axial load

$$V = -\frac{1}{2}\int_{-L/2}^{L/2} N_x^e(x)\left[\int_{-L/2}^{x}\left\{\left(\frac{\partial w_b}{\partial x}\right)^2 + 2\frac{\partial w_b}{\partial x}\frac{\partial w_s}{\partial x} + \left(\frac{\partial w_s}{\partial x}\right)^2\right\}dx\right]dx \quad \text{......Eq. 6}$$

Here, $N_x^e(x)$ = external axially variable in-plane load. The following polynomial axial load distribution is chosen in the present study:

$$N_x^e(x) = N_0\left[\alpha_1\left(x+\frac{L}{2}\right)^2 + \alpha_2\left(x+\frac{L}{2}\right) + \alpha_3\right] \quad \text{......Eq. 7a}$$

$$N_x^e(x) = N_0 P(x) \quad \text{......Eq. 7b}$$

Positive load is assumed compressive. The distributed external loads obtained based on the different values of i (i = 1, 2, 3) given in Table 1 are used for the numerical computations.

Load Type	*Load Symbol*	α_1	α_2	α_3
Constant Load	N_x^1	0	0	1
Linear Load-zero from left side	N_x^2	0	2	0
Linear Load-zero from right side	N_x^3	0	-2	2
Parabolic Load-zero from left side	N_x^4	3	0	0
Parabolic Load-zero from right side	N_x^5	3	-6	3
Symmetric Parabolic Load	N_x^6	-6	6	0

Table 1:- Coefficients of external axial distributed loads (Karamanli and Aydogdu [20])

The integral of the each axially variable in-plane load through the length of the beam is equal to integral of the uniformly distributed in-plane load. This may allow making the comparisons in a fair ground. Using Eqs. (1)–(7), the total potential energy can be obtained as follows:

$$\pi = u + V$$

Dimensionless critical buckling load (DCBLs) of isotropic beams subjected to various axially varying in plane loads for different BCs and fibre angles.

$$\pi = \frac{1}{2}\int_{-L/2}^{L/2}\left[A\left(\frac{\partial u}{\partial x}\right)^2 + D\left(\frac{\partial^2 w_b}{\partial x^2}\right)^2 + D_s\left(\frac{\partial^2 w_s}{\partial x^2}\right)^2 + 2H\left(\frac{\partial^2 w_b}{\partial x^2}\right)\left(\frac{\partial^2 w_s}{\partial x^2}\right) - 2B\left(\frac{\partial u}{\partial x}\right)\left(\frac{\partial^2 w_b}{\partial x^2}\right) - 2B_s\left(\frac{\partial u}{\partial x}\right)\left(\frac{\partial^2 w_s}{\partial x^2}\right) + 2X\left(\frac{\partial u}{\partial x}\right)w_z + Z{w_z}^2 + A_s\left\{\left(\frac{\partial w_s}{\partial x}\right)^2 + \left(\frac{\partial w_z}{\partial x}\right)^2 + 2\left(\frac{\partial w_s}{\partial x}\right)\left(\frac{\partial w_z}{\partial x}\right)\right\} - N_0 P(x)\left[\int_{-L/2}^{x}\left\{\left(\frac{\partial w_b}{\partial x}\right)^2 + 2\frac{\partial w_b}{\partial x}\frac{\partial w_s}{\partial x} + \left(\frac{\partial w_b}{\partial x}\right)^2\right\}dx\right]\right]dx \qquad \text{Eq. 8}$$

The stiffness coefficients of the present SNDBT are defined as:

$$(A, B, B_s, D, D_s, H) = \int_{-h/2}^{+h/2} Q_{11} b(1, z, f, z^2, f^2, fz, g^2)dz \qquad \text{Eq. 9a}$$

$$A_s = \int_{-h/2}^{+h/2} Q_{55} g^2 dz \qquad \text{Eq. 9b}$$

$$(X, Y, Y_s) = \int_{-h/2}^{+h/2} Q_{13} bg(1, z, f)dz \qquad \text{Eq. 9c}$$

$$Z = \int_{-h/2}^{+h/2} Q_{33} bg^2\, dz \qquad \text{Eq. 9d}$$

where A = the classical extensional
B = bending-stretching coupling
D = bending stiffness coefficients
Ds = shear bending stiffness coefficients
Z = normal bending stiffness coefficients
Bs = stretching-shear bending coupling coefficients
X = the and stretching-normal bending stiffness coefficients
H = shear-bending coupling,
Y = bending-normal bending coupling
Ys = shear bending-normal bending coupling stiffness coefficients respectively and
As= is the shear stiffness coefficient.

4. Ritz procedure

Since the differential equations derived for the buckling analysis of LCBs consist of variable coefficients, the exact solution cannot be obtained. Various numerical techniques can be employed for the solution of this complicated problem like FEM, DQM, meshless methods, Ritz method etc. Among them, the Ritz procedure has been found accurate and simple by the researchers [21-22]. In this study, the Ritz procedure is employed for the solution of the problem. The kinematic boundary conditions for the studied problem are given in Table 2.

BC	x = -L/2	x = L/2
SS	u = 0, w_b=0, $w_s = 0, w_z = 0$	w_b= 0, $w_s = 0, w_z = 0$
CS	u = 0, w_b= 0, $w_s = 0, w_z = 0$, w_b= 0, $w_s = 0$	u = 0, w_b= 0, $w_s = 0, w_z = 0$, w_b= 0, $w_s = 0$
CC	u = 0, w_b= 0, $w_s = 0, w_z = 0$, w_b= 0, $w_s = 0$	u = 0, w_b= 0, $w_s = 0, w_z = 0$, w_b= 0, $w_s = 0$
CF	u = 0, w_b= 0, $w_s = 0, w_z = 0$, w_b= 0, $w_s = 0$	

Table 2:- Kinematic boundary conditions used for the numerical computations

BC	Left end				Right end			
	P_u	P_{w_b}	P_{w_s}	P_{w_z}	q_u	q_{w_b}	q_{w_s}	q_{w_z}
SS	1	1	1	1	0	1	1	1
CS	1	2	2	1	0	1	1	1
CC	1	2	2	1	1	2	2	1
CF	1	2	2	1	0	1	0	0

Table 3:- Boundary exponents for various BCs

The following displacement functions u(x), w_b (x), w_s (x) and w_z (x) are defined for the present composite beam buckling problem:

$$u(x) = \sum_{j=1}^{m} A_j\theta_j(x),\ \ \theta_j(x) = \left(x + \frac{L}{2}\right)^{p_u}\left(x - \frac{L}{2}\right)^{q_u} x^{j-1} \quad \text{......Eq. 10a}$$

$$w_b(x) = \sum_{j=1}^{m} B_j\varphi_j(x),\ \ \varphi_j(x) = \left(x + \frac{L}{2}\right)^{p_{w_b}}\left(x - \frac{L}{2}\right)^{q_{w_b}} x^{j-1} \quad \text{......Eq. 10b}$$

$$w_s(x) = \sum_{j=1}^{m} C_j\zeta_j(x),\ \ \zeta_j(x) = \left(x + \frac{L}{2}\right)^{p_{w_s}}\left(x - \frac{L}{2}\right)^{q_{w_s}} x^{j-1} \quad \text{......Eq. 10c}$$

$$w_z(x) = \sum_{j=1}^{m} D_j\psi_j(x), \psi(x) = \left(x + \frac{L}{2}\right)^{p_{w_z}}\left(x - \frac{L}{2}\right)^{q_{w_z}} x^{j-1} \quad \text{......Eq. 10d}$$

where A_j, B_j, C_j and D_j are undetermined coefficients, j (x), j (x), j (x) and j (x) are the trial functions, p and q (ξ = u, wb, ws, wz) are the boundary exponents used to satisfy the BCs given in Table 3. The governing equations for the LCBs under the in-plane loads can be derived by substituting Eqs. (10a)–(10d) into Eq. (8) and then employing the minimum energy principle:

$$\frac{d\pi}{dq_j} = 0 \quad \text{......Eq. 11}$$

with q_j representing the undetermined coefficients, that leads to,

$$\left(\begin{bmatrix} [K_{11}] & [K_{12}] & [K_{13}] & [K_{14}] \\ [K_{12}]^T & [K_{22}] & [K_{23}] & [K_{241}] \\ [K_{31}]^T & [K_{32}]^T & [K_{33}] & [K_{34}] \\ [K_{41}]^T & [K_{42}]^T & [K_{43}]^T & [K_{44}] \end{bmatrix} - N_0 \begin{bmatrix} [0] & [0] & [0] & [0] \\ [0] & [M_{22}] & [M_{23}] & [0] \\ [0] & [M_{23}]^T & [M_{33}]^T & [0] \\ [0] & [0] & [0] & [0] \end{bmatrix}\right) \begin{Bmatrix} \{A\} \\ \{B\} \\ \{C\} \\ \{D\} \end{Bmatrix} = \begin{Bmatrix} \{0\} \\ \{0\} \\ \{0\} \\ \{0\} \end{Bmatrix} \quad \text{......Eq. 12}$$

where is the dimensionless critical buckling load (DCBL), $[K_{kl}]$ and $[M_{kl}]$ are the stiffness and geometry matrices, respectively. The above mentioned matrices are symmetric. The DCBL of the LCBs can be defined by:

$$\lambda = \frac{N_0 L^2}{E_1 b h^3} \quad \text{......Eq. 13}$$

The symmetric matrices are given by:

$$K_{11}(i,j) = A\int_{-L/2}^{L/2} \theta_{i,x}\theta_{j,x}dx \quad \text{......Eq. 14a}$$

$$K_{12}(i,j) = -B\int_{-L/2}^{L/2} \theta_{i,x}\varphi_{j,xx}dx \quad \text{......Eq. 14b}$$

$$K_{13}(i,j) = -B_s\int_{-L/2}^{L/2} \theta_{i,x}\zeta_{j,xx}dx \quad \text{......Eq. 14c}$$

$$K_{14}(i,j) = X\int_{-L/2}^{L/2} \theta_{i,x}\varphi_j dx \quad \text{......Eq. 14d}$$

$$K_{22}(i,j) = D\int_{-L/2}^{L/2} \varphi_{i,xx}\varphi_{j,xx}dx \quad \text{......Eq. 14e}$$

$$K_{23}(i,j) = H\int_{-L/2}^{L/2} \varphi_{i,xx}\zeta_{j,xx}dx \quad \text{......Eq. 14f}$$

$$K_{24}(i,j) = -Y\int_{-L/2}^{L/2} \varphi_{i,xx}\varphi_j dx \quad \text{......Eq. 14g}$$

$$K_{33}(i,j) = D_s\int_{-L/2}^{L/2} \zeta_{i,xx}\zeta_{j,xx}dx + A_s\int_{-L/2}^{L/2} \zeta_{i,x}\zeta_{j,x}dx \quad \text{......Eq. 14h}$$

$$K_{34}(i,j) = -Y_s\int_{-L/2}^{L/2} \zeta_{i,xx}\varphi_j dx + A_s\int_{-L/2}^{L/2} \zeta_{i,x}\varphi_j, xdx \quad \text{......Eq. 14i}$$

$$K_{44}(i,j) = Z\int_{-L/2}^{L/2} \psi_i\psi_j dx + A_s\int_{-L/2}^{L/2} \psi_{i,x}\psi_{j,x}dx \quad \text{......Eq. 14j}$$

$$M_{22}(i,j) = -N_0\int_{-L/2}^{L/2} P(x)\left(\int_{-L/2}^{x} \varphi_{i,x}\varphi_{j,x}dx\right)dx \quad \text{......Eq. 14k}$$

$$M_{23}(i,j) = -N_0 \int_{-L/2}^{L/2} P(x) \left(\int_{-L/2}^{x} \varphi_{i,x}\zeta_{j,x} dx\right) dx \ldots\ldots \text{Eq. 14l}$$

$$M_{33}(i,j) = -N_0 \int_{-L/2}^{L/2} P(x) \left(\int_{-L/2}^{x} \zeta_{j,x}\zeta_{j,x} dx\right) dx \ldots\ldots \text{Eq. 14m}$$

5. Buckling Modes

This section is devoted to the buckling mode-shapes analysis and its effect by loading type, fiber angle and boundary conditions. Influences of load functions and boundary conditions on buckling mode-shapes of laminated composite beams are presented through this section. To investigate the characteristics of buckling mode shapes corresponding to different load types, the first buckling mode shapes of a SS, CC, CF, CS composite laminated symmetric beams are strategized. Effects of the 6 loading types (N_x^1, N_x^2, N_x^3, N_x^4, N_x^5, N_x^6) on the first buckling mode-shapes on SS, CC, CF, CS and fiber angle (0,15,30,45,60,75,90) are presented. It can be concluded that, the load type has a significant effect on the buckling mode shapes. Further, it is observed that for all boundary conditions, the first critical buckling is high in CC and minimum in CF. It is important to note that, referring to all previous results, values of critical buckling for different loading types are ordered in the same sequence with highest buckling load.

N_x^1:- alpha1 = 0, alpha2 = 0, alpha3 = 1

Angle: 0/0/0

BL	0	0.25	0.5	0.75	0.99
SS	46.4351	40.5255	33.3678	23.8118	4.8084
CC	105.2815	92.1837	76.2098	54.6643	11.1094
CF	16.5445	14.3355	11.7113	8.2859	1.6581
CS	72.5537	63.2658	52.0513	37.1219	7.4938

Angle: 0/15/0

BL	0	0.25	0.5	0.75	0.99
SS	46.3326	40.4361	33.2942	23.7592	4.7978
CC	105.0489	91.9800	76.0414	54.5436	11.0849
CF	16.5080	14.3039	11.6855	8.2676	1.6545
CS	72.3934	63.1261	51.9363	37.0399	7.4773

Angle: 0/30/0

BL	0	0.25	0.5	0.75	0.99
SS	46.0907	40.2250	33.1204	23.6352	4.7728
CC	104.4993	91.4988	75.6436	54.2583	11.0270
CF	16.4217	14.2291	11.6244	8.2244	1.6458
CS	72.0149	62.7960	51.6648	36.8462	7.4382

Angle: 0/45/0

BL	0	0.25	0.5	0.75	0.99
SS	45.8505	40.0154	32.9478	23.5121	4.7479
CC	103.9514	91.0191	75.2471	53.9739	10.9694
CF	16.3359	14.1547	11.5637	8.1814	1.6372
CS	71.6378	62.4672	51.3943	36.6534	7.3993

Angle: 0/60/0

BL	0	0.25	0.5	0.75	0.99
SS	45.7124	39.8949	32.8486	23.4413	4.7336
CC	103.6343	90.7416	75.0178	53.8095	10.9361
CF	16.2864	14.1119	11.5286	8.1567	1.6323
CS	71.4200	62.2773	51.2382	36.5420	7.3768

Angle: 0/75/0

BL	0	0.25	0.5	0.75	0.99
SS	45.6749	39.8622	32.8217	23.4221	4.7297
CC	103.5463	90.6645	74.9541	53.7639	10.9268
CF	16.2729	14.1001	11.5190	8.1499	1.6309
CS	71.3598	62.2249	51.1950	36.5113	7.3706

Angle: 0/90/0

BL	0	0.25	0.5	0.75	0.99
SS	45.6745	39.8618	32.8213	23.4219	4.7297
CC	103.5442	90.6627	74.9526	53.7628	10.9262
CF	16.2726	14.0999	11.5189	8.1498	1.6309
CS	71.3585	62.2238	51.1942	36.5107	7.3705

N_x^2:- alpha1 = 0, alpha2 = 2, alpha3 = 0

Angle: 0/0/0

BL	0	0.25	0.5	0.75	0.99
SS	46.4351	40.4331	33.2098	23.6360	4.7598
CC	105.2815	92.0541	75.9746	54.3834	11.0249
CF	16.5445	14.3486	11.7332	8.3098	1.6646
CS	72.5537	63.1913	51.9179	36.9653	7.4477

Angle: 0/15/0

BL	0	0.25	0.5	0.75	0.99
SS	46.3326	40.3438	33.1364	23.5838	4.7493
CC	105.0489	91.8507	75.8068	54.2632	11.0006
CF	16.5080	14.3169	11.7073	8.2915	1.6610
CS	72.3934	63.0517	51.8033	36.8836	7.4312

Angle: 0/30/0

BL	0	0.25	0.5	0.75	0.99
SS	46.0907	40.0893	32.8881	23.3764	4.7010
CC	104.4993	91.3702	75.4102	53.9794	10.9430

Angle: 0/45/0

BL	0	0.25	0.5	0.75	0.99
SS	45.8505	39.8804	32.7166	23.2546	4.6765
CC	103.9514	90.8911	75.0149	53.6964	10.8857

CF	16.4217	14.2421	11.6462	8.2482	1.6523
CS	72.0149	62.7220	51.5324	36.6908	7.3924

CF	16.3359	14.1676	11.5853	8.2050	1.6436
CS	71.6378	62.3936	51.2626	36.4987	7.3537

Angle: 0/60/0

BL	0	0.25	0.5	0.75	0.99
SS	45.7124	39.7708	32.8486	23.2682	4.6857
CC	103.6343	90.5393	75.0178	53.8095	10.9361
CF	16.2864	14.1376	11.5286	8.1567	1.6323
CS	71.4200	62.1674	51.2382	36.5420	7.3768

Angle: 0/75/0

BL	0	0.25	0.5	0.75	0.99
SS	45.6749	39.7172	32.5653	23.1093	4.2654
CC	103.5463	90.4624	74.5886	53.3288	10.7961
CF	16.2729	14.1258	11.5622	8.1969	1.6437
CS	71.3598	62.1151	50.9992	36.2822	7.3034

Angle: 0/90/0

BL	0	0.25	0.5	0.75	0.99
SS	45.6745	39.7273	32.5653	23.1653	4.2654
CC	103.5442	90.4606	74.5886	53.3278	10.7961
CF	16.2726	14.1256	11.5622	8.1968	1.6437
CS	71.3585	62.1140	50.9992	36.2816	7.3034

N_x^3:- alpha1 = 0, alpha2 = -2, alpha3 =2

Angle: 0/0/0

BL	0	0.25	0.5	0.75	0.99
SS	46.4351	40.7162	33.6911	24.1681	4.9061
CC	105.2815	92.7290	77.1790	55.7937	11.4390
CF	16.5445	14.3583	11.7495	8.3275	1.6694
CS	72.5537	63.6423	52.7137	37.8846	7.7136

Angle: 0/15/0

BL	0	0.25	0.5	0.75	0.99
SS	46.3326	40.3438	33.1364	23.5838	4.7493
CC	105.0489	91.8507	75.8068	54.2632	11.0006
CF	16.5080	14.3169	11.7073	8.2915	1.6610
CS	72.3934	63.0517	51.8033	36.8836	7.4312

Angle: 0/30/0

BL	0	0.25	0.5	0.75	0.99
SS	46.0907	40.0893	32.8881	23.3764	4.7257
CC	104.4993	91.2949	75.4102	53.8194	13.1550
CF	16.4217	14.2550	11.6462	8.2719	1.6674
CS	72.0149	62.6852	51.5324	36.6151	7.4091

Angle: 0/45/0

BL	0	0.25	0.5	0.75	0.99
SS	45.8505	39.8804	32.7166	23.2546	4.6765
CC	103.9514	90.8162	74.8802	53.6964	10.8383
CF	16.3359	14.1805	11.6069	8.2050	1.6501
CS	71.6378	62.3570	51.1978	36.4987	7.3318

Angle: 0/60/0

BL	0	0.25	0.5	0.75	0.99
SS	45.8505	39.7708	39.7708	23.2406	4.6765
CC	103.9514	90.5393	90.5393	53.3741	10.8383
CF	16.3359	14.1376	14.1376	8.2038	1.6501
CS	71.6378	62.1674	62.1674	36.3128	7.3318

Angle: 0/75/0

BL	0	0.25	0.5	0.75	0.99
SS	45.6749	39.7172	32.8217	23.4219	4.2654
CC	103.5463	90.4624	74.9541	53.7628	10.7961
CF	16.2729	14.1258	11.5190	8.1498	1.6437
CS	71.3598	62.1151	51.1950	36.5107	7.3034

Angle: 0/90/0

BL	0	0.25	0.5	0.75	0.99
SS	45.6749	39.8618	32.8217	23.4219	4.6586
CC	103.5463	90.6627	74.9541	53.7628	10.7959
CF	16.2729	14.0999	11.5190	8.1498	1.6437
CS	71.3598	62.2238	51.1950	36.5107	7.3033

N_x^4:- alpha1 = 3, alpha2 = 0, alpha3 = 0

Angle: 0/0/0

BL	0	0.25	0.5	0.75	0.99
SS	46.4351	40.3888	33.1338	23.5510	4.7361
CC	105.2815	92.0421	75.9985	54.5714	13.1550
CF	16.5445	14.3616	11.7552	8.3338	1.6711
CS	72.5537	63.1542	51.8523	36.8891	7.4255

Angle: 0/15/0

BL	0	0.25	0.5	0.75	0.99
SS	46.3326	40.2997	33.0606	23.4990	4.7257
CC	105.2815	92.0421	75.9985	54.5714	13.1550
CF	16.5080	14.3299	11.7292	8.3154	1.6674
CS	72.3934	63.0147	51.7377	36.8076	7.4091

Angle: 0/30/0

BL	0	0.25	0.5	0.75	0.99

Angle: 0/45/0

BL	0	0.25	0.5	0.75	0.99

SS	46.0907	40.0893	32.8881	23.3764	4.7010
CC	104.4993	91.2949	75.2749	53.8194	10.8954
CF	16.4217	14.2550	11.6679	8.2719	1.6587
CS	72.0149	62.6852	51.4672	36.6151	7.3704

SS	45.8505	39.8804	32.7166	23.2546	4.6765
CC	103.9514	90.8162	74.8802	53.5372	10.8383
CF	16.3359	14.1805	11.6069	8.2287	1.6501
CS	71.6378	62.3570	51.1978	36.4235	7.3318

Angle: 0/60/0

BL	0	0.25	0.5	0.75	0.99
SS	45.7124	39.7708	39.7708	23.2406	5.0246
CC	103.6343	90.5393	90.5393	53.3741	10.8053
CF	16.2864	14.1376	14.1376	8.2038	1.6451
CS	71.4200	62.1674	62.1674	36.3128	7.3095

Angle: 0/75/0

BL	0	0.25	0.5	0.75	0.99
SS	45.6749	39.7172	32.5653	23.1093	4.2654
CC	103.5463	90.4624	74.5886	53.3288	10.7961
CF	16.2729	14.1258	11.5622	8.1969	1.6437
CS	71.3598	62.1151	50.9992	36.2822	7.3034

Angle: 0/90/0

BL	0	0.25	0.5	0.75	0.99
SS	45.6745	39.7273	32.5911	23.1653	4.6586
CC	103.5442	90.4606	74.5872	53.3278	10.7959
CF	16.2726	14.1256	11.5620	8.1968	1.6437
CS	71.3585	62.1140	50.9983	36.2816	7.3033

N_x^5:- alpha1 = 3, alpha2 = -6, alpha3 = 3

Angle: 0/0/0

BL	0	0.25	0.5	0.75	0.99
SS	46.44	45.05	43.62	42.14	40.67
CC	105.28	104.35	103.41	102.45	101.53
CF	16.54	14.38	11.78	8.36	1.68
CS	72.55	71.56	70.55	69.53	68.52

Angle: 0/15/0

BL	0	0.25	0.5	0.75	0.99
SS	46.33	44.95	43.53	42.05	40.58
CC	105.05	104.12	103.18	102.23	101.30
CF	16.51	14.34	11.75	8.34	1.67
CS	72.39	71.40	70.40	69.37	68.37

Angle: 0/30/0

BL	0	0.25	0.5	0.75	0.99
SS	46.09	44.72	43.30	41.83	40.37
CC	104.50	103.57	102.64	101.69	100.77
CF	16.42	14.27	11.69	8.30	1.67
CS	72.01	71.03	70.03	69.01	68.02

Angle: 0/45/0

BL	0	0.25	0.5	0.75	0.99
SS	45.850	44.451	43.002	41.498	39.997
CC	103.951	103.007	102.051	101.085	100.146
CF	16.336	14.195	11.632	8.255	1.657
CS	71.638	70.632	69.610	68.569	67.553

Angle: 0/60/0

BL	0	0.25	0.5	0.75	0.99
SS	45.712	44.352	42.946	41.488	40.034
CC	103.634	102.716	101.788	100.849	99.937
CF	16.286	14.152	11.596	8.230	1.652
CS	71.420	70.443	69.449	68.439	67.453

Angle: 0/75/0

BL	0	0.25	0.5	0.75	0.99
SS	45.675	43.890	42.019	40.049	38.051
CC	103.546	102.629	101.702	100.763	99.852
CF	16.273	14.141	11.587	8.224	1.651
CS	71.360	70.383	69.391	68.382	67.396

Angle: 0/90/0

BL	0	0.25	0.5	0.75	0.99
SS	45.674	44.316	42.910	41.453	40.001
CC	103.544	102.627	101.700	100.761	99.850
CF	16.273	14.140	11.587	8.223	1.651
CS	71.359	70.382	69.390	68.380	67.395

N_x^6:- alpha1 = -6, alpha2 = 6, alpha3 = 0

Angle: 0/0/0

BL	0	0.25	0.5	0.75	0.99
SS	46.4351	40.4135	33.1762	23.5985	4.7494
CC	105.2815	92.0173	75.9086	54.3055	11.0018
CF	16.5445	14.3539	11.7422	8.3196	1.6673
CS	72.5537	63.1723	51.8843	36.9263	7.4364

Angle: 0/15/0

BL	0	0.25	0.5	0.75	0.99
SS	46.3326	40.3243	33.1029	23.5464	4.7389
CC	105.0489	91.8140	75.7409	54.1855	10.9775
CF	16.5080	14.3222	11.7163	8.3012	1.6636
CS	72.3934	63.0327	51.7697	36.8448	7.4200

Angle: 0/30/0						**Angle: 0/45/0**					
BL	0	0.25	0.5	0.75	0.99	BL	0	0.25	0.5	0.75	0.99
SS	46.0907	40.1138	32.9301	23.4235	4.7141	SS	45.8505	39.9047	32.7585	23.3014	4.6896
CC	104.4993	91.3336	75.3447	53.9021	10.9201	CC	103.9514	90.8547	74.9497	53.6196	10.8629
CF	16.4217	14.2474	11.6550	8.2578	1.6549	CF	16.3359	14.1729	11.5941	8.2147	1.6463
CS	72.0149	62.7031	51.4990	36.6521	7.3812	CS	71.6378	62.3748	51.2294	36.4603	7.3425

Angle: 0/60/0						**Angle: 0/75/0**					
BL	0	0.25	0.5	0.75	0.99	BL	0	0.25	0.5	0.75	0.99
SS	45.7124	39.7846	32.6599	23.2313	4.6755	SS	45.6749	39.7520	32.6331	23.2122	4.6716
CC	103.6343	90.5777	74.7212	53.4562	10.8298	CC	103.5463	90.5008	74.6578	53.4109	10.8206
CF	16.2864	14.1300	11.5590	8.1898	1.6413	CF	16.2729	14.1182	11.5494	8.1830	1.6399
CS	71.4200	62.1852	51.0737	36.3495	7.3202	CS	71.3598	62.1328	51.0307	36.3189	7.3141

Angle: 0/90/0					
BL	0	0.25	0.5	0.75	0.99
SS	45.6745	39.7516	32.6328	23.2120	4.6716
CC	103.5442	90.4990	74.6563	53.4098	10.8204
CF	16.2726	14.1180	11.5492	8.1829	1.6399
CS	71.3585	62.1317	51.0298	36.3183	7.3140

Table 5:- Buckling Modes

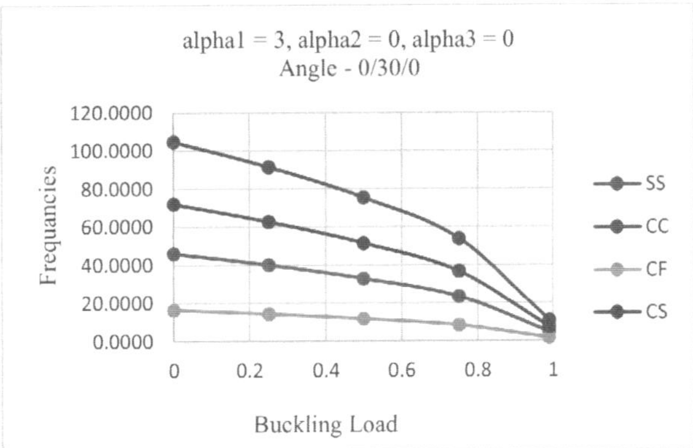

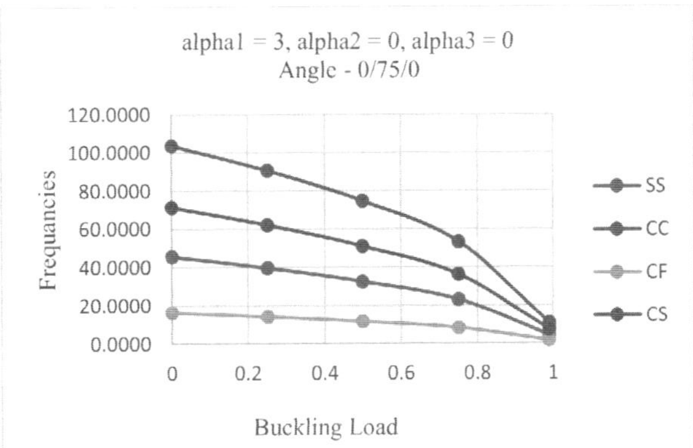

6. Numerical results

This section is concerned with DCBL and effects of in-plane distribution functions, aspect ratio, fiber orientation, boundary conditions, on SNDBT on LCBs structure.

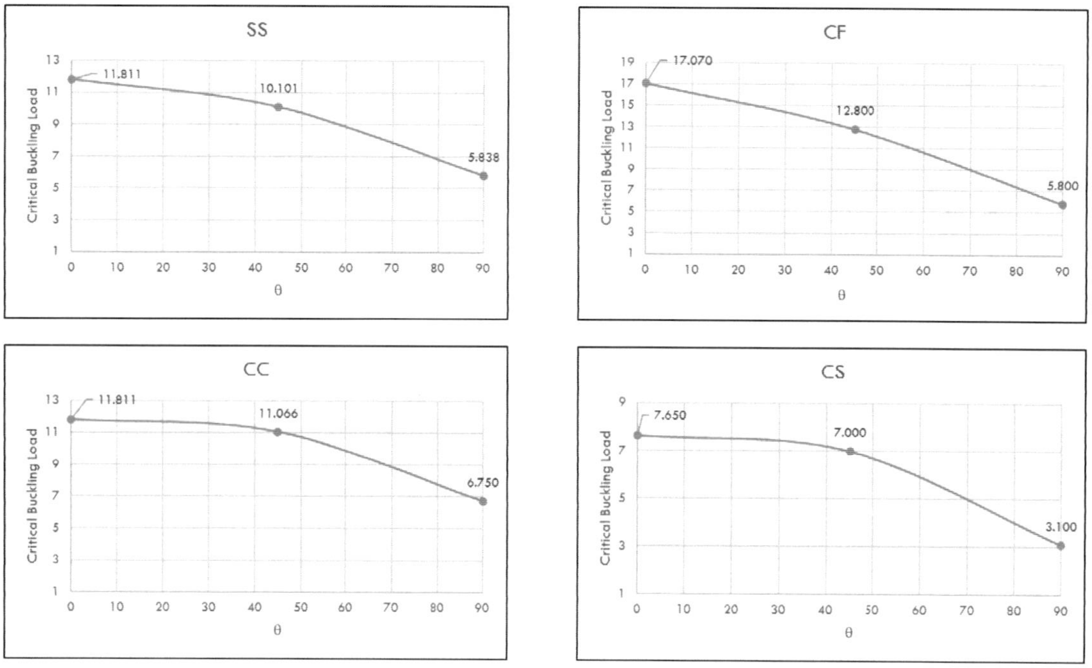

Figure 3:- Effects of the fiber angles on the variation of the CBLs of the symmetric [0°/ θ /0°] LCBs for various BCs L/h = 100

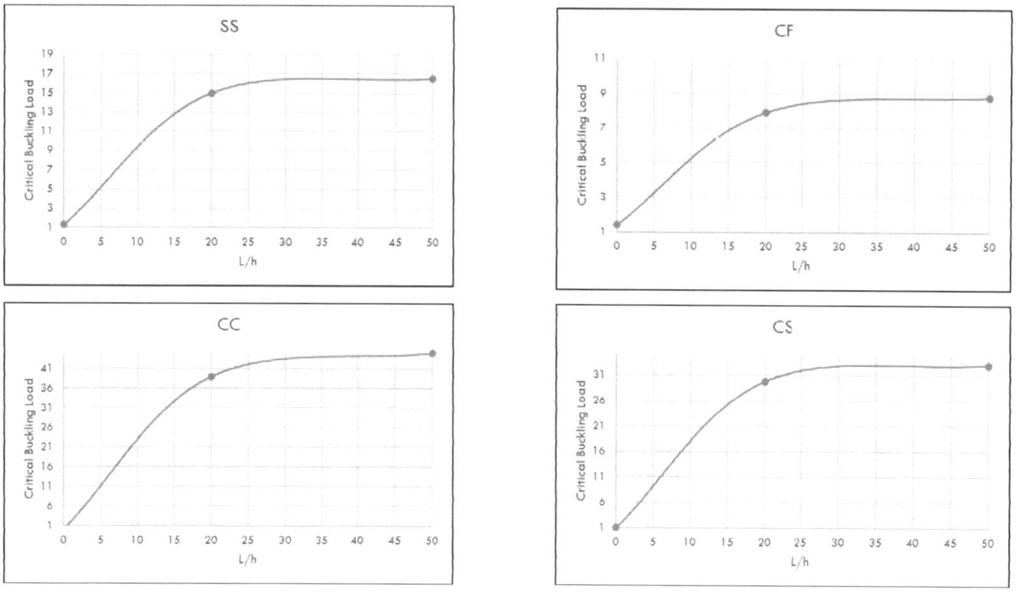

Figure 4:- The effects of the aspect ratio on the variation of the DCBLs of the symmetric [0°/90°/0°] LCSBs

The effect of shear deformation function on the critical buckling load of composite beam with different boundary conditions and fiber angle will be investigated firstly. After that, shear normal deformation beam theory (SNDBT), will be exploited in investigation of other effects. The buckling analysis of the symmetric [0°/θ/0°] LCBs under the variable in-plane loads are performed in this section. It is assumed that all the laminate of the composite beams has the same material properties and thickness. the DCBLs of the symmetric [0°/θ/0°] LCBs are given for various axially variable in plane loads, BCs, fiber angles (θ) and aspect ratios. It can be seen that the minimum DCBL is always obtained for the LCBs subjected to the N_x^4. Moreover, applying the N_x^5 always leads to have the maximum DCBL for all type of BCs. As expected, the DCBL increases with an increment on the aspect ratio. It is found that the DCBL is decreased by an increment in fiber angle for all types of BCs. The effect of aspect ratio variation on the DCBL is significant for CC LCBs. On the other hand, it is the least for CF LCBs. The DCBLs obtained for CC and CS LCBs is almost same while the aspect ratio is set to 100 for all fiber orientation angles. On the other hand, with the increasing of the aspect ratio, the difference between the DCBLs of CC and CS LCBs increases. It is seen that N_x^3 and N_x^5 types load distributions give higher critical buckling loads.

7. Conclusion:

In this article, the static critical buckling loads and mod-shapes of composite laminated beams are investigated under distributed in-plane axial load described by six functions. Unified higher order shear deformation theories are proposed to include the shear effects, extension bending, and rigidity of the beam structure. Ritz method is exploited to solve the govern equilibrium equations and derive the critical buckling loads and their mode-shapes.
The most finding are summarized as:

- The distributed axial in-plane load tends to distort the mode-shapes and deviate them from symmetry.
- CC is the most boundary condition significant for distortion of its mode rather than other boundaries under distributed axial load.
- Aspect ratio and fiber orientations have significant effects on the DCBLs.
- It is seen that N_x^3 and N_x^5 types load distributions give higher critical buckling loads.
- The type of axially variable in-plane load is significantly affecting the DCBL of the beam structures. The fiber angle value (θ) is important for the DCBL. Moreover, its effect on the DCBL is more pronounced especially for CS and CC symmetric LCBs.
- The buckling mode shapes of LCBs are significantly affected by the type of axially variable in-plane load for all BCs.
- The DCBL of the LCSBs depend on not only the aspect ratio but also the BCs and fiber orientation angles. The effect of the thickness stretching on the buckling mode shapes may vary depending on the type of the axially variable in-plane load.

References

1. Y.X. Zhang a,*, C.H. Yang Recent developments in finite element analysis for laminated composite plates Composite Structures 88 (2009) 147–157
2. Ghugal YM, Shimpi RP. A review of refined shear deformation theories for isotropic and anisotropic laminated beams. J Reinf Plast Compos 2001;20(3):255–72.
3. Aguiar R, Moleiro F, Soares CM. Assessment of mixed and displacement-based models for static analysis of composite beams of different cross-sections. Compos Struct 2012;94(2):601–16.
4. Zhen W, Wanji C. An assessment of several displacement-based theories for the vibration and stability analysis of laminated composite and sandwich beams. Compos Struct 2008;84(4):337–49.
5. Mathew TC, Singh G, Rao GV. Thermal buckling of cross-ply composite laminates. Comput Struct 1992;42(2):281–7.
6. S.K. Panda, L.S. Ramachandra, Buckling of rectangular plates with various boundary conditions loaded by non-uniform in-plane loads, Int. J. Mech. Sci. 52 (6) (2010) 819–828.
7. J.H. Kang, A.W. Leissa, Exact solutions for the buckling of rectangular plates having linearly varying in-plane loading on two opposite simply supported edges, Int. J. Solids Struct. 42 (14) (2005) 4220–4238.
8. N.I. Kim, D.K. Shin, Y.S. Park, Coupled stability analysis of thin-walled composite beams with closed cross-section, Thin-Walled Struct. 48 (8) (2010) 581–596.
9. A. Farajpour, A.R. Shahidi, M. Mohammadi, M. Mahzoon, Buckling of orthotropic micro/nanoscale plates under linearly varying in-plane load via nonlocal continuum mechanics, Compos. Struct. 94 (5) (2012) 1605–1615.
10. M.A. Eltaher a,b,*, S.A. Mohamed c, A. Melaibari a Thin–Walled Structures 147 (2020) 106488 Static stability of a unified composite beams under varying axial loads
11. Eisenberger M. Buckling loads for variable cross-section members with variable axial forces. Int J Solids Struct 1991;27(2):135–44.
12. Lee K. Buckling of fibers under distributed axial load. Fibers Polym 2008;9(2):200–2.
13. Wang CM, Wang CY, Reddy JN. Exact solutions for buckling of structural members. Boca Raton, Florida: CRC Press; 2004.
14. Li QS. Exact solutions for the generalized Euler's problem. J Appl Mech 2009;76(4):041015.
15. Vo TP, Thai HT, Nguyen TK, Lanc D, Karamanli A. Flexural analysis of laminated composite and sandwich beams using a four-unknown shear and normal deformation theory. Compos Struct 2017;176:388–97.

16. Vo TP, Thai HT, Aydogdu M. Free vibration of axially loaded composite beams using a four-unknown shear and normal deformation theory. Compos Struct 2017;178:406–14.
17. Karamanli A, Vo T. Size dependent bending analysis of two directional functionally graded microbeams via a quasi-3D theory and finite element method. Composites Part B 2018;144:171–83.
18. Trinh LC, Vo T, Thai HT, Nguyen TK. Size-dependent vibration of bi-directional functionally graded microbeams with arbitrary boundary conditions. Composites Part B 2018;134:225–45.
19. Pagano N. Exact solutions for composite laminates in cylindrical bending. J Compos Mater 1969;3(3):398–411.
20. A. Karamanli, M. Aydogdu, Buckling of laminated composite and sandwich beams due to axially varying in-plane loads, Compos. Struct. 210 (2019) 391–408.
21. Aydogdu M. Buckling analysis of cross-ply laminated beams with general boundary conditions by Ritz method. Compos Sci Technol 2006;66(10):1248–55.
22. Aydogdu M. Thermal buckling analysis of cross-ply laminated composite beams with general boundary conditions. Compos Sci Technol 2007;67(6):1096–104.

Kinematics and compliance analysis of active suspension system and Development of control algorithm to maximize ride comfort

Sandesh R Karande[1], Lenin Babu Mailan Chinnapandi[1, 2, *], P Jeyaraj[3] and Jeyanthi Subramanian[1]

[1]School of Mechanical Engineering, Vellore Institute of Technology, Chennai, India
[2]Electric Vehicles-Incubation, Testing and Research Centre, Vellore Institute of Technology, Chennai, India
[3]Department of Mechanical Engineering, NIT Surathkal, Mangalore

*Corresponding author's e-mail address: lenin.babu@vit.ac.in

Abstract. It is important to understand the interrelationship between steering stability and NVH to adjust stiffness of the suspension to satisfy the aim of maximum ride comfort. This work aims to perform kinematic and compliance analysis on comprehensive quarter car model of McPherson suspension system to validate its reliability with respect to various road profiles and development of control algorithm to control its stiffness. Despite the wide usage of PID controllers, controllers based on Adaptive Neuro Fuzzy Inference System (ANFIS) are proved to give better results as they consider the non-linearity of active car suspension models. The modelling of the comprehensive non-linear quarter car active McPherson suspension system is done using MSC ADAMS and the control algorithm is implemented using MATLAB. The approach used is to perform parametric sensitivity analysis for prime parameters such as toe angle, camber angle and caster angle. The approach used to design a better robust control algorithm is achieved using μ-synthesis function of MATLAB.

1. Introduction

Developing a reliable vehicle suspension system often involve compromises in various incongruous parameters. Vehicle handling and ride comfort are two such incongruous measures which need certain compromises among them. Further designing the car only for its ride comfort demands compromise in the space requirements those including the boot space and leg room space requirements [1]. Conventional suspension models have been used in the evaluation of active and semi-active suspension systems [2]. The wide use of McPherson Suspension system due to its features such as light weight and low-cost characteristics makes it the perfect candidate to carry out to be used for vehicle dynamic and kinematic analysis [3].

In general, using the conventional suspension models involve linear models, which often neglect the non-linearity introduced due to incorrect consideration of tyre to be rigid element or often only the vertical movement of tyre is considered. This only compensates for the spring stiffness of the tyre. Vertical damping [4] and lateral deflection [5] both should be taken into account to make the suspension system truly non-linear. To perform a similar kind of Kinematic and Dynamic analysis on a double wishbone suspension quarter car model system a comprehensive model is created by incorporating wheel hop, vertical longitudinal damping and compliance of bushing at joints [5]. Deriving useful data from kinematic analysis of suspension system Stenson et al. proposed a dynamic model of McPherson suspension. Though this study included the non-linearity in some sense it ignored the dynamics due to vibration induced in chassis and those due to tyre [6]. Considering rotation of control arm and longitudinal displacement of unsprung mass as generalized co-ordinates, two different models have been worked out in the previous studies. Model by Hong et. al. consists of strut mounted on control arm [7], while model by Fallah et. al. consists of strut mounted on wheel spindle [8]. The study also included the disturbances caused due kingpin angles, caster and camber. Previous work also included working on the three-dimensional models to study the kinematics of McPherson suspension system, but none included the variations caused due to tyre hop, its longitudinal damping and lateral deviation caused during the actual events [9] [10]. Kim et al. performed a similar work on a three-dimensional model in ADAMS Software [11]. Anderson et al. based on the readings obtained from the experimental McPherson suspension setup in the laboratory concluded that both linear and non-linear models are comparable if correct parameters are identified. The very little difference in these two models is due to small variation in strut angle [12]. It is very important to create a reliable model of suspension system taking into account all its non-linearities to save the time and resources spend on an incorrect model and realizing at a further time when making the changes in the template prepared in vehicle dynamic simulation software like MSC ADAMS. It is very tedious work to make the desired changes in the template at later time.

This work proposes to develop a non-linear McPherson suspension system comprising of strut mounted firmly on wheel spindle while considering the effects due to tyre hop and lateral tyre displacement. Kinematics and Compliance Analysis tests are performed in MSC ADAMS to ensure that the virtual suspension system developed is in agreement with the real suspension system. Thus, the suspension system developed is more reliable and more realistic and is used for the second part of this work comprising of the development of control algorithm that can truly increase the ride comfort by actuating a correct signal that develops precise amount of counter force by force actuators to reduce the disturbances caused due to road unevenness and various road profiles. Figure 1 shows typical McPherson Suspension system.

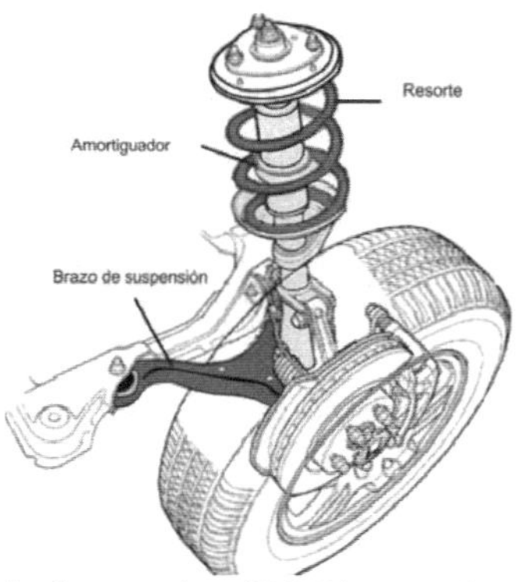

Figure 1: Typical example of McPherson Suspension system

With the continuous development of technology, passenger safety and ride comfort has become a topic of prime importance for the automotive industries. To increase the ride comfort it is very crucial to optimize stiffness of spring (k_t), various damping coefficient like tyre damping co-efficient and damping co-efficient related with sprung mass, unsprung mass spring stiffness of tyre. Sprung mass acceleration is inversely proportional to ride comfort. ISO 2631-1: 1997 defines ride comfort. Vibrations are caused due to various factors out of which vibrations due to road disturbances are predominant. Acceleration of sprung mass is the measure of Ride comfort.

Adverse effects of road excitations on the ride comfort has been studied by Griffin et. al. [12]. Added advantage of smooth transition of coefficient of damping of semi-active suspension makes it very effective in providing a better ride comfort when compared with passive suspension system [13]. Much research has been carried out on the semi-active suspension system that uses Magnetorheological damper. Although semi-active suspension system is the most commonly used suspension system, active suspension system turns to work out more effectively as it can perform better over wide range of frequencies [14]. Working space, ride comfort and handling have been determined as the objective function in the previous work involving 8 DOF model [15]. G. Verros et. al. presents the foundation to optimize parameters such as spring stiffness and damping co-efficient for non-linear suspensions models under random excitations [16].

Comparative study of Fuzzy Control system and Linear Quadratic Gaussian system has been carried out on Two Degree of Freedom suspension system [17]. It is seen that Adaptive-Network based Fuzzy Inference System (ANFIS) controllers performs better when optimising Active suspension system. The ANFIS controllers rely upon the data from FOPID control system to better compensate the non-linearity of suspension systems [18]. LQR control system has been suggested for Active suspension systems [19] [20]. This paper proposes a control algorithm with two approaches, H_∞ approach and μ-synthesis approach implemented using MATLAB on the SIMULINK model prepared after verifying the results of Kinematics and Compliance Analysis.

The following paper is organised as follows. The development of two DOF non-linear model of McPherson System is developed and Kinematics and Compliance (Dynamic) Analysis is performed in section 2. In section 3 based on the SIMULINK model developed in section 2, control Algorithm is implemented. Section 4 comprises of conclusion of work.

2. Modelling of the suspension and Kinematic and Compliance Analysis:

The model is built with sheer consideration that the unsprung mass of the suspension system which basically consists of wheel assembly experiences vertical translational motion as well as rotational motion and the chassis (sprung mass) undergoes only vertical translational motion. Figure 2 shows a typical example of non-linear model of McPherson suspension system.

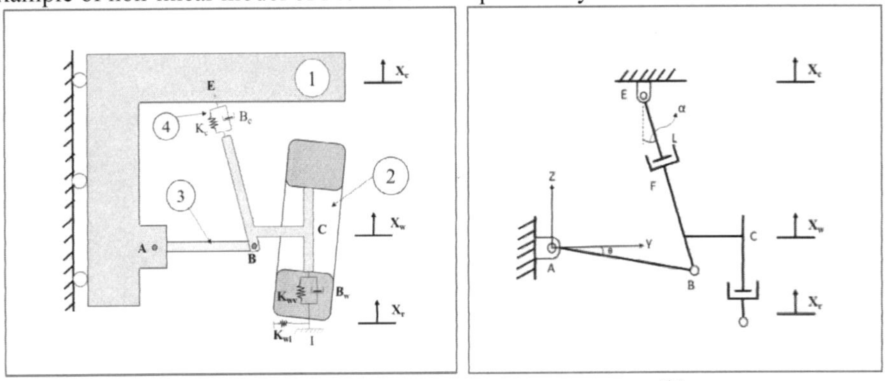

(a) (b)

Figure 2. McPherson suspension (a) Non-linear representation (b) kinematic representation

The system mainly consists of 4 main elements: (1) Car Body (Chassis)- sprung mass, (2) Wheel Assembly – Unsprung mass, (3) Control Arm and (4) Strut. The displacement associated with the sprung mass i.e., chassis is denoted by X_c, the displacement related with the unsprung mass is denoted X_w and the displacement due to road disturbances is denoted by X_r. The spring stiffness of the chassis is denoted by K_c and the vertical spring stiffness of the wheel is denoted by K_{wv} and the lateral spring stiffness of wheel (tyre) is denoted by K_{wl}, similarly the damping co-efficient associated with suspension is denoted as B_s and the damping co-efficient of the wheel is denoted as B_w. B_w thus represents the non-linearity due to tire hop. Considering the lateral tyre deflection which causes changes in caster and camber angles, the Moment of inertia of the wheel assembly at centre C is denoted by I_C. The key features or assumptions of these models are (1) The control arm and the strut are of insignificant masses so they can be ignored. (2) The tyre behaves as rigidless element (3) Springs and Dampers behave linearly.

Consider Y_C and X_C to be the coordinates of the wheel centre C and Y_{C0} and X_{C0} as their initial values and ϕ is the camber angle.

$$Y_F - Y_C = b\phi + a \tag{1}$$

$$X_F - a\phi = X_{F0} + X_W \tag{2}$$

$$Y_L - Y_C = d\phi + c \tag{3}$$

$$X_L + c\phi = X_{L0} + X_W \tag{4}$$

$$Y_B - Y_C = f\phi + e \tag{5}$$

$$X_P + e\phi = X_{B0} + X_W \tag{6}$$

here constants a, b, c, d, e and f are obtained from equilibrium conditions in initial conditions and their values are equal to: $a = Y_{F0} - Y_{C0}$, $b = X_{F0} - X_{C0}$, $c = Y_{L0} - Y_{C0}$, $d = X_{L0} - X_{C0}$, $e = Y_{B0} - Y_{C0}$ and $f = X_{B0} - X_{C0}$.

Equations (1)-(6) represent the kinematic behaviour of the unsprung mass and thus establish the relation between ϕ and X_w. Figure 3 shows motion of the system for the case of negative X_r. Due to the motion of the system, there is lateral as well as vertical displacement of wheel. The resultant lateral displacement of the wheel is denoted δY_{wl}. From the geometry of the system shown in Figure 3, we get three more equations as:

$$Y_{B0}Y_B - (2X_W - X_{B0})X_B = p - X_{B0}X_W \tag{7}$$

$$Y_B = -L\theta - q \tag{8}$$

$$X_P + q\theta = L + X_W \tag{9}$$

where $p = Y_{B0}^2 + X_{BO}^2$; $L = L_1 sin\theta_0$; $q = L_1 cos\theta_0$.

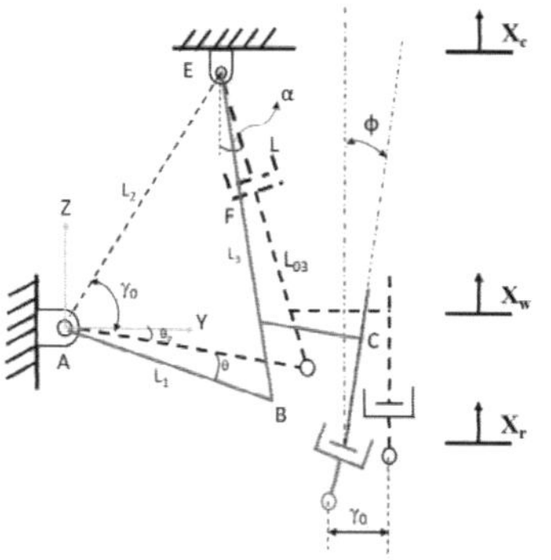

Figure 3. Suspension motion. Initial position is represented by dashed lines and solid lines represent position after motion

Equations (1)–(8), represents linear system with nine variables. These equations can further be evaluated in terms of X_c and X_w. We get series of expressions each of which is for different elements of suspension. Out of these expressions the three predominant key factors are Y_C, θ and ϕ. These are required for Compliance analysis:

$$Y_C = -e - \frac{f(X_{B0}+X_w)}{e} + sigma[\left(L + \frac{fq}{e}\right)(p + X_c - X_{B0}) + (q^2 + L^2 + LX_c)\left(\frac{fY_{B0}}{e} + 2X_c - X_{B0}\right)] \tag{10}$$

$$\phi = \frac{X_{B0}+X_W}{e} - sigma(\frac{Y_{B0}q^2}{e} + \frac{q(p-X_{B0}X_c)}{e} + \frac{LY_{B0}+X_c}{e}) \tag{11}$$

$$\theta = -sigma[p - X_{B0}X_c + Y_{B0}q - (L + X_c)(X_{B0} - 2X_c)] \tag{12}$$

where, $sigma = \frac{1}{LY_{B0}+qX_{B0}-2qX_c} = \frac{1}{G}$ and G denotes denominator equivalent.

For further simplicity in solving the equations, following terms are established:

$$d' = p(L + \frac{fq}{e})(q^2 + L^2)(\frac{fY_{B0}}{e} - X_{B0}) \qquad e' = X_{B0}\left(L + \frac{fq}{e}\right) + L\left(\frac{fY_{B0}}{e} + X_{B0}\right) + 2(q^2 + L^2)$$

$c' = -\frac{fX_{B0}}{e} - e$ $\quad i' = -qX_{B0} - LY_{B0}$ $\quad g' = \frac{X_{B0}}{e}$ $\quad h' = pq + (q^2 + L^2)Y_{B0}$

$f' = 2L$ $\quad t' = d' + \frac{Rh'}{e}$ $\quad u' = e' + \frac{Rh'}{e}$

Lateral deflection of tyre, δY_{wl} is then calculated using Equations (10) and (11) as:

$$\delta Y_{wl} = (Y_C - \phi R) - Y_{C0} = s' + \frac{f+R}{e} X_w + \frac{1}{G}(t' + u'X_c + f'X_c^2) \tag{13}$$

For the dynamic part of analysis, if D denotes the dissipation energy, m_c and m_w represent the masses of chassis (sprung mass) and wheel assembly (unsprung mass) and δ_{wv} represents the vertical displacement of spring and damper assembly of wheel, then value of δ_{wv} is given by:

$$\delta_{wv} = \sqrt{(L_{03})^2 + k\theta} - L_{03} \tag{14}$$

where, due to small angle:

$$(L_{03})^2 + k\theta = (L_3)^2 \tag{15}$$

And constant k is determined by:

$$k = -2L_1L_2\sin(\gamma_0)$$

The Lagrangian L to determine the dynamic response of the system is given by:

$$L = 0.5 \times [m_c\dot{X}_c^2 + I_C\dot{\phi}^2 + m_c(\dot{Y}_C^2 + \dot{X}_c^2) - K_w\delta_{wv}^2 - K_w(X_w - X_r)^2 - K_{wl}\delta Y_{wl}^2] \tag{16}$$

and

$$D = 0.5[B_c\dot{\delta l}^2 + B_w\dot{X}_t^2] \tag{17}$$

where,

$$X_t = X_w + X_r \tag{18}$$

combining value of L with X_w, we get:

$$\frac{d}{dt}\left[\frac{\partial L}{\partial \dot{X}_w}\right] - \frac{\partial L}{\partial X_c} + \frac{\partial D}{\partial \dot{X}_w} = 0 \tag{19}$$

Using values of equations (16) and (17) in equation (19), we get:

$$m_w\ddot{X}_w + m_w\ddot{Y}_C\frac{\partial \dot{Y}_C}{\partial \dot{X}_w} + I_C\ddot{\phi}\frac{\partial \phi}{\partial X_w} + B_w(\dot{X}_w - \dot{X}_r) + K_{wl}\delta Y_{wl}\frac{\partial(\delta Y_{wl})}{\partial X_w} + K_w(X_w - X_r) \tag{20}$$

Equation (20) represents non-linear differential equation as a function of X_w.

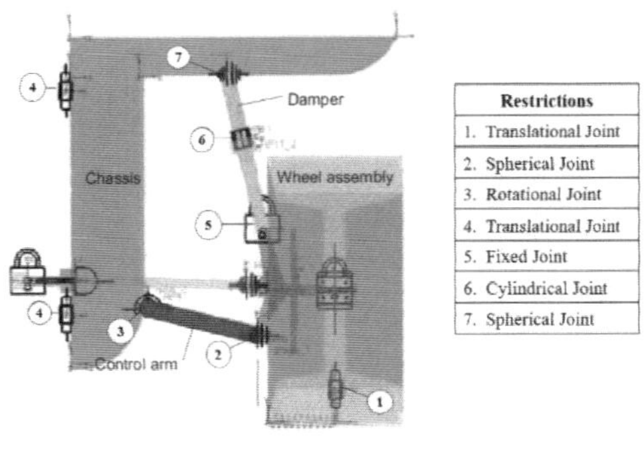

Figure 4: Non-linear McPherson suspension system developed in MSC ADAMS

With the help of above equations, a SIMULINK block model is derived for further use in MATLAB for developing control algorithm to optimize suspension. Figure 4 shows, equivalent quarter car model of non-linear McPherson Suspension system into consideration developed using MSC ADAMS for Kinematics and Compliance Analysis.

The modelled is carefully taking into account certain design parameters as hard point geometry the suspension model is subjected under vertical parallel movement test and vertical oppose movement test, with prime objective of optimizing camber angle, caster angle and track width as these three parameters play major role in providing ride comfort in vehicles with active McPherson suspension system. The results of these test on MSC ADAMS model are compared with the results obtained from solving mathematical model and similar evaluation on experimental model in previous research work.

2.1 Results and Discussion of Kinematics and Dynamic Analysis

Previous research work shows that similar kind of experimental evaluation has been done on a real McPherson suspension system in lab environment. The experimental setup underwent similar tests i.e., vertical parallel movement test and vertical oppose movement test. To make a comparative evaluation between experimental and virtual suspension system, certain values like vertical displacement (bound and rebound) travel value and steps.

A total of 6 graphs were obtained from the test data set. Figure 5 and figure 6 shows the comparison between the three non-linear models – experimental model, virtual model created using MSC ADAMS and mathematical model. Figure 5 particularly shows the result of Kinematic & Compliance (dynamic) analysis (K & C analysis) when the models are subjected to vertical oppose movement test and similarly Figure 6 shows results obtained from vertical parallel movement test.

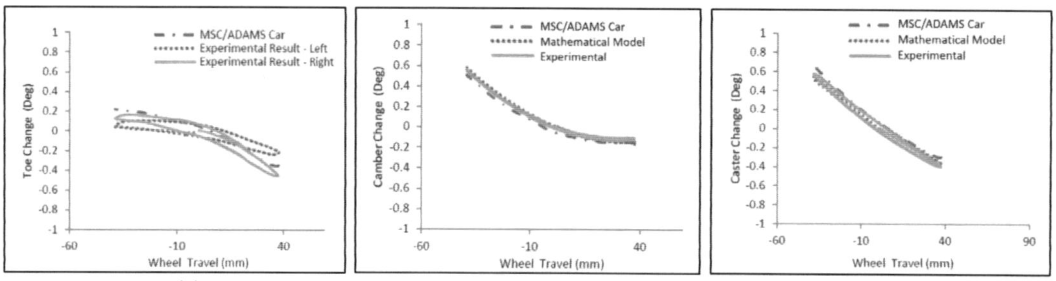

(a) (b) (c)

Figure 5. McPherson suspension virtual model subjected to vertical oppose movement test (a) Toe change (b) camber change (c) Caster change

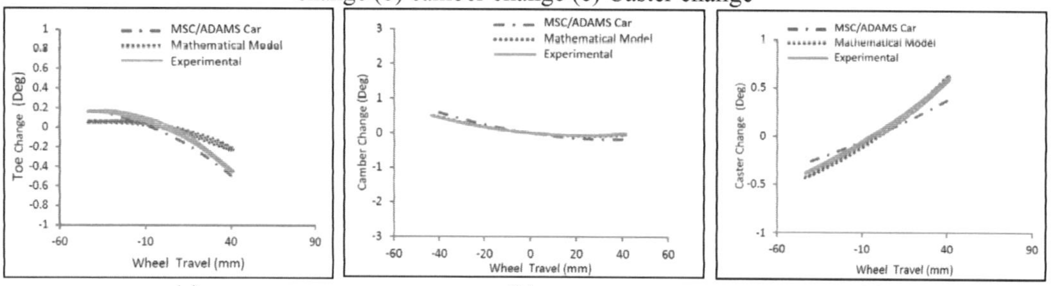

(a) (b) (c)

Figure 6. McPherson suspension virtual model subjected to vertical parallel movement test (a) Toe change (b) camber change (c) Caster change

From Figure 6 (a) and 6 (b) we can see there is negative slope of in the graph which denotes decrease in value of toe change and caster change with increase in wheel travel with time. This is advantageous as it allows better handling and good stability during cornering [21] [22]. From figure 5 and figure 6, it is evident that the non-linear virtual McPherson suspension system created in MSC ADAMS is in great agreement with the results of real experimental suspension system. Therefore this virtual suspension model can be worked with for developing control algorithm to control the input to force actuators that can counter act on the force disturbances that vehicle experience.

3. Development of Control Algorithm:

While in motion vehicle experiences Vibration due to various force acting on it. Some of the forces include vibrations due to irregularities in road like bumps and potholes, vibrations due to aerodynamic forces acting on it. When a car hits a bump or a pothole, the tire absorbs some of the impact force and compresses a little bit and the force is then transfer to the car body causing it to bounce. The desirable function of the suspension system is to dampen these vibrations to provide ride comfort to occupants and better vehicle handling. Vehicle nowadays are equipped with active suspension systems. Ride experience can be augmented using Active Suspension Systems. The Vehicle chassis is fitted with numerous sensors also including those to measure body acceleration.

In general, the ideal suspension system is one which work in accordance with Skyhook theory. Active suspension system has force actuator in addition to the elements of the passive suspension systems. Body acceleration of the vehicle is converted into a suitable form to act as an input signal to the force actuators.

In the development of this control algorithm, H-infinity synthesis was used to create a nominal plant model that although guarantees performance but is not necessarily robust in the sense it does not operate over a large range of frequencies. Later an uncertain model is created using mu-synthesis to design a controller that is robust. Mu-synthesis of unregulated controllers works similar to h-infinity synthesis only difference being that mu synthesis takes into account the uncertainty of the system.

Based on the calculation of mathematical model, SIMULINK Model is created and is further used for developing this control algorithm. Two approaches are used to develop the control algorithm. One is using H_∞ (H-infinity) approach and the second approach used is using μ-synthesis. The Robust control toolbox of MATLAB is used in the process.

The controller takes body acceleration (a_b) and suspension deflection (S_d) as inputs and produces hydraulic force (f_s) as output. This hydraulic force act as a counter force to the forces causing undesired vibrations in system. Thus, the force created by hydraulic force try to dampen these undesired vibrations. Whenever the car hits a bump or a goes through a pothole, the disturbances (undesired vibrations) are introduced in the system. These disturbances are amplified in some ways which cause the car body to hop (a_b) and the suspension to compress (S_d). The performance of active suspension systems is determined by how well the control algorithm is developed and implemented that it produces certainly the right amount of actuator force to isolate the vibrations away from passengers and give them better ride experience. The feature of a best controller is that it minimizes this amplification of the disturbances in the least amount of actuator force possible. To manipulate the signals and treat them unequally within the model we add weighting factors. Using H_∞ controllers This enables us to create different modes like comfort mode, balanced mode and Handling mode (Performance mode). We generally see these different modes that can be adjusted from car's dashboard on supercars. This way if we want a very smooth ride, we can choose comfort mode which decreases the weighting factor (β) of actuating energy of force actuator which dampen the vibrations slowly or other way if we are concerned more about the life of suspension and want a better handling feature, we can select handling mode which dampens the vibrations quickly increasing the weighting factor of actuating energy. A balanced mode is trade-off between ride comfort and vehicle handling capabilities. The gist of the control algorithm is to determine which factors we have access to (in this case a_b and S_d) and which variables can be controlled (f_s), then we need to choose the error signals and decide their cost functions by setting weight factors. The weighting functions (β) can be frequency dependent, so high pass filter is set up on actuator signal to penalize high frequency controller commands than low frequency commands, which allows to limit the bandwidth of operational frequencies. The tire-hop frequency and rattlespace frequency were calculated to be 84.56 rad/s and 34.46 rad/s respectively.

Setting up the model for H_∞ controller is easier as we only need to create an open loop function and declare the variables that it has access to (a_b and S_d) using the '*hinfsyn*' command of Robust control toolbox of MATLAB. '*hinfsyn*' command solves the optimization problem to find a controller 'K' that minimizes the amplification of energy from external input signals 'w' (vibration due to forces acting on the body) to error signals (a_b, S_d and f_s) of virtual suspension model. The gain factors of the three desired modes (gain sets) i.e., comfort, balanced and performance is declared. '*hinfsyn*' command is setup such that it calculated the optimized controller for each of the three gain sets. The simulation is carried out for time of 1 second and the road bump of 5cm.

From Figure 7 (a), we can observe that although the body travel for open loop suspension system i.e. in absence of force actuators and body travel in case of comfort mode and open loop response are same, the body returns to its original position slowly damping away the vibrations thus increasing the ride comfort for passengers. On the other hand, in handling mode the body quickly returns to its position enabling better handling of the vehicle. The green line represents the road bump. The blue line is representation of passive suspension system (open loop). As a result of high gain factor, β of performance mode the suspension creates a higher actuating force which quickly dampens the vibrations in less suspension deflection, The higher control force can be seen in figure 7 (b).

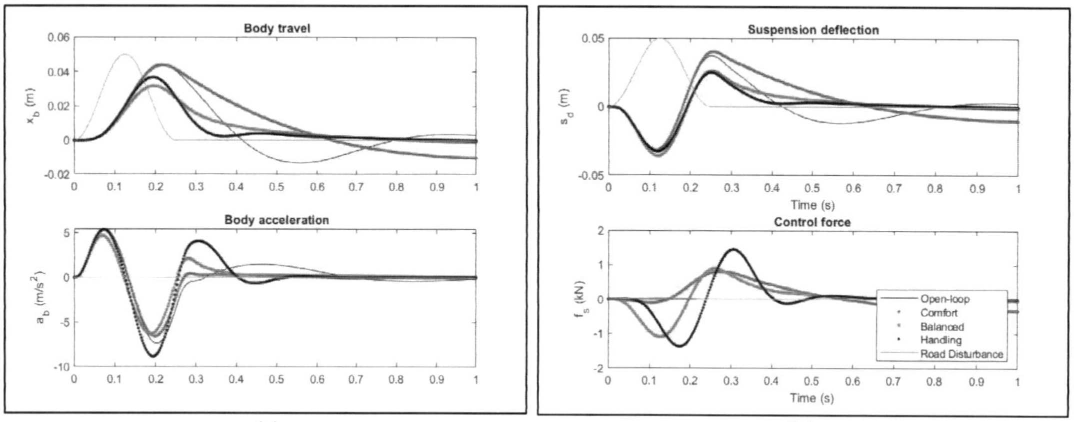

(a) (b)

Figure 7. Comparision of different modes using control algorithm (a) Body travel and Body acceleration (b) Suspension deflection and control force

The only problem with using this controller designed using H_∞-synthesis is it works on a nominal model and it fails to compensate for the noise (error) in the sensors value and dynamics of the actuator. Figure 8 shows comparision of simulations performed on Nominal H_∞ controller and μ-synthesis controllers. The models are simulated with uncertain model simulated for 5 cm road bump and 100 random actuator variations for balanced mode.

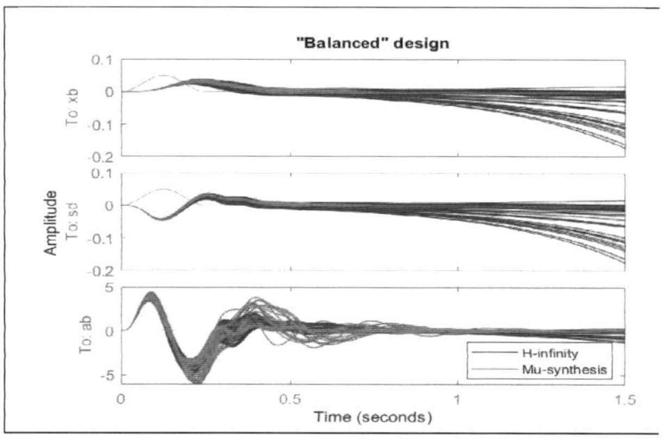

Figure 8. Simulation result of Nominal H_∞ controller vs μ-synthesis controller for 100 random actuator variations

It can be seen that out of the 100 actuator models, most of the actuators give a reasonably stable response. However some actuators produces unstable response. For a system with such high uncertainites, controllers developed using μ-synthesis provides better results at it takes into account all the uncertainities from beginning of simulation. μ-synthesis controllers work similar to H_∞ controllers except that it used decay iterative process to minimize the worst case scenario gain across range of uncertainities. It keeps calculating repeatedly the robustness of the nominal H_∞ controllers and scale each of them based on the uncertainity of the system. This process continues until the the robustness value (peak μ value) of the suspension attains some constant value. The robustness value for the Non-linear McPherson suspension system designed was calculated to be 0.946. The lesser this value, more robust is the suspension. It can be observed that μ-synthesis controller (representey red lines) produce stable response with less variability due to uncertainity. The wild acceleration is associated with the variations that deviate more from the nominal actuator. So we a suspension system is designed near the variation limit the system is stable but comes at the expense of rougher ride.

4. Conclusion:

The widely used McPherson suspension poses a challenging modelling problem of its nonlinear asymmetric behaviour. Certain important parameters are considered while designing the mathematical model of the suspension compensate the non-linearity in actual suspension system. The parameters include (1) rotation and translation for the unsprung mass (2) wheel mass and its inertia moment about the longitudinal axis and (3) tyre damping and lateral deflection due to the suspension mechanism. The proposed model is compared with ADAMS model and the SIMULINK model is prepared based on the mathematical model.

If actuator dynamics vary greatly, controllers developed using *H_∞-synthesis* is not a great solution.

References:

[1] Sharp, R.S., and Crolla, D.A., 1987, "Road Vehicle Suspension System Design-a Review", Vehicle System Dynamics, 16, pp. 167-192.

[2] Kim, C., and Ro, P.I., 1998, "A Sliding Mode Controller for Vehicle Active Suspension Systems with
Non linearities", Proceedings of IMechE, Journal of Automobile Engineering, 212 (1998), pp. 79-92.

[3] Chang, P., and Huang, A., 2006, "Adaptive Sliding Control of Active Suspension Systems with Uncertain Hydraulic Actuator Dynamics", Vehicle System Dynamics, 44(5), pp 357-368.

[4] H. Akcay and S. T¨urkay, "RMS performance limitations and constraints for quarter-car active suspensions," in *16th Mediterranean Conference on Control and Automation*, 2008, pp. 425 – 430.

[5] K. P. Balike, S. Rakheja, and I. Stiharu, "Development of kinetodynamic quarter-car model for synthesis of a double wishbone suspension," *Vehicle System Dynamics*, vol. 49, no. 1-2, pp. 107–128, 2011.

[6] Stenson, A., Asplund, C., and Karlsson, L., 1994, "The Nonlinear Behavior of a MacPherson Strut Wheel Suspension", Vehicle System Dynamics, 23, pp. 85-106.

[7] Hong, K., Jeon, D., Yoo, W., Sunwoo, H., Shin, S., Kim, C., and Park, B., 1999, "A New Model and an Optimal Pole-Placement Control of the MacPherson Suspension System", SAE Technical Paper, 1999-01-1331.

[8] Fallah, M.S., Bhat, R., and Xie, W.F., 2009, "New Model and Simulation of MacPherson Suspension System for Ride Control Applications", Vehicle System Dynamics, 47 (2), pp. 195-220.

[9] D. A. Mantaras, P. Luque, and C. Vera, "Development and validation of a three-dimensional kinematic model for the mcpherson steering and suspension mechanisms," *Mechanism and Machine Theory*, vol. 39, pp. 603–619, 2004.

[10] Y. A. Papegay, J.-P. Merlet, and D. Daney, "Exact kinematics analysis of car's suspension mechanisms using symbolic computation and interval analysis," *Mechanism and Machine Theory*, vol. 40, no. 4, pp. 395 – 413, 2005.

[11] Anderson, E., Sandu, C., and Southward, S., 2007, "Multibody Dynamic Modeling and System Identification of a Quarter-car Test Rig with McPherson Strut Suspension", SAE Technical Paper, 2007-01-4184.

[12] M. J. Griffin.: Handbook of Human Vibration. Academic Press, London Publication, 1996.

[13] Lee, C. H. Chen, C. L. Shen, J. Tyan F., 2010, Modelling and simulations of Half car Suspension system with a MR Damper Using RecurDyn and Simulink", Journal of Advanced Engineering, pp. 27-32

[14] Yao, G. Z., Yap F.F., Chen G. Li, W. H., Yeo S. H. 2002, "MR Damper and its application for semi active control of vehicle suspension system", Mechatronics 12, pp. 963-973.

[15] Mehrdad N., Khajavi, GolamhassanPaygane: A multi objective optimization Approach to Optimize Vehicle Ride and Handling Characteristics, International Scholarly and Scientific Research & Innovation. 4(2) (2010) 502-506

[16] G. Verros, S. Natsiavas C., Design Optimization of quarter car models with passive and semi active suspensions under random road excitations. Journal of Vibration and control, Sage publications, 11 (2005) 581-606

[17] A. J. Barr, J. L. Ray, Control of an active suspension using linear and fuzzy logic. Proceedings of the Fifth IEEE International Conference on Fuzzy Systems 1996, 1: 42-48

[18] A. Adlair, W. J. Wang, Adaptive Neuro Fuzzy Inference Controller for Full Vehicle Nonlinear Active Suspension systems. 2010 Ist International conference on Energy, Power and Control (EPC-IQ) 2010; 97-106

[19] M. Senthil Kumar, S Vijayarangan. Design of LQR controller for active suspension system. *Indian Journal of Engineering & Material Sciences 2006; 13: 173-179*

[20] M. Senthil Kumar, S Vijayarangan. Linear Quadratic regulator controller design for active suspension system subjected to random road surfaces. *Journal of Scientific & Industrial Research* 2006; 65: 213-226

[21] Milliken W, Milliken D. Rave car vehicle dynamics. SAE International ISBM 1-56098-526-9;1995

[22] Mitchell EC, Simons R, Sutherland T, Keena-Levin M. Suspension geometry: thepry vs. K&c measurement. SAE Technical Paper No. 2008-01-2948; 2008.

Impact of web and counterweight shape factors on crank shaft design for better strength and optimised geometry

G Mathan, J Daniel, P Vishnu and V Satyanarayana
Powertrain Division, Mahindra & Mahindra Ltd, Chennai, Tamilnadu, India

E-mail:g.mathan@mahindra.com

Abstract. Automotive industry is continuously focusing on developing weight efficient engines for better fuel economy without altering engine performance. To achieve a trade-off between performance and fuel economy, the design always demands downsizing of engine components with optimum geometry. Crankshaft is such an important component in deciding engine performance and overall weight. Torsional and bending stiffness of crankshaft plays a vital role in its structural integrity, particularly the Critical Cross Section (CCS) which connects main journal and crankpin. In the present study, the objective is to upgrade the crankshaft for higher torsional and bending stiffness and reduce weight without changing its performance. Three geometric shape factors of the crankshaft are identified and optimized in FEA and analyzed in AVL Excite® (1-D) simulation for its structural performance. Study shows the identification of shape factors and effect of their variation on torsional and bending stiffness of crankshaft. In addition, sensitivity analysis for various shape factors and their optimization has been done. The identified shape factors help designers to identify any scope to optimize the geometry of crankshaft and understand its role in structural strength. Using this methodology, specific regions defined by shape factors can be optimized for better strength and weight.

1.Introduction

An automotive Cranktrain assembly consists of crankshaft, connecting rod, pistons, piston pins and flywheel. Crankshaft transfers the combustion and inertial loads of cranktrain components to flywheel which in turn transfer it further to clutch and transmission. Diesel engines have high compression ratios, higher combustion pressures and hence high combustion loads. Crankshaft is one of the most stressed members in an engine and hence crankshaft optimization is rarely pursued in automotive industry. Due to increasing fuel efficiency requirements and impending CAFE (Combined Average Fleet Economy) regulations, weight reduction has become the main thrust area for OEMs.

Approximately 10 percentage of the total engine weight is contributed by crankshaft and hence crankshaft can be considered as one of the heavier component in an engine assembly. Crankshaft is a rotating component which rotates at 1000-4000 RPM. It is very important to select possible weight reduction regions of crankshaft. A study shows drastic variation in crankshaft stiffness for same amount of material removal from different regions (1).

Crankshaft contributes to 20-40% of frictional losses (FMEP) in an engine over different speeds. FMEP is directly proportional to contact area of running surfaces and weight of the component. Mass

reduction of rotating parts of engine provides advantage of lower power requirements during acceleration phase which measures drivability of vehicle (2).
In order to increase the power density of an engine, the requirement for higher cylinder pressures and hence higher torsional and bending stiffness of crankshaft becomes more relevant (3). Study shows about 10% increase in web width produces 15% increase in torsional strength whereas 10% increase in web thickness produces only 4% increase in torsion strength (4).
In this study, possibility of crankshaft weight reduction and stiffness optimization is analyzed. Three key regions are identified by respective shape factors and their effect on torsional and bending stiffness of crankshaft has been evaluated. Optimization of these shape factors have been done to achieve desired torsional and bending stiffness by DOE. 1-D simulation for the optimized crankshaft is done to confirm cranktrain design acceptance criteria. The present work consists of four major phases as shown as follow and as shown in figure 1

a.Identification of shape factors
b.Design of experiments and collection of data as per designs
c.Analysis of DOE results.
d.Optimization

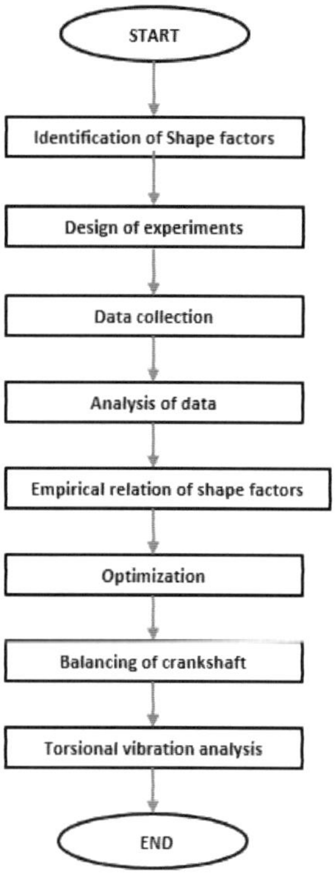

Figure 1: Process flow chart

2. Identification of Shape factors

Shape factors can be defined as dimensions of important features of a crankshaft. These factors should be selected in such a way that they don't affect overall dimensions of the crankshaft. The shape factors were also selected in such a way that it doesn't affect the critical cross section of crankshaft. Critical cross section (refer figure 2) is defined as the section between lowermost point of crankpin fillet and uppermost point of main journal fillet. This section comes in the area of pin overlap. Increase in pin overlap results in increase in dimensions of critical cross section. Hence, three areas were selected which do not interfere with this critical cross section. They are as given below.

a. Web Top Offset Factor
b. Counterweight Profile Factor
c. Side Width Factor

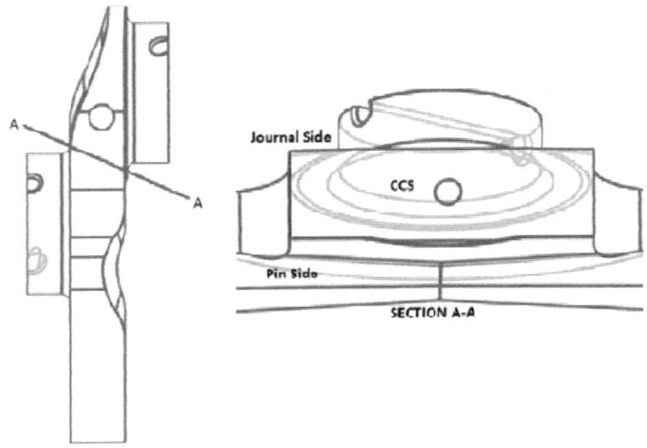

Figure 2: Critical Cross-Section (CCS) of crank web

2.1 Web top face offset factor

The top face thickness is varying over its height as shown in figure 3. A control point was selected on the centre of the collar. The geometrical dimension X is defined as the horizontal distance between the control point and collar machined face. This X dimension is controlled by forging and manufacturing requirements (minimum material width required for forging).When adjusting the control point, another geometrical dimension Y also to be considered. This is the distance between control point and crankshaft rotating axis. Y is also constrained indirectly due to the profile of face. The web top face offset factor of (A) = X × Y is defined for this region. There are three levels of profile modification as shown in figure 3.

- Low level-red
- Medium level-Blue
- High level-Black

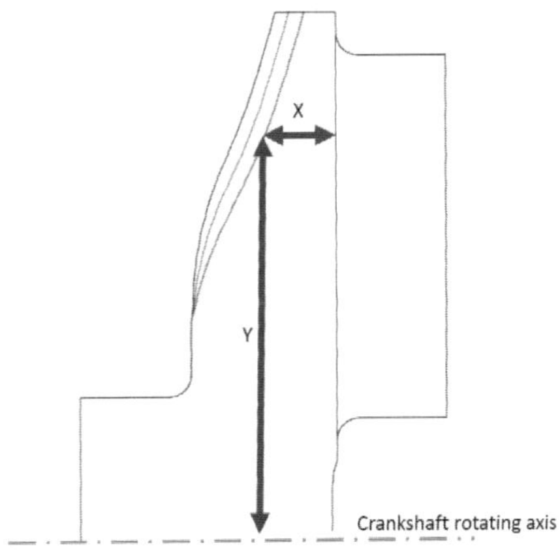

Figure 3: Web top face offset (A)

2.2 counter weight profile factor

Counterweight profile was chosen as it is just under the critical cross section. This profile is controlled by two geometric dimensions D and R. The factor D is defined as the distance between crankpin centre and centre of profile radius. It is constrained due to minimum distance requirement between end of crankpin collar and start of profile. This distance was fixed as 7mm to take care of minimum stiffness requirement of crankshaft. The factor R is defined as the radius of cut. R is constrained due to minimum thickness requirement of counter weight width which is 10 mm. Counterweight profile factor (B) = D/R has been defined for this region. There are three levels of profile modification as shown in figure 4.

- Low level-Blue
- Medium level-Red
- High level-Green

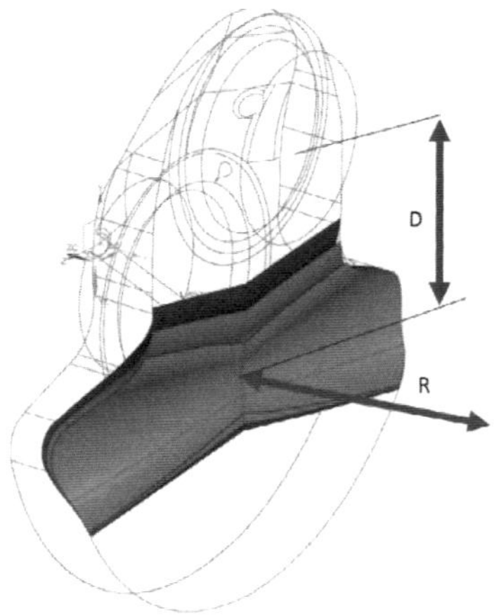

Figure 4: Counterweight profile (B)

2.3 Side width factor

The sides of web have been defined by a geometric parameter called as Z. The dimension Z is defined as distance between the two outermost faces of crank web sides. Z has a constraint with respect to manufacturing as the distance between side face and main journal collar cannot be less than 2 mm. The side width factor (C) = Z/2 has been defined for this region. There are three levels of profile modification as shown in figure 5.

- Low-Red
- Medium-Blue
- High-Black

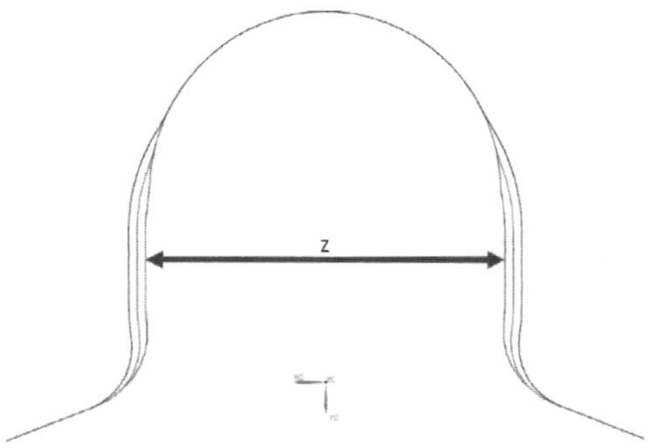

Figure 5: Side width (C)

3. Stiffness calculation

Crankshaft torsional and bending stiffness are important in overall cranktrain performance. The crankshaft stiffness can be calculated analytically or numerically (FEA). In FEA, the stiffness of crankshaft is calculated by evaluating deformation under specified loading condition. Stiffness can be calculated analytically with length and diameter, hence all empirical relations generates equivalent length with major crankshaft dimensions. Some of the empirical relations are BICEAA provisional formula, Carter's formula, Norman-Stinson's formula, Timoshenko's formula, Tuplin's formula(5).

Study shows considerable difference in stiffness values calculated by FEA and applicable empirical formula. Small change in dimension causes large deviations between empirical and real values hence stiffness estimation by empirical equations is not reliable (6). Meshing influence study with hexa and tetra elements shows less than 1% variation and hence second order tetrahedron elements are used for the ease of meshing and with reasonable computational time (7).

Since the crankshaft shape is continuously varying across the length, the stiffness of the various regions has been evaluated individually. Each region stiffness values are used in the 1-D simulation later in the paper. For shape factor evaluation and optimization a representative region (Single Web) of the crankshaft has been selected. The stiffness evaluation as described in the present section is used in the next section (DOE). The crankshaft 3D model is split into different webs by using the following method. The crankshaft crankpins and main bearings are split at the centre. This gives us different webs for calculating stiffness. Figure 6 gives the methodologies of splitting crankshafts.These individual webs are evaluated for their stiffness using FEA.

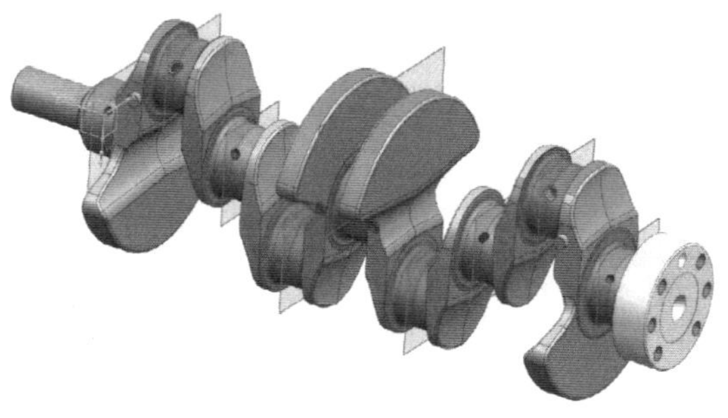

Fig.6: Splitting of crank shaft into individual webs

3.1 Torsional stiffness using moment

The 3D model of web is meshed using second order tetra elements. The nodes attached to the face of cut section are coupled to a single point at the centre of the section using rigid elements. Similarly the cut section at main bearing is also done. The centre of main bearing split section is also connected to the entire cut surface using rigid elements. The connected centre point of crankpin is constrained in all 6 degrees of freedom. A clockwise torsional moment of 1 N-mm is applied on the main bearing centre. The direction of moment application is parallel to crankshaft centre axis as shown in figure 7.

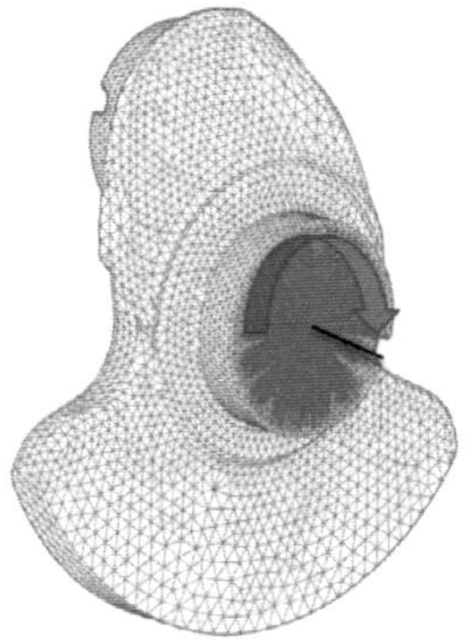

Figure 7: Boundary condition for torsional stiffness calculation by FEA

3.2 Bending stiffness using moment

The same finite element model which was used for torsional stiffness calculation is used for bending stiffness calculation. The connected centre point of crankpin is constrained in all 6 degrees of freedom. A bending moment of 1 N-mm is applied on the main bearing centre. The direction of moment application is perpendicular to crankshaft centre axis as shown in figure 8.

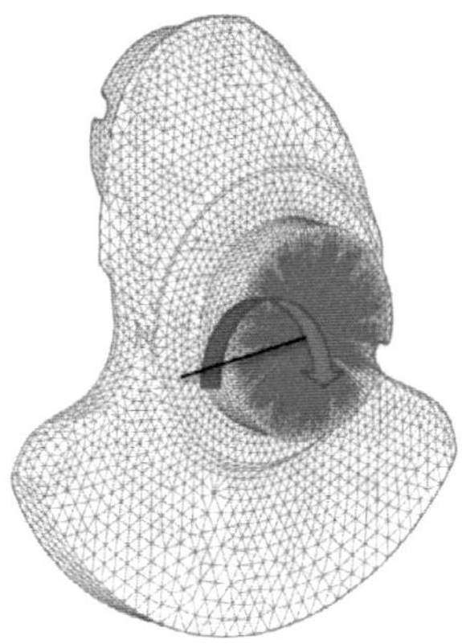

Figure 8: Boundary condition for bending stiffness calculation by FEA

IOP Conf. Series: Materials Science and Engineering 1128 (2021) 012045 doi:10.1088/1757-899X/1128/1/012045

4.Parametric study of shape factors(design of experiments)

For studying shape factors in crankshaft, the crankshaft of a 2.5Ltr 4 cylinder engine is selected. Design of experiments is a systematic method to determine the relationship between factors affecting a process and output of that process. Factorial designs allow for simultaneous study of the effects that several factors may have on output. While designing experiments, varying the levels of the factors simultaneously rather than One Factor at a Time (OFAT) is efficient in terms of time and cost. In full factorial experiment, responses are measured for all combinations of the experimental factor levels. These combinations represents conditions at which responses will be measured. Each experiment condition is called as “Run”, response measurement as “Observation” and entire set of runs is called as “Design” (8)

Number of runs are given by the following formula

$$\text{Number of runs} = \text{Level}^{\text{Factors}} \quad (1)$$

4.1 defining factors

In this experiment input parameters or factors of interest are as designated in Table 1.

- Web Top Face Offset (A)
- Counterweight Profile (B)
- Side Width (C)

4.2 defining responses

Crankshaft rigidity can be indicated with Torsional and Bending stiffness. These two responses are captured for each set of input conditions and evaluated.

4.3 DOE

Minitab® [8] is used to design experiments by defining factors and expected response parameters. The sequence and configuration of experiments is as follows in Table 2

Table 1: Crankshaft shape factors

Factors	Symbol	Level 1	Level 2	Level 3
Web Top Face Offset	A	1	2	3
Counterweight Profile	B	1	2	3
Side Width	C	1	2	3

Table 2: Design of Experiments

Iteration	A	B	C	Iteration	A	B	C
1	1	1	2	15	1	2	3
2	3	1	3	16	2	2	3
3	3	3	1	17	3	2	2
4	3	1	1	18	2	2	1
5	1	3	1	19	2	1	1
6	3	2	3	20	2	3	3
7	2	2	2	21	1	1	3
8	3	3	2	22	3	2	1
9	2	3	2	23	1	2	1
10	3	1	2	24	2	1	3
11	3	3	3	25	2	3	1
12	1	3	2	26	1	1	1
13	1	2	2	27	2	1	2
14	1	3	3				

4.4 Regression analysis

Regression is widely used method in statistics to discover the influence of more than one input variable on output response [9]. Linear multiple regression model is as shown in the equation 2

$$y = b_0 + b_1x_1 + b_2x_2 + \cdots \quad (2)$$

More detailed model can be defined according to predicted response and required accuracy. Following equation shows second order model with interaction effect

$$y = b_0 + b_1x_1 + b_2x_1x_2 + b_3x_1^2 + b_4x_2 \ldots \quad (3)$$

Using Minitab® regression analysis, torsional and bending stiffness can be expressed as,

$$\text{Torsional Stiffness} = f\{A^2, C^2, AC, B\} \quad (4)$$

$$\text{Bending Stiffness} = f\{A^2, C^2, B^2\} \quad (5)$$

Equations 4 and 5 indicates quadratic relation between torsional /bending stiffness and shape factors(Table 3).

Table 3: Quality of regression model

Equation	R^2	$R^2_{adjused}$	$R^2_{predicted}$	Model
Torsional Stiffness	99%	98.8%	98.51%	Quadratic
Bending Stiffness	94.6%	93.9%	92.77%	Quadratic

Correlation coefficient R^2 above 90% shows that it is stable model with reliable predictability indicated by high predictability values $R^2_{predicted}$ is above 90%.

Residual plots for above two regression models are shown in figure 9 and 10 and indicates goodness of model fit.

Normal probability plot: Points in this plot form approximate straight line hence residuals are normally distributed (refer figure 9 and 10).

Residuals versus fits: Random pattern of residuals on both sides of zero and hence no recognizable pattern (refer figure 9 and 10).

Residuals versus order: Plot of all residuals in an order that data was collected and used to find non-random error (refer figure 9 and 10).

Histogram of residuals: Indicates normally distributed residuals (refer figure 9 and 10).

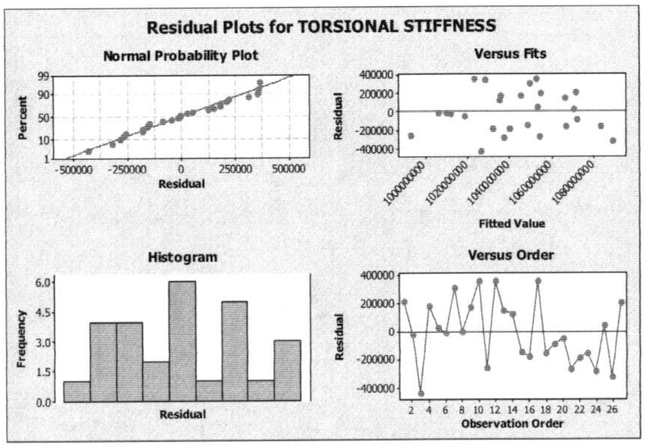

Figure 9: Residual plots for torsional stiffness model

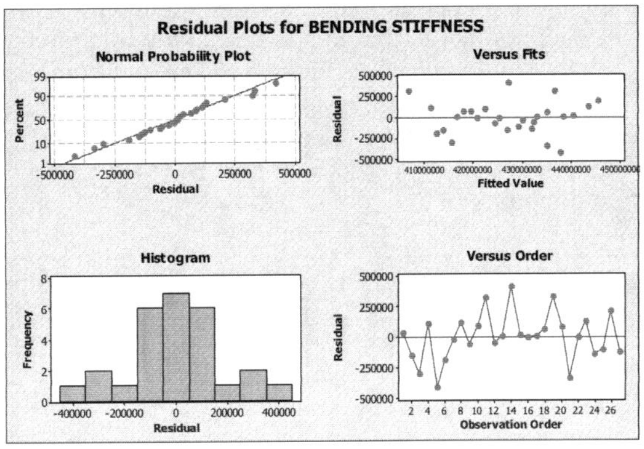

Figure 10: Residual plots for bending stiffness model

4.5 Interpretation of doe results

Significant and insignificant factors can be identified by plotting means of response simulated in experiments against each level of factors. These plots helps us to establish the trend in direction of variation of response for unit change in input factors.

4.5.1 Torsional stiffness. Main effects plot for torsional stiffness shows that web top face offset (A) and side width (C) has major effect on torsional stiffness and relation is quadratic in nature as shown in figure 11. Both these factors has "Higher the better" relation with torsional stiffness. Higher the factor means more material and increase in stiffness. Counterweight profile (B) is not very sensitive to torsional stiffness of the crankweb. Variation in shape factor 'B' is not altering CCS and hence effect on torsional stiffness is not that significant.

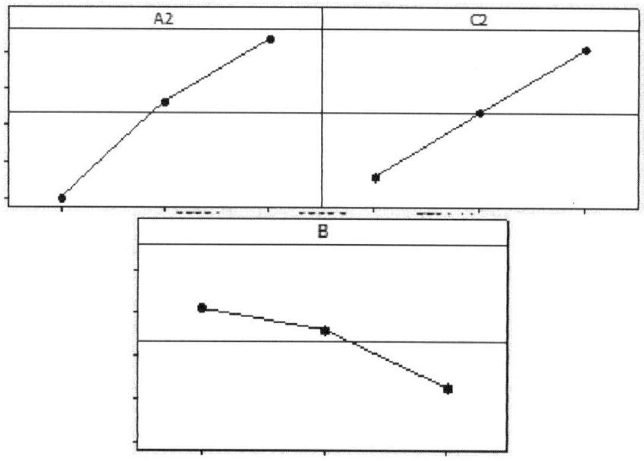

Figure 11: Main effect plots for torsional stiffness

4.5.2 Bending stiffness. Main effects plot for bending stiffness shows that web top face offset (A) and side width (C) has major effect on bending stiffness. Both these factors has "Higher the better" relation with bending stiffness and this relation is quadratic as shown in figure 12. Higher the factor means more material and increase in stiffness. Counterweight profile (B) is not very sensitive to bending stiffness of the crankweb. Variation in shape factor 'B' is not altering CCS and hence effect on bending stiffness is not that significant.

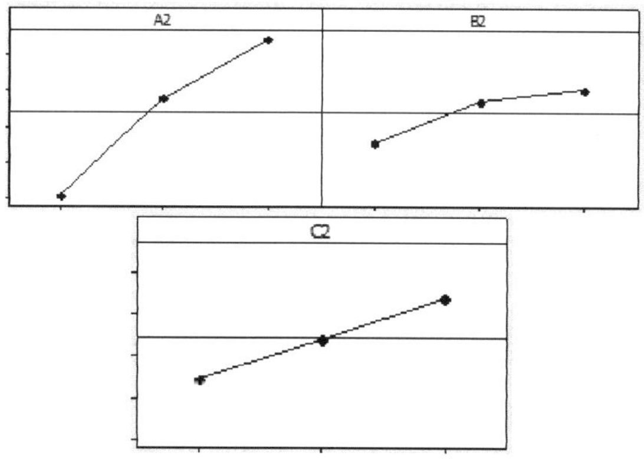

Figure 12: Main effect plots for bending stiffness

5.Optimization of shape factors

Statistical optimization of shape factors is done through "OptQuest" (Excel based tool) [11]. Shape factors are defined as 'Decision Variables' which can be varied in the defined range. Torsional and bending stiffness of crankshaft are defined as 'Forecast Variables' (Output variables) and minimum values of stiffness are kept as 'Constraints'. Solution gives desired values of shape factors to satisfy stiffness constraints. Slight variation of these values are done to meet balancing requirement and manufacturing constraints.

Crankshaft is designed with finalized values of shape factors and further used for 1-D simulation through AVL Excite®.

6.Analysis of strength and torsional vibrational performance

Strength and torsional vibration of the crankshaft cannot be determined without the whole Cranktrain assembly. A specialized commercial software called as AVL EXCITE® [10] is used to determine these parameters. 1-D simulation of the optimized crankshaft is done to confirm cranktrain design acceptance criteria.

Excite model of Cranktrain is a condensed representation of actual physical crank train system [10]. The geometry of various parts and their physical properties such as Mass moment of inertia and mass are given as inputs in this model. The various Cranktrain components such as piston, piston pin, connecting rod, flywheel, and damper are defined in terms of their mass moments of inertia and geometry.

The crankshaft is reduced to a simplified model of masses which are connected by springs with stiffness (spring-mass system).All the components are finally connected and loads on the Cranktraini.e. peak firing pressure in combustion chamber is applied to the system. Refer figure 13 for schematic representation of excite model

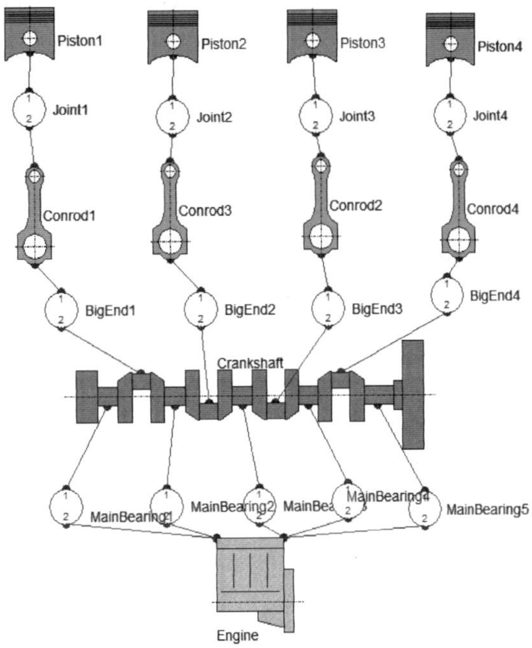

Figure 13: Cranktrain model in AVL Excite®

Above mentioned Multi-Body Dynamic (MBD) model is solved and following standard results as shown in figure 14-18 are obtained. It is observed that all these results are within acceptable limits. It shows that, crankshaft stiffness achieved by optimization of shape factors is sufficient to pass standard Cranktrain acceptance criteria.

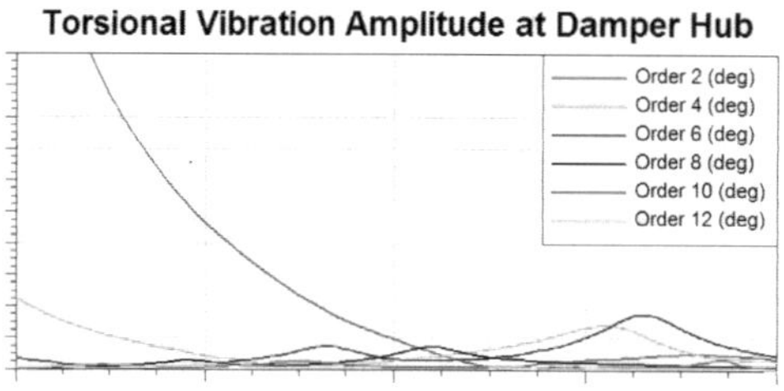

Figure 14: Torsional vibration amplitude at hub (degrees)

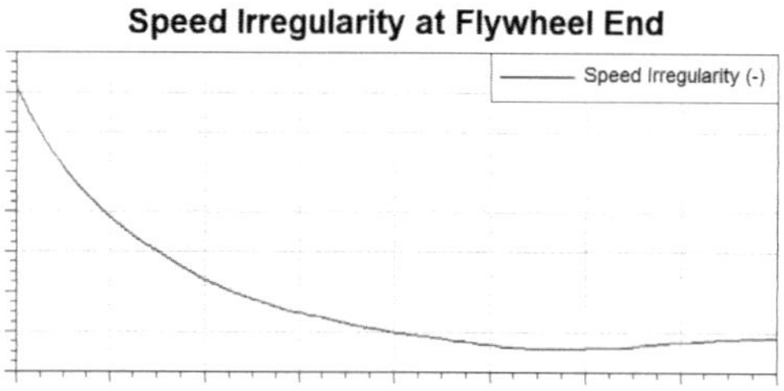

Figure 15: Speed irregularity at flywheel end

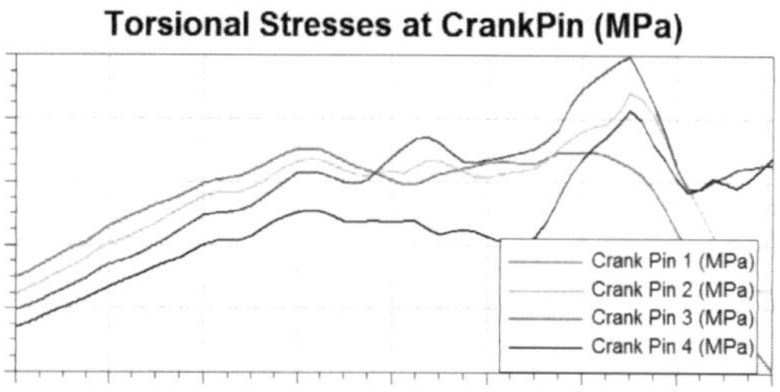

Figure 16: Torsional stresses at crankpin (MPa)

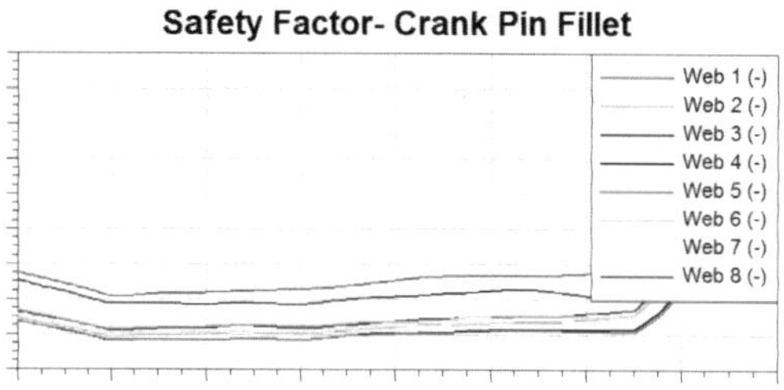

Figure 17: Crank pin fillet safety factor

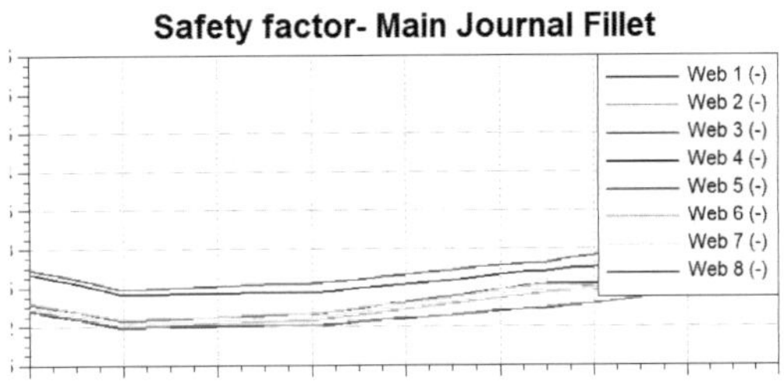

Figure 18: Main journal fillet safety factor

7. Summary

The shape factors governing the performance of crankshaft web have been identified. The methodology to optimize the crankshaft weight has been established. The following conclusions have been made from this work.

1. Top face offset and side width have major effect on torsional and bending stiffness.
2. Counterweight profile has less significant effect on torsional and bending stiffness
3. Crankshaft designed with optimized shape factors satisfies all design acceptance criteria.
4. This methodology can be implemented in early design phase to get weight and stiffness optimized crankshaft design.

8. References

[1] CagriCevik M, Rebbert M and Franz M 2009 Weight and friction optimized cranktrain design supported by coupled CAE tools SAE International USA 01 1452
[2] Druschitz A, Fitzgerald D and Hoegfeldt I 2006 Lightweight Crankshafts SAE International USA 01 0016
[3] Manoj U, Rajesh M and Shegavi G 2012 Optimization of crankshaft torsional rigidity for fatigue strength improvement using CAE SAE International USA 01 0016
[4] Wilson W 1967 Practical solution of torsional vibration problems (Chapman and Hall Ltd. London)
[5] Cevik M, Pischinger S, Rebbert M and Maassen F 2009 Boardline design of crankshafts based on hybrid simulation technology SAE International USA 01-1918.
[6] Milasinovic A, Filipovic I and Hribernik A 2009 Contribution to the definition of the torsional stiffness of the crankshaft of a diesel engine used in heavy-duty vehicles International journal of automobile engineering, 223, 921-929.
[7] User manual MiniTab 16®
[8] Edwards A 1979 Multiple regression and the analysis of variance and covariance", W.H. Freeman and company
[9] AVL Excite® hand book.
[10]Microsoft Excel OptQuest®

IOP Conf. Series: Materials Science and Engineering 1128 (2021) 012046 doi:10.1088/1757-899X/1128/1/012046

A Finite Element Analysis Based Study on Effects of Polyurea Coating for Blast Mitigation

R Dinesh[1], P Vasundra[2], D Davidson Jebaseelan[3]
School of mechanical engineering, VIT University, Chennai – 600127, India[1,3]
CVRDE AVADI, Chennai – 600054, India
dineshr9294@gmail.com[1]
vasundhra.p@cvrde.drdo.in[2]
Davidson.jd@vit.ac.in[3]

Abstract. One of main threats for Armoured fighting vehicles (AFVs) is anti-tank mines. Future war scenario insists on light tanks that can withstand land mines with higher order destruction. Hence a non-classical approach for blast mitigating structures has dealt with. In this sequence, there are two kinds of approach, one is to redesign the bottom plate structure and second is using special materials for coating. This work investigates the effect of special material Viz., Polyurea coating on the bottom structure of the AFV hull. Polyurea coating added to the Steel plate (considered as localized AFV vehicle hull structure) in such a way that the coating thickness is lesser than the thickness of base material. Explicit dynamic Finite Element (FE) based numerical simulations were performed using ANSYS AUTODYN. In this setup, coating is introduced to the Steel plate in two variations, namely, on non-blast face (opposite to the blast load) and blast face. The simulation is carried out with varying coating thickness (6mm, 8mm, 10mm 14mm) and with varying blast charges of TNT (2kg, 5kg, 7kg,and 12kg) in the stand-off distance (distance from the center of the explosive to the plate) of 450mm. The peak deformation of the plate caused by the blast load on the coated and uncoated homogenous plate is compared for their relative performances.

1. Introduction

Landmines can be defined as an explosive or other material, normally encased, designed to destroy or damage vehicles, boats or aircraft, or designed to wound, kill, or otherwise incapacitate personnel. It may be denoted by the action of its target, the passage of time or by controlled means [1]. Although the history of mines can be traced back to Roman times, the first modern mines were developed during first world war [2]. Based on the weight of explosives, mines can be categorized into anti-personnel mine and anti-tank mine. Anti-personnel mines are designed to harm humans, their quantity of explosive amount ranges from 0.003 to 0.25 kg [3]. Anti-tank(AT) mines are specifically designed for armoured vehicles, their explosive content ranges from 1.5kg to 12kg or more. Because of high explosive content AT mines possess a greater threat to vehicle and its crews. This work will focus only on the effects of AT mines to the AFV hull structure.

1.1. Blast Phenomena

An explosion is an extremely rapid release of energy in the form of light, heat, sound, and a shock wave. A shock wave consists of highly compressed air traveling radially outward from the source at supersonic velocities. The intensity of the wave fades with distance. Blast wave travels through the atmosphere at the speed of sound, which is approximately 1126 feet per second at 68°F in dry air, and at about 1100 fps at 45° F. In an ideal blast wave as seen in Figure 1, when the blast occurs, the pressure increases to a maximum value called as incident and reflected peak overpressure for incident and reflected blast waves, respectively. After this phase, the pressure starts decreasing gradually and reaches atmospheric pressure level, this phase is called as positive phase duration. After this phase, the pressure decreases further, dropping below the atmospheric level called as negative phase [4][5].

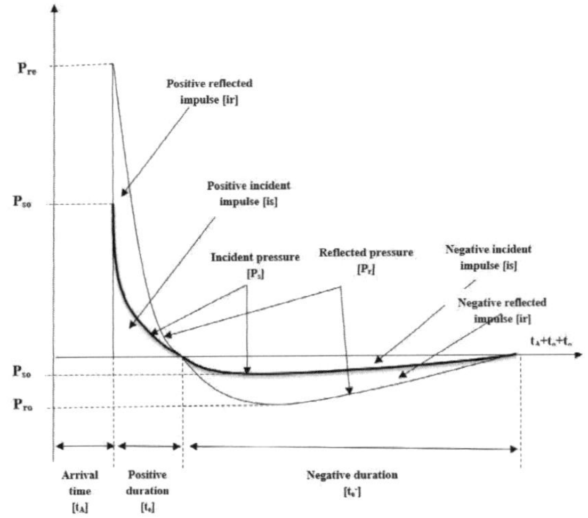

Figure 1. Blast phenomena

The decaying of pressure that occurs during a blast event can be characterized by Friedlander equation.

$$P(t) = P_{so}\left[1 - \frac{t - T_a}{T_0}\right]\exp\left[\frac{-A(t - T_a)}{T_0}\right] \tag{1}$$

Here P(t) is pressure at time t, P_{so} is the peak overpressure, T_0 is the positive phase duration, A is the decay coefficient, and Ta is arrival time [6].

1.1.2. Scaled distance The important parameter for blast loading computation is scaled distance. The peak pressure value of the blast wave, decrease rapidly by increasing the distance between the blast source and the target surface. The pressure is high when the distance from detonation point is less and the duration are longer whenever the distance from the detonation point increases. The variation of blast characteristics can be taken into account with the help of scaled distance or scaled laws. Scaled distance helps to scale parameters, which were defined through experiments. The experimental results are, in this way, generalized to include cases that are different from initial experimental setup. The most commonly used scaling law is introduced by Hopkinson-Cranz [7] and it is defined by

$$Z = \frac{R}{\sqrt[3]{W}} \tag{2}$$

Where R is the stand-off (m) distance (distance from the center of the mine to the target body), W is the TNT equivalent charge mass (kg).

1.3. Polyurea overview

Polyurea is a cross-linked amorphous isocyanate monomer or prepolymer and polyamine curative. Polyurea should contain up to 80% of polyamine . It is an elastomeric-thermoset polymer with high elongation property, in most cases they are capable of elongate up to 100%. Polyurea is often used as coating materials because it can be applied just by spraying. Also, it has good ductile behavior and able to reduce fragmentation during blast and impact situations. Polyurea is most commonly used to coat composite laminates, building structures because of their potential in resisting blast and impacts. Research indicates that adding polyurea to concrete structures are capable reducing the displacement caused by blast up to 40%. In Middle East countries where the terrorist activities are high, buildings and houses are reinforced with polyurea coating to protect from bomb blasts [8][9].

2. Literature Survey

Many research studies have been carried out to understand the effects of mine blast on various structures. Some of the most significant works are discussed below

Hrvoje Draganic et al (2012) studied the process of determination of blast load on structures, this paper reviews the analytical formulations to determine blast load. The peak over pressure has been calculated analytically and simulated on building model using SAP2000 software package. Loading was defined change of pressure over time. The determined blast load is used to find deformation at certain point of interest using the software. This paper provides a detailed description of blast wave. It shows how the TNT blast wave pressure varies with time, which is helpful in understanding the blast phenomena. [10].

T. Ngo et al (2011) express the nature of explosion and blast phenomenon. Here the classification of explosives and the material used for explosives has been studied. And also, the temperature rise during the detonation has been given. The blast wave propagation graph has been plotted by using the graphical technique. The next significant part of this paper is blast wave-structure interaction. The structural behaviour of an object or structure exposed to such blast wave may be analysed by dealing with two main issues. Firstly, blast-loading effects. Secondly, the structural response, or the expected damage criteria associated with such loading effects. The remaining part deals with material behaviours at high strain rate. Here the dynamic properties of Concrete and reinforcing steel have been studied under high-strain rates and stress-strain rate graphs have been plotted [11].

Jong Yil Park et al (2017) studied the dynamic behaviour of steel plate subjected to blast load. Steel plate was subjected to 50kg TNT at different stand-off distance. The study has been done both numerically and experimentally and studied for strain variable. Numerically it was modelled using AUTODYN software package in ALE method. A good agreement was observed between numerical and experimental value [12].

D.K Sahoo et al (2017) investigated blast resistance of layered plates using CONWEP method. Layered plate with different material combinations with varying thickness for the layers. This paper gives an idea about CONWEP method and method to use it in numerical simulation [13].

S N. Raman et al (2011) This paper provides a review on the status of research in using polyurea coating to enhance blast and impact resistance of structures. Application of coating on masonry, composite as well as RC structureshave been discussed. This paper helps to understand the application and effectiveness of polyurea coating around various fields [14].

Ackland et al (2011) studied blast mitigation of mild steel plates subjected to blast load of 0.5 Pentolite at close range blast with polyurea coating. Mile steel plates is applied with polyurea of thickness 350% and 575% more than the steel thickness. It was observed that the deformation reduction of steel plate coated with polyurea on non-blast face reduced deformation effectively than at the blast face [15].

Kathryn Ackland et al (2013) studied the effect of polyurea coating on the blast resistance of mild steel. Plates with and without polyurea coating applied to non-blast face is subjected to blast load of 0.5 pentolite at close range distance. Coating were applied in a way such that the bare plate density suits the polyurea applied plate by reducing the thickness of the steel plate. Bare steel plate of 6mm, 5mm steel plate of thin coating thickness and 4mm plate of thick coating were considered for investigation. It has been observed that bare plate dissipated more energy that polyurea coating plate with same aerial density. Also, residual deformation is low is bare plate in comparison to coated plate with same aerial density [16].

Ashan samiee et al (2013) performed numerical simulation of steel plates with polyurea bilayers, subject to blast like loads. Different cases of coating were investigated such as coating of front face and back face of 10mm steel with same thickness (10mm, 20mm, 30mm and 40mm), Bare steel and coated steel maintained with same aerial thickness in four configurations (11.4mm, 12.8mm, 14.2mm and 15.6mm). Pressure is applied in two types, direct application on bilayer system and application of pressure through a fluid medium such as polyurethane or water. It is observed that polyurea coated on non-blast face had less average plastic strain on central region of steel plate. It is concluded that in steel plate applying thin layer of coating when applied on either side, change in performance is negligible. As thickness is increased the advantages becomes clear. Regarding the effect of loading between direct and indirect pressure application to the bilayer system has same response but in indirect application of pressure plastic strain is lower. Overall cases it is identified that when polyurea is coated on non-blast face, the blast mitigation is more effective [17].

Li-hui Dai et al (2018) investigated the effect of polyurea coated steel plates subjected to underwater explosive load. Experimental and numerical investigation were conducted with thin steel plate of 2mm. Coating was applied at two configurations, thickness same as base plate and thickness double the base plate on blast and non-blast face. Blast load of 0.1kg charge of RDX at different standoff distance in close range, experiment was conducted underwater. It is observed that for thin steel plate, coating on front face provided significant blast resistant which contradicts other research works. Also, it is noted that increase in thickness provide better deformation resistance [18].

From the literature study, it is clear that the polyurea coating enhances blast resistance to building and concrete structures [14]. In the case of ductile materials like steel the coating is added equal to the thickness or more than the thickness of the base material [16] [17] [18], but not lesser than the thickness of base plate. This work investigates the effect blast mitigation of polyurea coating added to steel plate whose thickness is lesser than the base material.

3. Methods To Predict Mine blast

In numerical simulations, effect of mine blast can be predicted into two different methods Arbitrary Lagrange Euler (ALE), Conventional Weapon (CONWEP).

3.1. ALE method

ALE method involves geometric modeling of all the elements that involves in the mine blast numerical simulation. The structural elements are modeled using Lagrange elements. Lagrange is used to model the materials which have less distortion as Lagrange keeps the material in its initial element as there is no transport of material from element to element. The material with high distortion such as

IOP Conf. Series: Materials Science and Engineering 1128 (2021) 012046 doi:10.1088/1757-899X/1128/1/012046

Air, Explosives are modeled using Eulerian element as it allows the material from flow to cell to cell [2]. The Lagrange and Eulerian are given a coupling to interact with each other. The advantage of ALE method is that it allows visualization of blast phenomena.

3.2. CONWEP method

CONWEP is well accepted method for predicting mine blast. It is based on the experimental parameters from air burst and hemispherical surface blast done by Kingery and Bulmash for US army [19]. It contains over explosive charge data from 1kg to over 40,000kg. The experimental values are categorized based on scaled distance in graphical form by curve fitting technique. CONWEP is pure Lagrange model where in only the structure is modeled. It is valid for over a scaled distance range of 0.147 $m/kg^{1/3}$ to 40 $m/kg^{1/3}$. Since CONWEP method eliminates the need for modeling Air and explosives, it saves computational time.

3.3. Problem description

The AFV vehicle hull has been localized as a dimensioned 1000mm x1000mm x 15mm steel plate. The plate is kept at over 450mm stand-off distance (distance from the center of the explosive to the plate. The explosive charge is TNT. The explosive buried 100mm inside the sand as show in Figure 2. The TNT charge is varied (2kg, 5kg, 7kg, 10kg, 12kg). Then Polyurea coating of various thicknesses is applied to the steel plate on two cases, on non-blast face and blast face with varying coating thickness (6mm, 8mm, 10mm, and 14mm) and studied for the response under same blast load conditions.

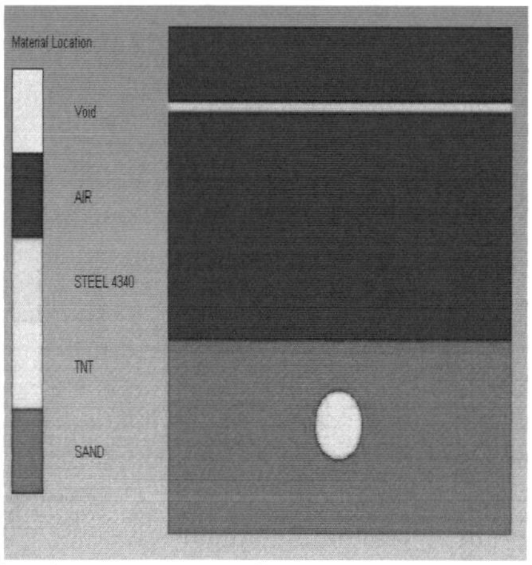

Figure 2. Visual representation of Problem description in ANSYS AUTODYN

3.4. Material Parameters

The material parameters involved in this study are as follows

3.4.1. AISI Steel 4340. AISI Steel 4340 is a medium carbon (carbon content is in between 0.3% to 0.6%), low alloy steel containing chromium, nickel and molybdenum. It is known for its toughness, strength and good fatigue resistance. The chemical composition of Steel 4340 can be seen in Table 1.

Table 1. The chemical composition of Steel 4340.

Element	Percentage [%]
Iron, Fe	95.195 - 96.33
Nickel, Ni	1.65 - 2.00
Chromium, Cr	0.70 - 0.90
Manganese, Mn	0.60 - 0.80
Carbon, C	0.37 - 0.43
Molybdenum, Mo	0.20 - 0.30
Silicon	0.15 - 0.30
Sulfur	0.04
Phosphorous	0.035

The behavior of Steel 4340 is defined by Johnson-Cook strength and failure model.Johnson-Cook model was used to define strength of the material Steel 4340. This constitutive model defines the strength behavior of materials subjected to large strains, high strain rates and high temperatures [20].

$$Y = \left[A + B\varepsilon_p^{\,n}\right]\left[1 + C\log\varepsilon_p^{\,n}\right]\left[1 - T_H^{\,m}\right] \quad (3)$$

Where
ε_p= effective plastic strain
ε_p^*= normalized effective plastic strain rate
T_H= homologous temperature = $(T - T_{room})/(T_{melt} - T_{room})$
The five material constants are A, B, C, n and m. The first set of brackets in above equation gives the stress as a function of strain, which can be found by quasi-static tensile testing ($\varepsilon_p = 1.0sec{-}1$ and T_H= 0). A is the basic yield stress at low strains whereas B and n define strain hardening. The second and third set of brackets represents the effects of strain rate hardening and temperature softening. With the thermal softening, the yield strength drops to zero at the melting temperature T_{melt}. The material constants can be obtained empirically via dynamic Split Hopkinson Bar tensile tests over a range of temperatures and strain rates. The obtained model constants were checked by calculations of Taylor tests of impacting metal cylinders on rigid metal targets which provided strain rates in excess of $10^5 sec^{-1}$ and strains in excess of 2.0.

Table 2. The Johnson-Cook parameters for AISI 4340 Steel.

Material	A [Mpa]	B [Mpa]	n	c	m	T_{melt} [K]	T_{room} [K]
Steel 4340	792	510	0.26	0.014	1.02	1793.15	293.15

3.4.2. Air. The material model of air used in this simulation is loaded from AUTODYN material library. Ideal equation of state (EOS) is used to describe air ***[9]***, which is described the form of

$$P = (\gamma - 1)\rho e + P_{shift} \quad (4)$$

Where p is pressure, is adiabatic constant (for air behaving as an ideal gas = 1.4). is the density, shift is a small pressure defined to dive zero starting pressure in order to avoid complications in problems with multiple material where initial small pressure in the gas would generate small unwanted velocities. e is the specific internal energy and can be defined by

$$e = c_v T \quad (5)$$

Where the constant C_v is the specific heat at constant volume.

The values for the variables in for ideal gas equation for air model in AUTODYN are listed in the Table 3.

Table 3. Air – Ideal Gas parameters.

r	ρ $[kg/m^3]$	C_v $[kJ/kg\ k]$	T [K]
1.4	1.225	0.000718	288.2

3.4.3. TNT (Explosive). The TNT is defined using Jones-Wilkins-Lee (JWL) form of equation of state

$$P = A\left[1 - \frac{\omega}{R_1 V}\right]e^{-R_1 V} + B\left[1 - \frac{\omega}{R_2 V}\right]e^{-R_2 V} + \frac{\omega E}{V} \tag{6}$$

Where A, B, R_1, R_2, are empirically derived constants varies with each explosive, V is the relative volume of expansion of explosive product and E is the detonation energy per initial unit volume [3][9]. The material property of TNT and variable for JWL equations listed in Table 4.

Table 4. TNT – JWL parameters.

Variable	Value
kgm3	1.225
A	373.77
B	374.71
R_1	4.15
R_2	0.9
ω	0.35
C-J Detonation velocity (m/ms)	6.93
C-J Energy/unit volume (MJ/m^3)	6000
C-J Pressure (Mpa)	21000

3.4.4. Polyurea. Material model for polyurea in most cases is defined using Mooney-Rivilin model [21]. Polyurea is observed to have higher elongation[22], As the deformation in the present study is within this elastic limit, linear elastic model is used to define the Polyurea material. Table 5 shows the material parameters for polyurea.

Table 5. Material parameters of Polyurea.

Material	ρ $[kg/m^3]$	E $[N/m^2]$	ν
Polyurea	1500	7×10^7	0.3

3.5. Numerical Simulation Of Bare Plate Subjected To 2KG TNT Blast Load

The plate of 1000mm x 1000mm x 18mm is subjected to 2kg blast load of TNT is to be studied with both ALE and CONWEP method using ANSYS AUTODYN solver.

3.5.1 ALE method. The plate is designed using Axisymmetric modeling using Lagrange elements. The air, sand and TNT are modeled by Eulerian element. Since 3D ALE method is time consuming, this study is made in 2d axisymmetric method. The plate is modeled with dimension 18mm x 500mm. Air is modeled with dimension 1000mm x 500mm, sand is modeled with dimension 383mm x 500mm and TNT is designed with semi-circular geometry with radius 66.42mm. The TNT is of 2kg using the relation m= ρV. The axisymmetric model is shown in Figure 3. Boundary condition, the plate is fixed at one end, and flow out condition is given at three edges for the air and sand. Gauge points are fixed at the edge of the plate. The mesh size for plate is 3mm and for Eulerian domains are 2.5mm, to allow interaction between Euler and Lagrange domain. The aim of this simulation is to capture the peak deformation of the plate.

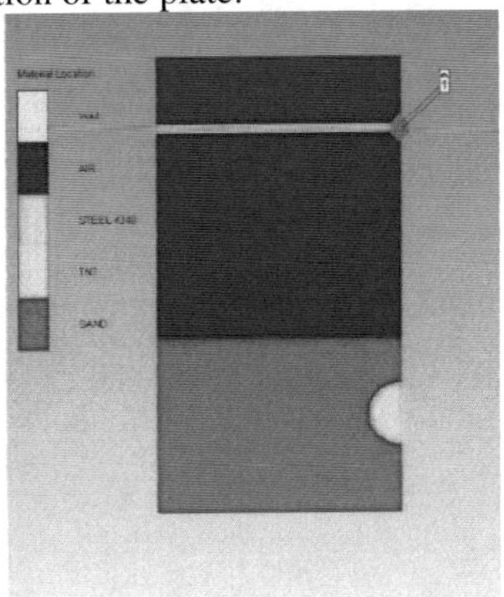

Figure 3. ALE setup – symmetric model before TNT explosion

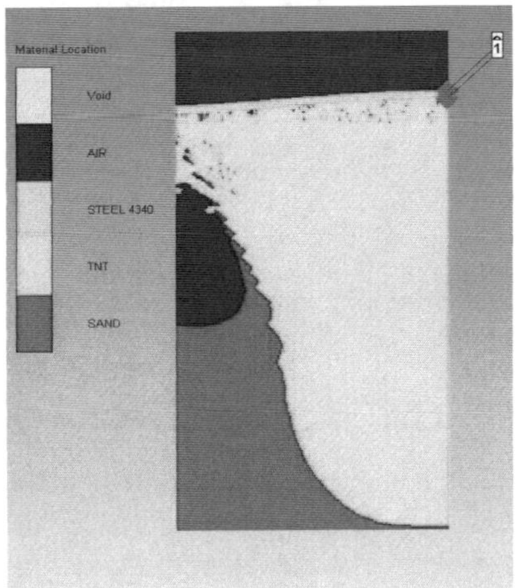

Figure 4. ALE setup – symmetric model after TNT explosion

3.5.2. CONWEP method. The same case is modeled in COWNEP method. A 3D plate with symmetric condition, the dimension of plate is 500mm x 500mm x 18mm. Since CONWEP is a pure Lagrange method there is no need to model Air, Sand and TNT. The CONWEP option should be given as boundary condition in AUODYN. Under analytical blast boundary condition, the weight of the TNT is entered as 2kg, stand-off distance is of 450mm and the type of blast is set to surface blast. And the pressure clearing is kept off because in symmetric method clearing does not work properly. The two edges of the plates are fixed, the CONWEP condition is given at the bottom face of the plate, and mesh size is 3 mm. Gauge point is kept at corner of the plate. The case setup in CONWEP method for steel plate is shown in Figure 5.

Figure 5. CONWEP model – case setup

The peak deformation for both the methods gives close values. The value obtained in both methods is shown in Table 6.

Table 5. ALE and CONWEP deformation result comparison

Peak deformation in ALE method	Peak deformation in CONWEP method
33.60 mm	32.676 mm

Since both the values are almost similar, the rest of the simulations are to be carried out in CONWEP method as it simplifies the model and saves computational time.

The same methodology is applied for blast charges of 5kg, 7kg, 10kg and 12kg. Only by altering the charge weight in the CONWEP boundary condition, the simulation is done.

3.6. Polyurea coated to plate

3.6.1 Numerical simulation of coated plate. In AUTODYN software the polyurea coating is modelled without any external gap between the plate and the polyurea coating. Based on experimental work done by Amini et al [23]. On steel plate with polyurea coating, it is found that while analyzing in numerical software, the bond strength of 100 Mpa gave most close agreement with experimental value. Hence a bonding strength of 100 Mpa is takes in this study.

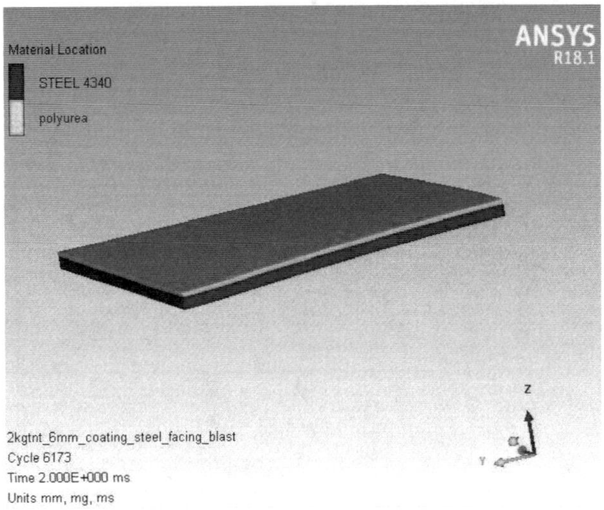

Figure 6. CONWEP model – case setup of 6mm coated plate

3.7.Nnumerical simulation of concrete with and without coating

Since it is clear from the research that polyurea has significant role along with concrete [14] for comparative study, concrete plate with same dimension as that of steel plate. Three cases are to be studied bare concrete, 6mm coating on blast face and non-blast face. The deformation resistance is studied with comparative to steel plate.

4. Results

4.1. Deformation observed on coated plate

The peak deformations of plate with coating on non-blast side and blast side subjected to mine blast of various masses of TNT are tabulated and compared.

4.1.1. Deformation results for 2kg TNT. The deformation obtained for bare (homogenous) and coated plate subjected to 2kg TNT blast is listed in Table 6.

Table 6. Deformation outputs comparison for 2kg TNT blast load

Coating thickness	Deformation (mm)	
Homogenous	32.188	
	Coating on non-blast side	Coating on blast side
6mm coating	31.464	31.375
8mm coating	31.229	31.283
10mm coating	30.969	30.982
14mm coating	30.262	30.458

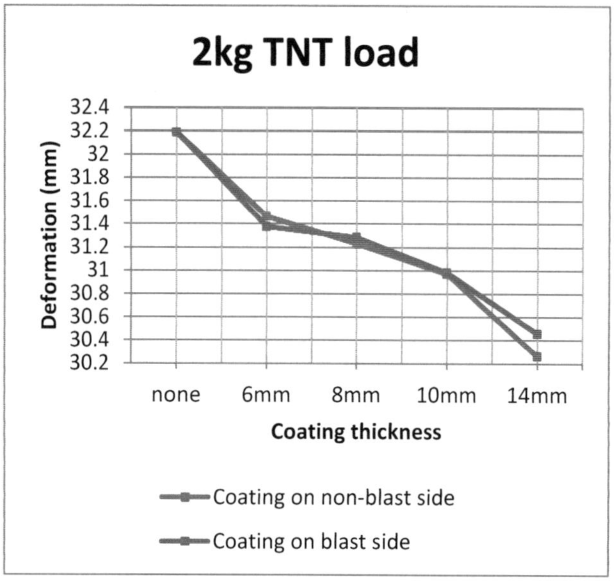

Figure 6. Deformation output of 2kg TNT load

4.1.2. Deformation results for 5kg TNT. The deformation obtained for bare and coated plate subjected to 5kg TNT blast is listed in Table 7.

Table 7. Deformation outputs comparison for 5kg TNT blast load.

Coating	Deformation (mm)	
Homogenous	55.735	
	Coating on non-blast side	Coating on blast side
6mm coating	54.504	54.940
8mm coating	54.112	54.642
10mm coating	53.787	54.428
14mm coating	52.811	53.841

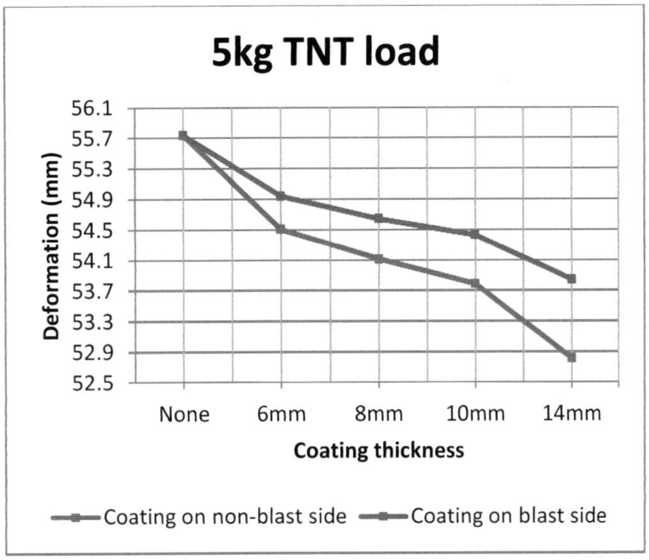

Figure 7. Deformation output of 5kg TNT load

4.1.3. Deformation results for 7kg TNT. The deformation obtained for bare and coated plate subjected to 7kg TNT blast is listed in Table 9.

Table 8. Deformation outputs comparison for 7kg TNT blast load.

Coating	Deformation (mm)	
Homogenous	69.619	
	Coating on non-blast side	Coating on blast side
6mm coating	68.004	68.176
8mm coating	67.483	67.782
10mm coating	66.966	67.628
14mm coating	65.690	65.69

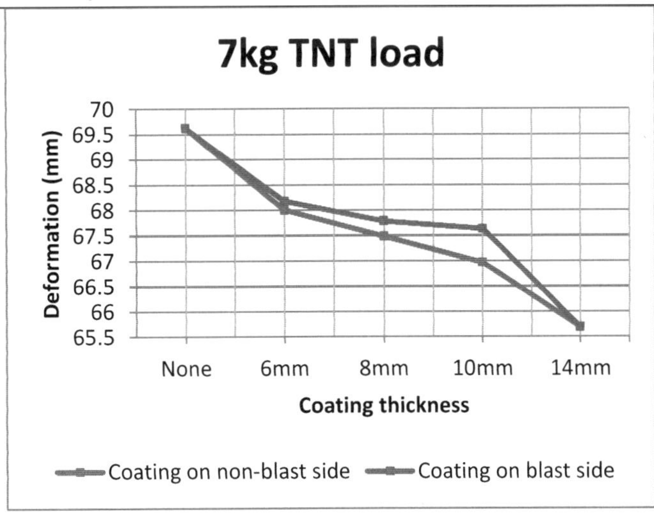

Figure 8. Deformation output of 7kg TNT load

The deformation obtained for bare and coated plate subjected to 10kg TNT blast is listed in Table 9.

Table 9. Deformation outputs comparison for 10kg TNT blast load.

Coating	Deformation (mm)	
Homogenous	90.156	
	Coating on non-blast side	Coating on blast side
6mm coating	87.833	88.315
8mm coating	87.081	87.328
10mm coating	86.359	86.727
14mm coating	84.582	85.786

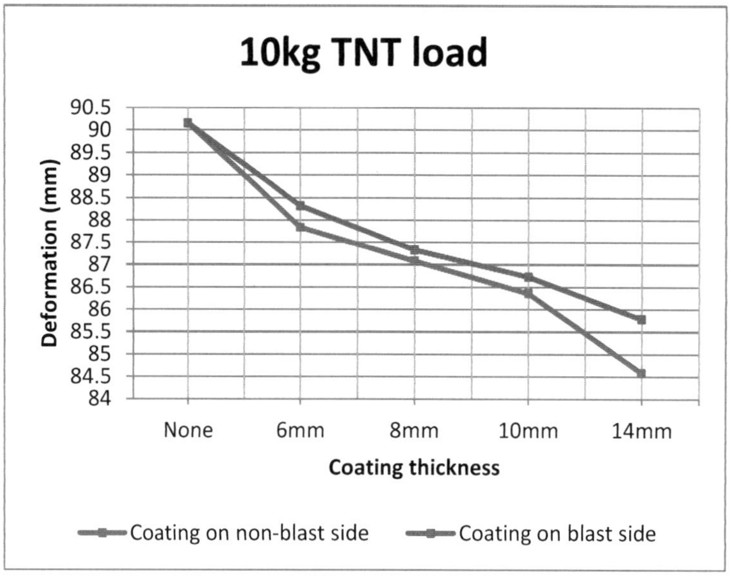

Figure 9. Deformation output of 10kg TNT load

4.1.5. Deformation results for 12kg TNT.

The deformation obtained for bare and coated plate subjected to 12kg TNT blast is listed in Table 10.

Table 10. Deformation outputs comparison for 12kg TNT blast load.

Coating	Deformation (mm)	
Homogenous	104.16	
	Coating on non-blast side	Coating on blast side
6mm coating	101.36	101.27
8mm coating	100.47	100.39
10mm coating	99.602	99.746
14mm coating	97.508	98.870

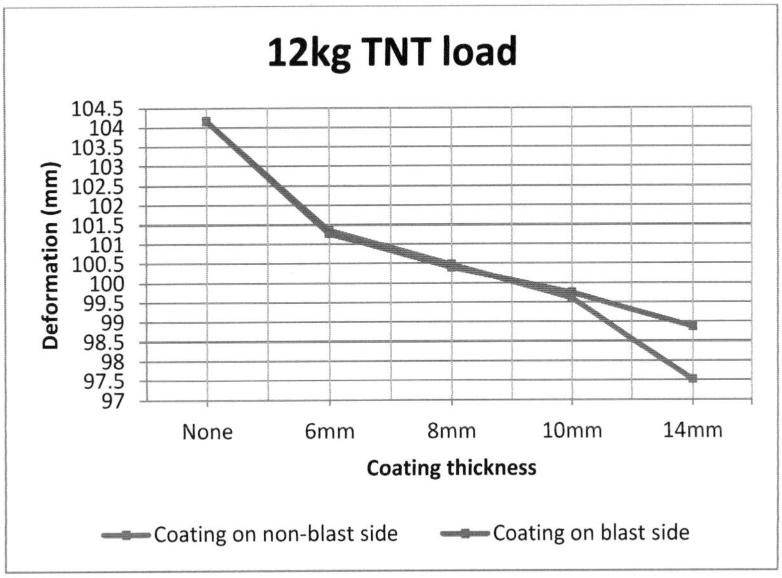

Figure 10. Deformation output of 12kg TNT load

4.2. Acceleration observed on coated plate

The acceleration obtained from plate bare plate and coated plate subjected to mine blast of various masses of TNT are tabulated and compared.

4.2.1. Acceleration results for 2kg TNT load. The acceleration obtained for bare (homogenous) and coated plate subjected to 2kg TNT blast is listed in Table 11.

Table 11. Acceleration outputs comparison for 2kg TNT blast load.

Coating	Acceleration (mm/ms^2)	
Homogenous	115.36	
	Coating on non-blast side	Coating on blast side
6mm coating	106.84	108.34
8mm coating	106.19	104.73
10mm coating	104.15	102.71
14mm coating	99.353	97.991

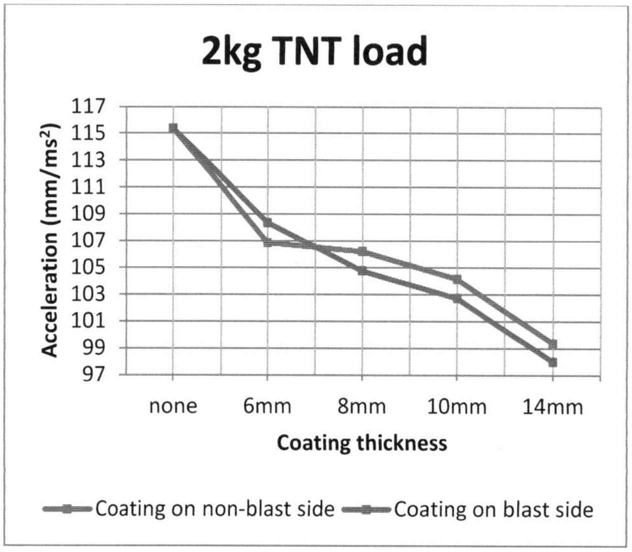

Figure 11. Deformation output of 2kg TNT load

4.2.2. Acceleration results for 5kg TNT load. The acceleration obtained for bare and coated plate subjected to 5kg TNT blast is listed in Table 12.

Table 12. Acceleration outputs comparison for 5kg TNT blast load.

Coating	Acceleration (in mm/ms^2)	
Homogenous	242.82	
	Coating on non-blast side	Coating on blast side
6mm coating	227.77	228.22
8mm coating	223.74	223.34
10mm coating	219.44	219.02
14mm coating	209.36	208.96

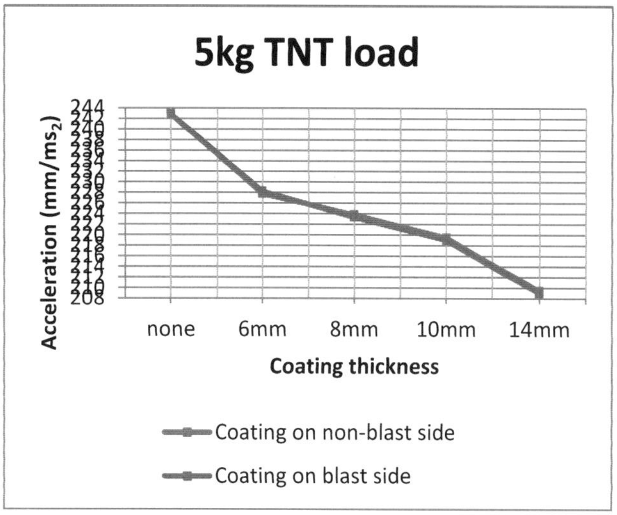

Figure 12. Acceleration output of 5kg TNT load

4.2.3. Acceleration results for 7kg TNT load. The acceleration obtained for bare and coated plate subjected to 7kg TNT blast is listed in Table 13.

Table 13. Acceleration outputs comparison for 7kg TNT blast load.

Coating	Acceleration (in mm/ms^2)	
Homogenous	311.03	
	Coating on non-blast side	Coating on blast side
6mm coating	292.74	292.37
8mm coating	286.63	287.04
10mm coating	281.12	281.56
14mm coating	268.22	268.622

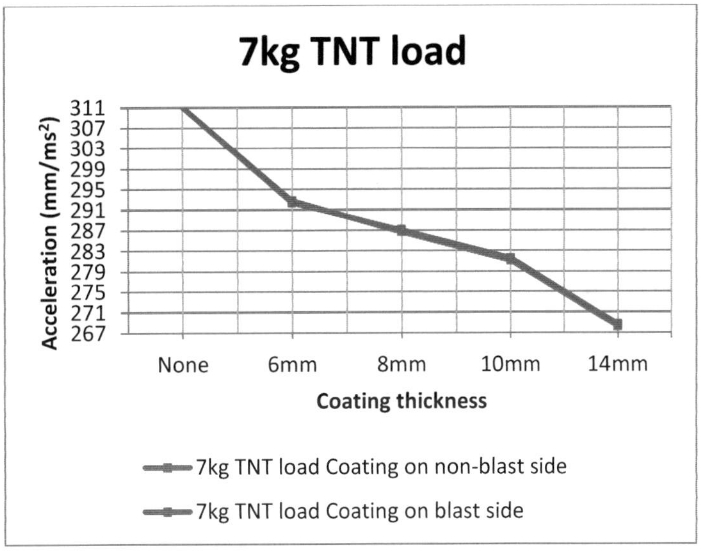

Figure 13. Acceleration output of 7kg TNT load

4.2.4. Acceleration results for 10kg TNT load. The acceleration obtained for bare and coated plate subjected to 10kg TNT blast is listed in Table 14.

Table 14. Acceleration outputs comparison for 10kg TNT blast load.

Coating	Acceleration (in mm/ms^2)	
Homogenous	399.07	
	Coating on non-blast side	Coating on blast side
6mm coating	377.40	378.13
8mm coating	367.77	370.09
10mm coating	360.70	360.30
14mm coating	344.15	346.32

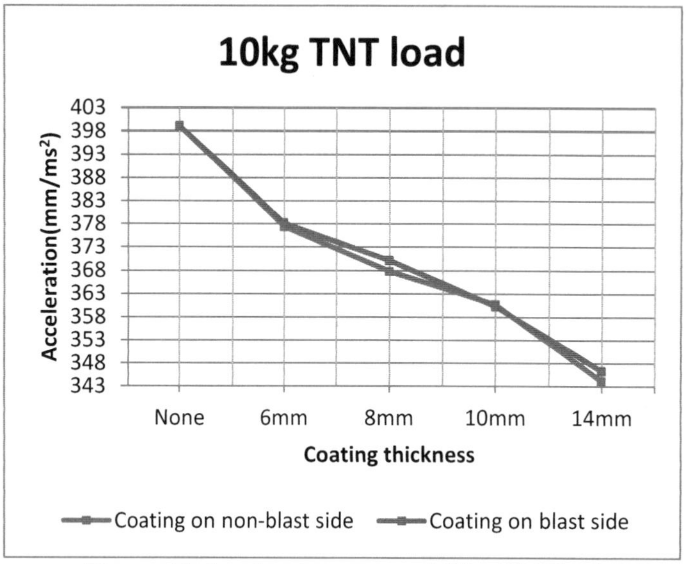

Figure 14. Acceleration output of 10kg TNT load

4.2.4. Acceleration results for 10kg TNT load. The acceleration obtained for bare and coated plate subjected to 12kg TNT blast is listed in Table 15.

Table 15. Acceleration outputs comparison for 10kg TNT blast load.

Coating	Acceleration (in mm/ms^2)	
Homogenous	450.94	
	Coating on non-blast side	Coating on blast side
6mm coating	426.84	423.88
8mm coating	415.57	419.07
10mm coating	407.57	419.01
14mm coating	388.67	392.80

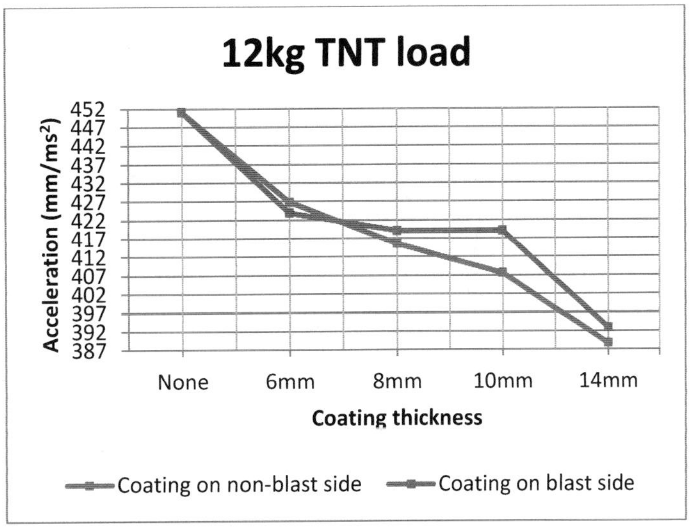

Figure 15. Acceleration output of 2kg TNT load

4.3. Concrete Results

The results obtained for concrete with 6mm coating, subjected to 2kg TNT explosion is listed in Table 16 and Table 17.

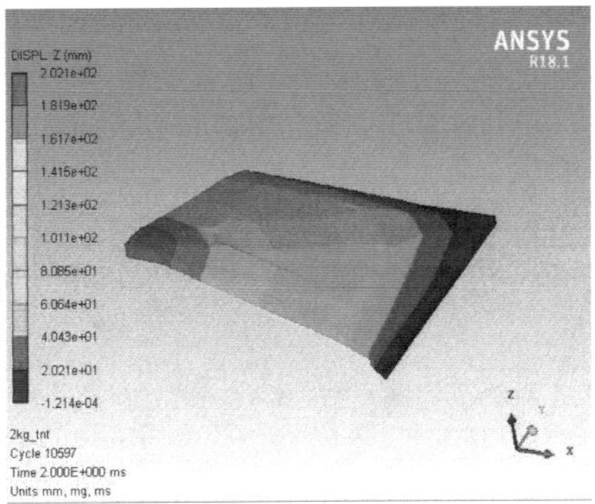

Figure 16. Deformation contour of bare Concrete plate

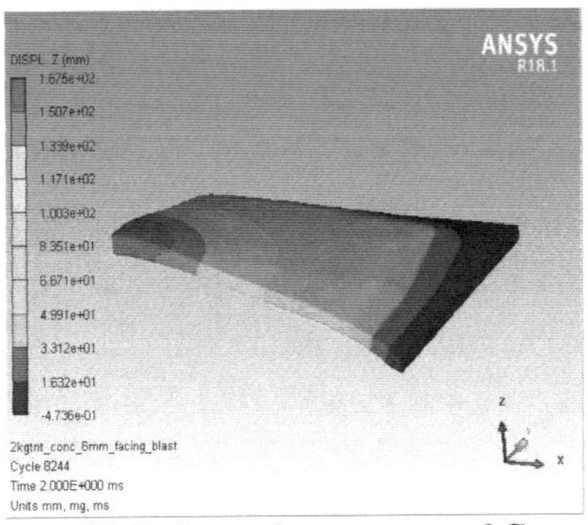

Figure 17. Deformation contour of Concrete coated plate on non-blast side

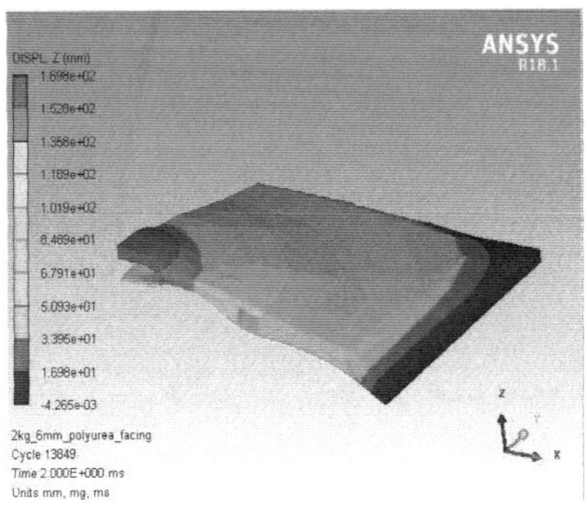

Figure 18. Deformation contour of Concrete coated plate on blast side

Table 16. Deformation outputs comparison for 2kg TNT blast load on Concrete.

Coating	Displacement (mm)	
Homogenous	202.13	
	Coating on non-blast side	Coating on blast side
6mm coating	167.46	169.79

Table 17. Acceleration outputs comparison for 2kg TNT blast load on Concrete.

Coating	Acceleration (mm/ms^2)	
Homogenous	354.80	
	Coating on non-blast side	Coating on blast side
6mm coating	295.86	294.65

4.4. De-bonding observation

De-bonding is observed in coating side facing on blast side in all coating thickness, for all charge masses. No de-bonding occurred in coating facing non-blast side. This is clear from the fact that in case on blast side there is excess energy available so as to shear the bonding between the two phases. In case of non-blast side coating, the excess energy is absorbed by the steel structure and hence there is no shear between the two materials. Fig 4.14 shows the debonding of polyurea of 14mm thickness subjected to 2kg TNT explosion.

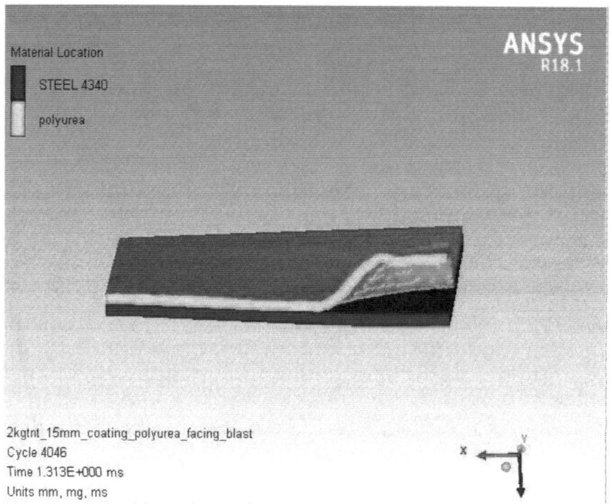

Figure 10. Debonding of polyurea coating

5. Conclusion

In all the case studied, it is clearly seen that there is decrease in the peak deformation and acceleration value. This is reduction is not significant when compared to the elastomeric properties of the material. In fact, the meager reduction can be attributed to the additional thickness that is added on to the structure. Further, it is noted from that there is significant reduction in peak deformation and acceleration on concrete structures coated just 6mm Polyurea. From these simulations, it is clear that the Polyurea coating has a significant role in case of blast mitigation where the base material. Whereas, in case of ductile materials like steel, the base material also has enough blast resistance and hence the elastomeric properties of the Polyurea coating becomes insignificant.

References

[1] Colin king 2007 Jane's Mines and Mine Clearance – 11th edition.
[2] Michel croll 1998 History of landmines.
[3] Dr. A Hameed 2008 Numerical analysis of vehicle bottom structures subjected to anti-tank mine explosions, PhD thesis, Cranfield University.
[4] Vasilis Karlos, George solomos, Martin larcher 2016, Analysis of blast parameters in the near-field for sphericalfree-air explosions, JRL technical reports - European commission.
[5] Jim ratcliff, Ed Sheehan, Keith carte 2011 Predictability Of air blast at surface coal mines in West Virginia, West Virginia department of environmental protection office of explosives and blasting.
[6] G. Le blanc, M. Adoum, V.Lapoujade, External blast load on structures – Empirical approach, CRIL technology, 5th European LS-DYNA users conference.
[7] Smith, P.D.; Hetherington 1994, J.G. Blast and Ballistic Loading of Structures; Butterworth-Heinemann: Oxford,UK.
[8] Srinivasan arjuntekalur, Arun Shukla, Kunigalshivakumar 2008 Blast resistance of polyurea based composite materials, *Composite structures Vol 84 Page 271-281.*
[9] Courtney E. Greene, John J. Myers, F. ASCE (2013). Flexural and shear behaviour of reinforced concrete members strengthened with a discrete fiber-reinforced polyurea system. Journal of composites for construction.
[10] Hrvoje Draganic, Vladmir Sigmund (2012). Blast loading on structures. *Technickivjesnik Vol.19, pg.643-652.*
[11] Tuan Duc Ngo, Anant Gupta, Priyan Mendis 2007 Blast loading and blast effects on structures *Electric journal of structural engineering –Special Edition.*
[12] Jong Yil Park, Eusun Jo, Seung Jae Lee, Young Hak Lee 2016 Dynamic behavior of a steel plate subjected to blast loading *Transaction of Canadian society for mechanical engineering, Vol.40, Pg.4.*
[13] D.K. Sahoo, Anirban Guha, AsimTewari, R.K. Singh 2017 Performance of monolithic plate and layered plates under blast load *Procedia engineering Vol.173.*
[14] S.N. Raman, T. Ngo, P. Mendis 2011 A review on the use of polymeric coatings for retrofitting of structural elements against blast effects *Electronic journal of structural engineering.*
[15] K. Ackland, N.St John 2011 Experimental and numerical investigations into polyurea coating for blast protection. Defense science and technology organization, Australian government department of defence.
[16] Kathryn Ackland, Christopher Anderson, Tuan Duc Ngo 2013 Deformation of polyurea-coated steel plates under localized blast loading *International journal of impact engineering, Vol.51, Pg.13-22.*
[17] Ashan samiee, Alireza V. Amirkhizi, Sia Nemat-Nasser 2013 Numerical study of the effect of polyurea on the performance of steel plates under blast loads. *Mechanics of materials, Vol.64, Pg.1-10.*
[18] Li-hui Dai, Cheng Wu, Feng-jiang An, Sha-sha Liao 2018 Experimental investigation of polyurea-coated steel plates at underwater explosive loading. *Advances in materials science and engineering, Vol 2018.*
[19] Charles N. Kingery, Gerald bulmash 1984 Air blast parameters from TNT spherical air burst and hemispherical surface burst. U.S Army ballistic research laboratory.
[20] Gordon R.Johnson, William H. Cook. A constitutive model and date for metals subjected to large strains, high strain rates and high temperatures.
[21] Yang Bai, Chunmei Liu, Guangyan Huang, Wei Li, Shunshan Feng 2016 A Hyper-Viscoelastic constitutive model for polyurea under uniaxial compressive loading *Polimers Vol.8, Issue – 4, Pg.133.*
[22] Nahid Iqbal, Manorama Tripathi, Surekha Parthasarathy, Devendra Kumar, Prasun Kumar Ro 2018 Turning the properties of polyurea by regulating soft segment length *Journal of Applied polymer science Vol.135, Issue 21.*

[23] M.R. Amini, J. Simon, S. Nemat-Nasser 2010 Numerical modelling of effect of polyurea on response of steel plates to impulsive loads in direct pressure-pulse experiments *Mechanics of materials, Vol.43, Pg.615.*

Analysis of Forces generated in Human Body segments for Real time Space flight

K A V Uday Kiran[1], D R S Raghuraman[2]
[1, 2] Department of Design and Automation, School of Mechanical Engineering, Vellore Institute of Technology, Vellore – 632014, INDIA
Corresponding author Email Id- ananta.udaykiran2019@vitstudent.ac.in

Abstract: Human body modelling for a space travel is a significant task to be considered for analyzing the forces generated on the human body segments. During the flight from earth's atmosphere due to high escape velocity and acceleration, huge reaction forces will be generated on body segments which are beyond human capacitance. Recent studies evaluated the forces generated in various body segments by equipping external sensor vest or strips around the body parts where the application is limited for several on ground operations like hopping, sitting etc. and few studies analyzed the theoretical forces by modelling the human body as a second order spring mass damper system. The main objective of this work is to analyze the forces experienced by the human body segments for like Head, Neck, Upper Torso, Central Torso and Lower Torso for real time space flight data. The analysis is done on the anthropometric data of 50th percentile male US citizens available in literature by considering the human body as a second order spring mass damper system and the real time space flight parameters are taken from STS-121 space shuttle launched from Kennedy Space centre on July 4th 2006. The obtained resultant forces are plotted against the time of flight and it is observed that high forces are generated on Central Torso compared to remaining body parts and are varying with altitude of the vehicle.

1. Introduction

Human body modelling is a prime tool to analyze the forces experiencing by the person under any condition. The basic modelling includes a two DOF spring mass damper system to evaluate the forces. During a space travel, the high altitudes and escape forces will adversely affect the human body parts [1]. Studies showed the space suits will restrict the motion and causes much forces during the flight [2]. Researchers used two types of models for evaluating the forces on various body segments such as a passive model in which the mass, damping coefficient and stiffness [3] of the body parts are analyzed and the forces are evaluated with the aid of a second order spring mass damper system and an Active model in which the real time forces are identified by attaching several sensor vests or strips around the body part. Most of the research works includes the analysis of forces in human body segments for various activities like bouncing [4], running [5], sitting in a vehicle [6] (impact forces) etc. where a multiple DOF lumped spring mass damper model is developed to analyze the forces. In [7] an active Head Neck model has been developed to analyze the effect of posture by equipping sensors at various points around the neck. Xiangyang Wang et.al [8] developed an ankle foot exoskeleton for walking assistance by attaching sensors around the leg of a human subject at various places to identify the forces. Few works [9-12] evaluated the anthropometric data of various nationals such as Indians, Americans etc. for several bio mechanical applications. In [13] a study on damping coefficients of a vibratory model of human body is done by S.C. Zanwar et.al where a 15 DOF human body is modeled using a spring mass damper system by considering the each body part as an ellipsoid. In [14], the work is done to analyze the motion and forces on the human Torso for various operations using an active sensor model, whose results stated that the torso plays a key role in carrying / bearing the loads. Though, not much bio mechanical analysis of human subjects is done in aerospace applications, based on the studies, it is understood that a theoretical analysis using a second order spring mass system is primarily sufficient to analyze the forces during a space travel. This work represents a five DOF spring mass damper system considered for the human body modeling which includes the body parts like Head, Neck, Upper Torso, Central Torso and Lower Torso and the values of mass, damping coefficient and stiffness for each body part are taken from [13]. The real time space flight trajectory values are taken from an open source online platform where the altitude, velocity and acceleration of the vehicle are taken from NASA STS-121 space shuttle launched on July 4th 2006 from Kennedy Space Center [15] so as to analyze the forces at every instant of time during the space travel. The evaluated forces are tabulated and plotted against the time of flight of 520 seconds with respect to the space flight data. The following subsections illustrate the human body modelling, SIMULINK model and the results of the work.

2. Methodology

The forces on each body part is evaluated using the second order spring mass damper system as shown in the equation 1 below using the mass, damping coefficients and stiffness of each body segment and the real time space flight trajectory data considered from the literature.

$$[M][A] + [C][V] + [K][X] = [F] \qquad [1]$$

2.1. Human Body Modeling

A 5 DOF spring mass damper based human body is modeled by considering the each body part as ellipsoids as shown in the figure 1, the 5 DOF includes the head, neck, upper torso, central torso and lower torso of 50th percentile U.S. Male citizens and the evaluated mass, damping coefficients and stiffness for each considered body segment are taken from [13] as shown in the table 1.

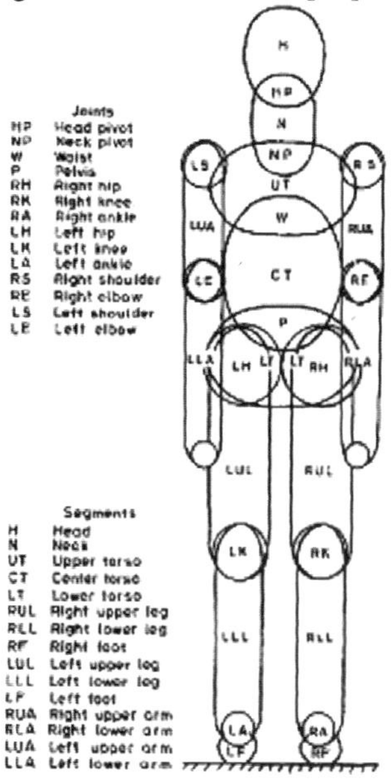

Figure 1. Anthropometric Model of Human Body [13]

Table 1. Mass, Damping Coefficients and Stiffness for various body segments taken form [13]

Body Segment	Mass (Kg)	Damping Coefficient	Stiffness (KN/mm)
Head	3.2	4.8	682.3
Neck	0.5	7.6	3572.3
Upper Torso	9.5	15.9	2246.1
Central Torso	13.41	55.55	785.8
Lower Torso	17.55	1204.02	1883.7

2.2. Real time Space Flight Data

The real time space flight trajectory data such as altitude, velocity and acceleration of the vehicle is taken from NASA's STS-121 space shuttle launched from Kennedy's space center on July 4th 2006. The table 2 below shows the real time ascent data of the vehicle up to a time of flight of 520 seconds which was taken from an open source online platform.

Table 2. Altitude, Velocity and Acceleration data of STS-121 space shuttle [15]

Time of Flight (sec)	Altitude (m)	Velocity (m/s)	Acceleration (m/s^2)
0	-8	0	2.45
20	1244	139	18.62
40	5377	298	16.37
60	11617	433	19.4
80	19872	685	24.5
100	31412	1026	24.01
120	44726	1279	8.72
140	57396	1373	9.7
160	67893	1490	10.19
180	77485	1634	10.68
200	85662	1800	11.17
220	92481	1986	11.86
240	98004	2191	12.45
260	102301	2417	13.23
280	105321	2651	13.92
300	107449	2915	14.9
320	108619	3203	15.97
340	108942	3516	17.15
360	108543	3860	18.62
380	107690	4216	20.29
400	106539	4630	22.34
420	105142	5092	24.89
440	103775	5612	28.09
460	102807	6184	29.01
480	102552	6760	29.3
500	103297	7327	29.02
520	105069	7581	0.1

3. Results and Discussion

A SIMULINK model has been developed using the above mentioned second order spring mass damper equation to evaluate the forces. The figure 2 shows a 1 DOF subsystem of the developed model for evaluating the forces on the head and the figure 3 shows the 5DOF model of the body segments. The obtained results are plotted against the time of flight of the vehicle where the figure 4 shows the forces generated on the Head, Neck, Upper Torso and Lower Torso and the figure 5 shows the forces generated on the Central Torso.

From the plots it is observed that the forces generated are high on each of the body segments and the Central Torso alone showed much higher forces compared to the other body segments which are severe for the human capacitance as stated in [16]. It is to be noted that, the body mass composition and strength of the Torso compared to other body parts is huge as shown in [17], which states that a major portion of the forces can be identified near the torso and F.J. Vera-Garcia et al [18] analyzed the forces on Torso due to sudden loading conditions and concluded the effect of forces are more on the trunk. Similarly, our study showed that the forces generated on the Torso are much higher during a space flight.

Though, this work evaluated the forces generated by the body segments' using a 5DOF model, the results are limited to theoretical analysis. More studies are needed to analyze the forces generated during a space travel such as an active model including a sensor strip around the body parts for improvising the human safety

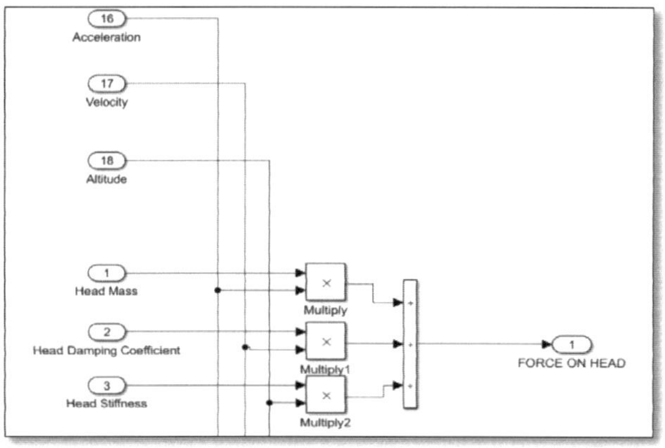

Figure 2. SIMULINK model for Individual body segment using second order spring mass damper equation

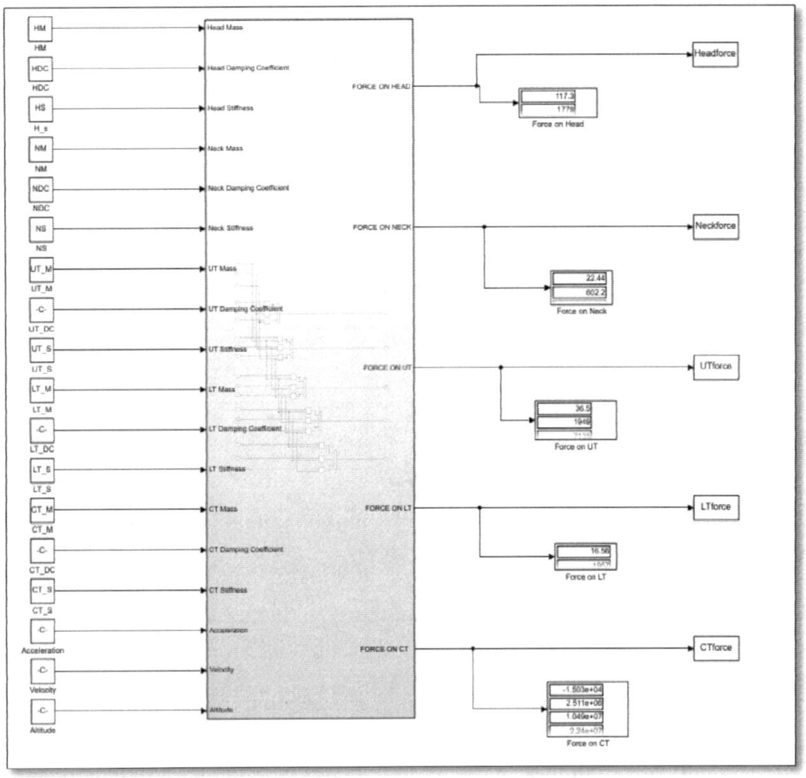

Figure 3. Developed 5 DOF SIMULINK model for various body segments (Head, Neck, Upper Torso, Central Torso and Lower Torso)

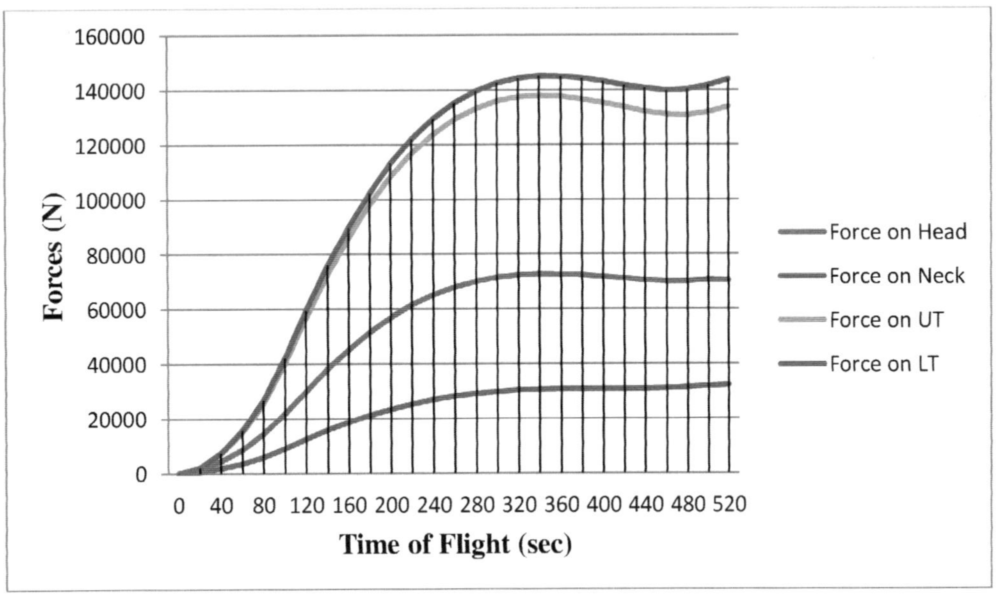

Figure 4. Resultant generated forces on Head, Neck, Upper and Lower Torso's against the time of flight

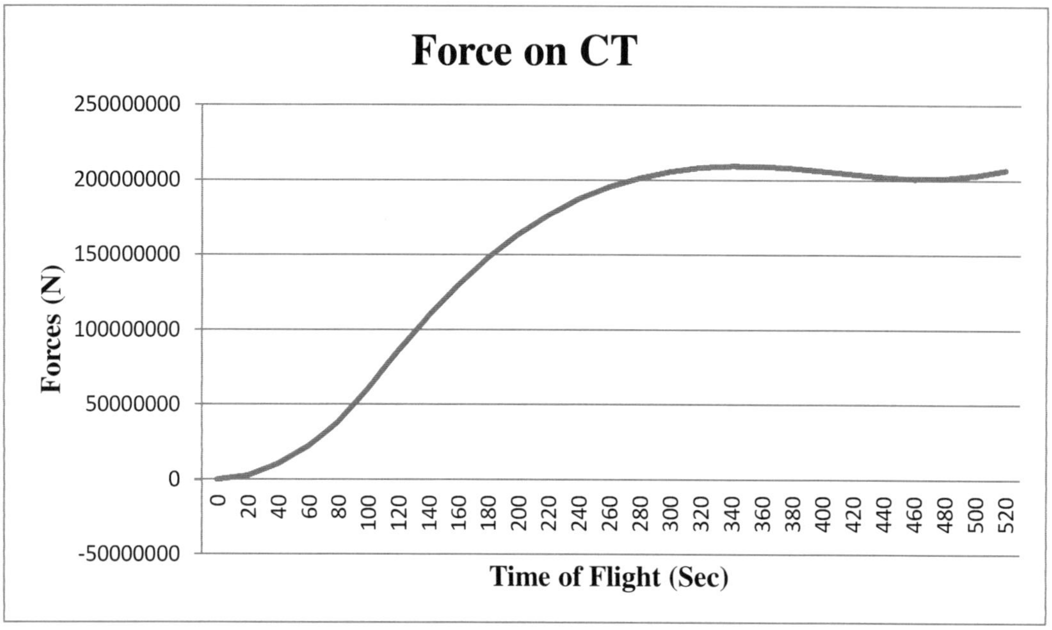

Figure 5. Resultant generated forces on the Central Torso against the time of flight

5. Conclusion

In conclusion, this work presents a theoretical analysis for evaluating the forces generated on human body segments for real time space flight using a second order spring mass damper system. A 5 DOF human body model is considered and the mass, damping coefficient and stiffness for each body segment is taken from the literature. The obtained forces on each body segment are high for the real time space flight data and the forces on Central Torso are much higher compared to the remaining body parts considered for the analysis.

References

[1] Sara Dina Feinzeig 2014 *The Science Journal of the Lander College of Arts and Sciences 8(1.)*
[2] Ana Diaz, Dava Newman 2014 *IEEE* **DOI:** 10.1109/AERO.2014.6836247.
[3] Nikooyan A A and Zadpoor A A 2011 *Proc. IMechE* **Vol. 225**.
[4] Haoqi Wang, Jun Chen and Tomonori Nagayama 2019 *Journal of Sound and Vibration* **456** pp 13-29.
[5] Andrew B Udofa, Laurence J Ryan and Peter G Weynad 2016 *IEEE* pp 371 – 376 DOI: 10.1109/BSN.2016.7516290.
[6] Bazil Basri and Michael J Griffin 2013 *Applied Ergonomics* **44** pp 423 – 434.
[7] Yang Wang and Salam Rahmatalla 2013 *Journal of Biomechanics* **46** pp 702 – 710.
[8] Xiangyang Wang, Sheng Guo, Haibo Qu and Majun Song 2019 *Sensors* **19** 3196.
[9] Joydeep Majumder 2014 *Euras J Anthropol* **5(2)** pp 45 – 53.
[10] GurleenKaur Luthra 2018 *IJRESM* **1(10).**
[11] Sandeep Singh Kharb, Belokar R M and Suman Kant 2018 *IJRASET* **6(V).**
[12] Sayeed Unsia and Kakoli Borkotoky 2017 *Demography India* **46(1)** pp 22 – 37.
[13] Zanwar S C, Mayank Mishra and Nigam S P 2008 *III rd NCME* 10.13140/RG.2.1.1807.6564.
[14] Daniele CAFOLLA, I-Ming CHEN, Marco CECCARELLI 2015 *Front. Mech. Eng.* **10(4.)**
[15] Space Shuttle Ascent 2006 (http://nasa.gov).
[16] Jose Jaime Esquivel Patricio 2019 *McNair Research Journal SJSU* **15(9).**
[17] Bihter Akinoglu, Tugba Kocahan 2019 *Progress in Nutrition* **21(4) : 1019-1028**.

[18] Francisco J Vera – Garcia, Stephen H.M. Brown, John R. Gray, Stuart M. McGill 2006 *Clinical Biomechanics* **21(2006) 443- 455**.

IOP Conf. Series: Materials Science and Engineering 1128 (2021) 012048 doi:10.1088/1757-899X/1128/1/012048

CT Guided Liver Tumor Biopsy Framework

G K Mourya[1], D Bhatia[1], M Gogoi[1] and A Handique[2]

[1]Department of Biomedical Engineering, North-Eastern Hill University, Shillong, Meghalaya, India
[2]Department of Radiology, North Eastern Indira Gandhi Regional Institute of Health and Medical Science, Shillong, Meghalaya, India

gkmourya@nehu.ac.in, bhatiadinesh@rediffmail.com, mgogoi@nehu.ac.in, drahandique@gmail.com

Abstract. Liver tumor biopsy is the first step of a liver cancer diagnosis. Accuracy of liver cancer diagnosis depends upon how precisely the sample is collected from the right location. The ultrasound-guided biopsy procedure is subjective and painful to patient. CT guided biopsy framework assists medical experts in planning biopsy needle path before performing the procedure. In this framework, first, CT image acquired, and 3D models are created for liver, skin, liver tumor, bone, and liver blood vessels using 3Dirac CT datasets. Second, an algorithm designed for biopsy needle insertion path recommendation. At the last 3D simulation generated. Needle insertion starts from the skin and ends at the tumor site. The recommended needle path does not carry bone and major blood vessels. The designed framework suggests two possible paths that medical experts can select and allow to make a further modification in that path. 3D simulation of needle insertion shown to medical experts for the evaluation. According to their comments, methods are producing promising results. The proposed framework may utilize for teaching-learning procedures for junior medical experts. Automatic biopsy needle path recommendations are perquisites for generating augmented reality tools and robotic biopsy systems.

1. Introduction

The liver is located below the thorax and on top of the pelvic region. The liver is the biggest internal organ in humans and performs plenty of body functions [1]. Liver cancer is one of the life-threatening diseases among other kinds of abnormality. Liver primary cancer (Hepatocellular Carcinoma) is initiated at the liver site and secondary cancer transferred by other organs, also known as metastasis. The liver is the first suspect organ to confirm the metastasis stage of different body parts cancer. Liver cancer incidence is increasing and its the fifth major cause of mortality worldwide [2],[3].

One of the significant causes of healthcare expenditure of patients with chronic liver disease is Hepatocellular Carcinoma (HCC). Computed Tomography (CT) guided liver biopsy procedure is performed by inserting a needle to the liver percutaneously, and the sample is collected to perform histopathology analysis for cancer diagnosis. Abdominal CT imaging is essential for the diagnosis and image-guided interventions of liver and liver tumors. Ultrasound and CT guided procedures are now the mainstays in managing liver diseases, avoiding unnecessary surgeries for obtaining biopsy specimen or removing liver tumors. Image-guided interventions fasten the workflow and reduce patient discomfort as daycare procedures are done under local anesthesia. Liver segmentation algorithms assist in diagnosing and prognosis of various liver disorders such as cirrhosis and liver tumors. Once the liver tumor is seen on CT imaging, the next step is to obtain a biopsy from the tumor for histopathological analysis to confirm the diagnosis. Liver tumor biopsy is mainly required to diagnose and determine the stage of cancer, and later for treatment evaluation. After the diagnosis of liver cancer is established, based on the staging (severity), it may be treated by various procedures such as surgery, radiation therapy, and thermal ablation, irreversible electroporation, etc. [4].

Ultrasound image-guided biopsy procedure is one of the favorable image-guided interventions. Some of the limitations of ultrasound-guided biopsies, like the lesion are not visualized as in obese patients; performance is very subjective, intervening bowel, etc. [5].US-guided biopsy procedures are very subjective, which needs to make objective to reduce operator dependency. Nowadays, most of the ultrasound-guided interventions are replacing with CT guided interventions to overcome the limitations of ultrasound. In this article, CT guided liver tumor biopsy framework is designed to counter the limitation of US-guided procedures.

2. Materials and methods

2.1 Data sets

The 3D Image Reconstruction for Comparison of Algorithm Database (3D-IRCAD) was utilized for this experiment. 3D-IRCAD Contains 22 subjects' 3D abdominal CT images along with liver and tumor ground truth. The image dimension is 512 x 512 pixels in the axial plane; the number of slices ranges from 74 to 225. The resolution in the X and Y directions ranges from 0.57 mm to 0.81 mm [6].

2.2 Methodology

We have designed a CT-guided biopsy framework comprising four steps; image acquisition, image segmentation, biopsy path recommendation, and 3D simulation **(Figure 1).** This framework will assist the clinician in planning the needle path in CT guided biopsy procedures. A biopsy gun needle is inserted from the skin to the tumor along the safest and shortest path avoiding bones and vital structures. Usually, a clinician selects an appropriate needle insertion path based on the imaging findings and personal experience.

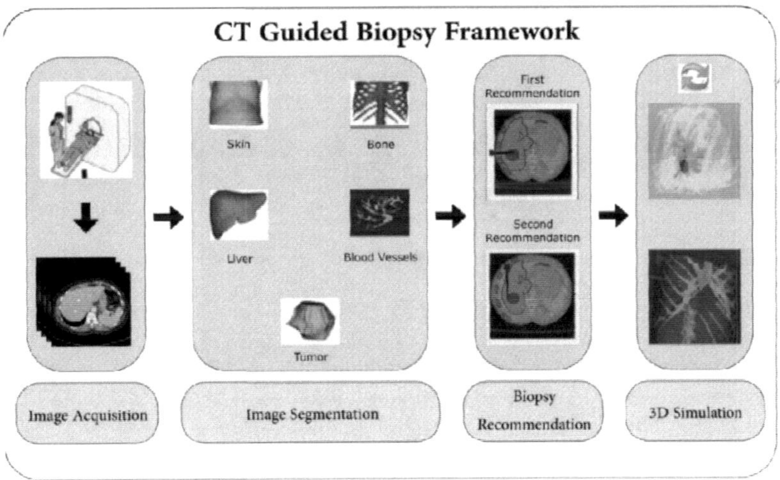

Figure 1. CT guided biopsy framework

2.2.1 Skin and Bone Segmentation

Skin is the body's outer surface; the liver biopsy procedure requires the outer shell to decide the entry point of needle insertion. In CT, the differences in Hounsfield units or gradient image, skin pixels have the highest gradient as it is surrounded by air. Skin caries a specific HU and gradient. A combination of thresholding and image gradient method is used to segment skin **(Figure 2).** Skin segmentation procedure further corrected by the morphological operations.

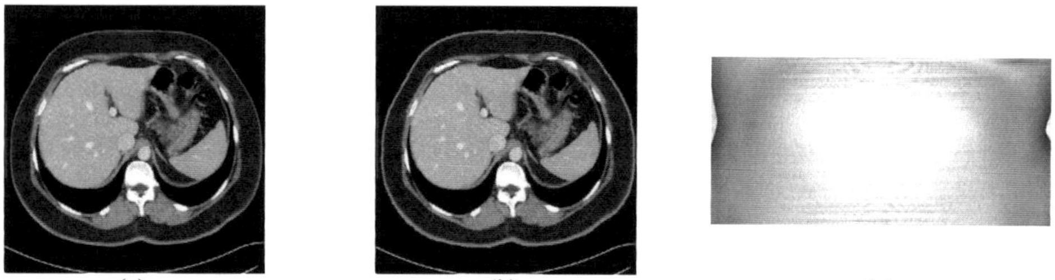

(a) (b) (c)

Figure 2. Result for skin segmentation (a) DICOM Image, (b) 2D Slice, (c) 3D Skin Segmentation

The liver is surrounded by the lower ribs of the right rib cage and situated below the diaphragm. Before performing the biopsy, it is essential to know the ribcage's exact location, which decides the track for needle insertion. Although bony structures show high contrast in CT images, their detection, identification, and correct segmentation remains a challenge [7]. The knowledge of the rib cage is very necessary for any CT guided biopsy procedure. A threshold-based approach has been used for the segmentation of ribcage. This method has performed an automatic and reliable image thresholding approach to segment the ribcage from CT images. We concluded the value of '**t** 'to be 50±2 on an average, which separates the ribs as the foreground from the background image [8]. Later, the segmented images are projected on a 3D platform, and the results obtained were found to be quite promising, as shown in **Figure 3**. Rib cage border outline marked accurately, which is a prerequisite for biopsy planning because bone parts should be avoided from needle path trajectory.

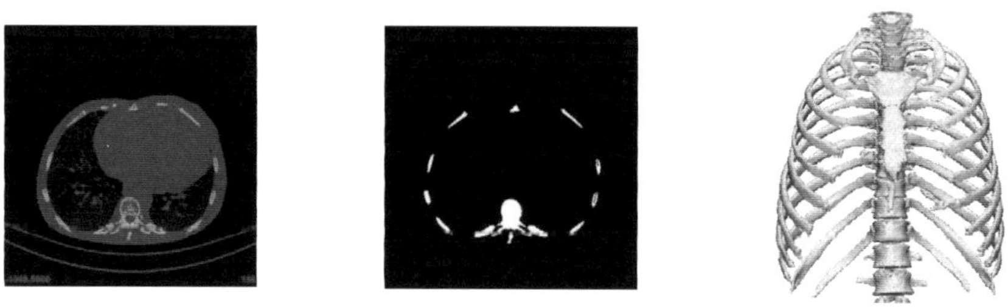

(a) (b) (c)

Figure 3. Rib cage segmentation (a) Reference Image (b) Segmented image (c) 3D view

2.2.2 Liver and Its Blood Vessels and Tumour Segmentation

Cascaded dilated deep residual network (DDRN) was utilized for segmenting liver, tumor, and blood vessels [9]. The abdominal CT image was segmented into three classes liver tissue, tumor, and blood vessels. After segmenting all body parts such as the rib cage, skin, liver, blood vessels, and liver tumor were superimposed in the same image **(Figure 4).**

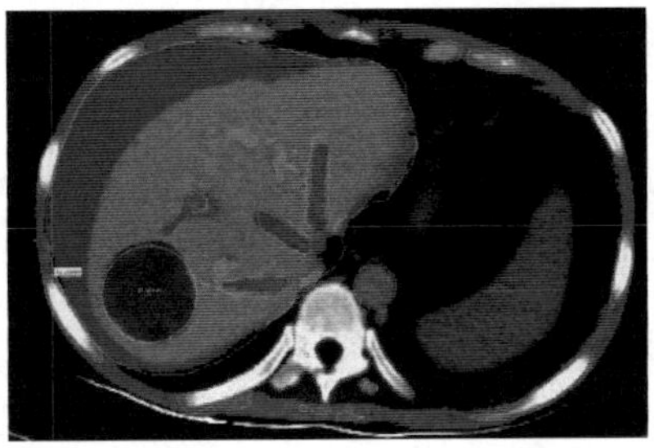

Figure 4. Result of liver segmentation and internal entity segmentation

2.2.3 Needle Path Planning
Accurate placement of the needle tip inside the tumor tissue is challenging, and critical structures and anatomical obstacles must be avoided. The first step is to select the tumor for biopsy if they are multiple. Criteria include viable tumor volume having viable (living) tissue (avoiding necrotic or dead tissue within) and distance from the skin entry site. Our implementation recommends the insertion point based on the shortest distance with the minimum number of obstacles. Needle path planning algorithm works based on cost function minimization concerning individual voxel within the path.

All steps are as follows:
Step 1: Selection of the target tumor if multiple tumors present within the liver.
Step 2: Selection of 2D axial slice, which carries the maximum area for the selected tumor in step one.
Step 3: Calculate the centroid of tumor selected in step two.
Step 4: Determine the centroid of the vertebra column present in the 2D axial view, which is selected in step two.
Step 5: Draw an arc at the concerning centroid of vertebra toward outer skin passed through tumor centroid width of the tumor's major axis.
Step 6: Consider all possible insertion points with respect to arc across the skin surface.
Step 7: Find the shortest and safest path with respect to the selected arc in the sixth step.
Step 8: Recommend a minimum two cost function path to a medical expert for their opinion and further corrections.

The same process may be performing for the skin arc, which covers the liver area's maximum size. The shortest distance with minimum cost function may be the selected possible safest path.
We use a 2D axial slice to determine the target point, making it easy to identify the location in Z-Direction. The next step is to choose the mask of possible entry points on the skin surface. We introduce a distance graph that, for every point of the pathway. This reduces the time needed for validating a possible path.

3. Results
For further enhancing the understanding of the whole biopsy procedure, a 3D simulation is also provided **(Figure 5).** This view helps in accurately locating the tumor and allows the inspection of the chosen needle pathway in a volumetric manner to provide a full spatial context for the biopsy needle. The simulation results were presented to three radiologists doing CT guided interventions in a tertiary care institute with such facilities for the validation in planning and performing CT guided biopsies and ablation of the liver tumors.

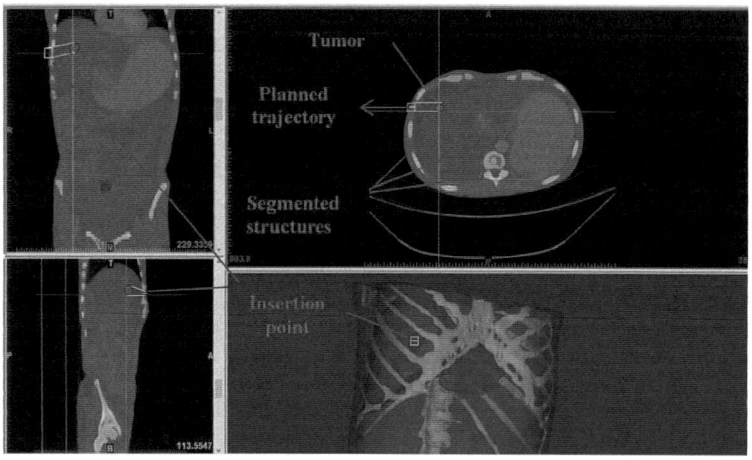

Figure 5. 3D simulation for the needle path planning

4. Conclusion

In this study, CT guided liver tumor biopsy planning is demonstrated. First, skin, bone, liver, blood vessels, and liver tumor were segmented using different segmentation algorithms. Skin was segmented using gradient operation followed by thresholding and morphological operation. After segmenting the rib cage, liver and inner components of the liver were segmented using DDRN. All blood vessels, including the portal vein, hepatic artery, and hepatic veins, were considered under the same classification. Our radiologists' team suggested that different phases of CT data should include and segment the portal vein and the hepatic artery separately. Blood pressure within the artery is greater than the veins, so the cost function weightage should be different for the artery and the veins to get exact planning. The needle path can be shifted due to the movement of the abdominal organs during respiration. Respiration gating should also be implemented along with an automated navigation system using augmented reality. Earlier, multiple CT scans were acquired to perform a single biopsy procedure; therefore, implementing the 3D simulations for tumor biopsy will reduce patient discomfort and excess x-ray radiation dose to the patient and the healthcare workers.

5. References

[1] J. L. Bollman and F. C. Mann, "The Physiology of the Liver," *Am. J. Physiol. Content*, vol. 96, no. 3, pp. 683–695, Mar. 1931.
[2] A. Forner, "Hepatocellular carcinoma surveillance with miRNAs," *Lancet Oncol.*, vol. 16, no. 7, pp. 743–745, Jul. 2015.
[3] S. K. Asrani, H. Devarbhavi, J. Eaton, and P. S. Kamath, "Burden of liver diseases in the world," *J. Hepatol.*, vol. 70, pp. 151–171, 2019.
[4] J. F. Cardella *et al.*, "Quality Improvement Guidelines for Image-guided Percutaneous Biopsy in Adults," *J. Vasc. Interv. Radiol.*, vol. 14, no. 9 (2), pp. 227–230, 2003.
[5] N. J. Khati, J. Gorodenker, and M. C. Hill, "Ultrasound-Guided Biopsies of the Abdomen," *Ultrasound Q.*, vol. 27, no. 4, pp. 255–268, 2011.
[6] IRCAD France, "3dircadb — ircad france." [Online]. Available: https://www.ircad.fr/research/3dircadb/. [Accessed: 13-Mar-2020].
[7] T. Klinder, C. Lorenz, J. von Berg, S. P. M. Dries, T. Bülow, and J. Ostermann, "Automated Model-Based Rib Cage Segmentation and Labeling in CT Images," in *Medical image computing and computer-assisted intervention : MICCAI ... International Conference on Medical Image Computing and Computer-Assisted Intervention*, vol. 10, no. Pt 2, 2007, pp.

195–202.

[8] G. K. Mourya, D. Bhatia, A. Handique, A. Bordoloi, P. S. Saha, and S. Swargiary, "Automatic segmentation of human ribcage from 3D medical dataset for diagnosis," in *2017 International Conference on Innovations in Electronics, Signal Processing and Communication (IESC)*, 2017, pp. 112–114.

[9] G. K. Mourya, M. Gogoi, S. N. Talbar, P. V. Dutande, and U. Baid, "Cascaded Dilated Deep Residual Network for Volumetric Liver Segmentation From CT Image," *Int. J. E-Health Med. Commun.*, vol. 12, no. 1, pp. 34–45, Jan. 2021.

CT Guided Diagnosis: Cascaded U-Net for 3D Segmentation of Liver and Tumor

G K Mourya[1], D Bhatia[1], M Gogoi[1] and A Handique[2]

[1]Department of Biomedical Engineering, North-Eastern Hill University, Shillong, Meghalaya, India
[2]Department of Radiology, North Eastern Indira Gandhi Regional Institute of Health and Medical Science, Shillong, Meghalaya, India

gkmourya@nehu.ac.in, bhatiadinesh@rediffmail.com, mgogoi@nehu.ac.in, drahandique@gmail.com

Abstract. Volumetric estimation of the liver tumor is the first step to identifying the liver's critical disorder. The liver and its tumor ratio prerequisite measures to select the therapeutic procedure. 3D printing and virtual reality platform require a segmented liver entity mask to evaluate the pre and post-treatment analysis. A cascaded U-Net model is proposed for automatic segmentation of liver and tumor in CT images. LiTS CT data set utilized for this study. The images were pre-processed using the windowing technique for contrast enhancement. Two U-Net models were modified for liver and tumor segmentation, respectively and connected in a cascaded manner. U-Net decoder end was modified in comparison to the original U-Net. The probability map of the first U-Net fed to the second U-Net and the input image to segment out the liver tumor. Eight subject volumetric CT datasets were utilized to test the cascaded U-Net performance and achieved average Dice coefficient for liver and tumor 0.95 and 0.69, respectively. Liver tumor diagnosis and treatment accuracy depend upon the precision of segmentation algorithms. Designed model segmented liver almost accurately and tumor segmented with limited accuracy. A further modification is required for the tumor segmentation cause of the occurrence of false negative.

1. Introduction

The liver is situated below the thorax and above the pelvic region. The liver is the largest internal organ in humans and performs ample body functions [1]. Liver cancer is one of the life threatening diseases among other kinds of abnormality. Liver primary cancer is initiated at the liver site and liver secondary cancer recived by other organs, also known as metastasis. The liver is the first suspect organ to confirm the metastasis stage of cancer. Liver cancer incidence increases and its is the fifth most cause of mortality worldwide [2], [3].

Liver and tumor segmentation from CT image is the first step for liver cancer diagnosis and therapeutic procedures. Liver and tumor segmentation is performed manually by medical experts for precise radiation therapy procedures, which is time-consuming and tedious. Liver tumor morphology, contrast, and location variability make segmentation challenging (Figure 1). This article's main objective is to devise a model for automatic segmentation of liver and tumor from abdominal CT images. These models reduced medical experts' workload and are treated as the second opinion for the segmentation results.

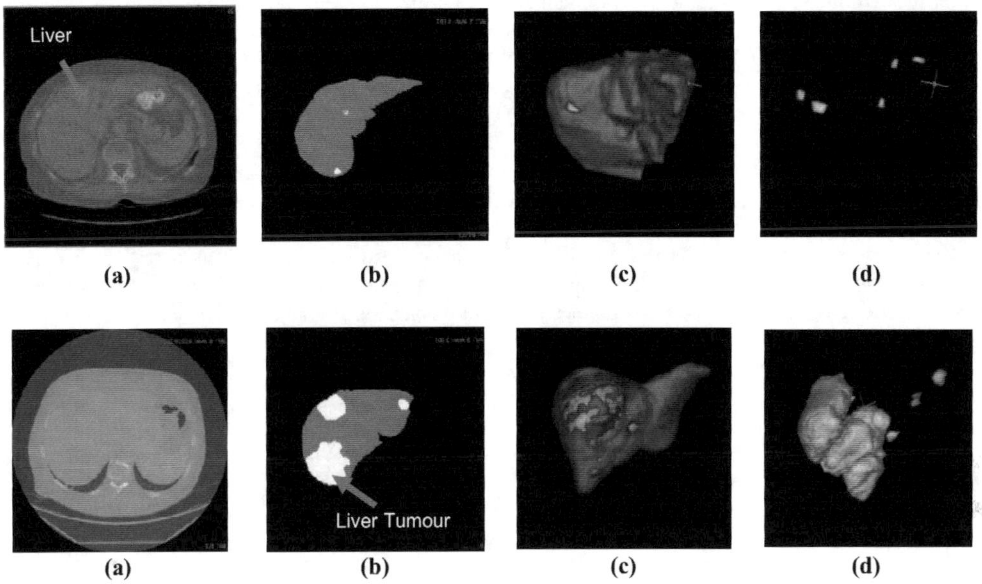

Figure 1. Heterogeneity in liver and tumour; shape, size, intensity profile (a) Raw Image, (b) Ground Truth [Gray: Liver; White: Tumour], (c) 3D Display [Red: Liver; Green: Tumour], (d) 3D Tumour [Green Tumour]

Many methods have been proposed for the segmentation of the liver and tumor. The segmentation technique is classified into two types first, semiautomatic segmentation and fully automatic segmentation methods [4]. Semiautomatic segmentation is mainly used in medical practice because results can be modified at any stage and need user interaction . Still, the fully automatic segmentation method not in practice because of less accuracy, and results are varied with respect to the source of data. Data drift still a challenging problem to get precise segmentation, mainly in medical image applications.

Various automatic segmentation technique studies like atlas-based, deformable model approach, but these techniques depend upon parameter optimization, which is rigid with respect to data and patient-specific [5], [6]. Fully automatic liver and tumor volume estimation is highly demanded and developed algorithm will be appreciated by medical experts if it will be error-free, more precise, and consistent. The research attempts to address the challenge of liver and tumor segmentation using graph cut[7],[8] statistical shape model[9], deformable model, level set technique, and machine learning [10], still limited by the generalizing capability of these approaches remains accepted for the healthy liver segmentation.

In the last decade, deep learning-based techniques have shown promising results for semantic segmentation of medical images. However, deep learning-based models have further classified a fully convolution network (FCN) most preferable model because of input and output as an image. FCN models overcome the limitation of machine learning techniques such as handcrafted feature extraction and selection of appropriate classification [11], [12]. In this article, the FCN model U-Net architecture is modified and connected in cascaded order for automatic segmentation of liver and tumor.

2. Materials and methods

2.1 Data sets

The Liver Tumor Segmentation (LiTS) Benchmark was utilized for this experiment. This data contains 201 volumetric abdominal CT of hyper and hypo contrast levels. This data is carrying ground truth for the liver and its tumor of 131 subjects. Image resolution ranges from 0.56 mm to 1.00 mm in axial and 0.45 mm to 6.0 mm in the Z-direction. The number of Z direction slices ranges from 42 to 1026 [13].

2.2 Methodology

The first input image is preprocess using window clipping [-100, 400], then fed to the model. The presented model was based on Cascaded U-Net, which was found to be optimal for segmentation. This model contains two U-Nets in a 'cascaded' order; one U-Net's output serves as the input for the second U-Net. The first U-Net input is the raw input data, and it produces a result as a probability map belongs to liver pixels. The probability map received from the first U-Net and original image serves as the second U-Net input. The second U-net's essential purpose is segmenting liver and tumor from a given input (Figure 2).

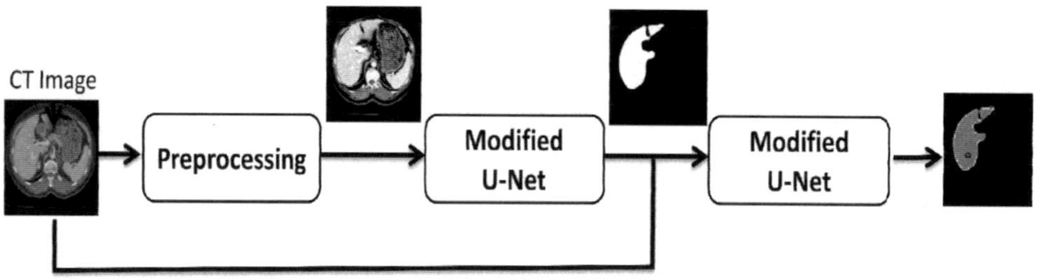

Figure 2 Block diagram of the proposed methodology

2.3 Network Architecture

The U-net architecture is synonymous with encoder-decoder architecture. The proposed architecture differs from the U-net architecture proposed by Ronneberger et al. [14]. We modified the decoder portion of architecture with unique upsampling feature sequences. In the original U-net, up-convolution was done with half a number of features in the previous layer, whereas in the proposed architecture, the numbers of features are kept the same as in the last layer. Due to this, the loss of information during up-convolution is prevented. The difference can be visualized in Figure 3. After up-convolution, features are concatenated with features from the corresponding layer in the encoder.

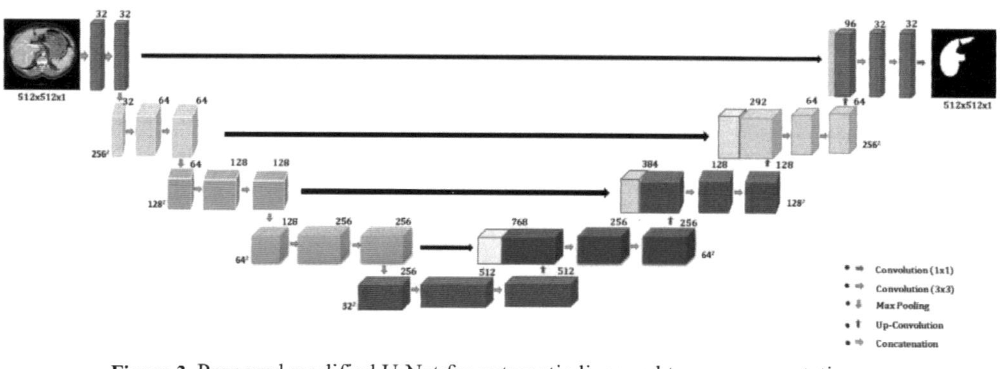

Figure 3. Proposed modified U-Net for automatic liver and tumor segmentation

Modified U-Net architectures are trained using a stochastic gradient descent optimizer. Categorical cross-entropy as loss function trained with learning rate 0.001 for the 40 epochs.

3. Results

Volumetric CT of 130 subjects data from LiTS Challenge data was used for this study. Training model results tested on eight subjects' data sets. The results quantify evaluated using DICE evaluation and methodology achieved average DICE 0.95 and 0.65 for liver and tumor. Qualitative results are demonstrated in Figure 4, which shows that liver and tumor segmented accurately (liver in yellow color and tumor is in red). We have considered the image slice where the liver shape has a concave boundary and multiple tumor regions in the same image. At the liver's concave boundary region, mostly algorithms fail to segment liver accurately; however, cascaded modified U-Net can segment the liver with the concave boundary. This method produces some false-negative results for the tumor segmentation (marked by the indigo square box) within the center tumor shown in Figure 4 (c).

(a) (b) (c)

Figure 4. Test results of Cascaded Modified U-Net (a) Original Image (b) Ground Truth Image (c) Predicted Image

The result of Cascaded modified U-Net compares with the exiting model with respect to DICE coefficient (Table 1). H-Dence U-Net performance is better among the existing model for segmentation of liver and tumor. The proposed model achieved an average Dice of 0.95 and 0.69 for liver and tumor segmentation. Cascaded modified U-Net needs to investigate for the segmentation of liver tumor because only 0.69 dice were achieved.

Table 1. Result comparison with other models

Method	Liver Dice	Tumor Dice
ResNet [15]	0.95	0.49
H-DenseUNet [16]	**0.98**	**0.94**
nnU-Net [17]	0.95	0.73
DeepX [18]	0.96	0.82
Cascaded ResNet [15]	0.95	0.50
Proposed	**0.95**	**0.69**

4. Conclusion

In this method, cascaded modified U-Net has proposed to segment liver and tumor from CT image automatically. The cascaded model can segment the liver precisely with the dice coefficient of 0.96. Proposed U-Net architecture has class in balancing problem, so two U-Net models connected in a cascaded manner. Further investigation is required for automatic segmentation of liver tumor cause of limited accuracy achieved. The ratio of tumor class to non-tumor class is less within the image, so class in balancing problem affects tumor segmentation results' accuracy. Accurate segmentation of liver tumors is a high need in clinical practice. In the future, architecture may changes with different connectivity for better results.

5. References

[1] J. L. Bollman and F. C. Mann, "The Physiology of the Liver," *Am. J. Physiol. Content*, vol. 96, no. 3, pp. 683–695, Mar. 1931.

[2] A. Forner, "Hepatocellular carcinoma surveillance with miRNAs," *Lancet Oncol.*, vol. 16, no. 7, pp. 743–745, Jul. 2015.

[3] S. K. Asrani, H. Devarbhavi, J. Eaton, and P. S. Kamath, "Burden of liver diseases in the world," *J. Hepatol.*, vol. 70, pp. 151–171, 2019.

[4] P. Campadelli, E. Casiraghi, and A. Esposito, "Liver segmentation from computed tomography scans: A survey and a new algorithm," *Artif. Intell. Med.*, vol. 45, no. 2–3, pp. 185–196, 2009.

[5] A. Schenk, "2012 - Liver Segmentation and its Application to Hepatic Interventions.pdf," Universitätsallee 29, 28357 Bremen, Germany, 2012.

[6] J.-B. Fasquel, V. Agnus, J. Moreau, L. Soler, and J. Marescaux, "An interactive medical image segmentation system based on the optimal management of regions of interest using topological medical knowledge.," *Comput. Methods Programs Biomed.*, vol. 82, no. 3, pp. 216–30, Jun. 2006.

[7] F. Malmberg, "Graph-based methods for interactive image segmentation," UPSALA UNIVERSITY, 2011.

[8] B. Peng, L. Zhang, D. Zhang, and J. Yang, "Image segmentation by iterated region merging with localized graph cuts $," *Pattern Recognit.*, pp. 1–12, 2011.

[9] H. Badakhshannoory and P. Saeedi, "A Model-Based Validation Scheme for Organ Segmentation in CT Scan Volumes," *IEEE Trans. Biomed. Eng.*, vol. 58, no. 9, pp. 2681–2693, Sep. 2011.

[10] F. A. Mohammed and S. Viriri, "Liver segmentation: A survey of the state-of-The-Art," in *Proceedings of: 2017 Sudan Conference on Computer Science and Information Technology, SCCSIT 2017*, 2018, vol. 2017-Novem, pp. 1–6.

[11] K. Munir, H. Elahi, A. Ayub, F. Frezza, and A. Rizzi, "Cancer diagnosis using deep learning: A bibliographic review," *Cancers (Basel).*, vol. 11, no. 9, pp. 1–36, 2019.

[12] F. Lateef and Y. Ruichek, "Survey on semantic segmentation using deep learning techniques," *Neurocomputing*, vol. 338, pp. 321–348, 2019.

[13] P. Bilic *et al.*, "The Liver Tumor Segmentation Benchmark (LiTS)," *arXiv:1901.04056v1*, vol. v1, pp. 1–43, 2019.

[14] O. Ronneberger, P. Fischer, and T. Brox, "U-net: Convolutional networks for biomedical image segmentation," in *International Conference on Medical image computing and computer-assisted intervention*, 2015, vol. 9351, pp. 234–241.

[15] L. Bi, J. Kim, A. Kumar, and D. Feng, "Automatic Liver Lesion Detection using Cascaded Deep Residual Networks," *arXiv ID 1704.02703*, Apr. 2017.

[16] X. Li, H. Chen, X. Qi, Q. Dou, C.-W. Fu, and P.-A. Heng, "H-DenseUNet: Hybrid Densely Connected UNet for Liver and Tumor Segmentation From CT Volumes," *IEEE Trans. Med. Imaging*, vol. 37, no. 12, pp. 2663–2674, 2018.

[17] F. Isensee *et al.*, "Abstract: nnU-Net: Self-adapting Framework for U-Net-Based Medical Image Segmentation," in *arXiv ID 1809.10486V1*, 2019, pp. 22–22.

[18] Y. Yuan, "Hierarchical Convolutional-Deconvolutional Neural Networks for Automatic Liver and Tumor Segmentation," *arXiv ID 1710.04540*, vol. i, pp. 3–6, Oct. 2017.

Finite element analysis of the slot milling of carbon fiber reinforced polymer composites

C Prakash[1,a], **K S Vijay Sekar**[2,b]

[1] Associate Professor, Department of Automobile Engineering, Surya Group of Institutions, Vikiravandi - 605 652, Tamilnadu, India.

[2]Associate Professor, Department of Mechanical Engineering, SSN College of Engineering, Kalavakkam - 603 110, Tamilnadu, India.

[a]prakashc@suryagroup.edu.in, [b]vijaysekarks@ssn.edu.in

Abstract. Carbon Fiber Reinforced Polymer (CFRP) composites are widely used for making the components and primary structures in the field of aircraft, aerospace, automobile, medical, construction and many essential items which we use every day. These composite laminates are light in weight without compromising its strength. Hence, CFRP composites attract the researchers and the modern industries to investigate more on it. Industries rely on conventional machining process to achieve required shape and size, but CFRP composites show intrinsic behaviour while machining because of its non homogeneous properties. Selecting the appropriate tool and the cutting conditions for machining the CFRP composite, influences more in achieving the desired shape, size with close tolerance. In this paper, a 3D Finite Element Model is developed to simulate the slot milling process and to predict the cutting forces for the different cutting conditions. The reliability and accuracy of 3D FE analysis depend on the friction coefficient and failure model used in the FE model. These FE input parameters were applied in a wider range and analysis was carried out. The cutting forces obtained from the FE results are correlated with the experimental results to validate and to evaluate the influence of the FEM parameters. The outcome of this investigation is to suggest a suitable Failure model and the Friction coefficient, which ensures the reliability of the FE Model.

Keywords: CFRP Laminate, Slot milling Process, Cutting forces, 3D FEM, Failure Criteria, Abaqus.

1. Introduction

The applications of Carbon Fiber Reinforced polymers (CFRP) are very extensive in the field of Automobile, Aerospace and Aircrafts since it is superior in strength to weight ratio, non corrosive and thermal resistance properties[1]. More than 60 % of the materials used in the construction of aircraft are CFRP composites. Moreover, the CFRP material is used in making tennis rackets, automobile components, domestic items such as chairs and tables, sports equipments, bridges, pressure vessels, pumps, electrical panels, wheel chairs, automobile bumpers, suspension springs and much more. Mostly these items are made in near net shape; however milling operation is required to achieve the required shape and size. The different types of failure occur during the milling process, which must be investigated to get defect free products [2]. Conducting experimental investigation require more resources and it is consumes more time; moreover it is not advisable for a wider machinability matrix [3, 4]. This requirements leads to the development of 3D FE Models to analyze the milling operation, which is very much suitable for studying the process at wider machinability matrix and effective even for elevated cutting parameter settings [5]. The 3D FE simulation result reveals the milling process very effectively and predicts the cutting forces during machining, chip formation process, chip characteristics and stress. However, the accurateness and consistency of the FE results rely on the FE model input parameter settings [6]. Hence, it is very much essential to investigate the influence of these FE model input model parameters to improve the quality of the 3D FE model and to get enhanced agreement with the experimental values [7]. The FE input parameters such as Friction coefficient, Element size, Failure model are vital parameters which influence more on FE output results. This paper aims in investigating the influence of Friction coefficient and the Failure models in the 3D slot milling simulation results and cutting force data. The validated FE model is used in this investigation and compared with the experimental result data from the literature [8]. The outcome of this investigation suggests a suitable failure model and friction coefficient for the researchers and industries to ensure good and consistent 3D FE model to simulate the slot milling process of CFRP laminate.

2. Milling experiment and results

The milling experimental studies are carried out in a NAMMIL vertical milling machine with the machining parameters given in the Table 1. The cutting parameters were selected based on the literature review and moreover, it helps to visualize and understand the chip formation, machining process effectively. The authors already have published a paper on the above mentioned experimental studies in detail [8]. The UD-CFRP composite laminates used in this experimental investigation measures 100 x 50 x 10 mm. The specifications and mechanical properties of the UD-CFRP composite laminate is given in the Table 2. A 10 mm diameter solid carbide end mill cutter coated with TiAlN is used for the slot milling operation [9]. The IEICOS mill tool dynamometer is used to obtain the cutting force data and the data is processed by an analogue to digital converter to plot the cutting force graph. The cutting force results obtained from experimental trials [8] are given in Table 3. The cutting force obtained from the experimental studies shows a fluctuating pattern which is due to the varying fiber orientation angle with the cutting edge of the tool (shown in Figure 3). The chips are mostly tiny, fragile and in the form of slightly curved flakes. The chips obtained from the experimental studies exposes the various composite failure modes such as fiber and matrix failure in tension, fiber and matrix failure in compression. The debonding of matrix and fibre, matrix cracking, interlaminar fibre failure is also observed in the milled chips.

Table 1. Machining parameters used in slot milling of UD-CFRP composite [8]

Machining parameters	Magnitude
Spindle speeds, N (rpm)	350, 660, 900, 1700
Cutting Speed, V_c (m/min)	10.99, 18.84, 28.26, 53.38
Feed rate, f (mm/min)	50, 150, 350
Depth of Cut (mm)	1
Fiber orientation	0^0 - Unidirectional

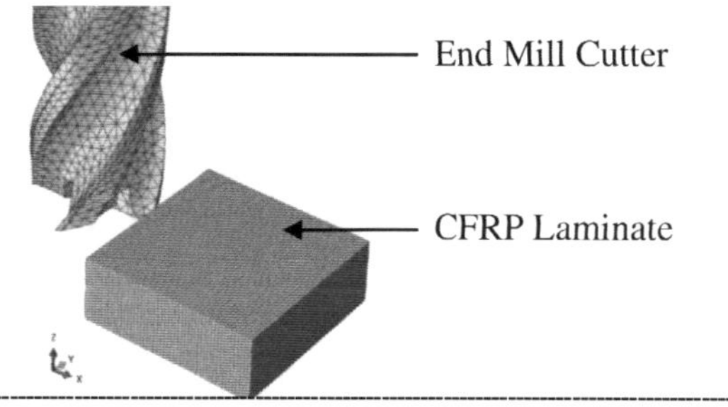

Figure 1. 3D FEA Model of slot milling CFRP laminate.

Table 2. Mechanical Properties and Specifications of UD - CFRP laminate [8]

Specifications	Value(s)
Modulus in Longitudinal Direction, E_x(GPa)	135
Modulus in Transverse Direction, E_y (GPa)	10
Modulus in Transverse Direction, E_z (GPa)	10
Shear Modulus along plane XY, G_{xy} (GPa)	5
Shear Modulus along plane YZ, G_{yz} (GPa)	3.7
Shear Modulus along plane XZ, G_{xz} (GPa)	5
Poisson's ratio, υ	0.27
Density (Kg/m^3)	1590
Resin / Fiber Ratio	35:65
Polymer	Epoxy Resin
Matrix	Carbon Fiber
Allowable Tensile Strength in fiber direction, X_t (MPa)	1500
Allowable Compressive Strength in fiber direction, X_c (MPa)	1200
Allowable Tensile Strength along Transverse direction (Y axis), Y_t (MPa)	50
Allowable Compressive Strength along Transverse direction (Y axis), Y_c (MPa)	250
Allowable Shear Strength in XY plane, S_{xy} (MPa)	70
Allowable Shear Strength in YZ plane, S_{yz} (MPa)	70
Allowable Shear Strength in XZ plane, S_{xz} (MPa)	70
Fiber Orientation (Degree)	0

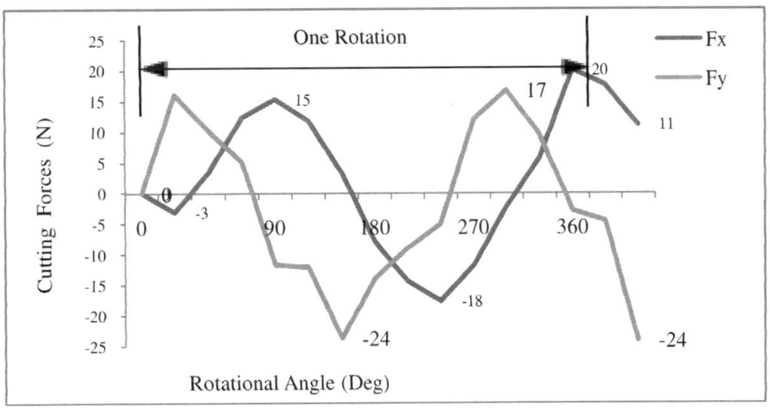

Figure 2. Changes in cutting force with cutting edge position during slot milling of CFRP Laminate [8, 13]. Cutting forces Vs Tool rotation angle in degrees.

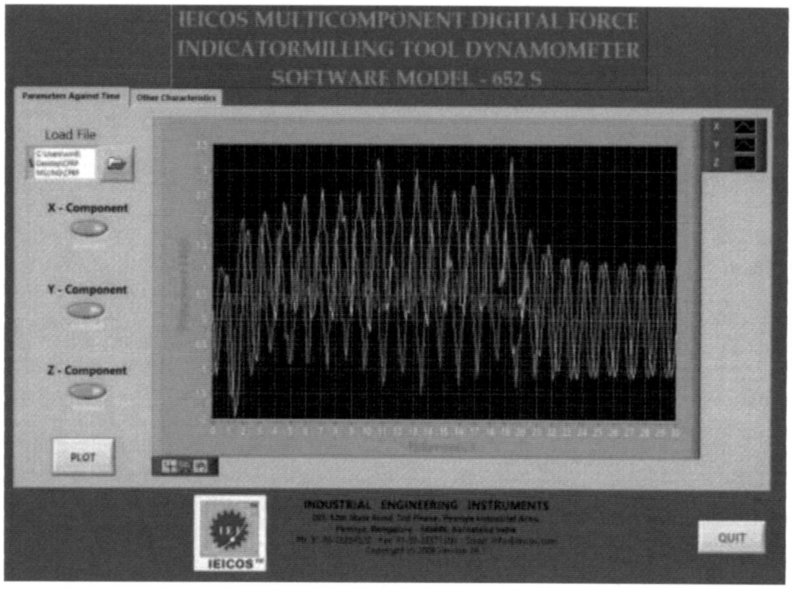

Figure 3. Experimental Cutting force Profile in slot milling of CFRP Laminate at V_c = 10.99 m/min, N = 350 rpm, Feed = 350 mm/min, Depth of cut = 1 mm.

3. FEA in slot milling of CFRP laminate

A 3D FE Model is developed (shown in Figure 1) in ABAQUS/Explicit to simulate the slot milling process and to understand the machining operation very effectively [10]. The End mill cutter is modeled in CATIA V5R14 [11] and workpiece (CFRP Laminate) is modeled in ABAQUS. The tool and the workpiece models are meshed with 3D elements and their material properties are incorporated with Autodesk Helius PFA plug-in [12]. The key FE input parameters are Failure model, friction coefficient, damage initiation and evolution criteria, output requirements, solvers, solving time, interaction properties between the tool and workpiece and the boundary conditions. A range of friction coefficients (0 to 1.0) and the Failure criteria such as Hashin, Christensen, LaRc02 and Puck were incorporated into the 3D FE model. The investigation was performed to report the influence of friction coefficients and to suggest a good failure model in order to enhance the accuracy of the 3D FE results and thereby ensuring the reliability of the FEM. The cutting force data obtained from 3D FE simulation results were compared with the experimental outcomes to evaluate the influence of the above mentioned FE model parameters.

4. FEA results and discussion

The Cutting force prediction is mandatory and essential to study the machinability of the CFRP laminate. Moreover, the larger fluctuation in the cutting force leads to poor quality of the milled surface and vibration of the machine. The composites are heterogeneous in nature and the cutting force varies with respect to fiber orientation and also with the cutting edge of the tool (shown in Figure 2) [13].

4.1. Comparison of cutting force data for various coefficient of Friction.

The interaction between the tool and the laminate is incorporated to the 3D FE model in order to define the surface erosion process of slot milling [14, 15]. The Coulomb friction theory with tangential behaviour is incorporated in the FE model to have a realistic machining approach. The friction coefficient used by the researchers for TiAlN coated tool and CFRP laminate is 0.7

approximately [2]. The maximum cutting force is predicted using 3D FE simulation results for different friction coefficients and it is compared with the Experimental data (given in Table 3). The results reveal that, Friction coefficient approximately around 0.7 provides comparable results with the experimental data and the literature [8, 9]. Hence, friction coefficient of 0.7 is recommended for further FE slot milling simulation studies which pocess similar input and output parameters.

Table 3. Predicted Cutting Force Data from 3D FE simulation for various Coefficient of Friction with HASHIN Failure Criteria.

Machining Parameter settings	Force Component (N)	Expt. data	Coefficient of Friction (μ)									
			0.1	0.2	0.3	0.4	0.5	0.6	0.7	0.8	0.9	1.0
			Maximum Cutting Force from FE results									
F = 50 mm/min, V_c = 10.99 m/min	F_x	35	24	24	23	24	23	25	38	24	25	24
	F_y	30	21	23	24	23	22	24	33	24	23	22
	F_z	11	1	2	1	2	1	3	7	1	2	1
F = 150 mm/min, V_c = 10.99 m/min	F_x	36	23	24	26	20	22	28	35	34	28	28
	F_y	28	24	23	22	21	20	24	26	24	22	24
	F_z	15	-4	-4	-4	4	3	8	10	5	3	1
F = 350 mm/min, V_c = 10.99 m/min	F_x	32	32	14	18	20	23	15	30	27	26	23
	F_y	19	-35	13	16	20	20	12	17	20	22	18
	F_z	12	30	10	9	5	7	3	10	8	-4	-5

4.2. Analysis of cutting force profile for various failure models

The selection of suitable failure criteria plays a vital responsibility in the accuracy and reliability of the machining simulation results. Moreover, the selection of failure criteria also depends upon the capability of the criteria in predicting the requisites (failure modes of the laminate, interlaminar failure). The cutting force profile and the data predicted from FE simulations must be in good agreement with the experimental results. Among the four failure models used in this investigation, Hashin failure criterion recognizes all four modes of failure such as, fiber and matrix failure in tension, fiber and matrix failure in compression and moreover the results are in better agreement with the experimental results [16]. Whereas, Christensen criterion make sure the fiber and matrix failure only.

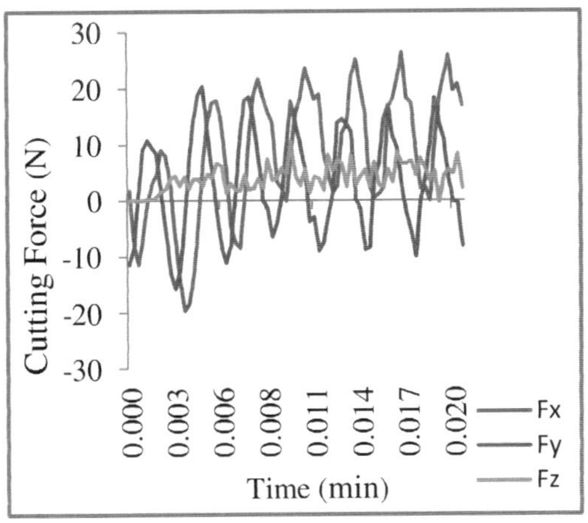

Figure 4. Hashin Criterion

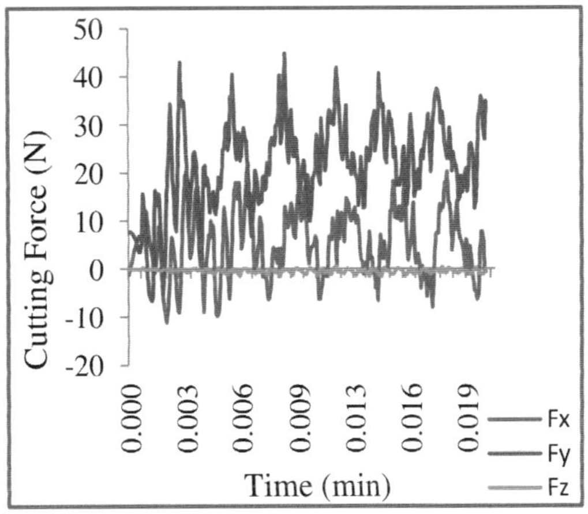

Figure 5. Christensen Criterion

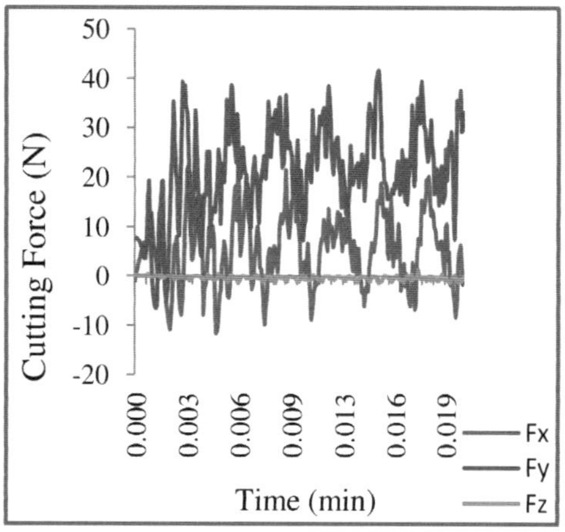

Figure 6. Puck Criterion

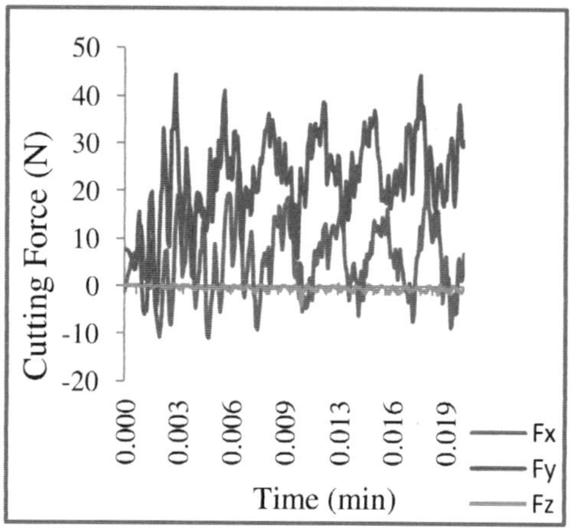

Figure 7. LaRc02 Criterion

Predicted Cutting force Profile from 3D FE Model in Slot milling CFRP Laminate at V_c = 10.99 m/min, N = 350 rpm, Feed = 350 mm/min, Depth of cut = 1 mm.

The Puck criterion predicts the fiber failure and inter fiber failure, LaRC02 criterion discloses the fiber failure and matrix cracking only in unidirectional composites. The FE analysis is carried out with all these damage models at V_c = 10.99 mm/min, Spindle speed, N = 350 rpm, Feed rate = 350 mm/min, Depth of cut = 1 mm. The cutting force profiles obtained were compared (shown in Figures 4, 5, 6, 7). Hashin failure model provides a unique cutting force profile, and whereas Puck, Christensen, LaRc02 failure models shows an entirely different kind of cutting force profile. Moreover, cutting forces F_x and F_y simulated by the Hashin failure model overlaps each other which is in good accordance with the experimental cutting force profile shown in figure 3 and the literature [2, 9, 13]. Hence, from this investigation it is confirmed that Hashin failure model is good enough in satisfying our requirements and it is suggested for further 3D FE slot milling machining simulation studies.

4.3. 3D Finite Element simulation of chip formation process.

The Chip formation can be visualized from Figure 8 which shows that, chip is formed due to plastic deformation and development of stress at the cutting zone. The Simulation results obtained from the developed 3D FE model shows the chip formation process.

a) Chip tend to curl b) Chip breaking c) Chip fragmented and released and Chip flows in air

Figure 8. Chip Formation Process in FEA Simulation at Cutting speed, V_c = 28.26 m/min and feed rate, f = 50 mm/min.

The chip have a propensity to curl slightly since the rake face of the tool slides over the composite laminate, however the inherent properties of the CFRP laminate does not allow the material to curl further, rather it breaks and forms a flake like chip. The chip obtained from FE simulation (shown in Figure 8) resembles the experimental chip, in size and shape. Since, the carbon fiber is brittle in nature the milled chip does not deform much, rather it is broken into tiny flakes which is understood from Figure 8.

5. Conclusion

The validated 3Dimensional FE model used in this investigation for slot milling simulating of UD-CFRP composite is trustworthy and good in predicting the cutting force profile and magnitude.

- The friction coefficient ranging from 0 to 1 is selected and 3D FE machining simulations studies are performed. The cutting force results predicted by the 3D FE simulation studies with the friction coefficient of 0.7 is in good agreement with the experimental results and the literature.

- Among the various composite failure models used in this research work, Hashin Failure model is suggested for further 3D FE investigations. Since, the cutting force profile and the data predicted by the Hashin damage criteria is in accordance with the literature and experimental results. Moreover, when compared to other failure models Hashin Failure model alone is capable of recognising all the mode of composite failures which is very much needed to predict at the earliest for avoiding premature component failures.

6. References

[1] R.El Alaiji, L. Lasri, A.Bouayad 2015 3D finite element modelling of chip formation and induced damage in machining of fiber reinforced composites *American Journal of Engineering Research* **4(7)** 123–132.

[2] Khaled Giasin, sabino Ayyar-Soberanis, Toby French, Vaibhav Padnis 2016 3D finite Element Modelling of cutting forces in drilling fibre Metal laminates and experimental hole quality analysis *Springer. Appl Compos mater. open access journal* DOI 10.1007/s10443-016-9517-0

[3] C.Prakash, KSV Sekar 2018 3D Finite Element Analysis of Drilling of Glass Fiber Reinforced Polymer Composites *Adv. Sci Eng. Med.* **10** 308–312 DOI:10.1166/asem.2018.2125

[4] Prakash.C, Vijaysekar.K.S 2018 3D Finite Element Analysis Simulation of Slot Milling Process for Titanium Alloy Ti6Al4V *IOSR Journal of Engineering (IOSRJEN)* 22–28

[5] Vaibhav A. Phadnis, Farrukh Makhdum, Anish Roy, Vadim V. Silberschmidt 2013 Drilling in carbon/epoxy composites: Experimental investigations and finite element implementation. *Composites. Part A.* **47** 41–51 http://dx.doi.org/10.1016/j.compositesa.2012.11.020

[6] Pedro J. Arrazola, Tugrul Ozel 2010 Investigations on the effects of friction modeling in finite element simulation of machining *International Journal of Mechanical Sciences* **52** 31–42

[7] Tugrul Ozel 2006 The influence of friction models on finite element simulations of machining. *Int. Journal of machine Tools & Manufacture* **46** 518–530

[8] Prakash.C, Vijaysekar.K.S 2017 3D Finite Element Analysis of Slot Milling of Carbon Fiber Reinforced Polymer Composite *Journal of Balkan Tribological Association* **23(3)** 497–514

[9] Prakash.C, Vijaysekar.K.S 2018 3D finite element analysis of slot milling of unidirectional glass fiber reinforced polymer composite *Journal of Brazilian Society of mechanical Sciences and Engineering* **40(6)** DOI : 10.1007/s40430-018-1195-4

[10] Simulia.D 2014 Abaqus 6.14 User's manual2014, *Dassault systems, Providence, RI* Available: http://50.16.225.63/v6.14/, last accessed 2019/07/30.

[11] CATIA V5R14 2014 User Manual2014, *Dassault systems, Providence, RI.*

[12] Autodesk., AUTODESK HELIUS PFA 2016 / HELP, *Autodesk Knowledge Network* http://help.autodesk.com/view/ACMPAN/2016/ENU/

[13] Sheikh-Ahmad J.Y 2009 Machining of Polymer Composites, *Springer, London*
[14] Nilanjan Das Chakladar, Surjya K. Pal, Parthasarathi Mandal 2012 Drilling of woven glass fiber-reinforced plastic - an experimental and finite element study *International Journal of Advanced Manufacturing Technology* **58** 267–278 DOI 10.1007/s00170-011-3386-3
[15] Kishore Debnath, Inderdeep singh 2015 Drilling of metal matrix composites: Experimental and Finite element analysis *Journal of Engineering Manufacture* **229(5)** 886–890
[16] Hashin Z 1980 Failure criteria for unidirectional fiber composites *Journal of Applied Mechanics* **47** 329–34

Experimental and Numerical Study on Ergonomic Evaluation of Automotive Seating Comfort

Anand V[1], Amit Dongale[2], Davidson Jebaseelan[3], M. Jagannath[4], Christo Michael[5]

[1,2,3,5]School of Mechanical Engineering, Vellore Institute of Technology (VIT), Chennai, Tamil Nadu, India
[4]School of Electronics Engineering, Vellore Institute of Technology (VIT), Chennai, Tamil Nadu, India

*Corresponding author: Davidson.jd@vit.ac.in

Abstract. Automotive car seats play a vital role in providing optimum comfort to the occupants when it comes to vehicle comfort. The objective of this study was to perform an experimental and numerical investigation of the ergonomic evaluation of automotive seating comfort. A questionnaire survey was carried out on car drivers driving two different cars namely Tata Indica and Maruti Swift, to understand the perception of drivers on seating comfort. Posture monitoring techniques using Kinect camera and pressure mapping systems were used to analyze the sitting posture and sitting angle. A multibody dynamics analysis technique was used to evaluate the human body exposed to vehicle vibration. The vertical vibration transmissibility was measured from different body segments using this model. Also, with the help of a multibody dynamics simulation model, the results were compared and validated with existing literature. This study suggests suitable sitting posture and sitting angle derived with the help of multibody dynamics simulation along with the experimental investigation, which would reduce the fatigue of the driver when exposed to prolonged driving.

Keywords: Seating comfort, Multibody dynamics, Vibration transmissibility, Fatigue, Prolonged driving, Kinect technology, Pressure distribution mapping.

1. Introduction

Globally automobile industries are growing fast, and comfort is the main approach of customers. One of the most important elements for comfort is the car seat since the human body is directly in contact with the seat and exposes to vibrations coming from the base to the top [1]. One of the methods to measure vibrations is to check transmissibility from the base to the human body through seat [2].

The interest to purchase vehicle is associated with various consumer specifications such as performance, mileage, safety, luxury, comfort and cost. These specifications vary based on the subjective perception of the customer. The seat and car manufacturers commonly highlight safety and seating comfort, to attract a larger number of customers and thereby beat the tough competition. As it is important to consider the safety of the occupants, the car manufacturers give extra care to the seats and sitting postures segment. The seating comfort is highly correlated to vehicle vibration due to varying road conditions, sitting postures, vehicle conditions, and so on [3]. Figure 1 shows the sitting postures (correct and bad) of the driver during driving. It is seen that each class of vehicle has a

IOP Conf. Series: Materials Science and Engineering 1128 (2021) 012051 doi:10.1088/1757-899X/1128/1/012051

different seating position according to the specific cabin design and dimensions. The number of vehicles on the road is increasing drastically every day. The number of light commercial vehicles (LCVs) is rather higher than heavy commercial vehicles (HCVs). Therefore the LCVs are considered for the research. It is seen that each class of vehicle has a different seating position according to the specific cabin design and dimensions. The number of vehicles on the road is increasing drastically every day. The number of light commercial vehicles (LCVs) is rather higher than heavy commercial vehicles (HCVs). Therefore the LCVs are considered for the research.

Figure 1. Sitting postures of driver in a car while driving.

The constant vibration, induced by vehicle structure due to traveling in different road conditions at different speeds is seen as a major factor for discomfort to the occupants in the car. These external vibrations are transmitted to occupants through the vehicle seat structure. Many studies show that the human body reaction to vehicle seat vibrations varies depending on the dynamic performance of the seat structure as well as the characteristics of the human body. Several studies have shown that the human body discomfort is significantly influenced by vibrations in frequencies of 0.5–80 Hz. Also, the resonance frequencies for the unoccupied and occupied vehicle seat structures lie mainly in the frequency range of 10–60 Hz [4]. It can be amplified depending on the frequency function of the combined human body and seat. Hence, it is essential to study the relationship between the vehicle seat and the occupant body at frequencies where the human body is highly susceptible to vibration since the dynamic characteristics of occupants are also highly susceptible to the seat structural dynamic responses. But the difficulty in analyzing the dynamics behaviour of the occupied seat is primarily due to the presence of non-linearity and the complex coupling between the seat of the vehicle and the human body.

The multibody dynamics is one of the reliable numerical modeling methods which can be used for analyzing the comfort of the seat and driver. The multibody dynamics simulation model is used to evaluate the effects of deformations in the rigid and flexible bodies with high computational efficiency. Therefore, for this research, the multibody dynamics model is used to simulate the effects on the seat and the occupants in real time conditions.

Kumbhar et al. [5, 6] proposed three-seat model hard seat, seat with cushion, seat with cushion, and base suspension. After numerical analysis for transmissibility between the car seat and human body, they concluded that optimizing cushion material and base suspension stiffness and damping coefficients can increase seating comfort. Kim et al. [7] established a human body sitting posture model for assessing the vibration transmissibility on head and apparent mass on vertical vibrations. For this, they tested various eight models and selected the best-suited model after putting experimental data.

Automobile seat plays an important role in any vehicle as it deals with the comfort and safety of the occupant during a long journey as it imposes a lot of physical and mental stress upon the driver; which would result in fatigue of the driver and gradually an accident. The exposure to vibration during prolonged driving in an automobile is associated with seating discomfort, increased prevalence of low back pain, and other musculoskeletal disorders [8]. This can be avoided by using a better seat model. In recent days, the preference of the customer is increased towards comfort. Since the field tests for assessing seating comfort with participants are time-consuming, the laboratory static and motion simulators can be utilized with suitable medical and ethical precautions. Computer based simulation models can also be used to assess human–seat interface.

Considering these issues faced by the occupant, the research focuses on determining the proper sitting postures and angles. The sitting posture and sitting angle were observed using Kinect technology and the pressure mapping system [9-10].

With regards to that, vibration analysis has to be done and frequency response function (FRF) is to be obtained by simulation software to find the induced vibration on the vehicle seat and the occupants. This has to be validated using an experimental setup consisting of a set of accelerometer pads fitted on different points of the seat. The above studies would provide enough data from which proper sitting postures and seating comfort are analyzed to reduce the fatigue induced on the driver due to prolonged driving.

2. Materials and Methods

2.1. Multibody Dynamics Analysis

Whole-body vibration in sitting posture on a car seat is taken up in this study. Since the human body is directly in contact with the car seat, vibrations are exposed from the seat to the human body. For this purpose, data from Kumbhar et al. [5, 6] and Kim et al. [6] were used which is based on experimental measurements and by using these data we run vibration analysis using multibody dynamics simulation software MSC ADAMS for the seat to body vibration transmissibility. For this study, vibration analysis is performed for three car seat models: hard seat, seat with a cushion but no seat suspension, and seat with cushion and base suspension. Assumptions include, all model parts are assumed as rigid bodies and their flexibility is defined through stiffness and damping coefficients. In the car seat, backrest support transfers a large amount of vibrations to the human body, so it is considered at an angle of 230° with a vertical plane. The biomechanical properties of the human body segments and car models are taken from Kumbhar et al. [5, 6]. The running condition of the vehicle is taken as ideal.

2.1.1 Hard Seat

In Figure 2, a human model is seated on a hard seat. The seat cushion and backrest cushion are also ignored in this model. There is nothing in this seat to protect the human body from vibrations. Thus, all vibrations are exposed to the human body directly [11]. The system (human model coupled with the seat model has 14 degrees of freedom) receives feedback directly from the floor.

2.1.2 Seat with a Cushion but No Seat Suspension

A human model is seated on a seat with cushion without suspension (Figure 3). In this case, any vibrations, as a consequence of cushion stiffness, are isolated. The stiffness of each cushion is extremely non-linear. Nevertheless, for convenience, they are assumed to be linear, in this study. With regard to its contact surface, each cushion has one horizontal motion and one vertical motion. The total number of degrees of freedom in this model is 18.

2.1.3 Seat with Cushion and Base Suspension

Figure 4 shows a human model placed on a seat with cushion, and suspension. Using such links, which serve as seat suspension, the seat is attached to the vehicle body floor. This seat has one horizontal and one vertical movement for which the vehicle body floor receives input. Thus, given the seat's two degrees of freedom, the cushions' two degrees of freedom, and the human model's 14 degrees of freedom, the total number of degrees of freedom in this model is 20.

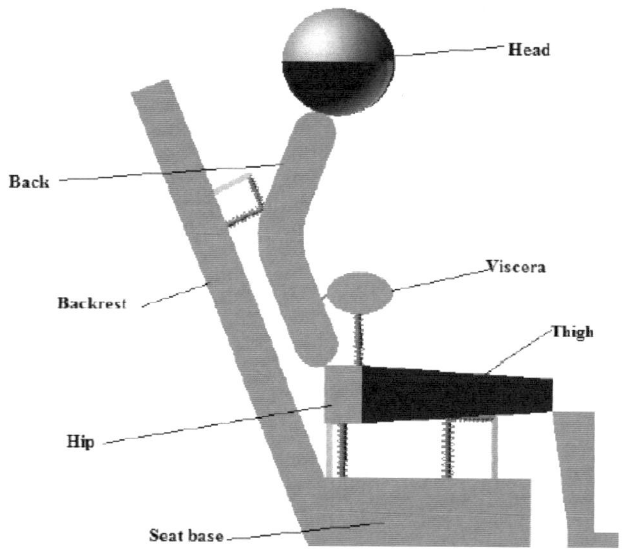

Figure 2. Human model sitting on a hard seat.

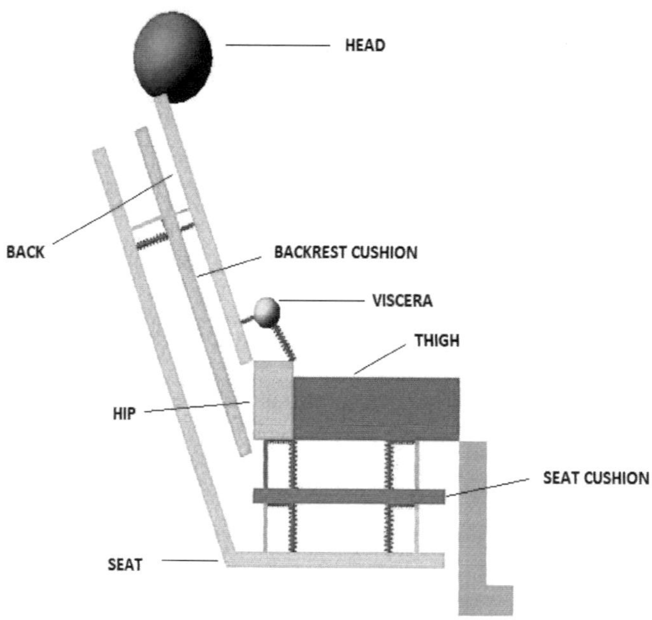

Figure 3. Seat with a cushion but no seat suspension.

2.2. Vibration Analysis

Vibration analysis is used to measure transmissibility from the vehicle floor to various human parts. The multibody dynamics simulation software MSC ADAMS was used to perform vibration analysis on three car seat models to get transmissibility magnitude and frequency. The seat parameters can be optimized with the maximum magnitude of transmissibility and its frequency to reduce discomfort [12, 13].

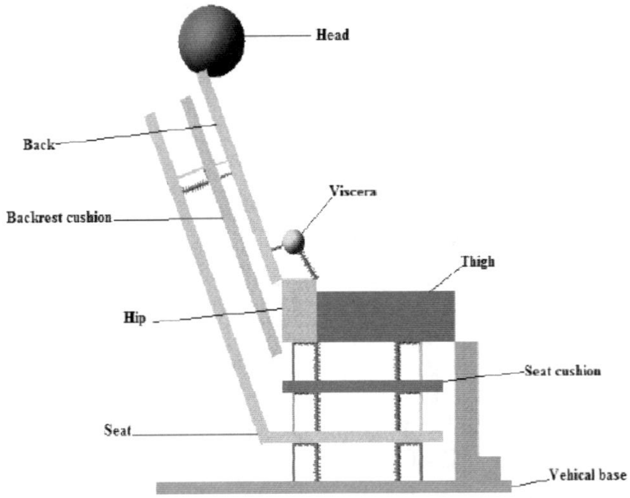

Figure 4. Seat with cushion and base suspension.

2.3. Questionnaire Survey

This questionnaire survey was conducted to estimate the driver discomfort in a different kind of passenger car driver seat and to describe the nature of sitting discomfort. The specifications of the seat have been chosen using anthropometric data such as thigh and buttocks, and seat design standards like seat length, seat width, etc. [14].

Table 1. Anthropometric data of the participant.

Variable	Range		Mean	
	Tata Indica Car	Maruti Swift Car	Tata Indica Car	Maruti Swift Car
Age (Years)	26 - 55	22 - 60	34.53	37.10
BMI (kg/m^2)	18.14 - 25.95	20.07 - 27.92	22.97	23.33
Height (m)	1.66 - 1.70	1.60 - 1.75	1.68	1.67
Weight (kg)	50 - 75	55 - 80	64.57	65.38

In this study, the brief subjective method of evaluating the driver's seat discomfort questionnaire is designed to cover all the major parameters on it. The survey was taken from 100 males, each professional passenger's car drivers in Chennai city. All the participants are from the Tata Indica and Maruti Swift car drivers alone. Table 1 shows anthropometric data obtained from the participants. The seats of Indica and Swift are shown in Figure 5. The participants were categorized into four groups based on their age groups. Group 1 consists of drivers whose age lies in between 26-30 years; Group 2 consists of drivers whose age lies in between 31-35 years; Group 3 consists of drivers whose age lies in between 36-40 years; and Group 4 consists of drivers whose age lies in between 41-55 years. Cronbach's alpha test is used to measure the internal consistency (reliability) of the questionnaire. In this questionnaire survey, the design is validated to be reliable with Cronbach Alpha score. (Cronbach's alpha > 0.070 is significant).

The scale selection is foremost important in any subjective method of evaluation. Based on that only the responses to be collected and analyzed. In this study, a 5 point Likert scale is used in this questionnaire survey. The driver's seat is to rate as [negative]-2 to +2 [positive] (Table 2).

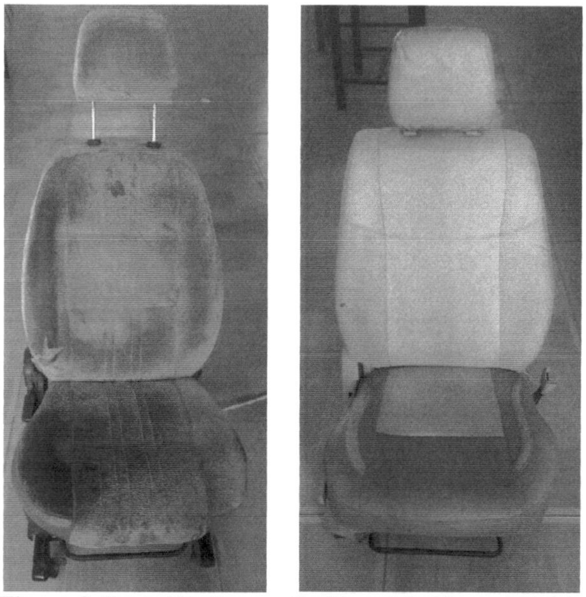

Figure 5. Seat of Indica car (Left) and Swift car (Right).

Table 2. Scale to measure seating comfort.

Discomfort	Little Discomfort	Neutral	Little Comfort	Comfort
-2	-1	0	+1	+2

2.4. Pressure Mapping System

The force sensitive resistor pressure sensor is used in this study (Figure 6). The sensors are arranged in a form of a 4x4 matrix (i.e., 16 sensors in all). They are joined with an Arduino Board and it is programmed and calibrated in such a way that with appropriate force, an appropriate reading is observed. Two different driver seats of Tata Indica and Maruti Swift were acquired for the study. Both seats are of cars that are extensively used in the range of commercial cars and by major Taxi aggregators, such as OLA and UBER.

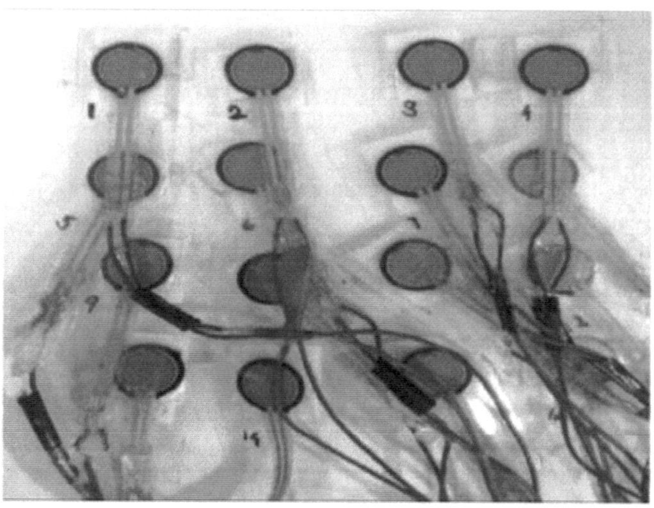

Figure 6. Pressure mapping system using force sensitive resistor pressure sensors.

In this pressure sensing system, the participants were asked to position themselves just above the seat having their back rested in a relaxed position. Before having them seated in the position, the arrangements of the apparatus had been done and they were asked to sit comfortably on the seat as prescribed with their hands resting on their laps. Ultimate care was taken so that the apparatuses were not moved while positioning the participants over the mat and throughout the whole reading procedure. No obstructions such as a wallet or a mobile phone are allowed to be kept on the buttock region while seated on the seat. Then the participant was asked to sit and then multiple reading was taken and recorded for less error and higher chances of convergence. The participants were asked to sit on the seat for about 30-40 seconds about 5 times. As the readings were a voluminous form of data, it was carefully settled down by taking an average at a particular instance. A total of 5 participants were considered for the study of different BMI ratings.

2.5. Kinect Sensor - Skeletal Mapping System
The Kinect is an RGB-D sensor providing skeletal mapping images with synchronized color and depth. This system uses IR camera that takes reading in the form of recordings and the snapshots of the recording are taken. A Kinect used here is in the virtue of obtaining the skeletal map of a body, and through it, can be obtained the stress levels that are exerted by the body after prolonged driving/sitting in a particular position [10]. The participants can use their arms at a distance, range of 1.2 m to 3.5 m.

As soon as the participant gets in front of the camera, the camera starts detecting the participant. The Kinect takes readings in the form of camera recordings and snapshots of the recording are taken. The angles of the participants are measured manually by KLONK image measuring software. All these processes are carried out by asking the participant to position on rigid car seats. These images of the skeletal tracked body were then used to find the sitting posture of those participants while driving the automobile. The participants were the only participants in the vicinity of the Kinect so that there are no miscalculations by the device. The entire process was conducted for approximately 60 seconds. Angles involving the lumbar region and the knees region were manually found out using KLONK Image measuring software.

3. Results and Discussion
Figure 7 shows the vibration analysis results of transmissibility for all body part segments from the hard seat model. Results show graphs for transmissibility magnitude versus frequency. We can obtain peak transmissibility value with its respective frequency from these results. For the hard seat model, where no vibration isolation happens, vibrations are transferred to the body directly through feet and seat contact with the body. Results show frequencies for all body segments are between 4 to 10 Hz. Figure 8 shows the vibration analysis results of transmissibility for all body part segment from the seat with a cushion but no base suspension. In the case of the second model, where only cushions are provided, vibrations isolated through cushions are not transferred to body parts directly. So frequency and magnitude are decreased as compared to the hard seat model. Results show transfer function frequencies for all body parts are approximately 2 to 3 Hz. The vibration analysis results of transmissibility for all body part segment from the seat with cushion and base suspension is depicted in Figure 9. In this model, where cushions and base suspension have been provided, vibrations are isolated through cushions and base suspension and are not transferred to body parts directly. So frequency and magnitude are decreased as compared to the hard seat model. Results show transfer function frequency for all body parts is approximately 2 Hz. Table 3 shows multibody dynamics simulation results for all three car seat models compared to results obtained from Kumbhar et al. [5]. The comparison for transmissibility magnitude and their respective frequencies are compared for three models; frequency values are approximately same and have some differences in transmissibility.

Table 3. Validation of results of the three car seat models.

Car Seat Model	Body Segment	Proposed Multibody Dynamics Simulation Model		Results from Kumbhar et al. [5]	
		Transmissibility Magnitude	Frequency (Hz)	Transmissibility Magnitude	Frequency (Hz)
Hard seat	Thigh	1	1.85	1.1	2.5
	Hip	1.59	5	1.5	5.2
	Back	1.7	5.1	1.15	5.2
	Head	1.58	5	1.2	5.2
	Viscera	3.6	5.97	2.9	5.7
Seat with a cushion without suspension	Thigh	1.62	3.07	1.35	3.45
	Hip	1.32	2.08	1.1	2
	Back	1.02	2.09	1.53	2
	Head	1.02	3.07	1.24	2.8
	Viscera	1.68	3.08	1.36	2.78
Seat with cushion and base suspension	Thigh	1.46	1.85	1.2	2.2
	Hip	1.9	2	1.25	2.2
	Back	1.6	2	1.2	2.2
	Head	1.53	2	1.2	2.2
	Viscera	2.35	2	1.7	2.2

Table 4. Comfort Scores of Head, Lumbar, Backrest, and Overall for Tata Indica and Maruti Swift.

Age Group (Years)	Car	Comfort Score			
		Head	Lumbar	Backrest	Overall
26-30	Tata Indica	-0.3235	-0.4412	-0.625	-0.8137
	Maruti Swift	0.1256	0.0272	-0.1033	0.1014
31-35	Tata Indica	-0.3547	-0.4231	-0.5385	-0.8846
	Maruti Swift	0.1634	0.0735	-0.1618	-0.0239
36-40	Tata Indica	-0.2731	-0.5021	-0.6508	-0.8583
	Maruti Swift	0.1471	0.0478	-0.1407	-0.0392
41-55	Tata Indica	-0.1589	-0.1641	-0.3047	-0.6667
	Maruti Swift	0.1068	0.0197	-0.1038	-0.0405

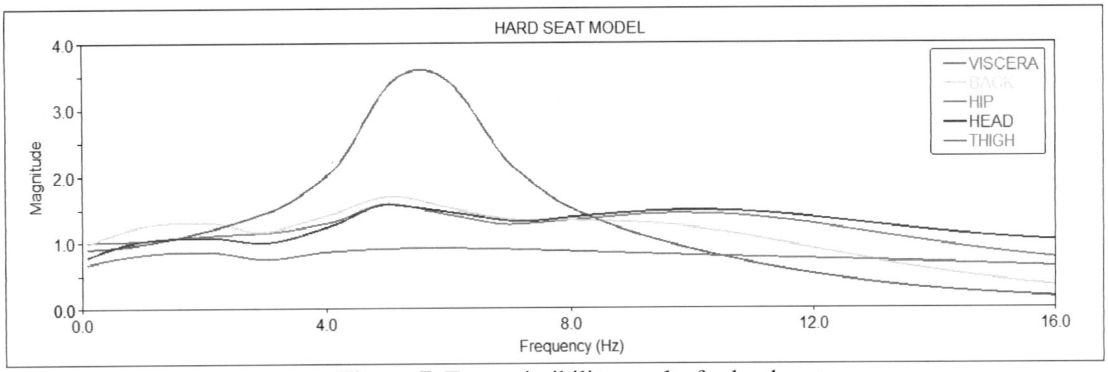

Figure 7. Transmissibility results for hard seat.

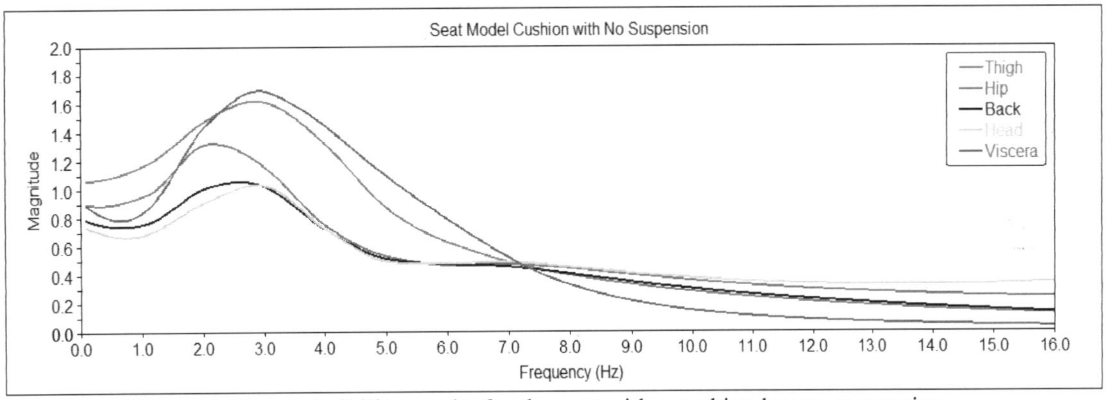

Figure 8. Transmissibility results for the seat with a cushion but no suspension.

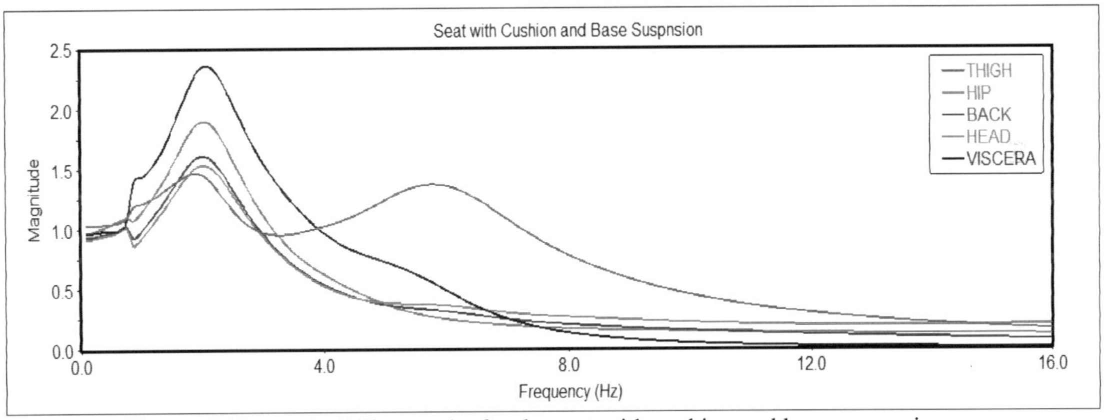

Figure 9. Transmissibility results for the seat with cushion and base suspension.

From the survey study, it can be seen that the comfort level of the Tata Indica seats was much lower compared to that of Maruti Swift seats. Figure 10 shows the relation between two seats with their respective comfort scores of different age groups. Table 4 shows the comfort levels of different seat segments concerning the different age groups. It is seen that the comfort levels of Tata Indica lack behind the comfort levels of the Maruti Swift for all age groups. The pressure mapping system gives the contour images of the seat pressure distribution obtained from Tata Indica and Maruti Swift cars. The contour images are taken for each seat on the following categories: Underweight participants; Healthy participants; and Over-weight participants (Figure 11).

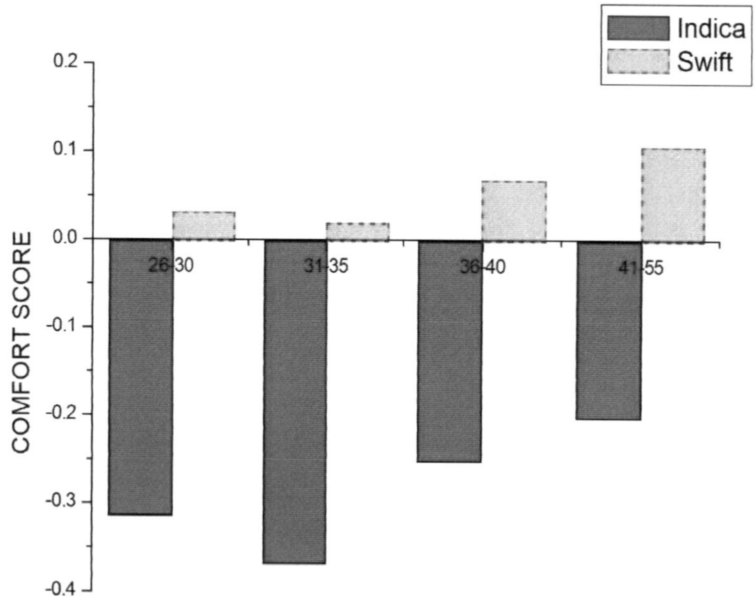

Figure 10. Comparison of comfort score of Tata Indica with Maruti Swift against the age groups.

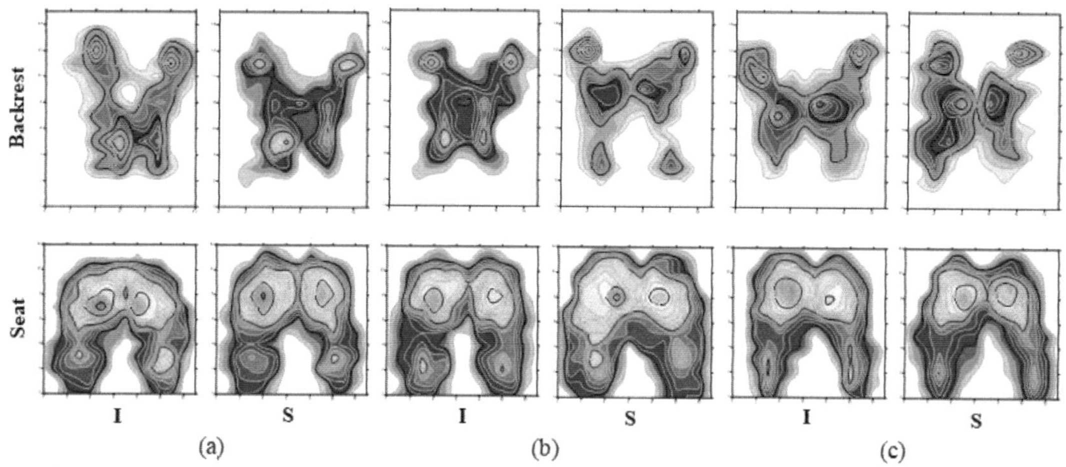

Figure 11. Pressure distribution mapping obtained from seat and backrest of Tata Indica (I) and Maruti Swift (S). (a) Underweight participants; (b) Healthy participants, and (c) Over-weight participants.

The final result that is obtained is that the buttock region has been stressed the most as corroborated with the study investigated by Rosaria et al. [15]. Also, a flexible back angle should be given to reduce the stress exerted by the participant. Moreover, there is no padding or support for the calves of the legs and the stress can be reduced if the padding supports the calves' region of the driver. All these lead to a conclusion that the comfort of the driver doesn't depend on a single parameter but multiple factors. The stresses in the buttock regions can be obviated by designing the seat appropriately and with proper material. Three participants (underweight participant, healthy participant, and over-weight participant) were made to sit on both the seats, and data were captured using the Kinect sensor – skeletal mapping system (Figure 12). The Kinect camera helps to find the angle that is made by all body parts of a participant driving the car. This method proved to enable us to find the optimal angle and where a body is strained the most and thus, posture can be improved. The result obtained by the Kinect camera is very close to the benchmark except for the hip joint angle and torso angle because all the procedure is carried out on a rigid car seat.

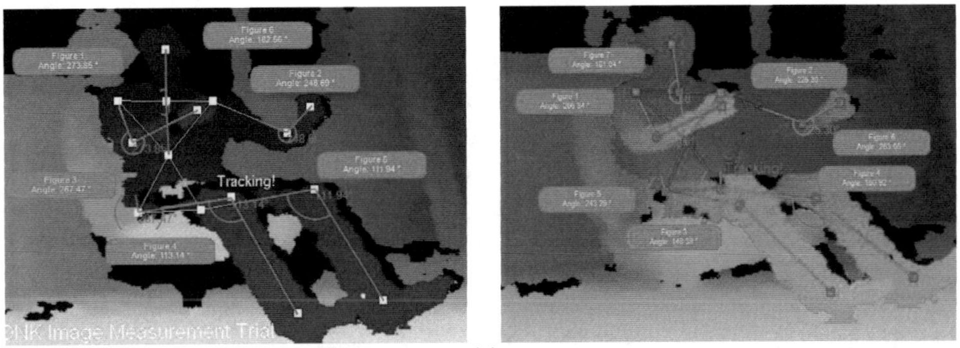

(a)

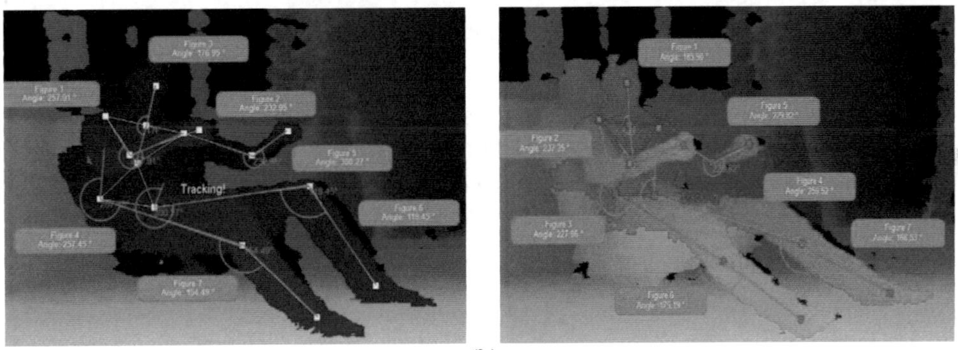

(b)

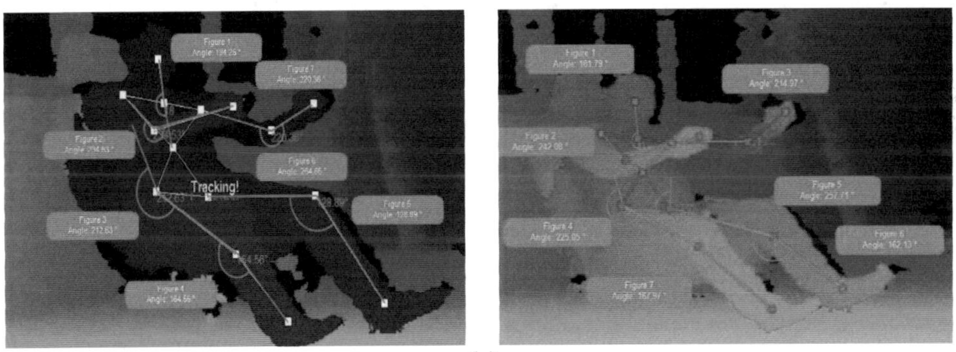

(c)

Figure 12. Skeletal image of participantccaptured using the Kinect camera system. (a) Underweight particcipant, (b) Healthy participant, and (c) Over-weight participant. (Blue – Tata Indica and Green – Maruti Swift).

4. Conclusion

In this study, the experimental and numerical investigation of human-car seating models was carried out and validated. The study reveals that by changing model parameters like spring stiffness, damping coefficients, and seat part masses, we can reduce vibration transmissibility can be reduced and thereby enhance more seating comfort. The pressure sensors produced accurate results of where is the most stressed part of the human body on the sea while the participant sits on it. It is seen that the buttock region has been stressed the most. The Kinect camera system helped to find the optimal angle and where a body is strained the most and thus, sitting posture can be improved. The stress felt on the buttock region of the participant while seated can be obviated by designing the seat with suitable cushion material. Also with thus, the seats could be given a smart option to adjust the seats with small degrees, as per the driver's requirement, manually or automatically.

5. References

[1] Fard M, Lo L, Subic A and Jazar R 2014 Effects of seat structural dynamics on current ride comfort criteria *Ergonomics*, 57(10), pp. 1549-1561.

[2] Lean Lo, Fard M, Subic A and Jazar R 2013 Structural dynamic characterization of a vehicle seat coupled with human occupant *Journal of Sound and Vibration*, 332(4), pp. 1141-1152.

[3] Tang A H, Tian J P and Liao Y H 2014 Analysis for ride comfort evaluation of passenger car traveling on roads with generalized road profiles and conventional speeds *Advanced Materials Research*, 926-930, Trans Tech Publications, Ltd., pp 877-880.

[4] Mansfield N J (2005), *Human Response to Vibration*, First Edition, CRC Press, Florida, pp. 1-244.

[5] Kumbhar P, Xu P and Yang J 2013 Evaluation of human body response for different vehicle seats using a multibody biodynamic model *SAE Technical Paper*, 2013-01-0994, pp. 1-14.

[6] Kumbhar P, Li N, Xu P and Yang J 2014 Optimal seat dynamic parameters determination for minimizing virtual driver's fatigue *SAE Technical Paper*, 2014-01-0877, pp. 1-13.

[7] Kim T H, Kim Y T and Yoon Y S 2005 Development of a biomechanical model of the human body in a sitting posture with vibration transmissibility in the vertical direction *International Journal of Industrial Ergonomics*, 35(9), pp. 817-829.

[8] Jagannath M and Balasubramanian V 2014 Assessment of early onset of driver fatigue using multimodal fatigue measures in a static simulator *Applied Ergonomics*, 45(4), pp. 1140 – 1147.

[9] Paul S, Basu S and Nasipuri M 2015 Microsoft Kinect in gesture recognition: A short review *I J C T A*, International Science Press, 8(5), pp. 2071-2076

[10] Napolia A, Glass S, Ward C, Tucker C and Obeid I 2017 Performance analysis of a generalized motion capture system uses microsoft kinect 2.0 *Biomedical Signal Processing and Control*, 38, pp. 265-280.

[11] Wang X, Savonnet L, Theodorakos I, Beurier G and Duprey S 2019 Biomechanical human models for seating discomfort assessment In: S. Scataglini and G. Paul (Eds.) *DHM and Posturography*, Academic Press, pp. 643-656.

[12] Ittianuwat R, Fard M and Kato K 2014 The transmission of vibration at various locations on vehicle seat to seated occupant body. In: *INTER-NOISE and NOISE-CON Congress and Conference Proceedings*, Melbourne, Australia, pp. 2062-2073.

[13] Toward M G R and Griffin M J 2011 The transmission of vertical vibration through seats: Influence of the characteristics of the human body *Journal of Sound and Vibration*, 330(26), pp. 6526–6543.

[14] Smith D R, Andrews D M and Wawrow P T 2006 Development and evaluation of the Automotive Seating Discomfort Questionnaire (ASDQ) *International Journal of Industrial Ergonomics*, 36(2), pp. 141-149.

[15] Rosaria C, Alessandro N and Chiara C 2020 Comfort seat design: Thermal sensitivity of human back and buttock *International Journal of Industrial Ergonomics*, 78, p. 102961.

Acknowledgement : Authors recognise the work of Late Prof.S.Kannan, Assistant Professor, VIT Chennai who initiated these studies but passed away before seeing the fruit of this work.

AUTHOR INDEX